T0328720

GEOMETRIC ALGEBRA FOR PHYSICISTS

Geometric algebra is a powerful mathematical language with applications across a range of subjects in physics and engineering. Written by two of the leading researchers in the field, this book is a complete guide to the current state of the subject.

Early chapters provide a self-contained development of geometric algebra and form the basis of an undergraduate lecture course. Topics covered include new techniques for handling rotations in arbitrary dimensions, and the links between rotations, bivectors and the structure of the Lie groups. Following chapters extend the concept of a complex analytic function theory to arbitrary dimensions. This has applications in quantum theory and electromagnetism. All four Maxwell equations are united into one single equation, and new techniques are discussed for its solution. Later chapters cover some advanced topics in physics, including non-Euclidean geometry, quantum entanglement and gauge theories. The final chapters describe the construction of a gauge theory of gravitation in Minkowski spacetime. Using the tools of geometric algebra, advanced applications such as black holes and cosmic strings are explored.

This book will be of interest to researchers working in the fields of geometry, relativity and quantum theory. It can also be used as a textbook for advanced undergraduate and graduate courses on the physical applications of geometric algebra.

CHRIS DORAN obtained his PhD from the University of Cambridge, having gained a distinction in Part III of his undergraduate degree. He was elected a Junior Research Fellow of Churchill College, Cambridge in 1993, was made a Lloyd's of London Fellow in 1996 and was the Schlumberger Interdisciplinary Research Fellow of Darwin College, Cambridge between 1997 and 2000. He is currently a Fellow of Sidney Sussex College, Cambridge and holds an EPSRC Advanced Fellowship. Dr Doran has published widely on aspects of mathematical physics and is currently researching applications of geometric algebra in engineering and computer science.

ANTHONY LASENBY is Professor of Astrophysics and Cosmology at the University of Cambridge, and is currently Head of the Astrophysics Group and the Mullard Radio Astronomy Observatory in the Cavendish Laboratory. He began his astronomical career with a PhD at Jodrell Bank, specialising in the Cosmic Microwave Background, which has been a major subject of his research ever since. After a brief period at the National Radio Astronomy Observatory in America, he moved from Manchester to Cambridge in 1984, and has been at the Cavendish since then. He is the author or coauthor of nearly 200 papers spanning a wide range of fields, from early universe cosmology to computer vision. His introduction to geometric algebra came in 1988, when he encountered the work of David Hestenes for the first time, and since then he has been developing geometric algebra techniques and employing them in his research in many areas.

GEOMETRIC ALGEBRA FOR PHYSICISTS

CHRIS DORAN
and
ANTHONY LASENBY
University of Cambridge

CAMBRIDGE
UNIVERSITY PRESS

University Printing House, Cambridge CB2 8BS, United Kingdom

Cambridge University Press is part of the University of Cambridge.

It furthers the University's mission by disseminating knowledge in the pursuit of education, learning and research at the highest international levels of excellence.

www.cambridge.org
Information on this title: www.cambridge.org/9780521715959

First published 2003
Reprinted 2003
First paperback edition published 2007
6th printing 2013

A catalogue record for this publication is available from the British Library

ISBN 978-0-521-48022-2 Hardback
ISBN 978-0-521-71595-9 Paperback

Contents

Preface

The ideas and concepts of physics are best expressed in the language of mathematics. But this language is far from unique. Many different algebraic systems exist and are in use today, all with their own advantages and disadvantages. In this book we describe what we believe to be the most powerful available mathematical system developed to date. This is *geometric algebra*, which is presented as a new mathematical tool to add to your existing set as either a theoretician or experimentalist. Our aim is to introduce the new techniques via their applications, rather than as purely formal mathematics. These applications are diverse, and throughout we emphasise the unity of the mathematics underpinning each of these topics.

The history of geometric algebra is one of the more unusual tales in the development of mathematical physics. William Kingdon Clifford introduced his geometric algebra in the 1870s, building on the earlier work of Hamilton and Grassmann. It is clear from his writing that Clifford intended his algebra to describe the geometric properties of vectors, planes and higher-dimensional objects. But most physicists first encounter the algebra in the guise of the Pauli and Dirac matrix algebras of quantum theory. Few then contemplate using these unwieldy matrices for practical geometric computing. Indeed, some physicists come away from a study of Dirac theory with the view that Clifford's algebra is inherently quantum-mechanical. In this book we aim to dispel this belief by giving a straightforward introduction to this new and fundamentally different approach to vectors and vector multiplication. In this language much of the standard subject matter taught to physicists can be formulated in an elegant and highly condensed fashion. And the portability of the techniques we discuss enables us to reach a range of advanced topics with little extra work.

This book is intended to be of interest to both students and researchers in physics. The early chapters grew out of an undergraduate lecture course that we have run for a number of years in the Physics Department at Cambridge University. We are indebted to the students who attended the early versions of this course, and helped to shape the material into a form suitable for undergraduate tuition. These early chapters require little more than a basic knowledge of linear algebra and vector geometry, and some exposure to classical mechanics. More advanced physical concepts are introduced as the book progresses.

A number of themes run throughout this book. The first is that geometric algebra enables us to express fundamental physics in a language that is free from coordinates or indices. Coordinates are only introduced later, when the geometry of a given problem is clear. This approach gives many equations a degree of clarity which is lost in tensor algebra. A second theme is the way in which rotations are handled in geometric algebra through the use of *rotors*. This approach extends to arbitrary spaces the idea of using a complex phase to rotate in a plane. Rotor techniques can be applied in spaces of arbitrary signature and are particularly well suited to formulating Lorentz and conformal transformations. The latter are central to our treatment of non-Euclidean geometry. Rotors also provide a framework for studying Lie groups and Lie algebras, and are essential to our discussion of gauge theories.

The third theme is the invertibility of the geometric product of vectors, which makes it possible to divide by a vector. This idea extends to the vector derivative, which has an inverse in the form a first-order Green's function. The vector derivative and its inverse enable us to extend complex analytic function theory to arbitrary dimensions. This theory is perfectly suited to electromagnetism, as all four Maxwell equations can be combined into a single spacetime equation involving the invertible vector derivative. The same vector derivative appears in the Dirac theory, and is central to the gauge treatment of gravitation which dominates the final two chapters of this book.

This book would not have been possible without the help and encouragement of a large number of people. We thank Stephen Gull for helping initiate much of the research described here, for his constant advice and criticism, and for use of a number of his figures. We also thank David Hestenes for all his work in shaping the modern subject of geometric algebra and for his constant encouragement. Special mention must be made of our many collaborators, in particular Joan Lasenby, Anthony Challinor, Leo Dorst, Tim Havel, Antony Lewis, Mark Ashdown, Frank Sommen, Shyamal Somaroo, Jeff Tomasi, Bill Fitzgerald, Youri Dabrowski and Mike Hobson. Special thanks also goes to Mike for his help with Latex and explaining the intricacies of the CUP style files. We thank the Physics Department of Cambridge University for the use of their facilities, and for the range of technical advice and expertise we regularly called on. Finally we thank everyone at Cambridge University Press who helped in the production of this book.

CD would also like to thank the EPSRC and Sidney Sussex College for their support, his friends and colleagues, all at Nomads HC, and above all Helen for not complaining about the lost evenings as I worked on this book. I promise to finish the decorating now it is complete.

AL thanks Joan and his children Robert and Alison for their constant enthusiasm and support, and their patience in the face of many explanations of topics from this book.

Cambridge
July 2002

C.J.L. Doran
A.N. Lasenby

Notation

The subject of vector geometry in general, and geometric algebra in particular, suffers from a profusion of notations and conventions. In short, there is no single convention that is perfectly suited to the entire range of applications of geometric algebra. For example, many of the formulae and results given in this book involve arbitrary numbers of vectors and are valid in vector spaces of arbitrary dimensions. These formulae invariably look neater if one does not embolden all of the vectors in the expression. For this reason we typically choose to write vectors in a lower case italic script, a, and more general multivectors in upper case italic script, M. But in some applications, particularly mechanics and dynamics, one often needs to reserve lower case italic symbols for coordinates and scalars, and in these situations writing vectors in bold face is helpful. This convention in adopted in chapter 3.

For many applications it is useful to have a notation which distinguishes frame vectors from general vectors. In these cases we write the former in an upright font as $\{\mathsf{e}_i\}$. But this notation looks clumsy in certain settings, and is not followed rigorously in some of the later chapters. In this book our policy is to ensure that we adopt a consistent notation within each chapter, and any new or distinct features are explained either at the start of the chapter or at their point of introduction.

Some conventions are universally adopted throughout this book, and for convenience we have gathered together a number of these here.

(i) The geometric (or Clifford) algebra generated by the vector space of signature (p, q) is denoted $\mathcal{G}(p, q)$. In the first three chapters we employ the abbreviations $\mathcal{G}_2$ and $\mathcal{G}_3$ for the Euclidean algebras $\mathcal{G}(2, 0)$ and $\mathcal{G}(3, 0)$. In chapter 4 we use $\mathcal{G}_n$ to denote all algebras $\mathcal{G}(p, q)$ of total dimension n.

(ii) The geometric product of A and B is denoted by juxtaposition, AB.

(iii) The inner product is written with a centred dot, $A \cdot B$. The inner product is only employed between homogeneous multivectors.

(iv) The outer (exterior) product is written with a wedge, $A \wedge B$. The outer product is also only employed between homogeneous multivectors.

(v) Inner and outer products are always performed before geometric products. This enables us to remove unnecessary brackets. For example, the expression $a \cdot b\, c$ is to be read as $(a \cdot b)c$.

(vi) Angled brackets $\langle M \rangle_p$ are used to denote the result of projecting onto the terms in M of grade p. The subscript zero is dropped for the projection onto the scalar part.

(vii) The reverse of the multivector M is denoted either with a dagger, $M^\dagger$, or with a tilde, $\tilde{M}$. The latter is employed for applications in spacetime.

(viii) Linear functions are written in an upright font as $\mathsf{F}(a)$ or $\mathsf{h}(a)$. This helps to distinguish linear functions from multivectors. Some exceptions are encountered in chapters 13 and 14, where caligraphic symbols are used for certain tensors in gravitation. The adjoint of a linear function is denoted with a bar, $\bar{\mathsf{h}}(a)$.

(ix) Lie groups are written in capital, Roman font as in $\mathrm{SU}(n)$. The corresponding Lie algebra is written in lower case, $\mathrm{su}(n)$.

Further details concerning the conventions adopted in this book can be found in sections 2.5 and 4.1.

1

Introduction

The goal of expressing geometrical relationships through algebraic equations has dominated much of the development of mathematics. This line of thinking goes back to the ancient Greeks, who constructed a set of geometric laws to describe the world as they saw it. Their view of geometry was largely unchallenged until the eighteenth century, when mathematicians discovered new geometries with different properties from the Greeks' *Euclidean* geometry. Each of these new geometries had distinct algebraic properties, and a major preoccupation of nineteenth century mathematicians was to place these geometries within a unified algebraic framework. One of the key insights in this process was made by W.K. Clifford, and this book is concerned with the implications of his discovery.

Before we describe Clifford's discovery (in chapter 2) we have gathered together some introductory material of use throughout this book. This chapter revises basic notions of vector spaces, emphasising pictorial representations of the underlying algebraic rules — a theme which dominates this book. The material is presented in a way which sets the scene for the introduction of Clifford's product, in part by reflecting the state of play when Clifford conducted his research. To this end, much of this chapter is devoted to studying the various products that can be defined between vectors. These include the scalar and vector products familiar from three-dimensional geometry, and the complex and quaternion products. We also introduce the *outer* or *exterior* product, though this is covered in greater depth in later chapters. The material in this chapter is intended to be fairly basic, and those impatient to uncover Clifford's insight may want to jump straight to chapter 2. Readers unfamiliar with the outer product are encouraged to read this chapter, however, as it is crucial to understanding Clifford's discovery.

1.1 Vector (linear) spaces

At the heart of much of geometric algebra lies the idea of vector, or linear spaces. Some properties of these are summarised here and assumed throughout this book. In this section we talk in terms of *vector* spaces, as this is the more common term. For all other occurrences, however, we prefer to use the term *linear* space. This is because the term '*vector*' has a very specific meaning within geometric algebra (as the grade-1 elements of the algebra).

1.1.1 Properties

Vector spaces are defined in terms of two objects. These are the vectors, which can often be visualised as directions in space, and the scalars, which are usually taken to be the real numbers. The vectors have a simple addition operation rule with the following obvious properties:

(i) Addition is *commutative*:

$$a + b = b + a. \tag{1.1}$$

(ii) Addition is *associative*:

$$a + (b + c) = (a + b) + c. \tag{1.2}$$

This property enables us to write expressions such as $a + b + c$ without ambiguity.

(iii) There is an identity element, denoted 0:

$$a + 0 = a. \tag{1.3}$$

(iv) Every element a has an inverse $-a$:

$$a + (-a) = 0. \tag{1.4}$$

For the case of directed line segments each of these properties has a clear geometric equivalent. These are illustrated in figure 1.1.

Vector spaces also contain a multiplication operation between the scalars and the vectors. This has the property that for any scalar λ and vector a, the product λa is also a member of the vector space. Geometrically, this corresponds to the dilation operation. The following further properties also hold for any scalars λ, μ and vectors a and b:

(i) $\lambda(a + b) = \lambda a + \lambda b$;
(ii) $(\lambda + \mu)a = \lambda a + \mu a$;
(iii) $(\lambda\mu)a = \lambda(\mu a)$;
(iv) if $1\lambda = \lambda$ for all scalars λ then $1a = a$ for all vectors a.

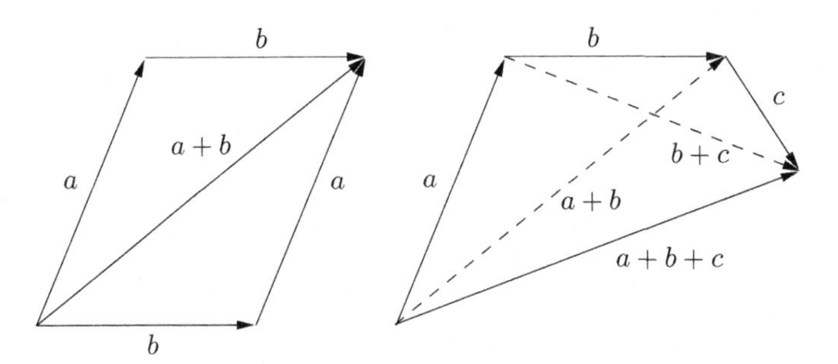

Figure 1.1 *A geometric picture of vector addition.* The result of $a + b$ is formed by adding the tail of b to the head of a. As is shown, the resultant vector $a + b$ is the same as $b + a$. This finds an algebraic expression in the statement that addition is commutative. In the right-hand diagram the vector $a + b + c$ is constructed two different ways, as $a + (b + c)$ and as $(a+b)+c$. The fact that the results are the same is a geometric expression of the associativity of vector addition.

The preceding set of rules serves to define a vector space completely. Note that the + operation connecting scalars is different from the + operation connecting the vectors. There is no ambiguity, however, in using the same symbol for both.

The following two definitions will be useful later in this book:

(i) Two vector spaces are said to be *isomorphic* if their elements can be placed in a one-to-one correspondence which preserves sums, and there is a one-to-one correspondence between the scalars which preserves sums and products.

(ii) If $\mathcal{U}$ and $\mathcal{V}$ are two vector spaces (sharing the same scalars) and all the elements of $\mathcal{U}$ are contained in $\mathcal{V}$, then $\mathcal{U}$ is said to form a *subspace* of $\mathcal{V}$.

1.1.2 Bases and dimension

The concept of dimension is intuitive for simple vector spaces — lines are one-dimensional, planes are two-dimensional, and so on. Equipped with the axioms of a vector space we can proceed to a formal definition of the dimension of a vector space. First we need to define some terms.

(i) A vector b is said to be a *linear combination* of the vectors $a_1, \dots, a_n$ if scalars $\lambda_1, \dots, \lambda_n$ can be found such that

$$b = \lambda_1 a_1 + \cdots + \lambda_n a_n = \sum_{i=1}^{n} \lambda_i a_i. \tag{1.5}$$

(ii) A set of vectors $\{a_1, \dots, a_n\}$ is said to be *linearly dependent* if scalars

$\lambda_1, \ldots, \lambda_n$ (not all zero) can be found such that

$$\lambda_1 a_1 + \cdots + \lambda_n a_n = 0. \tag{1.6}$$

If such a set of scalars cannot be found, the vectors are said to be *linearly independent.*

(iii) A set of vectors $\{a_1, \ldots, a_n\}$ is said to *span* a vector space $\mathcal{V}$ if every element of $\mathcal{V}$ can be expressed as a linear combination of the set.

(iv) A set of vectors which are both linearly independent and span the space $\mathcal{V}$ are said to form a *basis* for $\mathcal{V}$.

These definitions all carry an obvious, intuitive picture if one thinks of vectors in a plane or in three-dimensional space. For example, it is clear that two independent vectors in a plane provide a basis for all vectors in that plane, whereas any three vectors in the plane are linearly dependent. These axioms and definitions are sufficient to prove the *basis theorem*, which states that *all bases of a vector space have the same number of elements.* This number is called the *dimension* of the space. Proofs of this statement can be found in any textbook on linear algebra, and a sample proof is left to work through as an exercise. Note that any two vector spaces of the same dimension and over the same field are isomorphic.

The axioms for a vector space define an abstract mathematical entity which is already well equipped for studying problems in geometry. In so doing we are not compelled to interpret the elements of the vector space as displacements. Often different interpretations can be attached to isomorphic spaces, leading to different types of geometry (affine, projective, finite, *etc.*). For most problems in physics, however, we need to be able to do more than just add the elements of a vector space; we need to multiply them in various ways as well. This is necessary to formalise concepts such as angles and lengths and to construct higher-dimensional surfaces from simple vectors.

Constructing suitable products was a major concern of nineteenth century mathematicians, and the concepts they introduced are integral to modern mathematical physics. In the following sections we study some of the basic concepts that were successfully formulated in this period. The culmination of this work, Clifford's *geometric product*, is introduced separately in chapter 2. At various points in this book we will see how the products defined in this section can all be viewed as special cases of Clifford's geometric product.

1.2 The scalar product

Euclidean geometry deals with concepts such as lines, circles and perpendicularity. In order to arrive at Euclidean geometry we need to add two new concepts to our vector space. These are distances between points, which allow us to define a circle, and angles between vectors so that we can say that two lines are perpendicular. The introduction of a scalar product achieves both of these goals.

Given any two vectors a, b, the scalar product $a\cdot b$ is a rule for obtaining a number with the following properties:

(i) $a\cdot b = b\cdot a$;
(ii) $a\cdot(\lambda b) = \lambda(a\cdot b)$;
(iii) $a\cdot(b+c) = a\cdot b + a\cdot c$;
(iv) $a\cdot a > 0$, unless $a = 0$.

(When we study relativity, this final property will be relaxed.) The introduction of a scalar product allows us to define the length of a vector, $|a|$, by

$$|a| = \surd(a\cdot a). \tag{1.7}$$

Here, and throughout this book, the positive square root is always implied by the $\surd$ symbol. The fact that we now have a definition of lengths and distances means that we have specified a *metric space.* Many different types of metric space can be constructed, of which the simplest are the *Euclidean* spaces we have just defined.

The fact that for Euclidean space the inner product is positive-definite means that we have a Schwarz inequality of the form

$$|a\cdot b| \leq |a|\,|b|. \tag{1.8}$$

The proof is straightforward:

$$\begin{aligned} (a+\lambda b)\cdot(a+\lambda b) \geq 0 \qquad \forall\lambda \\ \Rightarrow a\cdot a + 2\lambda a\cdot b + \lambda^2 b\cdot b \geq 0 \qquad \forall\lambda \\ \Rightarrow (a\cdot b)^2 \leq a\cdot a\, b\cdot b, \end{aligned} \tag{1.9}$$

where the last step follows by taking the discriminant of the quadratic in λ. Since all of the numbers in this inequality are positive we recover (1.8). We can now define the *angle* θ between a and b by

$$a\cdot b = |a||b|\cos(\theta). \tag{1.10}$$

Two vectors whose scalar product is zero are said to be *orthogonal.* It is usually convenient to work with bases in which all of the vectors are mutually orthogonal. If all of the basis vectors are further normalised to have unit length, they are said to form an *orthonormal* basis. If the set of vectors $\{\mathsf{e}_1,\ldots,\mathsf{e}_n\}$ denote such a basis, the statement that the basis is orthonormal can be summarised as

$$\mathsf{e}_i\cdot\mathsf{e}_j = \delta_{ij}. \tag{1.11}$$

Here the δ_{ij} is the Kronecker delta function, defined by

$$\delta_{ij} = \begin{cases} 1 & \text{if } i = j, \\ 0 & \text{if } i \neq j. \end{cases} \tag{1.12}$$

We can expand any vector a in this basis as

$$a = \sum_{i=1}^{n} a_i \mathsf{e}_i = a_i \mathsf{e}_i, \tag{1.13}$$

where we have started to employ the *Einstein summation convention* that pairs of indices in any expression are summed over. This convention will be assumed throughout this book. The $\{a_i\}$ are the *components* of the vector a in the $\{\mathsf{e}_i\}$ basis. These are found simply by

$$a_i = \mathsf{e}_i \cdot a. \tag{1.14}$$

The scalar product of two vectors $a = a_i \mathsf{e}_i$ and $b = b_i \mathsf{e}_i$ can now written simply as

$$a \cdot b = (a_i \mathsf{e}_i) \cdot (b_j \mathsf{e}_j) = a_i b_j \, \mathsf{e}_i \cdot \mathsf{e}_j = a_i b_j \delta_{ij} = a_i b_i. \tag{1.15}$$

In spaces where the inner product is not positive-definite, such as Minkowski spacetime, there is no equivalent version of the Schwarz inequality. In such cases it is often only possible to define an 'angle' between vectors by replacing the cosine function with a cosh function. In these cases we can still introduce orthonormal frames and use these to compute scalar products. The main modification is that the Kronecker delta is replaced by η_{ij} which again is zero if $i \neq j$, but can take values ± 1 if $i = j$.

1.3 Complex numbers

The scalar product is the simplest product one can define between vectors, and once such a product is defined one can formulate many of the key concepts of Euclidean geometry. But this is by no means the only product that can be defined between vectors. In two dimensions a new product can be defined via complex arithmetic. A complex number can be viewed as an ordered pair of real numbers which represents a direction in the complex plane, as was realised by Wessel in 1797. Their product enables complex numbers to perform geometric operations, such as rotations and dilations. But suppose that we take the complex number $z = x + iy$ and square it, forming

$$z^2 = (x + iy)^2 = x^2 - y^2 + 2xyi. \tag{1.16}$$

In terms of vector arithmetic, neither the real nor imaginary parts of this expression have any geometric significance. A more geometrically useful product is defined instead by

$$zz^* = (x + iy)(x - iy) = x^2 + y^2, \tag{1.17}$$

which returns the square of the length of the vector. A product of two vectors in a plane, z and $w = u + vi$, can therefore be constructed as

$$zw^* = (x + iy)(u - iv) = xu + vy + i(uy - vx). \tag{1.18}$$

The real part of the right-hand side recovers the scalar product. To understand the imaginary term consider the polar representation

$$z = |z|\mathrm{e}^{i\theta}, \qquad w = |w|\mathrm{e}^{i\phi} \tag{1.19}$$

so that

$$zw^* = |z||w|\mathrm{e}^{i(\theta - \phi)}. \tag{1.20}$$

The imaginary term has magnitude $|z||w|\sin(\theta - \phi)$, where $\theta - \phi$ is the angle between the two vectors. The magnitude of this term is therefore the area of the parallelogram defined by z and w. The sign of the term conveys information about the *handedness* of the area element swept out by the two vectors. This will be defined more carefully in section 1.6.

We thus have a satisfactory interpretation for both the real and imaginary parts of the product zw^*. The surprising feature is that these are still both parts of a complex number. We thus have a second interpretation for complex addition, as a sum between scalar objects and objects representing plane segments. The advantages of adding these together are precisely the advantages of working with complex numbers as opposed to pairs of real numbers. This is a theme to which we shall return regularly in following chapters.

1.4 Quaternions

The fact that complex arithmetic can be viewed as representing a product for vectors in a plane carries with it a further advantage — it allows us to divide by a vector. Generalising this to three dimensions was a major preoccupation of the physicist W.R. Hamilton (see figure 1.2). Since a complex number $x + iy$ can be represented by two rectangular axes on a plane it seemed reasonable to represent directions in space by a triplet consisting of one real and two complex numbers. These can be written as $x+iy+jz$, where the third term jz represents a third axis perpendicular to the other two. The complex numbers i and j have the properties that $i^2 = j^2 = -1$. The norm for such a triplet would then be

$$(x + iy + jz)(x - iy - jz) = (x^2 + y^2 + z^2) - yz(ij + ji). \tag{1.21}$$

The final term is problematic, as one would like to recover the scalar product here. The obvious solution to this problem is to set $ij = -ji$ so that the last term vanishes.

Figure 1.2 *William Rowan Hamilton 1805–1865.* Inventor of quaternions, and one of the key scientific figures of the nineteenth century. He spent many years frustrated at being unable to extend his theory of couples of numbers (complex numbers) to three dimensions. In the autumn of 1843 he returned to this problem, quite possibly prompted by a visit he received from the young German mathematician Eisenberg. Among Eisenberg's papers was the observation that matrices form the elements of an algebra that was much like ordinary arithmetic except that multiplication was non-commutative. This was the vital step required to find the quaternion algebra. Hamilton arrived at this algebra on 16 October 1843 while out walking with his wife, and carved the equations in stone on Brougham Bridge. His discovery of quaternions is perhaps the best-documented mathematical discovery ever.

The anticommutative law $ij = -ji$ ensures that the norm of a triplet behaves sensibly, and also that multiplication of triplets in a plane behaves in a reasonable manner. The same is not true for the general product of triplets, however. Consider

$$\begin{aligned}(a + ib + jc)(x + iy + jz) = (ax - by - cz) + i(ay + bx)\\ + j(az + cx) + ij(bz - cy).\end{aligned} \tag{1.22}$$

Setting $ij = -ji$ is no longer sufficient to remove the ij term, so the algebra does not close. The only thing for Hamilton to do was to set $ij = k$, where k is some unknown, and see if it could be removed somehow. While walking along the Royal Canal he suddenly realised that if his triplets were instead made up of four terms he would be able to close the algebra in a simple, symmetric way.

To understand his discovery, consider

$$(a + ib + jc + kd)(a - ib - jc - kd)$$
$$= a^2 + b^2 + c^2 + d^2(-k^2) - bd(ik + ki) - cd(jk + kj), \quad (1.23)$$

where we have assumed that $i^2 = j^2 = -1$ and $ij = -ji$. The expected norm of the above product is $a^2 + b^2 + c^2 + d^2$, which is obtained by setting $k^2 = -1$ and $ik = -ki$ and $jk = -kj$. So what values do we use for jk and ik? These follow from the fact that $ij = k$, which gives

$$ik = i(ij) = (ii)j = -j \quad (1.24)$$

and

$$kj = (ij)j = -i. \quad (1.25)$$

Thus the multiplication rules for quaternions are

$$i^2 = j^2 = k^2 = -1 \quad (1.26)$$

and

$$ij = -ji = k, \quad jk = -kj = i, \quad ki = -ik = j. \quad (1.27)$$

These can be summarised neatly as $i^2 = j^2 = k^2 = ijk = -1$. It is a simple matter to check that these multiplication laws define a closed algebra.

Hamilton was so excited by his discovery that the very same day he obtained leave to present a paper on the quaternions to the Royal Irish Academy. The subsequent history of the quaternions is a fascinating story which has been described by many authors. Some suggested material for further reading is given at the end of this chapter. In brief, despite the many advantages of working with quaternions, their development was blighted by two major problems.

The first problem was the status of vectors in the algebra. Hamilton identified vectors with *pure quaternions*, which had a null scalar part. On the surface this seems fine — pure quaternions define a three-dimensional vector space. Indeed, Hamilton invented the word '*vector*' precisely for these objects and this is the origin of the now traditional use of i, j and k for a set of orthonormal basis vectors. Furthermore, the full product of two pure quaternions led to the definition of the extremely useful cross product (see section 1.5). The problem is that the product of two pure vectors does not return a new pure vector, so the vector part of the algebra does not close. This means that a number of ideas in complex analysis do not extend easily to three dimensions. Some people felt that this meant that the full quaternion product was of little use, and that the scalar and vector parts of the product should be kept separate. This criticism misses the point that the quaternion product is *invertible*, which does bring many advantages.

The second major difficulty encountered with quaternions was their use in describing rotations. The irony here is that quaternions offer the clearest way of handling rotations in three dimensions, once one realises that they provide a 'spin-1/2' representation of the rotation group. That is, if a is a vector (a pure quaternion) and R is a unit quaternion, a new vector is obtained by the *double-sided* transformation law

$$a' = RaR^*, \tag{1.28}$$

where the * operation reverses the sign of all three 'imaginary' components. A consequence of this is that each of the basis quaternions i, j and k generates rotations through π. Hamilton, however, was led astray by the analogy with complex numbers and tried to impose a single-sided transformation of the form $a' = Ra$. This works if the axis of rotation is perpendicular to a, but otherwise does not return a pure quaternion. More damagingly, it forces one to interpret the basis quaternions as generators of rotations through $\pi/2$, which is simply wrong!

Despite the problems with quaternions, it was clear to many that they were a useful mathematical system worthy of study. Tait claimed that quaternions 'freed the physicist from the constraints of coordinates and allowed thoughts to run in their most natural channels' — a theme we shall frequently meet in this book. Quaternions also found favour with the physicist James Clerk Maxwell, who employed them in his development of the theory of electromagnetism. Despite these successes, however, quaternions were weighed down by the increasingly dogmatic arguments over their interpretation and were eventually displaced by the hybrid system of vector algebra promoted by Gibbs.

1.5 The cross product

Two of the lasting legacies of the quaternion story are the introduction of the idea of a vector, and the cross product between two vectors. Suppose we form the product of two pure quaternions a and b, where

$$a = a_1 i + a_2 j + a_3 k, \qquad b = b_1 i + b_2 j + b_3 k. \tag{1.29}$$

Their product can be written

$$ab = -a_i b_i + c, \tag{1.30}$$

where c is the pure quaternion

$$c = (a_2 b_3 - a_3 b_2) i + (a_3 b_1 - a_1 b_3) j + (a_1 b_2 - a_2 b_1) k. \tag{1.31}$$

Writing $c = c_1 i + c_2 j + c_3 k$ the component relation can be written as

$$c_i = \epsilon_{ijk} a_j b_k, \tag{1.32}$$

where the alternating tensor ϵ_{ijk} is defined by

$$\epsilon_{ijk} = \begin{cases} 1 & \text{if } ijk \text{ is a cylic permutation of 123,} \\ -1 & \text{if } ijk \text{ is an anticylic permutation of 123,} \\ 0 & \text{otherwise.} \end{cases} \tag{1.33}$$

We recognise the preceding as defining the cross product of two vectors, $a \times b$. This has the following properties:

(i) $a \times b$ is perpendicular to the plane defined by a and b;
(ii) $a \times b$ has magnitude $|a||b|\sin(\theta)$;
(iii) the vectors a, b and $a \times b$ form a right-handed set.

These properties can alternatively be viewed as defining the cross product, and from them the algebraic definition can be recovered. This is achieved by starting with a right-handed orthonormal frame $\{\mathsf{e}_i\}$. For these we must have

$$\mathsf{e}_1 \times \mathsf{e}_2 = \mathsf{e}_3 \qquad \text{etc.} \tag{1.34}$$

so that we can write

$$\mathsf{e}_i \times \mathsf{e}_j = \epsilon_{ijk}\mathsf{e}_k. \tag{1.35}$$

Expanding out a vector in terms of this basis recovers the formula

$$\begin{aligned} a \times b &= (a_i\mathsf{e}_i) \times (b_j\mathsf{e}_j) \\ &= a_i b_j (\mathsf{e}_i \times \mathsf{e}_j) \\ &= (\epsilon_{ijk} a_i b_j)\mathsf{e}_k. \end{aligned} \tag{1.36}$$

Hence the geometric definition recovers the algebraic one.

The cross product quickly proved itself to be invaluable to physicists, dramatically simplifying equations in dynamics and electromagnetism. In the latter part of the nineteenth century many physicists, most notably Gibbs, advocated abandoning quaternions altogether and just working with the individual scalar and cross products. We shall see in later chapters that Gibbs was misguided in some of his objections to the quaternion product, but his considerable reputation carried the day and by the 1900s quaternions had all but disappeared from mainstream physics.

1.6 The outer product

The cross product has one major failing — it only exists in three dimensions. In two dimensions there is nowhere else to go, whereas in four dimensions the concept of a vector orthogonal to a pair of vectors is not unique. To see this, consider four orthonormal vectors $\mathsf{e}_1, \ldots, \mathsf{e}_4$. If we take the pair e_1 and e_2 and attempt

Figure 1.3 *Hermann Gunther Grassmann (1809–1877)*, born in Stettin, Germany (now Szczecin, Poland). A German mathematician and schoolteacher, Grassmann was the third of his parents' twelve children and was born into a family of scholars. His father studied theology and became a minister, before switching to teaching mathematics and physics at the Stettin Gymnasium. Hermann followed in his father's footsteps, first studying theology, classical languages and literature at Berlin. After returning to Stettin in 1830 he turned his attention to mathematics and physics. Grassmann passed the qualifying examination to win a teaching certificate in 1839. This exam included a written assignment on the tides, for which he gave a simplified treatment of Laplace's work based upon a new geometric calculus that he had developed. By 1840 he had decided to concentrate on mathematics research. He published the first edition of his geometric calculus, the 300 page *Lineale Ausdehnungslehre* in 1844, the same year that Hamilton announced the discovery of the quaternions. His work did not achieve the same impact as the quaternions, however, and it was many years before his ideas were understood and appreciated by other mathematicians. Disappointed by this lack of interest, Grassmann turned his attention to linguistics and comparative philology, with greater immediate impact. He was an expert in Sanskrit and translated the *Rig-Veda* (1876–1877). He also formulated the linguistic law (named after him) stating that in Indo-European bases, successive syllables may not begin with aspirates. He died before he could see his ideas on geometry being adopted into mainstream mathematics.

to find a vector perpendicular to both of these, we see that any combination of e_3 and e_4 will do.

A suitable generalisation of the idea of the cross product was constructed by

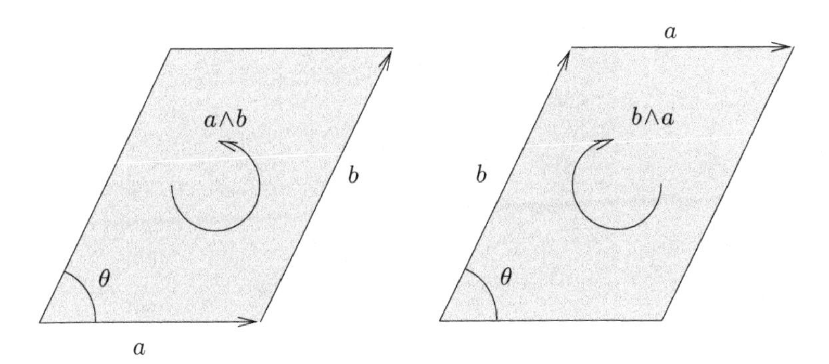

Figure 1.4 *The outer product.* The outer or wedge product of a and b returns a directed area element of area $|a||b|\sin(\theta)$. The orientation of the parallelogram is defined by whether the circuit a, b, $-a$, $-b$ is right-handed (anticlockwise) or left-handed (clockwise). Interchanging the order of the vectors reverses the orientation and introduces a minus sign in the product.

the remarkable German mathematician H.G. Grassmann (see figure 1.3). His work had its origin in the *Barycentrischer Calcul* of Möbius. There the author introduced expressions like AB for the line connecting the points A and B and ABC for the triangle defined by A, B and C. Möbius also introduced the crucial idea that the sign of the quantity should change if any two points are interchanged. (These *oriented* segments are now referred to as *simplices.*) It was Grassmann's leap of genius to realise that expressions like AB could actually be viewed as a product between vectors. He thus introduced the *outer* or *exterior product* which, in modern notation, we write as $a \wedge b$, or 'a wedge b'.

The outer product can be defined on any vector space and, geometrically, we are not forced to picture these vectors as displacements. Indeed, Grassmann was motivated by a *projective* viewpoint, where the elements of the vector space are interpreted as points, and the outer product of two points defines the line through the points. For our purposes, however, it is simplest to adopt a picture in which vectors represent directed line segments. The outer product then provides a means of encoding a plane, without relying on the notion of a vector perpendicular to it. The result of the outer product is therefore neither a scalar nor a vector. It is a new mathematical entity encoding an oriented plane and is called a *bivector.* It can be visualised as the parallelogram obtained by sweeping one vector along the other (figure 1.4). Changing the order of the vectors reverses the orientation of the plane. The magnitude of $a \wedge b$ is $|a||b|\sin(\theta)$, the same as the area of the plane segment swept out by the vectors.

The outer product of two vectors has the following algebraic properties:

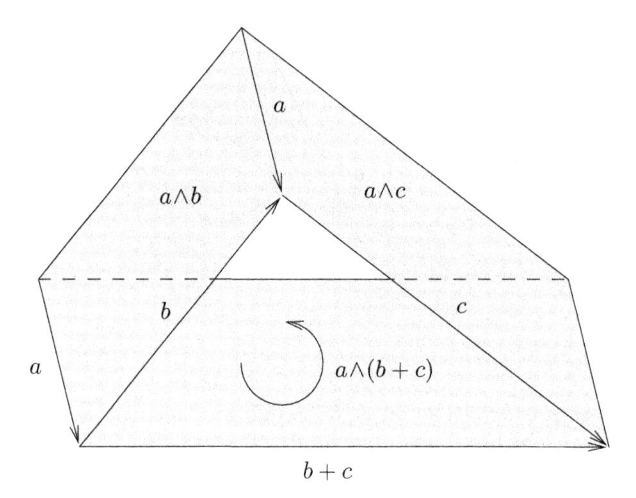

Figure 1.5 *A geometric picture of bivector addition.* In three dimensions any two non-parallel planes share a common line. If this line is denoted a, the two planes can be represented by $a \wedge b$ and $a \wedge c$. Bivector addition proceeds much like vector addition. The planes are combined at a common boundary and the resulting plane is defined by the initial and final edges, as opposed to the initial and final points for vector addition. The mathematical statement of this addition rule is the distributivity of the outer product over addition.

(i) The product is *antisymmetric*:

$$a\wedge b = -b\wedge a. \tag{1.37}$$

This has the geometric interpretation of reversing the orientation of the surface defined by a and b. It follows immediately that

$$a\wedge a = 0, \quad \text{for all vectors } a. \tag{1.38}$$

(ii) Bivectors form a linear space, the same way that vectors do. In two and three dimensions the addition of bivectors is easy to visualise. In higher dimensions this addition is not always so easy to visualise, because two planes need not share a common line.

(iii) The outer product is distributive over addition:

$$a\wedge(b + c) = a\wedge b + a\wedge c. \tag{1.39}$$

This helps to visualise the addition of bivectors which share a common line (see figure 1.5).

While it is convenient to visualise the outer product as a parallelogram, the actual shape of the object is not conveyed by the result of the product. This can be seen easily by defining $a' = a + \lambda b$ and forming

$$a' \wedge b = a \wedge b + \lambda b \wedge b = a \wedge b. \tag{1.40}$$

The same bivector can therefore be generated by many different pairs of vectors. In many ways it is better to replace the picture of a directed parallelogram with that of a directed circle. The circle defines both the plane and a handedness, and its area is equal to the magnitude of the bivector. This therefore conveys all of the information one has about the bivector, though it does make bivector addition harder to visualise.

1.6.1 Two dimensions

The outer product of any two vectors defines a plane, so one has to go to at least two dimensions to form an interesting product. Suppose then that $\{e_1, e_2\}$ are an orthonormal basis for the plane, and introduce the vectors

$$a = a_1 e_1 + a_2 e_2, \quad b = b_1 e_1 + b_2 e_2. \tag{1.41}$$

The outer product $a \wedge b$ contains

$$\begin{aligned} a \wedge b &= a_1 b_1 e_1 \wedge e_1 + a_1 b_2 e_1 \wedge e_2 + a_2 b_1 e_2 \wedge e_1 + a_2 b_2 e_2 \wedge e_2 \\ &= (a_1 b_2 - a_2 b_1) e_1 \wedge e_2, \end{aligned} \tag{1.42}$$

which recovers the imaginary part of the product of (1.18). The term therefore immediately has the expected magnitude $|a|\,|b| \sin(\theta)$. The coefficient of $e_1 \wedge e_2$ is positive if a and b have the same orientation as e_1 and e_2. The orientation is defined by traversing the boundary of the parallelogram defined by the vectors a, b, $-a$, $-b$ (see figure 1.4). By convention, we usually work with a right-handed set of reference axes (viewed from above). In this case the coefficient $a_1 b_2 - a_2 b_1$ will be positive if a and b also form a right-handed pair.

1.6.2 Three dimensions

In three dimensions the space of bivectors is also three-dimensional, because each bivector can be placed in a one-to-one correspondence with the vector perpendicular to it. Suppose that $\{e_1, e_2, e_3\}$ form a right-handed basis (see comments below), and the two vectors a and b are expanded in this basis as $a = a_i e_i$ and $b = b_i e_i$. The bivector $a \wedge b$ can then be decomposed in terms of an orthonormal frame of bivectors by

$$\begin{aligned} a \wedge b &= (a_i e_i) \wedge (b_j e_j) \\ &= (a_2 b_3 - b_3 a_2) e_2 \wedge e_3 + (a_3 b_1 - a_1 b_3) e_3 \wedge e_1 \\ &\quad + (a_1 b_2 - a_2 b_1) e_1 \wedge e_2. \end{aligned} \tag{1.43}$$

The components in this frame are therefore the same as those of the cross product. But instead of being the components of a vector perpendicular to a and b, they are the components of the bivector $a \wedge b$. It is this distinction which enables the outer product to be defined in any dimension.

1.6.3 Handedness

We have started to employ the idea of *handedness* without giving a satisfactory definition of it. The only space in which there is an unambiguous definition of handedness is three dimensions, as this is the space we inhabit and most of us can distinguish our left and right hands. This concept of 'left' and 'right' is a man-made convention adopted to make our life easier, and it extends to the concept of a frame in a straightforward way. Suppose that we are presented with three orthogonal vectors $\{e_1, e_2, e_3\}$. We align the 3 axis with the thumb of our right hand and then close our fist. If the direction in which our fist closes is the same as that formed by rotating from the 1 to the 2 axis, the frame is right-handed. If not, it is left-handed.

Swapping any pair of vectors swaps the handedness of a frame. Performing two such swaps returns us to the original handedness. In three dimensions this corresponds to a cyclic reordering, and ensures that the frames $\{e_1, e_2, e_3\}$, $\{e_3, e_1, e_2\}$ and $\{e_2, e_3, e_1\}$ all have the same orientation.

There is no agreed definition of a 'right-handed' orientation in spaces of dimensions other than three. All one can do is to make sure that any convention used is adopted consistently. In all dimensions the orientation of a set of vectors is changed if any two vectors are swapped. In two dimensions one does still tend to talk about right-handed axes, though the definition is dependent on the idea of looking down on the plane *from above*. The idea of above and below is not a feature of the plane itself, but depends on how we embed it in our three-dimensional world. There is no definition of left or right-handed which is intrinsic to the plane.

1.6.4 Extending the outer product

The preceding examples demonstrate that in arbitrary dimensions the components of $a \wedge b$ are given by

$$(a \wedge b)_{ij} = a_{[i}b_{j]} \tag{1.44}$$

where the $[\,]$ denotes antisymmetrisation. Grassmann was able to take this idea further by defining an outer product for any number of vectors. The idea is a simple extension of the preceding formula. Expressed in an orthonormal frame, the components of the outer product on n vectors are the totally antisymmetrised products of the components of each vector. This definition has the useful property that the outer product is *associative*,

$$a \wedge (b \wedge c) = (a \wedge b) \wedge c. \tag{1.45}$$

For example, in three dimensions we have

$$a \wedge b \wedge c = (a_i \mathrm{e}_i) \wedge (b_j \mathrm{e}_j) \wedge (c_k \mathrm{e}_k) = \epsilon_{ijk} a_i b_j c_k \mathrm{e}_1 \wedge \mathrm{e}_2 \wedge \mathrm{e}_3, \tag{1.46}$$

which represents a *directed volume* (see section 2.4).

A further feature of the antisymmetry of the product is that the outer product of any set of linearly dependent vectors vanishes. This means that statements like 'this vector lies on a given plane', or 'these two hypersurfaces share a common line' can be encoded algebraically in a simple manner. Equipped with these ideas, Grassmann was able to construct a system capable of handling geometric concepts in arbitrary dimensions.

Despite Grassmann's considerable achievement, the book describing his ideas, his *Lineale Ausdehnungslehre*, did not have any immediate impact. This was no doubt due largely to his relative lack of reputation (he was still a German schoolteacher when he wrote this work). It was over twenty years before anyone of note referred to Grassmann's work, and during this time Grassmann produced a second, extended version of the *Ausdehnungslehre*. In the latter part of the nineteenth century Grassmann's work started to influence leading figures like Gibbs and Clifford. Gibbs wrote a number of papers praising Grassmann's work and contrasting it favourably with the quaternion algebra. Clifford used Grassmann's work as the starting point for the development of his geometric algebra, the subject of this book.

Today, Grassmann's ideas are recognised as the first presentation of the abstract theory of vector spaces over the field of real numbers. Since his death, his work has given rise to the influential and fashionable areas of *differential forms* and *Grassmann variables*. The latter are anticommuting variables and are fundamental to the foundations of much of modern supersymmetry and superstring theory.

1.7 Notes

Descriptions of linear algebra and vector spaces can be found in most introductory textbooks of mathematics, as can discussions of the scalar and cross products and complex arithmetic. Quaternions, on the other hand, are much less likely to be mentioned. There is a large specialised literature on the quaternions, and a good starting point are the works of Altmann (1986, 1989). Altmann's paper on 'Hamilton, Rodriques and the quaternion scandal' (1989) is also a good introduction to the history of the subject.

The outer product is covered in most modern textbooks on geometry and physics, such as those by Nakahara (1990), Schutz (1980), and Gockeler & Schucker (1987). In most of these works, however, the exterior product is only treated in the context of differential forms. Applications to wider topics in geometry have been discussed by Hestenes (1991) and others. A useful summary in provided in the proceedings of the conference *Hermann Gunther Grassmann (1809–1877)*, edited by Schubring (1996). Grassmann's *Lineale Ausdehnungslehre* is also finally available in English translation due to Kannenberg (1995).

For those with a deeper interest in the history of mathematics and the development of vector algebra a good starting point is the set of books by Kline (1972). There are also biographies available of many of the key protagonists. Perhaps even more interesting is to return to their original papers and experience first hand the robust and often humorous language employed at the time. The collected works of J.W. Gibbs (1906) are particularly entertaining and enlightening, and contain a good deal of valuable historical information.

1.8 Exercises

1.1 Suppose that the two sets $\{a_1, \ldots, a_m\}$ and $\{b_1, \ldots, b_n\}$ form bases for the same vector space, and suppose initially that $m > n$. By establishing a contradiction, prove the *basis theorem* that all bases of a vector space have the same number of elements.

1.2 Demonstrate that the following define vector spaces:

   (a) the set of all polynomials of degree less than or equal to n;
   (b) all solutions of a given linear homogeneous ordinary differential equation;
   (c) the set of all $n \times m$ matrices.

1.3 Prove that in Euclidean space $|a + b| \leq |a| + |b|$. When does equality hold?

1.4 Show that the unit quaternions $\{\pm 1, \pm i, \pm j \pm k\}$ form a discrete group.

1.5 The unit quaternions i, j, k are generators of rotations about their respective axes. Are rotations through either π or $\pi/2$ consistent with the equation $ijk = -1$?

1.6 Prove the following:

   (a) $a\cdot(b\times c) = b\cdot(c\times a) = c\cdot(a\times b)$;
   (b) $a\times(b\times c) = a\cdot c\, b - a\cdot b\, c$;
   (c) $|a\times b| = |a|\,|b| \sin(\theta)$, where $a\cdot b = |a|\,|b| \cos(\theta)$.

1.7 Prove that the dimension of the space formed by the exterior product of m vectors drawn from a space of dimension n is

$$\frac{n(n-1)\cdots(n-m+1)}{1\cdot 2\cdots m} = \frac{n!}{(n-m)!m!}.$$

1.8 Prove that the n-fold exterior product of a set of n *dependent* vectors is zero.

1.9 A convex polygon in a plane is specified by the ordered set of points $\{x_0, x_1, \ldots, x_n\}$. Prove that the directed area of the polygon is given by

$$A = \tfrac{1}{2}(x_0 \wedge x_1 + x_1 \wedge x_2 + \cdots + x_n \wedge x_0).$$

What is the significance of the sign? Can you extend the idea to a triangulated surface in three dimensions?

2

Geometric algebra in two and three dimensions

Geometric algebra was introduced in the nineteenth century by the English mathematician William Kingdon Clifford (figure 2.1). Clifford appears to have been one of the small number of mathematicians at the time to be significantly influenced by Grassmann's work. Clifford introduced his *geometric algebra* by uniting the inner and outer products into a single *geometric* product. This is associative, like Grassmann's product, but has the crucial extra feature of being *invertible*, like Hamilton's quaternion algebra. Indeed, Clifford's original motivation was to unite Grassmann's and Hamilton's work into a single structure. In the mathematical literature one often sees this subject referred to as *Clifford algebra*. We have chosen to follow the example of David Hestenes, and many other modern researchers, by returning to Clifford's original choice of name — *geometric algebra*. One reason for this is that the first published definition of the geometric product was due to Grassmann, who introduced it in the second *Ausdehnungslehre*. It was Clifford, however, who realised the great potential of this product and who was responsible for advancing the subject.

In this chapter we introduce the basics of geometric algebra in two and three dimensions in a way that is intended to appear natural and geometric, if somewhat informal. A more formal, axiomatic approach is delayed until chapter 4, where geometric algebra is defined in arbitrary dimensions. The meaning of the various terms in the algebra we define will be illustrated with familiar examples from geometry. In so doing we will also uncover how Hamilton's quaternions fit into geometric algebra, and understand where it was that Hamilton and his followers went wrong in their treatment of three-dimensional geometry. One of the most powerful applications of geometric algebra is to rotations, and these are considered in some detail in this chapter. It is well known that rotations in a plane can be efficiently handled with complex numbers. We will see how to extend this idea to rotations in three-dimensional space. This representation has many applications in classical and quantum physics.

Figure 2.1 *William Kingdon Clifford 1845–1879.* Born in Exeter on 4 May 1845, his father was a justice of the peace and his mother died early in his life. After school he went to King's College, London and then obtained a scholarship to Trinity College, Cambridge, where he followed the likes of Thomson and Maxwell in becoming Second Wrangler. There he also achieved a reputation as a daring athlete, despite his slight frame. He was recommended for a fellowship at Trinity College by Maxwell, and in 1871 took the Professorship of Applied Mathematics at University College, London. He was made a Fellow of the Royal Society at the extremely young age of 29. He married Lucy in 1875, and their house became a fashionable meeting place for scientists and philosophers. As well as being one of the foremost mathematicians of his day, he was an accomplished linguist, philosopher and author of children's stories. Sadly, his insatiable appetite for physical and mental exercise was not matched by his physique, and in 1878 he was instructed to stop work and leave England for the Mediterranean. He returned briefly, only for his health to deteriorate further in the English climate. He left for Madeira, where he died on 3 March 1879 at the age of just 33. Further details of his life can be found in the book *Such Silver Currents* (Chisholm, 2002). Portrait by John Collier (©The Royal Society).

2.1 A new product for vectors

In chapter 1 we studied various products for vectors, including the symmetric scalar (or inner) product and the antisymmetric exterior (or outer) product. In two dimensions, we showed how to interpret the result of the complex product zw^* (section 1.3). The scalar term is the inner product of the two vectors representing the points in the complex plane, and the imaginary term records their directed area. Furthermore, the scalar term is symmetric, and the imaginary term is antisymmetric in the two arguments. Clifford's powerful idea was to generalise this product to arbitrary dimensions by replacing the imaginary term with the outer product. The result is the *geometric product* and is written simply as ab. The result is the sum of a scalar and a bivector, so

$$ab = a\cdot b + a\wedge b. \tag{2.1}$$

This sum of two distinct objects — a scalar and a bivector — looks strange at first and goes against the rule that one should only add like objects. This is the feature of geometric algebra that initially causes the greatest difficulty, in much the same way that $i^2 = -1$ initially unsettles most school children. So how is the sum on the right-hand side of equation (2.1) to be viewed? The answer is that it should be viewed in precisely the same way as the addition of a real and an imaginary number. The result is neither purely real nor purely imaginary — it is a mixture of two different objects which are combined to form a single complex number. Similarly, the addition of a scalar to a bivector enables us to keep track of the separate components of the product ab. The advantages of this are precisely the same as the advantages of complex arithmetic over working with the separate real and imaginary parts. This analogy between *multivectors* in geometric algebra and complex numbers is more than a mere pedagogical device. As we shall discover, geometric algebra encompasses both complex numbers and quaternions. Indeed, Clifford's achievement was to generalise complex arithmetic to spaces of arbitrary dimensions.

From the symmetry and antisymmetry of the terms on the right-hand side of equation (2.1) we see that

$$ba = b\cdot a + b\wedge a = a\cdot b - a\wedge b. \tag{2.2}$$

It follows that

$$a\cdot b = \tfrac{1}{2}(ab + ba) \tag{2.3}$$

and

$$a\wedge b = \tfrac{1}{2}(ab - ba). \tag{2.4}$$

We can thus define the inner and outer products in terms of the geometric product. This forms the starting point for an axiomatic development of geometric algebra, which is presented in chapter 4.

If we form the product of a and the parallel vector λa we obtain

$$a(\lambda a) = \lambda a\cdot a + \lambda a\wedge a = \lambda a\cdot a, \tag{2.5}$$

which is therefore a pure scalar. It follows similarly that a^2 is a scalar, so we can write $a^2 = |a|^2$ for the square of the length of a vector. If instead a and b

are perpendicular vectors, their product is

$$ab = a\cdot b + a\wedge b = a\wedge b \tag{2.6}$$

and so is a pure bivector. We also see that

$$ba = b\cdot a + b\wedge a = -a\wedge b = -ab, \tag{2.7}$$

which shows us that *orthogonal vectors anticommute.* The geometric product between general vectors encodes the relative contributions of both their parallel and perpendicular components, summarising these in the separate scalar and bivector terms.

2.2 An outline of geometric algebra

Clifford went further than just allowing scalars to be added to bivectors. He defined an algebra in which elements of any type could be added or multiplied together. This is what he called a *geometric algebra.* Elements of a geometric algebra are called *multivectors* and these form a linear space — scalars can be added to bivectors, and vectors, etc. Geometric algebra is a *graded* algebra, and elements of the algebra can be broken up into terms of different *grade.* The scalar objects are assigned grade-0, the vectors grade-1, the bivectors grade-2 and so on. Essentially, the grade of the object is the dimension of the hyperplane it specifies. The term 'grade' is preferred to 'dimension', however, as the latter is regularly employed for the size of a linear space. We denote the operation of projecting onto the terms of a chosen grade by $\langle\ \rangle_r$, so $\langle ab\rangle_2$ denotes the grade-2 (bivector) part of the geometric product ab. That is,

$$\langle ab\rangle_2 = a\wedge b. \tag{2.8}$$

The subscript 0 on the scalar term is usually suppressed, so we also have

$$\langle ab\rangle_0 = \langle ab\rangle = a\cdot b. \tag{2.9}$$

Arbitrary multivectors can also be multiplied together with the geometric product. To do this we first extend the geometric product of two vectors to an arbitrary number of vectors. This is achieved with the additional rule that the geometric product is *associative*:

$$a(bc) = (ab)c = abc. \tag{2.10}$$

The associativity property enables us to remove the brackets and write the product as abc. Arbitrary multivectors can now be written as sums of products of vectors. The geometric product of multivectors therefore inherits the two main properties of the product for vectors, which is to say it is associative:

$$A(BC) = (AB)C = ABC, \tag{2.11}$$

and distributive over addition:

$$A(B+C) = AB + AC. \tag{2.12}$$

Here $A, B, \ldots, C$ denote multivectors containing terms of arbitrary grade.

The associativity property ensures that it is now possible to divide by vectors, thus realising Hamilton's goal. Suppose that we know that $ab = C$, where C is some combination of a scalar and bivector. We find that

$$Cb = (ab)b = a(bb) = ab^2, \tag{2.13}$$

so we can define $b^{-1} = b/b^2$, and recover a from

$$a = Cb^{-1}. \tag{2.14}$$

This ability to divide by vectors gives the algebra considerable power.

As an example of these axioms in action, consider forming the square of the bivector $a\wedge b$. The properties of the geometric product allow us to write

$$\begin{aligned}(a\wedge b)(a\wedge b) &= (ab - a\cdot b)(a\cdot b - ba) \\ &= -ab^2a - (a\cdot b)^2 + a\cdot b(ab+ba) \\ &= (a\cdot b)^2 - a^2b^2 \\ &= -a^2b^2\sin^2(\theta),\end{aligned} \tag{2.15}$$

where we have assumed that $a\cdot b = |a|\,|b|\,\cos(\theta)$. The magnitude of the bivector $a\wedge b$ is therefore equal to the area of the parallelogram with sides defined by a and b. Manipulations such as these are commonplace in geometric algebra, and can provide simplified proofs of a number of useful results.

2.3 Geometric algebra of the plane

The easiest way to understand the geometric product is by example, so consider a two-dimensional space (a plane) spanned by two orthonormal vectors e_1 and e_2. These basis vectors satisfy

$$\mathsf{e}_1{}^2 = \mathsf{e}_2{}^2 = 1, \qquad \mathsf{e}_1\cdot\mathsf{e}_2 = 0. \tag{2.16}$$

The final entity present in the algebra is the bivector $\mathsf{e}_1\wedge\mathsf{e}_2$. This is the highest grade element in the algebra, since the outer product of a set of dependent vectors is always zero. The highest grade element in a given algebra is usually called the *pseudoscalar*, and its grade coincides with the dimension of the underlying vector space.

The full algebra is spanned by the basis set

$$\begin{matrix} 1 & \{\mathsf{e}_1, \mathsf{e}_2\} & \mathsf{e}_1\wedge\mathsf{e}_2 \\ \text{1 scalar} & \text{2 vectors} & \text{1 bivector} \end{matrix}. \tag{2.17}$$

We denote this algebra $\mathcal{G}_2$. Any multivector can be decomposed in this basis, and sums and products can be calculated in terms of this basis. For example, suppose that the multivectors A and B are given by

$$A = \alpha_0 + \alpha_1 e_1 + \alpha_2 e_2 + \alpha_3 e_1 \wedge e_2,$$
$$B = \beta_0 + \beta_1 e_1 + \beta_2 e_2 + \beta_3 e_1 \wedge e_2,$$

then their sum $S = A + B$ is given by

$$S = (\alpha_0 + \beta_0) + (\alpha_1 + \beta_1) e_1 + (\alpha_2 + \beta_2) e_2 + (\alpha_3 + \beta_3) e_1 \wedge e_2. \tag{2.18}$$

This result for the addition of multivectors is straightforward and unsurprising. Matters become more interesting, however, when we start forming products.

2.3.1 The bivector and its products

To study the properties of the bivector $e_1 \wedge e_2$ we first recall that for orthogonal vectors the geometric product is a pure bivector:

$$e_1 e_2 = e_1 \cdot e_2 + e_1 \wedge e_2 = e_1 \wedge e_2, \tag{2.19}$$

and that orthogonal vectors anticommute:

$$e_2 e_1 = e_2 \wedge e_1 = -e_1 \wedge e_2 = -e_1 e_2. \tag{2.20}$$

We can now form products in which $e_1 e_2$ multiplies vectors from the left and the right. First from the left we find that

$$(e_1 \wedge e_2) e_1 = (-e_2 e_1) e_1 = -e_2 e_1 e_1 = -e_2 \tag{2.21}$$

and

$$(e_1 \wedge e_2) e_2 = (e_1 e_2) e_2 = e_1 e_2 e_2 = e_1. \tag{2.22}$$

If we assume that e_1 and e_2 form a right-handed pair, we see that left-multiplication by the bivector rotates vectors 90° clockwise (i.e. in a negative sense). Similarly, acting from the right

$$e_1(e_1 e_2) = e_2, \qquad e_2(e_1 e_2) = -e_1. \tag{2.23}$$

So right multiplication rotates 90° anticlockwise — a positive sense.

The final product in the algebra to consider is the square of the bivector $e_1 \wedge e_2$:

$$(e_1 \wedge e_2)^2 = e_1 e_2 e_1 e_2 = -e_1 e_1 e_2 e_2 = -1. \tag{2.24}$$

Geometric considerations have led naturally to a quantity which squares to -1. This fits with the fact that two successive left (or right) multiplications of a vector by $e_1 e_2$ rotates the vector through 180°, which is equivalent to multiplying by -1. The fact that we now have a firm geometric picture for objects whose algebraic square is -1 opens up the possibility of providing a geometric interpretation for the unit imaginary employed throughout physics, a theme which will be explored further in this book.

2.3.2 Multiplying multivectors

Now that all of the individual products have been found, we can compute the product of the two general multivectors A and B of equation (2.18),

$$AB = M = \mu_0 + \mu_1 \mathrm{e}_1 + \mu_2 \mathrm{e}_2 + \mu_3 \mathrm{e}_1 \mathrm{e}_2, \tag{2.25}$$

where

$$\begin{aligned} \mu_0 &= \alpha_0\beta_0 + \alpha_1\beta_1 + \alpha_2\beta_2 - \alpha_3\beta_3, \\ \mu_1 &= \alpha_0\beta_1 + \alpha_1\beta_0 + \alpha_3\beta_2 - \alpha_2\beta_3, \\ \mu_2 &= \alpha_0\beta_2 + \alpha_2\beta_0 + \alpha_1\beta_3 - \alpha_3\beta_1, \\ \mu_3 &= \alpha_0\beta_3 + \alpha_3\beta_0 + \alpha_1\beta_2 - \alpha_2\beta_1. \end{aligned} \tag{2.26}$$

The full product shown here is actually rarely used, but writing it out explicitly does emphasise some of its key features. The product is always well defined, and the algebra is closed under it. Indeed, the product could easily be made an intrinsic part of a computer language, in the same way that complex arithmetic is already intrinsic to some languages. The basis vectors can also be represented with matrices, for example

$$\boldsymbol{E}_1 = \begin{pmatrix} 0 & 1 \\ 1 & 0 \end{pmatrix} \qquad \boldsymbol{E}_2 = \begin{pmatrix} 1 & 0 \\ 0 & -1 \end{pmatrix}. \tag{2.27}$$

(Verifying that these satisfy the required algebraic relations is left as an exercise.) Geometric algebras in general are associative algebras, so it is always possible to construct a matrix representation for them. The problem with this is that the matrices hide the geometric content of the elements they represent. Much of the mathematical literature does focus on matrix representations, and for this work the term *Clifford algebra* is appropriate. For the applications in this book, however, the underlying geometry is the important feature of the algebra and matrix representations are usually redundant. *Geometric algebra* is a much more appropriate name for this subject.

2.3.3 Connection with complex numbers

It is clear that there is a close relationship between geometric algebra in two dimensions and the algebra of complex numbers. The unit bivector squares to -1 and generates rotations through $90°$. The combination of a scalar and a bivector, which is formed naturally via the geometric product, can therefore be viewed as a complex number. We write this as

$$Z = u + v\mathrm{e}_1\mathrm{e}_2 = u + Iv, \tag{2.28}$$

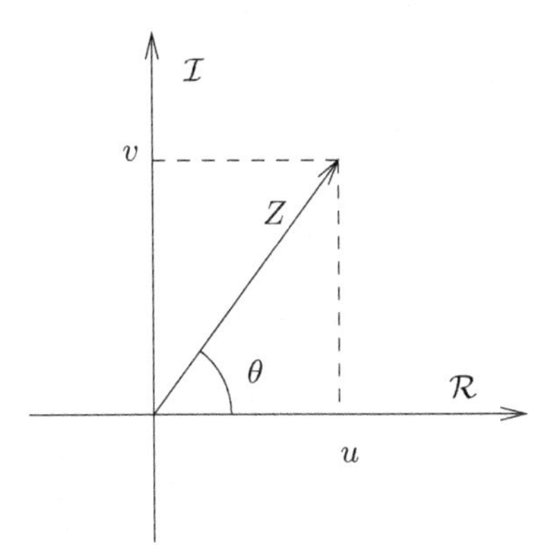

Figure 2.2 *The Argand diagram.* The complex number $Z = u + iv$ represents a vector in the complex plane, with Cartesian components u and v. The polar decomposition into $|Z|\exp(i\theta)$ can alternatively be viewed as an instruction to rotate 1 through θ and dilate by $|Z|$.

where

$$I = \mathsf{e}_1 \wedge \mathsf{e}_2, \qquad I^2 = -1. \tag{2.29}$$

Throughout we employ the symbol I for the pseudoscalar of the algebra of interest. That is why we have used it here, rather than the tempting alternative i. The latter is seen often in the literature, but the i symbol has the problem of suggesting an element which commutes with all others, which is not necessarily a property of the pseudoscalar.

Complex numbers serve a dual purpose in two dimensions. They generate rotations and dilations through their polar decomposition $|Z|\exp(i\theta)$, and they also represent vectors as points on the Argand diagram (see figure 2.2). But in the geometric algebra $\mathcal{G}_2$ complex numbers are replaced by scalar + bivector combinations, whereas vectors are grade-1 objects,

$$x = u\mathsf{e}_1 + v\mathsf{e}_2. \tag{2.30}$$

Is there a natural map between x and the multivector Z? The answer is simple — pre-multiply by e_1,

$$\mathsf{e}_1 x = u + v\mathsf{e}_1\mathsf{e}_2 = u + Iv = Z. \tag{2.31}$$

That is all there is to it! The role of the preferred vector e_1 is clear — it is the real axis. Using this product vectors in a plane can be interchanged with complex numbers in a natural manner.

If we now consider the complex conjugate of Z, $Z^\dagger = u - iv$, we see that

$$Z^\dagger = u + v\mathsf{e}_2\mathsf{e}_1 = x\mathsf{e}_1, \tag{2.32}$$

which has simply reversed the order of the geometric product of x and e_1. This operation of reversing the order of products is one of the fundamental operations performed in geometric algebra, and is called *reversion* (see section 2.5). Suppose now that we introduce a second complex number W, with vector equivalent y:

$$W = e_1 y. \tag{2.33}$$

The complex product $ZW^\dagger = W^\dagger Z$ now becomes

$$W^\dagger Z = y e_1 e_1 x = yx, \tag{2.34}$$

which returns the geometric product yx. This is as expected, as the complex product was used to suggest the form of the geometric product.

2.3.4 Rotations

Since we know how to rotate complex numbers, we can use this to find a formula for rotating vectors in a plane. We know that a positive rotation through an angle ϕ for a complex number Z is achieved by

$$Z \mapsto Z' = e^{i\phi} Z, \tag{2.35}$$

where i is the standard unit imaginary (see figure 2.3). Again, we now view Z as a combination of a scalar and a pseudoscalar in $\mathcal{G}_2$ and so replace i with I. The exponential of $I\phi$ is defined by power series in the normal way, so we still have

$$e^{I\phi} = \sum_{n=0}^{\infty} \frac{(I\phi)^n}{n!} = \cos\phi + I\,\sin\phi. \tag{2.36}$$

Suppose that Z' has the vector equivalent x',

$$x' = e_1 Z'. \tag{2.37}$$

We now have a means of rotating the vector directly by writing

$$x' = e_1 e^{I\phi} Z = e_1 e^{I\phi} e_1 x. \tag{2.38}$$

But

$$\begin{aligned} e_1 e^{I\phi} e_1 &= e_1(\cos\phi + I\,\sin\phi) e_1 \\ &= \cos\phi - I\,\sin\phi = e^{-I\phi}, \end{aligned} \tag{2.39}$$

where we have employed the result that I *anticommutes* with vectors. We therefore arrive at the formulae

$$x' = e^{-I\phi} x = x e^{I\phi}, \tag{2.40}$$

which achieve a rotation of the vector x in the I plane, through an angle ϕ. In section 2.7 we show how to extend this idea to arbitrary dimensions. The

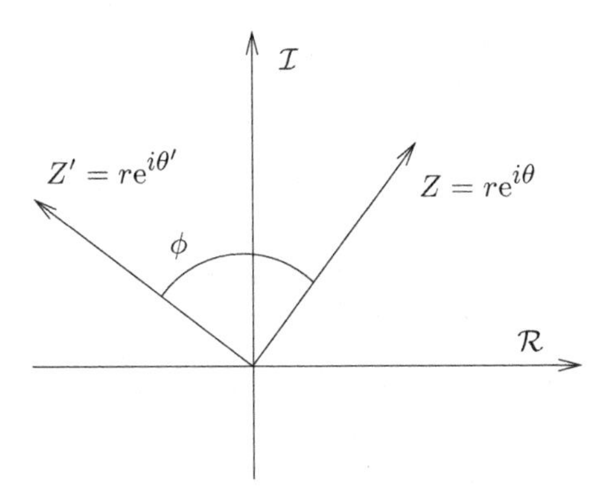

Figure 2.3 *A rotation in the complex plane.* The complex number Z is multiplied by the phase term $\exp(I\phi)$, the effect of which is to replace θ by $\theta' = \theta + \phi$.

change of sign in the exponential acting from the left and right of the vector x is to be expected. We saw earlier that left-multiplication by I generated left-handed rotations, and right-multiplication generated right-handed rotations. As the overall rotation is right-handed, the sign of I must be negative when acting from the left.

This should illustrate that geometric algebra fully encompasses complex arithmetic, and we will see later that complex analysis is fully incorporated as well. The beauty of the geometric algebra formulation is that it shows immediately how to extend the ideas of complex analysis to higher dimensions, a problem which had troubled mathematicians for many years. The key to this is the separation of the two roles of complex numbers by treating vectors as grade-1 objects, and the quantities acting on them (the complex numbers) as combinations of grade-0 and grade-2 objects. These two roles generalise differently in higher dimensions and, once one sees this, extending complex analysis becomes straightforward.

2.4 The geometric algebra of space

The geometric algebra of three-dimensional space is a remarkably powerful tool for solving problems in geometry and classical mechanics. It describes vectors, planes and volumes in a single algebra, which contains all of the familiar vector operations. These include the vector cross product, which is revealed as a disguised form of bivector. The algebra also provides a very clear and compact method for encoding rotations, which is considerably more powerful than working with matrices.

We have so far constructed the geometric algebra of a plane. We now add a third vector e_3 to our two-dimensional set $\{\mathsf{e}_1, \mathsf{e}_2\}$. All three vectors are assumed to be orthonormal, so they all *anticommute*. From these three basis vectors we generate the independent bivectors

$$\{\mathsf{e}_1\mathsf{e}_2, \mathsf{e}_2\mathsf{e}_3, \mathsf{e}_3\mathsf{e}_1\}.$$

This is the expected number of independent planes in space. There is one further term to consider, which is the product of all three vectors:

$$(\mathsf{e}_1\mathsf{e}_2)\mathsf{e}_3 = \mathsf{e}_1\mathsf{e}_2\mathsf{e}_3. \tag{2.41}$$

This results in a grade-3 object, called a *trivector*. It corresponds to sweeping the bivector $\mathsf{e}_1 \wedge \mathsf{e}_2$ along the vector e_3, resulting in a three-dimensional volume element (see section 2.4.3). The trivector represents the unique volume element in three dimensions. It is the highest grade element and is unique up to scale (or volume) and handedness (sign). This is again called the *pseudoscalar* for the algebra.

In three dimensions there are no further directions to add, so the algebra is spanned by

$$\begin{array}{cccc} 1 & \{\mathsf{e}_i\} & \{\mathsf{e}_i \wedge \mathsf{e}_j\} & \mathsf{e}_1\mathsf{e}_2\mathsf{e}_3 \\ \text{1 scalar} & \text{3 vectors} & \text{3 bivectors} & \text{1 trivector} \end{array} \tag{2.42}$$

This basis defines a graded linear space of total dimension $8 = 2^3$. We call this algebra $\mathcal{G}_3$. Notice that the dimensions of each subspace are given by the binomial coefficients.

2.4.1 Products of vectors and bivectors

Our expanded algebra gives us a number of new products to consider. We start by considering the product of a vector and a bivector. We have already looked at this in two dimensions, and found that a normalised bivector rotates vectors in its plane by 90°. Each of the basis bivectors in equation (2.42) shares the properties of the single bivector studied previously for two dimensions. So

$$(\mathsf{e}_1\mathsf{e}_2)^2 = (\mathsf{e}_2\mathsf{e}_3)^2 = (\mathsf{e}_3\mathsf{e}_1)^2 = -1 \tag{2.43}$$

and each bivector generates 90° rotations in its own plane.

The geometric product for vectors extends to all objects in the algebra, so we can form expressions such as aB, where a is a vector and B is a bivector. Now that our algebra contains a trivector $\mathsf{e}_1(\mathsf{e}_2 \wedge \mathsf{e}_3)$, we see that the result of the product aB can contain both vector and trivector terms, the latter arising if a does not lie fully in the B plane. To understand the properties of the product aB we first decompose a into terms in and out of the plane,

$$a = a_{\parallel} + a_{\perp}, \tag{2.44}$$

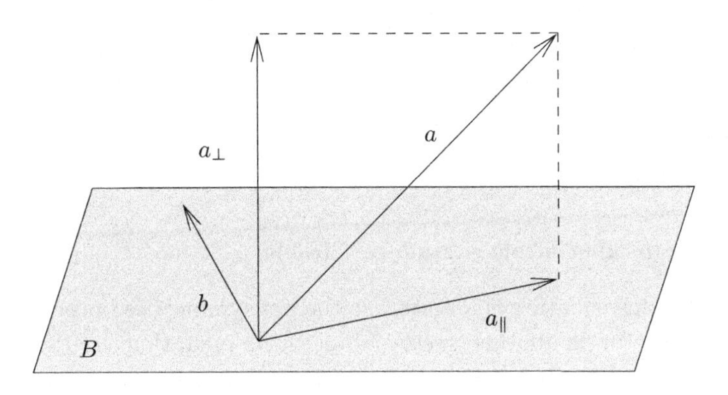

Figure 2.4 *A vector and a bivector.* The vector a can be written as the sum of a term in the plane B and a term perpendicular to the plane, so that $a = a_\parallel + a_\perp$. The bivector B can be written as $a_\parallel \wedge b$, where b is perpendicular to $a_\parallel$.

as shown in figure 2.4. We can now write $aB = (a_\parallel + a_\perp)B$. Suppose that we also write

$$B = a_\parallel \wedge b = a_\parallel b, \tag{2.45}$$

where b is orthogonal to $a_\parallel$ in the B plane. It is always possible to find such a vector b. We now see that

$$a_\parallel B = a_\parallel(a_\parallel b) = {a_\parallel}^2 b \tag{2.46}$$

and so is a vector. This is clear in that the product of a plane with a vector in the plane must remain in the plane. On the other hand

$$a_\perp B = a_\perp(a_\parallel \wedge b) = a_\perp a_\parallel b, \tag{2.47}$$

which is the product of three orthogonal (anticommuting) vectors and so is a trivector. As expected, the product of a vector and a bivector will in general contain vector and trivector terms.

To explore this further let us form the product of the vector a with the bivector $b \wedge c$. From the associative and distributive properties of the geometric product we have

$$a(b\wedge c) = a\tfrac{1}{2}(bc - cb) = \tfrac{1}{2}(abc - acb). \tag{2.48}$$

We now use the rearrangement

$$ab = 2a\cdot b - ba \tag{2.49}$$

to write

$$a(b\wedge c) = (a\cdot b)c - (a\cdot c)b - \tfrac{1}{2}(bac - cab)$$
$$= 2(a\cdot b)c - 2(a\cdot c)b + \tfrac{1}{2}(bc - cb)a, \tag{2.50}$$

so that

$$a(b\wedge c) - (b\wedge c)a = 2(a\cdot b)c - 2(a\cdot c)b. \tag{2.51}$$

The right-hand side of this equation is a vector, so the antisymmetrised product of a vector with a bivector is another vector. Since this operation is grade-lowering, we give it the dot symbol again and write

$$a\cdot B = \tfrac{1}{2}(aB - Ba), \tag{2.52}$$

where B is an arbitrary bivector. The preceding rearrangement means that we have proved one of the most useful results in geometric algebra,

$$a\cdot(b\wedge c) = a\cdot b\, c - a\cdot c\, b. \tag{2.53}$$

Returning to equation (2.46) we see that we must have

$$a\cdot B = a_{\parallel}B = a_{\parallel}\cdot B. \tag{2.54}$$

So the effect of taking the inner product of a vector with a bivector is to project onto the component of the vector in the plane, and then rotate this through 90° and dilate by the magnitude of B. We can also confirm that

$$a\cdot B = a_{\parallel}{}^2 b = -(a_{\parallel}b)a_{\parallel} = -B\cdot a, \tag{2.55}$$

as expected.

The remaining part of the product of a vector and a bivector returns a grade-3 trivector. This product is denoted with a wedge since it is grade-raising, so

$$a\wedge(b\wedge c) = \tfrac{1}{2}\big(a(b\wedge c) + (b\wedge c)a\big). \tag{2.56}$$

A few lines of algebra confirm that this outer product is associative,

$$\begin{aligned} a\wedge(b\wedge c) &= \tfrac{1}{2}\big(a(b\wedge c) + (b\wedge c)a\big) \\ &= \tfrac{1}{4}\big(abc - acb + bca - cba\big) \\ &= \tfrac{1}{4}\big(2(a\wedge b)c + bac + bca + 2c(a\wedge b) - cab - acb\big) \\ &= \tfrac{1}{2}\big((a\wedge b)c + c(a\wedge b) + b(c\cdot a) - (c\cdot a)b\big) \\ &= (a\wedge b)\wedge c, \end{aligned} \tag{2.57}$$

so we can unambiguously write the result as $a \wedge b \wedge c$. The product $a \wedge b \wedge c$ is therefore associative and antisymmetric on all pairs of vectors, and so is precisely Grassmann's exterior product (see section 1.6). This demonstrates that

Grassmann's exterior product sits naturally within geometric algebra. From equation (2.47) we have

$$a \wedge B = a_\perp B = a_\perp \wedge B, \tag{2.58}$$

so the effect of the exterior product with a bivector is to project onto the component of the vector perpendicular to the plane, and return a volume element (a trivector). We can confirm simply that this product is symmetric in its vector and bivector arguments:

$$a \wedge B = a_\perp \wedge a_\parallel \wedge b = -a_\parallel \wedge a_\perp \wedge b = a_\parallel \wedge b \wedge a_\perp = B \wedge a. \tag{2.59}$$

The full product of a vector and a bivector can now be written as

$$aB = a \cdot B + a \wedge B, \tag{2.60}$$

where the dot is generalised to mean the *lowest* grade part of the product, while the wedge means the *highest* grade part of the product. In a similar manner to the geometric product of vectors, the separate dot and wedge products can be written in terms of the geometric product as

$$\begin{aligned} a \cdot B &= \tfrac{1}{2}(aB - Ba), \\ a \wedge B &= \tfrac{1}{2}(aB + Ba). \end{aligned} \tag{2.61}$$

But pay close attention to the signs in these formulae, which are the opposite way round to the case of two vectors. The full product of a vector and a bivector wraps up the separate vector and trivector terms in the single product aB. The advantage of this is again that the full product is invertible.

2.4.2 The bivector algebra

Our three independent bivectors also give us another new product to consider. We already know that squaring a bivector results in a scalar. But if we multiply together two bivectors representing orthogonal planes we find that, for example,

$$(\mathsf{e}_1 \wedge \mathsf{e}_2)(\mathsf{e}_2 \wedge \mathsf{e}_3) = \mathsf{e}_1 \mathsf{e}_2 \mathsf{e}_2 \mathsf{e}_3 = \mathsf{e}_1 \mathsf{e}_3, \tag{2.62}$$

resulting in a third bivector. We also find that

$$(\mathsf{e}_2 \wedge \mathsf{e}_3)(\mathsf{e}_1 \wedge \mathsf{e}_2) = \mathsf{e}_3 \mathsf{e}_2 \mathsf{e}_2 \mathsf{e}_1 = \mathsf{e}_3 \mathsf{e}_1 = -\mathsf{e}_1 \mathsf{e}_3, \tag{2.63}$$

so the product of orthogonal bivectors is antisymmetric. The symmetric contribution vanishes because the two planes are perpendicular.

If we introduce the following labelling for the basis bivectors:

$$B_1 = \mathsf{e}_2 \mathsf{e}_3, \quad B_2 = \mathsf{e}_3 \mathsf{e}_1, \quad B_3 = \mathsf{e}_1 \mathsf{e}_2, \tag{2.64}$$

we find that their product satisfies

$$B_i B_j = -\delta_{ij} - \epsilon_{ijk} B_k. \tag{2.65}$$

There is a clear analogy with the geometric product of vectors here, in that the symmetric part is a scalar, whereas the antisymmetric part is a bivector. In higher dimensions it turns out that the symmetrised product of two bivectors can have grade-0 and grade-4 terms (which we will ultimately denote with the dot and wedge symbols). The antisymmetrised product is always a bivector, and bivectors form a closed algebra under this product.

The basis bivectors satisfy

$$B_1{}^2 = B_2{}^2 = B_3{}^2 = -1 \tag{2.66}$$

and

$$B_1 B_2 = -B_2 B_1, \quad \text{etc.} \tag{2.67}$$

These are the properties of the generators of the quaternion algebra (see section 1.4). This observation helps to sort out some of the problems encountered with the quaternions. Hamilton attempted to identify pure quaternions (null scalar part) with vectors, but we now see that they are actually *bivectors*. This causes problems when looking at how objects transform under reflections. Hamilton also imposed the condition $ijk = -1$ on his unit quaternions, whereas we have

$$B_1 B_2 B_3 = \mathsf{e}_2\mathsf{e}_3\mathsf{e}_3\mathsf{e}_1\mathsf{e}_1\mathsf{e}_2 = +1. \tag{2.68}$$

To set up an isomorphism we must flip a sign somewhere, for example in the y component:

$$i \leftrightarrow B_1, \quad j \leftrightarrow -B_2, \quad k \leftrightarrow B_3. \tag{2.69}$$

This shows us that the quaternions are a *left-handed* set of bivectors, whereas Hamilton and others attempted to view the i, j, k as a right-handed set of vectors. Not surprisingly, this was a potential source of great confusion and meant one had to be extremely careful when applying quaternions in vector algebra.

2.4.3 The trivector

Given three vectors, a, b and c, the trivector $a \wedge b \wedge c$ is formed by sweeping $a \wedge b$ along the vector c (see figure 2.5). The result can be represented pictorially as an oriented parallelepiped. As with bivectors, however, the picture should not be interpreted too literally. The trivector $a \wedge b \wedge c$ does not contain any shape information. It just records a volume and an orientation.

The various algebraic properties of trivectors have straightforward geometric interpretations. The same oriented volume is obtained by sweeping $a \wedge b$ along c or $b \wedge c$ along a. The mathematical expression of this is that the outer product is associative, $a \wedge (b \wedge c) = (a \wedge b) \wedge c$. The trivector $a \wedge b \wedge c$ changes sign under interchange of any pair of vectors, which follows immediately from the

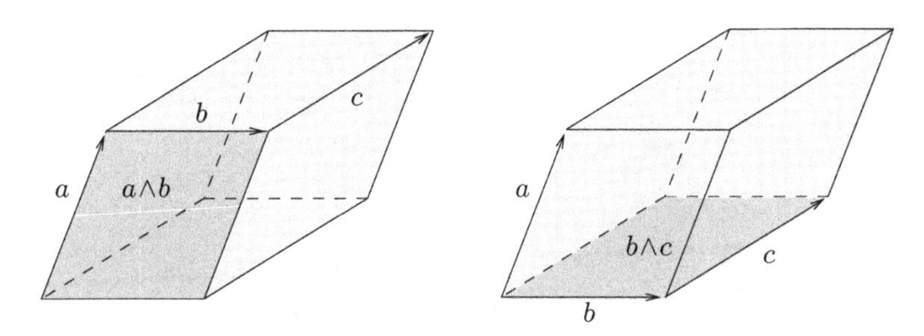

Figure 2.5 *The trivector.* The trivector $a \wedge b \wedge c$ can be viewed as the oriented parallelepiped obtained from sweeping the bivector $a \wedge b$ along the vector c. In the left-hand diagram the bivector $a \wedge b$ is swept along c. In the right-hand one $b \wedge c$ is swept along a. The result is the same in both cases, demonstrating the equality $a \wedge b \wedge c = b \wedge c \wedge a$. The associativity of the outer product is also clear from such diagrams.

antisymmetry of the exterior product. The geometric picture of this is that swapping any two vectors reverses the orientation by which the volume is swept out. Under two successive interchanges of pairs of vectors the trivector returns to itself, so

$$a \wedge b \wedge c = c \wedge a \wedge b = b \wedge c \wedge a. \tag{2.70}$$

This is also illustrated in figure 2.5.

The unit right-handed pseudoscalar for space is given the standard symbol I, so

$$I = \mathsf{e}_1 \mathsf{e}_2 \mathsf{e}_3, \tag{2.71}$$

where the $\{\mathsf{e}_1, \mathsf{e}_2, \mathsf{e}_3\}$ are any right-handed frame of orthonormal vectors. If a left-handed set of orthonormal vectors is multiplied together the result is $-I$. Given an arbitrary set of three vectors we must have

$$a \wedge b \wedge c = \alpha I, \tag{2.72}$$

where α is a scalar. It is not hard to show that $|\alpha|$ is the volume of the parallelepiped with sides defined by a, b and c. The sign of α encodes whether the set $\{a, b, c\}$ forms a right-handed or left-handed frame. In three dimensions this fully accounts for the information in the trivector.

Now consider the product of the vector e_1 and the pseudoscalar,

$$\mathsf{e}_1 I = \mathsf{e}_1(\mathsf{e}_1 \mathsf{e}_2 \mathsf{e}_3) = \mathsf{e}_2 \mathsf{e}_3. \tag{2.73}$$

This returns a bivector — the plane perpendicular to the original vector (see figure 2.6). The product of a grade-1 vector with the grade-3 pseudoscalar is therefore a grade-2 bivector. Multiplying from the left we find that

$$I \mathsf{e}_1 = \mathsf{e}_1 \mathsf{e}_2 \mathsf{e}_3 \mathsf{e}_1 = -\mathsf{e}_1 \mathsf{e}_2 \mathsf{e}_1 \mathsf{e}_3 = \mathsf{e}_2 \mathsf{e}_3. \tag{2.74}$$

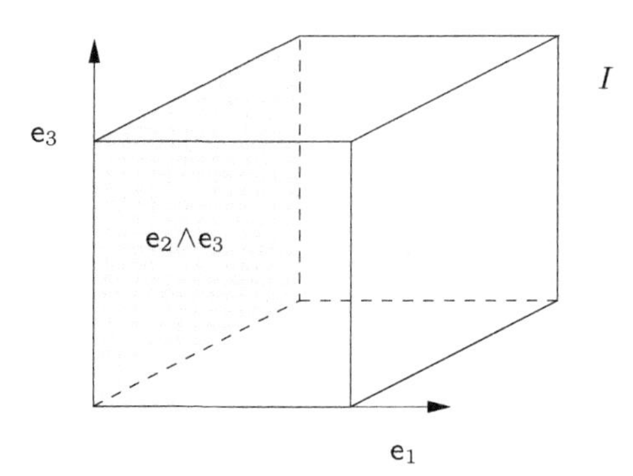

Figure 2.6 *A vector and a trivector.* The result of multiplying the vector e_1 by the trivector I is the plane $e_1(e_1e_2e_3) = e_2e_3$. This is the plane perpendicular to the e_1 vector.

The result is therefore independent of order, and this holds for any basis vector. It follows that the pseudoscalar commutes with all vectors in three dimensions:

$$Ia = aI. \tag{2.75}$$

This is always the case for the pseudoscalar in spaces of odd dimension. In even dimensions, the pseudoscalar anticommutes with all vectors, as we have already seen in two dimensions.

We can now express each of our basis bivectors as the product of the pseudoscalar and a *dual* vector:

$$e_1e_2 = Ie_3, \qquad e_2e_3 = Ie_1, \qquad e_3e_1 = Ie_2. \tag{2.76}$$

This operation of multiplying by the pseudoscalar is called a *duality* transformation and was originally introduced by Grassmann. Again, we can write

$$aI = a\cdot I \tag{2.77}$$

with the dot used to denote the lowest grade term in the product. The result of this can be understood as a projection — projecting onto the component of I perpendicular to a.

We next form the square of the pseudoscalar:

$$I^2 = e_1e_2e_3e_1e_2e_3 = e_1e_2e_1e_2 = -1. \tag{2.78}$$

So the pseudoscalar commutes with all elements and squares to -1. It is therefore a further candidate for a unit imaginary. In some physical applications this is the correct one to use, whereas for others it is one of the bivectors. The properties of I in three dimensions make it particularly tempting to replace it with the symbol i, and this is common practice in much of the literature. This convention can still lead to confusion, however, and is not adopted in this book.

Finally, we consider the product of a bivector and the pseudoscalar:

$$I(\mathsf{e}_1\wedge\mathsf{e}_2) = I\mathsf{e}_1\mathsf{e}_2\mathsf{e}_3\mathsf{e}_3 = II\mathsf{e}_3 = -\mathsf{e}_3. \tag{2.79}$$

So the result of the product of I with the bivector formed from e_1 and e_2 is $-\mathsf{e}_3$, that is, minus the vector perpendicular to the $\mathsf{e}_1\wedge\mathsf{e}_2$ plane. This provides a definition of the vector cross product as

$$a\times b = -I(a\wedge b). \tag{2.80}$$

The vector cross product is largely redundant now that we have the exterior product and duality at our disposal. For example, consider the result for the double cross product. We form

$$\begin{aligned} a\times(b\times c) &= -Ia\wedge(-I(b\wedge c)) \\ &= \tfrac{1}{2}I\big(aI(b\wedge c) - (b\wedge c)Ia\big) \\ &= -a\cdot(b\wedge c). \end{aligned} \tag{2.81}$$

We have already calculated the expansion of the final line, which turns out to be the first example of a much more general, and very useful, formula.

Equation (2.80) shows how the cross product of two vectors is a disguised bivector, the bivector being mapped to a vector by a duality operation. It is now clear why the product only exists in three dimensions — this is the only space for which the dual of a bivector is a vector. We will have little further use for the cross product and will rarely employ it from now on. This means we can also do away with the awkward distinction between polar and axial vectors. Instead we just talk in terms of vectors and bivectors. Both may belong to three-dimensional linear spaces, but they are quite different objects with distinct algebraic properties.

2.4.4 The Pauli algebra

The full geometric product for vectors can be written

$$\mathsf{e}_i\mathsf{e}_j = \mathsf{e}_i\cdot\mathsf{e}_j + \mathsf{e}_i\wedge\mathsf{e}_j = \delta_{ij} + I\epsilon_{ijk}\mathsf{e}_k. \tag{2.82}$$

This may be familiar to many — it is the Pauli algebra of quantum mechanics! The Pauli matrices therefore form a matrix representation of the geometric algebra of space. The Pauli matrices are

$$\sigma_1 = \begin{pmatrix} 0 & 1 \\ 1 & 0 \end{pmatrix}, \quad \sigma_2 = \begin{pmatrix} 0 & -i \\ i & 0 \end{pmatrix}, \quad \sigma_3 = \begin{pmatrix} 1 & 0 \\ 0 & -1 \end{pmatrix}. \tag{2.83}$$

These matrices satisfy

$$\sigma_i\sigma_j = \delta_{ij}\mathsf{I} + i\epsilon_{ijk}\sigma_k, \tag{2.84}$$

where I is the 2×2 identity matrix. Historically, these matrices were discovered by Pauli in his investigations of the quantum theory of spin. The link with geometric algebra ('Clifford algebra' in the quantum theory textbooks) was only made later.

Surprisingly, though the link with the geometric algebra of space is now well established, one seldom sees the Pauli matrices referred to as a representation for the algebra of a set of vectors. Instead they are almost universally referred to as the components of a single vector in 'isospace'. A handful of authors (most notably David Hestenes) have pointed out the curious nature of this interpretation. Such discussion remains controversial, however, and will only be touched on in this book. As with all arguments over interpretations of quantum mechanics, how one views the Pauli matrices has little effect on the predictions of the theory.

The fact that the Pauli matrices form a matrix representation of $\mathcal{G}_3$ provides an alternative way of performing multivector manipulations. This method is usually slower, but can sometimes be used to advantage, particularly in programming languages where complex arithmetic is built in. Working directly with matrices does obscure geometric meaning, and is usually best avoided.

2.5 Conventions

A number of conventions help to simplify expressions in geometric algebra. For example, expressions such as $(a \cdot b)c$ and $I(a \wedge b)$ demonstrate that it would be useful to have a convention which allows us to remove the brackets. We thus introduce the operator ordering convention that in the absence of brackets, *inner and outer products are performed before geometric products.* This can remove significant numbers of unnecessary brackets. For example, we can safely write

$$I(a \wedge b) = I\, a \wedge b. \tag{2.85}$$

and

$$(a \cdot b)c = a \cdot b\, c. \tag{2.86}$$

In addition, unless brackets specify otherwise, inner products are performed before outer products,

$$a \cdot b\, c \wedge d = (a \cdot b) c \wedge d. \tag{2.87}$$

A simple notation for the result of projecting out the elements of a multivector that have a given grade is also invaluable. We denote this with angled brackets $\langle\,\rangle_r$, where r is the grade onto which we want to project. With this notation we can write, for example,

$$a \wedge b = \langle a \wedge b \rangle_2 = \langle ab \rangle_2. \tag{2.88}$$

The final expression holds because $a \wedge b$ is the sole grade-2 component of the geometric product ab. This notation can be extremely useful as it often enables inner and outer products to be replaced by geometric products, which are usually simpler to manipulate. The operation of taking the scalar part of a product is often needed, and it is conventional for this to drop the subscript zero and simply write

$$\langle M \rangle = \langle M \rangle_0. \tag{2.89}$$

The scalar part of any pair of multivectors is symmetric:

$$\langle AB \rangle = \langle BA \rangle. \tag{2.90}$$

It follows that the scalar part satisfies the cyclic reordering property

$$\langle AB \cdots C \rangle = \langle B \cdots CA \rangle, \tag{2.91}$$

which is frequently employed in manipulations.

An important operation in geometric algebra is that of *reversion*, which reverses the order of vectors in any product. There are two conventions for this in common usage. One is the dagger symbol, $A^\dagger$, used for Hermitian conjugation in matrix algebra. The other is to use a tilde, $\tilde{A}$. In three-dimensional applications the dagger symbol is often employed, as the reverse operation returns the same result as Hermitian conjugation of the Pauli matrix representation of the algebra. In spacetime physics, however, the tilde symbol is the better choice as the dagger is reserved for a different (frame-dependent) operation in relativistic quantum mechanics. For the remainder of this chapter we will use the dagger symbol, as we will concentrate on applications in three dimensions.

Scalars and vectors are invariant under reversion, but bivectors change sign:

$$(\mathsf{e}_1\mathsf{e}_2)^\dagger = \mathsf{e}_2\mathsf{e}_1 = -\mathsf{e}_1\mathsf{e}_2. \tag{2.92}$$

Similarly, we see that

$$I^\dagger = \mathsf{e}_3\mathsf{e}_2\mathsf{e}_1 = \mathsf{e}_1\mathsf{e}_3\mathsf{e}_2 = -\mathsf{e}_1\mathsf{e}_2\mathsf{e}_3 = -I. \tag{2.93}$$

A general multivector in $\mathcal{G}_3$ can be written

$$M = \alpha + a + B + \beta I, \tag{2.94}$$

where a is a vector, B is a bivector and α and β are scalars. From the above we see that the reverse of M, $M^\dagger$, is

$$M^\dagger = \alpha + a - B - \beta I. \tag{2.95}$$

As stated above, this operation has the same effect as Hermitian conjugation applied to the Pauli matrices.

We have now introduced a number of terms, some of which have overlapping meaning. It is useful at this point to refer to multivectors which only contain terms of a single grade as *homogeneous*. The term *inner* product is reserved for the lowest grade part of the geometric product of two homogeneous multivectors. For two homogeneous multivectors of the same grade the inner product and scalar product reduce to the same thing. The terms *exterior* and *outer* products are interchangeable, though we will tend to prefer the latter for its symmetry with the inner product. The inner and outer products are also referred to colloquially as the *dot* and *wedge* products. We have followed convention in referring to the highest grade element in a geometric algebra as the *pseudoscalar*. This is a convenient name, though one must be wary that in tensor analysis the term can mean something subtly different. Both *directed volume element* and *volume form* are good alternative names, but we will stick with *pseudoscalar* in this book.

2.6 Reflections

The full power of geometric algebra begins to emerge when we consider reflections and rotations. We start with an arbitrary vector a and a unit vector n ($n^2 = 1$), and resolve a into parts parallel and perpendicular to n. This is achieved simply by forming

$$\begin{aligned} a &= n^2 a \\ &= n(n\cdot a + n\wedge a) \\ &= a_{\|} + a_{\perp}, \end{aligned} \tag{2.96}$$

where

$$a_{\|} = a\cdot n\, n, \qquad a_{\perp} = n\, n\wedge a. \tag{2.97}$$

The formula for $a_{\|}$ is certainly the projection of a onto n, and the remaining term must be the perpendicular component (sometimes called the rejection). We can check that $a_{\perp}$ is perpendicular to n quite simply:

$$n\cdot a_{\perp} = \langle nn\, n\wedge a\rangle = \langle n\wedge a\rangle = 0. \tag{2.98}$$

This is a simple example of how using the projection onto grade operator to replace inner and outer products with geometric products can simplify derivations.

The result of reflecting a in the plane orthogonal to n is the vector $a' = a_{\perp} - a_{\|}$ (see figure 2.7). This can be written

$$\begin{aligned} a' = a_{\perp} - a_{\|} &= n\, n\wedge a - a\cdot n\, n \\ &= -n\cdot a\, n - n\wedge a\, n \\ &= -nan. \end{aligned} \tag{2.99}$$

This formula is already more compact than can be written down without the geometric product. The best one can do with just the inner product is the

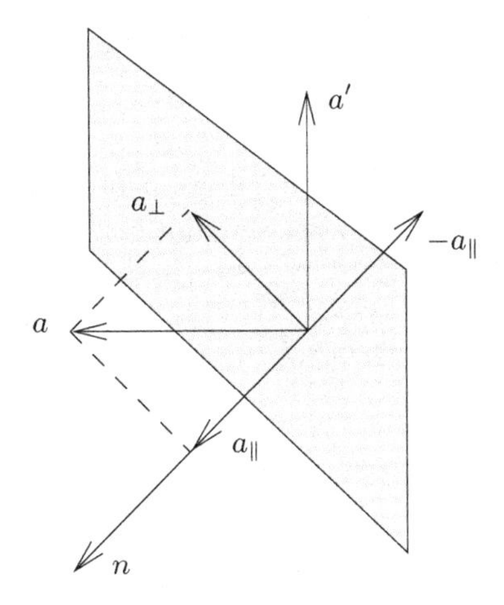

Figure 2.7 *A reflection.* The vector a is reflected in the (hyper)plane perpendicular to n. This is the way to describe reflections in arbitrary dimensions. The result a' is formed by reversing the sign of $a_\|$, the component of a in the n direction.

equivalent expression

$$a' = a - 2a \cdot n\, n. \tag{2.100}$$

The compression afforded by the geometric product becomes increasingly impressive as reflections are compounded together. The formula

$$a' = -nan \tag{2.101}$$

is valid is spaces of any dimension — it is a quite general formula for a reflection.

We should check that our formula for the reflection has the desired property of leaving lengths and angles unchanged. To do this we need only verify that the scalar product between vectors is unchanged if both are reflected, which is achieved with a simple rearrangement:

$$(-nan)\cdot(-nbn) = \langle(-nan)(-nbn)\rangle = \langle nabn\rangle = \langle abnn\rangle = a\cdot b. \tag{2.102}$$

In this manipulation we have made use of the cyclic reordering property of the scalar part of a geometric product, as defined in equation (2.91).

2.6.1 Complex conjugation

In two dimensions we saw that the vector x is mapped to a complex number Z by

$$Z = \mathsf{e}_1 x, \quad x = \mathsf{e}_1 Z. \tag{2.103}$$

The complex conjugate $Z^\dagger$ is the reverse of this, $Z^\dagger = x\mathsf{e}_1$, so maps to the vector

$$x' = \mathsf{e}_1 Z^\dagger = \mathsf{e}_1 x \mathsf{e}_1. \tag{2.104}$$

This can be converted into the formula for a reflection if we remember that the two-dimensional pseudoscalar $I = \mathsf{e}_1\mathsf{e}_2$ anticommutes with all vectors and squares to -1. We therefore have

$$x' = -\mathsf{e}_1 I I x \mathsf{e}_1 = -\mathsf{e}_1 I x \mathsf{e}_1 I = -\mathsf{e}_2 x \mathsf{e}_2. \tag{2.105}$$

This is precisely the expected relation for a reflection in the line perpendicular to e_2, which is to say a reflection in the real axis.

2.6.2 Reflecting bivectors

Now suppose that we form the bivector $B = a \wedge b$ and reflect both of these vectors in the plane perpendicular to n. The result is

$$B' = (-nan)\wedge(-nbn). \tag{2.106}$$

This simplifies as follows:

$$\begin{aligned}(-nan)\wedge(-nbn) &= \tfrac{1}{2}(nannbn - nbnnan) \\ &= \tfrac{1}{2}n(ab - ba)n \\ &= nBn.\end{aligned} \tag{2.107}$$

The effect of sandwiching a multivector between a vector, nMn, always preserves the grade of the multivector M. We will see how to prove this in general when we have derived a few more results for manipulating inner and outer products. The resulting formula nBn shows that bivectors are subject to the same transformation law as vectors, *except for a change in sign.* This is the origin of the conventional distinction between polar and axial vectors. Axial vectors are usually generated by the cross product, and we saw in section 2.4.3 that the cross product generates a bivector, and then dualises it back to a vector. But when the two vectors in the cross product are reflected, the bivector they form is reflected according to (2.107). The dual vector IB is subject to the same transformation law, since

$$I(nBn) = n(IB)n, \tag{2.108}$$

and so does not transform as a (polar) vector. In many texts this can be a source of much confusion. But now we have a much healthier alternative: banish all talk of axial vectors in favour of bivectors. We will see in later chapters that all of the main examples of 'axial' vectors in physics (angular velocity, angular momentum, the magnetic field etc.) are better viewed as bivectors.

2.6.3 Trivectors and handedness

The final object to try reflecting in three dimensions is the trivector $a \wedge b \wedge c$. We first write

$$
\begin{aligned}
(-nan)\wedge(-nbn)\wedge(-ncn) &= \langle(-nan)(-nbn)(-ncn)\rangle_3 \\
&= -\langle nabcn\rangle_3,
\end{aligned}
\tag{2.109}
$$

which follows because the only way to form a trivector from the geometric product of three vectors is through the exterior product of all three. Now the product abc can only contain a vector and trivector term. The former cannot give rise to an overall trivector, so we are left with

$$
(-nan)\wedge(-nbn)\wedge(-ncn) = -\langle na\wedge b\wedge cn\rangle_3. \tag{2.110}
$$

But any trivector in three dimensions is a multiple of the pseudoscalar I, which commutes with all vectors, so we are left with

$$
(-nan)\wedge(-nbn)\wedge(-ncn) = -a\wedge b\wedge c. \tag{2.111}
$$

The overall effect is simply to flip the sign of the trivector, which is a way of stating that reflections have determinant -1. This means that if all three vectors in a right-handed triplet are reflected in some plane, the resulting triplet is left handed (and vice versa).

2.7 Rotations

Our starting point for the treatment of rotations is the result that *a rotation in the plane generated by two unit vectors m and n is achieved by successive reflections in the (hyper)planes perpendicular to m and n.* This is illustrated in figure 2.8. Any component of a perpendicular to the $m\wedge n$ plane is unaffected, and simple trigonometry confirms that the angle between the initial vector a and the final vector c is twice the angle between m and n. (The proof of this is left as an exercise.) The result of the successive reflections is therefore to rotate through 2θ in the $m\wedge n$ plane, where $m\cdot n = \cos(\theta)$.

So how does this look using geometric algebra? We first form

$$
b = -mam \tag{2.112}
$$

and then perform a second reflection to obtain

$$
c = -nbn = -n(-mam)n = nmamn. \tag{2.113}
$$

This is starting to look extremely simple! We define

$$
R = nm, \tag{2.114}
$$

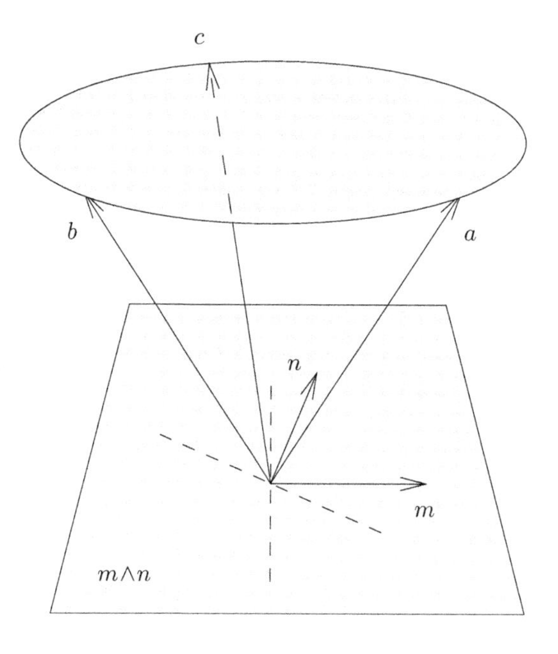

Figure 2.8 *A rotation from two reflections.* The vector b is the result of reflecting a in the plane perpendicular to m, and c is the result of reflecting b in the plane perpendicular to n.

so that we can now write the result of the rotation as

$$c = RaR^{\dagger}. \tag{2.115}$$

This transformation $a \mapsto RaR^{\dagger}$ is a totally general way of handling rotations. In deriving this transformation the dimensionality of the space of vectors was never specified, so the transformation law must work in all spaces, *whatever their dimension*. The rule also works for *any grade* of multivector!

2.7.1 Rotors

The quantity $R = nm$ is called a *rotor* and is one of the most important objects in applications of geometric algebra. Immediately, one can see the importance of the *geometric* product in both (2.114) and (2.115), which tells us that rotors provide a way of handling rotations that is unique to geometric algebra. To study the properties of the rotor R we first write

$$R = nm = n\cdot m + n\wedge m = \cos(\theta) + n\wedge m. \tag{2.116}$$

We already calculated the magnitude of the bivector $m \wedge n$ in equation (2.15), where we obtained

$$(n\wedge m)(n\wedge m) = -\sin^2(\theta). \tag{2.117}$$

We therefore define the *unit* bivector B in the $m\wedge n$ plane by

$$B = \frac{m\wedge n}{\sin(\theta)}, \qquad B^2 = -1. \tag{2.118}$$

The reason for this choice of orientation ($m \wedge n$ rather than $n \wedge m$) is to ensure that the rotation has the orientation specified by the generating bivector, as can be seen in figure 2.8. In terms of the bivector B we now have

$$R = \cos(\theta) - B\sin(\theta), \tag{2.119}$$

which is simply the polar decomposition of a complex number, with the unit imaginary replaced by the unit bivector B. We can therefore write

$$R = \exp(-B\theta), \tag{2.120}$$

with the exponential defined in terms of its power series in the normal way. (The power series for the exponential is absolutely convergent for any multivector argument.)

Now recall that our formula was for a rotation through 2θ. If we want to rotate through θ, the appropriate rotor is

$$R = \exp(-B\theta/2), \tag{2.121}$$

which gives the formula

$$a \mapsto a' = \mathrm{e}^{-B\theta/2} a \mathrm{e}^{B\theta/2} \tag{2.122}$$

for a rotation through θ in the B plane, with handedness determined by B (see figure 2.9). This description encourages us to think of rotations taking place *in a plane*, and as such gives equations which are valid in any dimension. The more traditional idea of rotations taking place around an axis is an entirely three-dimensional concept which does not generalise.

Since the rotor R is a geometric product of two unit vectors, we see immediately that

$$RR^\dagger = nm(nm)^\dagger = nmmn = 1 = R^\dagger R. \tag{2.123}$$

This provides a quick proof that our formula has the correct property of preserving lengths and angles. Suppose that $a' = RaR^\dagger$ and $b' = RbR^\dagger$, then

$$\begin{aligned} a'\cdot b' &= \tfrac{1}{2}(RaR^\dagger RbR^\dagger + RbR^\dagger RaR^\dagger) \\ &= \tfrac{1}{2}R(ab + ba)R^\dagger \\ &= a\cdot b\, RR^\dagger \\ &= a\cdot b. \end{aligned} \tag{2.124}$$

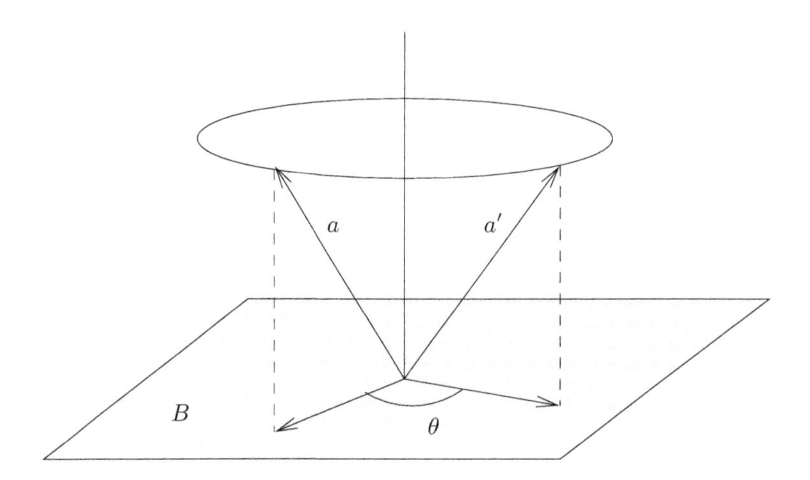

Figure 2.9 *A rotation in three dimensions.* The vector a is rotated to $a' = RaR^\dagger$. The rotor R is defined by $R = \exp(-B\theta/2)$, which describes the rotation directly in terms of the plane and angle. The rotation has the orientation specified by the bivector B.

We can also see that the inverse transformation is given by

$$a = R^\dagger a' R. \tag{2.125}$$

The proof is straightforward:

$$R^\dagger a' R = R^\dagger R a R^\dagger R = a. \tag{2.126}$$

The usefulness of rotors provides ample justification for adding up terms of different grades. The rotor R on its own has no geometric significance, which is to say that no meaning should be attached to the separate scalar and bivector terms. When R is written in the form $R = \exp(-B\theta/2)$, however, the bivector B has clear geometric significance, as does the vector formed from $RaR^\dagger$. This illustrates a central feature of geometric algebra, which is that both geometrically meaningful objects (vectors, planes etc.) and the elements that act on them (in this case rotors) are represented in the same algebra.

2.7.2 Constructing a rotor

Suppose that we wish to rotate the unit vector a into another unit vector b, leaving all vectors perpendicular to a and b unchanged. This is accomplished by a reflection perpendicular to the unit vector n half-way between a and b followed by a reflection in the plane perpendicular to b (see figure 2.10). The vector n is

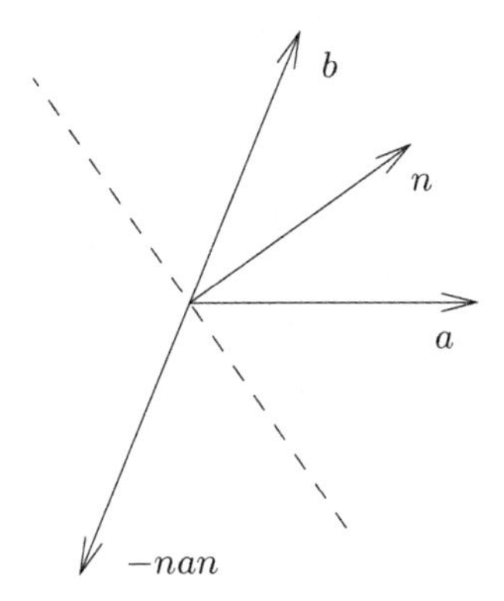

Figure 2.10 *A rotation from a to b.* The vector a is rotated onto b by first reflecting in the plane perpendicular to n, and then in the plane perpendicular to b. The vectors a, b and n all have unit length.

given by

$$n = \frac{(a+b)}{|a+b|}, \tag{2.127}$$

which reflects a into $-b$. Combining this with the reflection in the plane perpendicular to b we arrive at the rotor

$$R = bn = \frac{1+ba}{|a+b|} = \frac{1+ba}{\sqrt{2(1+b\cdot a)}}, \tag{2.128}$$

which represents a simple rotation in the $a \wedge b$ plane. This formula shows us that

$$Ra = \frac{a+b}{\sqrt{2(1+b\cdot a)}} = a\frac{1+ab}{\sqrt{2(1+b\cdot a)}} = aR^{\dagger}. \tag{2.129}$$

It follows that we can write

$$RaR^{\dagger} = R^2 a = a{R^{\dagger}}^2. \tag{2.130}$$

This is always possible for vectors in the plane of rotation. Returning to the polar form $R = \exp(-B\theta/2)$, where B is the $a \wedge b$ plane, we see that

$$R^2 = \exp(-B\theta), \tag{2.131}$$

so we can rotate a onto b with the formula

$$b = \mathrm{e}^{-B\theta} a = a\mathrm{e}^{B\theta}. \tag{2.132}$$

This is precisely the form found in the plane using complex numbers, and was the source of much of the confusion over the use of quaternions for rotations. Hamilton thought that a single-sided transformation law of the form $a \mapsto Ra$ should be the correct way to encode a rotation, with the full angle appearing in the exponential. He thought that this was the natural generalisation of the complex number representation. But we can see now that this formula only works for vectors *in the plane of rotation*. The correct formula for all vectors is the double-sided, half-angle formula $a \mapsto RaR^\dagger$. This formula ensures that given a vector c perpendicular to the $a \wedge b$ plane we have

$$Rc = c\frac{1+ba}{\sqrt{2(1+b\cdot a)}} = \frac{1+ba}{\sqrt{2(1+b\cdot a)}}c = Rc, \tag{2.133}$$

so that

$$RcR^\dagger = cRR^\dagger = c, \tag{2.134}$$

and the vector is unrotated. The single-sided law does not have this property. Correctly identifying the double-sided transformation law means that unit bivectors such as

$$\mathrm{e}_1\mathrm{e}_2 = \mathrm{e}^{\mathrm{e}_1\mathrm{e}_2\pi/2} \tag{2.135}$$

are generators of rotations through π, and not $\pi/2$. The fact that unit bivectors square to -1 is consistent with this because, acting double sidedly, the rotor -1 is the identity operation. More generally, R and $-R$ generate the same rotation, so there is a two-to-one map between rotors and rotations. (Mathematicians talk of the rotors providing a *double-cover* representation of the rotation group.)

2.7.3 Rotating multivectors

Suppose that the two vectors forming the bivector $B = a\wedge b$ are both rotated. What is the expression for the resulting bivector? To find this we form

$$\begin{aligned} B' = a'\wedge b' &= \tfrac{1}{2}(RaR^\dagger RbR^\dagger - RbR^\dagger RaR^\dagger) \\ &= \tfrac{1}{2}R(ab-ba)R^\dagger \\ &= Ra\wedge bR^\dagger \\ &= RBR^\dagger, \end{aligned} \tag{2.136}$$

where we have used the rotor normalisation formula $R^\dagger R = 1$. Bivectors are rotated using precisely the same formula as vectors! The same turns out to be true for all geometric multivectors, and this is one of the most attractive features of geometric algebra. In section 4.2 we prove that the transformation $A \mapsto RAR^\dagger$ preserves the grade of the multivector on which the rotors act. For applications in three dimensions we only need check this result for the trivector case, as we have already demonstrated it for vectors and bivectors. The pseudoscalar in three dimensions, I, commutes with all other terms in the algebra, so we have

$$RIR^\dagger = IRR^\dagger = I, \tag{2.137}$$

which is certainly grade-preserving. This result is one way of saying that rotations have determinant +1. We now have a means of rotating all geometric objects in three dimensions. In chapter 3 we will take full advantage of this when studying rigid-body dynamics.

2.7.4 Rotor composition law

Having seen how individual rotors are used to represent rotations, we now look at their composition law. Let the rotor R_1 transform the vector a into a vector b:

$$b = R_1 a R_1^\dagger. \tag{2.138}$$

Now rotate b into another vector c, using a rotor R_2. This requires

$$c = R_2 b R_2^\dagger = R_2 R_1 a R_1^\dagger R_2^\dagger = R_2 R_1 a (R_2 R_1)^\dagger, \tag{2.139}$$

so that if we write

$$c = R a R^\dagger, \tag{2.140}$$

then the composite rotor is given by

$$R = R_2 R_1. \tag{2.141}$$

This is the *group combination rule* for rotors. Rotors form a group because the product of two rotors is a third rotor, as can be checked from

$$R_2 R_1 (R_2 R_1)^\dagger = R_2 R_1 R_1^\dagger R_2^\dagger = R_2 R_2^\dagger = 1. \tag{2.142}$$

In three dimensions the fact that the multivector R contains only even-grade elements and satisfies $RR^\dagger = 1$ is sufficient to ensure that R is a rotor. The fact that rotors form a continuous group (called a *Lie group*) is a subject we will return to later in this book.

Rotors are the exception to the rule that all multivectors are subject to a double-sided transformation law. Rotors are already mixed-grade objects, so multiplying on the left (or right) by another rotor does not take us out of the space of rotors. All geometric entities, such as lines and planes, are single-grade objects, and their grades cannot be changed by a rotation. They are therefore all subject to a double-sided transformation law. Again, this brings us back to the central theme that both geometric objects and the operators acting on them are contained in a single algebra.

The composition rule (2.141) has a surprising consequence. Suppose that the rotor R_1 is kept fixed, and we set $R_2 = \exp(-B\theta/2)$. We now take the vector c on a 2π excursion back to itself. The final rotor R is

$$R = \mathrm{e}^{-B\pi} R_1 = -R_1. \tag{2.143}$$

The rotor has changed sign under a 2π rotation! This is usually viewed as a quantum-mechanical phenomenon related to the existence of fermions. But we can now see that the result is classical and is simply a consequence of our rotor description of rotations. (The relationship between rotors and fermion wavefunctions is discussed in chapter 8.) A geometric interpretation of the distinction between R and $-R$ is provided by the *direction* in which a rotation is performed. Suppose we want to rotate e_1 onto e_2. The rotor to achieve this is

$$R(\theta) = e^{-e_1 e_2 \theta/2}. \tag{2.144}$$

If we rotate in a positive sense through $\pi/2$ the final rotor is given by

$$R(\pi/2) = \frac{1}{\sqrt{2}}(1 - e_1 e_2). \tag{2.145}$$

If we rotate in the negative (clockwise) sense, however, the final rotor is

$$R(-3\pi/2) = -\frac{1}{\sqrt{2}}(1 - e_1 e_2) = -R(\pi/2). \tag{2.146}$$

So, while R and $-R$ define the same absolute rotation (and the same rotation matrix), their different signs can be employed to record information about the handedness of the rotation.

The rotor composition rule provides a simple formula for the compound effect of two rotations. Suppose that we have

$$R_1 = e^{-B_1\theta_1/2}, \quad R_2 = e^{-B_2\theta_2/2}, \tag{2.147}$$

where both B_1 and B_2 are unit bivectors. The product rotor is

$$\begin{aligned} R =& \big(\cos(\theta_2/2) - \sin(\theta_2/2)B_2\big)\big(\cos(\theta_1/2) - \sin(\theta_1/2)B_1\big) \\ =& \cos(\theta_2/2)\cos(\theta_1/2) - \big(\cos(\theta_2/2)\sin(\theta_1/2)B_1 + \cos(\theta_1/2)\sin(\theta_2/2)B_2\big) \\ &+ \sin(\theta_2/2)\sin(\theta_1/2)B_1B_2. \end{aligned} \tag{2.148}$$

So if we write $R = R_2R_1 = \exp(-B\theta/2)$, where B is a new unit bivector, we immediately see that

$$\cos(\theta/2) = \cos(\theta_2/2)\cos(\theta_1/2) + \sin(\theta_2/2)\sin(\theta_1/2)\langle B_1B_2\rangle \tag{2.149}$$

and

$$\begin{aligned} \sin(\theta/2)B =& \cos(\theta_2/2)\sin(\theta_1/2)B_1 + \cos(\theta_1/2)\sin(\theta_2/2)B_2 \\ &- \sin(\theta_2/2)\sin(\theta_1/2)\langle B_1B_2\rangle_2. \end{aligned} \tag{2.150}$$

These half-angle relations for rotations were first discovered by the mathematician Rodriguez, three years before the invention of the quaternions! It is well known that these provide a simple means of calculating the compound effect of two rotations. Numerically, it is usually even simpler to just multiply the rotors directly and not worry about calculating any trigonometric functions.

2.7.5 Euler angles

A standard way to parameterise rotations is via the three Euler angles $\{\phi, \theta, \psi\}$. These are defined to rotate an initial set of axes, $\{e_1, e_2, e_3\}$, onto a new set $\{e_1', e_2', e_3'\}$ (often denoted x, y, z and x', y', z' respectively). First we rotate about the e_3 axis — i.e. in the e_1e_2 plane — anticlockwise through an angle ϕ. The rotor for this is

$$R_\phi = \mathrm{e}^{-e_1e_2\phi/2}. \tag{2.151}$$

Next we rotate about the axis formed by the transformed e_1 axis through an amount θ. The plane for this is

$$IR_\phi e_1 R_\phi^\dagger = R_\phi e_2 e_3 R_\phi^\dagger. \tag{2.152}$$

The rotor is therefore

$$R_\theta = \exp\bigl(-R_\phi e_2 e_3 R_\phi^\dagger \theta/2\bigr) = R_\phi \mathrm{e}^{-e_2e_3\theta/2} R_\phi^\dagger. \tag{2.153}$$

The intermediate rotor is now

$$R' = R_\theta R_\phi = \mathrm{e}^{-e_1e_2\phi/2}\mathrm{e}^{-e_2e_3\theta/2}. \tag{2.154}$$

Note the order! Finally, we rotate about the transformed e_3 axis through an angle ψ. The appropriate plane is now

$$IR'e_3R'^\dagger = R'e_1e_2R'^\dagger \tag{2.155}$$

and the rotor is

$$R_\psi = \exp\bigl(-R'e_1e_2R'^\dagger \psi/2\bigr) = R'\mathrm{e}^{-e_1e_2\psi/2}R'^\dagger. \tag{2.156}$$

The resultant rotor is therefore

$$R = R_\psi R' = \mathrm{e}^{-e_1e_2\phi/2}\mathrm{e}^{-e_2e_3\theta/2}\mathrm{e}^{-e_1e_2\psi/2}, \tag{2.157}$$

which has decoupled very nicely and is really quite simple — it is much easier to visualise and work with than the equivalent matrix formula! Now that we have geometric algebra at our disposal we will, in fact, have little cause to use the Euler angles in calculations.

2.8 Notes

In this chapter we have given a lengthy introduction to geometric algebra in two and three dimensions. The latter algebra is generated entirely by three basis vectors $\{e_1, e_2, e_3\}$ subject to the rule that $e_ie_j + e_je_i = 2\delta_{ij}$. This simple rule generates an algebra of remarkable power and richness which we will explore in following chapters.

There is a large literature on the geometric algebra of three-dimensional space and its applications in physics. The most complete text is *New Foundations for Classical Mechanics* by David Hestenes (1999). Hestenes has also written many papers on the subject, most of which are listed in the bibliography at the end of this book. Other introductory papers have been written by Gull, Lasenby and Doran (1993a), Doran et al. (1996a) and Vold (1993a, 1993b). Clifford's *Mathematical Papers* (1882) are also of considerable interest. The use of geometric algebra for handling rotations is very common in the fields of engineering and computer science, though often purely in the guise of the quaternion algebra. Searching one of the standard scientific databases with the keyword 'quaternions' returns too many papers to begin to list here.

2.9 Exercises

2.1 From the properties of the geometric product, show that the symmetrised product of two vectors satisfies the properties of a scalar product, as listed in section 1.2.

2.2 By expanding the bivector $a\wedge b$ in terms of geometric products, prove that it anticommutes with both a and b, but commutes with any vector perpendicular to the $a\wedge b$ plane.

2.3 Verify that the $\boldsymbol{E}_1$ and $\boldsymbol{E}_2$ matrices of equation (2.27) satisfy the correct multiplication relations to form a representation of $\mathcal{G}_2$. Use these to verify equations (2.26).

2.4 Construct the multiplication table generated by the orthonormal vectors e_1, e_2 and e_3. Do these generate a (finite) group?

2.5 Prove that all of the following forms are equivalent expressions of the vector cross product:

$$a\times b = -Ia\wedge b = b\cdot(Ia) = -a\cdot(Ib).$$

Interpret each form geometrically. Hence establish that

$$a\times(b\times c) = -a\cdot(b\wedge c) = -(a\cdot b\, c - a\cdot c\, b)$$

and

$$a\cdot(b\times c) = [a, b, c] = a\wedge b\wedge c\, I^{-1}.$$

2.6 Prove that the effect of successive reflections in the planes perpendicular to the vectors m and n results in a rotation through twice the angle between m and n.

2.7 What is the reverse of $RaR^{\dagger}$, where a is a vector? Which objects in three dimensions have this property, and why must the result be another vector?

2.8 Show that the rotor

$$R = \frac{1+ba}{|a+b|}$$

can also be written as $\exp(-B\theta/2)$, where B is the unit bivector in the $a \wedge b$ plane and θ is the angle between a and b.

2.9 The Cayley–Klein parameters are a set of four real numbers α, β, γ and δ subject to the normalisation condition

$$\alpha^2 + \beta^2 + \gamma^2 + \delta^2 = 1.$$

These can be used to paramaterise an arbitrary rotation matrix as follows:

$$\boldsymbol{U} = \begin{pmatrix} \alpha^2 + \beta^2 - \gamma^2 - \delta^2 & 2(\beta\gamma + \alpha\delta) & 2(\beta\delta - \alpha\gamma) \\ 2(\beta\gamma - \alpha\delta) & \alpha^2 - \beta^2 + \gamma^2 - \delta^2 & 2(\gamma\delta + \alpha\beta) \\ 2(\beta\delta + \alpha\gamma) & 2(\gamma\delta - \alpha\beta) & \alpha^2 - \beta^2 - \gamma^2 + \delta^2 \end{pmatrix}.$$

Can you relate the Cayley–Klein parameters to the rotor description?

2.10 Show that the set of all rotors forms a continuous group. Can you identify the group manifold?

3

Classical mechanics

In this chapter we study the use of geometric algebra in classical mechanics. We will assume that readers already have a basic understanding of the subject, as a complete presentation of classical mechanics with geometric algebra would require an entire book. Such a book has been written, *New Foundations for Classical Mechanics* by David Hestenes (1999), which looks in detail at many of the topics discussed here. Our main focus in this chapter is to areas where geometric algebra offers some immediate benefits over traditional methods. These include motion in a central force and rigid-body rotations, both of which are dealt with in some detail. More advanced topics in Lagrangian and Hamiltonian dynamics are covered in chapter 12, and relativistic dynamics is covered in chapter 5.

Classical mechanics was one of the areas of physics that prompted the development of many of the mathematical techniques routinely used today. This is particularly true of vector analysis, and it is now common to see classical mechanics described using an abstract vector notation. Many of the formulae in this chapter should be completely familiar from such treatments. A key difference comes in adopting the outer product of vectors in place of the cross product. This means, for example, that angular momentum and torque both become bivectors. The outer product is clearer conceptually, but on its own it does not bring any calculational advantages. The main new computational tool we have at our disposal is the geometric product, and here we highlight a number of examples of its use.

In this chapter we have chosen to write all vectors in a bold font. This is conventional for three-dimensional physics and many of the formulae presented below look unnatural if this notation is not followed. Bivectors and other general multivectors are left in regular font, which helps to distinguish them from vectors.

3.1 Elementary principles

We start by considering a point particle with a trajectory $\boldsymbol{x}(t)$ described as a function of time. Here $\boldsymbol{x}$ is the position vector relative to some origin and the time t is taken as some absolute 'Newtonian' standard on which all observers agree. The particle has velocity

$$\boldsymbol{v} = \dot{\boldsymbol{x}} = \frac{d\boldsymbol{x}}{dt}, \tag{3.1}$$

where the overdot denotes differentiation with respect to time t. If the particle has mass m, then the momentum $\boldsymbol{p}$ is defined by $\boldsymbol{p} = m\boldsymbol{v}$. Newton's second law of motion states that

$$\dot{\boldsymbol{p}} = \boldsymbol{f}, \tag{3.2}$$

where the vector $\boldsymbol{f}$ is the force acting on the particle. Usually the mass m is constant and we recover the familiar expression $\boldsymbol{f} = m\boldsymbol{a}$, where $\boldsymbol{a}$ is the acceleration

$$\boldsymbol{a} = \frac{d^2\boldsymbol{x}}{dt^2}. \tag{3.3}$$

The case of constant mass is assumed throughout this chapter. The path for a single particle is then determined by a second-order differential equation (assuming $\boldsymbol{f}$ does not depend on higher derivatives).

The work done by the force $\boldsymbol{f}$ on a particle is defined by the line integral

$$W_{12} = \int_{t_1}^{t_2} \boldsymbol{f}\cdot\boldsymbol{v}\, dt = \int_1^2 \boldsymbol{f}\cdot d\boldsymbol{s}. \tag{3.4}$$

The final form here illustrates that the integral is independent of how the path is parameterised. From Newton's second law we have

$$W_{12} = m\int_{t_1}^{t_2} \dot{\boldsymbol{v}}\cdot\boldsymbol{v}\, dt = \frac{m}{2}\int_{t_1}^{t_2} \frac{d}{dt}(v^2)\, dt, \tag{3.5}$$

where $v = |\boldsymbol{v}| = \sqrt{(\boldsymbol{v}^2)}$. It follows that the work done is equal to the change in kinetic energy T, where

$$T = \tfrac{1}{2}mv^2. \tag{3.6}$$

In the case where the work is independent of the path from point 1 to point 2 the force is said to be *conservative*, and can be written as the gradient of a potential:

$$\boldsymbol{f} = -\boldsymbol{\nabla}V. \tag{3.7}$$

For conservative forces the work also evaluates to

$$W_{12} = -\int_1^2 d\boldsymbol{s}\cdot\boldsymbol{\nabla}V = V_1 - V_2 \tag{3.8}$$

and the total energy $E = T + V$ is conserved.

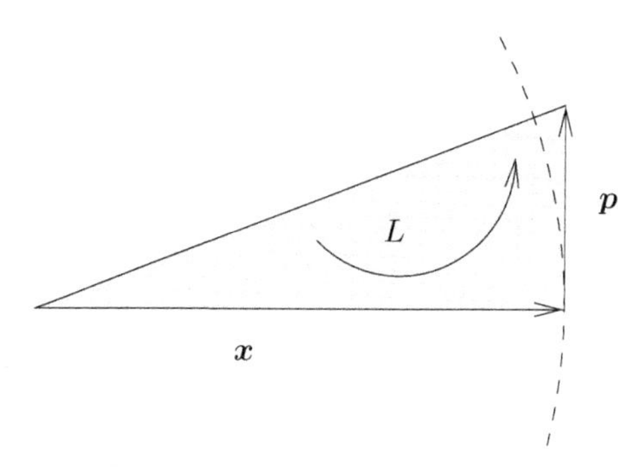

Figure 3.1 *Angular momentum.* The particle sweeps out the plane $L = \boldsymbol{x}\wedge\boldsymbol{p}$. The angular momentum should be directly related to the area swept out (cf. Kepler's second law), so is naturally encoded as a bivector. The position vector $\boldsymbol{x}$ depends on the choice of origin.

3.1.1 Angular momentum

Angular momentum is traditionally discussed in terms of the cross product, even though it is quite clear that what is required is a way of encoding the area swept out by a particle as it moves relative to some origin (see figure 3.1). We saw in chapter 2 that the exterior product provides this, and that the more traditional cross product is a derived concept based on the three-dimensional result that every directed plane has a unique normal. We therefore have no hesitation in dispensing with the traditional definition of angular momentum as an axial vector, and replace it with a bivector. So, if a particle has momentum $\boldsymbol{p}$ and position vector $\boldsymbol{x}$ from some origin, we define the angular momentum of the particle about the origin as the bivector

$$L = \boldsymbol{x}\wedge\boldsymbol{p}. \tag{3.9}$$

This definition does not alter the steps involved in computing L since the components are the same as those of the cross product. We will see, however, that the freedom we have to now use the geometric product can speed up derivations. The definition of angular momentum as a bivector maintains a clear distinction with vector quantities such as position and velocity, removing the need for the rather awkward definitions of polar and axial vectors. The definition of L as a bivector also fits neatly with the rotor description of rotations, as we shall see later in this chapter.

If we differentiate L we obtain

$$\frac{dL}{dt} = \boldsymbol{v}\wedge(m\boldsymbol{v}) + \boldsymbol{x}\wedge(m\boldsymbol{a}) = \boldsymbol{x}\wedge\boldsymbol{f}. \tag{3.10}$$

We define the torque N about the origin as the *bivector*

$$N = \boldsymbol{x} \wedge \boldsymbol{f}, \tag{3.11}$$

so that the torque and angular momentum are related by

$$\frac{dL}{dt} = N. \tag{3.12}$$

The idea of the torque being a bivector is also natural as torques act over a plane. The plane in question is defined by the vector $\boldsymbol{f}$ and the chosen origin, so both L and N depend on the origin. Recall also that bivectors are additive, much like vectors, so the result of applying two torques is found by adding the respective bivectors.

The angular momentum bivector can be written in an alternative way by first defining $r = |\boldsymbol{x}|$ and writing

$$\boldsymbol{x} = r\hat{\boldsymbol{x}}. \tag{3.13}$$

We therefore have

$$\dot{\boldsymbol{x}} = \frac{d}{dt}(r\hat{\boldsymbol{x}}) = \dot{r}\hat{\boldsymbol{x}} + r\dot{\hat{\boldsymbol{x}}}, \tag{3.14}$$

so that

$$L = m\boldsymbol{x} \wedge (\dot{r}\hat{\boldsymbol{x}} + r\dot{\hat{\boldsymbol{x}}}) = mr\hat{\boldsymbol{x}} \wedge (\dot{r}\hat{\boldsymbol{x}} + r\dot{\hat{\boldsymbol{x}}}) = mr^2\, \hat{\boldsymbol{x}} \wedge \dot{\hat{\boldsymbol{x}}}. \tag{3.15}$$

But since $\hat{\boldsymbol{x}}^2 = 1$ we must have

$$0 = \frac{d}{dt}\hat{\boldsymbol{x}}^2 = 2\hat{\boldsymbol{x}} \cdot \dot{\hat{\boldsymbol{x}}}. \tag{3.16}$$

We can therefore eliminate the outer product in equation (3.15) and write

$$L = mr^2\hat{\boldsymbol{x}}\dot{\hat{\boldsymbol{x}}} = -mr^2\dot{\hat{\boldsymbol{x}}}\hat{\boldsymbol{x}}, \tag{3.17}$$

which is useful in a number of problems.

3.1.2 Systems of particles

The preceding definitions generalise easily to systems of particles. For these it is convenient to distinguish between internal and external forces, so the force on the ith particles is

$$\sum_j \boldsymbol{f}_{ji} + \boldsymbol{f}_i^e = \dot{\boldsymbol{p}}_i. \tag{3.18}$$

Here $\boldsymbol{f}_i^e$ is the external force and $\boldsymbol{f}_{ij}$ is the force on the jth particle due to the ith particle. We assume that $\boldsymbol{f}_{ii} = 0$. Newton's third law (in its weak form) states that

$$\boldsymbol{f}_{ij} = -\boldsymbol{f}_{ji}. \tag{3.19}$$

This is not obeyed by all forces, but is assumed to hold for the forces considered in this chapter. Summing the force equation over all particles we find that

$$\sum_i m_i \boldsymbol{a}_i = \sum_i \boldsymbol{f}_i^e + \sum_{i,j} \boldsymbol{f}_{ij} = \sum_i \boldsymbol{f}_i^e. \tag{3.20}$$

All of the internal forces cancel as a consequence of the third law. We define the centre of mass $\boldsymbol{X}$ by

$$\boldsymbol{X} = \frac{1}{M} \sum_i m_i \boldsymbol{x}_i, \tag{3.21}$$

where M is the total mass

$$M = \sum_i m_i. \tag{3.22}$$

The position of the centre of mass is governed by the force law

$$M \frac{d^2 \boldsymbol{X}}{dt^2} = \sum_i \boldsymbol{f}_i^e = \boldsymbol{f}^e \tag{3.23}$$

and so only responds to the total external force on the system. The total momentum of the system is defined by

$$\boldsymbol{P} = \sum_i \boldsymbol{p}_i = M \frac{d\boldsymbol{X}}{dt} \tag{3.24}$$

and is conserved if the total external force is zero.

The total angular momentum about the chosen origin is found by summing the individual bivector contributions,

$$L = \sum_i \boldsymbol{x}_i \wedge \boldsymbol{p}_i. \tag{3.25}$$

The rate of change of L is governed by

$$\dot{L} = \sum_i \boldsymbol{x}_i \wedge \dot{\boldsymbol{p}}_i = \sum_i \boldsymbol{x}_i \wedge \boldsymbol{f}_i^e + \sum_{i,j} \boldsymbol{x}_i \wedge \boldsymbol{f}_{ji}. \tag{3.26}$$

The final term is a double sum containing pairs of terms going as

$$\boldsymbol{x}_i \wedge \boldsymbol{f}_{ji} + \boldsymbol{x}_j \wedge \boldsymbol{f}_{ij} = (\boldsymbol{x}_i - \boldsymbol{x}_j) \wedge \boldsymbol{f}_{ji}. \tag{3.27}$$

The *strong* form of Newton's third law states that the interparticle force $\boldsymbol{f}_{ij}$ is directed along the vector $\boldsymbol{x}_i - \boldsymbol{x}_j$ between the two particles. This law is obeyed by a sufficiently wide range of forces to make it a useful restriction. (The most notable exception to this law is electromagnetism.) Under this restriction the total angular momentum satisfies

$$\frac{dL}{dt} = N^e, \tag{3.28}$$

where N^e is the total external torque. If the applied external torque is zero, and the strong law of action and reaction is obeyed, then the total angular momentum is conserved.

A useful expression for the angular momentum is obtained by introducing a set of position vectors relative to the centre of mass. We write

$$\boldsymbol{x}_i = \boldsymbol{x}_i' + \boldsymbol{X}, \tag{3.29}$$

so that

$$\sum_i m_i \boldsymbol{x}_i' = 0. \tag{3.30}$$

The velocity of the ith particle is now

$$\boldsymbol{v}_i = \boldsymbol{v}_i' + \boldsymbol{v}, \tag{3.31}$$

where $\boldsymbol{v} = \dot{\boldsymbol{X}}$ is the velocity of the centre of mass. The total angular momentum contains four terms:

$$L = \sum_i \bigl(\boldsymbol{X}\wedge m_i\boldsymbol{v} + \boldsymbol{x}_i'\wedge m_i\boldsymbol{v}_i' + m_i\boldsymbol{x}_i'\wedge\boldsymbol{v} + \boldsymbol{X}\wedge m_i\boldsymbol{v}_i'\bigr). \tag{3.32}$$

The final two terms both contain factors of $\sum m_i \boldsymbol{x}_i'$ and so vanish, leaving

$$L = \boldsymbol{X}\wedge\boldsymbol{P} + \sum_i \boldsymbol{x}_i'\wedge\boldsymbol{p}_i'. \tag{3.33}$$

The total angular momentum is therefore the sum of the angular momentum of the centre of mass about the origin, plus the angular momentum of the system about the centre of mass. In many cases it is possible to chose the origin so that the centre of mass is at rest, in which case L is simply the total angular momentum about the centre of mass. Similar considerations hold for the kinetic energy, and it is straightforward to show that

$$T = \sum_i \tfrac{1}{2} m_i \boldsymbol{v}_i^2 = \tfrac{1}{2} M \boldsymbol{v}^2 + \tfrac{1}{2}\sum_i m_i \boldsymbol{v}_i'^2. \tag{3.34}$$

3.2 Two-body central force interactions

One of the most significant applications of the preceding ideas is to a system of two point masses moving under the influence of each other. The force acting between the particles is directed along the vector between them, and all external forces are assumed to vanish. It follows that both the total momentum $\boldsymbol{P}$ and angular momentum L are conserved.

We suppose that the particles have positions $\boldsymbol{x}_1$ and $\boldsymbol{x}_2$, and masses m_1 and m_2. Newton's second law for the central force problem takes the form

$$m_1\ddot{\boldsymbol{x}}_1 = \boldsymbol{f}, \tag{3.35}$$

$$m_2\ddot{\boldsymbol{x}}_2 = -\boldsymbol{f}, \tag{3.36}$$

where $\boldsymbol{f}$ is the interparticle force. We define the relative separation vector $\boldsymbol{x}$ by

$$\boldsymbol{x} = \boldsymbol{x}_1 - \boldsymbol{x}_2. \tag{3.37}$$

This vector satisfies

$$m_1 m_2 \ddot{\boldsymbol{x}} = (m_1 + m_2)\boldsymbol{f}. \tag{3.38}$$

We accordingly define the *reduced mass* μ by

$$\frac{1}{\mu} = \frac{1}{m_1} + \frac{1}{m_2}, \tag{3.39}$$

so that the final force equation can be written as

$$\mu \ddot{\boldsymbol{x}} = \boldsymbol{f}. \tag{3.40}$$

The two-body problem has now been reduced to an equivalent single-body equation. The strong form of the third law assumed here means that the force $\boldsymbol{f}$ is directed along $\boldsymbol{x}$, so we can write $\boldsymbol{f}$ as $f\hat{\boldsymbol{x}}$.

We next re-express the total angular momentum in terms of the centre of mass $\boldsymbol{X}$ and the relative vector $\boldsymbol{x}$. We start by writing

$$m_1 \boldsymbol{x}_1 = m_1 \boldsymbol{X} + \mu \boldsymbol{x}, \quad m_2 \boldsymbol{x}_2 = m_2 \boldsymbol{X} - \mu \boldsymbol{x}. \tag{3.41}$$

It follows that the total angular momentum L_t is given by

$$\begin{aligned} L_t &= m_1 \boldsymbol{x}_1 \wedge \dot{\boldsymbol{x}}_1 + m_2 \boldsymbol{x}_2 \wedge \dot{\boldsymbol{x}}_2 \\ &= M \boldsymbol{X} \wedge \dot{\boldsymbol{X}} + \mu \boldsymbol{x} \wedge \dot{\boldsymbol{x}}. \end{aligned} \tag{3.42}$$

We have assumed that there are no external forces acting, so both L_t and $\boldsymbol{P}$ are conserved. It follows that the internal angular momentum is also conserved and we write this as

$$L = \mu \boldsymbol{x} \wedge \dot{\boldsymbol{x}}. \tag{3.43}$$

Since L is constant, the motion of the particles is confined to the L plane. The trajectory of $\boldsymbol{x}$ must also sweep out area at a constant rate, since this is how L is defined. For planetary motion this is Kepler's second law, though he did not state it in quite this form. Kepler treated the sun as the origin, whereas L should be defined relative to the centre of mass.

The internal kinetic energy is

$$T = \tfrac{1}{2}\mu \dot{\boldsymbol{x}}^2 = \tfrac{1}{2}\mu(\dot{r}\hat{\boldsymbol{x}} + r\dot{\hat{\boldsymbol{x}}})^2 = \tfrac{1}{2}\mu \dot{r}^2 + \tfrac{1}{2}\mu r^2 \dot{\hat{\boldsymbol{x}}}^2. \tag{3.44}$$

From equation (3.17) we see that

$$L^2 = -\mu^2 r^4 \hat{\boldsymbol{x}} \dot{\hat{\boldsymbol{x}}} \dot{\hat{\boldsymbol{x}}} \hat{\boldsymbol{x}} = -\mu^2 r^4 \dot{\hat{\boldsymbol{x}}}^2. \tag{3.45}$$

We therefore define the constant l as the magnitude of L, so

$$l = \mu r^2 |\dot{\hat{\boldsymbol{x}}}|. \tag{3.46}$$

The kinetic energy can now be written as a function of r and $\dot{r}$ only:

$$T = \frac{\mu \dot{r}^2}{2} + \frac{l^2}{2\mu r^2}. \tag{3.47}$$

The force $\boldsymbol{f}$ is conservative and can be written in terms of a potential $V(r)$ as

$$\boldsymbol{f} = f\hat{\boldsymbol{x}} = -\boldsymbol{\nabla} V(r), \tag{3.48}$$

where

$$f = -\frac{dV}{dr}. \tag{3.49}$$

Since the force is conservative the total energy is conserved, so

$$E = \frac{\mu \dot{r}^2}{2} + \frac{l^2}{2\mu r^2} + V(r) \tag{3.50}$$

is a constant. For a given potential $V(r)$ this equation can be integrated to find the evolution of r. The full motion can then be recovered from L.

3.2.1 Inverse-square forces

The most important example of a two-body central force interaction is that described by an inverse-square force law. This case is encountered in gravitation and electrostatics and has been analysed in considerable detail by many authors (see the end of this chapter for suggested additional reading). In this section we review some of the key features of this system, highlighting the places where geometric algebra offers something new. An alternative approach to this problem is discussed in section 3.3.

Writing $f = -k/r^2$ the basic equation to solve is

$$\mu \ddot{\boldsymbol{x}} = -\frac{k}{r^2}\hat{\boldsymbol{x}} = -\frac{k}{r^3}\boldsymbol{x}. \tag{3.51}$$

The sign of k determines whether the force is attractive or repulsive (positive for attractive). This is a second-order vector differential equation, so we expect there to be two constant vectors in the solution — one for the initial position and one for the velocity. We already know that the angular momentum L is a constant of motion, and we can write this as

$$L = \mu r^2 \hat{\boldsymbol{x}}\dot{\hat{\boldsymbol{x}}} = -\mu r^2 \dot{\hat{\boldsymbol{x}}}\hat{\boldsymbol{x}}. \tag{3.52}$$

It follows that

$$L\dot{\boldsymbol{v}} = -\frac{k}{\mu r^2} L\hat{\boldsymbol{x}} = k\dot{\hat{\boldsymbol{x}}}, \tag{3.53}$$

which we can write in the form

$$\frac{d}{dt}(L\boldsymbol{v} - k\hat{\boldsymbol{x}}) = 0. \tag{3.54}$$

Eccentricity	Energy	Orbit
$e>1$	$E>0$	Hyperbola
$e=1$	$E=0$	Parabola
$e<1$	$E<0$	Ellipse
$e=0$	$E=-\mu k^2/(2l^2)$	Circle

Table 3.1 *Classification of orbits for an inverse-square force law.*

The motion is therefore described by the simple equation

$$Lv = k(\hat{x} + e), \tag{3.55}$$

where the *eccentricity vector* e is a second vector constant of motion. This vector is also known in various contexts as the Laplace vector and as the Runge–Lenz vector. From its definition we can see that e must lie in the L plane.

To find a direct equation for the trajectory we first write

$$Lvx = L(v\cdot x + v\wedge x) = \frac{1}{\mu}L\tilde{L} + v\cdot x\, L = k(r + ex). \tag{3.56}$$

The scalar part of this equation gives

$$r = \frac{l^2}{k\mu(1 + e\cdot\hat{x})}. \tag{3.57}$$

This equation specifies a conic surface in three dimensions with symmetry axis e. The surface is formed by rotating a two-dimensional conic about this axis. Since the motion takes place entirely within the L plane the motion is described by a conic. That is, the trajectory $x(t)$ is one of a hyperbola, parabola, ellipse or circle. The generic cases are ellipses for bound orbits and hyperbolae for free states. The cases of parabolic and circular orbits are exceptional as they require precise values of $|e|$ (table 3.1).

In L and e we have found five of the six constants of motion (we only have two arbitrary constants in e as it is constrained to lie in the L plane). The final constant specifies where on the conic we start at time $t = 0$. We know that the energy is also a constant of motion, so it should be possible to express the energy directly in terms of L and e. From equation (3.51) we see that the potential energy must go as k/r, provided we set the arbitrary constant so that $V = 0$ at infinity. The full energy is therefore given by

$$E = \frac{\mu}{2}v^2 - \frac{k}{r}. \tag{3.58}$$

To simplify this we first form

$$Lv v\tilde{L} = l^2 v^2 = k^2(\hat{x} + e)^2. \tag{3.59}$$

It follows that

$$E = \frac{\mu k^2}{2l^2}(e^2 + 1 + 2\hat{x}\cdot e) - \frac{k}{r} = \frac{\mu k^2}{2l^2}(e^2 - 1), \tag{3.60}$$

where $e = |e|$ is the eccentricity. The sign of the energy is governed entirely by e. Since the potential is set to zero at infinity, all bound states must have negative energy and hence an eccentricity $e < 1$. The limiting case of $e = 1$ describes a parabola (table 3.1).

3.2.2 Motion in time for elliptic orbits

Many methods can be used to find the trajectory as a function of time and these are discussed widely in the literature. Here we describe one of the simplest, which serves to highlight the essential difficulty of this problem. An alternative solution, which more fully exploits the techniques of geometric algebra, is described in section 3.3. From the energy equation we see that

$$\mu^2\dot{r}^2 = 2\mu E - \frac{l^2}{r^2} + \frac{2\mu k}{r}, \tag{3.61}$$

so t is given by

$$t = \mu \int_{r_0}^{r_1} \frac{r\,dr}{(2\mu k r + 2\mu E r^2 - l^2)^{1/2}}. \tag{3.62}$$

Evaluating this integral results in a rather complicated function of r, the general form of which is hard to invert and not very helpful. More useful formulae are obtained by specialising to one form of orbit. For bound problems we are interested in elliptic orbits for which E is negative. For these orbits it is useful to introduce the *semi-major axis* a defined by

$$a = \tfrac{1}{2}(r_1 + r_2) = -\frac{k}{2E}, \tag{3.63}$$

where r_1 and r_2 are the maximum and minimum values of r respectively. In terms of this we can write

$$2\mu k r + 2\mu E r^2 - l^2 = -\frac{\mu k}{a}(r^2 - 2ar) - l^2 = \frac{\mu k}{a}\big(a^2e^2 - (r-a)^2\big). \tag{3.64}$$

We now introduce a new variable Ψ, the *eccentric anomaly*, defined by

$$r = a\big(1 - e\cos(\Psi)\big). \tag{3.65}$$

In terms of this we find

$$t = \left(\frac{\mu a^3}{k}\right)^{1/2} \int_{\Psi_0}^{\Psi_1} \big(1 - e\cos(\Psi)\big)\,d\Psi, \tag{3.66}$$

so if we choose $t = 0$ to correspond to closest approach we have

$$\omega t = \Psi - e\sin(\Psi), \tag{3.67}$$

where

$$\omega^2 = \frac{k}{\mu a^3}. \tag{3.68}$$

Equations (3.65) and (3.67) provide a parametric solution relating r and t. This solution highlights the fact that the equation relating t and r is transcendental and does not have a simple closed form. The time taken for one orbit is $2\pi/\omega$, so the orbital period τ is related to the major axis a by

$$\tau^2 = \frac{4\pi^2\mu}{k}a^3. \tag{3.69}$$

This gives us the third of Kepler's three laws of planetary motion, that the square of the period is proportional to the cube of the major axis.

3.3 Celestial mechanics and perturbations

By far the most important application of the Newtonian theory of gravitation is to the motion of the planets in the solar system. This is a complicated subject of considerable historical and current importance, and we will only touch on a few applications. Detailed calculation of the motions of all of the planets in the solar system still represents a major computational challenge. Aside from the obvious problem of having to calculate the gravitational effects of every planet on every other planet, further effects must also be incorporated. These can include deviations of the shapes of the planets from spherical, the effects of tidal forces and ultimately general relativistic corrections.

A significant number of problems in celestial mechanics are best treated using perturbation theory. In this technique orbits are calculated as a series of ever smaller deviations from Kepler orbits. Since the Kepler orbit is specified entirely by L and $\boldsymbol{e}$, we should first form equations for these in the presence of a perturbing force. We modify the force law to read

$$\mu\ddot{\boldsymbol{x}} = -\frac{k}{r^3}\boldsymbol{x} + \boldsymbol{f}, \tag{3.70}$$

and assume that $\boldsymbol{f}$ is always small compared with the inverse-square term. The angular momentum L now satisfies

$$\dot{L} = \boldsymbol{x}\wedge\boldsymbol{f}, \tag{3.71}$$

so L is now only conserved if $\boldsymbol{f}$ is also a central force. With the eccentricity vector still defined by equation (3.55), we find that

$$k\dot{\boldsymbol{e}} = \dot{L}\cdot\boldsymbol{v} + \frac{1}{\mu}L\cdot\boldsymbol{f}. \tag{3.72}$$

Only five of the six equations for L and $\boldsymbol{e}$ are independent, as we always have $L\wedge\boldsymbol{e} = 0$.

For many problems the variation in L and $\boldsymbol{e}$ is slow compared to the orbital period. For these a useful approximation is obtained by finding the orbital average of $\boldsymbol{f}$ over one cycle, with L and $\boldsymbol{e}$ held constant. The quantities L and $\boldsymbol{e}$ are then assumed to vary slowly under the influence of the time-averaged force. Results for the orbital averages of numerous quantities can be found tabulated in many textbooks and are discussed in the exercises at the end of this chapter.

3.3.1 Example — general relativistic perturbations

Later in this book we will study how general relativity modifies the Newtonian view of gravity. For particles moving in a central potential, the modification is quite simple and can be handled efficiently using perturbation theory. The modified force law is

$$\ddot{\boldsymbol{x}} = -\frac{GM}{r^2}\left(1 + \frac{3l^2}{\mu^2 c^2 r^2}\right)\hat{\boldsymbol{x}}, \tag{3.73}$$

where c is the speed of light and we have replaced k by the gravitational expression $GM\mu$. (A small subtlety is that the derivatives here are with respect to proper time, but this does not affect our reasoning.) The force is still central, so the angular momentum L is still conserved. The eccentricity vector satisfies the simple equation

$$\dot{\boldsymbol{e}} = \frac{3l^2}{\mu^3 c^2 r^4} L \cdot \hat{\boldsymbol{x}}. \tag{3.74}$$

For bound orbits this gives rise to a precession of the major axis (see figure 3.2). The quantity of most interest is the amount $\boldsymbol{e}$ changes in one orbit. To get an approximate result for this we use the time-averaging idea and assume that the orbit is precisely elliptical. We therefore have

$$\Delta \boldsymbol{e} = -\frac{3l^2}{\mu^3 c^2} L \int_0^T dt \frac{\hat{\boldsymbol{x}}}{r^4}, \tag{3.75}$$

where T is the orbital period. Evaluating this integral is left as an exercise, and the final result is

$$\Delta \boldsymbol{e} = \frac{6\pi GM}{a(1-e^2)c^2} \boldsymbol{e} \cdot \hat{L}, \tag{3.76}$$

where $\hat{L} = L/l$. This gives a precession of $\boldsymbol{e}$ with the orientation of L, which corresponds to an advance (figure 3.2). For Mercury this gives rise to the famous advance in the perihelion of 43 arcseconds per century, which was finally explained by general relativity.

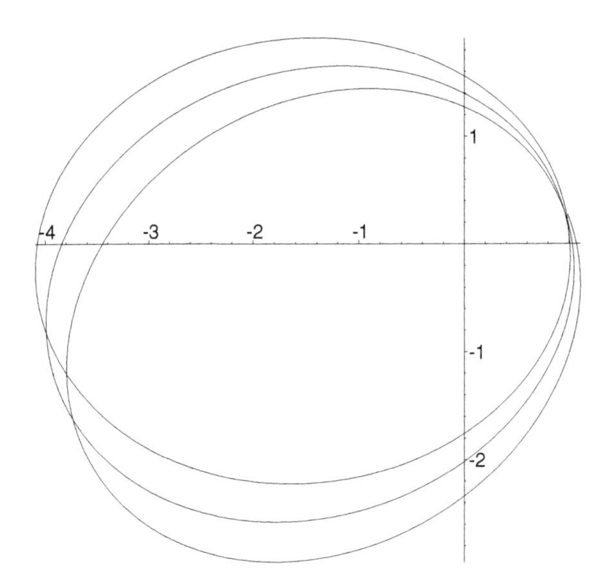

Figure 3.2 *Orbital precession.* The plot shows a modified orbit as predicted by general relativity. The ellipse precesses round in the same direction as the orbital motion. The parameters have been chosen to exaggerate the precession effect.

3.3.2 Spinor equations

An alternative method for analysing the Kepler problem is through the use of 'spinors'. These will be defined more carefully in later chapters, but in two and three dimensions they can be viewed as elements of the subalgebra of $\mathcal{G}_2$ and $\mathcal{G}_3$ consisting entirely of even elements. In two dimensions a spinor can therefore be identified with a complex number. The position vector $\boldsymbol{x}$ in two dimensions can be formed through a rotation and dilation via the polar decomposition

$$\boldsymbol{x} = \mathsf{e}_1 r \exp(\theta \mathsf{e}_1 \mathsf{e}_2) = r \exp(-\theta \mathsf{e}_1 \mathsf{e}_2) \mathsf{e}_1, \tag{3.77}$$

where $\{\mathsf{e}_1, \mathsf{e}_2\}$ denote a right-handed orthonormal frame and we assume that the vector lies in the $\mathsf{e}_1\mathsf{e}_2$ plane. We know from chapter 2 that the rotation formula only extends to higher dimensions if a double-sided prescription is adopted, so we write the vector $\boldsymbol{x}$ as

$$\boldsymbol{x} = U \mathsf{e}_1 U^\dagger = U^2 \mathsf{e}_1 = \mathsf{e}_1 {U^\dagger}^2. \tag{3.78}$$

In writing this we have placed all of the dynamics in the complex number U.

For the Kepler problem it turns out that the equation for U is considerably easier than that for $\boldsymbol{x}$. We assume that the plane of L is given by $\mathsf{e}_1\mathsf{e}_2$, and start by forming

$$r = |\boldsymbol{x}| = UU^\dagger. \tag{3.79}$$

(Recall that, for a scalar + bivector combination in two dimensions, the reverse operator is the same as complex conjugation.) On differentiating we find that

$$\dot{\boldsymbol{x}} = 2\dot{U}U\mathrm{e}_1, \tag{3.80}$$

hence

$$2r\dot{U} = \dot{\boldsymbol{x}}\mathrm{e}_1 U^\dagger = \dot{\boldsymbol{x}}U\mathrm{e}_1. \tag{3.81}$$

We now introduce the new variable s defined by

$$\frac{d}{ds} = r\frac{d}{dt}, \qquad \frac{dt}{ds} = r. \tag{3.82}$$

In terms of this

$$2\frac{dU}{ds} = \dot{\boldsymbol{x}}U\mathrm{e}_1 \tag{3.83}$$

and

$$2\frac{d^2U}{ds^2} = r\ddot{\boldsymbol{x}}U\mathrm{e}_1 + \dot{\boldsymbol{x}}\frac{dU}{ds}\mathrm{e}_1 = U\big(\ddot{\boldsymbol{x}}\boldsymbol{x} + \tfrac{1}{2}\dot{\boldsymbol{x}}^2\big). \tag{3.84}$$

Now suppose we have motion in a central inverse-square force:

$$\mu\ddot{\boldsymbol{x}} = -k\frac{\boldsymbol{x}}{r^3}. \tag{3.85}$$

The equation for U becomes

$$\frac{d^2U}{ds^2} = \frac{1}{2\mu}U\left(\tfrac{1}{2}\mu\dot{\boldsymbol{x}}^2 - \frac{k}{r}\right) = \frac{E}{2\mu}U, \tag{3.86}$$

which is simply the equation for harmonic motion! This has a number of advantages. First of all, the equation is easy to solve. If we set

$$\omega^2 = -\frac{E}{2\mu} \tag{3.87}$$

then the general solution is

$$U = A\exp(\hat{L}\omega s) + B\exp(-\hat{L}\omega s), \tag{3.88}$$

where A and B are constants and $\hat{L}$ is the unit bivector for the plane of motion. The motion is illustrated in figure 3.3. The particle trajectory maps out an ellipse with the origin at one focus, whereas U defines an ellipse with the origin at the centre. The particle completes two orbits for each full cycle of U.

Further advantages of formulating the dynamics in terms of U are that the equation for U is linear, so is better suited to perturbation theory, and that there is no singularity at $r = 0$, which provides better numerical stability. (Removing this singularity is called 'regularization'.) In addition, equation (3.86) is universal

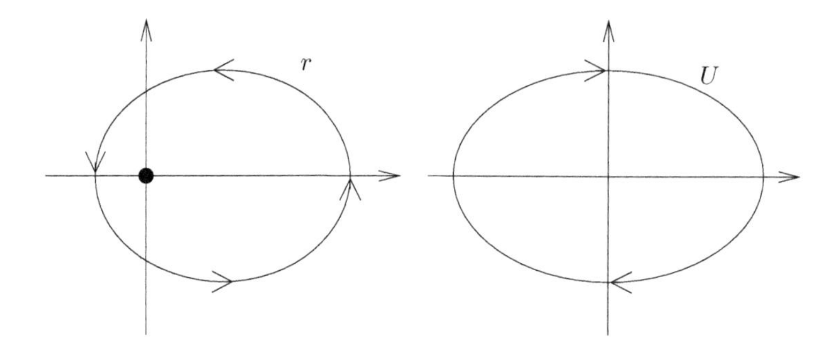

Figure 3.3 *Solution to the Kepler problem.* The particle orbit is shown on the left, and the corresponding spinor on the right. The particle completes two orbits every time U completes one cycle, since U and $-U$ describe the same position.

— it holds for $E > 0$ and $E < 0$. The solution when $E > 0$ simply has trigonometric functions replaced by exponentials. This universality is important, because perturbations can often send bound orbits into unbound ones.

For the method to be truly powerful, however, it must extend to three dimensions. The relevant formula in three dimensions is

$$\boldsymbol{x} = U\mathsf{e}_1U^\dagger, \tag{3.89}$$

where U is a general even element. This means that U has four degrees of freedom now, whereas only three are required to specify $\boldsymbol{x}$. We are therefore free to impose a further additional constraint on U, which we will use to ensure the equations take on a convenient form. The quantity $UU^\dagger$ is still a scalar in three dimensions, so we have

$$r = UU^\dagger = U^\dagger U. \tag{3.90}$$

We next form $\dot{\boldsymbol{x}}$:

$$\dot{\boldsymbol{x}} = \dot{U}\mathsf{e}_1U^\dagger + U\mathsf{e}_1\dot{U}^\dagger. \tag{3.91}$$

We would like this to equal $2\dot{U}\mathsf{e}_1U^\dagger$ for the preceding analysis to follow through. For this to hold we require

$$\dot{U}\mathsf{e}_1U^\dagger - U\mathsf{e}_1\dot{U}^\dagger = \dot{U}\mathsf{e}_1U^\dagger - (\dot{U}\mathsf{e}_1U^\dagger)^\dagger = 0. \tag{3.92}$$

The quantity $\dot{U}\mathsf{e}_1U^\dagger$ only contains odd grade terms (grade-1 and grade-3). If we subtract its reverse, all that remains is the trivector (pseudoscalar) term. We therefore require that

$$\langle \dot{U}\mathsf{e}_1U^\dagger\rangle_3 = 0, \tag{3.93}$$

which we adopt as our extra condition on U. With this condition satisfied we have

$$2\frac{dU}{ds} = \dot{\boldsymbol{x}}U\mathsf{e}_1 \tag{3.94}$$

and

$$2\frac{d^2U}{ds^2} = \left(\ddot{\boldsymbol{x}}\boldsymbol{x} + \tfrac{1}{2}\dot{\boldsymbol{x}}^2\right)U. \tag{3.95}$$

For an inverse-square force law we therefore recover the same harmonic oscillator equation. In the presence of a perturbing force we have

$$2\mu\frac{d^2U}{ds^2} - EU = \boldsymbol{f}\boldsymbol{x}U = r\boldsymbol{f}U\mathsf{e}_1. \tag{3.96}$$

This equation for U can be handled using standard techniques from perturbation theory. The equation was first found (in matrix form) by Kustaanheimo and Stiefel in 1964 . The analysis was refined and cast in its present form by Hestenes (1999).

3.4 Rotating systems and rigid-body motion

Rigid bodies can be viewed as another example of a system of particles, where now the effect of the internal forces is to keep all of the interparticle distances fixed. For such systems the internal forces can be ignored once one has found a set of dynamical variables that enforce the rigid-body constraint. The problem then reduces to solving for the motion of the centre of mass and for the angular momentum in the presence of any external forces or torques. Suitable variables are a vector $\boldsymbol{x}(t)$ for the centre of mass, and a set of variables to describe the attitude of the rigid body in space. Many forms exist for the latter variables, but here we will concentrate on parameterising the attitude of the rigid body with a *rotor*. Before applying this idea to rigid-body motion, we first look at the description of rotating frames with rotors.

3.4.1 Rotating frames

Suppose that the frame of vectors $\{\mathsf{f}_k\}$ is rotating in space. These can be related to a fixed orthonormal frame $\{\mathsf{e}_k\}$ by the time-dependent rotor $R(t)$:

$$\mathsf{f}_k(t) = R(t)\mathsf{e}_k R^\dagger(t). \tag{3.97}$$

The angular velocity vector $\boldsymbol{\omega}$ is traditionally defined by the formula

$$\dot{\mathsf{f}}_k = \boldsymbol{\omega}\times\mathsf{f}_k, \tag{3.98}$$

where the cross denotes the vector cross product. From section 2.4.3 we know that the cross product is related to the inner product with a bivector by

$$\boldsymbol{\omega}\times\mathsf{f}_k = (-I\boldsymbol{\omega})\cdot\mathsf{f}_k = \mathsf{f}_k\cdot(I\boldsymbol{\omega}). \tag{3.99}$$

We are now used to the idea that angular momentum is best viewed as a bivector, and we must expect the same to be true for angular velocity. We therefore define the angular velocity bivector Ω by

$$\Omega = I\boldsymbol{\omega}. \tag{3.100}$$

This choice ensures that the rotation has the orientation implied by Ω.

To see how Ω is related to the rotor R we start by differentiating equation (3.97):

$$\dot{\mathsf{f}}_k = \dot{R}\mathsf{e}_k R^\dagger + R\mathsf{e}_k\dot{R}^\dagger = \dot{R}R^\dagger\mathsf{f}_k + \mathsf{f}_k R\dot{R}^\dagger. \tag{3.101}$$

From the normalisation equation $RR^\dagger = 1$ we find that

$$0 = \frac{d}{dt}(RR^\dagger) = \dot{R}R^\dagger + R\dot{R}^\dagger. \tag{3.102}$$

Since differentiation and reversion are interchangeable operations we now have

$$\dot{R}R^\dagger = -R\dot{R}^\dagger = -(\dot{R}R^\dagger)^\dagger. \tag{3.103}$$

The quantity $\dot{R}R^\dagger$ is equal to minus its own reverse and has even grade, so must be a pure bivector. The equation for f_k now becomes

$$\dot{\mathsf{f}}_k = \dot{R}R^\dagger\mathsf{f}_k - \mathsf{f}_k\dot{R}R^\dagger = (2\dot{R}R^\dagger)\cdot\mathsf{f}_k. \tag{3.104}$$

Comparing this with equation (3.99) and equation (3.100) we see that $2\dot{R}R^\dagger$ must equal minus the angular velocity bivector Ω, so

$$2\dot{R}R^\dagger = -\Omega. \tag{3.105}$$

The dynamics is therefore contained in the single *rotor equation*

$$\dot{R} = -\tfrac{1}{2}\Omega R. \tag{3.106}$$

The reversed form of this is also useful:

$$\dot{R}^\dagger = \tfrac{1}{2}R^\dagger\Omega. \tag{3.107}$$

Equations of this type are surprisingly ubiquitous in physics. In the more general setting, rotors are viewed as elements of a *Lie group*, and the bivectors form their *Lie algebra*. We will have more to say about this in chapter 11.

3.4.2 Constant Ω

For the case of constant Ω equation (3.106) integrates immediately to give

$$R = \mathrm{e}^{-\Omega t/2}R_0, \tag{3.108}$$

which is the rotor for a constant frequency rotation in the positive sense in the Ω plane. The frame rotates according to

$$\mathsf{f}_k(t) = \mathrm{e}^{-\Omega t/2}R_0\mathsf{e}_k R_0^\dagger \mathrm{e}^{\Omega t/2}. \tag{3.109}$$

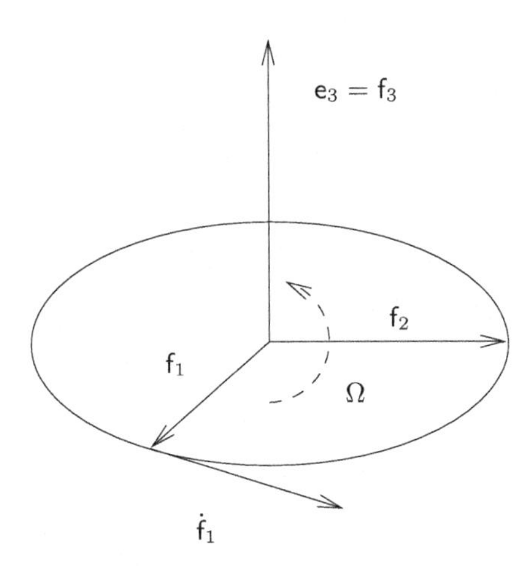

Figure 3.4 *Orientation of the angular velocity bivector.* Ω has the orientation of $\mathsf{f}_1 \wedge \dot{\mathsf{f}}_1$. It must therefore have orientation $+\mathsf{e}_1 \wedge \mathsf{e}_2$ when $\boldsymbol{\omega} = \mathsf{e}_3$.

The constant term R_0 describes the orientation of the frame at $t = 0$, relative to the $\{\mathsf{e}_k\}$ frame.

As an example, consider the case of motion about the e_3 axis (figure 3.4). We have

$$\Omega = \omega I \mathsf{e}_3 = \omega \mathsf{e}_1 \mathsf{e}_2, \tag{3.110}$$

and for convenience we set $R_0 = 1$. The motion is described by

$$\mathsf{f}_k(t) = \exp\left(-\tfrac{1}{2}\mathsf{e}_1\mathsf{e}_2\omega t\right)\mathsf{e}_k \exp\left(\tfrac{1}{2}\mathsf{e}_1\mathsf{e}_2\omega t\right), \tag{3.111}$$

so that the f_1 axis rotates as

$$\mathsf{f}_1 = \mathsf{e}_1 \exp(\mathsf{e}_1\mathsf{e}_2\omega t) = \cos(\omega t)\mathsf{e}_1 + \sin(\omega t)\mathsf{e}_2. \tag{3.112}$$

This defines a *right-handed* (anticlockwise) rotation in the $\mathsf{e}_1\mathsf{e}_2$ plane, as prescribed by the orientation of Ω.

3.4.3 Rigid-body motion

Suppose that a rigid body is moving through space. To describe the position in space of any part of the body, we need to specify the position of the centre of mass, and the vector to the point in the body from the centre of mass. The latter can be encoded in terms of a rotation from a fixed 'reference' body onto the body in space (figure 3.5). We let $\boldsymbol{x}_0$ denote the position of the centre of mass and $\boldsymbol{y}_i(t)$ denote the position (in space) of a point in the body. These are related by

$$\boldsymbol{y}_i(t) = R(t)\boldsymbol{x}_i R^\dagger(t) + \boldsymbol{x}_0(t), \tag{3.113}$$

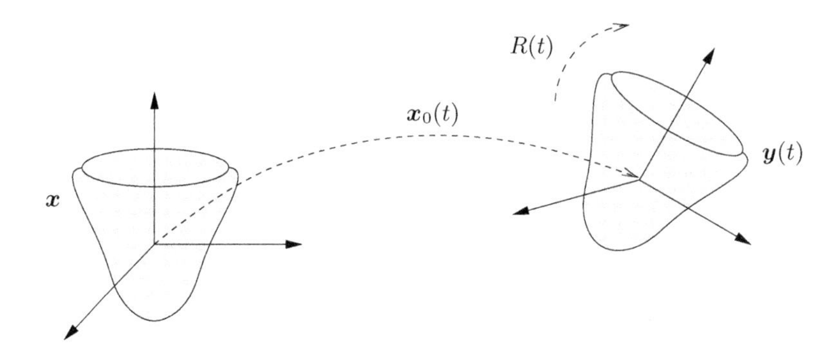

Figure 3.5 *Description of a rigid body.* The vector $\boldsymbol{x}_0(t)$ specifies the position of the centre of mass, relative to the origin. The rotor $R(t)$ defines the orientation of the body, relative to a fixed copy imagined to be placed at the origin. $\boldsymbol{x}$ is a vector in the reference body, and $\boldsymbol{y}$ is the vector in space of the equivalent point on the moving body.

where $\boldsymbol{x}_i$ is a fixed constant vector in the reference copy of the body. In this manner we have placed all of the rotational motion in the time-dependent rotor $R(t)$.

The velocity of the point $\boldsymbol{y} = R\boldsymbol{x}R^\dagger + \boldsymbol{x}_0$ is

$$\begin{aligned} \boldsymbol{v}(t) &= \dot{R}\boldsymbol{x}R^\dagger + R\boldsymbol{x}\dot{R}^\dagger + \dot{\boldsymbol{x}}_0 \\ &= -\tfrac{1}{2}\Omega R\boldsymbol{x}R^\dagger + \tfrac{1}{2}R\boldsymbol{x}R^\dagger\Omega + \boldsymbol{v}_0 \\ &= (R\boldsymbol{x}R^\dagger)\cdot\Omega + \boldsymbol{v}_0, \end{aligned} \tag{3.114}$$

where $\boldsymbol{v}_0$ is the velocity of the centre of mass. The bivector Ω defines the plane of rotation in space. This plane will lie at some orientation relative to the current position of the rigid body. For studying the motion it turns out to be extremely useful to transform the rotation plane back into the fixed, reference copy of the body. Since bivectors are subject to the same rotor transformation law as vectors we define the 'body' angular velocity Ω_B by

$$\Omega_B = R^\dagger \Omega R. \tag{3.115}$$

In terms of the body angular velocity the rotor equation becomes

$$\dot{R} = -\tfrac{1}{2}\Omega R = -\tfrac{1}{2}R\Omega_B, \qquad \dot{R}^\dagger = \tfrac{1}{2}\Omega_B R^\dagger. \tag{3.116}$$

The velocity of the body is now re-expressed as

$$\boldsymbol{v}(t) = R\,\boldsymbol{x}\cdot\Omega_B\,R^\dagger + \boldsymbol{v}_0, \tag{3.117}$$

which will turn out to be the more convenient form. (We have used the operator ordering conventions of section 2.5 to suppress unnecessary brackets in writing $R\,\boldsymbol{x}\cdot\Omega_B\,R^\dagger$ in place of $R(\boldsymbol{x}\cdot\Omega_B)R^\dagger$.)

To calculate the momentum of the rigid body we need the masses of each of the constituent particles. It is easier at this point to go to a continuum approximation and introduce a density $\rho = \rho(\boldsymbol{x})$. The position vector $\boldsymbol{x}$ is taken relative to the centre of mass, so we have

$$\int d^3x\, \rho = M \quad \text{and} \quad \int d^3x\, \rho \boldsymbol{x} = 0. \tag{3.118}$$

The momentum of the rigid body is simply

$$\int d^3x\, \rho \boldsymbol{v} = \int d^3x\, \rho (R\boldsymbol{x}\cdot\Omega_B\, R^\dagger + \boldsymbol{v}_0) = M\boldsymbol{v}_0, \tag{3.119}$$

so is specified entirely by the motion of the centre of mass. This is the continuum version of the result of section 3.1.2.

3.4.4 The inertia tensor

The next quantity we require is the angular momentum bivector L for the body about its centre of mass. We therefore form

$$\begin{aligned} L &= \int d^3x\, \rho (\boldsymbol{y} - \boldsymbol{x}_0)\wedge \boldsymbol{v} \\ &= \int d^3x\, \rho (R\boldsymbol{x}R^\dagger)\wedge(R\boldsymbol{x}\cdot\Omega_B\, R^\dagger + \boldsymbol{v}_0) \\ &= R\left(\int d^3x\, \rho \boldsymbol{x}\wedge(\boldsymbol{x}\cdot\Omega_B)\right)R^\dagger. \end{aligned} \tag{3.120}$$

The integral inside the brackets refers only to the fixed copy and so defines a time-independent function of Ω_B. This is the reason for working with Ω_B instead of the space angular velocity Ω. We define the *inertia tensor* $\mathcal{I}(B)$ by

$$\mathcal{I}(B) = \int d^3x\, \rho \boldsymbol{x}\wedge(\boldsymbol{x}\cdot B). \tag{3.121}$$

This is a linear function mapping bivectors to bivectors. This way of writing linear functions may be unfamiliar to those used to seeing tensors labelled with indices, but the notation is the natural extension to linear functions of the index-free approach advocated in this book. The linearity of the map is easy to check:

$$\begin{aligned} \mathcal{I}(\lambda A + \mu B) &= \int d^3x\, \rho \boldsymbol{x}\wedge\big(\boldsymbol{x}\cdot(\lambda A + \mu B)\big) \\ &= \int d^3x\, \rho \big(\lambda \boldsymbol{x}\wedge(\boldsymbol{x}\cdot A) + \mu \boldsymbol{x}\wedge(\boldsymbol{x}\cdot B)\big) \\ &= \lambda \mathcal{I}(A) + \mu \mathcal{I}(B). \end{aligned} \tag{3.122}$$

The fact that the inertia tensor maps bivectors to bivectors, rather than vectors to vectors, is also a break from tradition. This viewpoint is very natural given our earlier comments about the merits of bivectors over axial vectors, and provides a

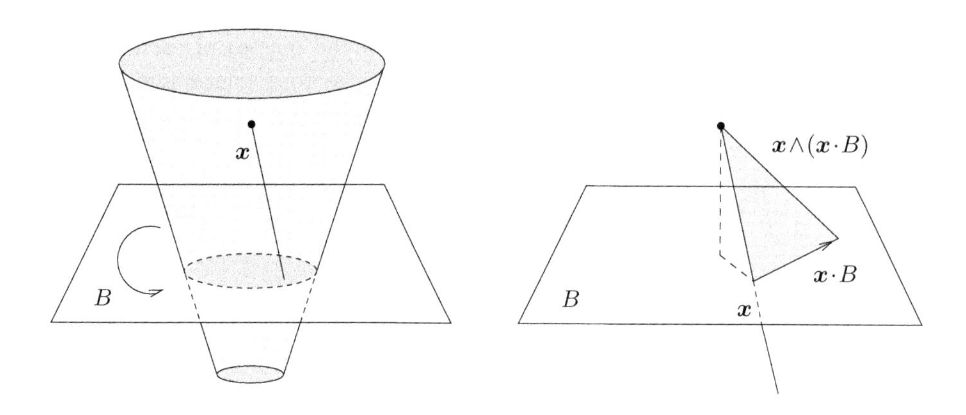

Figure 3.6 *The inertia tensor.* The inertia tensor $\mathcal{I}(B)$ is a linear function mapping its bivector argument B onto a bivector. It returns the total angular momentum about the centre of mass for rotation in the B plane.

clear geometric picture of the tensor (figure 3.6). Since both vectors and bivectors belong to a three-dimensional linear space, there is no additional complexity introduced in this new picture.

To understand the effect of the inertia tensor, suppose that the body rotates in the B plane at a fixed rate $|B|$, and we place the origin at the centre of mass (which is fixed). The velocity of the vector $\boldsymbol{x}$ is simply $\boldsymbol{x}\cdot B$, and the momentum density at this point is $\rho\boldsymbol{x}\cdot B$, as shown in figure 3.6. The angular momentum density bivector is therefore $\boldsymbol{x}\wedge(\rho\boldsymbol{x}\cdot B)$, and integrating this over the entire body returns the total angular momentum bivector for rotation in the B plane.

In general, the total angular momentum will not lie in the same plane as the angular velocity. This is one reason why rigid-body dynamics can often seem quite counterintuitive. When we see a body rotating, our eyes naturally pick out the angular *velocity* by focusing on the vector the body rotates around. Deciding the plane of the angular momentum is less easy, particularly if the internal mass distribution is hidden from us. But it is the angular momentum that responds directly to external torques, not the angular velocity, and this can have some unexpected consequences.

We have calculated the inertia tensor about the centre of mass, but bodies rotating around a fixed axis can be forced to rotate about any point. A useful theorem relates the inertia tensor about an arbitrary point to one about the centre of mass. Suppose that we want the inertia tensor relative to the point $\boldsymbol{a}$, where $\boldsymbol{a}$ is a vector taken from the centre of mass. Returning to the definition of equation (3.121) we see that we need to compute

$$\mathcal{I}_a(B) = \int d^3x\, \rho(\boldsymbol{x}-\boldsymbol{a})\wedge\big((\boldsymbol{x}-\boldsymbol{a})\cdot B\big). \tag{3.123}$$

This integral evaluates to give

$$\begin{aligned}\mathcal{I}_a(B) &= \int d^3x\,\rho\big(\boldsymbol{x}\wedge(\boldsymbol{x}\cdot B) - \boldsymbol{x}\wedge(\boldsymbol{a}\cdot B) - \boldsymbol{a}\wedge(\boldsymbol{x}\cdot B) + \boldsymbol{a}\wedge(\boldsymbol{a}\cdot B)\big)\\ &= \mathcal{I}(B) + M\boldsymbol{a}\wedge(\boldsymbol{a}\cdot B).\end{aligned} \tag{3.124}$$

The inertia tensor relative to $\boldsymbol{a}$ is simply the inertia tensor about the centre of mass, plus the tensor for a point mass M at position $\boldsymbol{a}$.

3.4.5 Principal axes

So far we have only given an abstract specification of the inertia tensor. For most calculations it is necessary to introduce a set of basis vectors fixed in the body. As we are free to choose the directions of these vectors, we should ensure that this choice simplifies the equations of motion as much as possible. To see how to do this, consider the $\{\mathsf{e}_i\}$ frame and define the matrix $\mathcal{I}_{ij}$ by

$$\mathcal{I}_{ij} = -(I\mathsf{e}_i)\cdot\mathcal{I}(I\mathsf{e}_j). \tag{3.125}$$

This defines a *symmetric* matrix, as follows from the result

$$A\cdot(\boldsymbol{x}\wedge(\boldsymbol{x}\cdot B)) = \langle A\boldsymbol{x}(\boldsymbol{x}\cdot B)\rangle = \langle (A\cdot\boldsymbol{x})\boldsymbol{x}B\rangle = B\cdot(\boldsymbol{x}\wedge(\boldsymbol{x}\cdot A)). \tag{3.126}$$

(This sort of manipulation, where one uses the projection onto grade to replace inner and outer products by geometric products, is very common in geometric algebra.) This result ensures that

$$\begin{aligned}\mathcal{I}_{ij} &= -\int d^3x\,\rho(I\mathsf{e}_i)\cdot\big(\boldsymbol{x}\wedge(\boldsymbol{x}\cdot(I\mathsf{e}_j))\big)\\ &= -\int d^3x\,\rho(I\mathsf{e}_j)\cdot\big(\boldsymbol{x}\wedge(\boldsymbol{x}\cdot(I\mathsf{e}_i))\big) = \mathcal{I}_{ji}.\end{aligned} \tag{3.127}$$

It follows that the matrix $\mathcal{I}_{ij}$ will be diagonal if the $\{\mathsf{e}_i\}$ frame is chosen to coincide with the eigendirections of the inertia tensor. These directions are called the *principal axes*, and we always choose our frame along these directions.

The matrix $\mathcal{I}_{ij}$ is also positive-(semi)definite, as can be seen from

$$\begin{aligned}a_i a_j \mathcal{I}_{ij} &= -\int d^3x\,\rho(I\boldsymbol{a})\cdot\big(\boldsymbol{x}\wedge(\boldsymbol{x}\cdot(I\boldsymbol{a}))\big)\\ &= \int d^3x\,\rho\big(\boldsymbol{x}\cdot(I\boldsymbol{a})\big)^2 \geq 0.\end{aligned} \tag{3.128}$$

It follows that all of the eigenvalues of $\mathcal{I}_{ij}$ must be positive (or possibly zero for the case of point or line masses). These eigenvalues are the principal moments of inertia and are crucial in specifying the properties of a rigid body. We denote these $\{i_1, i_2, i_3\}$, so that

$$\mathcal{I}_{jk} = \delta_{jk} i_k \quad \text{(no sum)}. \tag{3.129}$$

(It is more traditional to use a capital I for the moments of inertia, but this symbol is already employed for the pseudoscalar.) If two or three of the principal moments are the same the principal axes are not uniquely specified. In this case one simply chooses one orthonormal set of eigenvectors from the degenerate family of possibilities.

Returning to the index-free presentation, we see that the principal axes satisfy

$$\mathcal{I}(I\mathrm{e}_j) = I\mathrm{e}_k\mathcal{I}_{jk} = i_j I\mathrm{e}_j, \tag{3.130}$$

where again there is no sum implied between eigenvectors and their associated eigenvalue in the final expression. To calculate the effect of the inertia tensor on an arbitrary bivector B we decompose B in terms of the principal axes as

$$B = B_j I\mathrm{e}_j. \tag{3.131}$$

It follows that

$$\mathcal{I}(B) = \sum_{j=1}^{3} i_j B_j I\mathrm{e}_j = i_1 B_1 \mathrm{e}_2\mathrm{e}_3 + i_2 B_2 \mathrm{e}_3\mathrm{e}_1 + i_3 B_3 \mathrm{e}_1\mathrm{e}_2. \tag{3.132}$$

The fact that for most bodies the principal moments are not equal demonstrates that $\mathcal{I}(B)$ will not lie in the same plane as B, unless B is perpendicular to one of the principal axes.

A useful result for calculating the inertia tensor is that the principal axes of a body always coincide with symmetry axes, if any are present. This simplifies the calculation of the inertia tensor for a range of standard bodies, the results for which can be found in some of the books listed at the end of this chapter.

3.4.6 Kinetic energy and angular momentum

To calculate the kinetic energy of the body from the velocity of equation (3.114) we form the integral

$$\begin{aligned} T &= \tfrac{1}{2}\int d^3x\, \rho(R\boldsymbol{x}\cdot\Omega_B\, R^\dagger + \boldsymbol{v}_0)^2 \\ &= \tfrac{1}{2}\int d^3x\, \rho\big((\boldsymbol{x}\cdot\Omega_B)^2 + 2\boldsymbol{v}_0\cdot(R\boldsymbol{x}\cdot\Omega_B\, R^\dagger) + \boldsymbol{v}_0^2\big) \\ &= \tfrac{1}{2}\int d^3x\, \rho(\boldsymbol{x}\cdot\Omega_B)^2 + M\boldsymbol{v}_0^2. \end{aligned} \tag{3.133}$$

Again, there is a clean split into a rotational contribution and a term due to the motion of the centre of mass. Concentrating on the former, we use the manipulation

$$(\boldsymbol{x}\cdot\Omega_B)^2 = \langle \boldsymbol{x}\cdot\Omega_B \boldsymbol{x}\Omega_B\rangle = -\Omega_B\cdot\big(\boldsymbol{x}\wedge(\boldsymbol{x}\cdot\Omega_B)\big) \tag{3.134}$$

to write the rotational contribution as

$$-\tfrac{1}{2}\Omega_B\cdot\Big(\int d^3x\,\rho\boldsymbol{x}\wedge(\boldsymbol{x}\cdot\Omega_B)\Big) = -\tfrac{1}{2}\Omega_B\cdot\mathcal{I}(\Omega_B). \tag{3.135}$$

The minus sign is to be expected because bivectors all have negative squares. The sign can be removed by reversing one of the bivectors to construct a positive-definite product. The total kinetic energy is therefore

$$T = \tfrac{1}{2}M\boldsymbol{v}_0^2 + \tfrac{1}{2}\Omega_B^\dagger\cdot\mathcal{I}(\Omega_B). \tag{3.136}$$

The inertia tensor is constructed from the point of view of the fixed body. From equation (3.120) we see that the angular momentum in space is obtained by rotating the body angular momentum $\mathcal{I}(\Omega_B)$ onto the space configuration, that is,

$$L = R\mathcal{I}(\Omega_B)\,R^\dagger. \tag{3.137}$$

We can understand this expression as follows. Suppose that a body rotates in space with angular velocity Ω. At a given instant we carry out a fixed rotation to align everything back with the fixed reference configuration. This reference copy then has angular velocity $\Omega_B = R^\dagger\Omega R$. The inertia tensor (fixed in the reference copy) returns the angular momentum, given an input angular velocity. The result of this is then rotated forwards onto the body in space, to return L.

The space and body angular velocities are related by $\Omega = R\Omega_B R^\dagger$, so the kinetic energy can be written in the form

$$T = \tfrac{1}{2}M\boldsymbol{v}_0^2 + \tfrac{1}{2}\Omega^\dagger\cdot L. \tag{3.138}$$

We now introduce components $\{\omega_k\}$ for both Ω and Ω_B by writing

$$\Omega = \sum_{k=1}^{3}\omega_k I\mathsf{f}_k, \quad \Omega_B = \sum_{k=1}^{3}\omega_k I\mathsf{e}_k. \tag{3.139}$$

In terms of these we recover the standard expression

$$T = \tfrac{1}{2}M\boldsymbol{v}_0^2 + \sum_{k=1}^{3}\tfrac{1}{2}i_k\omega_k^2. \tag{3.140}$$

3.4.7 Equations of motion

The equations of motion are $\dot{L} = N$, where N is the external torque. The inertia tensor is time-independent since it only refers to the static 'reference' copy of the rigid body, so we find that

$$\begin{aligned}\dot{L} &= \dot{R}\mathcal{I}(\Omega_B)R^\dagger + R\mathcal{I}(\Omega_B)\dot{R}^\dagger + R\mathcal{I}(\dot{\Omega}_B)R^\dagger \\ &= R\big(\mathcal{I}(\dot{\Omega}_B) - \tfrac{1}{2}\Omega_B\mathcal{I}(\Omega_B) + \tfrac{1}{2}\mathcal{I}(\Omega_B)\Omega_B\big)R^\dagger.\end{aligned} \tag{3.141}$$

At this point it is extremely useful to have a symbol to denote one-half of the commutator of two bivectors. The standard symbol for this is the cross, $\times$, so we define the *commutator product* by

$$A\times B = \tfrac{1}{2}(AB - BA). \tag{3.142}$$

This notation does raise the possibility of confusion with the vector cross product, but as the latter is not needed any more this should not pose a problem. The commutator product is so ubiquitous in applications that it needs its own symbol, and the cross is particularly convenient as it correctly conveys the antisymmetry of the product. In section 4.1.3 we prove that the commutator of any two bivectors results in a third bivector. This is easily confirmed in three dimensions by expressing both bivectors in terms of their dual vectors.

With the commutator product at our disposal the equations of motion are now written concisely as

$$\dot{L} = R\big(\mathcal{I}(\dot{\Omega}_B) - \Omega_B\times\mathcal{I}(\Omega_B)\big)R^\dagger. \tag{3.143}$$

The typical form of the rigid-body equations is recovered by expanding in terms of components. In terms of these we have

$$\begin{aligned}\dot{L} &= R\left(\sum_{k=1}^{3} i_k\dot{\omega}_k I\mathsf{e}_k - \sum_{j,k=1}^{3} i_k\omega_j\omega_k (I\mathsf{e}_j)\times(I\mathsf{e}_k)\right)R^\dagger\\ &= \sum_{k=1}^{3}\dot{\omega}_k I\mathsf{f}_k + \sum_{j,k,l=1}^{3}\epsilon_{jkl} i_k\omega_j\omega_k I\mathsf{f}_l.\end{aligned} \tag{3.144}$$

If we let N_k denote the components of the torque N in the rotating f_k frame,

$$N = \sum_{k=1}^{3} N_k I\mathsf{f}_k, \tag{3.145}$$

we recover the Euler equations of motion for a rigid body:

$$\begin{aligned} i_1\dot{\omega}_1 - \omega_2\omega_3(i_2 - i_3) &= N_1,\\ i_2\dot{\omega}_2 - \omega_3\omega_1(i_3 - i_1) &= N_2,\\ i_3\dot{\omega}_3 - \omega_1\omega_2(i_1 - i_2) &= N_3.\end{aligned} \tag{3.146}$$

Various methods can be used to solve these equations and are described in most mechanics textbooks. Here we will simply illustrate some features of the equations, and describe a solution method which does not resort to the explicit coordinate equations.

3.4.8 Torque-free motion

The torque-free equation $\dot{L} = 0$ reduces to

$$\mathcal{I}(\dot{\Omega}_B) - \Omega_B \times \mathcal{I}(\Omega_B) = 0. \tag{3.147}$$

This is a first-order constant coefficient differential equation for the bivector Ω_B. Closed form solutions exist, but before discussing some of these it is useful to consider the conserved quantities. Throughout this section we ignore any overall motion of the centre of mass of the rigid body. Since $\dot{L} = 0$ both the kinetic energy and the magnitude of L are constant. To exploit this we introduce the components

$$L_k = i_k \omega_k, \quad L = \sum_{k=1}^{3} L_k I \mathsf{f}_k. \tag{3.148}$$

These are the components of L in the rotating f_k frame. So, even though L is constant, the components L_k are time-dependent. In terms of these components the magnitude of L is

$$LL^\dagger = L_1^2 + L_2^2 + L_3^3 \tag{3.149}$$

and the kinetic energy is

$$T = \frac{L_1^2}{2i_1} + \frac{L_2^2}{2i_2} + \frac{L_3^2}{2i_3}. \tag{3.150}$$

Both $|L|$ and T are constants of motion, which imposes two constraints on the three components L_k. A useful way to visualise this is to think in terms of a vector $\boldsymbol{l}$ with components L_k:

$$\boldsymbol{l} = \sum_{k=1}^{3} L_k \mathsf{e}_k = -IR^\dagger L R. \tag{3.151}$$

This is the vector perpendicular to $R^\dagger L R$ — a rotating vector in the fixed reference body. Conservation of $|L|$ means that $\boldsymbol{l}$ is constrained to lie on a sphere, and conservation of T restricts $\boldsymbol{l}$ to the surface of an ellipsoid. Possible paths for $\boldsymbol{l}$ for a given rigid body are therefore defined by the intersections of a sphere with a family of ellipsoids (governed by T). For the case of unequal principal moments these orbits are non-degenerate. Examples of these orbits are shown in figure 3.7. This figure shows that orbits around the axes with the smallest and largest principal moments are stable, whereas around the middle axis the orbits are unstable. Any small change in the energy of the body will tend to throw it into a very different orbit if the orbit of $\boldsymbol{l}$ approaches close to e_2.

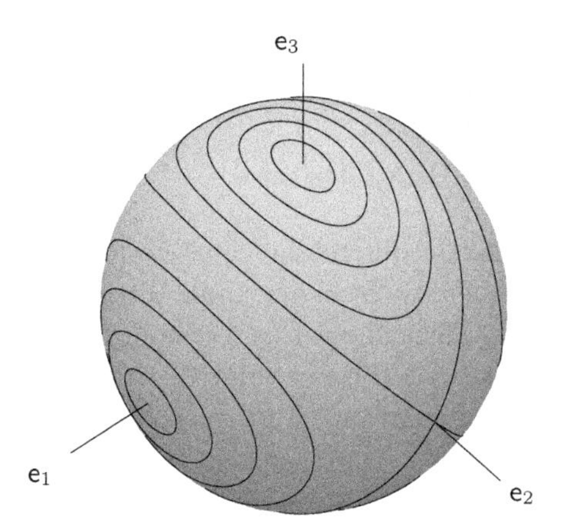

Figure 3.7 *Angular momentum orbits.* The point described by the vector $\boldsymbol{l}$ simultaneously lies on the surface of a sphere and an ellipse. The figure shows possible paths on the sphere for $\boldsymbol{l}$ in the case of $i_1 < i_2 < i_3$, with the 3 axis vertical.

3.4.9 The symmetric top

The full analytic solution for torque-free motion is complicated and requires elliptic functions. If the body has a single symmetry axis, however, the solution is quite straightforward. In this case the body has two equal moments of inertia, $i_1 = i_2$, and the third principal moment i_3 is assumed to be different. With this assignment e_3 is the symmetry axis of the body. The action of the inertia tensor on Ω_B is

$$\begin{aligned}\mathcal{I}(\Omega_B) &= i_1\omega_1\mathsf{e}_2\mathsf{e}_3 + i_1\omega_2\mathsf{e}_3\mathsf{e}_1 + i_3\omega_3\mathsf{e}_1\mathsf{e}_2 \\ &= i_1\Omega_B + (i_3 - i_1)\omega_3 I\mathsf{e}_3,\end{aligned} \tag{3.152}$$

so we can write $\mathcal{I}(\Omega_B)$ in the compact form

$$\mathcal{I}(\Omega_B) = i_1\Omega_B + (i_3 - i_1)(\Omega_B\wedge\mathsf{e}_3)\mathsf{e}_3. \tag{3.153}$$

(This type of expression offers many advantages over the alternative 'dyad' notation.) The torque-free equations of motion are now

$$\mathcal{I}(\dot{\Omega}_B) = \Omega_B\times\mathcal{I}(\Omega_B) = (i_3 - i_1)\Omega_B\times\big((\Omega_B\wedge\mathsf{e}_3)\mathsf{e}_3\big). \tag{3.154}$$

Since $\Omega_B\wedge\mathsf{e}_3$ is a trivector, we can dualise the final term and write

$$\mathcal{I}(\dot{\Omega}_B) = -(i_3 - i_1)\mathsf{e}_3\wedge\big((\Omega_B\wedge\mathsf{e}_3)\Omega_B\big). \tag{3.155}$$

It follows that

$$\mathsf{e}_3\wedge\mathcal{I}(\dot{\Omega}_B) = 0 = i_3\dot{\omega}_3 I, \tag{3.156}$$

which shows that ω_3 is a constant. This result can be read off directly from the Euler equations, but it is useful to see how it can be derived without dropping down to the individual component equations. The ability to do this becomes ever more valuable as the complexity of the equations increases.

Next we use the result that

$$\begin{aligned} i_1 \Omega_B &= \mathcal{I}(\Omega_B) - (i_3 - i_1)(\Omega_B \wedge \mathsf{e}_3)\mathsf{e}_3 \\ &= \mathcal{I}(\Omega_B) + (i_1 - i_3)\omega_3 I \mathsf{e}_3 \end{aligned} \tag{3.157}$$

to write

$$\Omega = R\Omega_B R^\dagger = \frac{1}{i_1} L + \frac{i_1 - i_3}{i_1} \omega_3 R I \mathsf{e}_3 R^\dagger. \tag{3.158}$$

Our rotor equation now becomes

$$\dot{R} = -\tfrac{1}{2}\Omega R = -\frac{1}{2i_1}(LR + R(i_1 - i_3)\omega_3 I \mathsf{e}_3). \tag{3.159}$$

The right-hand side of this equation involves two constant bivectors, one multiplying R to the left and the other to the right. We therefore define the two bivectors

$$\Omega_l = \frac{1}{i_1} L, \qquad \Omega_r = \omega_3 \frac{i_1 - i_3}{i_1} I \mathsf{e}_3, \tag{3.160}$$

so that the rotor equation becomes

$$\dot{R} = -\tfrac{1}{2}\Omega_l R - \tfrac{1}{2} R \Omega_r. \tag{3.161}$$

This equation integrates immediately to give

$$R(t) = \exp(-\tfrac{1}{2}\Omega_l t) R_0 \exp(-\tfrac{1}{2}\Omega_r t). \tag{3.162}$$

This fully describes the motion of a symmetric top. It shows that there is an 'internal' rotation in the $\mathsf{e}_1\mathsf{e}_2$ plane (the symmetry plane of the body). This is responsible for the precession of a symmetric top. The constant rotor R_0 defines the attitude of the rigid body at $t = 0$ and can be set to 1. The resultant body is then rotated in the plane of its angular momentum to obtain the final attitude in space.

3.5 Notes

Much of this chapter follows *New Foundations for Classical Mechanics* by David Hestenes (1999), which gives a comprehensive account of the applications to classical mechanics of geometric algebra in three dimensions. Readers are encouraged to compare the techniques used in this chapter with more traditional methods, a good description of which can be found in *Classical Mechanics* by Goldstein (1950), or *Analytical Mechanics* by Hand & Finch (1998). The standard reference for the Kustaanheimo–Stiefel equation is *Linear and Regular Celestial Mechanics* by Stiefel and Scheifele (1971). Many authors have explored this technique, particularly in the quaternionic framework. These include Hestenes' 'Celestial mechanics with geometric algebra' (1983) and the papers by Aramanovitch (1995) and Vrbik (1994, 1995).

3.6 Exercises

3.1 An elliptical orbit in an inverse-square force law is parameterised in terms of a scalar + pseudoscalar quantity U by $\boldsymbol{x} = U^2\mathsf{e}_1$. Prove that U can be written

$$U = A_0 \mathrm{e}^{I\omega s} + B_0 \mathrm{e}^{-I\omega s},$$

where $dt/ds = r$, $r = |\boldsymbol{x}| = UU^\dagger$ and I is the unit bivector for the plane. What is the value of ω? Find the conditions on A_0 and B_0 such that at time $t = 0$, $s = 0$ and the particle lies on the positive e_1 axis with velocity in the positive e_2 direction. For which value of s does the velocity point in the $-\mathsf{e}_1$ direction? Find the values for the shortest and longest diameters of the ellipse, and verify that we can write

$$U = \sqrt{a(1+e)}\cos(\omega s) - \sqrt{a(1-e)}I\sin(\omega s),$$

where e is the eccentricity and a is the semi-major axis.

3.2 For elliptical orbits the semi-major axis a is defined by $a = \frac{1}{2}(r_1 + r_2)$, where r_1 and r_2 are the distances of closest and furthest approach. Prove that

$$\frac{l^2}{k\mu} = a(1-e^2).$$

Hence show that we can write

$$r = \frac{a(1-e^2)}{1+e\cos(\theta)},$$

where $e\cos(\theta) = \boldsymbol{e}\cdot\hat{\boldsymbol{x}}$. The eccentricity vector points to the point of closest approach. Why would we expect the orbital average of $\hat{\boldsymbol{x}}/r^4$ to also point in this direction? Prove that

$$\int_0^T dt \frac{\hat{\boldsymbol{x}}}{r^4} = \hat{\boldsymbol{e}} \frac{\mu}{la^2(1-e^2)^2} \int_0^{2\pi} \bigl(1+e\cos(\theta)\bigr)^2 \cos(\theta)\, d\theta$$

and evaluate the integral.

3.3 A particle in three dimensions moves along a curve $\boldsymbol{x}(t)$ such that $|\boldsymbol{v}|$ is constant. Show that there exists a bivector Ω such that

$$\dot{\boldsymbol{v}} = \Omega\cdot\boldsymbol{v},$$

and give an explicit formula for Ω. Is this bivector unique?

3.4 Suppose that we measure components of the position vector $\boldsymbol{x}$ in a rotating frame $\{\mathsf{f}_i\}$. By referring this frame to a fixed frame, show that the components of $\boldsymbol{x}$ are given by

$$x_i = \mathsf{e}_i \cdot (R^\dagger \boldsymbol{x} R).$$

By differentiating this expression twice, prove that we can write

$$\mathsf{f}_i \ddot{x}_i = \ddot{\boldsymbol{x}} + \Omega \cdot (\Omega \cdot \boldsymbol{x}) + 2\Omega \cdot \dot{\boldsymbol{x}} + \dot{\Omega} \cdot \boldsymbol{x}.$$

Hence deduce expressions for the centrifugal, Coriolis and Euler forces in terms of the angular velocity bivector Ω.

3.5 Show that the inertia tensor satisfies the following properties:

$$\begin{aligned} \text{linearity:} \quad & \mathcal{I}(\lambda A + \mu B) = \lambda \mathcal{I}(A) + \mu \mathcal{I}(B) \\ \text{symmetry:} \quad & \langle A \mathcal{I}(B) \rangle = \langle \mathcal{I}(A) B \rangle. \end{aligned}$$

3.6 Prove that the inertia tensor $\mathcal{I}(B)$ for a solid cylinder of height h and radius a can be written

$$\mathcal{I}(B) = \frac{Mh^2}{12}(B - B \wedge \mathsf{e}_3 \, \mathsf{e}_3) + \frac{Ma^2}{4}(B + B \wedge \mathsf{e}_3 \, \mathsf{e}_3),$$

where e_3 is the symmetry axis.

3.7 For a torque-free symmetric top prove that the angular momentum, viewed back in the reference copy, rotates around the symmetry axis at an angular frequency ω, where

$$\omega = \omega_3 \frac{i_3 - i_1}{i_1}.$$

Show that the angle between the symmetry axis and the vector $\boldsymbol{l} = -IL$ is given by

$$\cos(\theta) = \frac{i_3 \omega}{l},$$

where $l^2 = \boldsymbol{l}^2 = LL^\dagger$. Hence show that the symmetry axis rotates in space in the L plane at an angular frequency ω', where

$$\omega' = \frac{i_3 \omega_3}{i_1 \cos(\theta)}.$$

4

Foundations of geometric algebra

In chapter 2 we introduced geometric algebra in two and three dimensions. We now turn to a discussion of the full, axiomatic framework for geometric algebra in arbitrary dimensions, with arbitrary signature. This will involve some duplication of material from chapter 2, but we hope that this will help reinforce some of the key concepts. Much of the material in this chapter is of primary relevance to those interested in the full range of applications of geometric algebra. Those interested solely in applications to space and spacetime may want to skip some of the material below, as both of these algebras are treated in a self-contained manner in chapters 2 and 5 respectively. The material on frames and linear algebra is important, however, and a knowledge of this is assumed for applications in gravitation.

The fact that geometric algebra can be applied in spaces of arbitrary dimensions is crucial to the claim that it is a mathematical tool of universal applicability. The framework developed here will enable us to extend geometric algebra to the study of relativistic dynamics, phase space, single and multiparticle quantum theory, Lie groups and manifolds. This chapter also highlights some of the new algebraic techniques we now have at our disposal. Many derivations can be simplified through judicious use of the geometric product at various intermediate steps. This is true even if the initial and final expressions contain only inner and outer products.

Many key relations in physics involve linear mappings between one space and another. In this chapter we also explore how geometric algebra simplifies the rich subject of linear transformations. We start with simple mappings between vectors in the same space and study their properties in a very general, basis-free framework. In later chapters this framework is extended to encompass functions between different spaces, and multilinear functions where the argument of the function can consist of one or more multivectors.

4.1 Axiomatic development

We should now have an intuitive feel for the elements of a geometric algebra — the multivectors — and some of their multiplicative properties. The next step is to define a set of axioms and conventions which enable us to efficiently manipulate them. Geometric algebra can be defined using a number of axiomatic frameworks, all of which give rise to the same final algebra. In the main we will follow the approach first developed by Hestenes and Sobczyk and raise the geometric product to primary status in the algebra. The properties of the inner and outer products are then inherited from the full geometric product, and this simplifies proofs of a number of important results.

Our starting point is the vector space from which the entire algebra will be generated. Vectors (i.e. grade-1 multivectors) have a special status in the algebra, as the grading of the algebra is determined by them. Three main axioms govern the properties of the geometric product for vectors.

(i) The geometric product is associative:

$$a(bc) = (ab)c = abc. \tag{4.1}$$

(ii) The geometric product is distributive over addition:

$$a(b + c) = ab + ac. \tag{4.2}$$

(iii) The square of any vector is a real scalar: $a^2 \in \Re$.

The final axiom is the key one which distinguishes a geometric algebra from a general associative algebra. We do not force the scalar to be positive, so we can incorporate Minkowski spacetime without modification of our axioms. Nothing is assumed about the commutation properties of the geometric product — matrix multiplication is one picture to keep in mind. Indeed, one can always represent the geometric product in terms of products of suitably chosen matrices, but this does not bring any new insights into the properties of the geometric product.

By successively multiplying together vectors we generate the complete algebra. Elements of this algebra are called multivectors and are usually written in upper-case italic font. The space of multivectors is *linear over the real numbers*, so if λ and μ are scalars and A and B are multivectors $\lambda A + \mu B$ is also a multivector. We only consider the algebra over the reals as most occurrences of complex numbers in physics turn out to have a geometric origin. This geometric meaning can be lost if we admit a scalar unit imaginary. Any multivector can be written as a sum of geometric products of vectors. They too can be multiplied using the geometric product and this product inherits properties (i) and (ii) above. So, for multivectors A, B and C, we have

$$(AB)C = A(BC) = ABC \tag{4.3}$$

and

$$A(B+C) = AB + AC. \tag{4.4}$$

If we now form the square of the vector $a+b$ we find that

$$(a+b)^2 = (a+b)(a+b) = a^2 + ab + ba + b^2. \tag{4.5}$$

It follows that the symmetrised product of two vectors can be written

$$ab + ba = (a+b)^2 - a^2 - b^2, \tag{4.6}$$

and so must also be a scalar, by axiom (iii). We therefore *define* the inner product for vectors by

$$a\cdot b = \frac{1}{2}(ab + ba). \tag{4.7}$$

The remaining, antisymmetric part of the geometric product is defined as the exterior product and returns a bivector,

$$a\wedge b = \frac{1}{2}(ab - ba). \tag{4.8}$$

These definitions combine to give the familiar result

$$ab = a\cdot b + a\wedge b. \tag{4.9}$$

In forming this decomposition we have defined both the inner and outer products of vectors in terms of the geometric product. This contrasts with the common alternative of defining the geometric product in terms of separate inner and outer products. Some authors prefer this alternative because the (less familiar) geometric product is defined in terms of more familiar objects. The main drawback, however, is that work still remains to establish the main properties of the geometric product. In particular, it is far from obvious that the product is associative, which is invaluable for its use.

4.1.1 The outer product, grading and bases

In the preceding we defined the outer product of two vectors and asserted that this returns a bivector (a grade-2 multivector). This is the key to defining the grade operation for the entire algebra. To do this we first extend the definition of the outer product to arbitrary numbers of vectors. The outer (exterior) product of the vectors $a_1, \ldots, a_r$ is denoted by $a_1 \wedge a_2 \wedge \cdots \wedge a_r$ and is defined as the totally antisymmetrised sum of all geometric products:

$$a_1\wedge a_2\wedge\cdots\wedge a_r = \frac{1}{r!}\sum (-1)^{\epsilon} a_{k_1} a_{k_2} \cdots a_{k_r}. \tag{4.10}$$

The sum runs over every permutation $k_1, \ldots, k_r$ of $1, \ldots, r$, and $(-1)^\epsilon$ is $+1$ or -1 as the permutation $k_1, \ldots, k_r$ is even or odd respectively. So, for example,

$$a_1 \wedge a_2 = \frac{1}{2!}(a_1 a_2 - a_2 a_1) \tag{4.11}$$

as required.

The antisymmetry of the outer product ensures that it vanishes if any two vectors are the same. It follows that the outer product vanishes if the vectors are linearly dependent, since in this case one vector can be written as a linear combination of the remaining vectors. The outer product therefore records the dimensionality of the object formed from a set of vectors. This is precisely what we mean by *grade*, so we define the outer product of r vectors as having grade r. Any multivector which can be written purely as the outer product of a set of vectors is called a *blade*. Any multivector can be expressed as a sum of blades, as can be verified by introducing an explicit basis. These blades all have definite grade and in turn define the grade or grades of the multivector.

We rarely need the full antisymmetrised expression when studying blades. Instead we can employ the result that *every blade can be written as a geometric product of orthogonal, anticommuting vectors.* The anticommutation of orthogonal vectors then takes care of the antisymmetry of the product. In Euclidean space this result is simple to prove using a form of Gram–Schmidt orthogonalisation. Given two vectors a and b we form

$$b' = b - \lambda a. \tag{4.12}$$

We then see that

$$a \wedge (b - \lambda a) = a \wedge b - \lambda a \wedge a = a \wedge b. \tag{4.13}$$

So the same bivector is obtained, whatever the value of λ (figure 4.1). The bivector encodes an oriented plane with magnitude determined by the area. Interchanging b and b' changes neither the orientation nor the magnitude, so returns the same bivector. We now form

$$a \cdot b' = a \cdot (b - \lambda a) = a \cdot b - \lambda a^2. \tag{4.14}$$

So if we set $\lambda = a \cdot b / a^2$ we have $a \cdot b' = 0$ and can write

$$a \wedge b = a \wedge b' = a b'. \tag{4.15}$$

One can continue in this manner and construct a complete set of orthogonal vectors generating the same outer product.

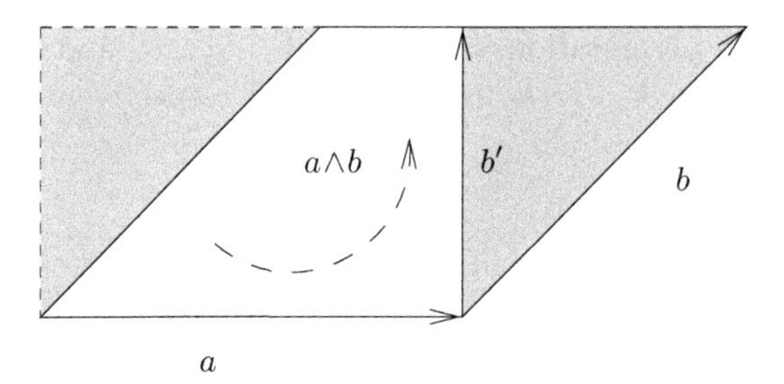

Figure 4.1 *The Gram–Schmidt process.* The outer product $a \wedge b$ is independent of shape of the parallelogram formed by a and b. The only information contained in $a \wedge b$ is the oriented plane and a magnitude. The vectors b and b' generate the same bivector, so we can choose b' orthogonal to a and write $a \wedge b = ab'$.

An alternative form for b' is quite revealing. We write

$$\begin{aligned} b' &= b - a^{-1} a \cdot b \\ &= a^{-1}(ab - a \cdot b) \\ &= a^{-1}(a \wedge b). \end{aligned} \tag{4.16}$$

This shows that b' is formed by rotating a through 90° in the $a \wedge b$ plane, and dilating by the appropriate amount. The algebraic form also makes it clear why $ab' = a \wedge b$, and gives a formula that extends simply to higher grades.

The above argument is fine for Euclidean space, but breaks down for spaces of mixed signature. The inverse $a^{-1} = a/a^2$ is not defined when a is null ($a^2 = 0$), so an alternative procedure is required. Fortunately this is a relatively straightforward exercise. We start with the set of r independent vectors $a_1, \ldots, a_r$ and form the $r \times r$ symmetric matrix

$$\mathsf{M}_{ij} = a_i \cdot a_j. \tag{4.17}$$

The symmetry of this matrix ensures that it can always be diagonalised with an orthogonal matrix R_{ij},

$$\mathsf{R}_{ik} \mathsf{M}_{kl} \mathsf{R}^{\mathrm{t}}_{lj} = \mathsf{R}_{ik} \mathsf{R}_{jl} \mathsf{M}_{kl} = \Lambda_{ij}. \tag{4.18}$$

Here Λ_{ij} is diagonal and, unless stated otherwise, the summation convention is employed. The matrix R_{ij} defines a new set of vectors via

$$e_i = \mathsf{R}_{ij} a_j. \tag{4.19}$$

These satisfy

$$\begin{aligned} e_i \cdot e_j &= (\mathsf{R}_{ik} a_k) \cdot (\mathsf{R}_{jl} a_l) \\ &= \mathsf{R}_{ik} \mathsf{R}_{jl} \mathsf{M}_{kl} \\ &= \Lambda_{ij}. \end{aligned} \tag{4.20}$$

The vectors $e_1, \ldots, e_r$ are therefore orthogonal and hence all anticommute. Their geometric product is therefore totally antisymmetric, and we have

$$\begin{aligned} e_1 e_2 \cdots e_r &= e_1 \wedge \cdots \wedge e_r \\ &= (\mathsf{R}_{1i} a_i) \wedge \cdots (\mathsf{R}_{rk} a_k) \\ &= \det\,(\mathsf{R}_{ij})\, a_1 \wedge a_2 \wedge \cdots \wedge a_r. \end{aligned} \tag{4.21}$$

The determinant appears here because of the total antisymmetry of the expression (see section 4.5.2). But since R_{ij} is an orthogonal matrix it has determinant ± 1, and by choosing the order of the $\{e_i\}$ vectors appropriately we can set the determinant of R_{ij} to 1. This ensures that we can always find a set of vectors such that

$$a_1 \wedge a_2 \wedge \cdots \wedge a_r = e_1 e_2 \cdots e_r. \tag{4.22}$$

This result will simplify the proofs of a number of results in this chapter.

For a given vector space, an orthonormal frame $\{\mathsf{e}_i\}, i = 1, \ldots, n$ provides a natural way to view the entire geometric algebra. We denote this algebra $\mathcal{G}_n$. Most of the results derived in this chapter are independent of signature, so in the following we let $\mathcal{G}_n$ denote the geometric algebra of a space of dimension n with arbitrary (non-degenerate) signature. One can also consider the degenerate case where some of the basis vectors are null, though we will not need such algebras in this book. The basis vectors build up to form a basis for the entire algebra as

$$1, \quad \mathsf{e}_i, \quad \mathsf{e}_i \mathsf{e}_j \ (i < j), \quad \mathsf{e}_i \mathsf{e}_j \mathsf{e}_k \ (i < j < k), \quad \ldots. \tag{4.23}$$

The fact that the basis vectors anticommute ensures that each product in the basis set is totally antisymmetric. The product of r distinct basis vectors is then, by definition, a grade-r multivector. The basis (4.23) therefore naturally defines a basis for each of the grade-r subspaces of $\mathcal{G}_n$. We denote each of these subspaces by $\mathcal{G}_n^r$. The size of each subspace is given by the number of distinct combinations of r objects from a set of n. (The order is irrelevant, because of the total antisymmetry.) These are given by the binomial coefficients, so

$$\dim(\mathcal{G}_n^r) = \binom{n}{r}. \tag{4.24}$$

For example, we have already seen that in two dimensions the algebra contains terms of grade $0, 1, 2$ with each space having dimension $1, 2, 1$ respectively. Similarly in three dimensions the separate graded subspaces have dimension $1, 3, 3, 1$. The binomial coefficients always exhibit a mirror symmetry between the r and $n - r$ terms. This gives rise to the notion of duality, which is explained in section 4.1.4 where we explore the properties of the highest grade element of the algebra — the pseudoscalar.

The total dimension of the algebra is

$$\dim(\mathcal{G}_n) = \sum_{r=0}^{n} \dim(\mathcal{G}_n^r) = \sum_{r=0}^{n} \binom{n}{r} = (1+1)^n = 2^n. \tag{4.25}$$

One can see that the total size of the algebra quickly becomes very large. If one wanted to find a matrix representation of the algebra, the matrices would have to be of the order of $2^{n/2} \times 2^{n/2}$. For all but the lowest values of n these matrices become totally impractical for computations. This is one reason why matrix representations do not help much with understanding and using geometric algebra.

We have now defined the grade operation for our linear space $\mathcal{G}_n$. An arbitrary multivector A can be decomposed into a sum of pure grade terms

$$A = \langle A \rangle_0 + \langle A \rangle_1 + \cdots = \sum_r \langle A \rangle_r. \tag{4.26}$$

The operator $\langle\ \rangle_r$ projects onto the grade-r terms in the argument, so $\langle A \rangle_r$ returns the grade-r components in A. Multivectors containing terms of only one grade are called *homogeneous*. They are often written as A_r, so

$$\langle A_r \rangle_r = A_r. \tag{4.27}$$

Take care not to confuse the grading subscript in A_r with frame indices in expressions like $\{\mathsf{e}_k\}$. The context should always make clear which is intended. The grade-0 terms in $\mathcal{G}_n$ are the real scalars and commute with all other elements. We continue to employ the useful abbreviation

$$\langle A \rangle = \langle A \rangle_0 \tag{4.28}$$

for the operation of taking the scalar part.

An important feature of a geometric algebra is that *not all homogeneous multivectors are pure blades.* This is confusing at first, because we have to go to four dimensions before we reach our first counterexample. Suppose that $\{\mathsf{e}_1, \ldots, \mathsf{e}_4\}$ form an orthonormal basis for the Euclidean algebra $\mathcal{G}_4$. There are six independent basis bivectors in this algebra, and from these we can construct terms like

$$B = \alpha \mathsf{e}_1 \wedge \mathsf{e}_2 + \beta \mathsf{e}_3 \wedge \mathsf{e}_4, \tag{4.29}$$

where α and β are scalars. B is a pure bivector, so is homogeneous, but it cannot be reduced to a blade. That is, we cannot find two vectors a and b such that $B = a \wedge b$. The reason is that $\mathsf{e}_1 \wedge \mathsf{e}_2$ and $\mathsf{e}_3 \wedge \mathsf{e}_4$ do not share a common vector. This is not possible in three dimensions, because any two planes with a common origin share a common line. A four-dimensional bivector like B is therefore hard for us to visualise. There is a way to visualise B in three dimensions, however, and it is provided by *projective geometry.* This is described in chapter 10.

4.1.2 Further properties of the geometric product

The decomposition of the geometric product of two vectors into a scalar term and a bivector term has a natural extension to general multivectors. To establish the results of this section we make repeated use of the formula

$$ab = 2a\cdot b - ba \tag{4.30}$$

which we use to reorder expressions. As a first example, consider the case of a geometric product of vectors. We find that

$$\begin{aligned}
aa_1a_2\cdots a_r &= 2a\cdot a_1\, a_2\cdots a_r - a_1aa_2\cdots a_r \\
&= 2a\cdot a_1\, a_2\cdots a_r - 2a\cdot a_2\, a_1a_3\cdots a_r + a_1a_2aa_3\cdots a_r \\
&= 2\sum_{k=1}^{r}(-1)^{k+1}a\cdot a_k\, a_1a_2\cdots \check{a}_k\cdots a_r + (-1)^r a_1a_2\cdots a_ra,
\end{aligned} \tag{4.31}$$

where the check on $\check{a}_k$ denotes that this term is missing from the series. We continue to follow the conventions introduced in chapter 2 so, in the absence of brackets, inner products are performed before outer products, and both are performed before geometric products.

Suppose now that the vectors $a_1, \ldots, a_r$ are replaced by a set of anticommuting vectors $e_1, \ldots, e_r$. We find that

$$\frac{1}{2}\Big(ae_1e_2\cdots e_r - (-1)^r e_1e_2\cdots e_ra\Big) = \sum_{k=1}^{r}(-1)^{k+1}a\cdot e_k\, e_1e_2\cdots \check{e}_k\cdots e_r. \tag{4.32}$$

The right-hand side contains a sum of terms formed from the product of $r-1$ anticommuting vectors, so has grade $r-1$. Since any grade-r multivector can be written as a sum of terms formed from anticommuting vectors, the combination on the left-hand side will always return a multivector of grade $r-1$. We therefore define the inner product between a vector a and a grade-r multivector A_r by

$$a\cdot A_r = \frac{1}{2}\Big(aA_r - (-1)^r A_ra\Big). \tag{4.33}$$

The inner product of a vector and a grade-r multivector results in a multivector with grade reduced by one.

The main work of this section is in establishing the properties of the remaining part of the product aA_r. For the case where A_r is a vector, the remaining term is the antisymmetric product, and so is a bivector. This turns out to be true in general — the remaining part of the geometric product returns the exterior product,

$$\frac{1}{2}\Big(a(a_1\wedge a_2\wedge\cdots\wedge a_r) + (-1)^r(a_1\wedge a_2\wedge\cdots\wedge a_r)a\Big) = a\wedge a_1\wedge a_2\wedge\cdots\wedge a_r. \tag{4.34}$$

We will prove this important result by induction. First, we write the blade as a geometric product of anticommuting vectors, so that the result we will establish becomes

$$\frac{1}{2}\Big(ae_1e_2\cdots e_r + (-1)^r e_1e_2\cdots e_r a\Big) = a\wedge e_1\wedge e_2\wedge\cdots\wedge e_r. \tag{4.35}$$

For $r=1$ the result is true as the right-hand side defines the bivector $a\wedge e_1$. For $r>1$ we proceed by writing

$$\begin{aligned} a\wedge e_1\wedge e_2\wedge\cdots\wedge e_r = {}& \frac{1}{r+1} ae_1e_2\cdots e_r \\ &+ \frac{1}{r+1}\sum_{k=1}^{r}(-1)^k e_k(a\wedge e_1\wedge\cdots\wedge\check{e}_k\wedge\cdots\wedge e_r). \end{aligned} \tag{4.36}$$

This result is easily established by writing out all terms in the full antisymmetric product and gathering together the terms which start with the same vector. Next we assume that equation (4.35) holds for the case of an $r-1$ blade, and expand the term inside the sum as follows:

$$\begin{aligned} &\sum_{k=1}^{r}(-1)^k e_k(a\wedge e_1\wedge\cdots\wedge\check{e}_k\wedge\cdots\wedge e_r) \\ &\qquad = \frac{1}{2}\sum_{k=1}^{r}(-1)^k e_k\Big(ae_1\cdots\check{e}_k\cdots e_r + (-1)^{r-1}e_1\cdots\check{e}_k\cdots e_r a\Big) \\ &\qquad = \frac{1}{2}\sum_{k=1}^{r}(-1)^k e_k a e_1\cdots\check{e}_k\cdots e_r + \frac{r}{2}(-1)^r e_1\cdots e_r a \\ &\qquad = \sum_{k=1}^{r}(-1)^k (e_k\cdot a) e_1\cdots\check{e}_k\cdots e_r + \frac{r}{2}\Big(ae_1\cdots e_r + (-1)^r e_1\cdots e_r a\Big) \\ &\qquad = \frac{r-1}{2} ae_1\cdots e_r + \frac{r+1}{2}(-1)^r e_1\cdots e_r a, \end{aligned} \tag{4.37}$$

where we have used equation (4.32). Substituting this result into equation (4.36) then proves equation (4.35) for a grade-r blade, assuming it is true for a blade of grade $r-1$. Since the result is already established for $r=1$, equation (4.34) holds for all blades and hence all multivectors.

We extend the definition of the wedge symbol by writing

$$a\wedge A_r = \frac{1}{2}\Big(aA_r + (-1)^r A_r a\Big). \tag{4.38}$$

With this definition we now have

$$aA_r = a\cdot A_r + a\wedge A_r, \tag{4.39}$$

which extends the decomposition of the geometric product in precisely the desired way. In equation (4.38) one can see how the geometric product can simplify many calculations. The left-hand side would, in general, require totally antisymmetrising all possible products. But the right-hand side only requires evaluating two products — an enormous saving! As we have established the grades of the separate inner and outer products, we also have

$$aA_r = \langle aA_r \rangle_{r-1} + \langle aA_r \rangle_{r+1}, \tag{4.40}$$

where

$$a \cdot A_r = \langle aA_r \rangle_{r-1}, \quad a \wedge A_r = \langle aA_r \rangle_{r+1}. \tag{4.41}$$

So, as expected, multiplication by a vector raises and lowers the grade of a multivector by 1.

A homogeneous multivector can be written as a sum of blades, and each blade can be written as a geometric product of anticommuting vectors. Applying the preceding decomposition, we establish that the product of two homogeneous multivectors decomposes as

$$A_r B_s = \langle A_r B_s \rangle_{|r-s|} + \langle A_r B_s \rangle_{|r-s|+2} + \cdots + \langle A_r B_s \rangle_{r+s}. \tag{4.42}$$

We retain the $\cdot$ and $\wedge$ symbols for the lowest and highest grade terms in this series:

$$\begin{aligned} A_r \cdot B_s &= \langle A_r B_s \rangle_{|r-s|}, \\ A_r \wedge B_s &= \langle A_r B_s \rangle_{r+s}. \end{aligned} \tag{4.43}$$

This is the most general use of the wedge symbol, and is consistent with the earlier definition as the antisymmetrised product of a set of vectors. We can check that the outer product is associative by forming

$$(A_r \wedge B_s) \wedge C_t = \langle A_r B_s \rangle_{r+s} \wedge C_t = \langle (A_r B_s) C_t \rangle_{r+s+t}. \tag{4.44}$$

Associativity of the outer product then follows from the fact that the geometric product is associative:

$$\langle (A_r B_s) C_t \rangle_{r+s+t} = \langle A_r B_s C_t \rangle_{r+s+t} = A_r \wedge B_s \wedge C_t. \tag{4.45}$$

In equation (4.32) we established a formula for the result for the inner product of a vector and a blade formed from orthogonal vectors. We now extend this to a more general result that is extremely useful in practice. We start by writing

$$a \cdot (a_1 \wedge a_2 \wedge \cdots \wedge a_r) = a \cdot \langle a_1 a_2 \cdots a_r \rangle_r, \tag{4.46}$$

where $a_1, \ldots, a_r$ are a general set of vectors. The geometric product $a_1 a_2 \cdots a_r$ can only contain terms of grade r, $r-2$, $\ldots$, so

$$\begin{aligned} &\frac{1}{2}\Big(a a_1 a_2 \cdots a_r - (-1)^r a_1 a_2 \cdots a_r a\Big) \\ &\qquad = a \cdot \langle a_1 a_2 \cdots a_r \rangle_r + a \cdot \langle a_1 a_2 \cdots a_r \rangle_{r-2} + \cdots . \end{aligned} \tag{4.47}$$

The term we are after is the $r-1$ grade part, so we have

$$a \cdot (a_1 \wedge a_2 \wedge \cdots \wedge a_r) = \frac{1}{2} \langle a a_1 a_2 \cdots a_r - (-1)^r a_1 a_2 \cdots a_r a \rangle_{r-1}. \tag{4.48}$$

We can now apply equation (4.31) inside the grade projection operator to form

$$a\cdot(a_1\wedge a_2\wedge\cdots\wedge a_r) = \sum_{k=1}^{r}(-1)^{k+1}a\cdot a_k\langle a_1\cdots\check{a}_k\cdots a_r\rangle_{r-1}$$
$$= \sum_{k=1}^{r}(-1)^{k+1}a\cdot a_k\, a_1\wedge\cdots\wedge\check{a}_k\wedge\cdots\wedge a_r. \tag{4.49}$$

The first two cases illustrate how the general formula behaves:

$$\begin{aligned} a\cdot(a_1\wedge a_2) &= a\cdot a_1\, a_2 - a\cdot a_2\, a_1, \\ a\cdot(a_1\wedge a_2\wedge a_3) &= a\cdot a_1\, a_2\wedge a_3 - a\cdot a_2\, a_1\wedge a_3 + a\cdot a_3\, a_1\wedge a_2. \end{aligned} \tag{4.50}$$

The first case was established in chapter 2, where it was used to replace the formula for the double cross product of vectors in three dimensions.

4.1.3 The reverse, the scalar and the commutator product

Now that the grading is established, we can establish some general properties of the reversion operator, which was first introduced in chapter 2. The reverse of a product of vectors is defined by

$$(ab\cdots c)^\dagger = c\cdots ba. \tag{4.51}$$

For a blade the reverse can be formed by a series of swaps of anticommuting vectors, each resulting in a minus sign. The first vector has to swap past $r-1$ vectors, the second past $r-2$, and so on. This demonstrates that

$$A_r^\dagger = (-1)^{r(r-1)/2}A_r. \tag{4.52}$$

If we now consider the scalar part of a geometric product of two grade-r multivectors we find that

$$\langle A_rB_r\rangle = \langle A_rB_r\rangle^\dagger = \langle B_r^\dagger A_r^\dagger\rangle = (-1)^{r(r-1)}\langle B_rA_r\rangle = \langle B_rA_r\rangle, \tag{4.53}$$

so, for general A and B,

$$\langle AB\rangle = \langle BA\rangle. \tag{4.54}$$

It follows that

$$\langle A\cdots BC\rangle = \langle CA\cdots B\rangle. \tag{4.55}$$

This cyclic reordering property is frequently useful for manipulating expressions. The product in equation (4.54) is sometimes given the symbol $*$, so we write

$$A*B = \langle AB\rangle. \tag{4.56}$$

A further product of considerable importance in geometric algebra is the commutator product of two multivectors. This is denoted with a cross, $\times$, and is defined by

$$A\times B = \frac{1}{2}(AB - BA). \tag{4.57}$$

Care must be taken to include the factor of one-half, which is different to the standard commutator of two operators in quantum mechanics. The commutator product satisfies the *Jacobi identity*

$$A\times(B\times C) + B\times(C\times A) + C\times(A\times B) = 0, \tag{4.58}$$

which is easily seen by expanding out the products.

The commutator arises most frequently in equations involving bivectors. Given a bivector B and a vector a we have

$$B\times a = \frac{1}{2}(Ba - aB) = B\cdot a, \tag{4.59}$$

which therefore results in a second vector. Now consider the product of a bivector and a blade formed from anticommuting vectors. We have

$$\begin{aligned} B(e_1e_2\cdots e_r) &= 2(B\times e_1)e_2\cdots e_r + e_1Be_2\cdots e_r \\ &= 2(B\times e_1)e_2\cdots e_r + \cdots + 2e_1\cdots(B\times e_r) + e_1e_2\cdots e_rB. \end{aligned} \tag{4.60}$$

It follows that

$$B\times(e_1e_2\cdots e_r) = \sum_{i=1}^{r} e_1\cdots(B\cdot e_i)\cdots e_r. \tag{4.61}$$

The sum involves a series of terms which can only contain grades r and $r-2$. But if we form the reverse of the commutator product between a bivector and a homogeneous multivector, we find that

$$\begin{aligned} (B\times A_r)^\dagger &= \frac{1}{2}(BA_r - A_rB)^\dagger \\ &= \frac{1}{2}(-A_r^\dagger B + BA_r^\dagger) \\ &= (-1)^{r(r-1)/2}B\times A_r. \end{aligned} \tag{4.62}$$

It follows that $B\times A_r$ has the same properties under reversion as A_r. But multivectors of grade r and $r-2$ always behave differently under reversion. The commutator product in equation (4.61) must therefore result in a grade-r multivector. Since this is true of any grade-r basis element, it must be true of any homogeneous multivector. That is,

$$B\times A_r = \langle B\times A_r\rangle_r. \tag{4.63}$$

The commutator of a multivector with a bivector therefore preserves the grade of the multivector. Furthermore, the commutator of two bivectors must result in a third bivector. This is the basis for incorporating the theory of Lie groups into geometric algebra.

A similar argument to the preceding one shows that the symmetric product with a bivector must raise or lower the grade by 2. We can summarise this by writing

$$\begin{aligned} BA_r &= \langle BA_r\rangle_{r-2} + \langle BA_r\rangle_r + \langle BA_r\rangle_{r+2} \\ &= B\cdot A_r + B\times A_r + B\wedge A_r, \end{aligned} \tag{4.64}$$

where

$$\frac{1}{2}(BA_r - A_rB) = B\times A_r \tag{4.65}$$

and

$$\frac{1}{2}(BA_r + A_rB) = B\cdot A_r + B\wedge A_r. \tag{4.66}$$

It is assumed in these formulae that A_r has grade $r > 1$.

4.1.4 Pseudoscalars and duality

The exterior product of n vectors defines a grade-n blade. For a given vector space the highest grade element is unique, up to a magnitude. The outer product of n vectors is therefore a multiple of the unique *pseudoscalar* for $\mathcal{G}_n$. This is denoted I, and has two important properties. The first is that I is normalised to

$$|I^2| = 1. \tag{4.67}$$

The sign of I^2 depends on the size of space and the signature. It turns out that the pseudoscalar squares to -1 for the three algebras of most use in this book — those of the Euclidean plane and space, and of spacetime. But this is in no way a general property.

The second property of the pseudoscalar I is that it defines an *orientation.* For any ordered set of n vectors, their outer product will either have the same sign as I, or the opposite sign. Those with the same sign are assigned a positive orientation, and those with opposite sign have a negative orientation. The orientation is swapped by interchanging any pair of vectors. In three dimensions we always choose the pseudoscalar I such that it has the orientation specified by a right-handed set of vectors. In other spaces one just asserts a choice of I and then sticks to that choice consistently.

The product of the grade-n pseudoscalar I with a grade-r multivector A_r is a grade $n-r$ multivector. This operation is called a *duality* transformation. If A_r is a blade, IA_r returns the *orthogonal complement* of A_r. That is, the blade formed from the space of vectors not contained in A_r. It is clear why this has grade $n-r$. Every blade acts as a pseudoscalar for the space spanned by its generating vectors. So, even if we are working in three dimensions, we can treat the bivector e_1e_2 as a pseudoscalar for any manipulation taking place entirely in the e_1e_2 plane. This is often a very helpful idea.

In spaces of odd dimension, I commutes with all vectors and so commutes with all multivectors. In spaces of even dimension, I anticommutes with vectors and so anticommutes with all odd-grade multivectors. In all cases the pseudoscalar commutes with all even-grade multivectors in its algebra. We summarise this by

$$IA_r = (-1)^{r(n-1)} A_r I. \tag{4.68}$$

An important use of the pseudoscalar is for interchanging inner and outer products. For example, we have

$$\begin{aligned} a\cdot(A_r I) &= \frac{1}{2}\Big(aA_r I - (-1)^{n-r} A_r I a\Big) \\ &= \frac{1}{2}\Big(aA_r I - (-1)^{n-r}(-1)^{n-1} A_r a I\Big) \\ &= \frac{1}{2}\Big(aA_r + (-1)^r A_r a\Big) I \\ &= a\wedge A_r\, I. \end{aligned} \tag{4.69}$$

More generally, we can take two multivectors A_r and B_s, with $r+s \leq n$, and form

$$\begin{aligned} A_r\cdot(B_s I) &= \langle A_r B_s I\rangle_{|r-(n-s)|} \\ &= \langle A_r B_s I\rangle_{n-(r+s)} \\ &= \langle A_r B_s\rangle_{r+s} I \\ &= A_r\wedge B_s\, I. \end{aligned} \tag{4.70}$$

This type of interchange is very common in applications. Note how simple this proof is made by the application of the geometric product in the intermediate steps.

4.2 Rotations and reflections

In chapter 2 we showed that in three dimensions a reflection in the plane perpendicular to the unit vector n is performed by

$$a \mapsto a' = -nan. \tag{4.71}$$

This formula holds in arbitrary numbers of dimensions. Provided $n^2 = 1$, we see that n is transformed to

$$n \mapsto -nnn = -n, \tag{4.72}$$

whereas any vector $a_\perp$ perpendicular to n is mapped to

$$a_\perp \mapsto -na_\perp n = a_\perp nn = a_\perp. \tag{4.73}$$

So, for a vector a, the component parallel to n has its sign reversed, whereas the component perpendicular to n is unchanged. This is what we mean by a reflection in the hyperplane perpendicular to n.

Two successive reflections in the hyperplanes perpendicular to m and n result in a rotation in the $m \wedge n$ plane. This is encoded in the rotor

$$R = nm = \exp(-\hat{B}\theta/2) \tag{4.74}$$

where

$$\cos(\theta/2) = n{\cdot}m, \quad \hat{B} = \frac{m \wedge n}{\sin(\theta/2)}. \tag{4.75}$$

The rotor R generates a rotation through the by now familiar formula

$$a \mapsto a' = RaR^\dagger. \tag{4.76}$$

Rotations form a group, as the result of combining two rotations is a third rotation. The same must therefore be true of rotors. Suppose that R_1 and R_2 generate two distinct rotations. The combined rotations take a to

$$a \mapsto R_2(R_1 a R_1^\dagger)R_2^\dagger = R_2 R_1 a R_1^\dagger R_2^\dagger. \tag{4.77}$$

We therefore define the product rotor

$$R = R_2 R_1, \tag{4.78}$$

so that the result of the composite rotation is described by $RaR^\dagger$, as usual. The product R is a new rotor, and in general it will consist of geometric products of an even number of unit vectors,

$$R = lk \cdots nm. \tag{4.79}$$

We will adopt this as our definition of a rotor. The reversed rotor is

$$R^\dagger = mn \cdots kl. \tag{4.80}$$

The result of the map $a \mapsto RaR^\dagger$ returns a vector for any vector a, since

$$RaR^\dagger = lk \cdots \big(n(mam)n\big) \cdots kl \tag{4.81}$$

and each successive sandwich between a vector returns a new vector.

We can immediately establish the normalisation condition

$$RR^\dagger = lk \cdots nmmn \cdots kl = 1 = R^\dagger R. \tag{4.82}$$

In Euclidean spaces, where every vector has a positive square, this normalisation is automatic. In mixed signature spaces, like Minkowski spacetime, unit vectors can have $n^2 = \pm 1$. In this case the condition $RR^\dagger = 1$ is taken as a further condition satisfied by a rotor. In the case where R is the product of two rotors we can easily confirm that

$$RR^\dagger = R_2 R_1 (R_2 R_1)^\dagger = R_2 R_1 R_1^\dagger R_2^\dagger = 1. \tag{4.83}$$

The set of rotors therefore forms a *group*, called a rotor group. This is similar to the group of rotation matrices, though not identical due to the two-to-one map between rotors and rotation matrices. We will have more to say about the group properties of rotors in chapter 11.

In Euclidean spaces every rotor can be written as the exponential of a bivector,

$$R = \exp(-B/2). \tag{4.84}$$

The bivector B defines the plane or planes in which the rotation takes place. The sign ensures that the rotation has the orientation defined by B. In mixed signature spaces one can always write a rotor as $\pm\exp(B)$. In either case the effect of the rotor R on the vector a is

$$a \mapsto \exp(-B/2)a\exp(B/2). \tag{4.85}$$

We can prove that the right-hand side always returns a vector by considering a Taylor expansion of

$$a(\lambda) = \exp(-\lambda B/2)a\exp(\lambda B/2). \tag{4.86}$$

Differentiating the expression on the right produces the power series expansion

$$a(\lambda) = a + \lambda a\cdot B + \frac{\lambda^2}{2!}(a\cdot B)\cdot B + \cdots. \tag{4.87}$$

Since the inner product of a vector and a bivector always results in a new vector, each term in this expansion is a vector. Setting $\lambda = 1$ then demonstrates that equation (4.85) results in a new vector, defined by

$$\exp(-B/2)a\exp(B/2) = a + a\cdot B + \frac{1}{2!}(a\cdot B)\cdot B + \cdots. \tag{4.88}$$

4.2.1 Multivector transformations

Suppose now that every vector in a blade undergoes the same rotation. This is the sort of transformation implied if a plane or volume element is to be rotated. The r-blade A_r can be written

$$A_r = a_1 \wedge \cdots \wedge a_r = \frac{1}{r!}\sum(-1)^\epsilon a_{k_1}a_{k_2}\cdots a_{k_r}, \tag{4.89}$$

with the sum running over all permutations. If each vector in a geometric product is rotated, the result is the multivector

$$\begin{aligned}(Ra_1R^\dagger)(Ra_2R^\dagger)\cdots(Ra_rR^\dagger) &= Ra_1R^\dagger Ra_2R^\dagger\cdots Ra_rR^\dagger \\ &= Ra_1a_2\cdots a_rR^\dagger. \end{aligned} \tag{4.90}$$

This holds for each term in the antisymmetrised sum, so the transformation law for the blade A_r is simply

$$A_r \mapsto A_r' = RA_rR^\dagger. \tag{4.91}$$

Blades transform with the same simple law as vectors! All multivectors share the same transformation law regardless of grade when each component vector is rotated. This is one reason why the rotor formulation is so powerful. The alternative, tensor form would require an extra matrix for each additional vector.

4.3 Bases, frames and components

Any set of linearly independent vectors form a basis for the vectors in a geometric algebra. Such a set is often referred to as a *frame*. Repeated use of the outer product then builds up a basis for the entire algebra. In this section we use the symbols $\mathsf{e}_1, \ldots, \mathsf{e}_n$ or $\{\mathsf{e}_k\}$ to denote a frame for n-dimensional space. We do not restrict the frame to be orthonormal, so the $\{\mathsf{e}_k\}$ do not necessarily anticommute. The reason for the change of font for frame vectors, as opposed to general sets of vectors, is that use of frames nearly always implies reference to coordinates. It is natural write the coordinates of the vector a as a_i or a^i so, to avoid confusion with a set of vectors, we write the frame vectors in a different font.

The volume element for the $\{\mathsf{e}_k\}$ frame is defined by

$$E_n \equiv \mathsf{e}_1 \wedge \mathsf{e}_2 \wedge \cdots \wedge \mathsf{e}_n. \tag{4.92}$$

The grade-n multivector E_n is a multiple of the pseudoscalar for the space spanned by the $\{\mathsf{e}_k\}$. The fact that the vectors are independent guarantees that $E_n \neq 0$. Associated with any arbitrary frame is a reciprocal frame $\{\mathsf{e}^k\}$ defined by the property

$$\mathsf{e}^i \cdot \mathsf{e}_j = \delta^i_j, \quad \forall i, j = 1 \ldots n. \tag{4.93}$$

The 'Kronecker δ', δ^i_j, has value $+1$ if $i = j$ and is zero otherwise. The reciprocal frame is constructed as follows:

$$\mathsf{e}^j = (-1)^{j-1} \mathsf{e}_1 \wedge \mathsf{e}_2 \wedge \cdots \wedge \check{\mathsf{e}}_j \wedge \cdots \wedge \mathsf{e}_n \, E_n^{-1}, \tag{4.94}$$

where as usual the check on $\check{\mathsf{e}}_j$ denotes that this term is missing from the expression. The formula for e^j has a simple interpretation. The vector e^j must be perpendicular to all the vectors $\{\mathsf{e}_i, i \neq j\}$. To find this we form the exterior product of the $n-1$ vectors $\{\mathsf{e}_i, i \neq j\}$. The dual of this returns a vector perpendicular to all vectors in the subspace, and this duality is achieved by the factor of E_n. All that remains is to fix up the normalisation. For this we recall the duality results of section 4.1.4 and form

$$\mathsf{e}_1 \cdot \mathsf{e}^1 = \mathsf{e}_1 \cdot (\mathsf{e}_2 \wedge \cdots \wedge \mathsf{e}_n \, E_n^{-1}) = (\mathsf{e}_1 \wedge \mathsf{e}_2 \wedge \cdots \wedge \mathsf{e}_n) E_n^{-1} = 1. \tag{4.95}$$

This confirms that the formula for the reciprocal frame is correct.

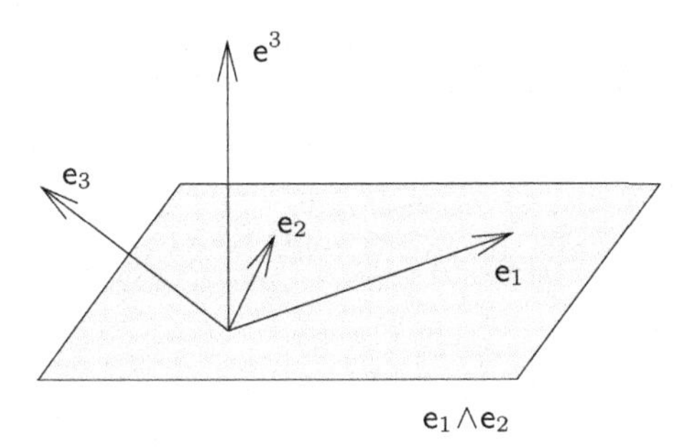

Figure 4.2 *The reciprocal frame.* The vectors e_1, e_2 and e_3 form a non-orthonormal frame for three-dimensional space. The vector e^3 is formed by constructing the $e_1 \wedge e_2$ plane, and forming the vector perpendicular to this plane. The length is fixed by demanding $e^3 \cdot e_3 = 1$.

4.3.1 Application — crystallography

An important application of the formula for a reciprocal frame is in crystallography. If a crystal contains some repeated structure defined by the vectors e_1, e_2, e_3, then constructive interference occurs for wavevectors whose difference satisfies

$$\Delta k = 2\pi(n_1 e^1 + n_2 e^2 + n_3 e^3), \tag{4.96}$$

where n_1, n_2, n_3 are integers. The reciprocal frame is defined by

$$e^1 = \frac{e_2 \wedge e_3}{e_1 \wedge e_2 \wedge e_3}, \quad e^2 = \frac{e_3 \wedge e_1}{e_1 \wedge e_2 \wedge e_3}, \quad e^3 = \frac{e_1 \wedge e_2}{e_1 \wedge e_2 \wedge e_3}. \tag{4.97}$$

If we write

$$e_1 \wedge e_2 \wedge e_3 = [e_1, e_2, e_3] I, \tag{4.98}$$

where I is the three-dimensional pseudoscalar and $[e_1, e_2, e_3]$ denotes the scalar triple product, we arrive at the standard formula

$$e^1 = \frac{(e_2 \wedge e_3) I^{-1}}{[e_1, e_2, e_3]} = \frac{e_2 \boldsymbol{\times} e_3}{[e_1, e_2, e_3]}, \tag{4.99}$$

with similar results holding for e^2 and e^3. Here the bold cross $\boldsymbol{\times}$ denotes the vector cross product, not to be confused with the commutator product. Figure 4.2 illustrates the geometry involved in defining the reciprocal frame.

4.3.2 Components

The basis vectors $\{e_k\}$ are linearly independent, so any vector a can be written uniquely in terms of this set as

$$a = a^i e_i = a_i e^i. \tag{4.100}$$

We continue to employ the summation convention and summed indices appear once as a superscript and once as a subscript. The set of scalars $(a^1, \dots, a^n)$ are the *components* of the vector a in the $\{\mathsf{e}_k\}$ frame. To find the components we form

$$a \cdot \mathsf{e}^i = a^j \mathsf{e}_j \cdot \mathsf{e}^i = a^j \delta_j^i = a^i \tag{4.101}$$

and

$$a \cdot \mathsf{e}_i = a_j \mathsf{e}^j \cdot \mathsf{e}_i = a_j \delta_i^j = a_i. \tag{4.102}$$

These formulae explain the labelling scheme for the components. In many applications we are only interested in orthonormal frames in Euclidean space. In this case the frame and its reciprocal are equivalent, and there is no need for the distinct subscript and superscript indices. The notation is unavoidable in mixed signature spaces, however, and is very useful in differential geometry, so it is best to adopt it at the outset.

Combining the equations (4.100), (4.101) and (4.102) we see that

$$a \cdot \mathsf{e}_i \, \mathsf{e}^i = a \cdot \mathsf{e}^i \, \mathsf{e}_i = a. \tag{4.103}$$

This holds for any vector a in the space spanned by the $\{\mathsf{e}_k\}$. This result generalises simply to arbitrary multivectors. First, for the bivector $a \wedge b$ we have

$$\mathsf{e}_i \, \mathsf{e}^i \cdot (a \wedge b) = \mathsf{e}_i \, \mathsf{e}^i \cdot a \, b - \mathsf{e}_i \, \mathsf{e}^i \cdot b \, a = ab - ba = 2a \wedge b. \tag{4.104}$$

This extends for an arbitrary grade-r multivector A_r to give

$$\mathsf{e}_i \, \mathsf{e}^i \cdot A_r = r A_r. \tag{4.105}$$

Since $\mathsf{e}_i \mathsf{e}^i = n$, we also see that

$$\mathsf{e}_i \, \mathsf{e}^i \wedge A_r = \mathsf{e}_i (\mathsf{e}^i A_r - \mathsf{e}^i \cdot A_r) = (n - r) A_r. \tag{4.106}$$

Subtracting the two preceding results we obtain,

$$\mathsf{e}_i A_r \mathsf{e}^i = (-1)^r (n - 2r) A_r. \tag{4.107}$$

The $\{\mathsf{e}_k\}$ basis extends easily to provide a basis for the entire algebra generated by the basis vectors. We can then decompose any multivector A into a set of components through

$$A_{i \cdots jk} = \langle (\mathsf{e}_k \wedge \mathsf{e}_j \cdots \wedge \mathsf{e}_i) A \rangle. \tag{4.108}$$

and

$$A = \sum_{i<j \cdots <k} A_{ij \cdots k} \mathsf{e}^i \wedge \cdots \wedge \mathsf{e}^j \wedge \mathsf{e}^k. \tag{4.109}$$

The components $A_{ij \cdots k}$ are totally antisymmetric on all indices and are usually referred to as the components of an *antisymmetric tensor*. We shall have more to say about tensors in following sections.

4.3.3 *Application — recovering a rotor*

As an application of the preceding results, suppose that we have two sets of vectors in three dimensions $\{\mathsf{e}_k\}$ and $\{\mathsf{f}_k\}$, $k = 1,2,3$. The vectors need not be orthonormal, but we know that the two sets are related by a rotation. The rotation is governed by the formula

$$\mathsf{f}_k = R\mathsf{e}_k R^\dagger \tag{4.110}$$

and we seek a simple expression for the rotor R. In three dimensions the rotor R can be written as

$$R = \exp(-B/2) = \alpha - \beta B, \tag{4.111}$$

where

$$\alpha = \cos(|B|/2), \quad \beta = \frac{\sin(|B|/2)}{|B|}. \tag{4.112}$$

The reverse is

$$R^\dagger = \exp(B/2) = \alpha + \beta B. \tag{4.113}$$

We therefore find that

$$\begin{aligned} \mathsf{e}_k R^\dagger \mathsf{e}^k &= \mathsf{e}_k(\alpha + \beta B)\mathsf{e}^k \\ &= 3\alpha - \beta B \\ &= 4\alpha - R^\dagger. \end{aligned} \tag{4.114}$$

We now form

$$\mathsf{f}_k\mathsf{e}^k = R\mathsf{e}_k R^\dagger \mathsf{e}^k = 4\alpha R - 1. \tag{4.115}$$

It follows that R is a scalar multiple of $1 + \mathsf{f}_k\mathsf{e}^k$. We therefore establish the simple formula

$$R = \frac{1 + \mathsf{f}_k\mathsf{e}^k}{|1 + \mathsf{f}_k\mathsf{e}^k|} = \frac{\psi}{\surd(\psi\tilde{\psi})}, \tag{4.116}$$

where $\psi = 1 + \mathsf{f}_k\mathsf{e}^k$. This compact formula recovers the rotor directly from the frame vectors. A problem arises if the rotation is through precisely 180°, in which case ψ vanishes. This case can be dealt with simply enough by considering the image of two of the three vectors.

4.4 Linear algebra

Many key relations in physics involve linear mappings between two, sometimes different, spaces. These are the subject of tensor analysis in the standard literature. Examples include the stress and strain tensors of elasticity, the conductivity tensor of electromagnetism and the inertia tensor of dynamics. If one has only met the study of linear transformations through tensor analysis, one could be forgiven for thinking that the subject cannot be discussed without a large dose of index notation. The indices refer to components of tensors in some frame, though the essence of tensor analysis is to establish a set of results which are independent of the choice of frame. In our opinion, this subject is much more simply dealt with if one can avoid specifying a frame until it is absolutely necessary. Perhaps unsurprisingly, it is geometric algebra that provides precisely the tools necessary to achieve such a development.

In this section we use capital, sans-serif symbols for linear functions. This helps to distinguish functions from their multivector argument. The dimension and signature of the vector space is arbitrary unless otherwise specified. We assume that readers are familiar with the basic properties of linear transformations in the guise of matrices. Suppose, then, that we are interested in a quantity F which maps vectors to vectors linearly in the same space. That is, if a is a vector in the space acted on by F, then $\mathsf{F}(a)$ lies in the same space. The linearity of F is expressed by

$$\mathsf{F}(\lambda a + \mu b) = \lambda \mathsf{F}(a) + \mu \mathsf{F}(b), \tag{4.117}$$

for scalars λ and μ and vectors a and b. Geometrically, we can think of F as an instruction to take a vector and rotate/dilate it to a new vector. No frame or components are required for such a picture. A simple example is provided by a rotation, which can be written as

$$\mathsf{R}(a) = RaR^\dagger, \tag{4.118}$$

where R is a rotor. It is a simple matter to confirm that this map is linear.

4.4.1 Extension to multivectors

Once one has formulated the action of a linear function on a vector, the obvious next step is to let the function act on a multivector. In this way we extend the action of a linear function to the full geometric algebra defined by the underlying vector space. Suppose that two vectors a and b are acted on by the linear function F. The bivector $a \wedge b$ then transforms to $\mathsf{F}(a) \wedge \mathsf{F}(b)$. We take this as the definition for the action of F on a bivector blade:

$$\mathsf{F}(a\wedge b) = \mathsf{F}(a)\wedge\mathsf{F}(b). \tag{4.119}$$

Since the right-hand side is the outer product of two vectors, it is also a bivector blade (see figure 4.3). The action on sums of blades is defined by the linearity of F:

$$\mathsf{F}(a\wedge b + c\wedge d) = \mathsf{F}(a\wedge b) + \mathsf{F}(c\wedge d). \tag{4.120}$$

Continuing in this manner, we define the action of F on an arbitrary blade by

$$\mathsf{F}(a\wedge b\wedge\cdots\wedge c) = \mathsf{F}(a)\wedge\mathsf{F}(b)\wedge\cdots\wedge\mathsf{F}(c). \tag{4.121}$$

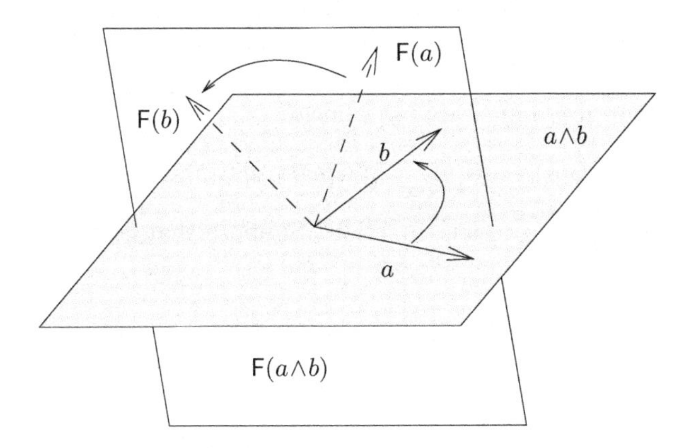

Figure 4.3 *The extended linear function.* The action of F on the bivector $a \wedge b$ results in the new plane $\mathsf{F}(a) \wedge \mathsf{F}(b)$. This is the definition of $\mathsf{F}(a \wedge b)$.

Extension by linearity then defines the action of F on arbitrary multivectors. By construction, F is both linear over multivectors,

$$\mathsf{F}(\lambda A + \mu B) = \lambda \mathsf{F}(A) + \mu \mathsf{F}(B), \tag{4.122}$$

and grade-preserving,

$$\mathsf{F}(A_r) = \langle \mathsf{F}(A_r) \rangle_r, \tag{4.123}$$

where A_r is a grade-r multivector. A simple example is provided by rotations. We have already established a formula for the result of rotating all of the vectors in a blade. For the extension of a rotation we therefore have

$$\begin{aligned} \mathsf{R}(a \wedge b \wedge \cdots \wedge c) &= (RaR^\dagger) \wedge (RbR^\dagger) \wedge \cdots \wedge (RcR^\dagger) \\ &= R\, a \wedge b \wedge \cdots \wedge c\, R^\dagger. \end{aligned} \tag{4.124}$$

It follows that acting on an arbitrary multivector A we have

$$\mathsf{R}(A) = RAR^\dagger. \tag{4.125}$$

Again, it is simple to confirm that this has the expected properties.

4.4.2 The product

The product of two linear functions is formed by letting a second function act on the result of the first function. Thus the action of the product of F and G is defined by

$$(\mathsf{FG})(a) = \mathsf{F}\big(\mathsf{G}(a)\big) = \mathsf{FG}(a). \tag{4.126}$$

The final expression enables us to remove some brackets without any ambiguity. A price to pay for removing indices is that brackets are often required to show

how calculations are ordered. Any convention that enables brackets to be systematically dropped is then well worth adopting. It is straightforward to show that FG is a linear function if F and G are both linear:

$$\mathsf{FG}(\lambda a + \mu b) = \mathsf{F}(\lambda \mathsf{G}(a) + \mu \mathsf{G}(b)) = \lambda \mathsf{FG}(a) + \mu \mathsf{FG}(b). \tag{4.127}$$

Next we form the extension of a product function. Suppose that H is given by the product of F and G:

$$\mathsf{H}(a) = \mathsf{F}\big(\mathsf{G}(a)\big) = \mathsf{FG}(a). \tag{4.128}$$

It follows that

$$\begin{aligned}\mathsf{H}(a \wedge b \wedge \cdots \wedge c) &= \mathsf{F}\big(\mathsf{G}(a)\big) \wedge \mathsf{F}\big(\mathsf{G}(b)\big) \wedge \cdots \wedge \mathsf{F}\big(\mathsf{G}(c)\big) \\ &= \mathsf{F}\big(\mathsf{G}(a) \wedge \mathsf{G}(b) \wedge \cdots \wedge \mathsf{G}(c)\big) \\ &= \mathsf{F}\big(\mathsf{G}(a \wedge b \wedge \cdots \wedge c)\big),\end{aligned} \tag{4.129}$$

so the multilinear action of the product of two linear functions is the product of their exterior actions. In dealing with combinations of linear functions we can therefore write

$$\mathsf{H}(A) = \mathsf{FG}(A), \tag{4.130}$$

since the meaning of the right-hand side is unambiguous.

4.4.3 The adjoint

Given a linear function F, the adjoint, or transpose, $\bar{\mathsf{F}}$ is defined so that

$$a \cdot \bar{\mathsf{F}}(b) = \mathsf{F}(a) \cdot b, \tag{4.131}$$

for all vectors a and b. If F is a mapping from one vector space to another, then the adjoint function maps from the second space back to the first. In terms of an arbitrary frame $\{\mathsf{e}_k\}$ we have

$$\mathsf{e}_i \cdot \bar{\mathsf{F}}(a) = a \cdot \mathsf{F}(\mathsf{e}_i), \tag{4.132}$$

so we can construct the adjoint using

$$\mathrm{ad}(\mathsf{F})(a) = \bar{\mathsf{F}}(a) = \mathsf{e}^i \, a \cdot \mathsf{F}(\mathsf{e}_i). \tag{4.133}$$

The notation of a bar for the adjoint, rather than a superscript T or †, is slightly unconventional, though it does agree with that of Hestenes & Sobczyk (1984). The notation is very useful in handwritten work, where it is also convenient to denote the linear function with an underline. Some formulae relating functions and their adjoints have a neat symmetry when this overbar/underbar convention is followed.

The operation of taking the adjoint of the adjoint of a function returns the original function. This is verified by forming

$$\text{ad}(\bar{\mathsf{F}})(a) = \mathsf{e}^i a \cdot \bar{\mathsf{F}}(\mathsf{e}_i) = \mathsf{e}^i\, \mathsf{e}_i \cdot \mathsf{F}(a) = \mathsf{F}(a). \tag{4.134}$$

The adjoint of a product of two functions is found as follows:

$$\begin{aligned}\text{ad}(\mathsf{FG})(a) &= \mathsf{e}^i\, a \cdot \mathsf{FG}(\mathsf{e}_i) = \bar{\mathsf{F}}(a) \cdot \mathsf{G}(\mathsf{e}_i)\, \mathsf{e}^i \\ &= \bar{\mathsf{G}}\bar{\mathsf{F}}(a) \cdot \mathsf{e}_i\, \mathsf{e}^i = \bar{\mathsf{G}}\bar{\mathsf{F}}(a).\end{aligned} \tag{4.135}$$

The operation of taking the adjoint of a product therefore reverses the order in which the linear functions act. A *symmetric* function is one which is equal to its own adjoint, $\bar{\mathsf{F}} = \mathsf{F}$. Two particularly significant examples of symmetric functions are the functions $\mathsf{F}\bar{\mathsf{F}}$ and $\bar{\mathsf{F}}\mathsf{F}$. To verify that these are symmetric we form

$$\text{ad}(\mathsf{F}\bar{\mathsf{F}}) = \text{ad}(\bar{\mathsf{F}})\text{ad}(\mathsf{F}) = \mathsf{F}\bar{\mathsf{F}}, \tag{4.136}$$

with a similar derivation holding for $\bar{\mathsf{F}}\mathsf{F}$. These functions will be met again later in this chapter.

The adjoint is still a linear function, so its extension to arbitrary multivectors is precisely as expected:

$$\bar{\mathsf{F}}(a \wedge b \wedge \cdots \wedge c) = \bar{\mathsf{F}}(a) \wedge \bar{\mathsf{F}}(b) \wedge \cdots \wedge \bar{\mathsf{F}}(c). \tag{4.137}$$

If we now consider two bivectors $a_1 \wedge a_2$ and $b_1 \wedge b_2$, we find that

$$\begin{aligned}(a_1 \wedge a_2) \cdot \mathsf{F}(b_1 \wedge b_2) &= a_1 \cdot \mathsf{F}(b_2)\, a_2 \cdot \mathsf{F}(b_1) - a_1 \cdot \mathsf{F}(b_1)\, a_2 \cdot \mathsf{F}(b_2) \\ &= \bar{\mathsf{F}}(a_1) \cdot b_2\, \bar{\mathsf{F}}(a_2) \cdot b_1 - \bar{\mathsf{F}}(a_1) \cdot b_1\, \bar{\mathsf{F}}(a_2) \cdot b_2 \\ &= \bar{\mathsf{F}}(a_1 \wedge a_2) \cdot (b_1 \wedge b_2).\end{aligned} \tag{4.138}$$

It follows that for two bivectors B_1 and B_2

$$B_1 \cdot \bar{\mathsf{F}}(B_2) = \mathsf{F}(B_1) \cdot B_2. \tag{4.139}$$

This result extends for arbitrary multivectors to give

$$\langle A \bar{\mathsf{F}}(B) \rangle = \langle \mathsf{F}(A) B \rangle. \tag{4.140}$$

This is a special case of an even more general and powerful result. Consider the expression

$$\begin{aligned}\mathsf{F}(a \wedge b) \cdot c &= \mathsf{F}(a)\, \mathsf{F}(b) \cdot c - \mathsf{F}(b)\, \mathsf{F}(a) \cdot c \\ &= \mathsf{F}\big(a\, b \cdot \bar{\mathsf{F}}(c) - b\, a \cdot \bar{\mathsf{F}}(c)\big) \\ &= \mathsf{F}\big((a \wedge b) \cdot \bar{\mathsf{F}}(c)\big).\end{aligned} \tag{4.141}$$

Building up in this way we establish the useful results:

$$\begin{aligned}A_r \cdot \bar{\mathsf{F}}(B_s) &= \bar{\mathsf{F}}\big(\mathsf{F}(A_r) \cdot B_s\big) \qquad r \le s, \\ \mathsf{F}(A_r) \cdot B_s &= \mathsf{F}\big(A_r \cdot \bar{\mathsf{F}}(B_s)\big) \qquad r \ge s.\end{aligned} \tag{4.142}$$

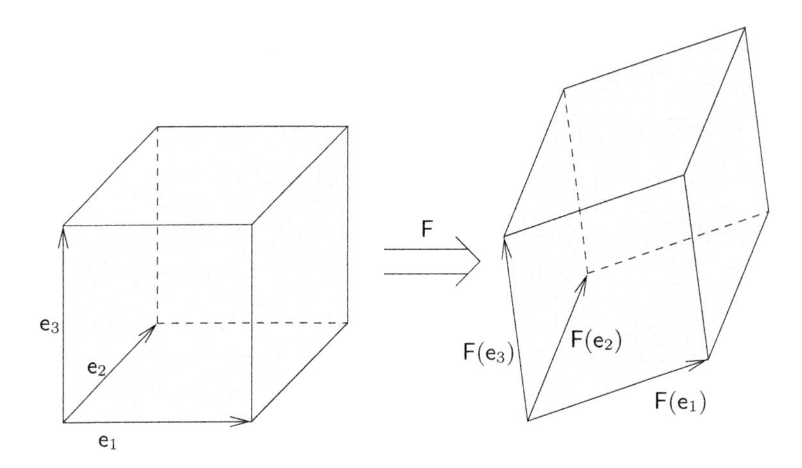

Figure 4.4 *The determinant.* The unit cube is transformed to a parallelepiped with sides $\mathsf{F}(\mathsf{e}_1)$, $\mathsf{F}(\mathsf{e}_2)$ and $\mathsf{F}(\mathsf{e}_3)$. The determinant is the volume scale factor, so is given by the volume of the parallelepiped, $\mathsf{F}(\mathsf{e}_1)\wedge\mathsf{F}(\mathsf{e}_2)\wedge\mathsf{F}(\mathsf{e}_3) = \mathsf{F}(I)$.

These reduce to equation (4.140) in the case when $r = s$. One way to think of these formulae is as follows. In the expression $\mathsf{F}(A_r) \cdot B_s$, with $r \geq s$, there are r separate applications of the function F on vectors. When the result is contracted with B_s, s of these applications are converted to adjoint functions $\bar{\mathsf{F}}$. The remaining $r - s$ applications act on the multivector $A_r \cdot \bar{\mathsf{F}}(B_s)$, which has grade $r - s$.

4.4.4 The determinant

Now that we have seen how a linear function defines an action on the entire geometric algebra, we can give a very compact definition of the determinant. The pseudoscalar for any space is unique up to scaling, and linear functions are grade-preserving, so we define

$$\mathsf{F}(I) = \det\,(\mathsf{F})\, I. \tag{4.143}$$

It should be immediately apparent that this definition of the determinant is much more compact and intuitive than the matrix definition (discussed later). The definition (4.143) shows clearly that the determinant is the volume scale factor for the operation F. In particular, acting on the unit hypercube, the result $\mathsf{F}(I)$ returns the directed volume of the resultant object (see figure 4.4).

As an example of the power of the geometric algebra definition, consider the product of two functions, F and G. From equation (4.130) it follows that

$$\det\,(\mathsf{FG})I = \mathsf{FG}(I) = \det\,(\mathsf{G})\,\mathsf{F}(I) = \det\,(\mathsf{F})\det\,(\mathsf{G})\,I, \tag{4.144}$$

which establishes that the determinant of the product of two functions is the product of their determinants. This is one of the key properties of the determinant, yet in conventional developments it is hard to prove. By contrast, the geometric algebra approach establishes the result in a few lines. Similarly, one can easily establish that the determinant of the adjoint is the same as that of the original function,

$$\det(\mathsf{F}) = \langle \mathsf{F}(I)I^{-1}\rangle = \langle I\bar{\mathsf{F}}(I^{-1})\rangle = \det(\bar{\mathsf{F}}). \tag{4.145}$$

Example 4.1

Consider the linear function

$$\mathsf{F}(a) = a + \alpha a\cdot f_1\, f_2, \tag{4.146}$$

where α is a scalar and f_1 and f_2 are a pair of arbitrary vectors. Construct the action of F on a general multivector and find its determinant.

We start by forming

$$\begin{aligned}\mathsf{F}(a\wedge b) &= (a + \alpha a\cdot f_1 f_2)\wedge(b + \alpha b\cdot f_1 f_2)\\ &= a\wedge b + \alpha(b\cdot f_1 a - a\cdot f_1 b)\wedge f_2\\ &= a\wedge b + \alpha\big((a\wedge b)\cdot f_1\big)\wedge f_2.\end{aligned} \tag{4.147}$$

It follows that

$$\mathsf{F}(A) = A + \alpha(A\cdot f_1)\wedge f_2. \tag{4.148}$$

The determinant is now calculated as follows:

$$\begin{aligned}\mathsf{F}(I) &= I + \alpha(I\cdot f_1)\wedge f_2\\ &= I + \alpha f_1\cdot f_2\, I,\end{aligned} \tag{4.149}$$

hence $\det(\mathsf{F}) = 1 + \alpha f_1\cdot f_2$.

4.4.5 The inverse

We now construct a simple, explicit formula for the inverse of a linear function. We start by considering a multivector B, lying entirely in the algebra defined by the pseudoscalar I. For these we have

$$\det(\mathsf{F})IB = \mathsf{F}(I)B = \mathsf{F}\big(I\bar{\mathsf{F}}(B)\big), \tag{4.150}$$

where we have used the adjoint formulae of equation (4.142). The inner product with a pseudoscalar is replaced with a geometric product, since no other grades are present in the full product. Replacing IB by A we find that

$$\det(\mathsf{F})A = \mathsf{F}\big(I\bar{\mathsf{F}}(I^{-1}A)\big) \tag{4.151}$$

with a similar result holding for the adjoint. It follows that

$$\begin{aligned} \mathsf{F}^{-1}(A) &= I\bar{\mathsf{F}}(I^{-1}A)\det(\mathsf{F})^{-1}, \\ \bar{\mathsf{F}}^{-1}(A) &= I\mathsf{F}(I^{-1}A)\det(\mathsf{F})^{-1}. \end{aligned} \tag{4.152}$$

These relations provide simple, explicit formulae for the inverse of a function. The derivation of these formulae is considerably quicker than anything available in traditional matrix/tensor analysis.

Example 4.2

Find the inverse of the function defined in equation (4.146).

With

$$\mathsf{F}(A) = A + \alpha(A\cdot f_1)\wedge f_2 \tag{4.153}$$

we have

$$\begin{aligned} \langle A_r \mathsf{F}(B_r)\rangle &= \langle A_r B_r\rangle + \alpha\langle A_r(B_r\cdot f_1)\wedge f_2\rangle \\ &= \langle A_r B_r\rangle + \alpha\langle f_2\cdot A_r B_r f_1\rangle, \end{aligned} \tag{4.154}$$

hence

$$\bar{\mathsf{F}}(A) = A + \alpha f_1\wedge(f_2\cdot A). \tag{4.155}$$

It follows that

$$\begin{aligned} \mathsf{F}^{-1}(A) &= \big(I^{-1}A + \alpha f_1\wedge(f_2\cdot(I^{-1}A))\big)(1+\alpha f_1\cdot f_2)^{-1} \\ &= (A + \alpha f_1\cdot(f_2\wedge A))(1+\alpha f_1\cdot f_2)^{-1} \\ &= A - \frac{\alpha}{1+\alpha f_1\cdot f_2} f_2\wedge(f_1\cdot A). \end{aligned} \tag{4.156}$$

Example 4.3

Find the inverse of the rotation

$$\mathsf{R}(a) = RaR^\dagger, \tag{4.157}$$

where R is a rotor.

We have already seen that the action of R on a general multivector is

$$\mathsf{R}(A) = RAR^\dagger \quad \text{and} \quad \bar{\mathsf{R}}(A) = R^\dagger AR \tag{4.158}$$

Hence

$$\det(\mathsf{R})I = RIR^\dagger = IRR^\dagger = I, \tag{4.159}$$

so $\det(\mathsf{R}) = 1$. It follows that

$$\mathsf{R}^{-1}(A) = IR^\dagger I^{-1}AR = R^\dagger AR = \bar{\mathsf{R}}(a), \tag{4.160}$$

so, as expected, the inverse of a rotation is the same as the adjoint. This is the definition of an orthogonal transformation.

4.4.6 Eigenvectors and eigenblades

We assume that readers are familiar with the concept of an eigenvalue and eigenvector of a matrix. All of the standard results for these have obvious counterparts in the geometric algebra framework. This subject will be explored more thoroughly in chapter 11. Here we give a simple outline, concentrating on the new concepts that geometric algebra offers. A linear function F has an eigenvector e if

$$\mathsf{F}(e) = \lambda e. \tag{4.161}$$

The scalar λ is the associated eigenvalue. It follows that

$$\det\,(\mathsf{F} - \lambda \mathsf{I}) = 0, \tag{4.162}$$

which defines a polynomial equation for λ. Techniques for finding eigenvalues and eigenvectors are discussed widely in the literature.

In general, the polynomial equation for λ will have complex roots. Traditional developments of the subject usually allow these and consider linear superpositions over the complex field. But if one starts with a real mapping between real vectors it is not clear that this formal complexification is useful. What one would like would be a more geometric classification of a general linear transformation. This is provided by the notion of an *eigenblade.* We extend the notion of an eigenvector to that of an eigenblade A_r satisfying

$$\mathsf{F}(A_r) = \lambda A_r, \tag{4.163}$$

where A_r is a grade-r blade and λ is real. One immediate example is the pseudoscalar, for which $\lambda = \det\,(\mathsf{F})$. More generally, each eigenblade determines an invariant subspace of the transformation.

As an example of the geometric clarity of the eigenblade concept, consider a function satisfying

$$\mathsf{F}(e_1) = \lambda e_2, \quad \mathsf{F}(e_2) = -\lambda e_1. \tag{4.164}$$

Traditionally, one might write that $e_1 \pm i e_2$ are eigenvectors with eigenvalues $\mp i\lambda$, where i is the unit imaginary. But the identity

$$\mathsf{F}(e_1 \wedge e_2) = \lambda^2 e_1 \wedge e_2 \tag{4.165}$$

identifies the plane $e_1 \wedge e_2$ as an eigenbivector of F. The role of the complex structure inherent in F is played by the unit bivector $e_1 \wedge e_2$. A linear function can have many distinct eigenbivectors, each acting as a distinct imaginary for its own plane. Replacing all of these by a single scalar imaginary throws away a considerable amount of useful information.

4.4.7 Symmetric and antisymmetric functions

An important aspect of the theory of linear functions is finding natural, *canonical*† expressions for a function. For symmetric functions in Euclidean space this form is via its spectral decomposition. If e_i and e_j are eigenvectors of a function, with eigenvalues λ_i and λ_j, we have (no sums implied)

$$e_i \cdot \mathsf{F}(e_j) = e_i \cdot (\lambda_j e_j) = \lambda_j e_i \cdot e_j. \tag{4.166}$$

But if F is symmetric, this also equals

$$\bar{\mathsf{F}}(e_i) \cdot e_j = \mathsf{F}(e_i) \cdot e_j = (\lambda_i e_i) \cdot e_j = \lambda_i e_i \cdot e_j. \tag{4.167}$$

It follows that

$$(\lambda_i - \lambda_j) e_i \cdot e_j = 0, \tag{4.168}$$

so eigenvectors of a symmetric function with distinct eigenvalues must be orthogonal.

If we admit the existence of complex eigenvectors and eigenvalues we also find that (no sums)

$$e^* \cdot \mathsf{F}(e) = \lambda e^* \cdot e = \mathsf{F}(e^*) \cdot e = \lambda^* e^* \cdot e. \tag{4.169}$$

So for any symmetric function we also have

$$(\lambda - \lambda^*) e^* \cdot e = 0. \tag{4.170}$$

Provided $e^* \cdot e \neq 0$ we can conclude that the eigenvalue, and hence the eigenvector, is real. In Euclidean space this inequality is always satisfied, and every symmetric function on an n-dimensional space has a spectral decomposition of the form

$$\mathsf{F}(a) = \lambda_1 \mathsf{P}_1(a) + \lambda_2 \mathsf{P}_2(a) + \cdots + \lambda_m \mathsf{P}_m(a). \tag{4.171}$$

Here $\lambda_1 < \lambda_2 < \cdots < \lambda_m$ are the m distinct eigenvalues ($m \leq n$) and the P_i are projections onto each of the invariant subspaces defined by the eigenvectors. For the case of a projection onto a one-dimensional space we have simply

$$\mathsf{P}_i(a) = a \cdot e_i \, e_i. \tag{4.172}$$

The eigenvectors form an orthonormal frame, which is the natural frame in which to study the linear function. If two eigenvalues are the same, it is always possible to choose the eigenvectors so that they remain orthogonal. In non-Euclidean spaces, such as spacetime, one has to be careful due to the possibility of complex null vectors. These can have $e^* \cdot e = 0$, so the above reasoning breaks down and

† The origin of the use of the word *canonical* is obscure — see for example the comments in Goldstein (1950). In mathematical physics, a canonical form usually refers to a standard way of simplifying an expression without altering its meaning.

one cannot guarantee the existence of an orthonormal frame of eigenvectors. We will encounter examples of this when we study gravitation.

Antisymmetric functions have $\bar{\mathsf{F}}(a) = -\mathsf{F}(a)$. It follows that

$$a\cdot\mathsf{F}(a) = \bar{\mathsf{F}}(a)\cdot a = -\mathsf{F}(a)\cdot a = 0. \tag{4.173}$$

The natural way to study antisymmetric functions is through the bivector

$$F = \frac{1}{2}\mathsf{e}^i\wedge\mathsf{F}(\mathsf{e}_i), \tag{4.174}$$

where the $\{\mathsf{e}_k\}$ are an arbitrary frame for the space acted on by F. The bivector F is independent of the choice of frame, so is an invariant quantity. One can easily confirm that the bivector F has the same number of degrees of freedom as F. If we now form $2a\cdot F$ we find that

$$\begin{aligned} 2a\cdot F &= a\cdot\big(\mathsf{e}^i\wedge\mathsf{F}(\mathsf{e}_i)\big) \\ &= a\cdot\mathsf{e}^i\,\mathsf{F}(\mathsf{e}_i) - \mathsf{e}^i a\cdot\mathsf{F}(\mathsf{e}_i) \\ &= \mathsf{F}(a\cdot\mathsf{e}^i\,\mathsf{e}_i) + \mathsf{e}^i\,\mathsf{e}_i\cdot\mathsf{F}(a) \\ &= 2\mathsf{F}(a). \end{aligned} \tag{4.175}$$

The action of an antisymmetric function therefore reduces to contracting with the *characteristic bivector* F:

$$\mathsf{F}(a) = a\cdot F. \tag{4.176}$$

The problem of reducing an antisymmetric function to its simplest form reduces to that of splitting F into a set of commuting blades:

$$F = \lambda_1\hat{F}_1 + \cdots + \lambda_k\hat{F}_k, \tag{4.177}$$

where $k \leq n/2$ and each of the $\hat{F}_i$ is a unit blade. This decomposition is always possible in Euclidean space, though the answer is only unique if the blades all have different magnitudes. Each component blade of F is an eigenblade of F and determines an invariant subspace. Within this subspace the effect of F is simply to rotate all vectors by $\pm 90°$, and to scale the result by the magnitude of the eigenblade. In non-Euclidean spaces such a decomposition is not always possible.

4.4.8 The singular value decomposition

For linear functions of no symmetry a number of alternative canonical forms can be found. Among these, perhaps the most useful is the singular value decomposition. We start with an arbitrary function F and restrict the discussion to the case where F acts on an n-dimensional Euclidean space. We also suppose that $\det(\mathsf{F}) \neq 0$; the case of $\det(\mathsf{F}) = 0$ is easily dealt with by separating out the space which is mapped onto the origin, and working with a reduced function acting in the subspace over which F is non-singular. We next form the function D by

$$\mathsf{D}(a) = \bar{\mathsf{F}}\mathsf{F}(a). \tag{4.178}$$

This function is symmetric and has n orthogonal eigenvectors with real, positive eigenvalues. The fact that the eigenvalues are positive follows from

$$\bar{\mathsf{F}}\mathsf{F}(e) = \lambda e \quad \Rightarrow \quad \mathsf{F}(e)\cdot\mathsf{F}(e) = \lambda e^2. \tag{4.179}$$

Since (in Euclidean space) the square of any vector is a positive scalar we see that λ must be positive. The assumption that det $(\mathsf{F}) \neq 0$ rules out the possibility of any eigenvalues being zero. It follows that we can write

$$\mathsf{D}(a) = \sum_{i=1}^{n} \lambda_i a\cdot\mathsf{e}_i\, \mathsf{e}_i, \tag{4.180}$$

where the $\{\mathsf{e}_i\}$ are the *orthonormal* frame of eigenvectors. Degenerate eigenvalues are dealt with by picking a set of arbitrary orthonormal vectors in the invariant subspace.

The linear function D has a simple (positive) square root,

$$\mathsf{D}^{1/2} = \sum_{i=1}^{n} \lambda_i^{1/2} a\cdot\mathsf{e}_i\, \mathsf{e}_i \tag{4.181}$$

and this is also invertible,

$$\mathsf{D}^{-1/2} = \sum_{i=1}^{n} \lambda_i^{-1/2} a\cdot\mathsf{e}_i\, \mathsf{e}_i. \tag{4.182}$$

We now set

$$\mathsf{S} = \mathsf{F}\mathsf{D}^{-1/2}. \tag{4.183}$$

This satisfies

$$\bar{\mathsf{S}}\mathsf{S} = \mathsf{D}^{-1/2}\bar{\mathsf{F}}\mathsf{F}\mathsf{D}^{-1/2} = \mathsf{D}^{-1/2}\mathsf{D}\,\mathsf{D}^{-1/2} = \mathsf{I}, \tag{4.184}$$

where I is the identity function. It follows that S is an orthogonal function. The function F can now be written

$$\mathsf{F} = \mathsf{S}\mathsf{D}^{1/2}. \tag{4.185}$$

This represents a series of dilations along the eigendirections of D, followed by a rotation.

If the linear function F is presented as an $n \times n$ matrix of components in some frame, then one usually includes a further rotation R to align this arbitrary frame with the frame of eigenvectors. In this case one writes

$$\mathsf{F} = \mathsf{S}\Lambda^{1/2}\bar{\mathsf{R}}, \tag{4.186}$$

where Λ is a diagonal matrix in the arbitrary coordinate frame. This writes a matrix as a dilation sandwiched between two rotations, and is called the singular value decomposition of the matrix. An arbitrary linear function in n dimensions has n^2 degrees of freedom. The singular value decomposition assigns $2 \times n(n-1)/2$ of these to the two orthogonal transformations R and S, with the remaining n degrees of freedom contained in the dilation Λ. The singular value decomposition appears frequently in subjects such as data analysis, where it is often used in connection with analysing non-square matrices.

4.5 Tensors and components

Many modern physics textbooks are written in the language of tensor analysis. In this approach one often works directly with the components of a vector, or linear function, in a chosen coordinate frame. The invariance of the laws under a change of frame can then be used to advantage to simplify the component equations. Since this approach is so ubiquitous it is important to establish the relationship between tensor analysis and the largely frame-free approach of the present chapter. We start by analysing Cartesian tensors, and then move onto the more general case of an arbitrary coordinate frame.

4.5.1 Cartesian tensors

The subject of Cartesian tensors arises when we restrict our frames to consist only of orthonormal vectors in Euclidean space. For these we have

$$\mathsf{e}_i \cdot \mathsf{e}_j = \delta_{ij}, \tag{4.187}$$

so there is no distinction between frames and their reciprocals. In this case we can drop all distinction between raised and lowered indices, and just work with all indices lowered. Provided both frames have the same orientation, a new frame is obtained from the $\{\mathsf{e}_k\}$ frame by a rotation,

$$\mathsf{e}_i' = R\mathsf{e}_i R^\dagger = \Lambda_{ij}\mathsf{e}_j. \tag{4.188}$$

Here R is a rotor and Λ_{ij} are the components of the rotation defined by R:

$$\Lambda_{ij} = (R\mathsf{e}_i R^\dagger)\cdot\mathsf{e}_j. \tag{4.189}$$

It follows that

$$\begin{aligned}\Lambda_{ij}\Lambda_{ik} &= (R\mathsf{e}_i R^\dagger)\cdot\mathsf{e}_j (R\mathsf{e}_i R^\dagger)\cdot\mathsf{e}_k \\ &= (R^\dagger\mathsf{e}_j R)\cdot(R^\dagger\mathsf{e}_k R) = \delta_{jk},\end{aligned} \tag{4.190}$$

and similarly

$$\Lambda_{ik}\Lambda_{jk} = \delta_{ij}. \tag{4.191}$$

A vector a has components $a_i = \mathsf{e}_i \cdot a$ and these transform under a change of frame in the obvious manner,

$$a_i' = \mathsf{e}_i' \cdot a = \Lambda_{ij} a_j. \tag{4.192}$$

It is important to realise here that it is only the components of a that change, not the underlying vector itself. The change in components is exactly cancelled by the change in the frame. Many equations in physics are invariant if the vector itself is transformed, but this is the result of an underlying symmetry in the equations, and not of the freedom to choose the coordinate system. These two concepts should not be confused!

Extending this idea, we define the components of the linear function F by

$$\mathsf{F}_{ij} = \mathsf{e}_i \cdot \mathsf{F}(\mathsf{e}_j). \tag{4.193}$$

The result of this decomposition is an $n \times n$ array of components, which can be stored and manipulated as a matrix. This definition ensures that the components of the vector $\mathsf{F}(a)$ are given by

$$\mathsf{e}_i \cdot \mathsf{F}(a) = \mathsf{e}_i \cdot \mathsf{F}(a_j \mathsf{e}_j) = \mathsf{F}_{ij} a_j, \tag{4.194}$$

which is the usual expression for a matrix acting on a column vector. Similarly, if F and G are a pair of linear functions, the components of the product function FG are given by

$$\begin{aligned}(\mathsf{FG})_{ij} &= \mathsf{FG}(\mathsf{e}_j) \cdot \mathsf{e}_i = \mathsf{G}(\mathsf{e}_j) \cdot \bar{\mathsf{F}}(\mathsf{e}_i) \\ &= \mathsf{G}(\mathsf{e}_j) \cdot \mathsf{e}_k \, \mathsf{e}_k \cdot \bar{\mathsf{F}}(\mathsf{e}_i) = \mathsf{F}_{ik} \mathsf{G}_{kj}.\end{aligned} \tag{4.195}$$

This recovers the familiar rule for multiplying matrices. If the frame is changed to a new rotated frame, the components of the tensor transform in the obvious way:

$$\mathsf{F}_{ij}' = \Lambda_{ik} \Lambda_{jl} \mathsf{F}_{kl}, \tag{4.196}$$

where the prime denotes the components in the new (primed) frame. Objects with two indices are referred to as rank-2 tensors. Rank-1 tensors are vectors, rank-3 tensors have three indices, and so on. Since rank-2 tensors appear regularly in physics they are often referred to simply as tensors. Also, it is usual to let the term tensor refer to either the component form F_{ij} or the abstract entity F.

For Cartesian tensors there are two important tensors which arise regularly in computations. These are the two *invariant* tensors. The first of these is the Kronecker δ, which transforms as

$$\delta_{ij}' = \Lambda_{ik} \Lambda_{jl} \delta_{kl} = \Lambda_{ik} \Lambda_{jk} = \delta_{ij}. \tag{4.197}$$

The components of the identity function are therefore the same in all orthonormal frames (and are those of the identity matrix in all cases). The second invariant is the alternating tensor $\epsilon_{ij\cdots k}$, where the number of indices matches the dimension of the space. This is totally antisymmetric and is defined as follows:

$$\epsilon_{ij\cdots k} = \begin{cases} 1 & i,\, j,\, \ldots,\, k = \text{even permutation of } 1, 2, \ldots, n \\ -1 & i,\, j,\, \ldots,\, k = \text{odd permutation of } 1, 2, \ldots, n \\ 0 & \text{otherwise} \end{cases} . \tag{4.198}$$

The order of a permutation is the number of pairwise swaps required to return to the original order $1, 2, \ldots, n$. If an even number of swaps is required the permutation is even, and similarly for the odd case. In three dimensions even permutations of $1, 2, 3$ coincide with cyclic orderings of the indices. The determinant of a matrix can be expressed in terms of the alternating tensor via

$$\mathsf{F}_{\alpha i}\mathsf{F}_{\beta j}\cdots\mathsf{F}_{\gamma k}\epsilon_{\alpha\beta\cdots\gamma} = \det(\mathsf{F})\,\epsilon_{ij\cdots k}. \tag{4.199}$$

Given this result, it is straightforward to prove the frame invariance of the alternating tensor under rotations:

$$\epsilon'_{ij\cdots k} = \Lambda_{i\alpha}\Lambda_{j\beta}\cdots\Lambda_{k\gamma}\epsilon_{\alpha\beta\cdots\gamma} = \det(\Lambda)\,\epsilon_{ij\cdots k}. \tag{4.200}$$

But since Λ_{ij} is a rotation matrix it has determinant $+1$, so the tensor is indeed invariant.

4.5.2 The determinant revisited

We should now establish that the definition of the determinant (4.199) agrees with our earlier definition (4.143). To prove this we first need the result that

$$\epsilon_{ij\cdots k} = \mathsf{e}_i\wedge\mathsf{e}_j\cdots\wedge\mathsf{e}_k\, I^\dagger, \tag{4.201}$$

where $I = \mathsf{e}_1\mathsf{e}_2\cdots\mathsf{e}_n$ and the $\{\mathsf{e}_k\}$ form an orthonormal frame. The right-hand side of (4.201) is zero if any of the indices are the same, because of the antisymmetry of the outer product. If the indices form an even permutation of $1, 2, \ldots, n$ we can reorder the vectors into the order $\mathsf{e}_1\mathsf{e}_2\cdots\mathsf{e}_n = I$, in which case the right-hand side of (4.201) returns $+1$. Similarly, any anticyclic combination of $1, 2, \ldots, n$ returns -1. Together these agree with the definition (4.198) of the alternating tensor $\epsilon_{ij\cdots k}$. We can now rearrange the left-hand side of (4.199) as follows:

$$\begin{aligned} \mathsf{F}_{\alpha i}\mathsf{F}_{\beta j}\cdots\mathsf{F}_{\gamma k}\epsilon_{\alpha\beta\cdots\gamma} &= \mathsf{F}_{\alpha i}\mathsf{F}_{\beta j}\cdots\mathsf{F}_{\gamma k}\,\mathsf{e}_\alpha\wedge\mathsf{e}_\beta\cdots\wedge\mathsf{e}_\gamma\, I^\dagger \\ &= \mathsf{F}(\mathsf{e}_i)\wedge\mathsf{F}(\mathsf{e}_j)\cdots\mathsf{F}(\mathsf{e}_k)\, I^\dagger \\ &= \det(\mathsf{F})\,\mathsf{e}_i\wedge\mathsf{e}_j\cdots\wedge\mathsf{e}_k\, I^\dagger \\ &= \det(\mathsf{F})\,\epsilon_{ij\cdots k}, \end{aligned} \tag{4.202}$$

which recovers the expected result.

We assume that most readers are familiar with the various techniques employed when computing the determinant of an $n \times n$ matrix. These can be found in most elementary textbooks on linear algebra. It is instructive to see how the same results arise in the geometric algebra treatment. We have already established that the determinant of the product of two functions is the product of the determinants, and that taking the adjoint does not change the determinant. To establish a further set of results we first introduce the (non-orthonormal) vectors $\{f_i\}$,

$$f_i \equiv \mathsf{F}(\mathsf{e}_i), \tag{4.203}$$

so that

$$\mathsf{F}_{ij} = \mathsf{e}_i \cdot f_j. \tag{4.204}$$

From equation (4.143) the determinant of F can be written

$$\det(\mathsf{F}) = (f_1 \wedge f_2 \wedge \cdots \wedge f_n) \cdot (\mathsf{e}_n \wedge \cdots \wedge \mathsf{e}_2 \wedge \mathsf{e}_1). \tag{4.205}$$

Expanding this product out in full recovers the standard expression for the determinant of a matrix. The first result we see is that swapping any two of the $\{f_i\}$ changes the sign of the determinant. This is the same as swapping two columns in the matrix F_{ij}. Since matrix transposition does not affect the result, the same is true for interchanging rows.

Next we single out one of the $\{\mathsf{e}_k\}$ vectors and write

$$\begin{aligned}\det(\mathsf{F}) &= (-1)^{j+1} (\mathsf{e}_n \wedge \cdots \check{\mathsf{e}}_j \cdots \wedge \mathsf{e}_1) \cdot \big(\mathsf{e}_j \cdot (f_1 \wedge \cdots \wedge f_n)\big) \\ &= \sum_{k=1}^{n} (-1)^{j+k}\, \mathsf{e}_j \cdot f_k\, (\mathsf{e}_n \wedge \cdots \check{\mathsf{e}}_j \cdots \wedge \mathsf{e}_1) \cdot (f_1 \wedge \cdots \check{f}_k \cdots \wedge f_n).\end{aligned} \tag{4.206}$$

The final part of each term in the sum corresponds to an $(n-1) \times (n-1)$ determinant, as can be seen by comparing with (4.205). This is equivalent to the familiar expression for the expansion of the determinant by the jth row. A further useful result is obtained from the identity

$$f_1 \wedge \cdots \wedge (f_j + \lambda f_k) \wedge \cdots \wedge f_n = f_1 \wedge \cdots \wedge f_j \wedge \cdots \wedge f_n \quad j \neq k. \tag{4.207}$$

This result means that any multiple of the kth row can be added to the jth row without changing the result. The same is true for columns. This is the key to the method of Gaussian elimination for finding a determinant. In this method the matrix is first transformed to upper (or lower) triangular form, so that the determinant is then simply the product of the entries down the leading diagonal. This is numerically a highly efficient method for calculating determinants. We can continue in this manner to give concise proofs of many of the key results for determinants. For a useful summary of these, see Turnbull (1960).

To see how these formulae also lead to the familiar expression for the inverse of a matrix, consider the decomposition:

$$\begin{aligned} \mathsf{F}^{-1}_{ij} &= \mathsf{e}_i \cdot \mathsf{F}^{-1}(\mathsf{e}_j) \\ &= \langle \mathsf{e}_i \, \mathsf{e}_1 \wedge \cdots \wedge \mathsf{e}_n \, \bar{\mathsf{F}}(\mathsf{e}_n \wedge \cdots \wedge \mathsf{e}_1 \, \mathsf{e}_j) \rangle \det(\mathsf{F})^{-1} \\ &= (-1)^{i+j} \langle \mathsf{F}(\mathsf{e}_1 \wedge \cdots \check{\mathsf{e}}_i \cdots \wedge \mathsf{e}_n) \, \mathsf{e}_n \wedge \cdots \check{\mathsf{e}}_j \cdots \wedge \mathsf{e}_1 \rangle \det(\mathsf{F})^{-1}. \end{aligned} \tag{4.208}$$

The term enclosed in angular brackets is the determinant of the $(n-1)\times(n-1)$ matrix obtained from F_{ij} by deleting the ith column and jth row. This is the definition of the i,j cofactor of F_{ij}. Equation (4.208) shows that the components of F^{-1}_{ij} are formed from the transposed matrix of cofactors, divided by the determinant $\det(\mathsf{F})$ — the familiar result. Similarly, all other matrix formulae have simple and often elegant counterparts in geometric algebra. Further examples of these are discussed in chapter 11.

4.5.3 General tensors

We now generalise the preceding treatment to the case of arbitrary basis sets in spaces of arbitrary (non-degenerate) signature. One reason for wanting to deal with non-orthonormal frames is that these regularly arise when working in curvilinear coordinate systems. In addition, in mixed signature spaces one has no option since it is impossible to identify a frame with its reciprocal. Suppose, then, that the vectors $\{\mathsf{e}_k\}$ constitute an arbitrary frame for n-dimensional space (of unspecified signature). The reciprocal frame is denoted $\{\mathsf{e}^k\}$ and the two frames are related by

$$\mathsf{e}^i \cdot \mathsf{e}_j = \delta^i_j. \tag{4.209}$$

Equation (4.94) for the reciprocal frame is general and still holds in mixed signature spaces.

As described in section 4.3.2, the vector a has components $(a^1, a^2, \ldots, a^n)$ in the $\{\mathsf{e}_k\}$ frame, and $(a_1, a_2, \ldots, a_n)$ in the $\{\mathsf{e}^k\}$ frame. When working with general coordinate frames we always ensure that upper and lower indices match separately on either side of an expression. Suppose we now form the inner product of two vectors a and b. We can write this as

$$a\cdot b = (a^i \mathsf{e}_i)\cdot(b_j \mathsf{e}^j) = a^i b_j \, \mathsf{e}_i \cdot \mathsf{e}^j = a^i b_j \delta^j_i = a^i b_i. \tag{4.210}$$

The general rule is that sums are only taken over pairs of indices where one is a superscript and the other a subscript. Another way to write an inner product is to introduce the *metric tensor* g_{ij}:

$$g_{ij} = \mathsf{e}_i \cdot \mathsf{e}_j. \tag{4.211}$$

In terms of its components g_{ij} is a symmetric $n \times n$ matrix. The inverse matrix is written as g^{ij} and is given by

$$g^{ij} = \mathsf{e}^i \cdot \mathsf{e}^j. \tag{4.212}$$

It is easily verified that this is the inverse of g_{ij}:

$$g^{ik} g_{kj} = \mathsf{e}^i \cdot \mathsf{e}^k \, \mathsf{e}_k \cdot \mathsf{e}_j = \mathsf{e}^i \cdot \mathsf{e}_j = \delta^i_j. \tag{4.213}$$

Employing the metric tensor we can write the inner product of two vectors in a number of equivalent forms:

$$a \cdot b = a^i b_i = a_i b^i = a^i b^j g_{ij} = a_i b_j g^{ij}. \tag{4.214}$$

Of course, all of these expressions encode the same thing and, unless there is a particular reason to introduce a frame, the index-free expression $a \cdot b$ is usually the simplest to use.

The same ideas extend to expressing the linear function F in a general non-orthonormal frame. We let F act on the frame vector e_j and find the components of the result in the reciprocal frame. The components are then given by

$$\mathsf{F}_{ij} = \mathsf{e}_i \cdot \mathsf{F}(\mathsf{e}_j). \tag{4.215}$$

Again, the set of numbers F_{ij} are referred to as the components of a rank-2 tensor and form an $n \times n$ matrix, the entries of which depend on the choice of frame. Similar expressions exist for combinations of frame vectors and reciprocal vectors, for example,

$$\mathsf{F}^{ij} = \mathsf{F}(\mathsf{e}^j) \cdot \mathsf{e}^i. \tag{4.216}$$

One use of the metric tensor is to interchange between these expressions:

$$\mathsf{F}^{ij} = \mathsf{e}^i \cdot \mathsf{F}(\mathsf{e}^j) = \mathsf{e}^i \cdot \mathsf{e}^k \, \mathsf{e}_k \cdot \mathsf{F}(\mathsf{e}_l \mathsf{e}^l \cdot \mathsf{e}^j) = g^{ik} g^{jl} \mathsf{F}_{kl}. \tag{4.217}$$

Again, we have at our disposal a variety of different ways of encoding the information in F. In terms of the abstract concept of a linear operator, the metric tensor g_{ij} is simply the identity operator expressed in a non-orthonormal frame.

If F_{ij} are the components of F in some frame then the components of $\bar{\mathsf{F}}$ are given by

$$\bar{\mathsf{F}}_{ij} = \bar{\mathsf{F}}(\mathsf{e}_j) \cdot \mathsf{e}_i = \mathsf{e}_j \cdot \mathsf{F}(\mathsf{e}_i) = \mathsf{F}_{ji}. \tag{4.218}$$

That is, viewed as a matrix, the components of $\bar{\mathsf{F}}$ are found from the components of F by matrix transposition. For mixed index tensors we have to be slightly more careful, as we now have

$$\mathsf{F}_i{}^j = \mathsf{F}(\mathsf{e}^j) \cdot \mathsf{e}_i = \mathsf{e}^j \cdot \bar{\mathsf{F}}(\mathsf{e}_i) = \bar{\mathsf{F}}^j{}_i. \tag{4.219}$$

If F is a symmetric function we have $\bar{\mathsf{F}} = \mathsf{F}$. In this case the component matrices satisfy

$$\mathsf{F}_{ij} = \mathsf{F}(\mathsf{e}_j) \cdot \mathsf{e}_i = \mathsf{F}(\mathsf{e}_i) \cdot \mathsf{e}_j = \mathsf{F}_{ji}, \tag{4.220}$$

so the components F_{ij} form a symmetric matrix. The same is true of $\mathsf{F}^{ij} = \mathsf{F}^{ji}$, but for the mixed tensor $\mathsf{F}_i{}^j$ we have $\mathsf{F}_i{}^j = \mathsf{F}^j{}_i$.

The components of the product function FG are found from the following rearrangement:

$$\begin{aligned}(\mathsf{FG})_{ij} &= \mathsf{FG}(\mathsf{e}_j)\cdot\mathsf{e}_i = \mathsf{G}(\mathsf{e}_j)\cdot\bar{\mathsf{F}}(\mathsf{e}_i) \\ &= \mathsf{G}(\mathsf{e}_j)\cdot\mathsf{e}_k\, \mathsf{e}^k\cdot\bar{\mathsf{F}}(\mathsf{e}_i) = \mathsf{F}_i{}^k\mathsf{G}_{kj}.\end{aligned} \tag{4.221}$$

Provided the correct combination of subscript and superscript indices is used, this can be viewed as a matrix product. Alternatively, one can work entirely with subscripted indices, and include suitable factors of the metric tensor,

$$(\mathsf{FG})_{ij} = \mathsf{F}_{ik}\mathsf{G}_{lj}g^{kl}. \tag{4.222}$$

Higher rank linear functions give rise to higher rank tensors. Suppose, for example, that $\phi(a_1, a_2, a_3)$ is a scalar function of three vectors, and is linear on each argument,

$$\phi(\lambda a_1 + \mu b, a_2, a_3) = \lambda\phi(a_1, a_2, a_3) + \mu\phi(b, a_2, a_3), \quad \text{etc.} \tag{4.223}$$

The components of this define a rank-3 tensor via

$$\phi_{ijk} = \phi(\mathsf{e}_i, \mathsf{e}_j, \mathsf{e}_k). \tag{4.224}$$

Using similar schemes it is a straightforward matter to set up a map between tensor equations and frame-free expressions in geometric algebra.

4.5.4 Coordinate transformations

If a second non-orthonormal frame $\{\mathsf{f}_\alpha\}$ is introduced we can relate the two frames via a transformation matrix $f_{\alpha i}$:

$$f_{\alpha i} = \mathsf{f}_\alpha\cdot\mathsf{e}_i, \quad f^{\alpha i} = \mathsf{f}^\alpha\cdot\mathsf{e}^i, \tag{4.225}$$

where Latin and Greek indices distinguish the components in one frame from the other. These matrices satisfy

$$f_{\alpha i}f^{\alpha j} = \mathsf{f}_\alpha\cdot\mathsf{e}_i\, \mathsf{f}^\alpha\cdot\mathsf{e}^j = \mathsf{e}_i\cdot\mathsf{e}^j = \delta_i^j \tag{4.226}$$

and

$$f_{\alpha i}f^{\beta i} = \mathsf{f}_\alpha\cdot\mathsf{e}_i\, \mathsf{f}^\beta\cdot\mathsf{e}^i = \mathsf{f}_\alpha\cdot\mathsf{f}^\beta = \delta_\alpha^\beta. \tag{4.227}$$

The decomposition of the vector a in terms of these frames gives

$$a = a^i\mathsf{e}_i = a^i\mathsf{f}^\alpha\mathsf{e}_i\cdot\mathsf{f}_\alpha = a^i f_{\alpha i}\mathsf{f}^\alpha. \tag{4.228}$$

If follows that the transformation law for the components is

$$a_\alpha = f_{\alpha i}a^i, \tag{4.229}$$

with similar expressions holding for the superscripted components.

These formulae extend simply to include linear functions. For example, we see that

$$\mathsf{F}_{\alpha\beta} = f_{\alpha i} f_{\beta j} \mathsf{F}^{ij}. \tag{4.230}$$

Again, similar expressions hold for superscripts and for mixtures of indices. In particular we have

$$\mathsf{F}_{\alpha}{}^{\beta} = f_{\alpha}{}^{i} f^{\beta}{}_{j} \mathsf{F}_{i}{}^{j}. \tag{4.231}$$

Expressed in terms of matrix multiplication, this would be an equivalence transformation. Of course, the abstract frame-free function F is unaffected by any change of basis. All that changes is the particular representation of the function in the chosen coordinate system. Any set of n^2 numbers with this transformation property are called the components of a rank 2 tensor, the implication being that the underlying function is frame-independent.

In conventional accounts, the subject of tensors is often built up by taking the transformation law as fundamental. That is, a vector (rank-1 tensor) is *defined* as a set of components which transform according to equation (4.229) under a change of basis. Once one has the tools available to treat vectors and linear operations in a frame-free manner, such an approach becomes entirely unnecessary. The defining property of a tensor is that it represents a genuine geometric object (or operation) and does not depend on a choice of frame. Given this, the transformation laws (4.229) and (4.231) follow automatically. In this book the name *tensor* is applied to any frame-independent linear function, such as F. We will encounter a variety of such objects in later chapters.

4.6 Notes

The realisation that geometric algebra is a universal tool for physics was a key point in the modern development of the subject, and was first strongly promoted by David Hestenes (figure 4.5). Before his work, physicists' sole interaction with geometric algebra was through the quantum theory of spin. The Pauli and Dirac matrices form representations of Clifford algebras, a fact that was realised as soon as they were introduced. But in the 50 years since Clifford's original idea, the geometry behind his algebra had been lost as mathematicians concentrated on its algebraic properties. This discovery of the Pauli and Dirac matrices thus gave rise to two mistaken beliefs. The first was that there was something intrinsically quantum-mechanical in the non-commutative properties of the matrices. This is clearly not the case. Clifford died long before quantum theory was first formulated and was motivated entirely by classical geometry, and his algebra is today routinely employed in a range of subjects far removed from quantum theory.

Figure 4.5 *David Hestenes.* Inventor of geometric calculus and first to draw attention to the universal nature of geometric algebra. He wrote the influential *Space-Time Algebra* in 1966, and followed this with a fully developed formalism in *Clifford Algebra to Geometric Calculus* (Hestenes & Sobczyk, 1984). This was followed by the (simpler) *New Foundations for Classical Mechanics*, first published in 1986 (second edition 1999). In a series of papers Hestenes and coworkers showed how geometric algebra could be applied in the study of classical and quantum mechanics, electrodynamics, projective and conformal geometry and Lie group theory. More recently, he has advocated the use of geometric algebra in the field of computer graphics.

The second widespread belief was that matrices were crucial to understanding the properties of Clifford algebras. This too is erroneous. The geometric algebra of a finite-dimensional vector space is an associative algebra, so always has a matrix representation. But these matrices add little, if anything, to understanding the properties of the algebra. Furthermore, an insistence on working with matrices deters one from applying geometric algebra to anything beyond the lowest dimensional spaces, because the size of the matrices increases exponentially with the dimension of the space. Working directly with the elements of the algebra imposes no such constraints, and one can easily apply the ideas to spaces of any dimension, including infinite-dimensional spaces.

Mathematicians had few such misconceptions, and Atiyah and others developed Clifford algebra as a powerful tool for geometry. Even in these developments, however, the emphasis was usually on Clifford algebra as an extra tool on top of the standard techniques for solving geometric problems. The algebra was seldom used as complete language for geometry. The picture first started to change when Hestenes recovered Clifford's original interpretation of the Pauli matrices. This led Hestenes to question whether the appearance of a Clifford algebra was telling us something about the underlying structure of quantum theory. Hestenes then went on to promote the universal nature of the algebra, which he publicised in a series of books and papers. Acceptance of this view is growing and, while not everyone is in full agreement, it is now hard to find an area of physics to which geometric algebra cannot or has not been applied without some degree of success.

4.7 Exercises

4.1 Prove that the outer product of a set of linearly dependent vectors vanishes.

4.2 In a Euclidean space, Gram–Schmidt orthogonalisation proceeds by successively replacing each vector in a set $\{a_i\}$ by one perpendicular to the preceding vectors. Prove that such a vector is given by

$$e_i = a_i - \sum_{j=1}^{i-1} \frac{a_i \cdot e_j}{e_j^2} e_j.$$

Prove that we can also write this as

$$e_i = a_i \wedge a_{i-1} \wedge \cdots \wedge a_1 (a_{i-1} \wedge \cdots \wedge a_1)^{-1}.$$

4.3 Prove that

$$(a \wedge b) \times (c \wedge d) = b \cdot c \, a \wedge d - a \cdot c \, b \wedge d + a \cdot d \, b \wedge c - b \cdot d \, a \wedge c.$$

4.4 The length of a vector in Euclidean space is defined by $|a| = \sqrt{(}a^2)$, and the angle θ between two vectors is defined by

$$\cos(\theta) = a \cdot b / (|a||b|).$$

Show that a linear transformation F which leaves lengths and angles unchanged must satisfy

$$\bar{\mathsf{F}} = \mathsf{F}^{-1}.$$

What does this imply for the determinant of F? A reflection in the (hyper)plane perpendicular to n is defined by

$$\mathsf{R}(a) = -nan,$$

where $n^2 = 1$. Show that $\bar{\mathsf{R}} = \mathsf{R}^{-1}$, and that R has determinant -1.

4.5 For the reflection in the preceding question introduce a suitable basis frame and express F in terms of a matrix F_{ij}. Verify the results for the determinant and inverse of this matrix. (Hint — align one of the basis vectors with n.)

4.6 A rotor R is defined by

$$R = \exp(-\lambda B/2).$$

By Taylor expanding in λ, prove that the operation

$$\mathsf{R}(A) = RAR^\dagger$$

preserves the grade(s) of the multivector A.

4.7 Show that the plane B is unchanged by the rotation defined by the rotor $R = \exp(B/2)$.

4.8 Analyse the properties of the matrix

$$\begin{pmatrix} 1 & 2\sinh(u) \\ 0 & 1 \end{pmatrix}.$$

To what geometric operation does this matrix correspond? Can this matrix be diagonalised, and does it have a sensible singular value decomposition?

4.9 Suppose that the linear transformation F has a complex eigenvector $e+if$ with associated eigenvector $\alpha + i\beta$. What is the effect of F on the $e\wedge f$ plane? How should one interpret the action of F in this plane?

4.10 Suppose that the vectors $\{\mathsf{e}_k\}$ form an orthonormal basis frame for n-dimensional Euclidean space. What is the effect of the transformation

$$\mathsf{T}(a) = a + \lambda a\cdot\mathsf{e}_1\, \mathsf{e}_2$$

on the rows of the matrix F_{ij} formed by decomposing F in the $\{\mathsf{e}_k\}$ frame? Use this result to prove that the determinant of a matrix is unchanged by adding a multiple of one row to another.

5

Relativity and spacetime

The geometric algebra of spacetime is called the *spacetime algebra*. Historically, the spacetime algebra was the first modern implementation of geometric algebra to gain widespread attention amongst the physics community. This is because it provides a *synthetic* framework for studying spacetime physics. There are two main approaches to the study of geometry, which can be loosely referred to as the algebraic and synthetic traditions. In the algebraic approach one works entirely with the components of a vector and manipulates these directly. Such an approach leads naturally to the subject of tensors, and places considerable emphasis on how coordinates transform under changes of frame. The synthetic approach, on the other hand, treats vectors as single, abstract entities x or a, and manipulates these directly. Geometric algebra follows in this tradition.

For much of modern physics the synthetic approach has come to dominate. The most obvious examples of this are classical mechanics and electromagnetism, both of which helped shape the development of abstract vector calculus. For these subjects, presentations typically perform all of the required calculations with the three-dimensional scalar and cross products. We have argued that geometric algebra provides extra efficiency and clarity, though it is not essential to a synthetic treatment of three-dimensional physics. But for spacetime calculations the cross product cannot be defined. Despite the obvious advantages of synthetic treatments, most relativity texts revert to a more basic, algebraic approach involving the components of 4-vectors and Lorentz-transform matrices. Such an approach has trouble encoding such basic notions as a plane in spacetime and, unsurprisingly, does a very poor job of handling the dynamics of extended bodies.

To develop a generally applicable algebra of vectors in spacetime one has little option but to use either geometric algebra, or the language of exterior forms (which is essentially a subset of geometric algebra which only employs the interior and exterior products). This is why relativistic physics still tends to dominate the literature of applications of geometric algebra. Many aspects of special relativity become clearer when viewed in the language of geometric algebra and, crucially, a wealth of new computational tools is provided which dramatically simplify relativistic problems.

5.1 An algebra for spacetime

It is not our intention in this chapter to give a fully self-contained introduction to relativity. Such an account can be found in the various books listed at the end of this chapter. In brief, a series of famous experiments conducted in the latter half of the nineteenth century showed that light did not appear to behave in quite the expected, Newtonian manner. This led Einstein to his 'second postulate', that the speed of light c is the same for all inertial (non-accelerating) observers. Combined with Einstein's 'first postulate', the principle of relativity, one is led inexorably to special relativity. The principle of relativity states simply that all inertial frames are equivalent for the purposes of physical experiment. An immediate consequence of these postulates is that the underlying geometry is no longer that of a (Euclidean) three-dimensional space, but instead the appropriate arena for physics is (Lorentzian) spacetime.

To understand why this is the case, suppose that a spherical flash of light is sent out from a source, and this event is described in two coordinate frames. We discuss the concept of a frame, as distinct from a single observer, later in this chapter. The frames are in relative motion, and their origins coincide with the location of the source at the moment the light is emitted. At this instant both frames also set their time measurements to zero. In the first frame the source is at rest and the light expands radially according to the equation

$$r = ct. \tag{5.1}$$

But the second frame must also record a radially expanding shell of light since the relative velocity of the source has no effect on the speed of light. The second frame therefore sees light expanding according to the equation

$$r' = ct'. \tag{5.2}$$

Since the two frames are in relative motion, points at a given fixed r cannot coincide with those at a fixed r'. So points reached at the same time in one frame are reached at *different* times in the second frame. But in both frames the light lies on a spherical expanding shell. So the one thing that is common to both frames is the value of

$$(ct)^2 - r^2 = (ct')^2 - (r')^2 = 0. \tag{5.3}$$

This defines the invariant interval of special relativity and is the fundamental algebraic concept we need to encode.

The preceding argument shows us that the algebra we need to construct is generated by four orthogonal vectors $\{\gamma_0, \gamma_1, \gamma_2, \gamma_3\}$ satisfying the algebraic relations

$$\gamma_0^2 = 1, \quad \gamma_0 \cdot \gamma_i = 0, \quad \gamma_i \cdot \gamma_j = -\delta_{ij}, \tag{5.4}$$

where i and j run from 1 to 3. These are summarised in relativistic notation as

$$\gamma_\mu \cdot \gamma_\nu = \eta_{\mu\nu} = \text{diag}(+ - - -), \quad \mu, \nu = 0, \ldots, 3. \tag{5.5}$$

The notation $\{\gamma_\mu\}$ for a spacetime frame is a widely adopted convention in the spacetime algebra literature. The notation is borrowed from Dirac theory and we continue to employ it in this book. We have also chosen the 'particle physics' choice of signature, which has spacelike vectors with negative norm. General relativists often work with the opposite signature and swap all of the signs in $\eta_{\mu\nu}$. Both choices have their advocates and all (known) physical laws are independent of the choice of signature. Throughout we use Latin indices to denote the range 1–3 and Greek for the full spacetime range 0–3.

The $\{\gamma_\mu\}$ vectors are dimensionless, as is clear from their squares. Since we are in a space of mixed signature, we must adopt the conventions of section 4.3 and distinguish between a frame and its reciprocal. For the $\{\gamma_\mu\}$ frame the reciprocal frame vectors, $\{\gamma^\mu\}$, have $\gamma^0 = \gamma_0$ and $\gamma^i = -\gamma_i$. A general vector in the spacetime algebra can be constructed from the $\{\gamma_\mu\}$ vectors. A spacetime event, for example, is encoded in the vector x, which has coordinates x^μ in the $\{\gamma_\mu\}$ frame. Explicitly, the vector x is

$$x = x^\mu \gamma_\mu = ct\gamma_0 + x^i \gamma_i, \tag{5.6}$$

which has dimensions of distance. From this point on it will be convenient to work in units where the speed of light c is 1. Factors of c can then be inserted in any final result if the answer is required in different units. The mixed signature means that the square of a vector (a, say) is no longer necessarily positive, and instead we have

$$a^2 = aa = \epsilon|a^2|. \tag{5.7}$$

ϵ is the signature of the vector and can be ± 1 or 0. The mixed signature does not affect the validity of the axiomatic development and results of chapter 4, which made no reference to the signature.

5.1.1 The bivector algebra

There are $4 \times 3/2 = 6$ bivectors in our algebra. These fall into two classes: those that contain a timelike component (e.g. $\gamma_i \wedge \gamma_0$), and those that do not (e.g. $\gamma_i \wedge \gamma_j$). For any pair of orthogonal vectors a and b, $a \cdot b = 0$, we have

$$(a \wedge b)^2 = abab = -abba = -a^2 b^2. \tag{5.8}$$

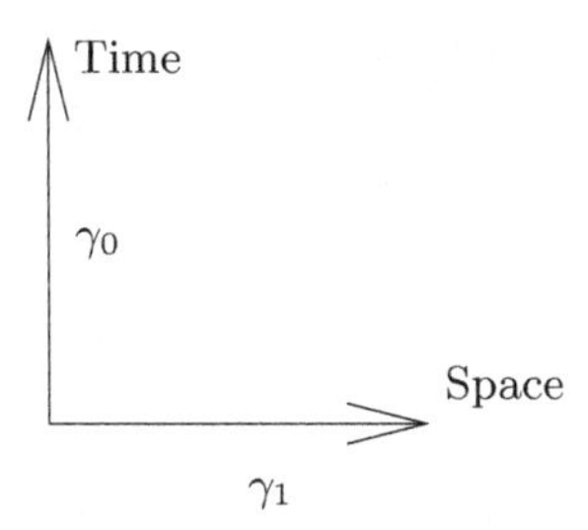

Figure 5.1 *A spacetime diagram.* Spacetime diagrams traditionally have the t axis vertical, so a suitable bivector for this plane is $\gamma_1\gamma_0$.

The two types of bivectors therefore have different signs of their squares. First, we have

$$(\gamma_i\wedge\gamma_j)^2 = -{\gamma_i}^2{\gamma_j}^2 = -1, \tag{5.9}$$

which is the familiar result for Euclidean bivectors. Each of these generates rotations in a plane. For bivectors containing a timelike component, however, we have

$$(\gamma_i\wedge\gamma_0)^2 = -{\gamma_i}^2{\gamma_0}^2 = +1. \tag{5.10}$$

Bivectors with positive square have a number of new properties. One immediate result we notice, for example, is that

$$\begin{aligned} e^{\alpha\gamma_1\gamma_0} &= 1 + \alpha\gamma_1\gamma_0 + \frac{\alpha^2}{2!} + \frac{\alpha^3}{3!}\gamma_1\gamma_0 + \cdots \\ &= \cosh(\alpha) + \sinh(\alpha)\gamma_1\gamma_0. \end{aligned} \tag{5.11}$$

This shows us that we are dealing with *hyperbolic geometry.* This will prove crucial to our treatment of Lorentz transformations. Traditionally, spacetime diagrams are drawn with the time axis vertical (see figure 5.1). For these diagrams the 'right-handed' bivector is, for example, $\gamma_1\gamma_0$. These bivectors do not generate 90° rotations, however, as we now have

$$\gamma_0\cdot(\gamma_1\gamma_0) = -\gamma_1, \quad \gamma_1\cdot(\gamma_1\gamma_0) = -\gamma_0. \tag{5.12}$$

5.1.2 The pseudoscalar

We define the (grade-4) pseudoscalar I by

$$I = \gamma_0\gamma_1\gamma_2\gamma_3. \tag{5.13}$$

In the literature the symbol i is often used for the pseudoscalar. We have departed from this practice to avoid confusion with the i of quantum theory. Using the latter symbol presents a potential problem because of the fact that the pseudoscalar anticommutes with vectors. The pseudoscalar defines an orientation for spacetime, and the reason for the above choice will emerge shortly. We still assume that $\{\gamma_1, \gamma_2, \gamma_3\}$ form a right-handed orthonormal set, as usual for a three-dimensional Cartesian frame. Since I is grade-4, it is equal to its own reverse:

$$\tilde{I} = \gamma_3\gamma_2\gamma_1\gamma_0 = I. \tag{5.14}$$

For relativistic applications we use the tilde $\tilde{}$ to denote the reverse operation. The problem with the alternative symbol, the dagger $\dagger$, is that it is usually reserved for a different role in relativistic quantum theory. The fact that $\tilde{I} = I$ makes it easy to compute the square of I :

$$I^2 = I\tilde{I} = (\gamma_0\gamma_1\gamma_2\gamma_3)(\gamma_3\gamma_2\gamma_1\gamma_0) = -1. \tag{5.15}$$

Multiplication of a bivector by I results in a multivector of grade $4 - 2 = 2$, so returns another bivector. This provides a map between bivectors with positive and negative squares, for example

$$I\gamma_1\gamma_0 = \gamma_1\gamma_0 I = \gamma_1\gamma_0\gamma_0\gamma_1\gamma_2\gamma_3 = -\gamma_2\gamma_3. \tag{5.16}$$

If we define $B_i = \gamma_i\gamma_0$ then the bivector algebra can be summarised by

$$\begin{aligned} B_i \times B_j &= \epsilon_{ijk}\, IB_k, \\ (IB_i)\times(IB_j) &= -\epsilon_{ijk}\, IB_k, \\ (IB_i)\times B_j &= -\epsilon_{ijk} B_k. \end{aligned} \tag{5.17}$$

These equations show that the pseudoscalar provides a natural complex structure for the set of bivectors. This in turn tells us that there is a complex structure hidden in the group of Lorentz transformations.

As well as the four vectors, we also have four trivectors in our algebra. The vectors and trivectors are interchanged by a duality transformation,

$$\gamma_1\gamma_2\gamma_3 = \gamma_0\gamma_0\gamma_1\gamma_2\gamma_3 = \gamma_0 I = -I\gamma_0. \tag{5.18}$$

The pseudoscalar I *anticommutes* with vectors and trivectors, as we are in a space of even dimensions. As always, I commutes with all even-grade multivectors.

5.1.3 The spacetime algebra

Combining the preceding results, we arrive at an algebra with 16 terms. The $\{\gamma_\mu\}$ define an explicit basis for this algebra as follows:

1	$\{\gamma_\mu\}$	$\{\gamma_\mu \wedge \gamma_\nu\}$	$\{I\gamma_\mu\}$	I
1 scalar	4 vectors	6 bivectors	4 trivectors	1 pseudoscalar

This is the *spacetime algebra*, $\mathcal{G}(1,3)$. The structure of this algebra tells us practically all one needs to know about (flat) spacetime and the Lorentz transformation group. A general element of the spacetime algebra can be written as

$$M = \alpha + a + B + Ib + I\beta, \tag{5.19}$$

where α and β are scalars, a and b are vectors and B is a bivector. The reverse of this element is

$$\tilde{M} = \alpha + a - B - Ib + I\beta. \tag{5.20}$$

The vector generators of the spacetime algebra satisfy

$$\gamma_\mu \gamma_\nu + \gamma_\nu \gamma_\mu = 2\eta_{\mu\nu}. \tag{5.21}$$

These are the defining relations of the Dirac matrix algebra, except for the absence of an identity matrix on the right-hand side. It follows that the Dirac matrices define a representation of the spacetime algebra. This also explains our notation of writing $\{\gamma_\mu\}$ for an orthonormal frame. But it must be remembered that the $\{\gamma_\mu\}$ are basis *vectors*, not a set of matrices in 'isospace'.

5.2 Observers, trajectories and frames

From a study of the literature on relativity one can easily form the impression that the subject is in the main concerned with transformations between frames. But it is the subject of relativistic dynamics that is of primary importance to us, and one aim of the spacetime algebra development is to minimise the use of coordinate frames. Instead, we aim to develop spacetime physics in a frame-free manner and, where necessary, then focus on the physics as seen from different observers. Developing relativistic physics in this manner has the added advantage of clarifying precisely which aspects of special relativity need modification to incorporate gravity.

5.2.1 Spacetime paths

Suppose that $x(\lambda)$ describes a curve in spacetime, where λ is some arbitrary, monotonically-increasing parameter along the curve. The tangent vector to the curve is

$$x' = \frac{dx(\lambda)}{d\lambda}. \tag{5.22}$$

Under a change of parameter from λ to τ the tangent vector becomes

$$\frac{dx}{d\tau} = \frac{d\lambda}{d\tau}\frac{dx}{d\lambda}. \tag{5.23}$$

It follows that

$$\left(\frac{dx}{d\tau}\right)^2 = \left(\frac{d\lambda}{d\tau}\right)^2 \left(\frac{dx}{d\lambda}\right)^2, \tag{5.24}$$

so the sign of $(x')^2$ is an invariant feature of the path. We assume for simplicity that this sign does not change along the path. As we are working in a space of mixed signature there are then three cases to consider.

The first possibility is that $(x')^2 > 0$, in which case the path is said to be *timelike*. Timelike trajectories are those followed by massive particles. For these paths we can define an invariant proper interval

$$\Delta\tau = \int_{\lambda_1}^{\lambda_2} \left(\frac{dx}{d\lambda}\cdot\frac{dx}{d\lambda}\right)^{1/2} d\lambda. \tag{5.25}$$

It is straightforward to check that this interval is independent of how the path is parameterised. If we consider the simplest case of a particle (or observer) at rest in the γ_0 system, its spacetime trajectory can be written as $x = t\gamma_0$. In this case it is clear that the interval defines the elapsed time in the observer's rest frame. This must be true for all possible paths, so the interval (5.25) defines the time as measured along the path. This is called the *proper time*, and is usually given the symbol τ. The proper time defines a preferred parameter along the curve with the unique property that the velocity v,

$$v = \frac{dx}{d\tau} = \dot{x}, \tag{5.26}$$

satisfies

$$v^2 = 1. \tag{5.27}$$

Throughout we use dots to denote differentiation with respect to proper time τ. The unit timelike vector v then defines the instantaneous rest frame. The definition of 'proper time' makes it clear that in relativity observers moving in relative motion measure different times.

The second case to consider is that $(x')^2 = 0$. In this case the trajectory is said to be *lightlike* or *null*. Null trajectories are followed by massless (point) particles and (in the geometric optics limit) they define possible photon paths. There is no preferred parameter along these curves, and the proper distance (or time) measured along the curve is 0. Photons do still carry an intrinsic clock, defined by their frequency, but this can tick at an arbitrary rate.

The third possibility is that $(x')^2 < 0$, in which case the trajectory is said to be *spacelike*. As with timelike paths there is a preferred (affine) parameter along the path such that $(x')^2 = -1$. In this case the parameter defines the *proper distance*. Spacelike curves cannot arise for the trajectories of (known) particles, which are constrained to move at less than (or equal to) the speed of light. Events which are separated by spacelike intervals cannot be in causal

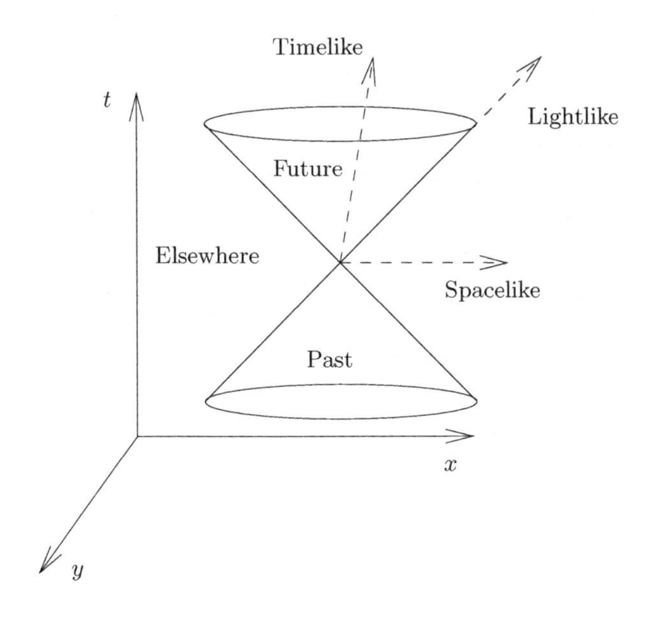

Figure 5.2 *Spacetime trajectories.* There are three different types of spacetime trajectory: timelike, lightlike and spacelike. The set of lightlike trajectories through a point separate spacetime into three regions: the past, the future and 'elsewhere'.

contact with each other and cannot exert any classical influence over each other. The three possibilities for spacetime trajectories are summarised in figure 5.2.

5.2.2 Spacetime frames

The subject of spacetime frames and coordinates dominates many discussions of the meaning of special relativity. The concept of a frame is distinct from that of an observer as it involves the notion of a coordinate lattice. We start with an inertial observer with constant velocity v. This velocity vector is then equated with the timelike vector e_0 from a spacetime frame $\{\mathsf{e}_\mu\}$. The remaining vectors e_i are chosen so that they form a right-handed set of orthonormal spacelike vectors perpendicular to $\mathsf{e}_0 = v$. The $\{\mathsf{e}_\mu\}$ then define a set of frame vectors satisfying

$$\mathsf{e}_\mu \cdot \mathsf{e}_\nu = \eta_{\mu\nu}. \tag{5.28}$$

So far these vectors are only defined at a single point on the observer's trajectory. We now assume that the vectors extend throughout all spacetime, so that any event can be given a set of spacetime coordinates

$$x^\mu = \mathsf{e}^\mu \cdot x. \tag{5.29}$$

Clearly these coordinates are a rather distinct concept from what an observer will actually measure, since the observer is constrained to remain in one place and only receives incoming photons. Frequently one sees discussions involving arrays of clocks all cleverly synchronised to read the time x^0 at each spatial location. But how such a frame is set up is not really the point. The assertion is that the coordinates as specified above are a reasonable model for the sort of distance and time measurements performed in a laboratory system using physical measuring devices. It is precisely this assertion that is challenged by general relativity, which insists that one talk entirely in terms of physically-defined coordinates, so that the x^μ defined above have no physical meaning. That said, for applications not involving gravity and for non-accelerating frames, we can safely identify the coordinates defined above with physical distances and times and will continue to do so in this chapter.

5.2.3 Relative vectors

Now suppose that we follow a timelike path with instantaneous velocity v, $v^2 = 1$. What sort of quantities do we measure? First we construct a frame of rest vectors $\{\mathsf{e}_i\}$ perpendicular to $v = \mathsf{e}_0$. We also take a point on the worldline as the spatial origin. Then a general event x can be decomposed in this frame as

$$x = t\mathsf{e}_0 + x^i \mathsf{e}_i, \tag{5.30}$$

where the time coordinate is

$$t = x \cdot \mathsf{e}_0 = x \cdot v \tag{5.31}$$

and spatial coordinates are

$$x^i = x \cdot \mathsf{e}^i. \tag{5.32}$$

Suppose now that the event is a point on the worldline of an object at rest in our frame. The three-dimensional vector to this object is

$$x^i \mathsf{e}_i = x \cdot \mathsf{e}^\mu \, \mathsf{e}_\mu - x \cdot \mathsf{e}^0 \, \mathsf{e}_0 = x - x \cdot v \, v = x \wedge v \, v. \tag{5.33}$$

Wedging with v projects onto the components of the vector x in the rest frame of v. The key quantity is the spacetime bivector $x \wedge v$. We call this the *relative* vector and write

$$\boldsymbol{x} = x \wedge v. \tag{5.34}$$

With these definitions we have

$$xv = x \cdot v + x \wedge v = t + \boldsymbol{x}. \tag{5.35}$$

The invariant distance now decomposes as

$$\begin{aligned} x^2 = xvvx &= (x\cdot v + x\wedge v)(x\cdot v + v\wedge x) \\ &= (t+\boldsymbol{x})(t-\boldsymbol{x}) = t^2 - \boldsymbol{x}^2, \end{aligned} \tag{5.36}$$

recovering the invariant interval. A second observer with a different velocity performs a different split of x into time and space components. But the interval x^2 is the same for all observers as it manifestly does not depend on the choice of frame.

5.2.4 The even subalgebra

Each observer sees a set of relative vectors, which we model as spacetime bivectors. What algebraic properties do these have? To simplify matters, we take the timelike velocity vector to be γ_0 and introduce a standard frame of relative vectors

$$\boldsymbol{\sigma}_i = \gamma_i\gamma_0. \tag{5.37}$$

These define a set of spacetime bivectors representing timelike planes. (The notation is again borrowed from quantum mechanics and is commonplace in the spacetime algebra literature.) The $\{\boldsymbol{\sigma}_i\}$ satisfy

$$\begin{aligned} \boldsymbol{\sigma}_i\cdot\boldsymbol{\sigma}_j &= \tfrac{1}{2}(\gamma_i\gamma_0\gamma_j\gamma_0 + \gamma_j\gamma_0\gamma_i\gamma_0) \\ &= \tfrac{1}{2}(-\gamma_i\gamma_j - \gamma_j\gamma_i) = \delta_{ij}. \end{aligned} \tag{5.38}$$

These act as vector generators for a three-dimensional algebra. This is the geometric algebra of the relative space in the rest frame defined by γ_0. Furthermore, the volume element of this algebra is

$$\boldsymbol{\sigma}_1\boldsymbol{\sigma}_2\boldsymbol{\sigma}_3 = (\gamma_1\gamma_0)(\gamma_2\gamma_0)(\gamma_3\gamma_0) = -\gamma_1\gamma_0\gamma_2\gamma_3 = I, \tag{5.39}$$

so the algebra of relative space shares the same pseudoscalar as spacetime. This was the reason for our earlier definition of I. Of course, we still have

$$\tfrac{1}{2}(\boldsymbol{\sigma}_i\boldsymbol{\sigma}_j - \boldsymbol{\sigma}_j\boldsymbol{\sigma}_i) = \epsilon_{ijk}I\boldsymbol{\sigma}_k, \tag{5.40}$$

so that both relative vectors and relative bivectors are spacetime bivectors.

The even-grade terms in the spacetime algebra define the *even subalgebra*. As we have just established, this algebra has precisely the properties of the algebra of three-dimensional (relative) space. The even subalgebra contains scalar and pseudoscalar terms, and six bivector terms. These are split into three timelike vectors and three spacelike vectors, which in turn become relative vectors and bivectors. This is called a *spacetime split*, and it is *observer-dependent*. Different velocity vectors generate different spacetime splits. Algebraically, this provides us with an extremely efficient tool for comparing physical effects in different frames.

Spacetime bivectors which are also used as relative vectors are written in bold. This conforms with our earlier usage of a bold face for vectors in three dimensions. There is a potential ambiguity here — how are we to interpret the expression $\boldsymbol{a}\wedge\boldsymbol{b}$? Our convention is that if all of the terms in an expression are bold, the dot and wedge symbols drop down to their three-dimensional meaning, otherwise they take their spacetime definition. This works pretty well in practice, though where necessary we will try to draw attention to the fact that this convention is in use.

5.2.5 Relative velocity

Suppose that an observer with constant velocity v measures the relative velocity of a particle with proper velocity $u(\tau) = \dot{x}(\tau)$, $u^2 = 1$. We have

$$uv = \frac{d}{d\tau}(x(\tau)v) = \frac{d}{d\tau}(t + \boldsymbol{x}), \tag{5.41}$$

where $t + \boldsymbol{x}$ is the description of the event x in the v frame. It follows that

$$\frac{dt}{d\tau} = u\cdot v, \quad \frac{d\boldsymbol{x}}{d\tau} = u\wedge v. \tag{5.42}$$

The relative velocity $\boldsymbol{u}$ as measured in the v frame is therefore

$$\boldsymbol{u} = \frac{d\boldsymbol{x}}{dt} = \frac{d\boldsymbol{x}}{d\tau}\frac{d\tau}{dt} = \frac{u\wedge v}{u\cdot v}. \tag{5.43}$$

This construction of the relative velocity is extremely elegant. It embodies the concept of relativity in its precise (anti)symmetry. If we interchange u and v the second observer measures precisely the same relative speed as the first, but in the opposite direction. Expressions like $u\wedge v/u\cdot v$ arise frequently in the subject of *projective geometry* (see section 10.1). The resulting bivector is homogeneous, which is to say we can rescale u and v and still recover the same result. So the choice of parameterisation of the two spacetime trajectories is irrelevant to their relative velocity. The relative velocity is determined solely by the spacetime trajectories themselves, and not by any evolution parameter.

The definition of the relative velocity ensures that the magnitude is

$$\frac{(u\wedge v)^2}{(u\cdot v)^2} = 1 - \frac{1}{(u\cdot v)^2} < 1, \tag{5.44}$$

so no two observers measure a relative velocity greater than the speed of light (which is 1 in our current choice of units). If we form the Lorentz factor γ using

$$\begin{aligned}\gamma^{-2} &= 1 - \boldsymbol{u}^2 \\ &= 1 + (u\cdot v)^{-2}[(uv - u\cdot v)(vu - v\cdot u)] = (u\cdot v)^{-2},\end{aligned} \tag{5.45}$$

we find that $\gamma = u \cdot v$. It follows that we can decompose the velocity as

$$u = uvv = (u \cdot v + u \wedge v)v = \gamma(1 + \boldsymbol{u})v, \tag{5.46}$$

which shows a neat split into a part $\gamma \boldsymbol{u} v$ in the rest space of v, and a part γv along v.

5.2.6 Momentum and wave vectors

The relativistic definitions of energy and momentum can be motivated in various ways. Perhaps the simplest is to consider photons with frequency ω and wavevector $\boldsymbol{k}$ measured in the γ_0 frame. From quantum theory, the energy and momentum are given by $\hbar\omega$ and $\hbar\boldsymbol{k}$ respectively. If we define the wavevector k by

$$k = \omega\gamma_0 + k^i\gamma_i, \tag{5.47}$$

then the energy-momentum vector for the photon is simply

$$p = \hbar k. \tag{5.48}$$

An observer with velocity v, as opposed to γ_0, measures energy and momentum given by

$$E = p \cdot v, \quad \boldsymbol{p} = p \wedge v. \tag{5.49}$$

We take this as the correct definition for massive particles as well. So a particle of rest mass m and velocity u has an energy-momentum vector $p = mu$. A spacetime split of this vector with the velocity vector v yields

$$pv = p \cdot v + p \wedge v = E + \boldsymbol{p}. \tag{5.50}$$

A significant feature of this definition is that the relative momentum is related to the velocity by

$$\boldsymbol{p} = mu \cdot v \, \boldsymbol{u} = \gamma m \boldsymbol{u}, \tag{5.51}$$

where again γ is the Lorentz factor. One sometimes sees this formula written in terms of a velocity-dependent mass $m' = \gamma m$, but we will not adopt this practice here.

From the definition of p we recover the invariant

$$m^2 = p^2 = pvvp = (E + \boldsymbol{p})(E - \boldsymbol{p}) = E^2 - \boldsymbol{p}^2. \tag{5.52}$$

Similarly, for a photon with wavevector k, $k^2 = 0$, we have

$$0 = kvvk = (\omega + \boldsymbol{k})(\omega - \boldsymbol{k}) = \omega^2 - \boldsymbol{k}^2. \tag{5.53}$$

This recovers the relation $|\boldsymbol{k}| = \omega$, which holds in all frames.

5.2.7 *Proper acceleration*

A final ingredient in the formulation of relativistic dynamics is the proper acceleration. A particle follows a trajectory $x(\tau)$, where τ is the proper time. The particle has velocity $v = \dot{x}$, $v^2 = 1$. The proper acceleration is simply

$$\dot{v} = \frac{dv}{d\tau}. \tag{5.54}$$

Since $v^2 = 1$, the velocity and acceleration are perpendicular

$$\frac{d}{d\tau}(v^2) = 0 = 2\dot{v}\cdot v. \tag{5.55}$$

In many physical phenomena it turns out that a more useful concept is provided by the *acceleration bivector*

$$B_v = \dot{v}\wedge v = \dot{v}v. \tag{5.56}$$

This bivector denotes the acceleration projected into the instantaneous rest frame of the particle. Typically this bivector multiplied by the rest mass is equated with a bivector encoding the forces acting on the particle. Any change in the parameter along the curve will rescale the velocity vector, so B_v can be written as

$$B_v = \frac{v'\wedge v}{(v\cdot v)^{3/2}}, \tag{5.57}$$

which is independent of the parameterisation of the trajectory.

Before applying the various preceding definitions to a range of dynamical problems, we turn to a discussion of the Lorentz transformations. This will pave the way for a powerful method for studying relativistic problems which is unique to geometric algebra.

5.3 Lorentz transformations

Lorentz transformations are usually expressed in the form of a coordinate transformation. We suppose that two inertial observers have set up 'coordinate lattices' in their own rest frames, as discussed in section 5.2.2. We denote these frames by S and S', and assume that they are set up such that their 1 and 2 axes coincide, but that S' moves at (scalar) velocity βc along the 3 axis as seen in the S frame. We denote the 0 and 3 components by t and z respectively. If the origins of the frames coincide at $t = t' = 0$, the coordinates of the same spacetime event as measured in the two frames are related by

$$t' = \gamma(t - \beta z), \quad x^{1'} = x^1, \quad x^{2'} = x^2, \quad z' = \gamma(z - \beta t), \tag{5.58}$$

where $\gamma = (1-\beta^2)^{-1/2}$ and β is the velocity in units of c ($\beta < 1$). The inverse relations are easily found to be

$$t = \gamma(t' + \beta z'), \quad x^1 = x^{1'}, \quad x^2 = x^{2'}, \quad z = \gamma(z' + \beta t'). \tag{5.59}$$

The arguments leading to these transformation laws are discussed in all introductory texts on relativity (see e.g. Rindler (1977) or French (1968)).

To get a clearer understanding of this transformation law we must first convert these relations into a transformation law for the frame vectors. The vector x has been decomposed in two frames, $\{\mathsf{e}_\mu\}$ and $\{\mathsf{e}'_\mu\}$, so that

$$x = x^\mu \mathsf{e}_\mu = x^{\mu'} \mathsf{e}'_\mu. \tag{5.60}$$

We then have, for example,

$$t = \mathsf{e}^0 \cdot x, \quad t' = \mathsf{e}^{0'} \cdot x. \tag{5.61}$$

Concentrating on the 0 and 3 components we have

$$t\mathsf{e}_0 + z\mathsf{e}_3 = t'\mathsf{e}'_0 + z'\mathsf{e}'_3, \tag{5.62}$$

and from this we derive the vector relations

$$\mathsf{e}'_0 = \gamma(\mathsf{e}_0 + \beta\mathsf{e}_3), \quad \mathsf{e}'_3 = \gamma(\mathsf{e}_3 + \beta\mathsf{e}_0). \tag{5.63}$$

These define the new frame in terms of the old. As a check the new frame vectors have the correct normalisation,

$$(\mathsf{e}'_0)^2 = \gamma^2(1-\beta^2) = 1, \quad (\mathsf{e}'_3)^2 = -1. \tag{5.64}$$

The geometry of this transformation is illustrated in figure 5.3.

We saw earlier that bivectors with positive square lead to hyperbolic geometry. This suggests that we introduce an 'angle' α with

$$\tanh(\alpha) = \beta \tag{5.65}$$

so that

$$\gamma = \big(1 - \tanh^2(\alpha)\big)^{-1/2} = \cosh(\alpha). \tag{5.66}$$

The vector e'_0 is now

$$\begin{aligned} \mathsf{e}'_0 &= \cosh(\alpha)\,\mathsf{e}_0 + \sinh(\alpha)\,\mathsf{e}_3 \\ &= \big(\cosh(\alpha) + \sinh(\alpha)\,\mathsf{e}_3\mathsf{e}_0\big)\mathsf{e}_0 \\ &= \exp(\alpha\,\mathsf{e}_3\mathsf{e}_0)\,\mathsf{e}_0, \end{aligned} \tag{5.67}$$

where we have expressed the scalar + bivector term as an exponential. Similarly, we have

$$\mathsf{e}'_3 = \cosh(\alpha)\,\mathsf{e}_3 + \sinh(\alpha)\,\mathsf{e}_0 = \exp(\alpha\,\mathsf{e}_3\mathsf{e}_0)\,\mathsf{e}_3. \tag{5.68}$$

Now recall that these are just two of four frame vectors, and the other pair

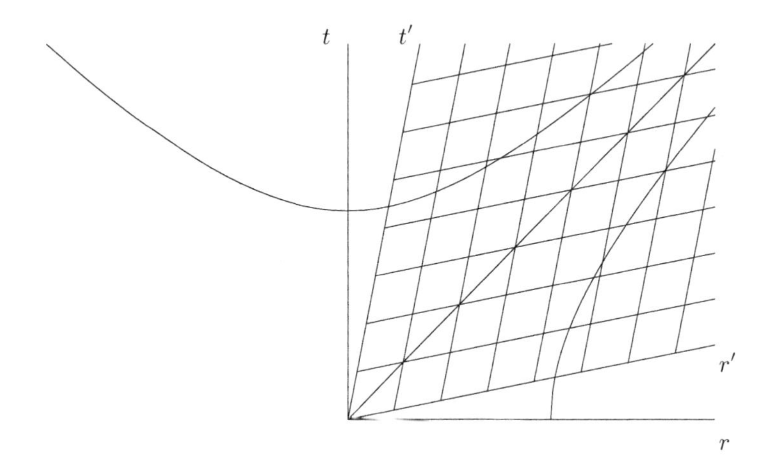

Figure 5.3 *A Lorentz transformation.* The transformation leaves the magnitude of a vector invariant. As the underlying geometry of a spacetime plane is Lorentzian, vectors of constant magnitude lie on hyperbolae, rather than circles. The transformed axes define a new coordinate grid.

are unchanged by the transformation. Since $\mathsf{e}_3\mathsf{e}_0$ anticommutes with e_0 and e_3, but commutes with e_1 and e_2, we can express the relationship between the two frames as

$$\mathsf{e}_\mu' = R\mathsf{e}_\mu \tilde{R}, \quad \mathsf{e}^{\mu\prime} = R\mathsf{e}^\mu \tilde{R}, \tag{5.69}$$

where

$$R = \mathrm{e}^{\alpha\, \mathsf{e}_3\mathsf{e}_0/2}. \tag{5.70}$$

The same rotor prescription introduced for rotations in Euclidean space also works for boosts in relativity! This is dramatically simpler than having to work with 4×4 Lorentz transform matrices.

5.3.1 Addition of velocities

As a simple example, suppose that we are in a frame with basis vectors $\{\gamma_\mu\}$. We observe two objects flying apart with 4-velocities

$$v_1 = \mathrm{e}^{\alpha_1\gamma_1\gamma_0/2}\gamma_0\mathrm{e}^{-\alpha_1\gamma_1\gamma_0/2} = \mathrm{e}^{\alpha_1\gamma_1\gamma_0}\gamma_0 \tag{5.71}$$

and

$$v_2 = \mathrm{e}^{-\alpha_2\gamma_1\gamma_0/2}\gamma_0\mathrm{e}^{\alpha_2\gamma_1\gamma_0/2} = \mathrm{e}^{-\alpha_2\gamma_1\gamma_0}\gamma_0. \tag{5.72}$$

What is the relative velocity they see for each other? We form

$$\frac{v_1 \wedge v_2}{v_1 \cdot v_2} = \frac{\langle e^{(\alpha_1+\alpha_2)\gamma_1\gamma_0}\rangle_2}{\langle e^{(\alpha_1+\alpha_2)\gamma_1\gamma_0}\rangle_0} = \frac{\sinh(\alpha_1+\alpha_2)\gamma_1\gamma_0}{\cosh(\alpha_1+\alpha_2)}. \tag{5.73}$$

Both observers therefore measure a relative velocity of

$$\tanh(\alpha_1+\alpha_2) = \frac{\tanh(\alpha_1)+\tanh(\alpha_2)}{1+\tanh(\alpha_1)\tanh(\alpha_2)}, \tag{5.74}$$

Addition of (collinear) velocities is achieved by adding hyperbolic angles, and not the velocities themselves. Replacing the tanh factors by the scalar velocities $u = c\tanh(\alpha)$ recovers the more familiar expression

$$u' = \frac{u_1+u_2}{1+u_1u_2/c^2}. \tag{5.75}$$

The surprising conclusion is that addition of velocities in spacetime is really a generalized rotation in a hyperbolic space! Quite dramatically different from the Newtonian prescription of simple vector addition of the velocities.

5.3.2 Photons, Doppler shifts and aberration

For many relativistic applications involving the properties of light it is sufficient to use a simplified model of a photon as a point particle following a null trajectory. The tangent vector to the path is the wavevector k. This provides for simple formulae for Doppler shifts and aberration. Suppose that two particles follow different worldlines and that particle 1 emits a photon which is received by particle 2 (see figure 5.4). The frequency seen by particle 1 is $\omega_1 = v_1 \cdot k$, and that by particle 2 is $\omega_2 = v_2 \cdot k$. The ratio of these describes the Doppler effect, often expressed as a redshift, z:

$$1+z = \frac{\omega_1}{\omega_2} = \frac{v_1 \cdot k}{v_2 \cdot k}. \tag{5.76}$$

This can be applied in many ways. For example, suppose that the emitter is receding in the γ_1 direction, and $v_2 = \gamma_0$. We have

$$k = \omega_2(\gamma_0+\gamma_1), \qquad v_1 = \cosh(\alpha)\,\gamma_0 - \sinh(\alpha)\,\gamma_1, \tag{5.77}$$

so that

$$1+z = \frac{\omega_2\big(\cosh(\alpha)+\sinh(\alpha)\big)}{\omega_2} = e^{\alpha}. \tag{5.78}$$

The velocity of the emitter in the γ_0 frame is $\tanh(\alpha)$, and it is easy to check that

$$e^{\alpha} = \left(\frac{1+\tanh(\alpha)}{1-\tanh(\alpha)}\right)^{1/2}. \tag{5.79}$$

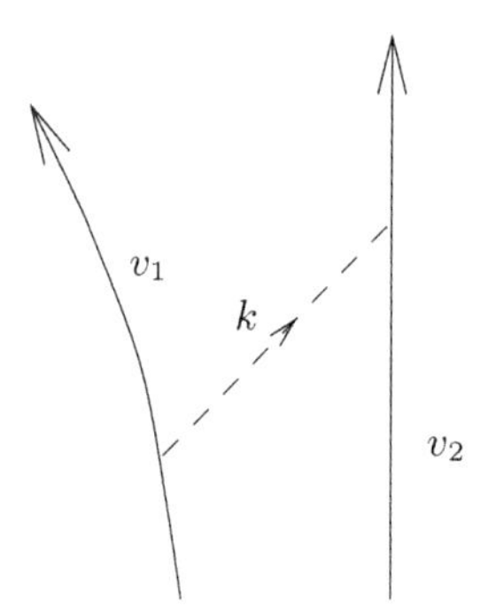

Figure 5.4 *Photon emission and absorption.* A photon is emitted by particle 1 and received by particle 2.

This formula recovers the standard expression for the relativistic Doppler effect:

$$\omega_2 = \left(\frac{1-\beta}{1+\beta}\right)^{1/2} \omega_1. \tag{5.80}$$

In its current form this formula is appropriate for a source and receiver moving away from each other at velocity βc. Had they been approaching each other the sign of β would be reversed, leading to an increased frequency at the receiver (a blueshift).

Aberration formulae can be obtained in a similar manner. Suppose that observer 1 has velocity γ_0, and that this observer receives photons at an angle θ to the 1 axis in the 12 plane. The photons are therefore on a null trajectory with tangent vector

$$n = \gamma_0 - \cos(\theta)\,\gamma_1 - \sin(\theta)\,\gamma_2, \tag{5.81}$$

and the γ_0 observer recovers the angle θ via

$$\tan(\theta) = \frac{n\cdot\gamma_2}{n\cdot\gamma_1}. \tag{5.82}$$

Suppose now that a second observer moves with velocity β relative to the first along the 1 axis. This observer's velocity is

$$v = \mathsf{e}_0 = \cosh(\alpha)\,\gamma_0 + \sinh(\alpha)\,\gamma_1 \tag{5.83}$$

and the frame vectors for this observer are

$$\mathsf{e}_1 = \cosh(\alpha)\,\gamma_1 + \sinh(\alpha)\,\gamma_0, \quad \mathsf{e}_2 = \gamma_2, \quad \mathsf{e}_3 = \gamma_3. \tag{5.84}$$

According to this observer the photons arrive at an angle

$$\tan(\theta') = \frac{n\cdot\mathsf{e}_2}{n\cdot\mathsf{e}_1} = \frac{\sin(\theta)}{\cosh(\alpha)\,\cos(\theta) + \sinh(\alpha)}. \tag{5.85}$$

A straightforward rearrangement gives

$$\cos(\theta') = \frac{\cosh(\alpha)\,\cos(\theta) + \sinh(\alpha)}{\cosh(\alpha) + \sinh(\alpha)\,\cos(\theta)} = \frac{\cos(\theta) + \beta}{1 + \beta\cos(\theta)}, \tag{5.86}$$

so observers in relative motion measure different angles to a fixed light source. This effect can be seen in observations of stars from the Earth. The Earth's orbital velocity around the sun has a β of roughly 10^{-4} so to a good approximation we have

$$\cos(\theta') \approx \cos(\theta) + \beta\sin^2(\theta). \tag{5.87}$$

The aberration angle $\phi = \theta - \theta'$ satisfies the approximate formula

$$\phi \approx \beta\sin(\theta), \tag{5.88}$$

which implies that the aberration varies over a year as θ varies through a complete cycle. This variation was first observed by James Bradley in 1727 and was explained in terms of a particle model of light. Bradley was able to use his data to give an improved estimate of the speed of light, though the full relativistic relation of (5.86) cannot be checked in this manner.

5.4 The Lorentz group

The full Lorentz group consists of the transformation group for vectors that preserves lengths and angles. These include reflections and rotations. A reflection in the hyperplane perpendicular to n is achieved by

$$a \mapsto -nan^{-1}. \tag{5.89}$$

The n^{-1} is necessary to accommodate both timelike $n^2 > 0$ and spacelike $n^2 < 0$ cases. We cannot have null n, as the inverse does not exist. A timelike n generates time-reversal transformations, whereas spacelike reflections preserve time-ordering. Pairs of either of these result in a transformation which preserves time-ordering. However, a combination of one spacelike and one timelike reflection does not preserve the time-ordering. The full Lorentz group therefore contains four sectors (table 5.1).

The structure of the Lorentz group is easily understood in the spacetime algebra. We concentrate on even numbers of reflections, which have determinant +1 and correspond to type I and type IV transformations. The remaining types are obtained from these by a single extra reflection. If we combine even numbers of reflections we arrive at a transformation of the form

$$a \mapsto \psi a\psi^{-1}, \tag{5.90}$$

where ψ is an even multivector. This expression is currently too general, as we

	Parity preserving	Space reflection
Time order preserving	I Proper orthochronous	II I with space reflection
Time reversal	III I with time reversal	IV I with $a \mapsto -a$

Table 5.1 *The full Lorentz group.* The group of Lorentz transformations falls into four disjoint sectors. Sectors I and IV have determinant $+1$, whereas II and III have determinant -1. Both I and II preserve time-ordering, and the proper orthochronous transformations (type I) are simply-connected to the identity.

have not ensured that the right-hand side is a vector. To see how to do this we decompose ψ into invariant terms. We first note that

$$\psi\tilde{\psi} = (\psi\tilde{\psi})^{\sim} \tag{5.91}$$

so $\psi\tilde{\psi}$ is even-grade and equal to its own reverse. It can therefore only contain a scalar and a pseudoscalar,

$$\psi\tilde{\psi} = \alpha_1 + I\alpha_2 = \rho e^{I\beta}, \tag{5.92}$$

where $\rho \neq 0$ in order for ψ^{-1} to exist. We can now define a rotor R by

$$R = \psi(\rho e^{I\beta})^{-1/2}, \tag{5.93}$$

so that

$$R\tilde{R} = \psi\tilde{\psi}(\rho e^{I\beta})^{-1} = 1, \tag{5.94}$$

as required. We now have

$$\psi = \rho^{1/2} e^{I\beta/2} R, \qquad \psi^{-1} = \rho^{-1/2} e^{-I\beta/2} \tilde{R} \tag{5.95}$$

and our general transformation becomes

$$a \mapsto e^{I\beta/2} R a e^{-I\beta/2} \tilde{R} = e^{I\beta} R a \tilde{R}. \tag{5.96}$$

The term $Ra\tilde{R}$ is necessarily a vector as it is equal to its own reverse, so we must restrict β to either 0 or π, leaving the transformation

$$a \mapsto \pm R a \tilde{R}. \tag{5.97}$$

The transformation $a \mapsto Ra\tilde{R}$ preserves causal ordering as well as parity. Transformations of this type are called 'proper orthochronous' transformations.

We can prove that transformations parameterised by rotors are proper orthochronous by starting with the velocity γ_0 and transforming it to $v = R\gamma_0\tilde{R}$. We require that the γ_0 component of v is positive, that is,

$$\gamma_0 \cdot v = \langle \gamma_0 R \gamma_0 \tilde{R} \rangle > 0. \tag{5.98}$$

Decomposing in the γ_0 frame we can write

$$R = \alpha + \boldsymbol{a} + I\boldsymbol{b} + I\beta \tag{5.99}$$

and we find that

$$\langle \gamma_0 R \gamma_0 \tilde{R} \rangle = \alpha^2 + \boldsymbol{a}^2 + \boldsymbol{b}^2 + \beta^2 > 0 \tag{5.100}$$

as required. Our rotor transformation law describes the group of proper orthochronous transformations, often called the *restricted Lorentz group*. These are the transformations of most physical relevance. The negative sign in equation (5.97) corresponds to $\beta = \pi$ and gives class-IV transformations.

5.4.1 Invariant decomposition and fixed points

Every rotor in spacetime can be written in terms of a bivector as

$$R = \pm \mathrm{e}^{B/2}. \tag{5.101}$$

(The minus sign is rarely required, and does not affect the vector transformation law.) We can understand many of the features of spacetime transformations and rotors through the properties of the bivector B. The bivector B can be decomposed in a Lorentz-invariant manner by first writing

$$B^2 = \langle B^2 \rangle_0 + \langle B^2 \rangle_4 = \rho \mathrm{e}^{I\phi}, \tag{5.102}$$

and we will assume that $\rho \neq 0$. (The case of a null bivector is treated slightly differently.) We now define

$$\hat{B} = \rho^{-1/2} \mathrm{e}^{-I\phi/2} B, \tag{5.103}$$

so that

$$\hat{B}^2 = \rho^{-1} \mathrm{e}^{-I\phi} B^2 = 1. \tag{5.104}$$

With this we can now write

$$B = \rho^{1/2} \mathrm{e}^{I\phi/2} \hat{B} = \alpha \hat{B} + \beta I \hat{B}, \tag{5.105}$$

which decomposes B into a pair of bivector blades, $\alpha\hat{B}$ and $\beta I\hat{B}$. Since

$$\hat{B}(I\hat{B}) = (I\hat{B})\hat{B} = I, \tag{5.106}$$

the separate bivector blades commute. The rotor R now decomposes into

$$R = \mathrm{e}^{\alpha\hat{B}/2} \mathrm{e}^{\beta I\hat{B}/2} = \mathrm{e}^{\beta I\hat{B}/2} \mathrm{e}^{\alpha\hat{B}/2}, \tag{5.107}$$

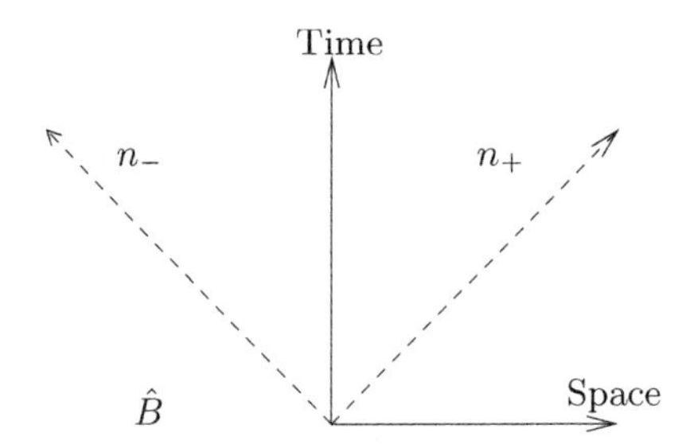

Figure 5.5 *A timelike plane.* Any timelike plane $\hat{B}$, $\hat{B}^2 = 1$, contains two null vectors n_+ and n_-. These can be normalised so that $n_+ \wedge n_- = 2\hat{B}$.

exhibiting an *invariant* split into a boost and a rotation. The boost is generated by $\hat{B}$ and the rotation by $I\hat{B}$.

For every timelike bivector $\hat{B}$, $\hat{B}^2 = 1$, we can construct a pair of null vectors $n_\pm$ satisfying

$$\hat{B} \cdot n_\pm = \pm n_\pm. \tag{5.108}$$

These are necessarily null, since

$$n_+ \cdot n_+ = (B \cdot n_+) \cdot n_+ = B \cdot (n_+ \wedge n_+) = 0, \tag{5.109}$$

with the same holding for n_-. The two null vectors can also be chosen so that

$$n_+ \wedge n_- = 2\hat{B}, \tag{5.110}$$

so that they form a null basis for the timelike plane defined by $\hat{B}$ (see figure 5.5).

The null vectors $n_\pm$ anticommute with $\hat{B}$ and therefore commute with $I\hat{B}$. The effect of the Lorentz transformation on $n_\pm$ is therefore

$$\begin{aligned} R n_\pm \tilde{R} &= \mathrm{e}^{\alpha \hat{B}/2} n_\pm \mathrm{e}^{-\alpha \hat{B}/2} \\ &= \cosh(\alpha)\, n_\pm + \sinh(\alpha)\, \hat{B} \cdot n_\pm \\ &= \mathrm{e}^{\pm\alpha} n_\pm. \end{aligned} \tag{5.111}$$

The two null directions are therefore just scaled — their direction is unchanged. It follows that every Lorentz transformation has two invariant null directions. The case where the bivector generator itself is null, $B^2 = 0$, corresponds to the special situation where these two null directions coincide.

5.4.2 The celestial sphere

One way to visualise the effect of Lorentz transformations is through their effect on the past light-cone (see figure 5.6). Each null vector on the past light-cone maps to a point on the sphere S^- — the *celestial sphere* for the observer. Suppose

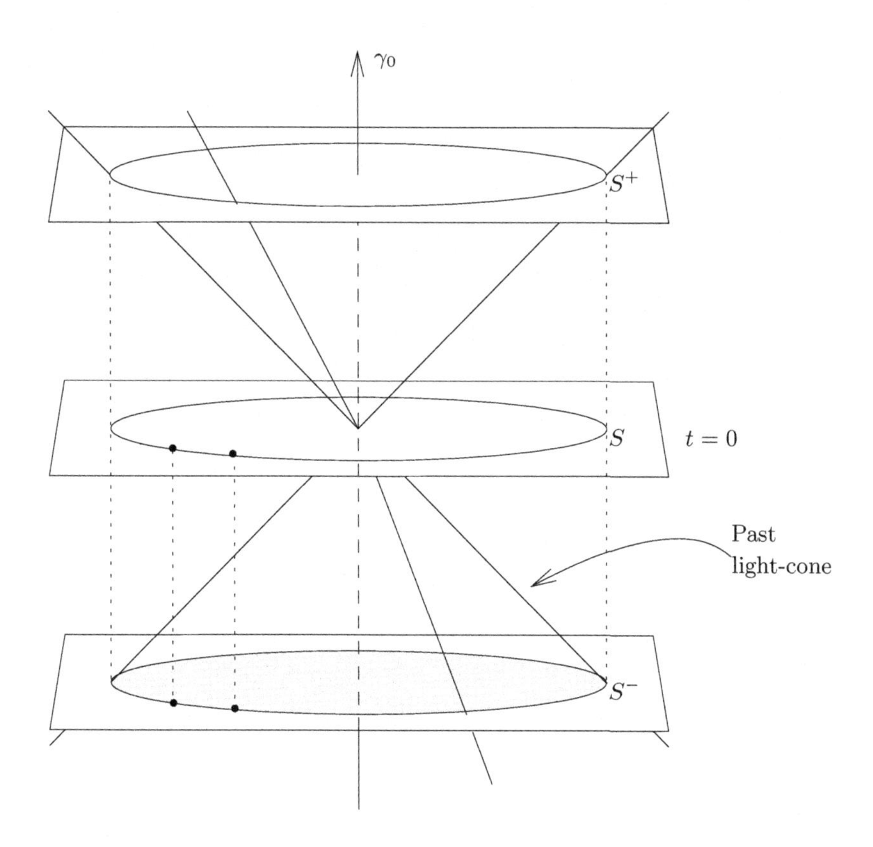

Figure 5.6 *The celestial sphere.* Each observer sees events in their past light-cone, which can be viewed as defining a sphere (shown here as a circle in a plane).

then that light is received along the null vector n, with the observer's velocity chosen to be γ_0. The relative vector in the γ_0 frame is $n \wedge \gamma_0$. This has magnitude

$$(n \wedge \gamma_0)^2 = (n \cdot \gamma_0)^2 - n^2 \gamma_0^2 = (n \cdot \gamma_0)^2. \tag{5.112}$$

We therefore define the unit relative vector $\boldsymbol{n}$ by the projective formula

$$\boldsymbol{n} = \frac{n \wedge \gamma_0}{n \cdot \gamma_0}. \tag{5.113}$$

Observers passing through the same spacetime point at different velocities see different celestial spheres. If a second observer has velocity $v = R\gamma_0\tilde{R}$, the unit relative vectors in this observer's frame are formed from $n \wedge v / n \cdot v$. These can be brought to the γ_0 frame for comparison by forming

$$\boldsymbol{n}' = \tilde{R}\frac{n \wedge v}{n \cdot v}R = \frac{n' \wedge \gamma_0}{n' \cdot \gamma_0}, \tag{5.114}$$

where $n' = \tilde{R}nR$. The effects of Lorentz transformations can be visualised simply by moving around points on the celestial sphere with the map $n \mapsto \tilde{R}nR$. We know immediately, then, that two directions remain invariant and so describe the same points on the celestial spheres of two observers.

5.4.3 Relativistic visualisation

We have endeavoured to separate the concept of a single observer from that of a coordinate lattice. A clear illustration of this distinction arises when one studies how bodies appear when seen by different observers. Concentrating purely on coordinates leads directly to the conclusion that there is a measurable Lorentz contraction in the direction of motion of a body moving relative to some coordinate system. But when we consider what two different observers actually *see*, the picture is rather different.

Suppose that two observers in relative motion observe a sphere. The sphere and one of the observers are both at rest in the γ_0 system. This observer sees the edge of the sphere as a circle defined by the unit vectors

$$\boldsymbol{n} = \sin(\theta)(\cos(\phi)\,\boldsymbol{\sigma}_1 + \sin(\phi)\,\boldsymbol{\sigma}_2) + \cos(\theta)\,\boldsymbol{\sigma}_3, \quad 0 \leq \phi < 2\pi. \tag{5.115}$$

The angle θ is fixed so the sphere subtends an angle 2θ on the sky and is centred on the 3 axis (see figure 5.7). The incoming photon paths from the sphere are defined by the family of null vectors

$$n = (1 - \boldsymbol{n})\gamma_0. \tag{5.116}$$

Now suppose that a second observer has velocity $\beta = \tanh(\alpha)$ along the 1 axis, so

$$v = \cosh(\alpha)\,\gamma_0 + \sinh(\alpha)\,\gamma_1 = R\gamma_0\tilde{R}, \tag{5.117}$$

where $R = \exp(\alpha\,\gamma_1\gamma_0/2)$. To compare what these two observers see we form

$$\begin{aligned} n' = \tilde{R}nR = {} & \cosh(\alpha)\big(1 + \beta\sin(\theta)\,\cos(\phi)\big)\gamma_0 - \cosh(\alpha)\big(\sin(\theta)\,\cos(\phi) + \beta\big)\gamma_1 \\ & - \sin(\theta)\,\sin(\phi)\,\gamma_2 - \cos(\theta)\,\gamma_3. \end{aligned} \tag{5.118}$$

And from this the new unit relative outward vector is

$$\boldsymbol{n}' = \frac{\cosh(\alpha)\big(\sin(\theta)\,\cos(\phi) + \beta\big)\boldsymbol{\sigma}_1 + \sin(\theta)\,\sin(\phi)\,\boldsymbol{\sigma}_2 + \cos(\theta)\,\boldsymbol{\sigma}_3}{\cosh(\alpha)\big(1 + \beta\sin(\theta)\,\cos(\phi)\big)}. \tag{5.119}$$

Now consider the vector

$$\boldsymbol{c} = \boldsymbol{\sigma}_3 + \sinh(\alpha)\,\cos(\theta)\,\boldsymbol{\sigma}_1. \tag{5.120}$$

This vector satisfies

$$\boldsymbol{c}\cdot\boldsymbol{n}' = \cosh(\alpha)\,\cos(\theta), \tag{5.121}$$

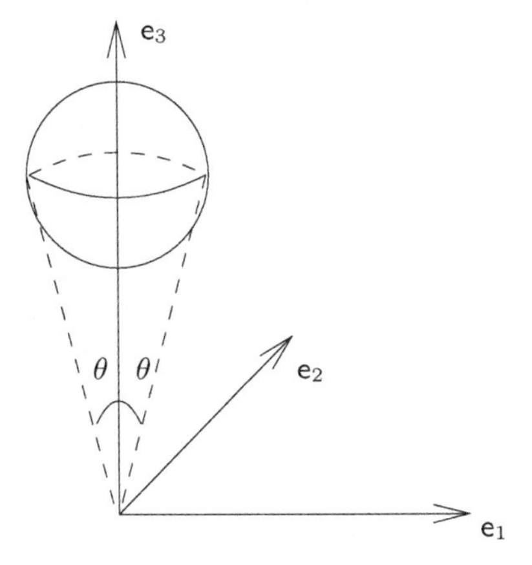

Figure 5.7 *Relativistic visualization of a sphere.* The sphere is at rest in the γ_0 frame with its centre a unit distance along the 3 axis. The sphere is simultaneously observed by two observers placed at the spatial origin. One observer is at rest in the γ_0 system, and the other is moving along the 1 axis.

which is independent of ϕ. It follows that, from the point of view of the second observer, all points on the edge of the sphere subtend the same angle to $\boldsymbol{c}$. So the vector $\boldsymbol{c}$ must lie at the centre of a circle, and the second observer still sees the edge of the sphere as circular. That is, both observers see the sphere as a sphere, and there is no observable contraction along the direction of motion. The only difference is that the moving observer sees the angular diameter of the sphere reduced from 2θ to $2\theta'$, where

$$\begin{aligned} \cos(\theta') &= \frac{\cos(\theta)\,\cosh(\alpha)}{\left(1+\sinh^2(\alpha)\,\cos^2(\theta)\right)^{1/2}}, \\ \tan(\theta') &= \frac{\tan(\theta)}{\gamma}. \end{aligned} \tag{5.122}$$

More generally, moving observers see solid objects as rotated, as opposed to contracted along their direction of motion. Visualising Lorentz transformations of solid objects has now been discussed by various authors (see Rau, Weiskopf & Ruder (1998)). But the original observation that spheres remain spheres for observers in relative motion had to wait until 1959 — more than 50 years after the development of special relativity! The first authors to point out this invisibility of the Lorentz contraction were Terrell (1959) and Penrose (1959). Both authors based their studies on the fact that the Lorentz group is isomorphic to the conformal group acting on the surface of a sphere. This type of geometry is discussed in chapter 10.

5.4.4 Pure boosts and observer splits

Suppose we are travelling with velocity u and want to boost to velocity v. We seek the rotor for this which contains no additional rotational factors. We have

$$v = Lu\tilde{L} \tag{5.123}$$

with $La_\perp\tilde{L} = a_\perp$ for any vector outside the $u\wedge v$ plane. It is clear that the appropriate bivector for the rotor is $u\wedge v$, and as this anticommutes with u and v we have

$$v = Lu\tilde{L} = L^2 u \quad \Rightarrow L^2 = vu. \tag{5.124}$$

The solution to this is

$$L = \frac{1+vu}{[2(1+u\cdot v)]^{1/2}} = \exp\left(\frac{\alpha}{2}\frac{v\wedge u}{|v\wedge u|}\right), \tag{5.125}$$

where the angle α is defined by $\cosh(\alpha) = u\cdot v$.

Now suppose that we start in the γ_0 frame and some arbitrary rotor R takes this to $v = R\gamma_0\tilde{R}$. We know that the pure boost for this transformation is

$$L = \frac{1+v\gamma_0}{[2(1+v\cdot\gamma_0)]^{1/2}} = \exp\left(\frac{\alpha}{2}\frac{v\wedge\gamma_0}{|v\wedge\gamma_0|}\right), \tag{5.126}$$

where $v\cdot\gamma_0 = \cosh(\alpha)$. Now define the further rotor U by

$$U = \tilde{L}R, \qquad U\tilde{U} = \tilde{L}R\tilde{R}L = 1. \tag{5.127}$$

This satisfies

$$U\gamma_0\tilde{U} = \tilde{L}vL = \gamma_0, \tag{5.128}$$

so $U\gamma_0 = \gamma_0 U$. We must therefore have $U = \exp(I\boldsymbol{b}/2)$, where $I\boldsymbol{b}$ is a relative bivector, and U generates a pure rotation in the γ_0 frame. We now have

$$R = LU, \tag{5.129}$$

which decomposes R into a relative rotation and boost. Unlike the invariant decomposition into a boost and rotation of equation (5.107), the boost L and rotation U will not usually commute. The fact that the LU decomposition initially singled out the γ_0 vector shows that the decomposition is frame-dependent. Both the invariant split of equation (5.107) and the frame-dependent split of equation (5.129) are useful in practice.

5.5 Spacetime dynamics

Dynamics in spacetime is traditionally viewed as a hard subject. This need not be the case, however. We have now established that Lorentz transformations which preserve parity and causal structure can be described with rotors. By parameterising the motion in terms of rotors many equations are considerably simplified, and can be solved in new ways. This provides a simple understanding of the Thomas precession, as well as a new formulation of the Lorentz force law for a particle in an electromagnetic field.

5.5.1 Rotor equations and Fermi transport

A spacetime trajectory $x(\tau)$ has a future-pointing velocity vector $\dot{x} = v$. This is normalised to $v^2 = 1$ by parameterising the curve in terms of the proper time. This suggests an analogy with rigid-body dynamics. We write

$$v = R\gamma_0\tilde{R}, \tag{5.130}$$

which keeps v future-pointing and normalised. This moves all of the dynamics into the rotor $R = R(\tau)$, and this is the key idea which simplifies much of relativistic dynamics. The next quantity we need to find is the acceleration

$$\dot{v} = \frac{d}{d\tau}(R\gamma_0\tilde{R}) = \dot{R}\gamma_0\tilde{R} + R\gamma_0\dot{\tilde{R}}. \tag{5.131}$$

But just as in three dimensions, $\dot{R}\tilde{R}$ is of even grade and is equal to minus its reverse, so can only contain bivector terms. We therefore have

$$\begin{aligned}\dot{v} &= \dot{R}\tilde{R}v - v\dot{R}\tilde{R} \\ &= 2(\dot{R}\tilde{R})\cdot v.\end{aligned} \tag{5.132}$$

This equation is consistent with the fact that $v\cdot\dot{v} = 0$, which follows from $v^2 = 1$.

If we now form the acceleration bivector we obtain

$$\dot{v}v = 2(\dot{R}\tilde{R})\cdot v\, v. \tag{5.133}$$

This determines the projection of the bivector into the instantaneous rest frame defined by v. In this frame the projected bivector is purely timelike and corresponds to a pure boost. The remaining freedom in $\dot{R}\tilde{R}$ corresponds to an additional rotation in R which does not change v.

For the purposes of determining the velocity and trajectory of a particle the component of $\dot{R}\tilde{R}$ perpendicular to v is of no relevance. In some applications, however, it is useful to attach physical significance to the comoving frame vectors $\{\mathsf{e}_\mu\}$,

$$\mathsf{e}_\mu = R\gamma_\mu\tilde{R}, \tag{5.134}$$

which have $\mathsf{e}_0 = v$. The spatial set of vectors $\{\mathsf{e}_i\}$ satisfy $\mathsf{e}_i\cdot v = 0$ and span the instantaneous rest space of v. In this case, the dynamics of the e_i can be used to determine the component of $\dot{R}\tilde{R}$ which is not fixed by v alone.

The vectors $\{\mathsf{e}_i\}$ are carried along the trajectory by the rotor R. They are said to be *Fermi-transported* if their transformation from one instant to the next is

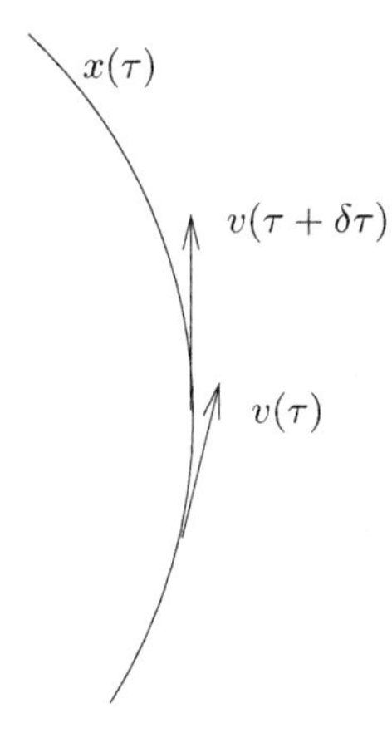

Figure 5.8 *The proper boost.* The change in velocity from τ to $\tau+\delta\tau$ should be described by a rotor solely in the $\dot{v}\wedge v$ plane.

a pure boost in the v frame. In this case the $\{\mathsf{e}_i\}$ vectors remain 'as constant as possible', subject to the constraint $\mathsf{e}_i\cdot v = 0$. For example, the direction defined by the angular momentum of an inertial guidance gyroscope (supported at its centre of mass so there are no torques) is Fermi-transported along the path of the gyroscope through spacetime.

To ensure Fermi-transport of $R\gamma_i\tilde{R}$ we need to ensure that the rotor describes pure boosts from one instant to the next (see figure 5.8). To first order in $\delta\tau$ we have

$$v(\tau+\delta\tau) = v(\tau) + \delta\tau\,\dot{v}. \tag{5.135}$$

The pure boost between $v(\tau)$ and $v(\tau+\delta\tau)$ is determined by the rotor

$$L = \frac{1+v(\tau+\delta\tau)v(\tau)}{[2(1+v(\tau+\delta\tau)\cdot v(\tau))]^{1/2}} = 1+\tfrac{1}{2}\delta\tau\,\dot{v}v, \tag{5.136}$$

to first order in $\delta\tau$. But since

$$R(\tau+\delta\tau) = R(\tau)+\delta\tau\dot{R}(\tau) = (1+\delta\tau\dot{R}\tilde{R})R(\tau), \tag{5.137}$$

the additional rotation that takes the $\{\mathsf{e}_i\}$ frame from τ to $\tau+\delta\tau$ is described by the rotor $1+\delta\tau\dot{R}\tilde{R}$. Equating this to the pure boost L of equation (5.136), we find that the correct expression to ensure Fermi-transport of the $\{\mathsf{e}_i\}$ is

$$\dot{R}\tilde{R} = \tfrac{1}{2}\dot{v}v. \tag{5.138}$$

This is as one would expect. The bivector describing the change in the rotor is simply the acceleration bivector, which is the acceleration seen in the instantaneous rest frame.

Under Fermi-transport the $\{\mathsf{e}_i\}$ frame vectors satisfy

$$\dot{\mathsf{e}}_i = 2(\dot{R}\tilde{R})\cdot\mathsf{e}_i = -\mathsf{e}_i\cdot(\dot{v}v). \tag{5.139}$$

This leads directly to the definition of the *Fermi derivative*

$$\frac{Da}{D\tau} = \dot{a} + a\cdot(\dot{v}v). \tag{5.140}$$

The Fermi derivative of a vector vanishes if the vector is Fermi-transported along the worldline. The derivative preserves both the magnitude a^2 and $a \cdot v$. The former holds because

$$\frac{d}{d\tau}(a^2) = -2a\cdot\big(a\cdot(\dot{v}\wedge v)\big) = 0. \tag{5.141}$$

Conservation of $a \cdot v$ is also straightforward to check:

$$\begin{aligned}\frac{d}{d\tau}(a\cdot v) &= -\big(a\cdot(\dot{v}v)\big)\cdot v + a\cdot\dot{v} \\ &= -a\cdot\dot{v} + a\cdot v\,\dot{v}\cdot v + a\cdot\dot{v} = 0.\end{aligned} \tag{5.142}$$

It follows that if a starts perpendicular to v it remains so. In the case where $a\cdot v = 0$ the Fermi derivative takes on the simple form

$$\frac{Da}{D\tau} = \dot{a} + a\cdot\dot{v}\,v = \dot{a} - \dot{a}\cdot v\,v = \dot{a}\wedge v\,v. \tag{5.143}$$

This is the projection of $\dot{a}$ perpendicular to v, as expected. The Fermi derivative extends simply to multivectors as follows:

$$\frac{DM}{D\tau} = \frac{dM}{d\tau} + M\times(\dot{v}v). \tag{5.144}$$

Derivatives of this type are important in gauge theories and gravity.

5.5.2 Thomas precession

As an application, consider a particle in a circular orbit (figure 5.9). The worldline is

$$x(\tau) = t(\tau)\gamma_0 + a(\cos(\omega t)\gamma_1 + \sin(\omega t)\gamma_2), \tag{5.145}$$

and the velocity is

$$v = \dot{x} = \dot{t}\big(\gamma_0 + a\omega(-\sin(\omega t)\gamma_1 + \cos(\omega t)\gamma_2)\big). \tag{5.146}$$

The relative velocity as seen in the γ_0 frame, $\boldsymbol{v} = v\wedge\gamma_0/v\cdot\gamma_0$, has magnitude $|\boldsymbol{v}| = a\omega$. We therefore introduce the hyperbolic angle α, with

$$\tanh(\alpha) = a\omega, \quad \dot{t} = \cosh(\alpha). \tag{5.147}$$

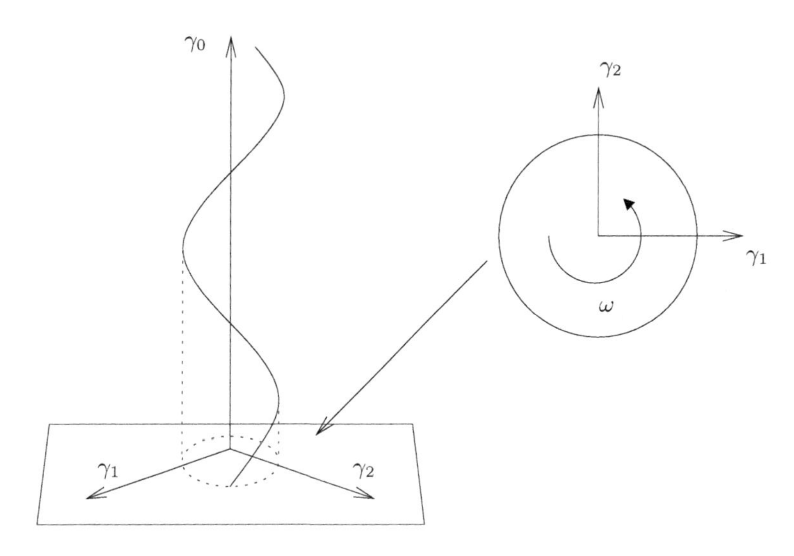

Figure 5.9 *Thomas precession.* The particle follows a helical worldline, rotating at a constant rate in the γ_0 frame.

The velocity is now

$$\begin{aligned} v &= \cosh(\alpha)\,\gamma_0 + \sinh(\alpha)\big(-\sin(\omega t)\gamma_1 + \cos(\omega t)\gamma_2\big) \\ &= \mathrm{e}^{\alpha \boldsymbol{n}/2}\gamma_0 \mathrm{e}^{-\alpha \boldsymbol{n}/2}, \end{aligned} \tag{5.148}$$

where

$$\boldsymbol{n} = -\sin(\omega t)\boldsymbol{\sigma}_1 + \cos(\omega t)\boldsymbol{\sigma}_2. \tag{5.149}$$

This form of time dependence in the rotor is inconvenient to work with. To simplify, we write

$$\boldsymbol{n} = \mathrm{e}^{-\omega t I \boldsymbol{\sigma}_3}\boldsymbol{\sigma}_2 = R_\omega \boldsymbol{\sigma}_2 \tilde{R}_\omega, \tag{5.150}$$

where $R_\omega = \exp(-\omega t I\boldsymbol{\sigma}_3/2)$. We now have

$$\mathrm{e}^{\alpha \boldsymbol{n}/2} = \exp(\alpha R_\omega \boldsymbol{\sigma}_2 \tilde{R}_\omega/2) = R_\omega R_\alpha \tilde{R}_\omega, \tag{5.151}$$

where

$$R_\alpha = \exp(\alpha \boldsymbol{\sigma}_2/2). \tag{5.152}$$

The velocity is now given by

$$v = R_\omega R_\alpha \tilde{R}_\omega \gamma_0 R_\omega \tilde{R}_\alpha \tilde{R}_\omega = R_\omega R_\alpha \gamma_0 \tilde{R}_\alpha \tilde{R}_\omega. \tag{5.153}$$

The final expression follows because R_ω commutes with γ_0.

We can now see that the rotor for the motion must have the form

$$R = R_\omega R_\alpha \Phi, \tag{5.154}$$

where Φ is a rotor that commutes with γ_0. We want R to describe Fermi transport of the $\{\mathsf{e}_i\}$, so we must have $\dot{v}v = 2\dot{R}\tilde{R}$. We begin by forming the acceleration bivector $\dot{v}v$. We can simplify this derivation by writing $v = R_\omega v_\alpha \tilde{R}_\omega$, where $v_\alpha = R_\alpha \gamma_0 \tilde{R}_\alpha$. We then find that

$$\begin{aligned}\dot{v}v &= R_\omega\big(2(\tilde{R}_\omega \dot{R}_\omega)\cdot v_\alpha \, v_\alpha\big)\tilde{R}_\omega \\ &= -\omega \cosh(\alpha)\, R_\omega\big((I\boldsymbol{\sigma}_3)\cdot v_\alpha \, v_\alpha\big)\tilde{R}_\omega \\ &= \omega \sinh(\alpha) \cosh(\alpha)\, R_\omega\big(-\cosh(\alpha)\,\boldsymbol{\sigma}_1 + \sinh(\alpha)\, I\boldsymbol{\sigma}_3\big)\tilde{R}_\omega. \end{aligned} \tag{5.155}$$

We also form the rotor equivalent, $2\dot{R}\tilde{R}$, which is

$$\begin{aligned}2\dot{R}\tilde{R} &= 2\dot{R}_\omega \tilde{R}_\omega + 2R_\omega R_\alpha \dot{\Phi}\tilde{\Phi}\tilde{R}_\alpha \tilde{R}_\omega \\ &= -\omega \cosh(\alpha)\, I\boldsymbol{\sigma}_3 + 2R_\omega R_\alpha \dot{\Phi}\tilde{\Phi}\tilde{R}_\alpha \tilde{R}_\omega. \end{aligned} \tag{5.156}$$

Equating the two preceding results we find that

$$\begin{aligned}2\dot{\Phi}\tilde{\Phi} &= \omega \cosh^2(\alpha)\, \tilde{R}_\alpha\big(-\sinh(\alpha)\,\boldsymbol{\sigma}_1 + \cosh(\alpha)\, I\boldsymbol{\sigma}_3\big)R_\alpha \\ &= \omega \cosh^2(\alpha)\, I\boldsymbol{\sigma}_3. \end{aligned} \tag{5.157}$$

The solution with $\Phi = 1$ at $t = 0$ is $\Phi = \exp(\omega \cosh(\alpha) t I\boldsymbol{\sigma}_3/2)$, so the full rotor is

$$R = \mathrm{e}^{-\omega t I\boldsymbol{\sigma}_3/2}\mathrm{e}^{\alpha\boldsymbol{\sigma}_2/2}\mathrm{e}^{\cosh(\alpha)\omega t I\boldsymbol{\sigma}_3/2}. \tag{5.158}$$

This form of the rotor ensures that the $\mathsf{e}_i = R\gamma_i\tilde{R}$ are Fermi transported. The fact that the 'internal' rotation rate $\omega\cosh(\alpha)$ differs from ω is due to the fact that the acceleration is formed in the instantaneous rest frame v and not the fixed γ_0 frame. This difference introduces a precession — the *Thomas precession.* We can see this effect by imagining the vector γ_1 being transported around the circle. The rotated vector is

$$\mathsf{e}_1 = R\gamma_1\tilde{R}. \tag{5.159}$$

In the low velocity limit $\cosh(\alpha) \mapsto 1$ the vector γ_1 continues to point in the γ_1 direction and the frame does not rotate, as we would expect. At larger velocities, however, the frame starts to precess. After time $t = 2\pi/\omega$, for example, the γ_1 vector is transformed to

$$\mathsf{e}_1(2\pi/\omega) = \mathrm{e}^{\alpha\boldsymbol{\sigma}_2/2}\mathrm{e}^{2\pi\cosh(\alpha) I\boldsymbol{\sigma}_3}\gamma_1\mathrm{e}^{-\alpha\boldsymbol{\sigma}_2/2}. \tag{5.160}$$

Dotting this with the initial vector $\mathsf{e}_1(0) = \gamma_1$ we see that the vector has precessed through an angle

$$\theta = 2\pi(\cosh(\alpha) - 1). \tag{5.161}$$

This shows that the effect is of order $|\boldsymbol{v}|^2/c^2$. The form of the Thomas precession justifies one of the relativistic corrections to the spin-orbit coupling in the Pauli theory of the electron.

5.5.3 The Lorentz force law

The non-relativistic form of the Lorentz force law for a particle of charge q is

$$\frac{d\boldsymbol{p}}{dt} = q(\boldsymbol{E} + \boldsymbol{v}\times\boldsymbol{B}), \tag{5.162}$$

where the $\times$ here denotes the vector cross product, and all relative vectors are expressed in some global Newtonian frame, which we will take to be the γ_0 frame. We seek a covariant relativistic version of this law. The quantity $\boldsymbol{p}$ on the left-hand side is the relative vector $p\wedge\gamma_0$. Since $dt = \gamma d\tau$, we must multiply through by $\gamma = v\cdot\gamma_0$ to convert the derivative into one with respect to proper time. The first term on the right-hand side then includes

$$\begin{aligned} v\cdot\gamma_0\, \boldsymbol{E} &= \tfrac{1}{4}\big(\boldsymbol{E}(v\gamma_0 + \gamma_0 v) + (v\gamma_0 + \gamma_0 v)\boldsymbol{E}\big) \\ &= \tfrac{1}{4}\big((\boldsymbol{E}v - v\boldsymbol{E})\gamma_0 - \gamma_0(\boldsymbol{E}v - v\boldsymbol{E})\big) \\ &= (\boldsymbol{E}\cdot v)\wedge\gamma_0. \end{aligned} \tag{5.163}$$

Recall at this point that $\boldsymbol{E}$ is a spacetime bivector built from the $\boldsymbol{\sigma}_k = \gamma_k\gamma_0$, so $\boldsymbol{E}$ *anticommutes* with γ_0.

For the magnetic term in equation (5.162) we first replace the cross product by the equivalent three-dimensional expression $(I\boldsymbol{B})\cdot\boldsymbol{v}$. Expanding out, and expressing in the full spacetime algebra, we obtain

$$\begin{aligned} \tfrac{1}{2}v\cdot\gamma_0(I\boldsymbol{B}\boldsymbol{v} - \boldsymbol{v}I\boldsymbol{B}) &= \tfrac{1}{4}\big(I\boldsymbol{B}(v\gamma_0 - \gamma_0 v) - (v\gamma_0 - \gamma_0 v)I\boldsymbol{B}\big) \\ &= \tfrac{1}{4}\big((I\boldsymbol{B}v - vI\boldsymbol{B})\gamma_0 - \gamma_0(I\boldsymbol{B}v - vI\boldsymbol{B})\big) \\ &= \big((I\boldsymbol{B})\cdot v\big)\wedge\gamma_0, \end{aligned} \tag{5.164}$$

where we use the fact that γ_0 *commutes* with $I\boldsymbol{B}$. Combining equations (5.163) and (5.164) we can now write the Lorentz force law (5.162) in the form

$$\frac{d\boldsymbol{p}}{d\tau} = \dot{p}\wedge\gamma_0 = q\big((\boldsymbol{E} + I\boldsymbol{B})\cdot v\big)\wedge\gamma_0. \tag{5.165}$$

We next define the *Faraday bivector* F by

$$F = \boldsymbol{E} + I\boldsymbol{B}. \tag{5.166}$$

This is the covariant form of the electromagnetic field strength. It unites the electric and magnetic fields into a single spacetime structure. We study this in greater detail in chapter 7. The Lorentz force law can now be written

$$\dot{p}\wedge\gamma_0 = q(F\cdot v)\wedge\gamma_0. \tag{5.167}$$

The rate of working on the particle is $q\boldsymbol{E}\cdot\boldsymbol{v}$, so

$$\frac{dp_0}{dt} = q\boldsymbol{E}\cdot\boldsymbol{v}. \tag{5.168}$$

Here, $p_0 = p \cdot \gamma_0$ is the particle's energy in the γ_0 frame. Multiplying through by $v \cdot \gamma_0$, we find

$$\dot{p} \cdot \gamma_0 = q\boldsymbol{E} \cdot (v \wedge \gamma_0) = q(F \cdot v) \cdot \gamma_0. \tag{5.169}$$

In the final step we have used $(I\boldsymbol{B}) \cdot (v \wedge \gamma_0) = 0$. Adding this equation to equation (5.167), and multiplying on the right by γ_0, we find

$$\dot{p} = qF \cdot v. \tag{5.170}$$

Recalling that $p = mv$, we arrive at the relativistic form of the *Lorentz force law*,

$$m\dot{v} = qF \cdot v. \tag{5.171}$$

This is *manifestly* Lorentz covariant, because no particular frame is picked out. The acceleration bivector is

$$\dot{v}v = \frac{q}{m} F \cdot v\, v = \frac{q}{m} (F \cdot v) \wedge v = \frac{q}{m} \boldsymbol{E}_v, \tag{5.172}$$

where $\boldsymbol{E}_v$ is the relative electric field in the v frame. A charged point particle only responds to the instantaneous electric field in its frame. Algebraically, this bivector is

$$\boldsymbol{E}_v = \tfrac{1}{2}(F - vFv). \tag{5.173}$$

So $\boldsymbol{E}_v$ is the component of the bivector F which anticommutes with v.

Now suppose that we parameterise the velocity with a rotor, so that $v = R\gamma_0\tilde{R}$. We have

$$\dot{v} = 2\dot{R}\tilde{R}v = 2(\dot{R}\tilde{R}) \cdot v = \frac{q}{m} F \cdot v. \tag{5.174}$$

The simplest form of the rotor equation comes from equating the projected terms:

$$\dot{R} = \frac{q}{2m} FR. \tag{5.175}$$

This is not the most general possibility as we could include an extra multiple of $F \wedge v\, v$. The rotor determined by equation (5.175) will not, in general, describe Fermi-transport of the $R\gamma_i\tilde{R}$ vectors. However, equation (5.175) is sufficient to determine the velocity of the particle, and is certainly the simplest form of rotor equation to work with. As we now demonstrate, the rotor equation (5.175) is remarkably efficient when it comes to solving the dynamical equations.

5.5.4 Constant field

Motion in a constant field is easy to solve for now. We can immediately integrate the rotor equation to give

$$R = \exp\left(\frac{q}{2m} F\tau\right) R_0. \tag{5.176}$$

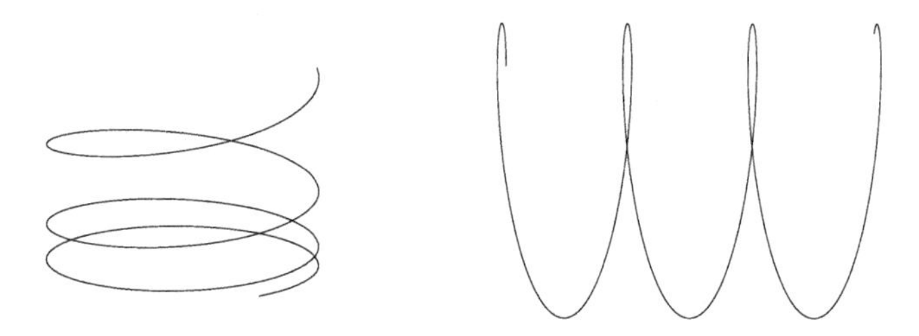

Figure 5.10 *Particle in a constant field.* The general motion is a combination of linear acceleration and circular motion. The plot on the left has $\boldsymbol{E}$ and $\boldsymbol{B}$ colinear. The plot on the right has $\boldsymbol{E}$ entirely in the $I\boldsymbol{B}$ plane, giving rise to cycloids.

To proceed and recover the trajectory we form the invariant decomposition of F. We first write

$$F^2 = \langle F^2 \rangle_0 + \langle F^2 \rangle_4 = \rho e^{I\theta}, \tag{5.177}$$

so that we can set

$$F = \rho^{1/2} e^{I\theta/2} \hat{F} = \alpha \hat{F} + I\beta \hat{F}, \tag{5.178}$$

where $\hat{F}^2 = 1$. (If F is null a slightly different procedure is followed.) We now have

$$R = \exp\left(\frac{q}{2m}\alpha \hat{F} \tau\right) \exp\left(\frac{q}{2m} I\beta \hat{F} \tau\right) R_0. \tag{5.179}$$

Next we decompose the initial velocity $v_0 = R_0 \gamma_0 \tilde{R}_0$ into components in and out of the $\hat{F}$ plane:

$$v_0 = \hat{F}^2 v_0 = \hat{F}\,\hat{F}\cdot v_0 + \hat{F}\,\hat{F}\wedge v_0 = v_{0\parallel} + v_{0\perp}. \tag{5.180}$$

Now $v_{0\parallel} = \hat{F}\,\hat{F}\cdot v_0$ anticommutes with $\hat{F}$, and $v_{0\perp}$ commutes with $\hat{F}$, so

$$\dot{x} = \exp\left(\frac{q}{m}\alpha \hat{F} \tau\right) v_{0\parallel} + \exp\left(\frac{q}{m} I\beta \hat{F} \tau\right) v_{0\perp}. \tag{5.181}$$

This integrates immediately to give the particle history

$$x - x_0 = \frac{e^{q\alpha \hat{F}\tau/m} - 1}{q\alpha/m} \hat{F}\cdot v_0 - \frac{e^{q\beta I \hat{F}\tau/m} - 1}{q\beta/m} (I\hat{F})\cdot v_0. \tag{5.182}$$

The first term gives linear acceleration and the second is periodic and drives rotational motion (see figure 5.10). One has to be slightly careful integrating the velocity equation in the case where either α or β is zero, which corresponds to perpendicular $\boldsymbol{E}$ and $\boldsymbol{B}$ fields.

5.5.5 Particle in a Coulomb field

As a further application we consider the case of a charged point particle moving in a central Coulomb field. If relativistic effects are ignored the problem reduces to the inverse-square force law described in section 3.2.1. We therefore expect that the relativistic description will add additional perturbative effects to the elliptic and hyperbolic orbits found in the inverse-square case. We assume for simplicity that the central charge has constant velocity γ_0 and is placed at the origin. The electromagnetic field is

$$F = \frac{Q\boldsymbol{x}}{4\pi\epsilon_0 r^3}, \tag{5.183}$$

where $\boldsymbol{x} = x\wedge\gamma_0$ and $r^2 = \boldsymbol{x}^2$. In this section all bold symbols denote relative vectors in the γ_0 frame. The question of how to generalise the non-relativistic definitions of centre of mass and relative separation turns out to be surprisingly complex and is not tackled here. Instead we will simply assume that the source of the Coulomb field is far heavier than the test charge so that the source's motion can be ignored.

There are two constants of motion for this force law. The first is the energy

$$E = mv\cdot\gamma_0 + \frac{qQ}{4\pi\epsilon_0 r}. \tag{5.184}$$

If the charges are opposite, qQ is negative and the potential is attractive. The force law can now be written in the γ_0 frame as

$$m\frac{d^2\boldsymbol{x}}{d\tau^2} = \frac{qQ\boldsymbol{x}}{4\pi\epsilon_0 r^3}\left(\frac{E}{m} - \frac{qQ}{4\pi\epsilon_0 m r}\right). \tag{5.185}$$

The second conserved quantity is the angular momentum, which is conserved for any central force, as is the case in equation (5.185). If we define the spacetime bivector $L = x\wedge p$ we find that

$$\dot{L} = qx\wedge(F\cdot v). \tag{5.186}$$

It follows that the trivector $L\wedge\gamma_0$ is conserved. Equivalently, we can define the relative bivector

$$I\boldsymbol{l} = L\wedge\gamma_0\,\gamma_0, \tag{5.187}$$

so that the relative vector $\boldsymbol{l}$ is conserved. This is the relative angular momentum vector and satisfies $\boldsymbol{x}\cdot\boldsymbol{l} = 0$. It follows that the test particle's motion takes place in a constant plane as seen from the source charge.

In order to integrate the rotor equation we need to find a way to express the field as a function of the particle's proper time. This is achieved by introducing an angular measure in the plane of motion. Suppose that we align the 3 axis with $\boldsymbol{l}$, so that we can write

$$\hat{\boldsymbol{x}}(\tau) = \boldsymbol{\sigma}_1 \exp\bigl(I\boldsymbol{\sigma}_3\theta(\tau)\bigr), \tag{5.188}$$

where $\hat{\boldsymbol{x}}$ is the unit relative vector $\boldsymbol{x}/r$. It follows that

$$\boldsymbol{l}^2 = m^2 r^4 \dot{\hat{\boldsymbol{x}}}^2 = m^2 r^4 \dot{\theta}^2. \tag{5.189}$$

If we set $l = |\boldsymbol{l}|$ we have $l = mr^2\dot{\theta}$, which enables us to express the Coulomb field as

$$F = \frac{Qm\dot{\theta}\boldsymbol{\sigma}_1 \exp\big(I\boldsymbol{\sigma}_3\theta(\tau)\big)}{4\pi\epsilon_0 l}. \tag{5.190}$$

If we now let

$$\kappa = \frac{qQ}{4\pi\epsilon_0 l} \tag{5.191}$$

the rotor equation takes on the simple form

$$\frac{dR}{d\theta} = \frac{\kappa}{2}\boldsymbol{\sigma}_1 \exp(I\boldsymbol{\sigma}_3\theta)R. \tag{5.192}$$

Re-expressing the differential equation in terms of θ is a standard technique for solving inverse-square problems in non-relativistic physics. But this technique fails to give a simple solution to the relativistic equation (5.185). Instead, we see that the technique gives a simple solution to the relativistic problem only if applied directly to the rotor equation.

To solve equation (5.192) we first set

$$R = \exp(-I\boldsymbol{\sigma}_3\theta/2)U. \tag{5.193}$$

It follows that

$$\frac{dU}{d\theta}\tilde{U} = \tfrac{1}{2}(\kappa\boldsymbol{\sigma}_1 + I\boldsymbol{\sigma}_3), \tag{5.194}$$

which integrates straightforwardly. The full rotor is then

$$R = \mathrm{e}^{-I\boldsymbol{\sigma}_3\theta/2}\mathrm{e}^{A\theta/2}R_0, \tag{5.195}$$

where

$$A = \kappa\boldsymbol{\sigma}_1 + I\boldsymbol{\sigma}_3. \tag{5.196}$$

The initial conditions can be chosen such that $\theta(0) = 0$, which tells us how to align the 1 axis. The rotor R_0 then specifies the initial velocity v_0. If we are not interested in transporting a frame, R_0 can be set equal to a pure boost from γ_0 to v_0.

With the rotor equation now solved, the velocity can be integrated to recover the trajectory. Clearly, different types of path are obtained for the different signs of $A^2 = \kappa^2 - 1$. The equation relating r and θ is found from the relation

$$-\frac{d}{d\theta}\left(\frac{1}{r}\right) = \frac{m}{l}\hat{\boldsymbol{x}}\cdot\dot{\boldsymbol{x}}. \tag{5.197}$$

To evaluate the right-hand side we need

$$\begin{aligned}\hat{\boldsymbol{x}}\cdot\dot{\boldsymbol{x}} &= \langle \mathrm{e}^{-I\sigma_3\theta/2}\boldsymbol{\sigma}_1\mathrm{e}^{I\sigma_3\theta/2}R\gamma_0\tilde{R}\gamma_0\rangle \\ &= -\langle\gamma_1\mathrm{e}^{A\theta/2}v_0\mathrm{e}^{-A\theta/2}\rangle \\ &= \langle\mathrm{e}^{-A\theta}\gamma^1 v_0\rangle. \end{aligned} \tag{5.198}$$

It follows that

$$-\frac{d}{d\theta}\left(\frac{1}{r}\right) = \frac{m}{l}\langle\mathrm{e}^{-A\theta}\gamma^1 v_0\rangle. \tag{5.199}$$

For a given l and v_0 this integrates to give the trajectory in the $\boldsymbol{Il}$ plane.

Suppose, for example, that we are interested in bound states. For these we must have $A^2 < 0$, which implies that $\kappa^2 < 1$. We write

$$|A| = (1-\kappa^2)^{1/2} \tag{5.200}$$

for the magnitude of A. To simplify the equations we will assume that $\tau = 0$ corresponds to a point on the trajectory where $\boldsymbol{v}$ is perpendicular to $\boldsymbol{x}$. In this case we have

$$v_0 = \cosh(\alpha_0)\,\gamma_0 + \sinh(\alpha_0)\,\gamma_2 \tag{5.201}$$

so that the trajectory is determined by

$$-\frac{d}{d\theta}\left(\frac{1}{r}\right) = \frac{m}{l|A|}\big(\kappa\cosh(\alpha_0)+\sinh(\alpha_0)\big)\sin(|A|\theta). \tag{5.202}$$

The magnitude of the angular momentum is given by $l = mr_0\sinh(\alpha_0)$, which can be used to write

$$m\big(\kappa\cosh(\alpha_0)+\sinh(\alpha_0)\big) = (E^2 - m^2|A|^2)^{1/2}. \tag{5.203}$$

The trajectory is then given by

$$\frac{l|A|^2}{r} = -\kappa E + (E^2 - m^2|A|^2)^{1/2}\cos(|A|\theta), \tag{5.204}$$

and since this represents a bound state, κ must be negative. The fact that the angular term goes as $\cos(|A|\theta)$ shows that this equation specifies a precessing ellipse (figure 5.11). The precession rate of the ellipse can be found simply using the technique of section 3.3.

5.5.6 The gyromagnetic moment

Particles with non-zero spin have a magnetic moment which is proportional to the spin. In non-relativistic physics we write this as $\boldsymbol{m} = \gamma\boldsymbol{s}$, where γ is the gyromagnetic ratio and $\boldsymbol{s}$ is the spin (which has units of angular momentum). The gyromagnetic ratio is usually written in the form

$$\gamma = g\frac{q}{2m}, \tag{5.205}$$

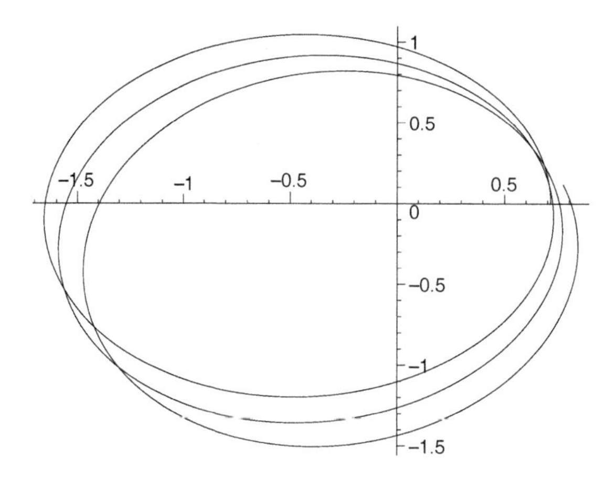

Figure 5.11 *Motion in a Coulomb field.* For bound orbits ($E < m$) the particle's motion is described by a precessing ellipse. The plot is for $|A| = 0.95$. The units are arbitrary.

where m is the particle mass, q is the charge and g is the (reduced) gyromagnetic ratio. The last is determined experimentally via the precession of the spin vector which, in classical physics, obeys

$$\dot{\boldsymbol{s}} = g\frac{q}{2m}(I\boldsymbol{B})\cdot\boldsymbol{s}. \tag{5.206}$$

We seek a relativistic extension of this equation. We start by introducing the relativistic spin vector s, which is perpendicular to the velocity v, so $s \cdot v = 0$. For a particle at rest in the γ_0 frame we have $s = \boldsymbol{s}\gamma_0$. The particle's spin will interact with the magnetic field only in the instantaneous rest frame, so we should regard equation (5.206) as referring to this frame.

Given that $\boldsymbol{s} = s\gamma_0$ we find that

$$\begin{aligned}(I\boldsymbol{B})\cdot\boldsymbol{s} &= \langle (F\wedge\gamma_0)\gamma_0 s\gamma_0\rangle_2 \\ &= (F\cdot s)\wedge\gamma_0.\end{aligned} \tag{5.207}$$

So, for a particle at rest in the γ_0 frame, equation (5.206) can be written

$$\frac{ds}{dt} = g\frac{q}{2m}(F\cdot s)\wedge\gamma_0\,\gamma_0. \tag{5.208}$$

To write down an equation which is valid for arbitrary velocity we must replace the two factors of γ_0 on the right-hand side with the velocity v. On the left-hand side we need the derivative of s which preserves $s \cdot v = 0$. This is the Fermi derivative of section 5.5.1, which tells us that the relativistic form of the spin precession equation is

$$\dot{s} + s\cdot(\dot{v}v) = g\frac{q}{2m}(F\cdot s)\wedge v\, v. \tag{5.209}$$

This equation tells us how much the spin vector rotates, relative to a Fermi-transported frame, which is physically sensible. We can eliminate the acceleration bivector $\dot{v}v$ by using the relativistic Lorentz force law to find

$$\begin{aligned}\dot{s} &= g\frac{q}{2m}(F\cdot s)\wedge v\, v - \frac{q}{m}s\cdot(F\cdot v\, v) \\ &= \frac{q}{2m}\big(g(F\cdot s)\wedge v + 2(F\cdot s)\cdot v\big)v \\ &= \frac{q}{m}F\cdot s + (g-2)\frac{q}{2m}(F\cdot s)\wedge v\, v. \end{aligned} \tag{5.210}$$

This is called the Bargmann–Michel–Telegdi equation.

For the value $g = 2$, the Bargmann–Michel–Telegdi equation reduces to

$$\dot{s} = \frac{q}{m}F\cdot s, \tag{5.211}$$

which has the same form as the Lorentz force law. In this sense, $g = 2$ is the most natural value of the gyromagnetic ratio of a point particle in relativistic physics. Ignoring quantum corrections, this is indeed found to be the value for an electron. Quantum corrections tell us that for an electron $g = 2(1+\alpha/2\pi+\cdots)$. The corrections are due to the fact that the electron is never truly isolated and constantly interacts with virtual particles from the quantum vacuum.

Given a velocity v and a spin vector s, with $v\cdot s = 0$ and s normalised to $s^2 = -1$, we can always find a rotor R such that

$$v = R\gamma_0\tilde{R}, \quad s = R\gamma_3\tilde{R}. \tag{5.212}$$

For these we have

$$\dot{v} = 2(\dot{R}\tilde{R})\cdot v, \quad \dot{s} = 2(\dot{R}\tilde{R})\cdot s. \tag{5.213}$$

For a particle with $g = 2$, this pair of equations reduces to the single rotor equation (5.175). The simple form of this equation further justifies the claim that $g = 2$ is the natural, relativistic value of the gyromagnetic ratio. This also means that once we have solved the rotor equation, we can simultaneously compute both the trajectory and the spin precession of a classical relativistic particle with $g = 2$.

5.6 Notes

There are many good introductions to special relativity. Standard references include the books by French (1968), Rindler (1977) and d'Inverno (1992). Practically all introductory books make heavy use of coordinate geometry. Geometric algebra was first systematically applied to the study of relativistic physics in the book *Space-Time Algebra* by Hestenes (1966). Since this book was published in 1966 many authors have applied spacetime algebra techniques to relativistic physics. The two most significant papers are again by Hestenes, 'Proper particle mechanics' and 'Proper dynamics of a rigid point particle' (1974a,b). These papers detail the use of rotor equations for solving problems in electrodynamics, and much of section 5.5 follows their presentation.

5.7 Exercises

5.1 Suppose that the spacetime bivector $\hat{B}$ satisfies $\hat{B}^2 = 1$. By writing $\hat{B} = \boldsymbol{a} + I\boldsymbol{b}$ in the γ_0 frame, show that we can write

$$\hat{B} = \cosh(u)\hat{\boldsymbol{a}} + \sinh(u)I\hat{\boldsymbol{b}} = \mathrm{e}^{uI\hat{\boldsymbol{b}}\hat{\boldsymbol{a}}}\hat{\boldsymbol{a}},$$

where $\hat{\boldsymbol{a}}^2 = \hat{\boldsymbol{b}}^2 = 1$. Hence explain why we can write $\hat{B} = R\boldsymbol{\sigma}_3\tilde{R}$. By considering the null vectors $\gamma_0 \pm \gamma_3$, prove that we can always find two null vectors satisfying

$$\hat{B}\cdot n_\pm = \pm n_\pm.$$

5.2 The boost L from velocity u to velocity v satisfies

$$v = Lu\tilde{L} = L^2 u,$$

with $L\tilde{L} = 1$. Prove that a solution to this equation is

$$L = \frac{1 + vu}{[2(1 + v\cdot u)]^{1/2}}.$$

Is this solution unique? Show further that this solution can be written in the form

$$L = \exp\left(\frac{\alpha}{2}\frac{v\wedge u}{|v\wedge u|}\right),$$

where $\alpha > 0$ satisfies $\cosh(\alpha) = u\cdot v$.

5.3 *Compton scattering* occurs when a photon scatters off an electron. If we ignore quantum effects this can be modelled as a relativistic collison process. The incident photon has wavelength λ_0 in the frame in which the electron is initially stationary. Show that the wavelength after scattering, λ, satisfies

$$\lambda - \lambda_0 = \frac{2\pi\hbar}{mc}\big(1 - \cos(\theta)\big),$$

where θ is the angle through which the photon scatters.

5.4 A relativistic particle has velocity $v = R\gamma_0\tilde{R}$. Show that v satisfies the Lorentz force equation $m\dot{v} = qF\cdot v$ if R satisfies

$$\dot{R} = \frac{q}{2m}FR.$$

Show that the solution to this for a constant field is

$$R = \exp(qF\tau/2m)R_0.$$

Given that F is *null*, $F^2 = 0$, show that v is given by the polynomial

$$v = v_0 + \tau\frac{q}{m}F\cdot v_0 - \tau^2\frac{q^2}{4m^2}Fv_0F.$$

Suppose now that $F = \boldsymbol{\sigma}_1 + I\boldsymbol{\sigma}_2$ and the particle is initially at rest in the γ_0 frame. Sketch the resultant motion in the $\gamma_1\gamma_3$ plane.

5.5 One way to construct the Fermi derivative of a vector a is to argue that we should 'de-boost' the vector at proper time $\tau + \delta\tau$ before comparing it with $a(\tau)$. Explain why this leads us to evaluate

$$\lim_{\delta\tau\to 0}\frac{1}{\delta\tau}\big(\tilde{L}a(\tau+\delta\tau)L - a(\tau)\big),$$

and confirm that this evaluates to $\dot{a} + a\cdot(\dot{v}v)$.

5.6 A frame is Fermi-transported along the worldline of a particle with velocity $v = R\gamma_0\tilde{R}$. The rotor R is decomposed into a rotation and boost in the γ_0 frame as $R = LU$. Show that the rotation U satisfies

$$2\dot{U}\tilde{U} = -(\tilde{L}\dot{L} + \gamma_0\tilde{L}\dot{L}\gamma_0).$$

What is the interpretation of the right-hand side in terms of the γ_0 frame?

5.7 The bivector $B = a\wedge b$ is Fermi-transported along a worldline by Fermi-transporting the two vectors a and b. Show that B remains a blade, and that the bivector satisfies

$$\frac{dB}{d\tau} + B\times(\dot{v}v) = 0.$$

5.8 A point particle with a gyromagnetic ratio $g = 2$ is in a circular orbit around a central Coulomb field. Show that in one complete orbit the spin vector rotates in the plane $A = \kappa\boldsymbol{\sigma}_1 + I\boldsymbol{\sigma}_3$ by an amount $2\pi|A|$, where

$$\kappa = \frac{qQ}{4\pi\epsilon_0 l},$$

and l is the angular momentum.

5.9 Show that the Bargmann–Michel–Telegdi equation of (5.210) for a relativistic point particle with spin vector s can be written

$$\dot{s} = \frac{q}{m}\big(F + \tfrac{1}{2}(g-2)F\wedge v\, v\big)\cdot s.$$

Given that $v = R\gamma_0\tilde{R}$ and $s = R\gamma_3\tilde{R}$, show that the rotor R satisfies the equation

$$\dot{R} = \frac{q}{2m}FR + \frac{q}{4m}(g-2)RI\boldsymbol{B}_0,$$

where

$$I\boldsymbol{B}_0 = (\tilde{R}FR)\wedge\gamma_0\,\gamma_0.$$

Assuming that the electromagnetc field F is constant, prove that $\boldsymbol{B}_0$ is also constant. Hence study the precession of s for a particle with a gyromagnetic ratio $g \neq 2$.

6

Geometric calculus

Geometric algebra provides us with an invertible product for vectors. In this chapter we investigate the new insights this provides for the subject of vector calculus. The familiar gradient, divergence and curl operations all result from the action of the vector operator, ∇. Since this operator is vector-valued, we can now form its geometric product with other multivectors. We call this the *vector derivative*. Unlike the separate divergence and curl operations, the vector derivative has the important property of being invertible. That is to say, Green's functions exist for ∇ which enable initial conditions to be propagated off a surface.

The synthesis of vector differentiation and geometric algebra described in this chapter is called '*geometric calculus*'. We will see that geometric calculus provides new insights into the subject of complex analysis and enables the concept of an analytic function to be extended to arbitrary dimensions. In three dimensions this generalisation gives rise to the angular eigenstates of the Pauli theory, and the spacetime generalisation of an analytic function defines the wavefunction for a massless spin-1/2 particle. Clearly there are many insights to be gained from a unified treatment of calculus based around the geometric product.

The early sections of this chapter discuss the vector derivative, and its associated Green's functions, in flat spaces. This way we can quickly assemble a number of results of central importance in later chapters. The generalisations to embedded surfaces and manifolds are discussed in the final section. This is a large and important subject, which has been widely discussed elsewhere. Our presentation here is kept brief, focusing on the key results which are required later in this book.

6.1 The vector derivative

The vector derivative is denoted with the symbol ∇ (or $\boldsymbol{\nabla}$ in two and three dimensions). Algebraically, this has all of the properties of a vector (grade-1) object in a geometric algebra. The operator properties of ∇ are contained in the definition that the inner product of ∇ with any vector a results in the *directional derivative* in the a direction. That is,

$$a\cdot\nabla F(x) = \lim_{\epsilon\mapsto 0} \frac{F(x+\epsilon a) - F(x)}{\epsilon}, \tag{6.1}$$

where we assume that this limit exists and is well defined. Suppose that we now define a constant coordinate frame $\{\mathsf{e}_k\}$ with reciprocal frame $\{\mathsf{e}^k\}$. Spatial coordinates are defined by $x^k = \mathsf{e}^k\cdot x$, and the summation convention is assumed except where stated otherwise. The vector derivative can be written

$$\nabla = \sum_k \mathsf{e}^k \frac{\partial}{\partial x^k} = \mathsf{e}^k\partial_k, \tag{6.2}$$

where we introduce the useful abbreviation

$$\partial_i = \frac{\partial}{\partial x^i}. \tag{6.3}$$

The frame decomposition $\nabla = \mathsf{e}^k\partial_k$ shows clearly how the the vector derivative combines the algebraic properties of a vector with the operator properties of the partial derivatives. It is a straightforward exercise to confirm that the definition of ∇ is independent of the choice of frame.

6.1.1 Scalar fields

As a first example, consider the case of a scalar field $\phi(x)$. Acting on ϕ, the vector derivative ∇ returns the *gradient*, $\nabla\phi$. This is the familiar grad operation. The result is a vector whose components in the $\{\mathsf{e}^k\}$ frame are the partial derivatives with respect to the x^k coordinates. The simplest example of a scalar field is the quantity $a\cdot x$, where a is a constant vector. We write $a\cdot x = x^j a_j$, so that the gradient becomes

$$\nabla(x\cdot a) = \mathsf{e}^i \frac{\partial x^j}{\partial x^i} a_j = \mathsf{e}^i a_j \delta^j_i. \tag{6.4}$$

But the right-hand side simply expresses the vector a in the $\{\mathsf{e}^k\}$ frame, so we are left with the frame-free result

$$\nabla(x\cdot a) = a. \tag{6.5}$$

This result is independent of both the dimensions and signature of the vector space. Many formulae for the vector derivative can be built up by combining this primitive result with the chain and product rules for differentiation. A particular application of this result is to the coordinates themselves,

$$\nabla x^k = \nabla(x \cdot \mathsf{e}^k) = \mathsf{e}^k, \tag{6.6}$$

a formula which generalises to curvilinear coordinate systems.

As a second example, consider the derivative of the scalar x^2. We first derive the result in coordinates before discussing a more elegant, frame-free derivation. We form

$$\begin{aligned} \nabla(x^2) &= \mathsf{e}^i \partial_i (x^j x^k) \mathsf{e}_j \cdot \mathsf{e}_k \\ &= \mathsf{e}^i \left(\frac{\partial x^j}{\partial x^i} x^k + \frac{\partial x^k}{\partial x^i} x^j \right) \mathsf{e}_j \cdot \mathsf{e}_k \\ &= x^k \mathsf{e}_k + x^j \mathsf{e}_j \\ &= 2x, \end{aligned} \tag{6.7}$$

which recovers the expected result. It is extremely useful to be able to perform such manipulations without reference to any coordinate frame. This requires a notation to keep track of which terms are being differentiated in a given expression. A suitable convention is to use overdots to define the scope of the vector derivative. With this notation we can write

$$\nabla(x^2) = \dot{\nabla}(\dot{x} \cdot x) + \dot{\nabla}(x \cdot \dot{x}) = 2\dot{\nabla}(\dot{x} \cdot x). \tag{6.8}$$

In the final term it is only the first factor of x which is differentiated, while the second is held constant. We can therefore apply the result of equation (6.5), which immediately gives $\nabla(x^2) = 2x$. More complex results can be built up in a similar manner.

In Euclidean spaces $\nabla\phi$ points in the direction of steepest increase of ϕ. This is illustrated in equation (6.5). To get the biggest increase in $a \cdot x$ for a given step size you must clearly move in the positive a direction, since moving in any orthogonal direction does not change the value. More generally, suppose $\nabla\phi = J$ and consider the contraction of this equation with the unit vector n,

$$n \cdot \nabla\phi = n \cdot J. \tag{6.9}$$

We seek the direction of n which maximises this value. Clearly in a Euclidean space this must be the J direction, so J points in the direction of greatest increase of ϕ. Also, setting n in the J direction shows that the magnitude of J is simply the derivative in the direction of steepest increase.

In mixed signature spaces, such as spacetime, this simple geometric picture can break down. As a simple example, consider a timelike plane defined by orthogonal basis vectors $\{\gamma_0, \gamma_1\}$, with $\gamma_0^2 = 1$ and $\gamma_1^2 = -1$. We introduce the scalar field

$$\phi = \langle x\gamma_0 x\gamma_0 \rangle = (x^0)^2 + (x^1)^2. \tag{6.10}$$

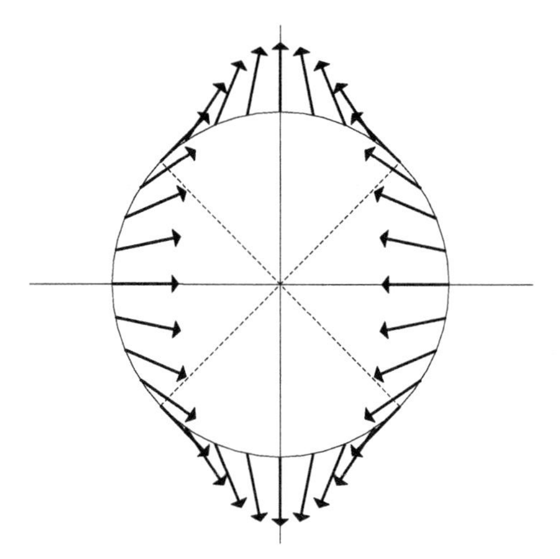

Figure 6.1 *Spacetime gradients.* The contours of the scalar field $\phi = \langle x\gamma_0 x\gamma_0 \rangle$ define circles in spacetime. But the direction of the vector derivative is only in the outward normal direction along the 0 axis. Along the 1 axis the gradient points inwards, which reflects the opposite signature. Around the circle the gradient interpolates between these two extremes. At points where x is null the gradient vector is tangential to the circle.

Contours of constant ϕ are circles in the spacetime plane, so the direction of steepest increase points radially outwards. But if we form the gradient of ϕ we obtain

$$\nabla\phi = 2\dot{\nabla}\langle \dot{x}\gamma_0 x\gamma_0 \rangle = 2\gamma_0 x\gamma_0. \tag{6.11}$$

Figure 6.1 shows the direction of this vector for various points on the unit circle. Clearly the vector does not point in the direction of steepest increase of ϕ. Instead, $\nabla\phi$ points in a direction 'normal' to tangent vectors in the circle. In mixed signature spaces, the 'normal' does not point in the direction our Euclidean intuition is used to. This example should be borne in mind when we consider directed integration in spaces of mixed signature. (This example may appear esoteric, but closed spacetime curves of this type are of considerable importance in some modern attempts to construct a quantum theory of gravity.)

6.1.2 Vector fields

Suppose now that we have a vector field $J(x)$. The full vector derivative ∇J contains two terms, a scalar and a bivector. The scalar term is the *divergence* of

$J(x)$. In terms of the constant frame vectors $\{\mathsf{e}_k\}$ we can write

$$\nabla\cdot J = \frac{\partial}{\partial x^k} e^k \cdot J = \frac{\partial J^k}{\partial x^k} = \partial_k J^k. \tag{6.12}$$

The divergence can also be defined in terms of the geometric product as

$$\nabla\cdot J = \tfrac{1}{2}(\nabla J + \dot{J}\dot{\nabla}). \tag{6.13}$$

The simplest example of the divergence is for the vector x itself, for which we find

$$\nabla\cdot x = \frac{\partial x^k}{\partial x^k} = n, \tag{6.14}$$

where n is the dimension of the space.

The remaining, antisymmetric, term defines the exterior derivative of the vector field. In terms of coordinates this can be written

$$\nabla\wedge J = \mathsf{e}^i\wedge(\partial_i J) = \mathsf{e}^i\wedge\mathsf{e}^j\,\partial_i J_j. \tag{6.15}$$

The components are the antisymmetrised terms in $\partial_i J_j$. In three dimensions these are the components of the curl, though $\boldsymbol{\nabla}\wedge\boldsymbol{J}$ is a bivector, rather than an (axial) vector. (In this chapter we write vectors in two and three dimensions in bold face.) The three-dimensional curl requires a duality operation to return a vector,

$$\mathrm{curl}(\boldsymbol{J}) = -I\,\boldsymbol{\nabla}\wedge\boldsymbol{J}. \tag{6.16}$$

The exterior derivative generalises the curl to arbitrary dimensions.

As an example, consider the exterior derivative of the position vector x. We find that

$$\nabla\wedge x = \mathsf{e}^i\wedge\mathsf{e}_i = \mathsf{e}^i\wedge\mathsf{e}^j\,(\mathsf{e}_i\cdot\mathsf{e}_j) = 0, \tag{6.17}$$

which follows because $\mathsf{e}^i\wedge\mathsf{e}^j$ is antisymmetric on i and j, whereas $\mathsf{e}_i\cdot\mathsf{e}_j$ is symmetric. Again, we can give an algebraic definition of the exterior derivative in terms of the geometric product as

$$\nabla\wedge J = \tfrac{1}{2}(\nabla J - \dot{J}\dot{\nabla}). \tag{6.18}$$

Equations (6.13) and (6.18) combine to give the familiar decomposition of a geometric product:

$$\nabla J = \nabla\cdot J + \nabla\wedge J. \tag{6.19}$$

So, for example, we have $\nabla x = n$.

6.1.3 Multivector fields

The preceding definitions extend simply to the case of the vector derivative acting on a multivector field. We have

$$\nabla A = \mathsf{e}^k \partial_k A, \tag{6.20}$$

and for an r-grade multivector field A_r we write

$$\nabla \cdot A_r = \langle \nabla A_r \rangle_{r-1}, \tag{6.21}$$

$$\nabla \wedge A_r = \langle \nabla A_r \rangle_{r+1}. \tag{6.22}$$

These define the interior and exterior derivatives respectively. The interior derivative is often referred to as the divergence, and the exterior derivative is sometimes called the curl. This latter name conflicts with the more familiar meaning of 'curl' in three dimensions, however, and we will avoid this name where possible.

An important result for the vector derivative is that the exterior derivative of an exterior derivative always vanishes,

$$\begin{aligned}\nabla \wedge (\nabla \wedge A) &= \mathsf{e}^i \wedge \partial_i (\mathsf{e}^j \wedge \partial_j A) \\ &= \mathsf{e}^i \wedge \mathsf{e}^j \wedge (\partial_i \partial_j A) = 0.\end{aligned} \tag{6.23}$$

This follows because $\mathsf{e}^i \wedge \mathsf{e}^j$ is antisymmetric on i, j, whereas $\partial_i \partial_j A$ is symmetric, due to the fact that partial derivatives commute. Similarly, the divergence of a divergence vanishes,

$$\nabla \cdot (\nabla \cdot A) = 0, \tag{6.24}$$

which is proved in the same way, or by using duality. (By convention, the inner product of a vector and a scalar is zero.)

Because ∇ is a vector, it does not necessarily commute with other multivectors. We therefore need to be careful in describing the scope of the operator. We use the following series of conventions to clarify the scope:

(i) In the absence of brackets, ∇ acts on the object to its immediate right.
(ii) When the ∇ is followed by brackets, the derivative acts on all of the terms in the brackets.
(iii) When the ∇ acts on a multivector to which it is not adjacent, we use overdots to describe the scope.

The 'overdot' notation was introduced in the previous section, and is invaluable when differentiating products of multivectors. For example, with this notation we can write

$$\nabla(AB) = \nabla AB + \dot{\nabla} A \dot{B}, \tag{6.25}$$

which encodes a version of the product rule. If necessary, the overdots can be replaced with partial derivatives by writing

$$\dot{\nabla} A \dot{B} = \mathsf{e}^k A \, \partial_k B. \tag{6.26}$$

Later in this chapter we also employ the overdot notation for linear functions. Suppose that $\mathsf{f}(a)$ is a position-dependent linear function. We write

$$\dot{\nabla} \dot{\mathsf{f}}(a) = \nabla \mathsf{f}(a) - \mathsf{e}^k \mathsf{f}(\partial_k a), \tag{6.27}$$

so that $\dot{\nabla} \dot{\mathsf{f}}(a)$ only differentiates the position dependence in the linear function, and not in its argument.

We can continue to build up a series of useful basic results by differentiating various multivectors that depend linearly on x. For example, consider

$$\nabla \, x \cdot A_r = \mathsf{e}^k \, \mathsf{e}_k \cdot A_r, \tag{6.28}$$

where A_r is a grade-r multivector. Using the results of section 4.3.2 we find that

$$\begin{aligned} \nabla \, x \cdot A_r &= r A_r, \\ \nabla \, x \wedge A_r &= (n - r) A_r, \\ \dot{\nabla} A_r \dot{x} &= (-1)^r (n - 2r) A_r, \end{aligned} \tag{6.29}$$

where n is the dimension of the space.

6.2 Curvilinear coordinates

So far we have only expressed the vector derivative in terms of a fixed coordinate frame (which is usually chosen to be orthonormal). In many applications, however, it is more convenient to work in a *curvilinear* coordinate system, where the frame vectors vary from point to point. A general set of coordinates consist of a set of scalar functions $\{x^i(x)\}$, $i = 1, \ldots, n$, defined over some region. In this region we can equally write $x(x^i)$, expressing the position vector x parametrically in terms of the coordinates. If one of the coordinates is varied and all of the others are held fixed we specify an associated coordinate curve. The derivatives along these curves specify a set of frame vectors by

$$\mathsf{e}_i(x) = \frac{\partial x}{\partial x^i} = \lim_{\epsilon \mapsto 0} \frac{x(x^1, \ldots, x^i + \epsilon, \ldots, x^n) - x}{\epsilon}, \tag{6.30}$$

where the ith coordinate is varied and all others are held fixed. The derivative in the e_i direction, $\mathsf{e}_i \cdot \nabla$, is found by moving a small amount along e_i. But this is precisely the same as varying the x^i coordinate with all others held fixed. We therefore have

$$\mathsf{e}_i \cdot \nabla = \frac{\partial}{\partial x^i} = \partial_i. \tag{6.31}$$

In order that the coordinate system be valid over a given region we require that throughout this region

$$\mathsf{e}_1\wedge\mathsf{e}_2\wedge\cdots\wedge\mathsf{e}_n \neq 0. \tag{6.32}$$

As this quantity can never pass through zero it follows that the frame has the same orientation throughout the valid region.

We can construct a second frame directly from the coordinate functions by defining

$$\mathsf{e}^i = \nabla x^i. \tag{6.33}$$

From their construction we see that the $\{\mathsf{e}^i\}$ vectors have vanishing exterior derivative:

$$\nabla\wedge\mathsf{e}^i = \nabla\wedge(\nabla x^i) = 0. \tag{6.34}$$

As the notation suggests, the two frames defined above are reciprocal to one another. This is straightforward to check:

$$\mathsf{e}_i\cdot\mathsf{e}^j = \mathsf{e}_i\cdot\nabla x^j = \frac{\partial x^j}{\partial x^i} = \delta_i^j. \tag{6.35}$$

This result is very useful because, when working with curvilinear coordinates, one usually has simple expressions for either $x^i(x)$ or $x(x^i)$, but rarely both. Fortunately, only one is needed to construct a set of frame vectors, and the reciprocal frame can then be constructed algebraically (see section 4.3). This construction provides a simple geometric picture for the gradient in a general space. Suppose we view the coordinate $x^1(x)$ as a scalar field. The contours of constant x^1 are a set of $(n-1)$-dimensional surfaces. The remaining coordinates $x^2, \dots, x^n$ define a set of directions in this surface. At each point on the surface of constant x^1 the vector ∇x^1 is orthogonal to all of the directions in the surface. In Euclidean spaces this vector is necessarily orthogonal (normal) to the surface. In other spaces this construct defines what we mean by normal.

Now suppose we have a function $F(x)$ that is expressed in terms of the coordinates as $F(x^i)$. A simple application of the chain rule gives

$$\nabla F = \nabla x^i\,\partial_i F = \mathsf{e}^i\partial_i F. \tag{6.36}$$

This is consistent with the decomposition

$$\nabla = \mathsf{e}^i\frac{\partial}{\partial x^i} = \mathsf{e}^i\partial_i = \mathsf{e}^i\mathsf{e}_i\cdot\nabla, \tag{6.37}$$

which holds as the $\{\mathsf{e}_i\}$ and $\{\mathsf{e}^i\}$ are reciprocal frames.

6.2.1 Tensor analysis

A consequence of curvilinear frame vectors is that one has to be careful when working entirely in terms of coordinates, as is the case in tensor analysis. The problem is that for a vector, for example, we have $J = J^i \mathsf{e}_i$. If we just keep the coordinates J^i we lose the information about the position dependence in the coordinate frame. When formulating the derivative of J in tensor analysis we must introduce *connection coefficients* to keep track of the derivatives of the frame vectors. This can often complicate derivations.

There are two cases of the vector derivative in curvilinear coordinates that do not require connection coefficients. The first is the exterior derivative, for which we can write

$$\nabla\wedge J = \nabla\wedge(J_i \mathsf{e}^i) = (\nabla J_i)\wedge \mathsf{e}^i. \tag{6.38}$$

It follows that the exterior derivative has coordinates $\partial_i J_j - \partial_j J_i$ regardless of chosen coordinate system. The second exception is provided by the divergence of a vector. We have

$$\nabla\cdot J = \nabla\cdot(J^i \mathsf{e}_i). \tag{6.39}$$

If we define the volume factor V by

$$\mathsf{e}_1\wedge\mathsf{e}_2\wedge\cdots\wedge\mathsf{e}_n = IV, \tag{6.40}$$

where I is the unit pseudoscalar, we can write (following section 4.3)

$$\mathsf{e}_i = (-1)^{i-1}\mathsf{e}^n\wedge\mathsf{e}^{n-1}\wedge\cdots\wedge\check{\mathsf{e}}^i\wedge\cdots\wedge\mathsf{e}^1\, IV. \tag{6.41}$$

Recalling that each of the e^i vectors has vanishing exterior derivative, one can quickly establish that

$$\nabla\cdot J = \frac{1}{V}\frac{\partial}{\partial x^i}(VJ^i). \tag{6.42}$$

Similarly, the Laplacian ∇^2 can be written as

$$\nabla^2\phi = \frac{1}{V}\frac{\partial}{\partial x^i}\left(Vg^{ij}\frac{\partial\phi}{\partial x^j}\right), \tag{6.43}$$

where $g^{ij} = \mathsf{e}^i\cdot\mathsf{e}^j$.

6.2.2 Orthogonal coordinates in three dimensions

A number of the most useful coordinate systems are orthogonal systems of coordinates in three dimensions. For these systems a number of special results hold. We define a set of orthonormal vectors by first introducing the magnitudes

$$h_i = |\mathsf{e}_i| = (\mathsf{e}_i\cdot\mathsf{e}_i)^{1/2}. \tag{6.44}$$

In terms of these we can write (no sums implied)

$$\mathsf{e}_i = h_i\hat{\mathsf{e}}_i, \quad \mathsf{e}^i = \frac{1}{h_i}\hat{\mathsf{e}}_i. \tag{6.45}$$

We now use the $\{\hat{\mathsf{e}}_i\}$ as our coordinate frame and, since this frame is orthonormal, we can work entirely with lowered indices. For a vector $\boldsymbol{J}$ we have

$$\boldsymbol{J} = J_i \hat{\mathsf{e}}_i = \sum_{i=1}^{3} \frac{J_i}{h_i} \mathsf{e}_i. \tag{6.46}$$

It follows that we can write

$$\boldsymbol{\nabla}\cdot\boldsymbol{J} = \frac{1}{h_1 h_2 h_3}\left(\frac{\partial}{\partial x_1}(h_2 h_3 J_1) + \frac{\partial}{\partial x_2}(h_3 h_1 J_2) + \frac{\partial}{\partial x_3}(h_1 h_2 J_3)\right). \tag{6.47}$$

A compact formula for the Laplacian is obtained by replacing each J_i term with $1/h_i\, \partial_i \phi$,

$$\begin{aligned}\boldsymbol{\nabla}^2\phi - \frac{1}{h_1 h_2 h_3}\Bigg(&\frac{\partial}{\partial x_1}\left(\frac{h_2 h_3}{h_1}\frac{\partial \phi}{\partial x_1}\right) + \frac{\partial}{\partial x_2}\left(\frac{h_3 h_1}{h_2}\frac{\partial \phi}{\partial x_2}\right) \\ &+ \frac{\partial}{\partial x_3}\left(\frac{h_1 h_2}{h_3}\frac{\partial \phi}{\partial x_3}\right)\Bigg).\end{aligned} \tag{6.48}$$

The components of the curl can be found in a similar manner. A number of useful curvilinear coordinate systems are summarised below.

Cartesian coordinates

These are the basic starting point for all other coordinate systems. We introduce a constant, right-handed orthonormal frame $\{\boldsymbol{\sigma}_i\}$, $\boldsymbol{\sigma}_1\boldsymbol{\sigma}_2\boldsymbol{\sigma}_3 = I$. This notation for a Cartesian frame is borrowed from quantum theory and is very useful in practice. The coordinates in the $\{\boldsymbol{\sigma}_i\}$ frame are written, following standard notation, as (x, y, z). To avoid confusion between the scalar coordinate x and the three-dimensional position vector we write the latter as $\boldsymbol{r}$. That is,

$$\boldsymbol{r} = x\boldsymbol{\sigma}_1 + y\boldsymbol{\sigma}_2 + z\boldsymbol{\sigma}_3. \tag{6.49}$$

Since the frame vectors are orthonormal we have $h_1 = h_2 = h_3 = 1$, so the divergence and Laplacian take on their simplest forms.

Cylindrical polar coordinates

These are denoted (ρ, ϕ, z) with ρ and ϕ the standard two-dimensional polar coordinates

$$\rho = \left(x^2 + y^2\right)^{1/2}, \qquad \tan\phi = \frac{y}{x}. \tag{6.50}$$

The coordinates lie in the ranges $0 \leq r < \infty$ and $0 \leq \phi < 2\pi$. The coordinate vectors are

$$\begin{aligned}\hat{\mathsf{e}}_\rho &= \cos(\phi)\,\boldsymbol{\sigma}_1 + \sin(\phi)\,\boldsymbol{\sigma}_2, \\ \hat{\mathsf{e}}_\phi &= -\sin(\phi)\,\boldsymbol{\sigma}_1 + \cos(\phi)\,\boldsymbol{\sigma}_2, \\ \hat{\mathsf{e}}_z &= \boldsymbol{\sigma}_3.\end{aligned} \tag{6.51}$$

We have adopted the common convention of labelling the frame vectors with the associated coordinate. The magnitudes are $h_\rho = 1$, $h_\phi = \rho$ and $h_z = 1$, and the frame vectors satisfy

$$\hat{\mathsf{e}}_\rho \hat{\mathsf{e}}_\phi \hat{\mathsf{e}}_z = \boldsymbol{\sigma}_1\boldsymbol{\sigma}_2\boldsymbol{\sigma}_3 = I \tag{6.52}$$

and so form a right-handed set in the order (ρ, ϕ, z).

Spherical polar coordinates

Spherical polar coordinates arise in many problems in physics, particularly quantum mechanics and field theory. They are typically labelled (r, θ, ϕ) and are defined by

$$r = |\boldsymbol{r}| = (\boldsymbol{r}\cdot\boldsymbol{r})^{1/2}, \quad r\cos(\theta) = z, \quad \tan(\phi) = \frac{y}{x}. \tag{6.53}$$

The coordinate ranges are $0 \leq r < \infty$, $0 \leq \theta \leq \pi$ and $0 \leq \phi < 2\pi$. The ϕ coordinate is ill defined along the z axis — a reflection of the fact that it is impossible to construct a global coordinate system over the surface of a sphere. The inverse relation giving $\boldsymbol{r}(r, \theta, \phi)$ is often useful,

$$\boldsymbol{r} = r\sin(\theta)(\cos(\phi)\,\boldsymbol{\sigma}_1 + \sin(\phi)\,\boldsymbol{\sigma}_2) + r\cos(\theta)\,\boldsymbol{\sigma}_3. \tag{6.54}$$

This expression makes it a straightforward exercise to compute the orthonormal frame vectors, which are

$$\begin{aligned}
\hat{\mathsf{e}}_r &= \sin(\theta)(\cos(\phi)\,\boldsymbol{\sigma}_1 + \sin(\phi)\,\boldsymbol{\sigma}_2) + \cos(\theta)\,\boldsymbol{\sigma}_3 = r^{-1}\boldsymbol{r}, \\
\hat{\mathsf{e}}_\theta &= \cos(\theta)(\cos(\phi)\,\boldsymbol{\sigma}_1 + \sin(\phi)\,\boldsymbol{\sigma}_2) - \sin(\theta)\,\boldsymbol{\sigma}_3, \\
\hat{\mathsf{e}}_\phi &= -\sin(\phi)\,\boldsymbol{\sigma}_1 + \cos(\phi)\,\boldsymbol{\sigma}_2.
\end{aligned} \tag{6.55}$$

The associated normalisation factors are

$$h_r = 1, \quad h_\theta = r, \quad h_\phi = r\sin(\theta). \tag{6.56}$$

The orthonormal vectors satisfy $\hat{\mathsf{e}}_r\hat{\mathsf{e}}_\theta\hat{\mathsf{e}}_\phi = I$ so that $\{\hat{\mathsf{e}}_r, \hat{\mathsf{e}}_\theta, \hat{\mathsf{e}}_\phi\}$ form a right-handed orthonormal frame. This frame can be obtained from the $\{\mathsf{e}_i\}$ frame through the application of a position-dependent rotor, so that $\hat{\mathsf{e}}_r = R\boldsymbol{\sigma}_3\tilde{R}$, $\hat{\mathsf{e}}_\theta = R\boldsymbol{\sigma}_1\tilde{R}$ and $\hat{\mathsf{e}}_\phi = R\boldsymbol{\sigma}_2\tilde{R}$. The rotor is then given by

$$R = \exp(-I\boldsymbol{\sigma}_3\phi/2)\exp(-I\boldsymbol{\sigma}_2\theta/2). \tag{6.57}$$

Spheroidal coordinates

These coordinates turn out to be useful in a number of problems in gravitation and electromagnetism involving rotating sources. We introduce a vector $\boldsymbol{a}$, so that $\pm\boldsymbol{a}$ denote the foci of a family of ellipses. The distances from the foci are given by

$$r_1 = |\boldsymbol{r} + \boldsymbol{a}|, \quad r_2 = |\boldsymbol{r} - \boldsymbol{a}|. \tag{6.58}$$

From these we define the orthogonal coordinates

$$u = \tfrac{1}{2}(r_1 + r_2), \quad v = \tfrac{1}{2}(r_1 - r_2). \tag{6.59}$$

The coordinate system is completed by rotating the ellipses around the $\boldsymbol{a}$ axis. This defines an oblate spheroidal coordinate system. Prolate spheroidal coordinates are formed by starting in a plane, defining (u_1, u_2) as above, and rotating this system around the minor axis.

If we define

$$\hat{\boldsymbol{r}}_1 = \frac{\boldsymbol{r} + \boldsymbol{a}}{r_1}, \quad \hat{\boldsymbol{r}}_2 = \frac{\boldsymbol{r} - \boldsymbol{a}}{r_2}, \tag{6.60}$$

we see that

$$\mathsf{e}^u = \tfrac{1}{2}(\hat{\boldsymbol{r}}_1 + \hat{\boldsymbol{r}}_2), \quad \mathsf{e}^v = \tfrac{1}{2}(\hat{\boldsymbol{r}}_1 - \hat{\boldsymbol{r}}_2), \tag{6.61}$$

which are clearly orthogonal. The normalisation factors are found from

$$h_u^2 = \frac{u^2 - v^2}{u^2 - a^2}, \quad h_v^2 = \frac{u^2 - v^2}{a^2 - v^2}. \tag{6.62}$$

If we align $\boldsymbol{a}$ with the 3 axis and let ϕ take its spherical-polar meaning, the coordinate frame is completed with the vector $\hat{\mathsf{e}}_\phi$, and

$$h_\phi^2 = (u^2 - a^2)(a^2 - v^2). \tag{6.63}$$

The frame vectors satisfy $\hat{\mathsf{e}}_u \hat{\mathsf{e}}_\phi \hat{\mathsf{e}}_v = I$. The hyperbolic nature of the coordinate system is often best expressed by redefining the u and v coordinates as $a\cosh(w)$ and $a\cos(\vartheta)$ respectively.

6.3 Analytic functions

The vector derivative combines the algebraic properties of geometric algebra with vector calculus in a simple and natural way. In this section we show how the vector derivative can be used to extend the definition of an analytic function to arbitrary dimensions. We start by considering the vector derivative in two dimensions to establish the link with complex analysis.

6.3.1 Analytic functions in two dimensions

Suppose that $\{\mathsf{e}_1, \mathsf{e}_2\}$ define an orthonormal frame in two dimensions. This is identified with the Argand plane by singling out e_1 as the real axis. We denote coordinates by (x, y) and write the position vector as $\boldsymbol{r}$:

$$\boldsymbol{r} = x\mathsf{e}_1 + y\mathsf{e}_2. \tag{6.64}$$

With this notation the vector derivative is

$$\boldsymbol{\nabla} = \mathsf{e}_1 \frac{\partial}{\partial x} + \mathsf{e}_2 \frac{\partial}{\partial y}. \tag{6.65}$$

In section 2.3.3 we showed that complex numbers sit naturally within the geometric algebra of the plane. The pseudoscalar is the bivector $I = \mathsf{e}_1\mathsf{e}_2$, which satisfies $I^2 = -1$. Complex numbers therefore map directly onto even-grade elements in the algebra by identifying the unit imaginary i with I. The position vector $\boldsymbol{r}$ is mapped onto a complex number by pre-multiplying by the vector representing the real axis:

$$z = x + Iy = \mathsf{e}_1\boldsymbol{r}. \tag{6.66}$$

Now suppose we introduce the complex field $\psi = u + Iv$. The vector derivative applied to ψ yields

$$\boldsymbol{\nabla}\psi = \left(\frac{\partial u}{\partial x} - \frac{\partial v}{\partial y}\right)\mathsf{e}_1 + \left(\frac{\partial v}{\partial x} + \frac{\partial u}{\partial y}\right)\mathsf{e}_2. \tag{6.67}$$

The terms in brackets are precisely the ones that vanish in the Cauchy–Riemann equations. The statement that ψ is an *analytic function* (a function that satisfies the Cauchy–Riemann equations) reduces to the equation

$$\boldsymbol{\nabla}\psi = 0. \tag{6.68}$$

This is the fundamental equation which can be generalised immediately to higher dimensions. These generalisations invariably turn out to be of mathematical and physical importance, and it is is no exaggeration to say that equations of the type of equation (6.68) are amongst the most studied in physics.

To complete the link with complex analysis we recall that the complex partial derivative ∂_z is defined by the properties

$$\frac{\partial z}{\partial z} = 1, \qquad \frac{\partial z^\dagger}{\partial z} = 0 \tag{6.69}$$

with the complex conjugate satisfying

$$\frac{\partial z}{\partial z^\dagger} = 0, \qquad \frac{\partial z^\dagger}{\partial z^\dagger} = 1. \tag{6.70}$$

From these we see that

$$\frac{\partial}{\partial z} = \frac{1}{2}\left(\frac{\partial}{\partial x} - I\frac{\partial}{\partial y}\right), \qquad \frac{\partial}{\partial z^\dagger} = \frac{1}{2}\left(\frac{\partial}{\partial x} + I\frac{\partial}{\partial y}\right). \tag{6.71}$$

An analytic function is one that depends on z alone. That is, we can write $\psi(x + Iy) = \psi(z)$. The function is therefore independent of $z^\dagger$, and we have

$$\frac{\partial\psi(z)}{\partial z^\dagger} = 0. \tag{6.72}$$

This summarises the content of the Cauchy–Riemann equations, though this fact is often obscured by the complex limiting argument favoured in many textbooks. Comparing the preceding forms, we see that this equation is equivalent to

$$\frac{1}{2}\left(\frac{\partial}{\partial x} + I\frac{\partial}{\partial y}\right)\psi = \tfrac{1}{2}\mathsf{e}_1\boldsymbol{\nabla}\psi = 0, \tag{6.73}$$

recovering our earlier equation.

It is instructive to see why solutions to $\boldsymbol{\nabla}\psi = 0$ can be constructed as power series in z. We first see that

$$\boldsymbol{\nabla} z = \boldsymbol{\nabla}(\mathsf{e}_1 \boldsymbol{r}) = 2\mathsf{e}_1 \cdot \boldsymbol{\nabla} \boldsymbol{r} - \mathsf{e}_1 \boldsymbol{\nabla} \boldsymbol{r} = 2\mathsf{e}_1 - 2\mathsf{e}_1 = 0. \tag{6.74}$$

This little manipulation drives most of analytic function theory! It follows immediately, for example, that

$$\boldsymbol{\nabla}(z - z_0)^n = n\boldsymbol{\nabla}(\mathsf{e}_1 \boldsymbol{r} - z_0)(z - z_0)^{n-1} = 0, \tag{6.75}$$

so a Taylor series expansion in z about z_0 automatically returns an analytic function. We will delay looking at poles until we have introduced the subject of directed integration.

6.3.2 Generalized analytic functions

There are two problems with the standard presentation of complex analytic function theory that prevent a natural generalisation to higher dimensions:

(i) Both the vector operator $\boldsymbol{\nabla}$ and the functions it operates on are mapped into the same algebra by picking out a preferred direction for the real axis. This only works in two dimensions.

(ii) The 'complex limit' argument does not generalise to higher dimensions. Indeed, one can argue that it is not wholly satisfactory in two dimensions, as it confuses the concept of a directional derivative with the concept of being independent of $z^\dagger$.

These problems are solved by keeping the derivative operator $\boldsymbol{\nabla}$ as a vector, while letting it act on general multivectors. The analytic requirement is then replaced with the equation $\boldsymbol{\nabla}\psi = 0$. Functions satisfying this equation are said to be *monogenic*. If ψ contains all grades it is clear that both the even-grade and odd-grade components must satisfy this equation independently. Without loss of generality, we can therefore assume that ψ has even grade.

We can construct monogenic functions by following the route which led to the conclusion that z is analytic in two dimensions. We recall that $\boldsymbol{\nabla}\boldsymbol{r} = 3$ and

$$\boldsymbol{\nabla}(\boldsymbol{a}\boldsymbol{r}) = -\boldsymbol{a}. \tag{6.76}$$

It follows that

$$\psi = \boldsymbol{r}\boldsymbol{a} + 3\boldsymbol{a}\boldsymbol{r} \tag{6.77}$$

is a monogenic for any constant vector $\boldsymbol{a}$. The main difference with complex analysis is that we cannot derive new monogenics simply from power series in this solution, due to the lack of commutativity. One can construct monogenic functions from series of geometric products, but a more instructive route is to classify monogenics via their angular properties.

First we assume that Ψ is a monogenic containing terms which scale uniformly with $\boldsymbol{r}$. If we introduce polar coordinates we can then write

$$\Psi(\boldsymbol{r}) = r^l \psi(\theta, \phi). \tag{6.78}$$

The function $\psi(\theta, \phi)$ then satisfies

$$l r^{l-1} \mathrm{e}_r \psi + r^l \boldsymbol{\nabla} \psi(\theta, \phi) = 0. \tag{6.79}$$

It follows that ψ satisfies the angular eigenvalue equation

$$-\boldsymbol{r} \wedge \boldsymbol{\nabla} \psi = l \psi. \tag{6.80}$$

These angular eigenstates play a key role in the Pauli and Dirac theories of the electron. Since Ψ satisfies $\boldsymbol{\nabla}\Psi = 0$, it follows that

$$\boldsymbol{\nabla}^2 \Psi = 0. \tag{6.81}$$

So each component of Ψ (in a constant basis) satisfies Laplace's equation. It follows that each component of ψ is a spherical harmonic, and hence that l is an integer. We can construct a monogenic by starting with the function $(x + yI\boldsymbol{\sigma}_3)^l$, which is the three-dimensional extension of the complex analytic function z^l. In terms of polar coordinates

$$(x + yI\boldsymbol{\sigma}_3)^l = r^l \sin^l(\theta)\, \mathrm{e}^{l\phi I\boldsymbol{\sigma}_3}, \tag{6.82}$$

which gives us our first angular monogenic function

$$\psi_l^l = \sin^l(\theta)\, \mathrm{e}^{l\phi I\boldsymbol{\sigma}_3}. \tag{6.83}$$

The remaining monogenic functions are constructed from this by acting with an operator which, in quantum terms, lowers the eigenvalue of the angular momentum around the z axis. These are discussed in more detail in section 8.4.1.

6.3.3 The spacetime vector derivative

To construct the vector derivative in spacetime suppose that we introduce the orthonormal frame $\{\gamma_\mu\}$ with associated coordinates x^μ. We can then write

$$\nabla = \gamma^\mu \frac{\partial}{\partial x^\mu} = \gamma_0 \frac{\partial}{\partial t} + \gamma^i \frac{\partial}{\partial x^i}. \tag{6.84}$$

This derivative is the key operator in all relativistic field theories, including electromagnetism and Dirac theory. If we post-multiply by γ_0 we see that

$$\nabla \gamma_0 = \partial_t + \gamma^i \gamma_0 \partial_i = \partial_t - \boldsymbol{\nabla}, \tag{6.85}$$

where $\boldsymbol{\nabla} = \boldsymbol{\sigma}_i \partial_i$ is the vector derivative in the relative space defined by the γ_0 vector. Similarly,

$$\gamma_0 \nabla = \partial_t + \boldsymbol{\nabla}. \tag{6.86}$$

These equations are consistent with

$$\nabla x = \nabla(\gamma_0 \gamma_0 x) = (\partial_t - \boldsymbol{\nabla})(t - \boldsymbol{r}) = 4, \tag{6.87}$$

where x is the spacetime position vector. The spacetime vector derivative satisfies

$$\nabla^2 = \frac{\partial^2}{\partial t^2} - \boldsymbol{\nabla}^2, \tag{6.88}$$

which is the fundamental operator describing waves travelling at the speed of light. The spacetime monogenic equation $\nabla\psi = 0$ is discussed in detail in chapters 7 and 8. We only note here that, if ψ is an even-grade element of the spacetime algebra, the monogenic equation is precisely the wave equation for a massless spin-1/2 particle.

6.3.4 Characteristic surfaces and propagation

The fact that ∇^2 can give rise to either elliptic or hyperbolic operators, depending on signature, suggests that the propagator theory for ∇ will depend strongly on the signature. This is confirmed by a simple argument which can be modified to apply to most first-order differential equations. Suppose we have a generic equation of the type

$$\nabla\psi = f(\psi, x), \tag{6.89}$$

where ψ is some multivector field, $f(\psi, x)$ is a known function and x is the position vector in an n-dimensional space. We are presented with data on some $(n-1)$-dimensional surface, and wish to propagate these initial conditions away from the surface. If surfaces exist for which this is not possible they are known as *characteristic surfaces*. Suppose that we construct a set of independent tangent vectors in the surface, $\{e_1, \dots, e_{n-1}\}$. Knowledge of ψ on the surface enables us to calculate each of the directional derivatives $e_i \cdot \nabla\psi$, $i = 1, \dots, n-1$. We now form the normal vector

$$n = I\, e_1 \wedge e_2 \wedge \cdots \wedge e_{n-1}, \tag{6.90}$$

where I is the pseudoscalar for the space. Pre-multiplying equation (6.89) with n we obtain

$$n \cdot \nabla\psi = -n \wedge \nabla\psi + n f(\psi, x). \tag{6.91}$$

But we have

$$
\begin{aligned}
n\wedge\nabla\psi &= I(e_1\wedge e_2\wedge\cdots\wedge e_{n-1})\cdot\nabla\psi \\
&= I\sum_{i=1}^{n-1}(-1)^{i+1-n}(e_1\wedge\cdots\wedge\check{e}_i\wedge\cdots\wedge e_{n-1})\, e_i\cdot\nabla\psi,
\end{aligned}
\tag{6.92}
$$

which is constructed entirely from known derivatives of ψ. Equation (6.91) then tells us how to propagate ψ in the n direction. The only situation in which we can fail to propagate ψ is when n still lies in the surface. This happens if n is linearly dependent on the surface tangent vectors. If this is the case we have

$$
n\wedge(e_1\wedge e_2\wedge\cdots\wedge e_{n-1}) = 0. \tag{6.93}
$$

But this implies that

$$
(I^{-1}n)\wedge n = I^{-1}n\cdot n = 0. \tag{6.94}
$$

We therefore only fail to propagate when $n^2 = 0$, so characteristic surfaces are always null surfaces. This possibility can only arise in mixed signature spaces, and unsurprisingly the propagators in these spaces can have quite different properties to their Euclidean counterparts.

6.4 Directed integration theory

The true power of geometric calculus begins to emerge when we study directed integration theory. This provides a very general and powerful integral theorem which enables us to construct Green's functions for the vector derivative in various spaces. These in turn can be used to generalise the many powerful results from complex function theory to arbitrary spaces.

6.4.1 Line integrals

The simplest integrals to start with are line integrals. The line integral of a multivector field $F(x)$ along a line $x(\lambda)$ is defined by

$$
\int F(x)\frac{dx}{d\lambda}\, d\lambda = \int F\, dx = \lim_{n\mapsto\infty}\sum_{i=1}^{n}\bar{F}^i\Delta x^i. \tag{6.95}
$$

In the final expression a set of successive points along the curve $\{x_i\}$ are introduced, with x_0 and x_n the endpoints, and

$$
\Delta x^i = x_i - x_{i-1}, \quad \bar{F}^i = \tfrac{1}{2}\big(F(x_{i-1}) + F(x_i)\big). \tag{6.96}
$$

If the curve is closed then $x_0 = x_n$. The result of the integral is independent of the way we choose to parameterise the curve, provided the parameterisation respects the required ordering of points along the curve. Curves that double back on themselves are handled by referring to the parameterised form $x(\lambda)$, which tells us how the curve is traversed.

The definition of the integral (6.95) looks so standard that it is easy to overlook the key new feature, which is that dx is a *vector-valued measure*, and the product $F\,dx$ is a geometric product between multivectors. This small extension to scalar integration is sufficient to bring a wealth of new features. We refer to dx, and its multivector-valued extensions, as a *directed measure*. The fact that dx is no longer a scalar means that equation (6.95) is not the most general line integral we can form. We can also consider integrals of the form

$$\int F(x)\frac{dx}{d\lambda}G(x)\,d\lambda = \int F(x)\,dx\,G(x), \tag{6.97}$$

and more generally we can consider sums of terms like these. The most general form of line integral can be written

$$\int \mathsf{L}(\partial_\lambda x;x)\,d\lambda = \int \mathsf{L}(dx), \tag{6.98}$$

where $\mathsf{L}(a) = \mathsf{L}(a;x)$ is a multivector-valued linear function of a. The position dependence in L can often be suppressed to streamline the notation.

Suppose now that the field F is replaced by the vector-valued function $v(x)$. We have

$$\int v\,dx = \int v\cdot dx + \int v\wedge dx, \tag{6.99}$$

which separates the directed integral into scalar and bivector-valued terms. If v is the unit tangent vector along the curve then the scalar integral returns the arc length. In many applications the scalar and bivector integrals are considered separately. But to take advantage of the most powerful integral theorems in geometric calculus we need to use the combined form, containing a geometric product with the directed measure.

6.4.2 Surface integrals

The natural extension of a line integral is to a directed surface integral. Suppose now that the the multivector-valued field F is defined over a two-dimensional surface embedded in some larger space. If the surface is parameterised by two coordinates $x(x^1,x^2)$ we define the directed measure by the bivector

$$dX = \frac{\partial x}{\partial x^1}\wedge\frac{\partial x}{\partial x^2}\,dx^1\,dx^2 = e_1\wedge e_2\,dx^1\,dx^2, \tag{6.100}$$

where $e_i = \partial_i x$. This measure is independent of how the surface is parameterised, provided we orient the coordinate vectors in the desired order. Sometimes more than one coordinate patch will be needed to parameterise the entire surface, but

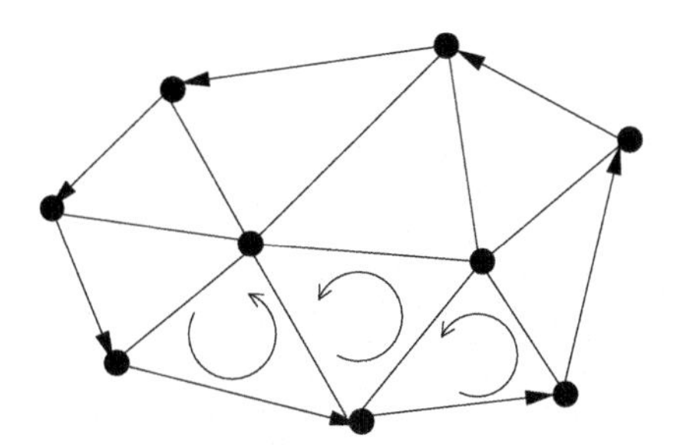

Figure 6.2 *A triangulated surface.* The surface is represented by a series of points, and each set of three adjacent points defines a triangle, or simplex. As more points are added the simplices become a closer fit to the true surface. Each simplex is given the same orientation by ensuring that for adjacent simplices, the common edge in traversed in opposite directions.

the directed measure dX is still defined everywhere. A directed surface integral then takes the form

$$\int F\,dX = \int F e_1 \wedge e_2\,dx^1\,dx^2, \tag{6.101}$$

or a sum of such terms if more than one coordinate patch is required. Again, we form the geometric product between the integrand and the measure. As in the case of a line integral, this is not the most general surface integral that can be considered, as the integrand can multiply the measure from the left or the right, giving rise to different integrals.

As an example of a surface integral, consider a closed surface in three dimensions, with unit outward normal $\boldsymbol{n}$. We let F be given by the bivector-valued function $\phi \boldsymbol{n} I^{-1}$, where ϕ is a scalar field. The surface integral is then

$$\oint \phi \boldsymbol{n} I^{-1}\,dX = \oint \phi |dS|. \tag{6.102}$$

Here $|dS| = I^{-1}\boldsymbol{n}\,dX$ is the scalar-valued measure over the surface. The directed measure is usually chosen so that $\boldsymbol{n}\,dX$ has the same orientation as I. As a second example, suppose that $F = 1$. In this case we can show that

$$\oint dX = 0, \tag{6.103}$$

which holds for any closed surface (see later). If the surface is open, the result of the directed surface integral depends entirely on the boundary, since all the internal simplices cancel out. This result is sometimes called the vector area, though in geometric algebra the result is a bivector.

In order to construct proofs of some of the more important results it is necessary to express the surface integral (6.101) in terms of a limit of a sum. This involves the idea of a triangulated surface (figure 6.2). A set of points are chosen

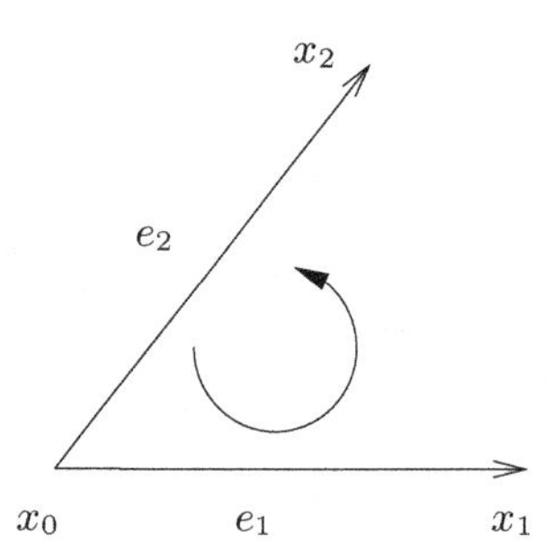

Figure 6.3 *A planar simplex.* The points x_0, x_1, x_2 define a triangle. The order specifies how the boundary is traversed, which defines an orientation for the simplex.

on the surface, and adjacent sets of three points define a series of planar triangles, or *simplices*. As more points are added these triangles become smaller and are an ever better model for the surface. (In computer graphics programs this is precisely how 'smooth' surfaces are represented internally.) Each simplex has an orientation attached such that, for a pair of adjacent simplices, the common edge is traversed in opposite directions. In this way an initial simplex builds up to define an orientation for the entire surface. For some surfaces, such as the Mobius strip, it is not possible to define a consistent orientation over the entire surface. For these it is not possible to define a directed integral, so our presentation is restricted to *orientable* surfaces.

Suppose now that the three points x_0, x_1, x_2 define the corners of a simplex, with orientation specified by traversing the edges in the order $x_0 \mapsto x_1 \mapsto x_2$ (see figure 6.3). We define the vectors

$$e_1 = x_1 - x_0, \qquad e_2 = x_2 - x_0. \tag{6.104}$$

The surface measure is then defined by

$$\Delta X = \tfrac{1}{2} e_1 \wedge e_2 = \tfrac{1}{2}(x_1 \wedge x_2 + x_2 \wedge x_0 + x_0 \wedge x_1). \tag{6.105}$$

ΔX has the orientation defined by the boundary, and an area equal to that of the simplex. The final expression makes it clear that ΔX is invariant under even permutations of the vertices. With this definition of ΔX we can express the surface integral (6.101) as the limit:

$$\int F\, dX = \lim_{n \mapsto \infty} \sum_{k=1}^{n} \bar{F}^k \Delta X^k. \tag{6.106}$$

The sum here runs over all simplices making up the surface, and for each simplex $\bar{F}$ is the average value of F over the simplex. For well-behaved integrals the value in the limit is independent of the precise nature of the limiting process.

6.4.3 *n-dimensional surfaces*

The simplex structure introduced in the previous section provides a means of defining a directed integral for any dimension of surface. We discretise the surface by considering a series of points, and adjacent sets of points are combined to define a simplex. Suppose that we have an n-dimensional surface, and that one simplex for the discretised surface has vertices $x_0, \ldots, x_n$, with the order specifying the desired orientation. For this simplex we define vectors

$$e_i = x_i - x_0, \quad i = 1, \ldots, n, \tag{6.107}$$

and the directed volume element is

$$\Delta X = \frac{1}{n!} e_1 \wedge \cdots \wedge e_n. \tag{6.108}$$

A point in the simplex can be described in terms of coordinates $\lambda^1, \ldots, \lambda^n$ by writing

$$x = x_0 + \sum_{i=1}^{n} \lambda^i e_i. \tag{6.109}$$

Each coordinate lies in the range $0 \leq \lambda^i \leq 1$, and the coordinates also satisfy

$$\sum_{i=1}^{n} \lambda^i \leq 1. \tag{6.110}$$

Now suppose we have a multivector field $F(x)$ defined over the surface. We denote the value at each vertex by $F_i = F(x_i)$. A new function $f(x)$ is then introduced which linearly interpolates the F_i over the simplex. This can be written

$$f(x) = F_0 + \sum_{i=1}^{n} \lambda^i (F_i - F_0). \tag{6.111}$$

As the number of points increases and the simplices grow smaller, $f(x)$ becomes an ever better approximation to $F(x)$, and the triangulated surface approaches the true surface.

The directed integral of F over the surface is now approximated by the integral of f over each simplex in the surface. To evaluate the integral over each simplex we use the λ^i as coordinates, so that

$$dX = e_1 \wedge \cdots \wedge e_n \, d\lambda^1 \cdots d\lambda^n. \tag{6.112}$$

It is then a straightforward exercise in integration to establish that

$$\int dX = \Delta X \tag{6.113}$$

and

$$\int \lambda^i \, dX = \frac{1}{n+1} \Delta X, \quad \forall \lambda^i. \tag{6.114}$$

Combining these two results we find that the integral of $f(x)$ over a single simplex evaluates to

$$\int f \, dX = \frac{1}{n+1} \left(\sum_{i=0}^{n} F_i \right) \Delta X. \tag{6.115}$$

The function is therefore replaced by its average value over the simplex. We write this as $\bar{F}$. Summing over all the simplices making up the surface we can now define

$$\int F \, dX = \lim_{n \mapsto \infty} \sum_{k=1}^{n} \bar{F}^k \, \Delta X^k, \tag{6.116}$$

where k runs over all of the simplices in the surface. More generally, suppose that $\mathsf{L}(A_n)$ is a position-dependent linear function of a grade-n multivector A_n. We can then write

$$\int \mathsf{L}(dX) = \lim_{n \mapsto \infty} \sum_{k=1}^{n} \bar{\mathsf{L}}^k(\Delta X^k), \tag{6.117}$$

with $\bar{\mathsf{L}}^k(\Delta X^k)$ the average value of $\mathsf{L}(\Delta X^k)$ over the vertices of each simplex.

6.4.4 The fundamental theorem of geometric calculus

Most physicists are familiar with a number of integral theorems, including the divergence and Stokes' theorems, and the Cauchy integral formula of complex analysis. We will now show that these are all special cases of a more general theorem in geometric calculus. In this section we will sketch of proof of this important theorem. Readers who are not interested in the details of the proof may want to jump straight to the following section, where some applications are discussed. The proof given here uses simplices and triangulated surfaces, which means that it is relevant to methods of discretising integrals for numerical computation.

We start by introducing a notation for simplices which helps clarify the nature of the boundary operator. We let $(x_0, x_1, \ldots, x_k)$ denote the k-simplex defined by the $k+1$ points $x_0, \ldots, x_k$. This is abbreviated to

$$(x)_{(k)} = (x_0, x_1, \ldots, x_k). \tag{6.118}$$

The order of points is important, as it specifies the orientation of the simplex. If any two adjacent points are swapped then the simplex changes sign. The boundary operator for a simplex is denoted by ∂ and is defined by

$$\partial(x)_{(k)} = \sum_{i=0}^{k}(-1)^i(x_0, \ldots, \check{x}_i, \ldots, x_k)_{(k-1)}, \tag{6.119}$$

where the check denotes that the term is missing from the product. So, for example,

$$\partial(x_0, x_1) = (x_1) - (x_0), \tag{6.120}$$

which returns the two points at the end of a line segment. The boundary of a boundary vanishes,

$$\partial\partial(x)_{(k)} = 0. \tag{6.121}$$

Proofs of this can be found in most differential geometry textbooks.

So far we have dealt only with ordered lists of points, not geometric sums or products. To add some geometry we introduce the operator Δ which returns the directed content of a simplex,

$$\Delta(x)_{(k)} = \frac{1}{k!}(x_1 - x_0)\wedge(x_2 - x_0)\wedge\cdots\wedge(x_k - x_0). \tag{6.122}$$

This is the result of integrating the directed measure over a simplex

$$\int_{(x)_{(k)}} dX = \Delta(x)_{(k)} = \Delta X. \tag{6.123}$$

The directed content of a boundary vanishes,

$$\Delta(\partial(x)_{(k)}) = 0. \tag{6.124}$$

As an example, consider a planar simplex consisting of three points. We have

$$\partial(x_0, x_1, x_2) = (x_1, x_2) - (x_0, x_2) + (x_0, x_1). \tag{6.125}$$

So the directed content of the boundary is

$$\Delta(\partial(x_0, x_1, x_2)) = (x_2 - x_1) - (x_2 - x_0) + (x_1 - x_0) = 0. \tag{6.126}$$

The general result of equation (6.124) can be established by induction from the case of a triangle. These results are sufficient to establish that the directed integral over the surface of a simplex is zero:

$$\oint_{\partial(x)_{(k)}} dS = \sum_{i=0}^{k}(-1)^i \int_{(\check{x}_i)_{(k-1)}} dX = \Delta(\partial(x)_{(k)}) = 0. \tag{6.127}$$

A general volume is built up from a chain of simplices. Simplices in the chain are defined such that, at any common boundary, the directed areas of the bounding faces of two simplices are equal and opposite. It follows that the surface integrals over two simplices cancel out over their common face. The surface integral over the boundary of the volume can therefore be replaced by the sum of the surface integrals over each simplex in the chain. If the boundary is closed we establish that

$$\oint dS = \lim_{n\mapsto\infty} \sum_{a=1}^{n} \oint dS^a = 0. \tag{6.128}$$

The sum runs over each simplex in the surface, with a labeling the simplex. It is implicit in this proof that the surface bounds a volume which can be filled by a connected set of simplices. So, as well as being oriented, the surface must be closed and simply connected.

Next, we return to equation (6.114) and introduce a constant vector b. If we define $b_i = b\cdot e_i$ we see that

$$\sum_{i=1}^{k} b_i\lambda^i = b\cdot(x - x_0), \tag{6.129}$$

which is valid for all vectors x in the simplex of interest. Multiplying equation (6.114) by b_i and summing over i we obtain

$$\int_{(x)_{(k)}} b\cdot(x - x_0)\, dX = \frac{1}{k+1}\sum_{i=1}^{k} b\cdot e_i\, \Delta X, \tag{6.130}$$

where the integral runs over a simplex defined by $k+1$ vertices. A simple re-ordering yields

$$\begin{aligned}\int b\cdot x\, dX &= \frac{1}{k+1}\left(\sum_{i=1}^{k} b\cdot(x_i - x_0) + (k+1)b\cdot x_0\right)\Delta X \\ &= b\cdot\bar{x}\,\Delta X,\end{aligned} \tag{6.131}$$

where $\bar{x}$ is the vector representing the (geometric) centre of the simplex,

$$\bar{x} = \frac{1}{k+1}\sum_{i=0}^{k} x_i. \tag{6.132}$$

Now suppose we have a k-simplex specified by the $k+1$ points $(x_0, \ldots, x_k)$ and we form the directed surface integral of $b\cdot x$. We obtain

$$\oint_{\partial(x)_{(k)}} b\cdot x\, dS = \frac{1}{k+1}\sum_{i=0}^{k}(-1)^i b\cdot(x_0 + \cdots \check{x}_i \cdots + x_n)\Delta(\check{x}_i)_{(k-1)}. \tag{6.133}$$

To evaluate the final sum we need the result that

$$\sum_{i=0}^{k}(-1)^i b\cdot(x_0 + \cdots \check{x}_i \cdots + x_n)\Delta(\check{x}_i)_{(k-1)} = \frac{1}{k!} b\cdot(e_1\wedge\cdots\wedge e_n). \tag{6.134}$$

The proof of this result is purely algebraic and is left as an exercise. We have now established the simple result that

$$\oint_{\partial(x)_{(k)}} b \cdot x \, dS = b \cdot (\Delta X), \tag{6.135}$$

where $\Delta X = \Delta((x)_{(k)})$. The order and orientations in this result are important. The simplex $(x)_{(k)}$ is oriented, and the order of points specifies how the boundary is traversed. With dS the oriented element over each boundary, and ΔX the volume element for the simplex, we find that the correct expression for the surface integral is $b \cdot (\Delta X)$.

We are now in a position to apply these results to the interpolated function $f(x)$ of equation (6.111). Suppose that we are working in a (flat) n-dimensional space and consider a simplex with points $(x_0, \ldots, x_n)$. The simplex is chosen such that its volume is non-zero, so the n vectors $e_i = x_i - x_0$ define a (non-orthonormal) frame. We therefore write

$$\mathsf{e}_i = x_i - x_0, \tag{6.136}$$

and introduce the reciprocal frame $\{\mathsf{e}^i\}$. These vectors satisfy

$$\mathsf{e}^i \cdot (x - x_0) = \lambda^i. \tag{6.137}$$

It follows that the surface integral of $f(x)$ over the simplex is given by

$$\begin{aligned} \oint_{\partial(x)_{(k)}} f(x) dS &= \sum_{i=1}^{n} (F_i - F_0) \oint \mathsf{e}^i \cdot (x - x_0) dS \\ &= \sum_{i=1}^{n} (F_i - F_0) \mathsf{e}^i \cdot (\Delta X). \end{aligned} \tag{6.138}$$

But if we consider the directional derivatives of $f(x)$ we find that

$$\frac{\partial f(x)}{\partial \lambda^i} = F_i - F_0. \tag{6.139}$$

The result of the surface integral can therefore be written

$$\begin{aligned} \oint_{\partial(x)_{(k)}} f(x) dS &= \sum_{i=1}^{n} (F_i - F_0) \mathsf{e}^i \cdot (\Delta X) \\ &= \sum_{i=1}^{n} \frac{\partial f}{\partial \lambda^i} \mathsf{e}^i \cdot (\Delta X) = \dot{f} \dot{\nabla} \cdot (\Delta X). \end{aligned} \tag{6.140}$$

Here we have used the result that $\nabla = \mathsf{e}^i \partial_i$, which follows from using the λ^i as a set of coordinates.

We now consider a chain of simplices, and add the result of equation (6.140) over each simplex in the chain. The interpolated function $f(x)$ takes on the same value over the common boundary of two adjacent simplices, since $f(x)$ is only defined by the values at the common vertices. In forming a sum over a chain, all of the internal faces cancel and only the surface integral over the boundary remains. We therefore arrive at

$$\oint f(x)\, dS = \sum_a \dot{f}\dot{\nabla}\cdot(\Delta X^a), \tag{6.141}$$

with the sum running over all of the simplices in the chain. Taking the limit as more points are added and each simplex is shrunk in size we arrive at our first statement of the fundamental theorem,

$$\oint_{\partial V} F\, dS = \int_V \dot{F}\dot{\nabla}\, dX. \tag{6.142}$$

We have replaced the interpolated function f with F, which is obtained in the limit as more points are added. We have also used the fact that ∇ lies entirely within the space defined by the pseudoscalar measure dX to remove the contraction on the right-hand side and write a geometric product.

The above proof is easily adapted for the case where the function sits to the right of the measure, giving

$$\oint_{\partial V} dS\, G = \int_V \dot{\nabla}\, dX\, \dot{G}. \tag{6.143}$$

Since ∇ is a vector, the commutation properties with dX will depend on the dimension of the space. A yet more general statement of the fundamental theorem can be constructed by introducing a linear function $\mathsf{L}(A_{n-1}) = \mathsf{L}(A_{n-1}; x)$. This function takes a multivector A_{n-1} of grade $n-1$ as its linear argument, and returns a general multivector. L is also position-dependent, and its linear interpolation over a simplex is defined by

$$L(A) = \mathsf{L}(A; x_0) + \sum_{i=1}^{n} \lambda^i \big(\mathsf{L}(A; x_i) - \mathsf{L}(A, x_0)\big). \tag{6.144}$$

The linearity of $L(A)$ means that sums and integrals can be moved inside the argument, and we establish that

$$\begin{aligned}\oint L(dS) &= \mathsf{L}\Big(\oint dS; x_0\Big) + \sum_{i=1}^{n} \mathsf{L}\Big(\oint \lambda^i dS; x_i\Big) - \sum_{i=1}^{n} \mathsf{L}\Big(\oint \lambda^i dS; x_0\Big) \\ &= \sum_{i=1}^{n} \mathsf{L}(\mathsf{e}^i \Delta X; x_i) - \mathsf{L}(\mathsf{e}^i \Delta X; x_0) \\ &= \dot{L}(\dot{\nabla}\Delta X).\end{aligned} \tag{6.145}$$

There is no position dependence in the final term as the derivative is constant over the simplex. Building up a chain of simplices and taking the limit we prove the general result

$$\oint_{\partial V} \mathsf{L}(dS) = \int_V \dot{\mathsf{L}}(\dot{\nabla} dX). \tag{6.146}$$

This holds for any linear function $\mathsf{L}(A_{n-1})$ integrated over a closed region of an n-dimensional flat space. This is still not the most general statement of the fundamental theorem, as we will later prove a version valid for surfaces embedded in a curved space, but equation (6.146) is sufficient to make contact with the main integral theorems of vector calculus.

6.4.5 The divergence and Green's theorems

To see the fundamental theorem of geometric calculus in practice, first consider the scalar-valued function

$$\mathsf{L}(A) = \langle JAI^{-1} \rangle. \tag{6.147}$$

Here J is a vector, and I is the (constant) unit pseudoscalar for the n-dimensional space. The argument A is a multivector of grade $n-1$. Equation (6.146) gives

$$\int_V \langle \dot{J} \dot{\nabla} dX I^{-1} \rangle = \int_V \nabla \cdot J \, |dX| = \oint_{\partial V} \langle J dS I^{-1} \rangle, \tag{6.148}$$

where $|dX| = I^{-1} dX$ is the scalar measure over the volume of interest. The normal to the surface, n is defined by

$$n|dS| = dS\, I^{-1}, \tag{6.149}$$

where $|dS|$ is the scalar-valued measure over the surface. This definition ensures that, in Euclidean spaces, $n\, dS$ has the orientation defined by I, and in turn that n points outwards. With this definition we arrive at

$$\int_V \nabla \cdot J \, |dX| = \oint_{\partial V} n \cdot J \, |dS|, \tag{6.150}$$

which is the familiar divergence theorem. This way of writing the theorem hides the fact that $n|dS|$ should be viewed as a single entity, which can be important in spaces of mixed signature.

Now return to the fundamental theorem in the form of equation (6.143), and let G equal the vector $\boldsymbol{J}$ in two-dimensional Euclidean space. We find that

$$\oint_{\partial V} dS\, \boldsymbol{J} = \int_V \dot{\boldsymbol{\nabla}}\, dX\, \dot{\boldsymbol{J}} = -\int_V \boldsymbol{\nabla} \boldsymbol{J}\, dX, \tag{6.151}$$

where we have used the fact that dX is a pseudoscalar, so it anticommutes with vectors in two dimensions. Introducing Cartesian coordinates we have $dX = I dx\, dy$, so

$$\oint_{\partial V} dS\, \boldsymbol{J} = -\int_V \boldsymbol{\nabla} \boldsymbol{J} I\, dx\, dy. \tag{6.152}$$

If we let $\boldsymbol{J} = P\mathsf{e}_1 + Q\mathsf{e}_2$ and take the scalar part of both sides, we prove Green's theorem in the plane

$$\oint P dx + Q dy = \int \left(\frac{\partial Q}{\partial x} - \frac{\partial P}{\partial y} \right) dx\, dy. \tag{6.153}$$

The line integral is taken around the perimeter of the area in a positive sense, as specified by $I = \mathsf{e}_1\mathsf{e}_2$.

6.4.6 Cauchy's integral formula

The fundamental theorem of geometric calculus enables us to view the Cauchy integral theorem of complex variable theory in a new light. We let ψ denote an even-grade multivector, which therefore commutes with dX, so we can write

$$\int \boldsymbol{\nabla} \psi\, dX = \oint d\boldsymbol{s}\, \psi = \oint \frac{\partial \boldsymbol{r}}{\partial \lambda} \psi\, d\lambda. \tag{6.154}$$

In the final expression λ is a parameter along the (closed) curve. Now recall from section 6.3.1 that we form the complex number z by $z = \mathsf{e}_1 \boldsymbol{r}$. We therefore have

$$\oint \psi dz = \int \mathsf{e}_1 \boldsymbol{\nabla} \psi\, dX, \tag{6.155}$$

where the term on the left is now a complex line integral. The condition that ψ is analytic can be written $\boldsymbol{\nabla}\psi = 0$ so we have immediately proved that the line integral of an analytic function around a closed curve always vanishes.

Cauchy's integral formula states that, for an analytic function,

$$f(a) = \frac{1}{2\pi i} \oint_C \frac{f(z)}{z - a} dz, \tag{6.156}$$

where the contour C encloses the point a and is traversed in a positive sense. The precise form of the contour is irrelevant, because the difference between two contour integrals enclosing a is a contour integral around a region not enclosing a (see figure 6.4). In such a region $f(z)/(z - a)$ is analytic so the difference has zero contribution.

To understand Cauchy's theorem in terms of geometric calculus we need to focus on the properties of the Cauchy kernel $1/(z - a)$. We first write

$$\frac{1}{z - a} = \frac{(z - a)^\dagger}{|(z - a)|^2} = \frac{\boldsymbol{r} - \boldsymbol{a}}{(\boldsymbol{r} - \boldsymbol{a})^2} \mathsf{e}_1, \tag{6.157}$$

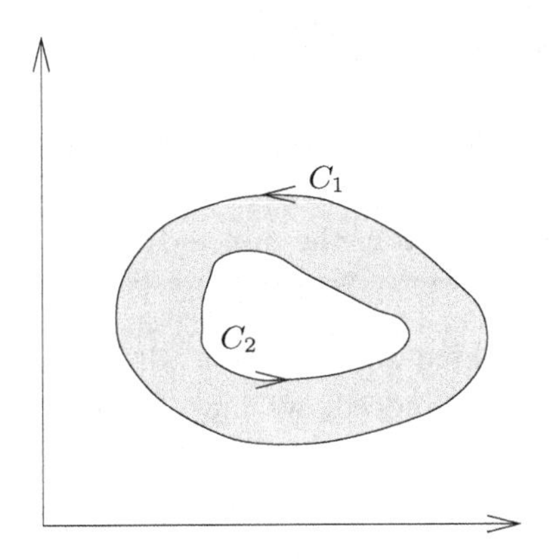

Figure 6.4 *Contour integrals in the complex plane.* The two contours C_1 and C_2 can be deformed into one another, provided the function to be integrated has no singularities in the intervening region. In this case the difference vanishes, by Cauchy's theorem.

where $\boldsymbol{a} = \mathsf{e}_1 a$ is the vector corresponding to the complex number a. The essential quantity here is the vector $(\boldsymbol{r}-\boldsymbol{a})/(\boldsymbol{r}-\boldsymbol{a})^2$, which we can write as

$$\frac{\boldsymbol{r}-\boldsymbol{a}}{(\boldsymbol{r}-\boldsymbol{a})^2} = \boldsymbol{\nabla} \ln |\boldsymbol{r}-\boldsymbol{a}|. \tag{6.158}$$

But $\ln |\boldsymbol{r}-\boldsymbol{a}|$ is the Green's function for the Laplacian operator in two dimensions,

$$\boldsymbol{\nabla}^2 \ln |\boldsymbol{r}-\boldsymbol{a}| = 2\pi\delta(\boldsymbol{r}-\boldsymbol{a}). \tag{6.159}$$

It follows that the vector part of the Cauchy kernel satisfies

$$\boldsymbol{\nabla} \frac{\boldsymbol{r}-\boldsymbol{a}}{(\boldsymbol{r}-\boldsymbol{a})^2} = 2\pi\delta(\boldsymbol{r}-\boldsymbol{a}). \tag{6.160}$$

The Cauchy kernel is the Green's function for the two-dimensional vector derivative! The existence of this Green's function proves that the vector derivative is invertible, which is not true of its separate divergence and curl components.

The Cauchy integral formula now follows from the fundamental theorem of geometric calculus in the form of equation (6.155),

$$\begin{aligned}\oint \frac{f(z)}{z-a} dz &= \mathsf{e}_1 \int \boldsymbol{\nabla} \left(\frac{\boldsymbol{r}-\boldsymbol{a}}{(\boldsymbol{r}-\boldsymbol{a})^2} \mathsf{e}_1 f(x) \right) dX \\ &= \mathsf{e}_1 \int \left(2\pi\delta(x-\boldsymbol{a}) \mathsf{e}_1 f(z) + \boldsymbol{\nabla} f(z) \frac{\boldsymbol{r}-\boldsymbol{a}}{(\boldsymbol{r}-\boldsymbol{a})^2} \mathsf{e}_1 \right) I|dX| \\ &= 2\pi I f(a), \end{aligned} \tag{6.161}$$

where we have assumed that f is analytic, $\boldsymbol{\nabla} f(z) = 0$. We can now understand precisely the roles of each term in the theorem:

(i) The dz encodes the tangent vector and forms a *geometric* product in the integrand.
(ii) The $(z-a)^{-1}$ is the Green's function for the vector derivative $\boldsymbol{\nabla}$ and ensures that the area integral only picks up the value at a.
(iii) The I (which replaces i) comes from the directed volume element $dX = I\,dx\,dy$.

Much of this is hidden in conventional accounts, but all of these insights are crucial to generalising the theorem. Indeed, we have already proved a more general theorem in two dimensions applying to non-analytic functions. For these we can now write, following section 6.3.1,

$$2\pi I f(a) = \oint \frac{f}{z-a} dz - 2\int \frac{\partial f}{\partial z^\dagger}\frac{1}{z-a} I|dX|. \tag{6.162}$$

A second key ingredient in complex analysis is the series expansion of a function. In particular, if $f(z)$ is analytic apart from a pole of order n at $z=a$, the function has a Laurent series of the form

$$f(z) = \frac{a_{-n}}{(z-a)^n} \cdots \frac{a_{-1}}{z-a} + \sum_{i=0}^{\infty} a_i (z-a)^i. \tag{6.163}$$

The powerful residue theorem states that for such a function

$$\oint_C f(z)\,dz = 2\pi i a_{-1}. \tag{6.164}$$

We now have a new interpretation for the residue term in a Laurent expansion — it is a weighted Green's function. The residue theorem just recovers the weight! Geometric calculus unifies the theory of poles and residues, supposedly unique to complex analysis, with that of Green's functions and δ-functions.

We now have an alternative picture of complex variable theory in terms of Green's functions and surface data. Suppose, for example, that we start with a function $f(x)$ on the real axis. We seek to propagate this function into the upper half-plane, subject to the boundary conditions that f falls to zero as $|z| \mapsto \infty$. The Cauchy formula tells us that we should propagate according to the formula

$$f(a) = \frac{1}{2\pi i}\int_{-\infty}^{\infty} \frac{f(x)}{x-a}\, dx. \tag{6.165}$$

But suppose now that we form the Fourier transform of the initial function $f(x)$,

$$f(x) = \int_{-\infty}^{\infty} \frac{dk}{2\pi} \bar{f}(k) \mathrm{e}^{ikx}. \tag{6.166}$$

We now have

$$f(a) = \frac{1}{2\pi i}\int_{-\infty}^{\infty} \frac{dk}{2\pi} \bar{f}(k) \int_{-\infty}^{\infty} \frac{\mathrm{e}^{ikx}}{x-a}\, dx. \tag{6.167}$$

Now we only close the x integral in the upper half-plane for positive k. For negative k there is no residue term, since a lies in the the upper half-plane. The Cauchy integral formula now returns

$$f(a) = \int_0^\infty \frac{dk}{2\pi} \bar{f}(k) e^{ika}. \tag{6.168}$$

This shows that only the part of the function consistent with the desired boundary conditions is propagated in the positive y direction. The remaining part of the function propagates in the $-y$ direction, if similar boundary conditions are imposed in the lower half plane. In this way the boundary conditions and the Green's function between them specify precisely which parts of a function are propagated in the desired direction. No restrictions are placed on the boundary values $f(x)$, which need not be part of an analytic function.

A second example, which generalises nicely, is the unit circle. Suppose we have initial data $f(\theta)$ defined over the unit circle. We write $f(\theta)$ as

$$f(\theta) = \sum_{-\infty}^{\infty} f_n e^{in\theta}. \tag{6.169}$$

The terms in $\exp(in\theta)$ are replaced by z^n over the unit circle, and we then choose whether to evaluate in interior or exterior closure of the Cauchy integral. The result is that only the negative powers are propagated outwards from the circle, resulting in the function

$$f(z) = \sum_{n=1}^{\infty} f_{-n} z^{-n}, \qquad |z| > 1. \tag{6.170}$$

(The constant component f_0 is technically propagated as well, but this can be removed trivially.) These observations are simple from the point of view of complex variable theory, but are considerably less obvious in propagator theory.

6.4.7 Green's functions in Euclidean spaces

The extension of complex variable theory to arbitrary Euclidean spaces is now straightforward. The analogue of an analytic function is a multivector ψ satisfying $\nabla\psi = 0$. We choose to work with even-grade multivectors to simplify matters. The fundamental theorem states that

$$\oint_{\partial V} dS\, \psi = \int \nabla\psi\, dX = 0. \tag{6.171}$$

where we have used the fact that ψ commutes with the pseudoscalar measure dX. For any *monogenic* function ψ, the directed integral of ψ over a closed surface must vanish.

The Green's function for the vector derivative in n dimensions is simply

$$G(x;y) = \frac{1}{S_n}\frac{x-y}{|x-y|^n}, \tag{6.172}$$

where x and y are vectors and S_n is the surface area of the unit ball in n-dimensional space. The Green's function satisfies

$$\nabla G(x;y) = \nabla\cdot G(x;y) = \delta(x-y). \tag{6.173}$$

In order to allow for the lack of commutativity between G and ψ we use the fundamental theorem in the form

$$\begin{aligned}\oint_{\partial V} G\,dS\,\psi &= \int_V (\dot{G}\dot{\nabla}\,\psi + G\,\nabla\psi)\,dX \\ &= \int_V \dot{G}\dot{\nabla}\psi\,dX,\end{aligned} \tag{6.174}$$

where we have used the fact that ψ is a monogenic function. Setting G equal to the Green's function of equation (6.172) we find that Cauchy's theorem in n dimensions can be written in the form

$$\psi(y) = \frac{1}{IS_n}\oint_{\partial V}\frac{x-y}{|x-y|^n}\,dS\,\psi(x). \tag{6.175}$$

This relates the value of a monogenic function at a point to the value of a surface integral over a region surrounding the point.

One consequence of equation (6.175) is that a generalisation of Liouville's theorem applies to monogenic functions in Euclidean spaces. We define the modulus function

$$|M| = \langle MM^{\dagger}\rangle^{1/2}, \tag{6.176}$$

which is a well-defined positive-definite function for all multivectors M in a Euclidean algebra. The modulus function is easily shown to satisfy Schwarz inequality in the form

$$|A+B| \leq |A| + |B|. \tag{6.177}$$

If we let a denote a unit vector and let ∇_y denote the derivative with respect to the vector y we find that

$$a\cdot\nabla_y\psi(y) = -\frac{1}{IS_n}\oint_{\partial V}\frac{a(x-y)^2 + na\cdot(x-y)\,(x-y)}{|x-y|^{n+2}}\,dS\,\psi(x). \tag{6.178}$$

It follows that

$$|a\cdot\nabla_y\psi(y)| \leq \frac{1}{S_n}\oint_{\partial V}\frac{n+1}{|x-y|^n}|dS|\,|\psi(x)|. \tag{6.179}$$

But if ψ is bounded, $|\psi(x)|$ never exceeds some given value. Taking the surface of integration out to large radius $r = |x|$, we find that the right-hand side falls off as $1/r$. This is sufficient to prove that the directional derivative of ψ must vanish in all directions, and the only monogenic function that is bounded over all space is constant ψ.

Equation (6.175) enables us to propagate a function off an initial surface in Euclidean space, subject to suitable boundary conditions. Suppose, for example, that we wish to propagate ψ off the surface of the unit ball, subject to the condition that the function falls to zero at large distance. Much like the two-dimensional case, we can write

$$\psi = \sum_{l=-\infty}^{\infty} \alpha_l \psi_l, \tag{6.180}$$

where the ψ_l are angular monogenics, satisfying

$$x \wedge \nabla \psi = -l\psi. \tag{6.181}$$

Each angular monogenic is multiplied by r^l to yield a full monogenic function, and only the negative powers have their integral closed over the exterior region. The result is the function

$$\psi = \sum_{l=1}^{\infty} \alpha_{-l} r^{-l} \psi_{-l}, \quad r > 1. \tag{6.182}$$

Similarly, the positive powers are picked up if we solve the interior problem.

6.4.8 Spacetime propagators

Propagation in mixed signature spaces is somewhat different to the Euclidean case. There is no analogue of Liouville's theorem to call on, so one can easily construct bounded solutions to the monogenic equation which are non-singular over all space. Plane wave solutions to the massless Dirac equation are an example of such functions. Furthermore, the existence of characteristic surfaces has implications for the how boundary values are specified. To see this, consider a two-dimensional Lorentzian space with basis vectors $\{\gamma_0, \gamma_1\}$, $\gamma_0^2 = -\gamma_1^2 = 1$, and pseudoscalar $I = \gamma_1\gamma_0$. The monogenic equation is $\nabla\psi = 0$, where ψ is an even-grade multivector built from a scalar and pseudoscalar terms. We define the null vectors

$$n_{\pm} = \gamma_0 \pm \gamma_1. \tag{6.183}$$

Pre-multiplying the monogenic equation by n_+ we find that

$$n_+ \cdot \nabla \psi = -n_+ \wedge \nabla \psi = I\,(n_+ I) \cdot \nabla \psi = -I n_+ \cdot \nabla \psi. \tag{6.184}$$

where we have used the result that $In_+ = n_+$. It follows that

$$(1 + I) n_+ \cdot \nabla \psi = 0, \tag{6.185}$$

and similarly,

$$(1-I)n_- \cdot \nabla \psi = 0. \tag{6.186}$$

If we take ψ and decompose it into $\psi = \psi_+ + \psi_-$,

$$\psi_\pm = \tfrac{1}{2}(1 \pm I)\psi, \tag{6.187}$$

we see that the values of the separate $\psi_\pm$ components have vanishing derivatives along the respective null vectors $n_\pm$. Propagation of ψ from an initial surface is therefore quite straightforward. The function is split into $\psi_\pm$, and the values of these are transported along the respective null vectors. That is, ψ_+ has the same value along each vector in the n_+ direction, and the same for ψ_-. There is no need for a complicated contour integral.

The fact that the values of ψ are carried along the characteristics illustrates a key point. Any surface on which initial values are specified can cut a characteristic surface only once. Otherwise the initial values are unlikely to be consistent with the differential equation. For the monogenic equation, $\nabla\psi = 0$, suitable initial conditions consist of specifying ψ along the γ_1 axis, for example. But the fundamental theorem involves integrals around closed loops. The theorem is still valid in a Lorentzian space, so it is interesting to see what happens to the boundary data if we attempt to construct an interior solution with arbitrary surface data. The first step is to construct the Lorentzian Green's function. This can be found routinely via its Fourier transformation. With $x = x^0\gamma_0 + x^1\gamma_1$ we find

$$\begin{aligned} G(x) &= i \int \frac{d\omega}{2\pi}\frac{dk}{2\pi}\,\frac{\omega\gamma_0 + k\gamma_1}{\omega^2 - k^2}\,\mathrm{e}^{i(kx^1 - \omega x^0)} \\ &= \frac{i}{2}\int \frac{d\omega}{2\pi}\frac{dk}{2\pi}\left(\frac{\gamma_0+\gamma_1}{\omega - k} + \frac{\gamma_0 - \gamma_1}{\omega + k}\right)\mathrm{e}^{i(kx^1 - \omega x^0)} \\ &= \frac{\epsilon(x^0)}{4}\big(\delta(x^1 - x^0)(\gamma_0+\gamma_1) + \delta(x^1 + x^0)(\gamma_0 - \gamma_1)\big). \end{aligned} \tag{6.188}$$

The function $\epsilon(x^0)$ takes the value $+1$ or -1, depending on whether x^0 is positive or negative respectively.

To apply the fundamental theorem, suppose we take the contour of figure 6.5, which runs along the γ_1 axis for two different times $t_i < t_f$ and is closed at spatial infinity. We assume that the function we are propagating, ψ, falls off at large spatial distance, and write $\psi(x)$ as $\psi(x^0, x^1)$. The fundamental theorem

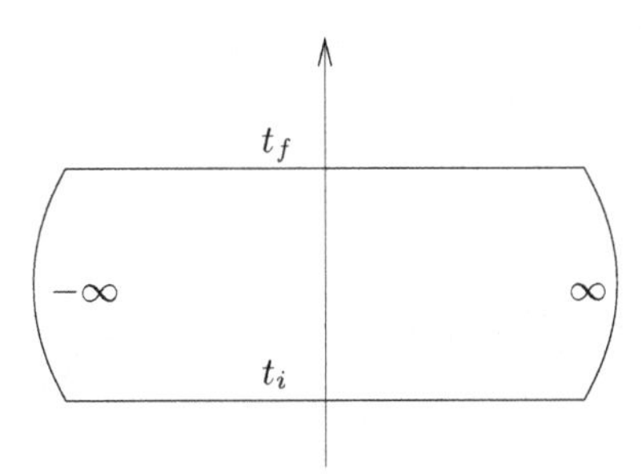

Figure 6.5 *A spacetime contour.* The contour is closed at spatial infinity.

then gives

$$\begin{aligned}\psi(y) =& I \int_{-\infty}^{\infty} d\lambda\, G(t_i\gamma_0 + \lambda\gamma_1 - y)\gamma_1\psi(t_i,\lambda) \\ &- I \int_{-\infty}^{\infty} d\lambda\, G(t_f\gamma_0 + \lambda\gamma_1 - y)\gamma_1\psi(t_f,\lambda) \\ =& \frac{1}{4}(1+I)\big(\psi(t_i, y^1 - y^0 + t_i) + \psi(t_f, y^1 - y^0 + t_f)\big) \\ &- \frac{1}{4}(1-I)\big(\psi(t_i, -y^1 + y^0 + t_i) + \psi(t_f, -y^1 + y^0 + t_f)\big). \end{aligned} \tag{6.189}$$

The construction of $\psi(y)$ in the interior region has a simple interpretation. For the function $\psi_+(y)$, for example, we form the null vector n_+ through y. The value at y is then the average value at the two intersections with the boundary. A similar construction holds for ψ_-. Much like the Euclidean case, only the part of the function on the boundary that is consistent with the monogenic equation is propagated to the interior.

These insights hold in other Lorentzian spaces, such as four-dimensional spacetime. The Green's functions become more complicated, and typically involve derivatives of δ-functions. These are more usefully handled via their Fourier transforms, and are discussed in more detail in section 8.5. In addition, the lack of a Liouville's theorem means that any monogenic function can be added to a Green's function to generate a new Green's function. This has no consequences if one rigorously applies surface integral formulae. In quantum theory, however, this is not usually the case. Rather than a rigorous application of the generalised Green's theorem, it is common instead to talk about propagators which transfer initial data from one timeslice to a later one. Used in this role, the Green's functions we have derived are referred to as *propagators*. As we are not specifying data over a closed surface, adding further terms to our Green's function can have an effect. These effects are related to the desired boundary conditions and are crucial to the formulation of a relativistic quantum field theory. There one is led to employ the complex-valued Feynman propagator, which ensures that positive frequency modes are propagated forwards in time, and negative frequency modes are propagated backwards in time. We will meet this object in greater detail in section 8.5.

6.5 Embedded surfaces and vector manifolds

We now seek a generalisation of the preceding results where the volume integral is taken over a curved surface. We will do this in the setting of the *vector manifold* theory developed by Hestenes and Sobczyk (1984). The essential concept is to treat a manifold as a surface embedded in a larger, flat space. Points in the manifold are then treated as vectors, which simplifies a number of derivations. Furthermore, we can exploit the coordinate freedom of geometric algebra to derive a set of general results without ever needing to specify the dimension of the background space. The price we pay for this approach is that we are working with a more restrictive concept of a manifold than is usually the case in mathematics. For a start, the surface naturally inherits a metric from the embedding space, so we are already restricting to Riemannian manifolds. We will also insist that a pseudoscalar can be uniquely defined throughout the surface, making it orientable.

While this may all appear quite restrictive, in fact these criteria rule out hardly any structures of interest in physics. This approach enables us to quickly prove a number of key results in Riemannian geometry, and to unite these with results for the exterior geometry of the manifold, achieving a richer general theory. We are not prevented from discussing topological features of surfaces either. Rather than build up a theory of topology which makes no reference to the metric, we instead build up results that are unaffected if the embedding is (smoothly) transformed.

We define a vector manifold as a set of points labelled by vectors lying in a geometric algebra of arbitrary dimension and signature. If we consider a path in the surface $x(\lambda)$, the tangent vector is defined in the obvious way by

$$x' = \left.\frac{\partial x(\lambda)}{\partial \lambda}\right|_{\lambda_0} = \lim_{\epsilon \mapsto 0} \frac{x(\lambda_0 + \epsilon) - x(\lambda_0)}{\epsilon}. \tag{6.190}$$

An advantage of the embedding picture is that the meaning of the limit is well defined, since the numerator exists for all ϵ. This is true even if, for finite epsilon, the difference vector does not lie entirely in the tangent space and only becomes a tangent vector in the limit. Standard formulations of differential geometry avoid any mention of an embedding, however, so have to resort to a more abstract definition of a tangent vector.

An immediate consequence of this approach is that we can define the path length as

$$s = \int_{\lambda_1}^{\lambda_2} |x' \cdot x'|^{1/2} \, d\lambda. \tag{6.191}$$

The embedded surface therefore inherits a metric from the 'ambient' background space. All finite-dimensional Riemannian manifolds can be studied in this way since, given a manifold, a natural embedding in a larger flat space can always be found. In applications such as general relativity one is usually not interested in the properties of the embedding, since they are physically unmeasurable. But in many other applications, particularly those involving constrained systems, the embedding arises naturally and useful information is contained in the extrinsic geometry of a manifold.

6.5.1 The pseudoscalar and projection

Suppose that we next introduce a set of paths in the surface all passing through the same point x. The paths define a set of tangent vectors $\{\mathsf{e}_1, \ldots, \mathsf{e}_n\}$. We assume that these are independent, so that they form a basis for the n-dimensional tangent space at the point x. The exterior product of the tangent vectors defines the pseudoscalar for the tangent space $I(x)$:

$$I(x) \equiv \mathsf{e}_1 \wedge \mathsf{e}_2 \wedge \cdots \wedge \mathsf{e}_n / |\mathsf{e}_1 \wedge \mathsf{e}_2 \wedge \cdots \wedge \mathsf{e}_n|. \tag{6.192}$$

The modulus in the denominator is taken as a positive number, so that I has the orientation specified by the tangent vectors. The pseudoscalar will satisfy

$$I^2 = \pm 1, \tag{6.193}$$

with the sign depending on dimension and signature. Clearly, to define I in this manner requires that the denominator in (6.192) is non-zero. This provides a restriction on the vector manifolds we consider here, and rules out certain structures in mixed signature spaces. The unit circle in the Lorentzian plane (figure 6.1), for example, falls outside the class of surfaces of studied here, as the tangent space has vanishing norm where the tangent vectors become null. Of course, there is no problem in referring to a closed spacetime curve as a vector manifold. The problem arises when attempting to generalise the integral theorems of the previous sections to such spaces.

The pseudoscalar $I(x)$ contains all of the geometric information about the surface and unites both its intrinsic and extrinsic properties. As well as assuming that $I(x)$ can be defined globally, we will also assume that $I(x)$ is continuous and differentiable over the entire surface, that it has the same grade everywhere, and that it is single-valued. The final assumption implies that the manifold is *orientable*, and rules out objects such as the Mobius strip, where the pseudoscalar is double-valued. Many of the restrictions on the pseudoscalar mentioned above can be relaxed to construct a more general theory, but this is only achieved at some cost to the ease of presentation. We will follow the simpler route, as the results developed here are sufficiently general for our purposes in later chapters.

The pseudoscalar $I(x)$ defines an operator which projects from an arbitrary multivector onto the component that is intrinsic to the manifold. This operator is

$$\mathsf{P}(A_r(x), x) = \begin{cases} A_r(x)\cdot I(x)\, I^{-1}(x) = A_r\cdot I\, I^{-1}, & r \le n \\ 0 & r > n \end{cases}. \tag{6.194}$$

which defines an operator at every point x on the manifold. It is straightforward to prove that P satisfies the essential requirement of a projection operator, that is,

$$\mathsf{P}^2(A) = \mathsf{P}\big(\mathsf{P}(A)\big) = \mathsf{P}(A). \tag{6.195}$$

The effect of P on a vector a is to project onto the component of a that lies entirely in the tangent space at the point x. Such vectors are said to be *intrinsic* to the manifold. The complement,

$$\mathsf{P}_\perp(a) = a - \mathsf{P}(a), \tag{6.196}$$

lies entirely outside the tangent space, and is said to be extrinsic to the manifold.

Suppose now that $A(x)$ is a multivector field defined over some region of the manifold. We do not assume that A is intrinsic to the manifold. Given a vector a in the tangent space, the directional derivative along a is defined in the obvious manner:

$$a\cdot\nabla A(x) = \lim_{\epsilon\mapsto 0} \frac{A(x+\epsilon a) - A(x)}{\epsilon}. \tag{6.197}$$

Again, the presence of the embedding enables us to write this limit without ambiguity. The derivative operator $a\cdot\nabla$ is therefore simply the vector derivative in the ambient space contracted with a vector in the tangent space. Given a set of linearly independent tangent vectors $\{\mathsf{e}_i\}$, we can now define a vector derivative ∂ intrinsic to the manifold by

$$\partial = \mathsf{e}^i\, \mathsf{e}_i\cdot\nabla = \mathsf{P}(\nabla). \tag{6.198}$$

This is simply the ambient space vector derivative projected onto the tangent space. The use of the ∂ symbol should not cause confusion with the boundary operator introduced in section 6.4.4. The definition of ∂ requires the existence of the reciprocal frame $\{\mathsf{e}^i\}$, which is why we restricted to manifolds over which I is globally defined. The projection of the vector operator ∂ satisfies

$$\mathsf{P}(\partial) = \partial. \tag{6.199}$$

The contraction of ∂ with a tangent vector a satisfies $a\cdot\partial = a\cdot\nabla$, which is simply the directional derivative in the a direction.

6.5.2 Directed integration for embedded surfaces

Now that we have defined the ∂ operator it is a straightforward task to write down a generalized version of the fundamental theorem of calculus appropriate for embedded surfaces. We can essentially follow through the derivation of section 6.4.4 with little modification. The volume to be integrated over is again triangulated into a chain of simplices. The only difference now is that the pseudoscalar for each simplex varies from one simplex to another. This changes very little. For example we still have

$$\oint dS = 0, \tag{6.200}$$

which holds for the directed integral over the closed boundary of any simply-connected vector manifold.

The linear interpolation results used in deriving equation (6.138) are all valid, because we can again fall back on the embedding picture. In addition, the assumption that the pseudoscalar $I(x)$ is globally defined means that the reciprocal frame required in equation (6.138) is well defined. The only change that has to be made is that the ambient derivative ∇ is replaced by its projection into the manifold, because we naturally assemble the inner product of ∇ with the pseudoscalar. The most general statement of the fundamental theorem can now be written as

$$\oint_{\partial V} \mathsf{L}(dS) = \int_V \dot{\mathsf{L}}(\dot{\partial} dX) = \int_V \dot{\mathsf{L}}(\dot{\nabla}\cdot dX). \tag{6.201}$$

The form of the volume integral involving ∂ is generally more useful as it forms a geometric product with the volume element. The function L can be any multivector-valued function in this equation — it is not restricted to lie in the tangent space. An important feature of this more general theorem is that if we write $dX = I|dX|$ we see that the directed element dX is position-dependent. But this position dependence is *not* differentiated in equation (6.201). It is only the integrand that is differentiated.

There are two main applications of the general theorem derived here. The first is a generalisation of the divergence theorem to curved spaces. We again write

$$\mathsf{L}(A) = \langle JAI^{-1}\rangle, \tag{6.202}$$

where J is a vector field in the tangent space, and I is the unit pseudoscalar for the n-dimensional curved space. Equation (6.201) now gives

$$\oint_{\partial V} n\cdot J|dS| = \int_V (\partial\cdot J + \langle J\dot{\partial}\dot{I}^{-1}I\rangle)|dX|, \tag{6.203}$$

where $|dX| = I^{-1}dX$ and $n|dS| = dS\,I^{-1}$. The final term in the integral vanishes, as can be shown by first writing $I^{-1} = \pm I$ and using

$$\langle J\dot{\partial}\dot{I}I\rangle = \tfrac{1}{2}\langle J\dot{\partial}(\dot{I}I + I\dot{I})\rangle = \tfrac{1}{2}\langle J\partial(I^2)\rangle = 0. \tag{6.204}$$

It follows that the divergence theorem in curved space is essentially unchanged from the flat-space version, so

$$\int_V \partial\cdot J\,|dX| = \oint_{\partial V} n\cdot J\,|dS|. \tag{6.205}$$

As a second application we derive Stokes' theorem in three dimensions. Suppose that σ denotes an open, connected surface in three dimensions, with boundary $\partial\sigma$. The linear function L takes a vector as its linear argument and we define

$$\mathsf{L}(\boldsymbol{a}) = \boldsymbol{J}\cdot\boldsymbol{a}. \tag{6.206}$$

Equation (6.201) now gives

$$\oint_{\partial\sigma} \boldsymbol{J}\cdot d\boldsymbol{l} = \int_\sigma \langle \dot{\boldsymbol{J}}\,\dot{\boldsymbol{\nabla}}\cdot dX\rangle = -\int_\sigma (\boldsymbol{\nabla}\wedge\boldsymbol{J})\cdot dX, \tag{6.207}$$

where the line integral is taken around the boundary of the surface, and since the embedding is specified we have chosen a form of the integral theorem involving the three-dimensional derivative $\boldsymbol{\nabla}$. We now define the normal vector to the surface by

$$dX = I\boldsymbol{n}|dX|, \tag{6.208}$$

where I is the three-dimensional (right-handed) pseudoscalar. This equation defines the vector $\boldsymbol{n}$ normal to the surface. The direction in which this points depends on the orientation of dX. Around the boundary, for example, we can denote the tangent vector at the boundary by $\boldsymbol{l}$, and the vector pointing into the surface as $\boldsymbol{m}$. Then dX has the orientation specified by $\boldsymbol{l}\wedge\boldsymbol{m}$, and from equation (6.208) we see that $\boldsymbol{l}, \boldsymbol{m}, \boldsymbol{n}$ must form a right-handed set. This extends inwards to define the normal vector $\boldsymbol{n}$ over the surface (see figure 6.6). We now have

$$\oint_{\partial\sigma} \boldsymbol{J}\cdot d\boldsymbol{l} = \int_\sigma -(I\boldsymbol{\nabla}\wedge\boldsymbol{J})\cdot\boldsymbol{n}\,|dX| = \int_\sigma (\mathrm{curl}\,\boldsymbol{J})\cdot\boldsymbol{n}\,|dX|, \tag{6.209}$$

which is the familiar Stokes' theorem in three dimensions. This is only the scalar part of a more general (and less familiar) theorem which holds in three dimensions. To form this result we remove the projection onto the scalar part, to obtain

$$\oint_{\partial\sigma} d\boldsymbol{l}\,\boldsymbol{J} = -I\int_\sigma \boldsymbol{n}\wedge\boldsymbol{\nabla}\,\boldsymbol{J}\,|dX|. \tag{6.210}$$

A version of this result holds for any open n-dimensional surface embedded in a flat space of dimension $n+1$.

6.5.3 Intrinsic and extrinsic geometry

Suppose now that the directional derivative $a\cdot\partial$ acts on a tangent vector field $b(x) = \mathsf{P}(b(x))$. There is no guarantee that the resulting vector also lies entirely

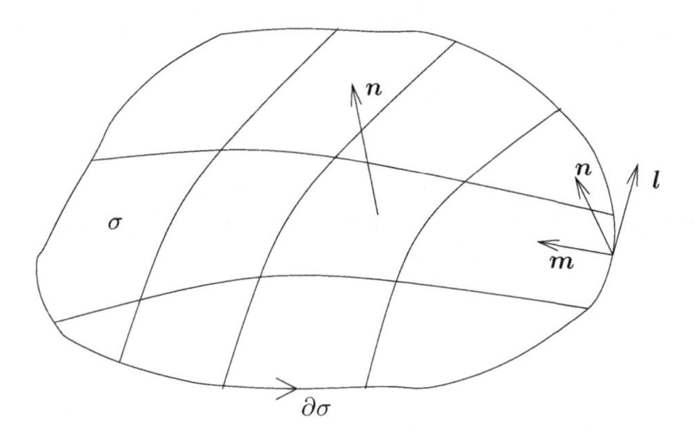

Figure 6.6 *Orientations for Stokes' theorem.* The bivector measure dX defines an orientation over the surface and at the boundary. With $\boldsymbol{l}$ and $\boldsymbol{m}$ the tangent and inward directions at the boundary, the normal $\boldsymbol{n}$ is defined so that $\boldsymbol{l}, \boldsymbol{m}, \boldsymbol{n}$ form a right-handed set.

in the tangent space, even if a does. For example, consider the simple case of a circle in the plane. The derivative of the tangent vector around the circle is a radial vector, which is entirely extrinsic to the manifold. In order to restrict to quantities intrinsic to the manifold we define a new derivative — the covariant derivative D — as follows:

$$a\cdot DA(x) = \mathsf{P}(a\cdot\partial A(x)). \tag{6.211}$$

The operator $a\cdot D$ acts on multivectors in the tangent space, returning a new multivector field in the tangent space. Since the $a\cdot\partial$ operator satisfies Leibniz's rule, the covariant derivative $a\cdot D$ must as well,

$$a\cdot D(AB) = \mathsf{P}\big(a\cdot\partial(AB)\big) = (a\cdot DA)B + A\,a\cdot DB. \tag{6.212}$$

The vector operator D is then defined in the obvious way from the covariant directional derivatives,

$$D = \mathsf{e}^i\,\mathsf{e}_i\cdot D. \tag{6.213}$$

So, for example, we can write

$$DA_r = \mathsf{e}^i(\mathsf{e}_i\cdot DA_r) = \mathsf{P}(\partial A_r). \tag{6.214}$$

The result decomposes into grade-raising and grade-lowering terms, so we write

$$\begin{aligned} D\cdot A_r &= \langle DA_r\rangle_{r-1}, \\ D\wedge A_r &= \langle DA_r\rangle_{r+1}. \end{aligned} \tag{6.215}$$

So, like ∂, D has the algebraic properties of a vector in the tangent space. Acting on a scalar function $\alpha(x)$ defined over the manifold the two derivatives coincide, so

$$\partial\alpha(x) = D\alpha(x). \tag{6.216}$$

Suppose now that a is a tangent vector to the manifold, and we look at how the pseudoscalar changes along the a direction. It should be obvious, from considering a 2-sphere for example, that the resulting quantity must lie at least partly outside the manifold. We let $\{\mathsf{e}_i\}$ denote an orthonormal frame, so

$$I = \mathsf{e}_1\mathsf{e}_2\cdots\mathsf{e}_n. \tag{6.217}$$

It follows that

$$\begin{aligned} a\cdot\partial\, I\, I^{-1} &= \sum_{i=1}^{n} \mathsf{e}_1 \cdots \big(a\cdot D\mathsf{e}_i + \mathsf{P}_\perp(a\cdot\partial\,\mathsf{e}_i)\big)\cdots\mathsf{e}_n\, I^{-1} \\ &= a\cdot D\, I\, I^{-1} + \mathsf{P}_\perp(a\cdot\partial\,\mathsf{e}_i)\wedge\mathsf{e}^i. \end{aligned} \tag{6.218}$$

The final term is easily shown to be independent of the choice of frame. But $a\cdot DI$ must remain in the tangent space, so it can only be a multiple of the pseudoscalar I. It follows that

$$(a\cdot D\, I)I = \langle (a\cdot D\, I)I\rangle = \tfrac{1}{2}\langle a\cdot D(I^2)\rangle = 0, \tag{6.219}$$

so

$$a\cdot D\, I = 0. \tag{6.220}$$

That is, the (unit) pseudoscalar is a covariant constant over the manifold. Equation (6.218) now simplifies to give

$$a\cdot\partial\, I = \mathsf{P}_\perp(a\cdot\partial\,\mathsf{e}_i)\wedge\mathsf{e}^i\, I = -S(a)I, \tag{6.221}$$

which defines the *shape tensor* $S(a)$. This is a bivector-valued, linear function of its vector argument a, where a is a tangent vector. Since the result of $a\cdot\partial\, I$ has the same grade as I, we can write

$$a\cdot\partial I = I\times S(a) \tag{6.222}$$

with

$$S(a)\cdot I = S(a)\wedge I = 0. \tag{6.223}$$

The fact that $S(a)\cdot I = 0$ confirms that $S(a)$ lies partly outside the manifold, so that $\mathsf{P}(S(a)) = 0$.

The shape tensor $S(a)$ unites the intrinsic and extrinsic geometry of the manifold in a single quantity. It can be thought of as the 'angular momentum' of $I(x)$ as it slides over the manifold. The shape tensor provides a compact relation between directional and covariant derivatives. We first form

$$b\cdot S(a) = b^i\mathsf{P}_\perp(a\cdot\partial\,\mathsf{e}_i) = \mathsf{P}_\perp(a\cdot\partial\, b), \tag{6.224}$$

where a and b are tangent vectors. It follows that

$$a\cdot\partial\, b = \mathsf{P}(a\cdot\partial\, b) + \mathsf{P}_\perp(a\cdot\partial\, b) = a\cdot D\, b + b\cdot S(a), \tag{6.225}$$

which we can rearrange to give the neat result

$$a\cdot D\, b = a\cdot\partial\, b + S(a)\cdot b. \tag{6.226}$$

Applying this result to the geometric product bc we find that

$$\begin{aligned} a\cdot D(bc) &= (a\cdot\partial\, b)c + S(a)\cdot b\, c + b(a\cdot\partial\, c) + b\, S(a)\cdot c \\ &= a\cdot\partial(bc) + S(a)\times(bc), \end{aligned} \tag{6.227}$$

where $\times$ is the commutator product, $A\times B = (AB - BA)/2$. It follows that for any multivector field A taking its values in the tangent space we have

$$a\cdot DA = a\cdot\partial A + S(a)\times A. \tag{6.228}$$

The fact that $S(a)$ is bivector-valued ensures that $S(a)\times A$ does not alter the grade of A. As a check, setting $A = I$ recovers equation (6.222). If we now write

$$a\cdot\partial\, b = a\cdot\partial\, \mathsf{P}(b) = a\cdot\dot{\partial}\,\dot{\mathsf{P}}(b) + \mathsf{P}(a\cdot\partial b) = a\cdot\dot{\partial}\,\dot{\mathsf{P}}(b) + a\cdot Db \tag{6.229}$$

we establish the further relation

$$a\cdot\dot{\partial}\,\dot{\mathsf{P}}(b) = b\cdot S(a). \tag{6.230}$$

This holds for any pair of tangent vectors a and b.

6.5.4 Coordinates and derivatives

A number of important results can be derived most simply by introducing a coordinate frame. In a region of the manifold we introduce local coordinates x^i and define the frame vectors

$$\mathsf{e}_i = \frac{\partial x}{\partial x^i}. \tag{6.231}$$

From the definition of ∂ it follows that $\mathsf{e}^i = \partial x^i$. The $\{\mathsf{e}_i\}$ are usually referred to as *tangent* vectors and the reciprocal frame $\{\mathsf{e}^i\}$ as *cotangent* vectors (or 1-forms). The fact that the space is curved implies that it may not be possible to construct a global coordinate system. The 2-sphere is the simplest example of this. In this case we simply patch together a series of local coordinate systems. The covariant derivative along a coordinate vector, $\mathsf{e}_i\cdot D$, satisfies

$$\mathsf{e}_i\cdot DA = D_i A = \mathsf{e}_i\cdot\partial A + S(\mathsf{e}_i)\times A = \partial_i A + S_i\times A, \tag{6.232}$$

which defines the D_i and S_i symbols.

The tangent frame vectors satisfy

$$\partial_i \mathsf{e}_j - \partial_j \mathsf{e}_i = (\partial_i\partial_j - \partial_j\partial_i)x = 0. \tag{6.233}$$

Projecting this result into the manifold establishes that

$$D_i \mathsf{e}_j - D_j \mathsf{e}_i = 0. \tag{6.234}$$

Projecting out of the manifold we similarly establish the result

$$\mathsf{e}_i \cdot S_j = \mathsf{e}_j \cdot S_i. \tag{6.235}$$

In terms of arbitrary tangent vectors a and b this can be written as

$$a \cdot S(b) = b \cdot S(a). \tag{6.236}$$

The shape tensor can be written in terms of the coordinate vectors as

$$S(a) = \mathsf{e}^k \wedge \mathsf{P}_\perp(a \cdot \partial \mathsf{e}_k). \tag{6.237}$$

It follows that

$$S_i = \mathsf{e}^k \wedge \mathsf{P}_\perp(\partial_i \mathsf{e}_k) = \mathsf{e}^k \wedge \mathsf{P}_\perp(\partial_k \mathsf{e}_i). \tag{6.238}$$

The tangent vectors therefore satisfy

$$\partial \wedge \mathsf{e}_i = \mathsf{e}^k \wedge \big(\mathsf{P}(\partial_k \mathsf{e}_i) + \mathsf{P}_\perp(\partial_k \mathsf{e}_i)\big) = D \wedge \mathsf{e}_i + S_i. \tag{6.239}$$

If we decompose a vector in the tangent space as $a = a^i \mathsf{e}_i$ we establish the general result that

$$\partial \wedge a = D \wedge a + S(a). \tag{6.240}$$

This gives a further interpretation to the shape tensor. It is the object which picks up the component of the curl of a tangent vector which lies outside the tangent space. As we can write

$$\partial \wedge a = \partial \wedge \big(\mathsf{P}(a)\big) = \dot{\partial} \wedge \dot{\mathsf{P}}(a) + \mathsf{P}(\partial \wedge a) = D \wedge a + \dot{\partial} \wedge \dot{\mathsf{P}}(a), \tag{6.241}$$

we establish the further result

$$\dot{\partial} \wedge \dot{\mathsf{P}}(a) = S(a). \tag{6.242}$$

This is easily seen to be consistent with the definition of the shape tensor in terms of the derivative of pseudoscalar.

If we now apply the preceding to the case of the curl of a gradient of a scalar, we find that

$$\partial \wedge \partial \phi = \mathsf{P}(\nabla) \wedge \mathsf{P}(\nabla \phi) = \mathsf{P}(\nabla \wedge \nabla \phi) + \dot{\partial} \wedge \dot{\mathsf{P}}(\nabla \phi). \tag{6.243}$$

But the ambient derivative satisfies the integrability condition $\nabla \wedge \nabla = 0$. It follows that we have

$$\partial \wedge \partial \phi = S(\nabla \phi), \tag{6.244}$$

which lies outside the manifold. The covariant derivative therefore satisfies

$$D \wedge (D\phi) = 0. \tag{6.245}$$

An important application of this result is to the coordinate scalars themselves. We find that

$$D\wedge(Dx^i) = D\wedge \mathsf{e}^i = 0, \tag{6.246}$$

which can also be proved directly from equation (6.234). Applying this result to an arbitrary vector $a = a_i \mathsf{e}^i$ we find that

$$D\wedge a = D\wedge(a_j \mathsf{e}^j) = \mathsf{e}^i\wedge\mathsf{e}^j(\partial_i a_j) = \tfrac{1}{2}\mathsf{e}^i\wedge\mathsf{e}^j(\partial_i a_j - \partial_j a_i). \tag{6.247}$$

This demonstrates that the $D\wedge$ operator is precisely the *exterior derivative* of differential geometry.

6.5.5 Riemannian geometry

To understand further how the shape tensor can specify the intrinsic geometry of a surface, we now make contact with Riemannian geometry. In Riemannian geometry one focuses entirely on the intrinsic properties of a manifold. It is customary to formulate the subject using the metric tensor as the starting point. In terms of the $\{\mathsf{e}_i\}$ coordinate frame the metric tensor is defined in the expected manner:

$$g_{ij} = \mathsf{e}_i \cdot \mathsf{e}_j. \tag{6.248}$$

In what follows we will not place any restriction on the signature of the tangent space. Some texts prefer to use the adjective 'Riemannian' to refer to extensions of Euclidean geometry to curved spaces (as Riemann originally intended). But in the physics literature it is quite standard now to refer to general relativity as a theory of Riemannian geometry, despite the Lorentzian signature.

After the metric, the next main object in Riemannian geometry is the Christoffel connection. The directional covariant derivative, D_i, restricts the result of its action to the tangent space. The result of its action on one of the $\{\mathsf{e}_i\}$ vectors can therefore be decomposed uniquely in the $\{\mathsf{e}_i\}$ frame. The coefficients of this define the Christoffel connection by

$$\Gamma^i_{jk} = (D_j \mathsf{e}_k)\cdot \mathsf{e}^i. \tag{6.249}$$

The components of the connection are clearly dependent on the choice of coordinate system, as well as the underlying geometry. It follows that a connection is necessary even when working in a curvilinear coordinate system in a flat space. A connection on its own does not imply that a space is curved. A typical use of the Christoffel connection is in finding the components in the $\{\mathsf{e}^i\}$ frame of a covariant derivative $a\cdot D\, b$, for example. We form

$$(a\cdot D\, b)\cdot \mathsf{e}^i = a^j \big(D_j(b^k \mathsf{e}_k)\big)\cdot \mathsf{e}^i = a^j(\partial_j b^i + \Gamma^i_{jk} b^k), \tag{6.250}$$

which shows how the connection accounts for the position dependence in the coordinate frame.

The components of the Christoffel connection can be found directly from the metric without referring to the frame vectors themselves. To achieve this we first establish a pair of results. The first is that the connection Γ^i_{jk} is symmetric on the jk indices. This follows from

$$\Gamma^i_{jk} - \Gamma^i_{kj} = (D_j \mathsf{e}_k - D_k \mathsf{e}_j)\cdot \mathsf{e}^i = 0, \tag{6.251}$$

where we have used equation (6.234). The second result is for the curl of a frame vector,

$$D\wedge \mathsf{e}_i = D\wedge (g_{ij}\mathsf{e}^j) = (Dg_{ij})\wedge \mathsf{e}^j. \tag{6.252}$$

We can now write

$$\begin{aligned}\Gamma^i_{jk} &= \tfrac{1}{2}\mathsf{e}^i\cdot(D_j\mathsf{e}_k + D_k\mathsf{e}_j)\\ &= \tfrac{1}{2}\mathsf{e}^i\cdot\big(\mathsf{e}_j\cdot(Dg_{kl}\wedge\mathsf{e}^l) + \mathsf{e}_k\cdot(Dg_{jl}\wedge\mathsf{e}^l) + Dg_{jk}\big)\\ &= \tfrac{1}{2}\mathsf{e}^i\cdot(\partial_j g_{kl}\mathsf{e}^l + \partial_k g_{jl}\mathsf{e}^l - Dg_{jk})\\ &= \tfrac{1}{2}g^{il}(\partial_j g_{kl} + \partial_k g_{jl} - \partial_l g_{jk}),\end{aligned} \tag{6.253}$$

which recovers the familiar definition of the Christoffel connection.

We now seek a method of encoding the intrinsic curvature of a Riemannian manifold. Suppose we form the commutator of two covariant derivatives

$$\begin{aligned}[D_i, D_j]A &= \partial_i(\partial_j A + S_j\times A) + S_i\times(\partial_j A + S_j\times A)\\ &\quad -\partial_j(\partial_i A + S_i\times A) - S_j\times(\partial_i A + S_i\times A)\\ &= (\partial_i S_j - \partial_j S_i)\times A + (S_i\times S_j)\times A,\end{aligned} \tag{6.254}$$

where we have used the Jacobi identity of section 4.1.3. Remarkably, all derivatives of the multivector A have cancelled out and what remains is a commutator with a bivector. To simplify this we form

$$\begin{aligned}\partial_i S_j - \partial_j S_i &= -\partial_i(\partial_j I\, I^{-1}) + \partial_j(\partial_i I\, I^{-1})\\ &= -S_j I S_i I^{-1} + S_i I S_j I^{-1}\\ &= -2S_i\times S_j,\end{aligned} \tag{6.255}$$

where we have used the fact that $S(a)$ anticommutes with I. On substituting this result in equation (6.254) we obtain the simple result

$$[D_i, D_j]A = -(S_i\times S_j)\times A. \tag{6.256}$$

The commutator of covariant derivatives defines the *Riemann tensor*. We denote this by $\mathsf{R}(a\wedge b)$, where

$$\mathsf{R}(\mathsf{e}_i\wedge\mathsf{e}_j)\times A = [D_i, D_j]A. \tag{6.257}$$

$\mathsf{R}(a\wedge b)$ is a bivector-valued linear function of its bivector argument. In terms of the shape tensor we have

$$\mathsf{R}(a\wedge b) = \mathsf{P}\big(S(b)\wedge S(a)\big). \tag{6.258}$$

The projection is required here because the Riemann tensor is defined to be entirely intrinsic to the manifold. The Riemann tensor (and its derivatives) fully encodes all of the local intrinsic geometry of a manifold. Since it can be derived easily from the shape tensor, it follows that the shape tensor also captures all of the intrinsic geometry. In addition to this, the shape tensor tells us about the extrinsic geometry — how the manifold is embedded in the larger ambient space.

The Riemann tensor can also be expressed entirely in terms of intrinsic quantities. To achieve this we first write

$$\mathsf{R}(\mathsf{e}_i \wedge \mathsf{e}_j) \cdot \mathsf{e}_k = [D_i, D_j]\mathsf{e}_k = D_i(\Gamma^a_{jk}\mathsf{e}_a) - D_j(\Gamma^a_{ik}\mathsf{e}_a). \tag{6.259}$$

It follows that

$$\begin{aligned} \mathsf{R}_{ijk}{}^l &= \mathsf{R}(\mathsf{e}_i \wedge \mathsf{e}_j) \cdot (\mathsf{e}_k \wedge \mathsf{e}^l) \\ &= \partial_i \Gamma^l_{jk} - \partial_j \Gamma^l_{ik} + \Gamma^a_{jk}\Gamma^l_{ia} - \Gamma^a_{ik}\Gamma^l_{ja}, \end{aligned} \tag{6.260}$$

recovering the standard definition of Riemannian geometry. An immediate advantage of the geometric algebra route is that many of the symmetry properties of $\mathsf{R}_{ijk}{}^l$ follow immediately from the fact that $\mathsf{R}(a \wedge b)$ is a bivector-valued linear function of a bivector. This immediately reduces the number of degrees of freedom to $n^2(n-1)^2/4$.

A further symmetry of the Riemann tensor can be found as follows:

$$\begin{aligned} \mathsf{R}(\mathsf{e}_i \wedge \mathsf{e}_j) \cdot \mathsf{e}_k &= D_i D_j \mathsf{e}_k - D_j D_i \mathsf{e}_k \\ &= D_i D_k \mathsf{e}_j - D_j D_k \mathsf{e}_i \\ &= [D_i, D_k]\mathsf{e}_j - [D_j, D_k]\mathsf{e}_i + D_k(D_i \mathsf{e}_j - D_j \mathsf{e}_i) \\ &= \mathsf{R}(\mathsf{e}_i \wedge \mathsf{e}_k) \cdot \mathsf{e}_j - \mathsf{R}(\mathsf{e}_j \wedge \mathsf{e}_k) \cdot \mathsf{e}_i. \end{aligned} \tag{6.261}$$

It follows that

$$a \cdot \mathsf{R}(b \wedge c) + c \cdot \mathsf{R}(a \wedge b) + b \cdot \mathsf{R}(c \wedge a) = 0, \tag{6.262}$$

for any three vectors a, b, c in the tangent space. This equation tells us that a vector quantity vanishes for all trivectors $a \wedge b \wedge c$, which provides a set of $n^2(n-1)(n-2)/6$ scalar equations. The number of independent degrees of freedom in the Riemann tensor is therefore reduced to

$$\frac{1}{4}n^2(n-1)^2 - \frac{1}{6}n^2(n-1)(n-2) = \frac{1}{12}n^2(n^2-1). \tag{6.263}$$

This gives the values 1, 6 and 20 for two, three and four dimensions respectively. Further properties of the Riemann tensor are covered in more detail in later chapters, where in particular we are interested in its relevance to gravitation.

The fact that Riemannian geometry is founded on the covariant derivative D, as opposed to the projected vector derivative ∂ limits the application of the integral theorem of equation (6.201). If one attempts to add multivectors from different points in the surface, there is no guarantee that the result remains intrinsic. The only quantities that can be combined from different points on the surface are scalars, or functions taking their values in a different space (such as a Lie group). The most significant integral theorem that remains is a generalization of Stokes' theorem, applicable to a grade-r multivector A_r and an open surface σ of dimension $r+1$. For this case we have

$$\oint_{\partial\sigma} A_r \cdot dS = \int_\sigma (\dot{A}_r \wedge \dot{\partial}) \cdot dX = (-1)^r \int_\sigma (D \wedge A_r) \cdot dX, \tag{6.264}$$

which only features intrinsic quantities. A particular case of this is when $r = n-1$, which recovers the divergence theorem. This is important for constructing conservation theorems in curved spaces.

6.5.6 Transformations and maps

The study of maps between vector manifolds helps to clarify some of the relationships between the structures defined in this chapter and more standard formulations of differential geometry. Suppose that $f(x)$ defines a map from one vector manifold to another. We denote these $\mathcal{M}$ and $\mathcal{M}'$, so that

$$x' = f(x) \tag{6.265}$$

associates a point in the manifold $\mathcal{M}'$ with one in $\mathcal{M}$. We will only consider smooth, differentiable, invertible maps between manifolds. In the mathematics literature these are known as *diffeomorphisms*. These are a subset of the more general concept of a *homeomorphism*, which maps continuously between spaces without the restriction of smoothness. Somewhat surprisingly, these two concepts are not equivalent. It is possible for two manifolds to be homeomorphic, but not admit a diffeomorphism between them. This implies that it is possible for a single topological space to admit more than one differentiable structure. The first example of this to be discovered was the sphere S^7, which admits 28 distinct differentiable structures! In 1983 Donaldson proved the even more striking result that four-dimensional space R^4 admits an infinite number of differentiable structures.

A path in $\mathcal{M}$, $x(\lambda)$, maps directly to a path in $\mathcal{M}'$. The map accordingly induces a map between tangent vectors, as seen by forming

$$\frac{\partial x'(\lambda)}{\partial \lambda} = \frac{\partial f(x(\lambda))}{\partial \lambda} = \mathsf{f}(v), \tag{6.266}$$

where v is the tangent vector in $\mathcal{M}$, $v = \partial_\lambda x(\lambda)$ and the linear function f is defined by

$$\mathsf{f}(a) = a \cdot \partial f(x) = \mathsf{f}(a; x). \tag{6.267}$$

The function $\mathsf{f}(a)$ takes a tangent vector in $\mathcal{M}$ as its linear argument, and returns the image tangent vector in $\mathcal{M}'$. If we denote the latter by a', and write out the position dependence explicitly, we have

$$a'(x') = \mathsf{f}(a(x);x). \tag{6.268}$$

This map is appropriate for tangent vectors, so applies to the coordinate frame vectors $\{\mathsf{e}_i\}$. These map to an equivalent frame for the tangent space to $\mathcal{M}'$,

$$\mathsf{e}_i' = \mathsf{f}(\mathsf{e}_i). \tag{6.269}$$

The reciprocal frame in the transformed space is therefore given by

$$\mathsf{e}^{i\prime} = \bar{\mathsf{f}}^{-1}(\mathsf{e}^i). \tag{6.270}$$

The fact that the map $x \mapsto f(x)$ is assumed to be invertible ensures that the adjoint function $\bar{\mathsf{f}}(a)$ is also invertible.

Under transformations, therefore, vectors in one space can transform in two different ways. If they are tangent vectors they transform under the action of $\mathsf{f}(a)$. If they are cotangent vectors they transform under action of $\bar{\mathsf{f}}^{-1}(a)$. In differential geometry it is standard practice to maintain a clear distinction between these types of vectors, so one usually thinks of tangent and cotangent vectors as lying in separate linear spaces. The contraction relation $\mathsf{e}^i \cdot \mathsf{e}_j = \delta^i_j$ identifies the spaces as dual to each other. This relation is metric-independent and is preserved by arbitrary diffeomorphisms. These maps relate differentiable manifolds, and two diffeomorphic spaces are usually viewed as the same manifold.

A metric is regarded as an additional construct on a differentiable manifold, which maps between the tangent and cotangent spaces. In the vector manifold picture this map is achieved by constructing the reciprocal frame using equation (4.94). In using this relation we are implicitly employing a metric in the contraction with the pseudoscalar. For the theory of vector manifolds it is therefore useful to distinguish objects and operations that transform simply under diffeomorphisms. These will define the metric-independent features of a vector manifold. Metric-dependent quantities, like the Riemann tensor, invariably have more complicated transformation laws.

The exterior product of a pair of tangent vectors transforms as

$$\mathsf{e}_i \wedge \mathsf{e}_j \mapsto \mathsf{f}(\mathsf{e}_i) \wedge \mathsf{f}(\mathsf{e}_j) = \mathsf{f}(\mathsf{e}_i \wedge \mathsf{e}_j). \tag{6.271}$$

For example, if I' is the unit pseudoscalar for $\mathcal{M}'$ we have

$$\mathsf{f}(I) = \det(\mathsf{f}) I' \tag{6.272}$$

and for invertible maps we must have $\det(\mathsf{f}) \neq 0$. Similarly, for cotangent vectors we see that

$$\mathsf{e}^i \wedge \mathsf{e}^j \mapsto \bar{\mathsf{f}}^{-1}(\mathsf{e}^i) \wedge \bar{\mathsf{f}}^{-1}(\mathsf{e}^j) = \bar{\mathsf{f}}^{-1}(\mathsf{e}^i \wedge \mathsf{e}^j). \tag{6.273}$$

So exterior products of like vectors give rise to higher grade objects in a manner that is unchanged by diffeomorphisms. Metric invariants are constructed from inner products between tangent and cotangent vectors. Since the derivative of a scalar field is

$$\partial\phi = \mathsf{e}^i \partial_i \phi, \tag{6.274}$$

we see that $\partial\phi$ is a cotangent vector, and we can write

$$\partial' = \bar{\mathsf{f}}^{-1}(\partial). \tag{6.275}$$

A similar result holds for the covariant derivative D. If a is a tangent vector the directional derivative of a scalar field $a\cdot\partial\phi$ is therefore an invariant,

$$a'\cdot\partial'\phi' = \mathsf{f}(a)\cdot\bar{\mathsf{f}}^{-1}(\partial)\phi = a\cdot\partial\phi, \tag{6.276}$$

where $\phi'(x') = \phi(x)$.

In constructing the covariant derivative in section 6.5.3, we made use of the projection operation $\mathsf{P}(a)$. This *is* a metric operation, as it relies on a contraction with I. Hence the covariant derivatives $D_i \mathsf{e}_j$ do depend on the metric (via the connection). To establish a metric-independent operation we let a and b represent tangent vectors and form

$$\begin{aligned} a\cdot\partial b - b\cdot\partial a &= a\cdot Db - b\cdot Da + a\cdot S(b) - b\cdot S(a) \\ &= a\cdot Db - b\cdot Da. \end{aligned} \tag{6.277}$$

The shape terms cancel, so the result is intrinsic to the manifold. Under a diffeomorphism the result transforms to

$$a\cdot\partial\mathsf{f}(b) - b\cdot\partial\mathsf{f}(a) = \mathsf{f}(a\cdot\partial b - b\cdot\partial a) + a\cdot\dot{\partial}\dot{\mathsf{f}}(b) - b\cdot\dot{\partial}\dot{\mathsf{f}}(a). \tag{6.278}$$

But $\mathsf{f}(a)$ is the differential of the map $f(x)$, so we have

$$(\partial_i\partial_j - \partial_j\partial_i)f(x) = \partial_i\mathsf{f}(\mathsf{e}_j) - \partial_j\mathsf{f}(\mathsf{e}_i) = \dot{\partial}_i\dot{\mathsf{f}}(\mathsf{e}_j) - \dot{\partial}_j\dot{\mathsf{f}}(\mathsf{e}_i) = 0. \tag{6.279}$$

It follows that, for tangent vectors a and b,

$$a\cdot\dot{\partial}\dot{\mathsf{f}}(b) - b\cdot\dot{\partial}\dot{\mathsf{f}}(a) = 0. \tag{6.280}$$

We therefore define the *Lie derivative* $\mathcal{L}_a b$ by

$$\mathcal{L}_a b = a\cdot\partial b - b\cdot\partial a. \tag{6.281}$$

This results in a new tangent vector, and transforms under diffeomorphisms as

$$\mathcal{L}_a b \mapsto \mathcal{L}'_{a'} b' = \mathsf{f}(\mathcal{L}_a b). \tag{6.282}$$

Relations between tangent vectors constructed from the Lie derivative will therefore be unchanged by diffeomorphisms.

A similar construction is possible for cotangent vectors. If we contract equation (6.279) with $\bar{\mathsf{f}}^{-1}(\mathsf{e}^k)$ we obtain

$$\mathsf{f}(\mathsf{e}_j)\cdot\big(\partial_i\bar{\mathsf{f}}^{-1}(\mathsf{e}^k)\big) - \mathsf{f}(\mathsf{e}_i)\cdot\big(\partial_j\bar{\mathsf{f}}^{-1}(\mathsf{e}^k)\big) = 0. \tag{6.283}$$

Now multiplying by $\bar{\mathsf{f}}^{-1}(\mathsf{e}^i \wedge \mathsf{e}^j)$ and summing we find that

$$\mathsf{P}'\big(\bar{\mathsf{f}}^{-1}(\partial)\wedge\bar{\mathsf{f}}^{-1}(\mathsf{e}^k)\big) = 0. \tag{6.284}$$

This result can be summarised simply as

$$D'\wedge \mathsf{e}^{k'} = D'\wedge\bar{\mathsf{f}}^{-1}(\mathsf{e}^k) = 0. \tag{6.285}$$

This is sufficient to establish that the exterior derivative of a cotangent vector results in a cotangent bivector (equivalent to a 2-form). The result transforms in the required manner:

$$D\wedge A \mapsto D'\wedge A' = \bar{\mathsf{f}}^{-1}(D\wedge A). \tag{6.286}$$

This is the result that makes the exterior algebra of cotangent vectors so powerful for studying the topological features of manifolds. This algebra is essentially that of differential forms, as is explained in section 6.5.7. For example, a form is said to be closed if its exterior derivative is zero, and to be exact if it can be written as the exterior derivative of a form of one degree lower. Both of these properties are unchanged by diffeomorphisms, so the size of the space of functions that are closed but not exact is a topological feature of a space. This is the basis of de Rham cohomology.

It is somewhat less common to see diffeomorphisms discussed when studying Riemannian geometry. More usually one focuses attention on the restricted class of *isometries*, which are diffeomorphisms that preserve the metric. These define symmetries of a Riemannian space. In the vector manifold setting, however, it is natural to study the effect of maps on metric-dependent quantities. The reason being that vector manifolds inherit their metric structure from the embedding, and if the embedding is changed by a diffeomorphism, the natural metric is changed as well. One does not have to inherit the metric from an embedding. One can easily impose a metric on a vector manifold by defining a linear transformation over the manifold. This takes us into the subject of induced geometries, which is closer to the spirit of the approach to gravity adopted in chapter 14. Similarly, when transforming a vector manifold, one need not insist that the transformed metric is that inherited by the new embedding. One can instead simply define a new metric on the transformed space directly from the original one.

The simplest example of a diffeomorphism inducing a new geometry is to consider a flat plane in three dimensions. If the plane is distorted in the third direction, and the new metric taken as that implied by the embedding, the surface clearly becomes curved. Formulae for the effects of such transformations are generally quite complex. Most can be derived from the transformation properties of the projection operation,

$$\mathsf{P}' = \mathsf{f}\mathsf{P}\mathsf{f}^{-1}. \tag{6.287}$$

This identity ensures that the projection and transformation formulae can be applied in either order. If we now form

$$\begin{aligned} \mathsf{e}'_i \cdot S'_j &= \mathsf{P}'_\perp\big(\partial_j \mathsf{f}(\mathsf{e}_i)\big) \\ &= \mathsf{f}(\mathsf{e}_i \cdot S_j) + \mathsf{P}'_\perp\big(\dot{\partial}_j \dot{\mathsf{f}}(\mathsf{e}_i)\big), \end{aligned} \tag{6.288}$$

we see that the shape tensor transforms according to

$$a' \cdot S'(b') = \mathsf{f}(a \cdot S(b)) + \mathsf{P}'_\perp\big(b \cdot \dot{\partial} \dot{\mathsf{f}}(a)\big). \tag{6.289}$$

Further results can be built up from this. For example, the new Riemann tensor is constructed from the commutator of the transformed shape tensor.

6.5.7 Differential geometry and forms

So far we have been deliberately loose in relating objects in vector manifold theory to those of modern differential geometry texts. In this section we clarify the relations and distinctions between the viewpoints. In the subject of differential geometry it is now common practice to identify directional derivatives as tangent vectors, so that the tangent vector a is the scalar operator

$$a = a^i \frac{\partial}{\partial x^i}. \tag{6.290}$$

Tangent vectors form a linear space, denoted $T_x\mathcal{M}$, where x labels a point in the manifold $\mathcal{M}$. This notion of a tangent vector is slightly different from that adopted in the vector manifold theory, where we explicitly let the directional derivative act on the vector x. As explained earlier, the limit implied in writing $\partial x/\partial x^i$ is only well defined if an embedding picture is assumed. The reason for the more abstract definition of a tangent vector in the differential geometry literature is to remove the need for an embedding, so that a topological space can be viewed as a single distinct entity. There are arguments in favour, and against, both viewpoints. For all practical purposes, however, the philosophies behind the two viewpoints are largely irrelevant, and calculations performed in either scheme will return the same results.

The dual space to $T_x\mathcal{M}$ is called the cotangent space and is denoted $T^*_x\mathcal{M}$. Elements of $T^*_x\mathcal{M}$ are called cotangent vectors, or 1-forms. The inner product between a tangent and cotangent vector can be written as $\langle \omega, a\rangle$. A basis for the dual space is defined by the coordinate differentials dx^i, so that

$$\langle dx^i, \partial/\partial x^j \rangle = \delta^i_j. \tag{6.291}$$

A 1-form therefore implicitly contains a directed measure on a manifold. So, if α is a 1-form we have

$$\alpha = \alpha_i dx^i = A \cdot (dx), \tag{6.292}$$

where A is a grade-1 multivector in the vector manifold sense. Similarly, if dX is a directed measure over a two-dimensional surface, we have

$$dX = e_i \wedge e_j \, dx^i \, dx^j, \tag{6.293}$$

so that

$$(e^j \wedge e^i) \cdot dX = dx^i \, dx^j - dx^j \, dx^i. \tag{6.294}$$

An arbitrary 2-form can be written as

$$\alpha_2 = \frac{1}{2!} \alpha_{ij} (dx^i \, dx^j - dx^j \, dx^i) = A_2^\dagger \cdot dX. \tag{6.295}$$

Here A_2 is the multivector

$$A_2 = \frac{1}{2!} \alpha_{ij} \, e^i \wedge e^j, \tag{6.296}$$

which has the same components as the differential form. More generally, an r-form α_r can be written as

$$\alpha_r = A_r^\dagger \cdot dX_r = A_r \cdot dX_r^\dagger. \tag{6.297}$$

Clearly there is little difference in working with the r-form α_r or the equivalent multivector A_r. So, for example, the outer product of two 1-forms results in the 2-form

$$\alpha_1 \wedge \beta_1 = \alpha_i \beta_i (e^i \wedge e^j) \cdot dX_2^\dagger = (A_1 \wedge B_1) \cdot dX_2^\dagger, \tag{6.298}$$

where dX_2 is a two-dimensional surface measure and A_1, B_1 are the grade-1 multivectors with components α_i and β_i respectively. Similarly, the exterior derivative of an r-form is given by

$$d\alpha_r = (D \wedge A_r) \cdot dX_{r+1}^\dagger. \tag{6.299}$$

The fact that forms come packaged with an implicit measure allows for a highly compact statement of Stokes' theorem, as given in equation (6.264). In ultra-compact notation this says that

$$\int_{\sigma_r} d\alpha = \oint_{\partial \sigma_r} \alpha, \tag{6.300}$$

where α is an $(r-1)$-form integrated over an open r-surface σ_r. This is entirely equivalent to equation (6.264), as can be seen by writing

$$\int_{\sigma_r} d\alpha = \int_{\sigma_r} (\dot{A}_{r-1}^\dagger \wedge \dot{D}) \cdot dX_r = \oint_{\partial \sigma_r} (A_{r-1}^\dagger) \cdot dS_{r-1} = \oint_{\partial \sigma_r} \alpha. \tag{6.301}$$

One can proceed in this manner to establish a direct translation scheme between the languages of differential forms and vector manifolds. Many of the expressions are so similar that there is frequently little point in maintaining a distinction.

If the language of differential forms is applied in a metric setting, an important additional concept is that of a duality transformation, also known as the *Hodge* $*$ (star) operation. To define this we first introduce the volume form

$$\Omega = \sqrt{|g|}dx^1\wedge dx^2\wedge\cdots\wedge dx^n = \sqrt{|g|}(e^n\wedge e^{n-1}\wedge\cdots\wedge e^1)\cdot dX. \tag{6.302}$$

The pseudoscalar for a vector manifold, given a coordinate frame with the specified orientation, is given by

$$I = \frac{1}{\sqrt{|g|}}(e_1\wedge e_2\wedge\cdots\wedge e_n). \tag{6.303}$$

This definition was chosen earlier to ensure that $I^2 = \pm 1$ and that I keeps the orientation specified by the frame. It follows that

$$\Omega = I^{-1}\cdot dX, \tag{6.304}$$

so that the equivalent multivector is $I^{-1\dagger}$. This will equal $\pm I$, depending on signature. The Hodge $*$ of an r-form α_r is the $(n-r)$-form

$$*\alpha_r = \frac{\sqrt{|g|}}{r!(m-r)!}\omega_{i_1,\ldots,i_r}\epsilon^{i_1,\ldots,i_r}{}_{j_{r+1},\ldots,j_n}\, dx^{j_{r+1}}\wedge\cdots\wedge dx^{j_n}, \tag{6.305}$$

where $\epsilon_{i_1,\ldots,i_n}$ denotes the alternating tensor. If A_r is the multivector equivalent of α_r, the Hodge $*$ takes on the rather simpler expression

$$*A_r = (I^{-1}A_r)^\dagger = (I^{-1}\cdot A_r)^\dagger. \tag{6.306}$$

In effect, we are multiplying by the pseudoscalar, as one would expect for a duality relation. Applied twice we find that

$$**A_r = \big(I^{-1}(I^{-1}\cdot A_r)^\dagger\big)^\dagger = (-1)^{r(m-r)}A_r(I^\dagger I). \tag{6.307}$$

In spaces with Euclidean signature, $I^\dagger I = +1$. In spaces of mixed signature the sign depends on whether there are an even or odd number of basis vectors with negative norm. It is a straightforward exercise to prove the main results for the Hodge $*$ operation, given equation (6.307) and the fact that I is covariantly conserved.

6.6 Elasticity

As a more extended application of some of the ideas developed in this chapter, we discuss the foundations of the subject of elasticity. The behaviour of a solid object is modelled by treating the object as a continuum. Locally, the strains in the object will tend to be small, but these can build up to give large global displacements. As such, it is important to treat the full, non-linear theory of elasticity. Only then can one be sure about the validity of various approximation schemes, such as assuming small deflections.

Our discussion is based on a generalisation of the ideas employed in the treatment of a rigid body. We first introduce an undeformed, *reference* configuration, with points in this labelled with the vector x. This is sometimes referred to as the material configuration. Points in the *spatial* configuration, y, are obtained by a non-linear displacement f of the reference configuration, so that

$$y = y(x,t) = f(x,t). \tag{6.308}$$

We use non-bold vectors to label points in the body, and bold to label tangent vectors in either the reference or spatial body. We assume that the background space is flat, three-dimensional Euclidean space.

6.6.1 Body strains

To calculate the strains in the body, consider the image of the vector between two nearby points in the reference configuration,

$$(x + \epsilon \boldsymbol{a}) - x \mapsto y(x + \epsilon \boldsymbol{a}) - y(x) = \epsilon \mathsf{f}(\boldsymbol{a}) + O(\epsilon^2), \tag{6.309}$$

where f is the deformation gradient,

$$\mathsf{f}(\boldsymbol{a}) = \boldsymbol{a}\cdot\boldsymbol{\nabla} y = \boldsymbol{a}\cdot\boldsymbol{\nabla} f(x,t). \tag{6.310}$$

The function f maps a tangent vector in the reference configuration to the equivalent vector in the spatial configuration. That is, if $x(\lambda)$ is a curve in the reference configuration with tangent vector

$$\boldsymbol{x}' = \frac{\partial x(\lambda)}{\partial \lambda}, \tag{6.311}$$

then the spatial curve has tangent vector $\mathsf{f}(\boldsymbol{v})$. The length of the curve $\boldsymbol{x}(\lambda)$ in the reference configuration is

$$\int \left| \frac{\partial x}{\partial \lambda} \right| d\lambda = \int |\boldsymbol{x}'| \, d\lambda. \tag{6.312}$$

The length of the induced curve in the spatial configuration is therefore

$$\int d\lambda \big(\mathsf{f}(\boldsymbol{x}')^2\big)^{1/2} = \int d\lambda \big(\boldsymbol{x}'\cdot\bar{\mathsf{f}}\mathsf{f}(\boldsymbol{x}')\big)^{1/2}. \tag{6.313}$$

We define the (right) *Cauchy–Green tensor* C, by

$$\mathsf{C}(\boldsymbol{a}) = \bar{\mathsf{f}}\mathsf{f}(\boldsymbol{a}). \tag{6.314}$$

This tensor is a symmetric, positive-definite map between vectors in the reference configuration. It describes a set of positive dilations along the principal directions in the reference configuration. The eigenvalues of C can be written as $(\lambda_1^2, \lambda_2^2, \lambda_3^2)$, where the λ_i define the *principal stretches*. The deviations of these from unity measure the strains in the material.

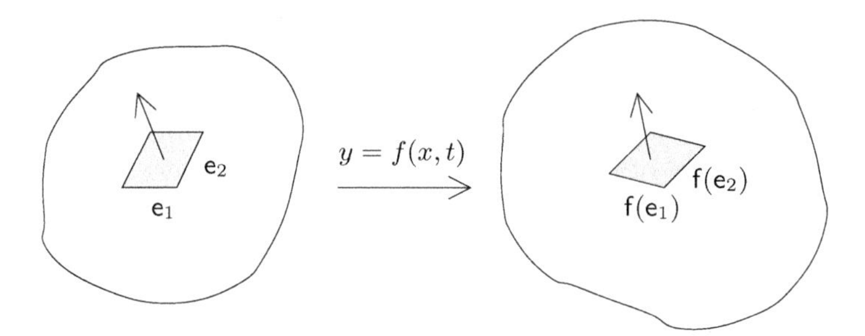

Figure 6.7 *An elastic body.* The function $f(x,t)$ maps points in the reference configuration to points in the spatial configuration. Coordinate curves e_1 and e_2 map to $\mathsf{f}(\mathsf{e}_1)$ and $\mathsf{f}(\mathsf{e}_2)$. The normal vector in the spatial configuration therefore lies in the $\bar{\mathsf{f}}^{-1}(\mathsf{e}^3)$ direction.

6.6.2 Body stresses

If we take a cut through the body then the contact force between the surfaces will be a function of the normal to the surface (and position in the body). Cauchy showed that, under reasonable continuity conditions, this force must be a linear function of the normal, which we write $\boldsymbol{\sigma}(\boldsymbol{n}) = \boldsymbol{\sigma}(\boldsymbol{n}; x)$. The tensor $\boldsymbol{\sigma}(\boldsymbol{n})$ maps a vector normal to a surface in the spatial configuration onto the force vector, also in the spatial configuration. We will verify shortly that $\boldsymbol{\sigma}$ is symmetric.

The total force on a volume segment in the body involves integrating $\boldsymbol{\sigma}(\boldsymbol{n})$ over the surface of the volume. But, as with the rigid body, it is simpler to perform all calculations back in the reference copy. To this end we let x^i denote a set of coordinates for position in the reference body. The associated coordinate frame is $\{\mathsf{e}_i\}$, with reciprocal frame $\{\mathsf{e}^i\}$. Suppose now that x^1 and x^2 are coordinates for a surface in the reference configuration. The equivalent normal in the spatial configuration is (see figure 6.7)

$$\boldsymbol{n} = \mathsf{f}(\mathsf{e}_1)\wedge\mathsf{f}(\mathsf{e}_2)\, I^{-1} = \det\,(\mathsf{f})\,\bar{\mathsf{f}}^{-1}(\mathsf{e}^3). \tag{6.315}$$

The force over this surface is found by integrating the quantity

$$\boldsymbol{\sigma}\big(\mathsf{f}(\boldsymbol{e}_1\wedge\boldsymbol{e}_2)I^{-1}\big)dx^1\,dx^2 = \det\,(\mathsf{f})\boldsymbol{\sigma}(\bar{\mathsf{f}}^{-1}(\boldsymbol{e}^3))dx^1\,dx^2. \tag{6.316}$$

We therefore define the *first Piola–Kirchoff stress tensor* T by

$$\mathsf{T}(\boldsymbol{a}) = \det\,(\mathsf{f})\boldsymbol{\sigma}\bar{\mathsf{f}}^{-1}(\boldsymbol{a}). \tag{6.317}$$

The stress tensor T takes as its argument a vector normal to a surface in the reference configuration, and returns the contact force in the spatial body. The force balance equation tells us that, for any sub-body, we have

$$\frac{d}{dt}\int d^3x\,\rho\boldsymbol{v} = \oint \mathsf{T}(d\boldsymbol{s}) + \int d^3x\,\rho\boldsymbol{b}, \tag{6.318}$$

where ρ is the density in the reference configuration, $\boldsymbol{v} = \dot{y}$ is the spatial velocity, and $\boldsymbol{b}$ is the applied *body force*. The fundamental theorem immediately converts this to the local equation

$$\rho\dot{\boldsymbol{v}} = \check{\mathsf{T}}(\check{\boldsymbol{\nabla}}) + \rho\boldsymbol{b}. \tag{6.319}$$

The check symbol is used for the scope of the derivative, to avoid confusion with time derivatives (denoted with an overdot). This equation is sensible as $\boldsymbol{\nabla}$ is the vector derivative in the reference configuration, and $\check{\mathsf{T}}(\check{\boldsymbol{\nabla}})$ is a vector in the spation configuration.

The total torque on a volume element, centred on y_0, is (ignoring body forces)

$$M = \oint (y - y_0)\wedge\mathsf{T}(d\boldsymbol{s}). \tag{6.320}$$

This integral runs over the reference body, and returns a torque in the spatial configuration. This must be equated with the rate of change of angular momentum, which is

$$\begin{aligned}\frac{d}{dt}\int d^3x\,\rho(y - y_0)\wedge\dot{y} &= \int d^3x\,(y - y_0)\wedge\check{\mathsf{T}}(\check{\boldsymbol{\nabla}}) \\ &= \oint (y - y_0)\wedge\mathsf{T}(d\boldsymbol{s}) - \int d^3x\,\check{y}\wedge\mathsf{T}(\check{\boldsymbol{\nabla}}).\end{aligned} \tag{6.321}$$

Equating this with M we see that

$$\check{y}\wedge\mathsf{T}(\check{\boldsymbol{\nabla}}) = (\partial_i f(x))\wedge\mathsf{T}(\mathsf{e}^i) = \mathsf{f}(\mathsf{e}_i)\wedge\mathsf{T}(\mathsf{e}^i) = 0. \tag{6.322}$$

It follows that

$$\mathsf{f}(\mathsf{e}_i)\wedge\mathsf{T}(\mathsf{e}^i) = \det\,(\mathsf{f})\,\mathsf{f}(\mathsf{e}_i)\wedge\boldsymbol{\sigma}\bar{\mathsf{f}}^{-1}(\mathsf{e}^i) = 0, \tag{6.323}$$

and we see that $\boldsymbol{\sigma}$ must be a symmetric tensor in order for angular momentum to be conserved.

It is often convenient to work with a version of T that is symmetric and defined entirely in the material frame. We therefore define the *second Piola–Kirchoff stress tensor* $\mathcal{T}$ by

$$\mathcal{T}(\boldsymbol{a}) = \mathsf{f}^{-1}\mathsf{T}(\boldsymbol{a}). \tag{6.324}$$

It is meaningless to talk about symmetries of T, since it maps between different spaces, whereas $\mathcal{T}$ is defined entirely in the reference configuration and, by construction, is symmetric.

The equations of motion for an elastic material are completed by defining a constitutive relation. This relates the stresses to the strains in the body. These relations are most easily expressed in the reference copy as a relationship between

$\mathcal{T}$ and C. There is no universal definition of the strain tensor $\mathcal{E}$, though for certain applications a useful definition is

$$\mathcal{E}(\boldsymbol{a}) = \mathsf{C}^{1/2}(\boldsymbol{a}) - \boldsymbol{a}. \tag{6.325}$$

This tensor is zero if the material is undeformed. Linear materials have the property that $\mathcal{T}$ and $\mathcal{E}$ are linearly related by a rank-4 tensor. This can, in principle, have 36 independent degrees of freedom, all of which may need to be determined experimentally. If the material is homogeneous then the components of the rank-4 tensor are constants. If the material is also isotropic then the 36 degrees of freedom reduce to two. These are usually given in terms of the bulk modulus B and shear modulus G, with $\mathcal{T}$ and $\mathcal{E}$ related by an expression of the form

$$\mathcal{T}(\boldsymbol{a}) = 2G\mathcal{E}(\boldsymbol{a}) + (B - \tfrac{2}{3}G)\mathrm{tr}(\mathcal{E})\boldsymbol{a}. \tag{6.326}$$

In many respects this is the simplest material one can consider, though even in this case the non-linearity of the force law makes the full equations very hard to analyse. The analysis can be aided by the fact that these materials are described by an action principle, as discussed in section 12.4.1.

6.7 Notes

The treatment of vector manifolds presented here is a condensed version of the theory developed by Hestenes & Sobczyk in the book *Clifford Algebra to Geometric Calculus* (1984) and in a series of papers by Garret Sobczyk. There are a number of differences in our presentation, however. Most significant is our definition of the orientations in the fundamental theorem of integral calculus. Our definition of the boundary operator ensures that a boundary inherits its orientation from the directed volume measure. Hestenes & Sobczyk used the opposite specification for their boundary operator, which gives rise to a number of (fairly trivial) differences. A significant advantage of our conventions is that in two dimensions the pseudoscalar has the correct orientation implied by the imaginary in the Cauchy integral formula.

A further difference is that from the outset we have emphasised both the implied embedding of a vector manifold, and the fact that this gives rise to a metric. A vector manifold thus has greater structure than a differentiable manifold in the sense of differential geometry. For applications to finite-dimensional Riemannian geometry the different approaches are entirely equivalent, as any finite-dimensional Riemannian manifold can be embedded in a larger dimensional flat space in such a way that the metric is generated by the embedding. This result was proved by John Nash in 1956. His remarkable story is the subject of the book *A Beautiful Mind* by Sylvia Nasar (1998) and, more recently, a film of the same name. In other applications of differential geometry the full range of validity of the vector manifold approach has yet to be fully established. The approach certainly does give streamlined proofs of a number of key results. But whether this comes with some loss of generality is an open question.

A final, small difference in our approach here to the original one of Hestenes & Sobczyk is our definition of the shape tensor. We have only considered the shape tensor $S(a)$ taking intrinsic vectors as its linear argument. This concept can be generalised to define a function that can act linearly on general vectors. One of the most interesting properties of this generalized version of the shape tensor is that it provides a natural square root of the Ricci tensor. This theory is developed in detail in chapter 5 of *Clifford Algebra to Geometric Calculus*, to which readers are referred for further information. There is no shortage of good textbooks on modern differential geometry. The books by Nakahara (1990), Schutz (1980) and Göckeler & Schucker (1987) are particularly strong on emphasising physical applications. Elasticity is described in the books by Marsden & Hughes (1994) and Antman (1995).

6.8 Exercises

6.1 Confirm that the vector derivative is independent of the choice of coordinate system.

6.2 If we denote the curl of a vector field $\boldsymbol{J}$ in three dimensions by $\boldsymbol{\nabla}\times\boldsymbol{J}$, show that

$$\boldsymbol{\nabla}\times\boldsymbol{J} = -I\nabla\wedge J.$$

Hence prove that

$$\boldsymbol{\nabla}\cdot(\boldsymbol{\nabla}\times\boldsymbol{J}) = 0,$$
$$\boldsymbol{\nabla}\times(\boldsymbol{\nabla}\times\boldsymbol{J}) = \boldsymbol{\nabla}(\boldsymbol{\nabla}\cdot\boldsymbol{J}) - \boldsymbol{\nabla}^2\boldsymbol{J}.$$

6.3 An oblate spheroidal coordinate system can be defined by

$$a\cosh(u)\sin(v) = \surd(x^2+y^2),$$
$$a\sinh(u)\cos(v) = z,$$
$$\tan(\phi) = y/x,$$

where (x, y, z) denote standard Cartesian coordinates and a is a scalar. Prove that

$$\mathsf{e}_u^2 = \mathsf{e}_v^2 = a^2\big(\sinh^2(u)+\cos^2(v)\big) = \rho^2,$$

which defines the quantity ρ. Hence prove that the Laplacian becomes

$$\boldsymbol{\nabla}^2\psi = \frac{1}{\rho^2\cosh(u)}\frac{\partial}{\partial u}\left(\cosh(u)\frac{\partial\psi}{\partial u}\right) + \frac{1}{\rho^2\sin(v)}\frac{\partial}{\partial v}\left(\sin(v)\frac{\partial\psi}{\partial v}\right) + \frac{1}{a^2\cosh^2(u)\sin^2(v)}\frac{\partial^2\psi}{\partial\phi^2},$$

and investigate the properties of separable solutions in oblate spheroidal coordinates.

6.4 Prove that over the surface of a tetrahedron the directed surface integral satisfies

$$\oint dS = 0.$$

By considering pairs of adjacent tetrahedra, prove that this integral vanishes for all orientable, connected closed surfaces.

6.5 For a circle in a plane confirm that the line integral around the perimeter satisfies

$$\oint b\cdot x\, dl = b\cdot A,$$

where A is the oriented area of the circle.

6.6 Prove that

$$\sum_{i=0}^{k}(-1)^i b\cdot(x_0 + \cdots \check{x}_i \cdots + x_n)\Delta(\check{x}_i)_{(k-1)} = \frac{1}{k!} b\cdot(e_1\wedge\cdots\wedge e_n),$$

where the notation follows section 6.4.4.

6.7 Suppose that σ is an n-dimensional surface embedded in a flat space of dimensions $n+1$ with (constant) unit pseudoscalar I. Prove that

$$\oint_{\partial\sigma} dS J = -I \int_{\sigma} l\wedge\nabla\, J\, |dX|,$$

where the normal l is defined by $dX = Il\,|dX|$.

6.8 The shape tensor is defined by

$$a\cdot\partial I = IS(a) = I\times S(a).$$

Prove that the shape tensor satisfies

$$a\cdot S(b) = b\cdot S(a)$$

and

$$\dot{\partial}\wedge\dot{\mathsf{P}}(a) = S(a),$$

where P projects into the tangent space, and a and b are tangent vectors.

6.9 An open two-dimensional surface in three-dimensional space is defined by

$$\boldsymbol{r}(x, y) = x\mathsf{e}_1 + y\mathsf{e}_2 + \alpha(r)\mathsf{e}_3,$$

where $r = (x^2 + y^2)^{1/2}$ and the $\{\mathsf{e}_i\}$ are a standard Cartesian frame. Prove that the Riemann tensor can be written

$$\mathsf{R}(a\wedge b) = \frac{\alpha'\alpha''}{r(1+\alpha'^2)^2}\, a\wedge b,$$

where the primes denote differentiation with respect to r. The scalar factor κ in $\mathsf{R}(a\wedge b) = \kappa a\wedge b$ is called the *Gaussian curvature.*

6.10 A linear, isotropic, homogeneous material is described by a bulk modulus B and shear modulus G. By linearising the elasticity equations, show that the longitudinal and transverse sound speeds v_l and v_t are given by

$$v_l^2 = \frac{1}{3\rho}(3B + 4G), \qquad v_t^2 = \frac{G}{\rho}.$$

6.11 Consider an infinite linear, isotropic, homogeneous material containing a spherical hole into which air is pumped. Show that, in the linearised theory, the radial stress τ_r is related to the radius of the hole r by $\tau_r \propto r^{-3}$. Discuss how the full non-linear theory might modify this result.

7

Classical electrodynamics

Geometric algebra offers a number of new techniques for studying problems in electromagnetism and electrodynamics. These are described in this chapter. We will not attempt a thorough development of electrodynamics, which is a vast subject with numerous specialist areas. Instead we concentrate on a number of selected applications which highlight the advantages that geometric algebra can bring. There are two particularly significant new features that geometric algebra adds to traditional formulations of electrodynamics. The first is that, through employing the spacetime algebra, all equations can be studied in the appropriate spacetime setting. This is much more transparent than the more traditional approach based on a $3+1$ formulation involving retarded times. The spacetime algebra simplifies the study of how electromagnetic fields appear to different observers, and is particularly powerful for handling accelerated charges and radiation. These results build on the applications of spacetime algebra described in section 5.5.3.

The second major advantage of the geometric algebra treatment is a new, compact formulation of Maxwell's equations. The spacetime vector derivative and the geometric product enable us to unite all four of Maxwell's equations into a single equation. This is one of the most impressive results in geometric algebra. And, as we showed in chapter 6, this is more than merely a cosmetic exercise. The vector derivative is invertible directly, without having to pass via intermediate, second-order equations. This has many implications for scattering and propagator theory. Huygen's principle is encoded directly, and the first-order theory is preferable for numerical computation of diffraction effects. In addition, the first-order formulation of electromagnetism means that plane waves are easily handled, as are their polarisation states.

7.1 Maxwell's equations

Before writing down the Maxwell equations, we remind ourselves of the notation introduced in chapter 5. We denote an orthonormal spacetime frame by $\{\gamma_\mu\}$, with coordinates $x_\mu = \gamma_\mu \cdot x$. The spacetime vector derivative is

$$\nabla = \gamma^\mu \partial_\mu, \qquad \partial_\mu = \frac{\partial}{\partial x^\mu}. \tag{7.1}$$

The spacetime split of the vector derivative is

$$\nabla \gamma_0 = (\gamma^0 \partial_t + \gamma^i \partial_i)\gamma_0 = \partial_t - \boldsymbol{\sigma}_i \partial_i = \partial_t - \boldsymbol{\nabla}, \tag{7.2}$$

where the $\boldsymbol{\sigma}_i = \gamma_i \gamma_0$ denote a right-handed orthonormal frame for the relative space defined by the timelike vector γ_0. The three-dimensional vector derivative operator is

$$\boldsymbol{\nabla} = \boldsymbol{\sigma}_i \frac{\partial}{\partial x^i} = \boldsymbol{\sigma}_i \partial_i, \tag{7.3}$$

and all relative vectors are written in bold.

The four Maxwell equations, in SI units, are

$$\begin{aligned} \boldsymbol{\nabla} \cdot \boldsymbol{D} &= \rho, & \boldsymbol{\nabla} \cdot \boldsymbol{B} &= 0, \\ -\boldsymbol{\nabla} \times \boldsymbol{E} &= \frac{\partial}{\partial t} \boldsymbol{B}, & \boldsymbol{\nabla} \times \boldsymbol{H} &= \frac{\partial}{\partial t} \boldsymbol{D} + \boldsymbol{J}, \end{aligned} \tag{7.4}$$

where

$$\begin{aligned} \boldsymbol{D} &= \epsilon_0 \boldsymbol{E} + \boldsymbol{P}, \\ \boldsymbol{H} &= \frac{1}{\mu_0} \boldsymbol{B} - \boldsymbol{M}, \end{aligned} \tag{7.5}$$

and the $\times$ symbol denotes the vector cross product. The cross product is ubiquitous in electromagnetic theory, and it will be encountered at various points in this chapter. To avoid any confusion, the commutator product (denoted by $\times$) will not be employed in this chapter.

The first step in simplifying the Maxwell equations is to assume that we are working in a vacuum region outside isolated sources and currents. We can then remove the polarisation and magnetisation fields $\boldsymbol{P}$ and $\boldsymbol{M}$. We also replace the cross product with the exterior product, and revert to natural units ($c = \epsilon_0 = \mu_0 = 1$), so that the equations now read

$$\begin{aligned} \boldsymbol{\nabla} \cdot \boldsymbol{E} &= \rho, & \boldsymbol{\nabla} \cdot \boldsymbol{B} &= 0, \\ \boldsymbol{\nabla} \wedge \boldsymbol{E} &= -\partial_t (I\boldsymbol{B}), & \boldsymbol{\nabla} \wedge \boldsymbol{B} &= I(\boldsymbol{J} + \partial_t \boldsymbol{E}). \end{aligned} \tag{7.6}$$

We naturally assemble equations for the separate divergence and curl parts of the vector derivative. We know that there are many advantages in uniting these into a single equation involving the vector derivative. First we take the two equations for $\boldsymbol{E}$ and combine them into the single equation

$$\boldsymbol{\nabla} \boldsymbol{E} = \rho - \partial_t(I\boldsymbol{B}). \tag{7.7}$$

A similar manipulation combines the $\boldsymbol{B}$-field equations into

$$\boldsymbol{\nabla}(I\boldsymbol{B}) = -\boldsymbol{J} - \partial_t \boldsymbol{E}, \tag{7.8}$$

where we have multiplied through by I. This equation is a combination of (spatial) bivector and pseudoscalar terms, whereas equation (7.7) contains only scalar and vector parts. It follows that we can combine all of these equations into the single multivector equation

$$\boldsymbol{\nabla}(\boldsymbol{E} + I\boldsymbol{B}) + \partial_t(\boldsymbol{E} + I\boldsymbol{B}) - \rho - \boldsymbol{J}. \tag{7.9}$$

This is already a significant compactification of the original equations. We have not lost any information in writing this, since each of the separate Maxwell equations can be recovered by picking out terms of a given grade.

In section 5.5.3 we introduced the Faraday bivector F. This represents the *electromagnetic field strength* and is defined by

$$F = \boldsymbol{E} + I\boldsymbol{B}. \tag{7.10}$$

The combination of relative vectors and bivectors tells us that this quantity is a spacetime bivector. Many authors have noticed that the Maxwell equations can be simplified if expressed in terms of the complex quantity $\boldsymbol{E} + i\boldsymbol{B}$. The reason is that the spacetime pseudoscalar has negative square, so can be represented by the unit imaginary for certain applications. It is important, however, to work with I in the full spacetime setting, as I anticommutes with spacetime vectors.

In terms of the field strength the Maxwell equations reduce to

$$\boldsymbol{\nabla} F + \partial_t F = \rho - \boldsymbol{J}. \tag{7.11}$$

We now wish to convert this to manifestly Lorentz covariant form. We introduce the spacetime current J, which has

$$\rho = J \cdot \gamma_0, \qquad \boldsymbol{J} = J \wedge \gamma_0. \tag{7.12}$$

It follows that

$$\rho - \boldsymbol{J} = \gamma_0 \cdot J + \gamma_0 \wedge J = \gamma_0 J. \tag{7.13}$$

But we know that $\partial_t + \boldsymbol{\nabla} = \gamma_0 \nabla$. We can therefore pre-multiply equation (7.11) by γ_0 to assemble the covariant equation

$$\nabla F = J. \tag{7.14}$$

This unites all four Maxwell equations into a single spacetime equation based on the *geometric* product with the vector derivative. An immediate consequence is seen if we multiply through by ∇, giving

$$\nabla^2 F = \nabla J = \nabla\cdot J + \nabla\wedge J. \tag{7.15}$$

Since ∇^2 is a scalar operator, the left-hand side can only contain bivector terms. It follows that the current J must satisfy the conservation equation

$$\nabla\cdot J = \frac{\partial \rho}{\partial t} + \boldsymbol{\nabla}\cdot\boldsymbol{J} = 0. \tag{7.16}$$

This equation tells us that the total charge generating the fields must be conserved.

The equation $\nabla F = J$ separates into a pair of spacetime equations for the vector and trivector parts,

$$\nabla\cdot F = J, \qquad \nabla\wedge F = 0. \tag{7.17}$$

In tensor language, these correspond to the pair of spacetime equations

$$\partial_\mu F^{\mu\nu} = J^\nu, \qquad \epsilon^{\mu\nu\rho\sigma}\partial_\nu F_{\rho\sigma} = 0. \tag{7.18}$$

These two tensor equations are as compact a formulation of the Maxwell equations as tensor algebra can achieve, and the same is true of differential forms. Only geometric algebra enables us to combine the Maxwell equations (7.17) into the single equation $\nabla F = J$.

7.1.1 The vector potential

The fact that $\nabla\wedge F = 0$ tells us that we can introduce a vector field A such that

$$F = \nabla\wedge A. \tag{7.19}$$

The equation $\nabla\wedge F = \nabla\wedge\nabla\wedge A = 0$ then follows automatically. The field A is known as the *vector potential.* We shall see in later chapters that the vector potential is key to the quantum theory of how matter interacts with radiation. The vector potential is also the basis for the Lagrangian treatment of electromagnetism, described in chapter 12.

The remaining source equation tells us that the vector potential satisfies

$$\nabla\cdot(\nabla\wedge A) = \nabla^2 A - \nabla(\nabla\cdot A) = J. \tag{7.20}$$

There is some residual freedom in A beyond the restriction of equation (7.19). We can always add the gradient of a scalar field to A, since

$$\nabla\wedge(A + \nabla\lambda) = \nabla\wedge A + \nabla\wedge(\nabla\lambda) = F. \tag{7.21}$$

For historical reasons, this ability to alter A is referred to as a *gauge* freedom.

Before we can solve the equations for A, we must therefore specify a gauge. A natural way to absorb this freedom is to impose the *Lorentz condition*

$$\nabla\cdot A = 0. \tag{7.22}$$

This does not totally specify A, as the gradient of a solution of the wave equation can still be added, but this remaining freedom can be removed by imposing appropriate boundary conditions. The Lorentz gauge condition implies that $F = \nabla A$. We then recover a wave equation for the components of A, since

$$\nabla F = \nabla^2 A = J. \tag{7.23}$$

One route to solving the Maxwell equations is to solve the associated wave equation $\nabla^2 A = J$, with appropriate boundary conditions applied, and then compute F at the end. In this chapter we explore alternative, more direct routes.

The fact that a gauge freedom exists in the formulation in terms of A suggests that some *conjugate* quantity should be conserved. This is the origin of the current conservation law derived in equation (7.16). Conservation of charge is therefore intimately related to gauge invariance. A more detailed understanding of this will be provided by the Lagrangian framework.

7.1.2 The electromagnetic field strength

In uniting the Maxwell equations we introduced the *electromagnetic field strength* $F = \boldsymbol{E}+I\boldsymbol{B}$. This is a covariant spacetime bivector. Its components in the $\{\gamma^\mu\}$ frame give rise to the tensor

$$F^{\mu\nu} = \gamma^\nu\cdot(\gamma^\mu\cdot F) = (\gamma^\nu\wedge\gamma^\mu)\cdot F. \tag{7.24}$$

These are the components of a rank-2 antisymmetric tensor which, written out as a matrix, has entries

$$F^{\mu\nu} = \begin{pmatrix} 0 & -E_x & -E_y & -E_z \\ E_x & 0 & -B_z & B_y \\ E_y & B_z & 0 & -B_x \\ E_z & -B_y & B_x & 0 \end{pmatrix}. \tag{7.25}$$

This matrix form of the field strength is often presented in textbooks on relativistic electrodynamics. It has a number of disadvantages. Amongst these are that Lorentz transformations cannot be handled elegantly and the natural complex structure is hidden.

Writing $F = \boldsymbol{E} + I\boldsymbol{B}$ decomposes F into the sum of a relative vector $\boldsymbol{E}$ and a relative bivector $I\boldsymbol{B}$. The separate $\boldsymbol{E}$ and $I\boldsymbol{B}$ fields are recovered from

$$\begin{aligned} \boldsymbol{E} &= \tfrac{1}{2}(F - \gamma_0 F \gamma_0), \\ I\boldsymbol{B} &= \tfrac{1}{2}(F + \gamma_0 F \gamma_0). \end{aligned} \tag{7.26}$$

This shows clearly how the split into $\boldsymbol{E}$ and $I\boldsymbol{B}$ fields depends on the observer velocity (γ_0 here). Observers in relative motion see different fields. For example, suppose that a second observer has velocity $v = R\gamma_0\tilde{R}$ and constructs the rest frame basis vectors

$$\gamma'_\mu = R\gamma_\mu\tilde{R}. \tag{7.27}$$

This observer measures components of an electric field to be

$$E'_i = (\gamma'_i\gamma'_0)\cdot F = (R\boldsymbol{\sigma}_i\tilde{R})\cdot F = \boldsymbol{\sigma}_i\cdot(\tilde{R}FR). \tag{7.28}$$

The effect of a Lorentz transformation can therefore be seen by taking F to $\tilde{R}FR$. The fact that bivectors are subject to the same rotor transformation law as vectors is extremely useful for computations.

Suppose now that two observers measure the F-field at a point. One has 4-velocity γ_0, and the other is moving at relative velocity $\boldsymbol{v}$ in the γ_0 frame. This observer has 4-velocity

$$v = R\gamma_0\tilde{R}, \qquad R = \exp(\alpha\hat{\boldsymbol{v}}/2), \tag{7.29}$$

where $\boldsymbol{v} = \tanh(\alpha)\hat{\boldsymbol{v}}$. The second observer measures the $\{\gamma_\mu\}$ components of $\tilde{R}FR$. To find these we decompose F into terms parallel and perpendicular to $\boldsymbol{v}$,

$$F = F_\parallel + F_\perp, \tag{7.30}$$

where

$$\boldsymbol{v}F_\parallel = F_\parallel\boldsymbol{v}, \qquad \boldsymbol{v}F_\perp = -F_\perp\boldsymbol{v}. \tag{7.31}$$

We quickly see that the parallel components are unchanged, but the perpendicular components transform to

$$\tilde{R}F_\perp R = \exp(-\alpha\hat{\boldsymbol{v}})F_\perp = \gamma(1-\boldsymbol{v})F_\perp, \tag{7.32}$$

where γ is the Lorentz factor $(1-\boldsymbol{v}^2)^{-1/2}$. This result is sufficient to immediately establish the transformation law

$$\begin{aligned} \boldsymbol{E}'_\perp &= \gamma(\boldsymbol{E} + \boldsymbol{v}\times\boldsymbol{B})_\perp, \\ \boldsymbol{B}'_\perp &= \gamma(\boldsymbol{B} - \boldsymbol{v}\times\boldsymbol{E})_\perp. \end{aligned} \tag{7.33}$$

Here the primed vectors are formed from $\boldsymbol{E}' = E'_i\boldsymbol{\sigma}_i$, for example. These have the components of F in the new frame, but combined with the original basis vectors.

Further useful information about the F field is contained in its square, which defines a pair of Lorentz-invariant terms. We form

$$F^2 = \langle FF\rangle + \langle FF\rangle_4 = a_0 + Ia_4, \tag{7.34}$$

which is easily seen to be Lorentz-invariant,

$$(\tilde{R}FR)(\tilde{R}FR) = \tilde{R}FFR = a_0 + Ia_4. \tag{7.35}$$

Both the scalar and pseudoscalar terms are independent of the frame in which they are measured. In the γ_0 frame these are

$$\alpha = \langle(\boldsymbol{E} + I\boldsymbol{B})(\boldsymbol{E} + I\boldsymbol{B})\rangle = \boldsymbol{E}^2 - \boldsymbol{B}^2 \tag{7.36}$$

and

$$\beta = -\langle I(\boldsymbol{E} + I\boldsymbol{B})(\boldsymbol{E} + I\boldsymbol{B})\rangle = 2\boldsymbol{E}\cdot\boldsymbol{B}. \tag{7.37}$$

The former yields the Lagrangian density for the electromagnetic field. The latter is seen less often. It is perhaps surprising that $\boldsymbol{E}\cdot\boldsymbol{B}$ is a full Lorentz invariant, rather than just being invariant under rotations.

7.1.3 Dielectric and magnetic media

The Maxwell equations inside a medium, with polarisation and magnetisation fields $\boldsymbol{P}$ and $\boldsymbol{M}$, were given in equation (7.4). These separate into a pair of spacetime equations. We introduce the spacetime bivector field G by

$$G = \boldsymbol{D} + I\boldsymbol{H}. \tag{7.38}$$

Maxwell's equations are now given by the pair of equations

$$\begin{aligned} \nabla\wedge F &= 0, \\ \nabla\cdot G &= J. \end{aligned} \tag{7.39}$$

The first tells us that F has vanishing curl, so can still be obtained from a vector potential, $F = \nabla\wedge A$. The second equation tells us how the $\boldsymbol{D}$ and $\boldsymbol{H}$ fields respond to the presence of free sources. These equations on their own are insufficient to fully describe the behaviour of electromagnetic fields in matter. They must be augmented by constitutive relations which relate F and G. The simplest examples of these are for linear, isotropic, homogeneous materials, in which case the constitutive relations amount to specifying a relative permittivity ϵ_r and permeability μ_r. The fields are then related by

$$\boldsymbol{D} = \epsilon_r\boldsymbol{E}, \qquad \boldsymbol{B} = \mu_r\boldsymbol{H}. \tag{7.40}$$

More complicated models for matter can involve considering responses to different frequencies, and the presence of preferred directions on the material. The subject of suitable constitutive relations is one of heuristic model building. We are, in effect, seeking models which account for the quantum properties of matter in bulk, without facing the full multiparticle quantum equations.

7.2 Integral and conservation theorems

A number of important integral theorems exist in electromagnetism. Indeed, the subject of integral calculus was largely shaped by considering applications to electromagnetism. Here the results are all derived as examples of the fundamental theorem of integral calculus, derived in chapter 6.

7.2.1 Static fields

We start by deriving a number of results for static field configurations. When the fields are static the Maxwell equations reduce to the pair

$$\nabla \boldsymbol{E} = \frac{\rho}{\epsilon_0}, \qquad \nabla \boldsymbol{B} = \mu_0 I \boldsymbol{J}, \tag{7.41}$$

where (for this section) we have reinserted the constants ϵ_0 and μ_0. A current $\boldsymbol{J}$ is static if the charge flows at a constant rate. The fact that $\nabla \wedge \boldsymbol{E} = 0$ implies that around any closed path

$$\oint_{\partial \sigma} \boldsymbol{E} \cdot d\boldsymbol{l} = 0, \tag{7.42}$$

which applies for all static configurations. We can therefore introduce a potential ϕ such that

$$\boldsymbol{E} = -\nabla \phi. \tag{7.43}$$

The potential ϕ is the timelike component of the vector potential A, $\phi = \gamma_0 \cdot A$. One can formulate many of the main results of electrostatics directly in terms of ϕ. Here we adopt a different approach and work directly with the $\boldsymbol{E}$ and $\boldsymbol{B}$ fields.

An extremely important integral theorem is a straightforward application of Gauss' law (indeed this *is* Gauss' original law)

$$\oint_{\partial V} \boldsymbol{E} \cdot \boldsymbol{n} \, |dA| = \frac{1}{\epsilon_0} \int_V \rho \, |dX| = \frac{Q}{\epsilon_0}, \tag{7.44}$$

where Q is the enclosed charge. In this formula $\boldsymbol{n}$ is the outward pointing normal, formed from $d\boldsymbol{A} = I\boldsymbol{n}|dA|$, where $d\boldsymbol{A}$ is the directed measure over the surface, and the scalar measure $|dX|$ is simply

$$|dX| = dx \, dy \, dz. \tag{7.45}$$

For the next application, recall from section 6.4.7 the form of the Green's function for the vector derivative in three dimensions,

$$\boldsymbol{G}(\boldsymbol{r}; \boldsymbol{r}') = \frac{1}{4\pi} \frac{\boldsymbol{r} - \boldsymbol{r}'}{|\boldsymbol{r} - \boldsymbol{r}'|^3}. \tag{7.46}$$

An application of the fundamental theorem tells us that

$$\int_V (\dot{G}\dot{\nabla} E + G\nabla E)|dX| = -I\oint_{\partial V} G\,dA\,E. \tag{7.47}$$

If we assume that the sources are localised, so that E falls off at large distance, we can take the integral over all space and the right-hand side will vanish. Replacing G by the Green's function above we find that the field from a static charge distribution is given by

$$E(r) = \frac{1}{4\pi\epsilon_0}\int \frac{\rho(r')(r-r')}{|r-r'|^3}\,|dX'|. \tag{7.48}$$

If ρ is a single δ-function source, $\rho = Q\delta(r' - r_0)$, we immediately recover the Coulomb field

$$E(r) = \frac{Q}{4\pi\epsilon_0}\frac{(r-r_0)}{|r-r_0|^3}. \tag{7.49}$$

Unsurprisingly, this is simply a weighted Green's function.

For the magnetic field B, the absence of magnetic monopoles is encoded in the integral equation

$$\oint B\cdot dA = 0. \tag{7.50}$$

This tells us that the integral curves of B always form closed loops. This is true both inside and outside matter, and holds in the time-dependent case as well. Next we apply the integral theorem of equation (7.47) with E replaced by B. If we again assume that the fields are produced by localised charges and fall off at large distances, we derive

$$IB(r) = -\frac{\mu_0}{4\pi}\int \frac{(r-r')}{|r-r'|^3} J(r')\,|dX'|. \tag{7.51}$$

The scalar term in the integrand vanishes as a consequence of the static conservation law $\nabla\cdot J = 0$. The bivector term gives the magnetic field bivector IB. Now suppose that the current is carried entirely in an 'ideal' wire. This is taken as an infinitely thin wire carrying a current J,

$$J = J\int d\lambda\,\frac{dy(\lambda)}{d\lambda}\delta\big(r - y(\lambda)\big) = J\int dl\,\delta\big(r - y(\lambda)\big). \tag{7.52}$$

We have little option but to use J for the current as the more standard symbol I is already taken for the pseudoscalar. The result is that the B-field is determined by a line integral along the wire. This is the Biot–Savart law, which can be written

$$B(r) = \frac{\mu_0 J}{4\pi}\int \frac{dl'\times(r-r')}{|r-r'|^3}, \tag{7.53}$$

where r' is the position vector to the line element dl'.

A further integral theorem for magnetic fields is found if we consider the integral around a loop enclosing a surface σ. We have

$$\oint_{\partial\sigma} \boldsymbol{B}\cdot d\boldsymbol{l} = \int_\sigma (\dot{\boldsymbol{B}}\wedge\dot{\boldsymbol{\nabla}})\cdot d\boldsymbol{A} = \mu_0 \int_\sigma \boldsymbol{J}\cdot(-I\,d\boldsymbol{A}). \tag{7.54}$$

Again, we write $d\boldsymbol{A} = I\boldsymbol{n}|dA|$, where $\boldsymbol{n}$ is the unit right-handed normal. That is, if we grip the surface in our right hands in the manner specified by the line integral, our thumbs point in the normal direction. The result is that we integrate $\boldsymbol{J}\cdot\boldsymbol{n}$ over the surface. This returns the total current through the loop, J, recovering Ampère's law,

$$\oint_{\partial\sigma} \boldsymbol{B}\cdot d\boldsymbol{l} = \mu_0 J. \tag{7.55}$$

This is routinely used for finding the magnetic fields surrounding electrical circuits.

7.2.2 Time-varying fields

If the fields vary in time, some of the preceding formulae remain valid, and others only require simple modifications. The two applications of Gauss' law, equations (7.44) and (7.50), remain unchanged. The two applications of Stokes' theorem acquire an additional term. For the $\boldsymbol{E}$-field we have

$$\oint_{\partial\sigma} \boldsymbol{E}\cdot d\boldsymbol{l} = \frac{d}{dt}\int_\sigma (I\boldsymbol{B})\cdot d\boldsymbol{A} = -\frac{d\Phi}{dt}, \tag{7.56}$$

where Φ is the linked magnetic flux. The flux is the integral of $\boldsymbol{B}\cdot\boldsymbol{n}$ over the area enclosed by the loop, with $\boldsymbol{n}$ the unit normal. Magnetic flux is an important concept for understanding inductance in circuits.

For the magnetic field we can derive a similar formula,

$$\oint_{\partial\sigma} \boldsymbol{B}\cdot d\boldsymbol{l} = \mu_0 J + \epsilon_0\mu_0\frac{d}{dt}\int_\sigma \boldsymbol{E}\cdot\boldsymbol{n}\,|dA|. \tag{7.57}$$

This is useful when studying boundary conditions at surfaces of media carrying time-varying currents. The equations involving the Euclidean Green's function are no longer valid when the sources vary with time. In section 7.5 we discuss an alternative Green's function suitable for the important case of electromagnetic radiation.

7.2.3 The energy-momentum tensor

The energy density contained in a vacuum electromagnetic field, measured in the γ_0 frame, is

$$\varepsilon = \tfrac{1}{2}(\boldsymbol{E}^2 + \boldsymbol{B}^2), \tag{7.58}$$

where we have reverted to natural units. In section 7.1.2 we saw that the quantity $\boldsymbol{E}^2 - \boldsymbol{B}^2$ is Lorentz-invariant. This is not true of the energy density, which should clearly depend on the observer performing the measurement. The total energy in a volume V is found by integrating ε over the volume. If we look at how this varies in time, assuming no sources are present, we find that

$$\begin{aligned}\frac{d}{dt}\int_V |dX|\,\tfrac{1}{2}(\boldsymbol{E}^2+\boldsymbol{B}^2) &= \int_V |dX|\,\langle -\boldsymbol{E}\boldsymbol{\nabla}(I\boldsymbol{B}) + I\boldsymbol{B}\boldsymbol{\nabla}\boldsymbol{E}\rangle \\ &= \oint_{\partial V} |dA|\,\boldsymbol{n}\cdot\big(\boldsymbol{E}\cdot(I\boldsymbol{B})\big).\end{aligned} \tag{7.59}$$

We therefore establish that the field momentum is described by the Poynting vector

$$\boldsymbol{P} = -\boldsymbol{E}\cdot(I\boldsymbol{B}) = \boldsymbol{E}\times\boldsymbol{B}. \tag{7.60}$$

The energy and momentum should be the components of a spacetime 4-vector P, so we form

$$\begin{aligned}P = (\varepsilon + \boldsymbol{P})\gamma_0 &= \tfrac{1}{2}(\boldsymbol{E}^2+\boldsymbol{B}^2)\gamma_0 + \tfrac{1}{2}(I\boldsymbol{B}\boldsymbol{E} - \boldsymbol{E}I\boldsymbol{B})\gamma_0 \\ &= \tfrac{1}{2}(\boldsymbol{E}+I\boldsymbol{B})(\boldsymbol{E}-I\boldsymbol{B})\gamma_0 \\ &= \tfrac{1}{2}F(-\gamma_0 F\gamma_0)\gamma_0 = -\tfrac{1}{2}F\gamma_0 F.\end{aligned} \tag{7.61}$$

This quantity is still observer-dependent as it contains a factor of γ_0. We have in fact constructed the *energy-momentum tensor* of the electromagnetic field. We write this as

$$\mathsf{T}(a) = -\tfrac{1}{2}FaF = \tfrac{1}{2}Fa\tilde{F}. \tag{7.62}$$

This is clearly a linear function of a and, since it is equal to its own reverse, the result is automatically a vector. It is instructive to contrast our neat form of the energy-momentum tensor with the tensor formula

$$\mathsf{T}^{\mu}{}_{\nu} = \tfrac{1}{4}\delta^{\mu}_{\nu}F^{\alpha\beta}F_{\alpha\beta} + F^{\mu\alpha}F_{\alpha\nu}. \tag{7.63}$$

The geometric algebra form of equation (7.62) does a far better job of capturing the geometric content of the electromagnetic energy-momentum tensor.

The energy-momentum tensor $\mathsf{T}(a)$ returns the flux of 4-momentum across the hypersurface perpendicular to a. This is the relativistic extension of the stress tensor, and it is as fundamental to field theory as momentum is to the mechanics of point particles. All relativistic fields, classical or quantum, have an associated energy-momentum tensor that contains information about the distribution of energy in the fields, and acts as a source of gravitation. The electromagnetic energy-momentum tensor demonstrates a number of properties that turn out to be quite general. The first is that the energy-momentum tensor is (usually) symmetric. For example, we have

$$a\cdot\mathsf{T}(b) = -\tfrac{1}{2}\langle aFbF\rangle = -\tfrac{1}{2}\langle FaFb\rangle = \mathsf{T}(a)\cdot b. \tag{7.64}$$

The reason for qualifying the above statement is that quantum spin gives rise to an antisymmetric contribution to the (matter) energy-momentum tensor. This will be discussed in more details when we look at Dirac theory.

A second property of the electromagnetic energy-momentum tensor is that the energy density $v \cdot \mathsf{T}(v)$ is positive for any timelike vector v. This is clear from the definition of ε in equation (7.58). The expression for ε is appropriate the γ_0 frame, but the sign of ε cannot be altered by transforming to a different frame. The reason is that

$$\langle vFvF \rangle = \langle R\gamma_0 \tilde{R} F R \gamma_0 \tilde{R} F \rangle = \langle \gamma_0 F' \gamma_0 F' \rangle, \tag{7.65}$$

where $F' = \tilde{R} F R$. Transforming to a different velocity is equivalent to back-transforming the fields in the γ_0 frame, so keeps the energy density positive. Matter which does not satisfy the inequality $v \cdot \mathsf{T}(v) \geq 0$ is said to be '*exotic*', and has curious properties when acting as a source of gravitational fields.

The third main property of energy-momentum tensors is that, in the absence of external sources, they give rise to a set of conserved vectors. This is because we have

$$\nabla \cdot \mathsf{T}(a) = 0 \quad \forall \text{ constant } a. \tag{7.66}$$

Equivalently, we can use the symmetry of $\mathsf{T}(a)$ to write

$$\dot{\mathsf{T}}(\dot{\nabla}) \cdot a = 0, \quad \forall \, a, \tag{7.67}$$

which implies that

$$\dot{\mathsf{T}}(\dot{\nabla}) = 0. \tag{7.68}$$

For the case of electromagnetism, this result is straightforward to prove:

$$\dot{\mathsf{T}}(\dot{\nabla}) = -\tfrac{1}{2}[\dot{F}\dot{\nabla}F + F\nabla F] = 0, \tag{7.69}$$

which follows since $\nabla F = \dot{F}\dot{\nabla} = 0$ in the absence of sources.

Conservation of the energy-momentum tensor implies that the total flux of energy-momentum over a closed hypersurface is zero:

$$\int_{\partial V} |dA| \, \mathsf{T}(n) = 0, \tag{7.70}$$

where ∂V is a closed 3-surface with directed measure $dA = nI \, |dA|$. That the flux vanishes is a simple application of the fundamental theorem of integral calculus (in flat spacetime),

$$\int_{\partial V} \mathsf{T}(n \, |dA|) = \int_{\partial V} \mathsf{T}(dAI^{-1}) = \int_{V} \dot{\mathsf{T}}(\dot{\nabla}) \, dX \, I^{-1} = 0. \tag{7.71}$$

Given that $\mathsf{T}(\gamma_0)$ is the energy-momentum density in the γ_0 frame, the total 4-momentum is

$$P_{tot} = \int |dX| \, \mathsf{T}(\gamma_0). \tag{7.72}$$

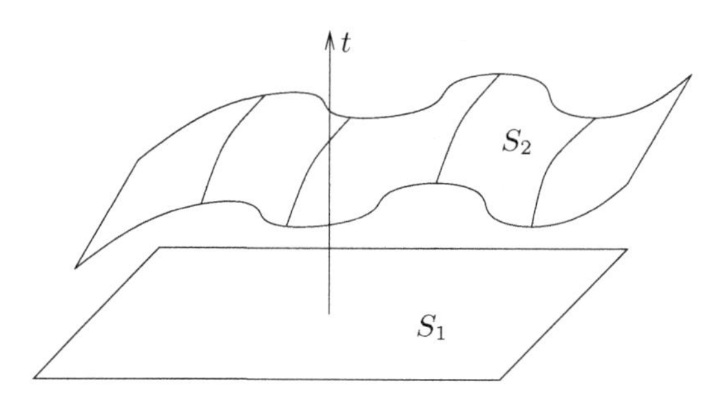

Figure 7.1 *Hypersurface integration.* The integral over a hypersurface of a (spacetime) conserved current is independent of the chosen hypersurface. The two surfaces S_1 and S_2 can be joined at spatial infinity (provided the fields vanish there). The difference is therefore the integral over a closed 3-surface, which vanishes by the divergence theorem.

The conservation equation (7.68) guarantees that, in the absence of charges, the total energy-momentum is conserved. We see that

$$\frac{d}{dt}P_{tot} = \int |dX|\, \partial_t \mathsf{T}(\gamma_0) = \int |dX|\, \dot{\mathsf{T}}(\dot{\boldsymbol{\nabla}}\gamma_0), \tag{7.73}$$

where we have used the fact that $\nabla = \gamma_0\partial_t - \boldsymbol{\nabla}\gamma_0$. The final integral here is a total derivative and so gives rise to a boundary term, which vanishes provided the fields fall off sufficiently fast at large distances. Similarly, we can also see that P_{tot} is independent of the chosen timelike axis. It is a covariant (non-local) property of the field configuration. The proof comes from considering the integral over two distinct spacelike hypersurfaces (figure 7.1). If the integrals are joined at infinity (which introduces zero contribution) we form a closed integral of $\mathsf{T}(n)$. This vanishes from the conservation equation, so the total energy-momentum is independent of the choice of hypersurface.

In the presence of additional sources the electromagnetic energy-momentum tensor is no longer conserved. The total energy-momentum tensor, including both the matter and electromagnetic content will be conserved, however. This is a general feature of field theory in a flat spacetime, though the picture is altered somewhat if gravitational fields are present. The extent to which the separate tensors for each field are not conserved contains useful information about the flow of energy-momentum. For example, suppose that an external current is present, so that

$$\dot{\mathsf{T}}(\dot{\nabla}) = -\tfrac{1}{2}(-JF + FJ) = J\cdot F. \tag{7.74}$$

An expression of the form $J\cdot F$ was derived in the Lorentz force law, discussed in section 5.5.3. In the γ_0 frame, $J\cdot F$ decomposes into

$$J\cdot F = \langle(\rho + \boldsymbol{J})\gamma_0(\boldsymbol{E} + I\boldsymbol{B})\rangle_1 = -\big(\boldsymbol{J}\cdot\boldsymbol{E} + \rho\boldsymbol{E} + \boldsymbol{J}\times\boldsymbol{B}\big)\gamma_0. \tag{7.75}$$

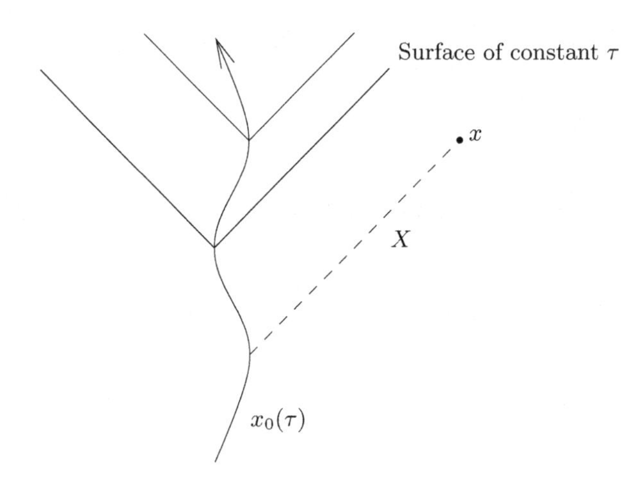

Figure 7.2 *Field from a moving point charge.* The charge follows the trajectory $x_0(\tau)$, and $X = x - x_0(\tau)$ is the retarded null vector connecting the point x to the worldline. The time τ can be viewed as a scalar field with each value of τ extended out over the forward null cone.

The timelike component, $\boldsymbol{J}\cdot\boldsymbol{E}$, is the work done — the rate of change of energy density. The relative vector term is the rate of change of field momentum, and so is closely related to the force on a point particle.

7.3 The electromagnetic field of a point charge

We now derive a formula for the electromagnetic fields generated by a radiating charge. This is one of the most important results in classical electromagnetic theory. Suppose that a charge q moves along a worldline $x_0(\tau)$, where τ is the proper time along the worldline (see figure 7.2). An observer at spacetime position x receives an electromagnetic influence from the point where the charge's worldline intersects the observer's past light-cone. The vector

$$X = x - x_0(\tau) \tag{7.76}$$

is the separation vector down the light-cone, joining the observer to this intersection point. Since this vector must be null, we can view the equation

$$X^2 = 0 \tag{7.77}$$

as defining a map from spacetime position x to a value of the particle's proper time τ. That is, for every spacetime position x there is a unique value of the (retarded) proper time along the charge's worldline for which the vector connecting x to the worldline is null. In this sense, we can write $\tau = \tau(x)$, and treat τ as a scalar field.

The Liénard–Wiechert potential for the retarded field from a point charge moving with an arbitrary velocity $v = \dot{x}_0$ is

$$A = \frac{q}{4\pi} \frac{v}{|X \cdot v|}. \tag{7.78}$$

This solution is obtained from the wave equation $\nabla^2 A = J$ using the appropriate retarded Green's function

$$G_{ret}(\boldsymbol{r}, t) = \frac{1}{4\pi|\boldsymbol{r}|} \delta(|\boldsymbol{r}| - t). \tag{7.79}$$

A similar solution exists if the advanced Green's function is used. The question of which is the correct one to use is determined experimentally by the fact that no convincing detection of an advanced (acausal) field has ever been reported. A deeper understanding of these issues is provided by the quantum treatment of radiation.

If the charge is at rest in the γ_0 frame, we have

$$x_0(\tau) = \tau\gamma_0 = (t - r)\gamma_0, \tag{7.80}$$

where r is the relative 3-space distance from the observer to the charge. The null vector X is therefore

$$X = r(\gamma_0 + e_r). \tag{7.81}$$

For this simple case the 4-potential A is a pure $1/r$ electrostatic field:

$$A = \frac{q}{4\pi} \frac{\gamma_0}{|X \cdot \gamma_0|} = \frac{q}{4\pi r} \gamma_0. \tag{7.82}$$

The same result is obtained if the advanced Green's function is used. The difference between the advanced and retarded solutions is only seen when the charge radiates. We know that radiation is not handled satisfactorily in the classical theory because it predicts that atoms are not stable and should radiate. Issues concerning the correct Green's function cannot be fully resolved without a quantum treatment.

7.3.1 The field strength

The aim now is to differentiate the potential of equation (7.78) to find the field strength. First, we differentiate the equation $X^2 = 0$ to obtain

$$\begin{aligned} 0 = \gamma^\mu (\partial_\mu X) \cdot X &= \dot{\nabla}\, \dot{x} \cdot X - \nabla\tau \, (\partial_\tau x_0) \cdot X \\ &= X - \nabla\tau \, (v \cdot X). \end{aligned} \tag{7.83}$$

It follows that

$$\nabla\tau = \frac{X}{X \cdot v}. \tag{7.84}$$

The gradient of τ points along X, which is the direction of constant τ. This is a peculiarity of null surfaces that was first encountered in chapter 6. In finding an expression for $\nabla\tau$ we have demonstrated how the particle proper time can be treated as a spacetime scalar field. Fields of this type are known as *adjunct* fields — they carry information, but do not exist in any physical sense.

To differentiate A we need an expression for $\nabla(X\cdot v)$. We find that

$$\begin{aligned}\nabla(X\cdot v) &= \dot{\nabla}(\dot{X})\cdot v + \nabla\tau\, X\cdot(\partial_\tau v)\\ &= v - \nabla\tau + \nabla\tau\, X\cdot\dot{v},\end{aligned} \tag{7.85}$$

where $\dot{v} = \partial_\tau v$. Provided X is defined in terms of the retarded time, $X\cdot v$ will always be positive and there is no need for the modulus in the denominator of equation (7.78). We are now in a position to evaluate ∇A. We find that

$$\begin{aligned}\nabla A &= \frac{q}{4\pi}\left(\frac{\nabla v}{X\cdot v} - \frac{1}{(X\cdot v)^2}\nabla(X\cdot v)v\right)\\ &= \frac{q}{4\pi}\left(\frac{X\dot{v}}{(X\cdot v)^2} - \frac{1}{(X\cdot v)^2} - \frac{(X\,X\cdot\dot{v} - X)v}{(X\cdot v)^3}\right)\\ &= \frac{q}{4\pi}\left(\frac{X\wedge\dot{v}}{(X\cdot v)^2} + \frac{X\wedge v - X\cdot\dot{v}\,X\wedge v}{(X\cdot v)^3}\right).\end{aligned} \tag{7.86}$$

The result is a pure bivector, so $\nabla\cdot A = 0$ and the A field of equation (7.78) is in the Lorentz gauge. This is to be expected, since the solution is obtained from the wave equation $\nabla^2 A = J$.

We can gain some insight into the expression for F by writing

$$X\cdot v\, X\wedge\dot{v} - X\cdot\dot{v}\, X\wedge v = -X\big(X\cdot(\dot{v}\wedge v)\big) = \tfrac{1}{2}X\dot{v}\wedge vX, \tag{7.87}$$

which uses the fact that $X^2 = 0$. Writing $\Omega_v = \dot{v}\wedge v$ for the acceleration bivector of the particle, we arrive at the compact formula

$$F = \frac{q}{4\pi}\frac{X\wedge v + \frac{1}{2}X\Omega_v X}{(X\cdot v)^3}. \tag{7.88}$$

One can proceed to show that, away from the worldline, F satisfies the free-field equation $\nabla F = 0$. The details are left as an exercise. The solution (7.88) displays a clean split into a velocity term proportional to $1/(\text{distance})^2$ and a long-range radiation term proportional to $1/(\text{distance})$. The term representing the distance is simply $X\cdot v$. This is just the distance between the events x and $x_0(\tau)$ as measured in the rest frame of the charge at its retarded position. The first term in equation (7.88) is the Coulomb field in the rest frame of the charge. The second, radiation, term:

$$F_{rad} = \frac{q}{4\pi}\frac{\frac{1}{2}X\Omega_v X}{(X\cdot v)^3}, \tag{7.89}$$

is proportional to the rest frame acceleration projected down the null vector X.

The fact that this term falls of as 1/(distance) implies that the energy-momentum tensor contains a term which falls of as the inverse square of distance. This gives a non-vanishing surface integral at infinity in equation (7.73) and describes how energy is carried away from the source.

7.3.2 Constant velocity

A charge with constant velocity v has the trajectory

$$x_0(\tau) = v\tau, \tag{7.90}$$

where we have chosen an origin so that the particle passes through this point at $\tau = 0$. The intersection of $x_0(\tau)$ with the past light-cone through x is determined by

$$(x - v\tau)^2 = 0 \quad \Rightarrow \tau = v\cdot x - \left((v\cdot x)^2 - x^2\right)^{1/2}. \tag{7.91}$$

We have chosen the earlier root to ensure that the intersection lies on the past light-cone. We now form $X\cdot v$ to find

$$X\cdot v = (x - v\tau)\cdot v = \left((v\cdot x)^2 - x^2\right)^{1/2}. \tag{7.92}$$

We can write this as $|x\wedge v|$ since

$$|x\wedge v|^2 = x\cdot\big(v\cdot(x\wedge v)\big) = (x\cdot v)^2 - x^2. \tag{7.93}$$

The acceleration bivector vanishes since v is constant, and $X\wedge v = x\wedge v$. It follows that the Faraday bivector is simply

$$F = \frac{q}{4\pi}\frac{x\wedge v}{|x\wedge v|^3}. \tag{7.94}$$

This is the Coulomb field solution with the velocity γ_0 replaced by v. This solution could be obtained by transforming the Coulomb field via

$$F \mapsto F' = RF(\tilde{R}xR)\tilde{R}, \tag{7.95}$$

where $v = R\gamma_0\tilde{R}$. Covariance of the field equations ensures that this process generates a new solution.

We next decompose F into electric and magnetic fields in the γ_0 frame. This requires the spacetime split

$$x\wedge v = \langle x\gamma_0\gamma_0 v\rangle_2 = \gamma\langle(t + \boldsymbol{r})(1 - \boldsymbol{v})\rangle_2 = \gamma(\boldsymbol{r} - \boldsymbol{v}t) - \gamma\boldsymbol{r}\wedge\boldsymbol{v}, \tag{7.96}$$

where $\boldsymbol{v}$ is the relative velocity and γ is the Lorentz factor. We now have

$$\boldsymbol{E} = \frac{q\gamma}{4\pi d^3}(\boldsymbol{r} - \boldsymbol{v}t), \qquad \boldsymbol{B} = \frac{q\gamma}{4\pi d^3} I\boldsymbol{r}\wedge\boldsymbol{v}. \tag{7.97}$$

Here, the effective distance d can be written

$$d^2 = \gamma^2(|\boldsymbol{v}|t - \boldsymbol{v}\cdot\boldsymbol{r}/|\boldsymbol{v}|)^2 + \boldsymbol{r}^2 - (\boldsymbol{r}\cdot\boldsymbol{v})^2/\boldsymbol{v}^2. \tag{7.98}$$

The electric field points towards the actual position of the charge at time t, and not its retarded position at time τ. The same is true of the advanced field, hence the retarded and advanced solutions are equal for charges with constant velocity.

7.3.3 Linear acceleration

Suppose that an accelerating charged particle follows the trajectory

$$x_0(\tau) = a\big(\sinh(g\tau)\gamma_0 + \cosh(g\tau)\gamma_3\big), \tag{7.99}$$

where $a = g^{-1}$ (see figure 7.3). The velocity is given by

$$v(\tau) = \cosh(g\tau)\gamma_0 + \sinh(g\tau)\gamma_3 = e^{g\tau\boldsymbol{\sigma}_3}\gamma_0 \tag{7.100}$$

and the acceleration bivector is simply

$$\dot{v}v = g\boldsymbol{\sigma}_3. \tag{7.101}$$

The charge has constant (relativistic) acceleration in the γ_3 direction. We again seek the retarded solution of $X^2 = 0$. This is more conveniently expressed in a cylindrical polar coordinate system, with

$$\boldsymbol{r} = \rho(\cos(\phi)\,\boldsymbol{\sigma}_1 + \sin(\phi)\,\boldsymbol{\sigma}_2) + z\boldsymbol{\sigma}_3, \tag{7.102}$$

so that $r^2 = \rho^2 + z^2$. We then find the following equivalent expressions for the retarded proper time:

$$\begin{aligned} e^{g\tau} &= \frac{1}{2a(z-t)}\Big(a^2 + r^2 - t^2 - \big((a^2+r^2-t^2)^2 - 4a^2(z^2-t^2)\big)^{1/2}\Big),\\ e^{-g\tau} &= \frac{1}{2a(z+t)}\Big(a^2 + r^2 - t^2 + \big((a^2+r^2-t^2)^2 - 4a^2(z^2-t^2)\big)^{1/2}\Big). \end{aligned} \tag{7.103}$$

These equations have a solution provided $z + t > 0$. As the trajectory assumes that the charge has been accelerating for ever, a *horizon* is formed beyond which no effects of the charge are felt (figure 7.3). Constant eternal acceleration of this type is unphysical and in practice we only consider the acceleration taking place for a short period.

We can now calculate the radiation from the charge. First we need the effective distance

$$X\cdot v = \frac{\big((a^2+r^2-t^2)^2 - 4a^2(z^2-t^2)\big)^{1/2}}{2a}. \tag{7.104}$$

This vanishes on the path of the particle ($\rho = 0$ and $z^2 - t^2 = a^2$), as required.

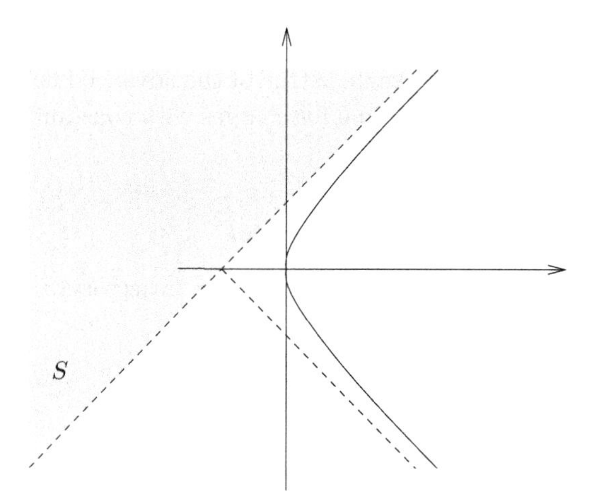

Figure 7.3 *Constant acceleration.* The spacetime trajectory of a particle with constant acceleration is a hyperbola. The asymptotes are null vectors and define future and past horizons. Any signal sent from within the shaded region S will never be received by the particle.

The remaining factor in F is

$$\begin{aligned}X \wedge v + \tfrac{1}{2} X \dot{v} v X &= x \wedge v - a\boldsymbol{\sigma}_3 + \frac{1}{2a}(x - x_0)\boldsymbol{\sigma}_3(x - x_0) \\ &= \frac{1}{2a} x \boldsymbol{\sigma}_3 x - \frac{a}{2}\boldsymbol{\sigma}_3 \\ &= \frac{1}{2a}(z^2 - \rho^2 - t^2 - a^2)\boldsymbol{\sigma}_3 + \frac{z\rho}{a}\boldsymbol{\sigma}_\rho + \frac{t\rho}{a} I \boldsymbol{\sigma}_\phi, \end{aligned} \tag{7.105}$$

where $\boldsymbol{\sigma}_\rho$ and $\boldsymbol{\sigma}_\phi$ are the unit spatial axial and azimuthal vectors respectively. An instructive way to display the information contained in the expression for F is to plot the field lines of $\boldsymbol{E}$ at a fixed time. We assume that the charge starts accelerating at $t = t_1$, and stops again at $t = t_2$. There are then discontinuities in the electric field line directions on the two appropriate light-spheres. In figure 7.4 the acceleration takes place for a short period of time, so that a pulse of radiation is sent outwards. In figure 7.5 the charge began accelerating from rest at $t = -10a$. The pattern is well developed, and shows clearly the refocusing of the field lines onto the 'image charge'. The image position corresponds to the place the charge would have reached had it not started accelerating. Of course, the image charge is not actually present, and the field lines diverge after they cross the light-sphere corresponding to the start of the acceleration.

For many applications we are only interested in the fields a long way from the source. In this region the fields can usually be approximated by simple dipole or higher order multipole fields. Suppose that the charge accelerates for a short

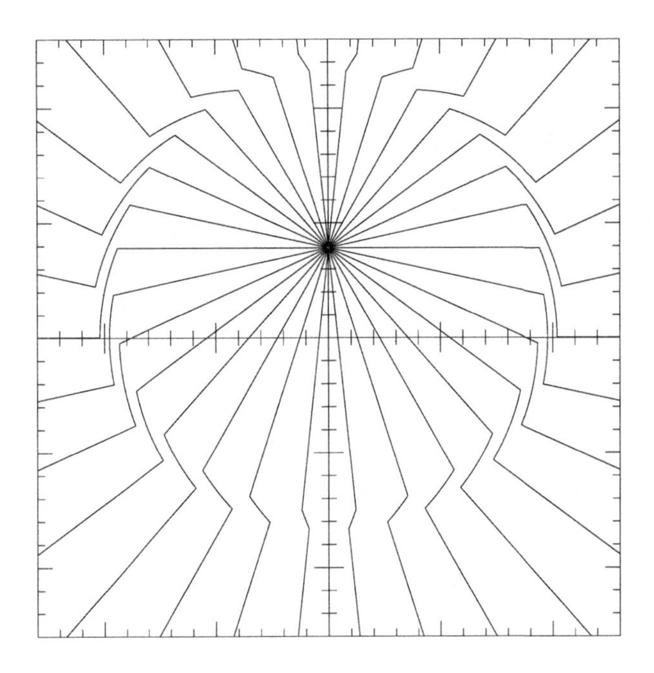

Figure 7.4 *Field lines from an accelerated charge I.* The charge accelerated for $-0.2a < t < 0.2a$, leaving an outgoing pulse of transverse radiation field. The field lines were computed at $t = 5a$.

period and emits a pulse of radiation. In the limit $r \gg a$ the pulse will arrive at some time which, to a good approximation, is centred around the time that minimises $X \cdot v$. This time is given by

$$t_0 = \sqrt{r^2 - a^2}. \tag{7.106}$$

At $t = t_0$ the proper distance $X \cdot v$ evaluates to ρ, the distance from the z axis. The point on the axis ρ away from the observer is where the charge would appear to be if it were not accelerating. For the large distance approximation to be valid we therefore also require that ρ is large, so that the proper distance from the source is large. (For small ρ and $z > a$ a different procedure can be used.) We can now obtain an approximate formula for the radiation field at a fixed location $\boldsymbol{r}$, with $r, \rho \gg a$, around $t = t_0$. For this we define

$$\delta_t = t - t_0 \tag{7.107}$$

so that the proper distance is approximated by

$$X \cdot v \approx \left(\rho^2 + r^2 \delta_t^2 / a^2\right)^{1/2}. \tag{7.108}$$

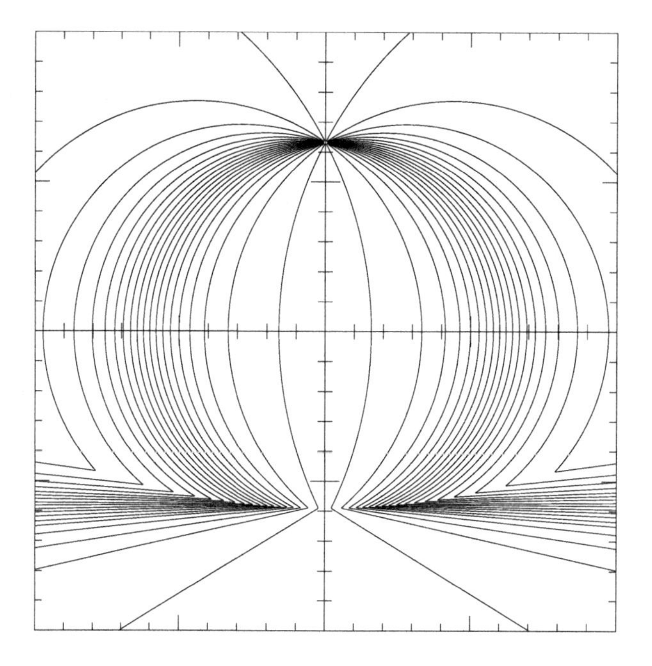

Figure 7.5 *Field lines from an accelerated charge II.* The charge began its acceleration at $t_1 = -10a$ and has thereafter accelerated uniformly. The field lines are plotted at $t = 3a$.

The remaining terms in F become

$$X \wedge v + \tfrac{1}{2} X \dot{v} v X \approx \frac{r\rho}{a} \left(\boldsymbol{\sigma}_\theta + I \boldsymbol{\sigma}_\phi \right), \tag{7.109}$$

where $\boldsymbol{\sigma}_\theta$ and $\boldsymbol{\sigma}_\phi$ are unit spherical-polar basis vectors. The final formula is

$$F \approx \frac{q}{4\pi} \frac{r\rho}{a} \left(\rho^2 + \frac{r^2 \delta_t^2}{a^2} \right)^{-3/2} \left(\boldsymbol{\sigma}_\theta + I \boldsymbol{\sigma}_\phi \right), \tag{7.110}$$

which describes a pure, outgoing radiation field a large distance from a linearly accelerating source. The magnitude of the acceleration is controlled by $g = a^{-1}$.

7.3.4 Circular orbits and synchrotron radiation

As a further application, consider a charge moving in a circular orbit. The worldline is defined by

$$x_0 = \tau \cosh(\alpha)\, \gamma_0 + a\big(\cos(\omega\tau)\gamma_1 + \sin(\omega\tau)\gamma_2\big), \tag{7.111}$$

where $a = \omega^{-1} \sinh(\alpha)$. The particle velocity is

$$v = \cosh(\alpha)\, \gamma_0 + \sinh(\alpha)\big(-\sin(\omega\tau)\gamma_1 + \cos(\omega\tau)\gamma_2\big) = R\gamma_0 \tilde{R}, \tag{7.112}$$

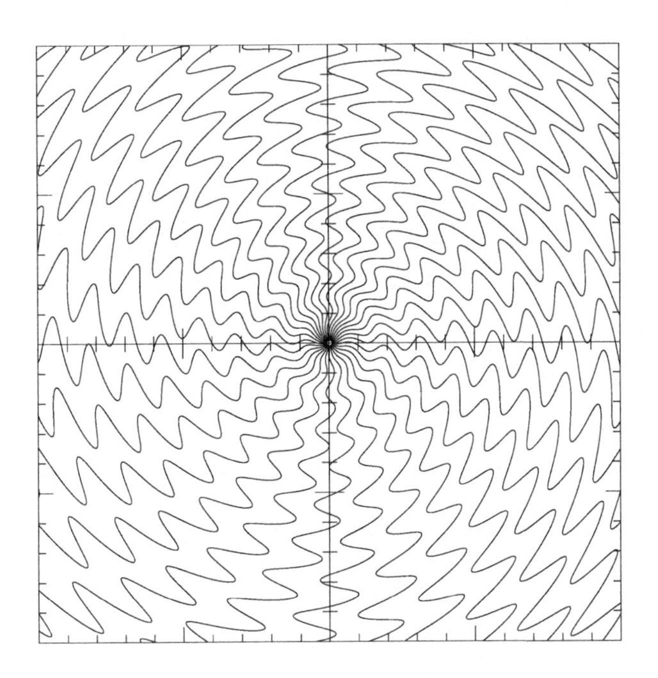

Figure 7.6 *Field lines from a rotating charge I.* The charge has $\alpha = 0.1$, which gives rise to a smooth, wavy pattern.

where the rotor R is given by

$$R = e^{-\omega\tau I\boldsymbol{\sigma}_3/2} e^{\alpha\boldsymbol{\sigma}_2/2}. \tag{7.113}$$

We must first locate the retarded null vector X. The equation $X^2 = 0$ reduces to

$$t = \tau\cosh(\alpha) + \left(r^2 + a^2 - 2a\rho\cos(\omega\tau - \phi)\right)^{1/2}, \tag{7.114}$$

which is an implicit equation for $\tau(x)$. No simple analytic solution exists, but a numerical solution is easy to achieve. This is aided by the observation that, for fixed $\boldsymbol{r}$, the mapping between t and τ is monotonic and τ is bounded by the conditions

$$t - \left(r^2 + 2a\rho + a^2\right)^{1/2} < \tau\cosh(\alpha) < t - \left(r^2 - 2a\rho + a^2\right)^{1/2}. \tag{7.115}$$

Once we have a satisfactory procedure for locating τ on the retarded light-cone, we can straightforwardly employ the formula for F in numerical simulations. The first term required is the effective distance $X \cdot v$, which is given by

$$X \cdot v = \cosh(\alpha)\left(r^2 + a^2 - 2a\rho\cos(\omega\tau - \phi)\right)^{1/2} + \rho\sinh(\alpha)\,\sin(\omega\tau - \phi). \tag{7.116}$$

The remaining term to compute, $X \wedge v + X\dot{v}vX/2$, is more complicated, as can be seen from the behaviour shown in figures 7.6, 7.7 and 7.8. They show the

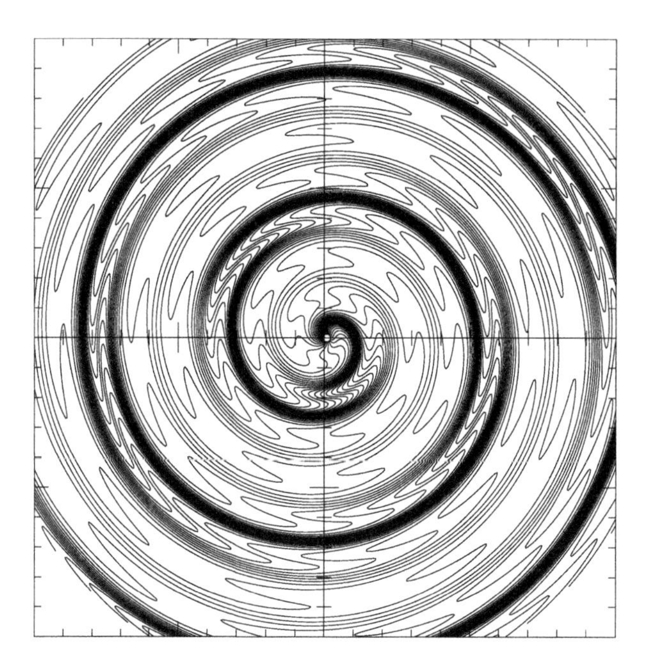

Figure 7.7 *Field lines from a rotating charge II.* The charge has an intermediate velocity, with $\alpha = 0.4$. Bunching of the field lines is clearly visible.

field lines in the equatorial plane of a rotating charge with $\omega = 1$. For 'low' speeds we get the gentle, wavy pattern of field lines shown in figure 7.6. The case displayed in figure 7.7 is for an intermediate velocity ($\alpha = 0.4$), and displays many interesting features. By $\alpha = 1$ (figure 7.8) the field lines have concentrated into synchrotron pulses, a pattern which continues thereafter.

Synchrotron radiation is important in many areas of physics, from particle physics through to radioastronomy. Synchrotron radiation from a radiogalaxy, for example, has $a \approx 10^8$ m and $r \approx 10^{25}$ m. A power-series expansion in a/r is therefore quite safe! Typical values of $\cosh(\alpha)$ are 10^4 for electrons producing radio emission. In the limit $r \gg a$, the relation between t and τ simplifies to

$$t - r \approx \tau \cosh(\alpha) - a \sin(\theta) \cos(\omega\tau - \phi). \tag{7.117}$$

The effective distance reduces to

$$X \cdot v \approx r \cosh(\alpha) \big(1 + \tanh(\alpha) \sin(\theta) \sin(\omega\tau - \phi)\big), \tag{7.118}$$

and the null vector X given by the simple expression

$$X \approx r(\gamma_0 + e_r). \tag{7.119}$$

In the expression for F of equation (7.88) we can ignore the $X \wedge v$ (Coulomb)

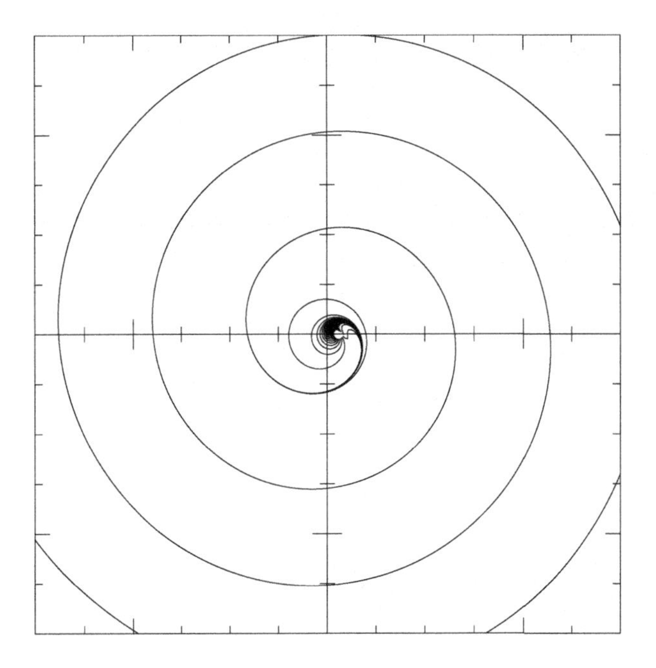

Figure 7.8 *Field lines from a rotating charge III.* The charge is moving at a highly relativistic velocity, with $\alpha = 1$. The field lines are concentrated into a series of synchrotron pulses.

term, which is negligible compared with the long-range radiation term. For the radiation term we need the acceleration bivector

$$\dot{v}v = -\omega \sinh(\alpha)\cosh(\alpha)\big(\cos(\omega\tau)\boldsymbol{\sigma}_1 + \sin(\omega\tau)\boldsymbol{\sigma}_2\big) + \omega \sinh^2(\alpha) I\boldsymbol{\sigma}_3. \quad (7.120)$$

The radiation term is governed by $X\Omega_v X/2$, which simplifies to

$$\begin{aligned}\tfrac{1}{2}X\dot{v}vX \approx\; &\omega r^2 \cosh(\alpha)\sinh(\alpha)(\cos(\theta)\cos(\omega\tau - \phi)\boldsymbol{\sigma}_\theta(1 - \boldsymbol{\sigma}_r)\\ &+ \omega r^2 \sinh(\alpha)\big(\cosh(\alpha)\sin(\omega\tau - \phi) + \sinh(\alpha)\sin(\theta)\big)\boldsymbol{\sigma}_\phi(1 - \boldsymbol{\sigma}_r). \quad (7.121)\end{aligned}$$

These formulae are sufficient to initiate studying synchrotron radiation. They contain a wealth of physical information, but a detailed study is beyond the scope of this book.

7.4 Electromagnetic waves

For many problems in electromagnetic theory it is standard practice to adopt a complex representation of the electromagnetic field, with the implicit assumption that only the real part represents the physical field. This is particularly convenient when discussing electromagnetic waves and diffraction, as studied in this and the following section. We have seen, however, that the field strength

F is equipped with a natural complex structure through the pseudoscalar I. We should therefore not be surprised to find that, in certain cases, the formal imaginary i plays the role of the pseudoscalar. This is indeed the case for circularly-polarised light. But one cannot always identify i with I, as is clear when handling plane-polarised light. The formal complexification retains its usefulness in such applications and we accordingly adopt it here. It is important to remember that this is a formal exercise, and that real parts must be taken before forming bilinear objects such as the energy-momentum tensor. The study of electromagnetic waves is an old and well-developed subject. Unfortunately, it suffers from the lack of a single, universal set of conventions. As far as possible, we have followed the conventions of Jackson (1999).

We seek vacuum solutions to the Maxwell equations which are purely oscillatory. We therefore start by writing

$$F = \mathrm{Re}\big(F_0 \mathrm{e}^{-ik\cdot x}\big). \tag{7.122}$$

The vacuum equation $\nabla F = 0$ then reduces to the algebraic equation

$$kF_0 = 0. \tag{7.123}$$

Pre-multiplying by k we immediately see that $k^2 = 0$, as expected of the wavevector. The constant bivector F_0 must contain a factor of k, as nothing else totally annihilates k. We therefore must have

$$F_0 = k\wedge n = kn, \tag{7.124}$$

where n is some vector satisfying $k\cdot n = 0$. We can always add a further multiple of k to n, since

$$k(n + \lambda k) = kn + \lambda k^2 = k\wedge n. \tag{7.125}$$

This freedom in n can be employed to ensure that n is perpendicular to the velocity vector of some chosen observer.

As an example, consider a wave travelling in the γ_3 direction with frequency ω as measured in the γ_0 frame. This implies that $\gamma_0\cdot k = \omega$, so the wavevector is given by

$$k = \omega(\gamma_0 + \gamma_3), \tag{7.126}$$

and the phase term is

$$-ik\cdot x = -i\omega(t - z). \tag{7.127}$$

The vector n can be chosen to just contain γ_1 and γ_2 components, so we can write

$$\begin{aligned} F &= -(\gamma_0 + \gamma_3)(\alpha_1\gamma_1 + \alpha_2\gamma_2)\cos(k\cdot x) \\ &= (1 + \boldsymbol{\sigma}_3)(\alpha_1\boldsymbol{\sigma}_1 + \alpha_2\boldsymbol{\sigma}_2)\cos(k\cdot x). \end{aligned} \tag{7.128}$$

This solution represents plane-polarised light, as both the $\boldsymbol{E}$ and $\boldsymbol{B}$ fields lie in fixed planes, 90° apart, and only their magnitudes oscillate in time.

An arbitrary phase can be added to the cosine term, so the most general solution for a wave travelling in the $+z$ direction is

$$F = (1+\boldsymbol{\sigma}_3)\big((\alpha_1\boldsymbol{\sigma}_1 + \alpha_2\boldsymbol{\sigma}_2)\cos(k\cdot x) + (\beta_1\boldsymbol{\sigma}_1 + \beta_2\boldsymbol{\sigma}_2)\sin(k\cdot x)\big), \tag{7.129}$$

where the constants α_i and β_i, are all real. This general solution can describe all possible states of polarisation. A convenient representation is to introduce the complex coefficients

$$c_1 = \alpha_1 + i\beta_1, \quad c_2 = \alpha_2 + i\beta_2. \tag{7.130}$$

These form the components of the complex *Jones vector* (c_1, c_2). In terms of these components we can write

$$F = \mathrm{Re}\big((1+\boldsymbol{\sigma}_3)(c_1\boldsymbol{\sigma}_1 + c_2\boldsymbol{\sigma}_2)\mathrm{e}^{-ik\cdot x}\big), \tag{7.131}$$

and it is a straightforward matter to read off the separate $\boldsymbol{E}$ and $\boldsymbol{B}$ fields.

The multivector $(1+\boldsymbol{\sigma}_3)$ has a number of interesting properties. It absorbs factors of $\boldsymbol{\sigma}_3$, as can be seen from

$$\boldsymbol{\sigma}_3(1+\boldsymbol{\sigma}_3) = 1+\boldsymbol{\sigma}_3. \tag{7.132}$$

In addition, $(1+\boldsymbol{\sigma}_3)$ squares to give a multiple of itself,

$$(1+\boldsymbol{\sigma}_3)^2 = 1 + 2\boldsymbol{\sigma}_3 + \boldsymbol{\sigma}_3^2 = 2(1+\boldsymbol{\sigma}_3). \tag{7.133}$$

This property implies that $(1+\boldsymbol{\sigma}_3)$ does not have an inverse, so in a multivector expression it acts as a projection operator. The combination $(1+\boldsymbol{\sigma}_3)/2$ has the particular property of squaring to give itself back again. Multivectors with this property are said to be *idempotent* and are important in the general classification of Clifford algebras and their spinor representations. In spacetime applications idempotents invariably originate from a null vector, in the manner that $(1+\boldsymbol{\sigma}_3)$ originates from a spacetime split of $\gamma_0 + \gamma_3$.

7.4.1 Circularly-polarised light

Many problems are more naturally studied using a basis of circularly-polarised states, as opposed to plane-polarised ones. These arise when c_1 and c_2 are $\pi/2$ out of phase. One form is given by $\alpha_1 = -\beta_2 = E_0$ and $\alpha_2 = \beta_1 = 0$, where E_0 denotes the magnitude of the electric field. For this solution we can write

$$\begin{aligned} F &= E_0(1+\boldsymbol{\sigma}_3)\big(\boldsymbol{\sigma}_1\cos(k\cdot x) - \boldsymbol{\sigma}_2\sin(k\cdot x)\big) \\ &= E_0(1+\boldsymbol{\sigma}_3)\boldsymbol{\sigma}_1\mathrm{e}^{-I\boldsymbol{\sigma}_3\omega(t-z)}. \end{aligned} \tag{7.134}$$

In a plane of constant z (a wavefront) the $\boldsymbol{E}$ field rotates in a clockwise (negative) sense, when viewed looking back towards the source (figure 7.9). In the optics

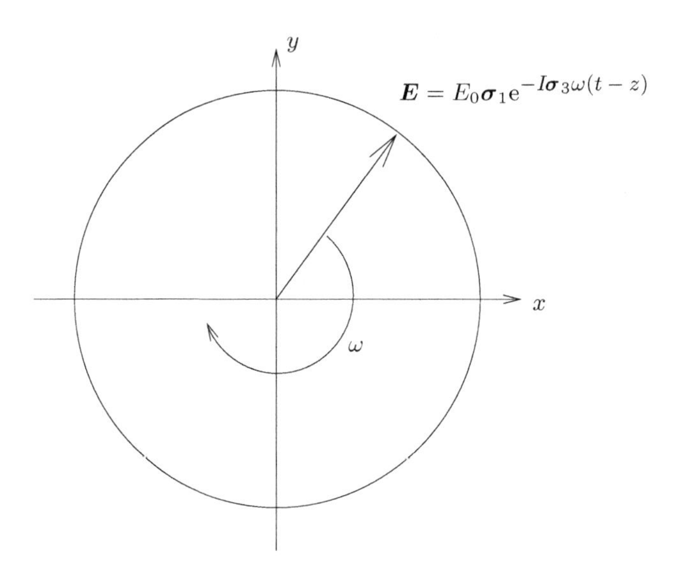

Figure 7.9 *Right-circularly-polarised light.* In the $z = 0$ plane the $\boldsymbol{E}$ vector rotates clockwise, when viewed from above. The wave vector points out of the page. In space, at constant time, the $\boldsymbol{E}$ field sweeps out a right-handed helix.

literature this is known as *right-circularly-polarised* light. The reason for this is that, at constant time, the $\boldsymbol{E}$ field sweeps out a helix in space which defines a right-handed screw. If you grip the helix in your right hand, your thumb points in the direction in which the helix advances if tracked along in the sense defined by your grip. This definition of handedness for a helix is independent of which way round you chose to grip it.

Left-circularly-polarised light has the $\boldsymbol{E}$ field rotating with the opposite sense. The general form of this solution is

$$F = (1 + \boldsymbol{\sigma}_3)(\alpha_1\boldsymbol{\sigma}_1 + \alpha_2\boldsymbol{\sigma}_2)e^{I\boldsymbol{\sigma}_3 k\cdot x}. \tag{7.135}$$

Particle physicists prefer an alternative labelling scheme for circularly-polarised light. The scheme is based, in part, on the quantum definition of angular momentum. In the quantum theory, the total angular momentum consists of a spatial part and a spin component. Photons, the quanta of electromagnetic radiation, have spin-1. The spin vector for these can either point in the direction of propagation, or against it, depending on the orientation of rotation of the $\boldsymbol{E}$ field. It turns out that for *right*-circularly-polarised light the spin vector points against the direction of propagation, which is referred to as a state of *negative helicity*. Conversely, *left*-circularly-polarised light has *positive* helicity.

Equation (7.132) enables us to convert phase rotations with the bivector $I\boldsymbol{\sigma}_3$

into duality rotations governed by the pseudoscalar I. This relies on the relation

$$\begin{aligned}(1+\boldsymbol{\sigma}_3)\mathrm{e}^{I\boldsymbol{\sigma}_3\phi} &= (1+\boldsymbol{\sigma}_3)\big(\cos(\phi)+I\boldsymbol{\sigma}_3\sin(\phi)\big) \\ &= (1+\boldsymbol{\sigma}_3)\big(\cos(\phi)+I\sin(\phi)\big) = (1+\boldsymbol{\sigma}_3)\mathrm{e}^{I\phi}. \end{aligned} \tag{7.136}$$

The general solution for right-circularly-polarised light can now be written

$$\begin{aligned}F &= (1+\boldsymbol{\sigma}_3)\mathrm{e}^{I\boldsymbol{\sigma}_3 k\cdot x}(\alpha_1\boldsymbol{\sigma}_1+\alpha_2\boldsymbol{\sigma}_2) \\ &= (1+\boldsymbol{\sigma}_3)(\alpha_1\boldsymbol{\sigma}_1+\alpha_2\boldsymbol{\sigma}_2)\mathrm{e}^{Ik\cdot x}. \end{aligned} \tag{7.137}$$

In this case the complex structure is now entirely geometric, generated by the pseudoscalar. This means that there is no longer any need to take the real part of the solution, as the bivector is already entirely real. A similar trick can be applied to write the constant terms as

$$(1+\boldsymbol{\sigma}_3)(\alpha_1\boldsymbol{\sigma}_1+\alpha_2\boldsymbol{\sigma}_2) = (1+\boldsymbol{\sigma}_3)\boldsymbol{\sigma}_1(\alpha_1 - I\alpha_2), \tag{7.138}$$

so that the coefficient also becomes 'complex' on the pseudoscalar. The general form for right-hand circularly-polarised light solution can now be written

$$F = (1+\boldsymbol{\sigma}_3)\boldsymbol{\sigma}_1\alpha_R\mathrm{e}^{Ik\cdot x}, \tag{7.139}$$

where α_R is a scalar + pseudoscalar combination. Left-hand circularly-polarised light is described by reversing the sign of the exponent to $-Ik\cdot x$. General polarisation states can be built up as linear combinations of these circularly polarised modes, so we can write

$$F = (1+\boldsymbol{\sigma}_3)\boldsymbol{\sigma}_1\big(\alpha_R\mathrm{e}^{Ik\cdot x}+\alpha_L\mathrm{e}^{-Ik\cdot x}\big). \tag{7.140}$$

Here both the coefficients α_L and α_R are scalar + pseudoscalar combinations. The complexification is now based on the pseudoscalar, and we can use α_R and α_L as alternative, geometrically meaningful, complex coefficients for describing general polarisation states. For completeness, the α_L and α_R parameters are related to the earlier plane-polarised coefficients α_i and β_i by

$$\begin{aligned}\alpha_R &= \tfrac{1}{2}(\alpha_1-\beta_2)+\tfrac{1}{2}(\alpha_2+\beta_1)I, \\ \alpha_L &= \tfrac{1}{2}(\alpha_1+\beta_2)+\tfrac{1}{2}(\alpha_2-\beta_1)I. \end{aligned} \tag{7.141}$$

The preceding solutions all assume that the wave vector is entirely in the $\boldsymbol{\sigma}_3$ direction. More generally, we can introduce a right-handed coordinate frame $\{\mathsf{e}_i\}$, with e_3 pointing along the direction of propagation. The solutions then all generalise straightforwardly. In more covariant notation the circularly-polarised modes can also be written

$$F = kn\big(\alpha_R\mathrm{e}^{Ik\cdot x}+\alpha_L\mathrm{e}^{-Ik\cdot x}\big), \tag{7.142}$$

where $k\cdot n = 0$.

7.4.2 Stokes parameters

A useful way of describing the state of polarisation in light emitted from some source is through the Stokes parameters. The general definition of these involves time averages of the fields, which we denote here with an overbar. To start with we assume that the light is coherent, so that all modes are in the same state. We first define the Stokes parameters in terms of the plane-polarised coefficients. The electric field is given by

$$\boldsymbol{E} = \mathrm{Re}\big((c_1\boldsymbol{\sigma}_1 + c_2\boldsymbol{\sigma}_2)\mathrm{e}^{-ik\cdot x}\big) = \mathrm{Re}(\mathcal{E}), \tag{7.143}$$

where $\mathcal{E}$ denotes the complex amplitude. The first Stokes parameter gives the magnitude of the electric field,

$$s_0 = 2\overline{\boldsymbol{E}^2} = \langle \mathcal{E}\mathcal{E}^* \rangle, \tag{7.144}$$

where the star denotes complex conjugation. This evaluates straightforwardly to

$$s_0 = |c_1|^2 + |c_2|^2. \tag{7.145}$$

The remaining three Stokes parameters describe the relative amounts of radiation present in various polarisation states. If we denote the real components of $\boldsymbol{E}$ by E_x and E_y the parameters are defined by

$$\begin{aligned} s_1 &= 2(\overline{E_x^2} - \overline{E_y^2}) = |c_1|^2 - |c_2|^2 \\ s_2 &= 4\overline{E_x E_y} = 2\mathrm{Re}(c_1 c_2^*) \\ s_3 &= 4\overline{E_x(t)E_y(t + \pi/(2\omega))} = -2\mathrm{Im}(c_1 c_2^*). \end{aligned} \tag{7.146}$$

The Stokes parameters can equally well be written in terms of the α_L and α_R coefficients of circularly-polarised modes:

$$\begin{aligned} s_0 &= 2(|\alpha_L|^2 + |\alpha_R|^2), \\ s_1 &= 4\langle \alpha_L \alpha_R \rangle, \\ s_2 &= -4\langle I\alpha_L \alpha_R \rangle, \\ s_3 &= 2(|\alpha_L^2| - |\alpha_R|^2). \end{aligned} \tag{7.147}$$

For coherent light the Stokes parameters are related by

$$s_0^2 = s_1^2 + s_2^2 + s_3^2. \tag{7.148}$$

The s_μ can therefore be viewed algebraically as the components of a null vector, though its direction in space has no physical significance. This representation for 'observables' in terms of a null vector is typical of a two-state quantum system. We can bring this out neatly in the spacetime algebra by introducing the three-dimensional rotor

$$\kappa = \langle \alpha_L \rangle + \langle I\alpha_L \rangle \boldsymbol{I\sigma}_3 - \langle \alpha_R \rangle \boldsymbol{I\sigma}_2 - \langle I\alpha_R \rangle \boldsymbol{I\sigma}_1. \tag{7.149}$$

The (quantum) origin of this object is explained in section 8.1. The rotor κ satisfies

$$\kappa\kappa^\dagger = \tfrac{1}{2}s_0, \qquad \kappa\boldsymbol{\sigma}_3\kappa^\dagger = \tfrac{1}{2}s_i\boldsymbol{\sigma}_i. \tag{7.150}$$

It follows that in spacetime

$$2\kappa(\gamma_0+\gamma_3)\tilde{\kappa} = 2\kappa(1+\boldsymbol{\sigma}_3)\kappa^\dagger\gamma_0 = s_0\gamma_0 + s_i\gamma_i, \tag{7.151}$$

and since we have rotated a null vector we automatically obtain a null vector. The unit spatial vector

$$\hat{\boldsymbol{s}} = \frac{\boldsymbol{s}}{s_0}, \qquad \boldsymbol{s} = s_i\boldsymbol{\sigma}_i \tag{7.152}$$

can be represented by a point on a sphere. For light polarisation states this is called the *Poincaré sphere*. For spin-1/2 systems the equivalent construction is known as the *Bloch sphere*. The construction is also useful for describing partially coherent light. In this case the light can be viewed as originating from a set of discrete (incoherent) sources. The single null vector is replaced by an average over the sources,

$$s = \sum_{k=1}^{n} s_k \tag{7.153}$$

and the unit vector $\hat{\boldsymbol{s}}$ is replaced by

$$\boldsymbol{s} = \frac{s\wedge\gamma_0}{s\cdot\gamma_0} = \sum_{k=1}^{n}\frac{\omega_k}{\omega}\hat{\boldsymbol{s}}_k, \qquad \omega = \sum_{k=1}^{n}\omega_k. \tag{7.154}$$

The resulting polarisation vector $\boldsymbol{s}$ has $\boldsymbol{s}^2 \leq 1$, so now defines a vector inside the Poincaré sphere. The length of this vector directly encodes the relative amounts of coherent and incoherent light present.

The preceding discussion also makes it a simple matter to compute how the Stokes parameters appear to observers moving at different velocities. Suppose that a second observer with velocity $v = \mathsf{e}_0$ sets up a frame $\{\mathsf{e}_\mu\}$. This is done in such a way that the wave vector still travels in the e_3 direction, which requires that

$$\mathsf{e}_3 = \frac{k - k\cdot v\, v}{k\cdot v}. \tag{7.155}$$

If the old and new frames are related by a rotor, $\mathsf{e}_\mu = R\gamma_\mu\tilde{R}$, then equation (7.155) restricts R to satisfy

$$Rk\tilde{R} = \lambda k. \tag{7.156}$$

Rather than work in the new frame, it is simpler to back-transform the field F and work in the original $\{\gamma_\mu\}$ frame. We define

$$F' = \tilde{R}F(Rx\tilde{R})R = \frac{1}{\lambda}kn'\left(\alpha_R e^{Ik\cdot x/\lambda} + \alpha_L e^{-Ik\cdot x/\lambda}\right), \tag{7.157}$$

where $n' = \tilde{R}nR$ and $k = \omega(\gamma_0+\gamma_3)$. We can again choose n' to be perpendicular to γ_0 by adding an appropriate multiple of k. It follows that the only change to the final vector n can be a rotation in the $I\boldsymbol{\sigma}_3$ plane. Performing a spacetime split on γ_0, and assuming that the original n was $-\gamma_1$, we obtain

$$F' = \frac{1}{\lambda}(1+\boldsymbol{\sigma}_3)\boldsymbol{\sigma}_1 \mathrm{e}^{-\phi I\boldsymbol{\sigma}_3}\left(\alpha_R \mathrm{e}^{Ik\cdot x/\lambda} + \alpha_L \mathrm{e}^{-Ik\cdot x/\lambda}\right), \tag{7.158}$$

where ϕ is the angle of rotation in the $I\boldsymbol{\sigma}_3$ plane. The rotation can again be converted to a phase factor on I, so the overall change is that α_R and α_L are multiplied by $\lambda^{-1}\exp(I\phi)$. The rescaling has no effect on the unit vector on the Poincaré sphere, so the only change is a rotation through 2ϕ in the $I\boldsymbol{\sigma}_3$ plane. This implies that the $\boldsymbol{\sigma}_3$ component of the vector on the Poincaré sphere is constant, which is sensible. This component determines the relative amounts of left and right-circularly-polarised light present, and this ratio is independent of which observer measures it. Similar arguments apply to the case of partially coherent light.

7.5 Scattering and diffraction

We turn now to the related subjects of the scattering and diffraction of electromagnetic waves. This is an enormous subject and our aim here is to provide little more than an introduction, highlighting in particular a unified approach based on the free-space multivector Green's function. This provides a first-order formulation of the scattering problem, which is valuable in numerical computation. We continue to adopt a complex representation for the electromagnetic field, and will concentrate on waves of a single frequency. The time dependence is then expressed via

$$F(x) = F(\boldsymbol{r})\mathrm{e}^{-i\omega t}, \tag{7.159}$$

so that the Maxwell equations reduce to

$$\boldsymbol{\nabla} F - i\omega F = 0. \tag{7.160}$$

This is the first-order equivalent of the vector Helmholtz equation. Throughout this section we work with the full, complex quantities, and suppress all factors of $\exp(i\omega t)$. All quadratic quantities are assumed to be time averaged.

If sources are present the Maxwell equations become

$$(\boldsymbol{\nabla} - i\omega)F = \rho - \boldsymbol{J}. \tag{7.161}$$

Current conservation tells us that the (complex) current satisfies

$$i\omega\rho = \boldsymbol{\nabla}\cdot\boldsymbol{J}. \tag{7.162}$$

Provided that all the sources are localised in some region in space, there can be no electric monopole term present. This follows because

$$Q = \int |dX| \rho = \frac{1}{i\omega} \oint \boldsymbol{J} \cdot \boldsymbol{n} \, |dA|, \tag{7.163}$$

where $\boldsymbol{n}$ is the outward normal. Taking the surface to totally enclose the sources, so that $\boldsymbol{J}$ vanishes over the surface of integration, we see that $Q = 0$.

7.5.1 First-order Green's function

The main result we employ in this section is Green's theorem in three dimensions in the general form

$$\int_V (\dot{G}\dot{\boldsymbol{\nabla}} F + G\, \boldsymbol{\nabla} F) \, |dX| = \oint_{\partial V} G n F \, dA \tag{7.164}$$

where $\boldsymbol{n}$ is the outward-pointing normal vector over the surface ∂V. If F satisfies the vacuum Maxwell equations, we have

$$\oint_{\partial V} G n F \, dA = \int_V (\dot{G}\dot{\boldsymbol{\nabla}} + i\omega G) F \, |dX|. \tag{7.165}$$

We therefore seek a Green's function satisfying

$$\dot{G}\dot{\boldsymbol{\nabla}} + i\omega G = \delta(\boldsymbol{r}). \tag{7.166}$$

It will turn out that G only contains (complex) scalar and vector terms, so (by reversing both sides) this equation is equivalent to

$$(\boldsymbol{\nabla} + i\omega) G = \delta(\boldsymbol{r}). \tag{7.167}$$

The Green's function is easily found from the Green's function for the (scalar) Helmholtz equation,

$$\phi(\boldsymbol{r}) = -\frac{1}{4\pi r} \mathrm{e}^{i\omega r}. \tag{7.168}$$

This is appropriate for *outgoing* radiation. Choosing the outgoing Green's function is equivalent to imposing causality by working with retarded fields. The function ϕ satisfies

$$(\boldsymbol{\nabla}^2 + \omega^2)\phi = \delta(\boldsymbol{r}) = (\boldsymbol{\nabla} + i\omega)(\boldsymbol{\nabla} - i\omega)\phi. \tag{7.169}$$

We therefore see that the required first-order Green's function is

$$\begin{aligned} G(\boldsymbol{r}) &= (\boldsymbol{\nabla} - i\omega)\phi \\ &= \frac{\mathrm{e}^{i\omega r}}{4\pi} \left(\frac{i\omega}{r}(1 - \boldsymbol{\sigma}_r) + \frac{\boldsymbol{r}}{r^3} \right), \end{aligned} \tag{7.170}$$

where $\boldsymbol{\sigma}_r = \boldsymbol{r}/r$ is the unit vector in the direction of $\boldsymbol{r}$. This Green's function is the key to much of scattering theory. With a general argument it satisfies

$$(\boldsymbol{\nabla} + i\omega) G(\boldsymbol{r} - \boldsymbol{r}') = \delta(\boldsymbol{r} - \boldsymbol{r}') \tag{7.171}$$

or, equivalently,

$$(\boldsymbol{\nabla}' - i\omega)G(\boldsymbol{r} - \boldsymbol{r}') = -\delta(\boldsymbol{r} - \boldsymbol{r}'), \tag{7.172}$$

where $\boldsymbol{\nabla}'$ denotes the vector derivative with respect to $\boldsymbol{r}'$.

7.5.2 Radiation and multipole fields

As a first application, suppose that a localised system of charges in free space, with sinusoidal time dependence, generates outgoing radiation fields. We could find these by generalising our point source solutions of section 7.3, but here we wish to exploit our new Green's function. We can now immediately write down the solution

$$F(\boldsymbol{r}) = -\int_V G(\boldsymbol{r}' - \boldsymbol{r})\big(\rho(\boldsymbol{r}') - \boldsymbol{J}(\boldsymbol{r}')\big)\,|dX'|, \tag{7.173}$$

where the integral is over a volume enclosing all of the sources. Equation (7.172) guarantees that this equation solves the Maxwell equations (7.161), subject to the boundary condition that only outgoing waves are present at large distances. It is worth stressing that the geometric algebra formulation is crucial to the way we have a single integral yielding both the electric and magnetic fields.

Often, one is mainly interested in the radiation fields present at large distances from the source. These are the contributions to F which fall off as $1/r$. To isolate these terms we use the expansion

$$\mathrm{e}^{i\omega|\boldsymbol{r} - \boldsymbol{r}'|} = \mathrm{e}^{i\omega r}\mathrm{e}^{-i\omega\boldsymbol{\sigma}_r\cdot\boldsymbol{r}'} + \mathrm{O}(r^{-1}), \tag{7.174}$$

so that the Green's function satisfies

$$\lim_{r\mapsto\infty} G(\boldsymbol{r}' - \boldsymbol{r}) = \frac{i\omega}{4\pi r}\mathrm{e}^{i\omega r}(1 + \boldsymbol{\sigma}_r)\mathrm{e}^{-i\omega\boldsymbol{\sigma}_r\cdot\boldsymbol{r}'}. \tag{7.175}$$

We therefore find that the limiting form of F can be written

$$F(\boldsymbol{r}) = -\frac{i\omega}{4\pi r}\mathrm{e}^{i\omega r}(1 + \boldsymbol{\sigma}_r)\int \mathrm{e}^{-i\omega\boldsymbol{\sigma}_r\cdot\boldsymbol{r}'}\big(\rho(\boldsymbol{r}') - \boldsymbol{J}(\boldsymbol{r}')\big)\,|dX'|. \tag{7.176}$$

As expected, the multivector is controlled by the idempotent term $(1 + \boldsymbol{\sigma}_r) = (\gamma_0 + e_r)\gamma_0$, appropriate for outgoing radiation.

A multipole expansion of the radiation field is achieved by expanding (7.176) in a series in ωd, where d is the dimension of the source. To leading order, and recalling that no monopole term is present, we find that

$$\begin{aligned}\int \mathrm{e}^{-i\omega\boldsymbol{\sigma}_r\cdot\boldsymbol{r}'}\big(\rho(\boldsymbol{r}') - \boldsymbol{J}(\boldsymbol{r}')\big)\,|dX'| &\approx \int(-\boldsymbol{J} - i\omega\rho\boldsymbol{\sigma}_r\cdot\boldsymbol{r}')|dX'| \\ &= \int(-\boldsymbol{J} + \boldsymbol{\sigma}_r\cdot\boldsymbol{J})\,|dX'|,\end{aligned} \tag{7.177}$$

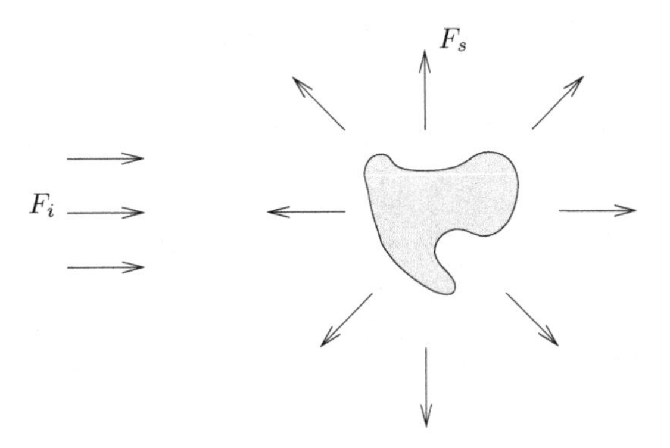

Figure 7.10 *Scattering by a localised object.* The incident field F_i sets up oscillating currents in the object, which generate an outgoing radiation field F_s.

where we have integrated by parts to obtain the final expression. This result is more commonly expressed in terms of the *electric dipole moment* $\boldsymbol{p}$, via

$$\int \boldsymbol{J}\,|dX| = -\int \boldsymbol{r}\,\boldsymbol{\nabla}\cdot\boldsymbol{J}\,|dX| = -i\omega\int \boldsymbol{r}\rho(\boldsymbol{r})\,|dX| = -i\omega\boldsymbol{p}. \tag{7.178}$$

The result is that the F field is given by

$$F(\boldsymbol{r}) = \frac{\omega^2}{4\pi r}\mathrm{e}^{i\omega r}(1+\boldsymbol{\sigma}_r)(\boldsymbol{p}-\boldsymbol{\sigma}_r\cdot\boldsymbol{p}). \tag{7.179}$$

An immediate check is that the scalar term in F vanishes, as it must. The electric and magnetic dipole fields can be read off easily now as

$$\boldsymbol{E} = \frac{\omega^2}{4\pi r}\mathrm{e}^{i\omega r}\boldsymbol{\sigma}_r\,\boldsymbol{\sigma}_r\wedge\boldsymbol{p}, \qquad I\boldsymbol{B} = \frac{\omega^2}{4\pi r}\mathrm{e}^{i\omega r}\boldsymbol{\sigma}_r\wedge\boldsymbol{p}. \tag{7.180}$$

These formulae are quite general for any (classical) radiating object.

7.6 Scattering

The geometry of a basic scattering problem is illustrated in figure 7.10. A (known) field F_i is incident on a localised object. Usually the incident radiation is taken to be a plane wave. This radiation sets up oscillating currents in the scatterer, which in turn generate a scattered field F_s. The total field F is given by

$$F = F_i + F_s, \tag{7.181}$$

and both F_i and F_s satisfy the vacuum Maxwell equations away from the scatterers.

The essential difficulty is how to solve for the currents set up by the incident radiation. This is extremely complex and a number of distinct approaches are described in the literature. One straightforward result is for scattering from a small uniform dielectric sphere. For this situation we have

$$\boldsymbol{p} = 4\pi a^3 \frac{\epsilon_r - 1}{\epsilon_r + 2} \boldsymbol{E}_i, \tag{7.182}$$

where a is the radius of the sphere. From equation (7.180) we see that the ratio of incident to scattered radiation is controlled by ω^2. This ratio determines the differential cross section via

$$\frac{d\sigma}{d\Omega} = r^2 \frac{|\boldsymbol{e}^* \cdot \boldsymbol{E}_s|^2}{|\boldsymbol{e}^* \cdot \boldsymbol{E}_i|^2}, \tag{7.183}$$

where the complex vector $\boldsymbol{e}$ determines the polarisation. The cross section clearly depends of the polarisation of the incident wave. Summing over polarisations the differential cross section is

$$\frac{d\sigma}{d\Omega} = \omega^4 a^6 \left(\frac{\epsilon_r - 1}{\epsilon_r + 2} \right)^2 \frac{1 + \cos^2(\theta)}{2}. \tag{7.184}$$

The factor of $\omega^4 = \lambda^{-4}$ is typical of Rayleigh scattering. These results are central to Rayleigh's explanation of blue skies and red sunsets.

Suppose now that we know the fields over a closed surface enclosing a volume V. Provided that F satisfies the vacuum Maxwell equations throughout V we can compute F_s directly from

$$F_s(\boldsymbol{r}') = \oint_{\partial V} G(\boldsymbol{r} - \boldsymbol{r}') \boldsymbol{n} F_s(\boldsymbol{r}) \, |dS|. \tag{7.185}$$

We take the volume V to be bounded by two surfaces, S_1 and S_2, as shown in figure 7.11. The surface S_1 is assumed to lie just outside the scatterers, so that $J = 0$ over S_1. The surface S_2 is assumed to be spherical, and is taken out to infinity. In this limit only the $1/r$ terms in G and F can contribute to the surface integral over S_2. But from equation (7.175) we know that

$$\lim_{r \mapsto \infty} G(\boldsymbol{r} - \boldsymbol{r}') = \frac{i\omega}{4\pi r} \mathrm{e}^{i\omega r} (1 - \boldsymbol{\sigma}_r) \mathrm{e}^{-i\omega \boldsymbol{\sigma}_r \cdot \boldsymbol{r}'}, \tag{7.186}$$

whereas F_s contains a factor of $(1 + \boldsymbol{\sigma}_r)$. It follows that the integrand $G\boldsymbol{n}F_s$ contains the term

$$(1 - \boldsymbol{\sigma}_r)\boldsymbol{\sigma}_r(1 + \boldsymbol{\sigma}_r) = 0. \tag{7.187}$$

This is identically zero, so there is no contribution from the surface at infinity. The result is that the scattered field is given by

$$F_s(\boldsymbol{r}) = \frac{1}{4\pi} \oint_{S_1} \mathrm{e}^{i\omega d} \left(\frac{i\omega}{d} + \frac{i\omega(\boldsymbol{r} - \boldsymbol{r}')}{d^2} - \frac{\boldsymbol{r} - \boldsymbol{r}'}{d^3} \right) \boldsymbol{n}' F_s(\boldsymbol{r}') \, |dS(\boldsymbol{r}')|, \tag{7.188}$$

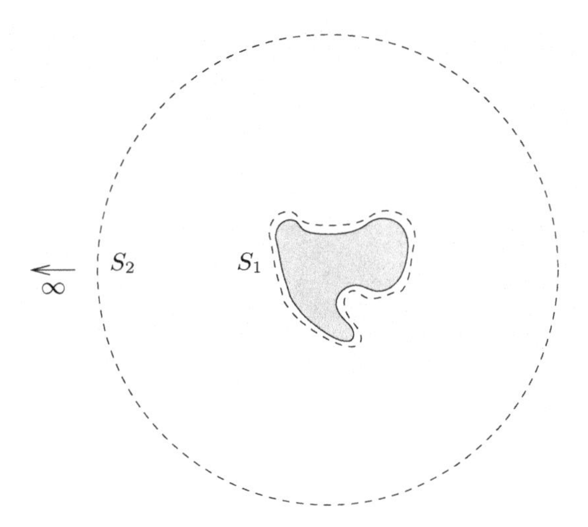

Figure 7.11 *Surfaces for Green's theorem.* The surface S_2 can be taken out to infinity, and S_1 lies just outside the scattering surface.

where

$$d = |\boldsymbol{r} - \boldsymbol{r}'|. \tag{7.189}$$

Since $\boldsymbol{n}$ is the outward pointing normal to the volume, this points *into* the scatterers. This result contains all the necessary polarisation and obliquity factors, often derived at great length in standard optics texts.

A significant advantage of this first-order approach is that it clearly embodies Huygens' principle. The scattered field F_s is propagated into the interior simply by multiplying it by a Green's function. This accords with Huygen's original idea of reradiation of wavelets from any given wavefront. Two significant problems remain, however. The first is how to specify F_s over the surface of integration. This requires detailed modelling of the polarisation currents set up by the incident radiation. A subtlety here is that we do not have complete freedom to specify F over the surface. The equation $\boldsymbol{\nabla} F = i\omega F$ implies that the components of $\boldsymbol{E}$ and $\boldsymbol{B}$ perpendicular to the boundary surface are determined by the derivatives of the components in the surface. This reduces the number of degrees of freedom in the problem from six to four, as is required for electromagnetism.

A further problem is that, even if F_s has been found, the integrals in equation (7.188) cannot be performed analytically. One can approximate to the large r regime and, after various approximations, recover Fraunhofer and Fresnel optics. Alternatively, equation (7.188) can be used as the basis for numerical simulations of scattered fields. Figure 7.12 shows the type of detailed patterns that can emerge. The plot was calculated using the two-dimensional equivalent of equation (7.188). The total energy density is shown, where the scattering

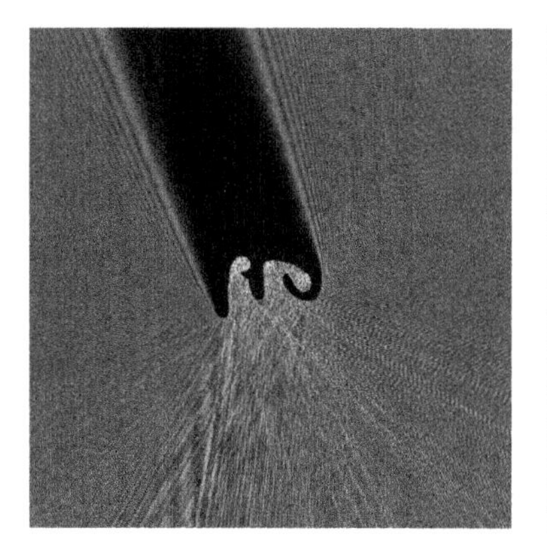
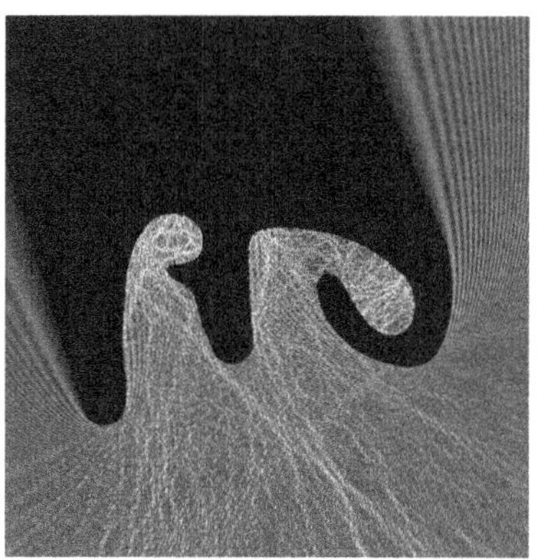

Figure 7.12 *Scattering in two dimensions.* The plots show the intensity of the electric field, with higher intensity coloured lighter. The incident radiation enters from the bottom right of the diagram and scatters off a conductor with complicated surface features. The conductor is closed in the shadow region. Various diffraction effects are clearly visible. The right-hand plot is a close-up near the surface and shows the complicated pattern of hot and cold regions that can develop.

is performed by a series of perfect conductors. A good check that the calculations have been performed correctly is that all the expected shadowing effects are present.

7.7 Notes

There is a vast literature on electromagnetism and electrodynamics. For this chapter we particularly made use of the classic texts by Jackson (1999) and Schwinger et al. (1998), both entitled *Classical Electrodynamics.* The former of these also contains an exhaustive list of further references. Applications of geometric algebra to electromagnetism are discussed in the book *Multivectors and Clifford Algebra in Electrodynamics* by Jancewicz (1989). This is largely an introductory text and stops short of tackling the more advanced applications.

We are grateful to Stephen Gull for producing the figures in section 7.3 and for stimulating much of the work described in this chapter. Further material can be found in the Banff series of lectures by Doran et al (1996a). Readers interested in the action at a distance formalism of Wheeler and Feynman can do no better than return to their original 1949 paper. It is a good exercise to convert their arguments into a more streamlined geometric algebra notation!

7.8 Exercises

7.1 A circular current loop has radius a and lies in the $z = 0$ plane with its centre at the origin. The loop carries a current J. Write down an integral expression for the $\boldsymbol{B}$ field, and show that on the z axis,

$$\boldsymbol{B} = \frac{\mu_0 J a^2}{2(a^2+z^2)^{3/2}}\boldsymbol{\sigma}_3.$$

7.2 An extension to the Maxwell equations which is regularly discussed is how they are modified in the presence of magnetic monopoles. If ρ_m and $\boldsymbol{J}_m$ denote magnetic charges and currents, the relevant equations are

$$\boldsymbol{\nabla}\cdot\boldsymbol{D} = \rho_e, \qquad \boldsymbol{\nabla}\cdot\boldsymbol{B} = \rho_m,$$
$$-\boldsymbol{\nabla}\times\boldsymbol{E} = \frac{\partial}{\partial t}\boldsymbol{B} + \boldsymbol{J}_m, \qquad \boldsymbol{\nabla}\times\boldsymbol{H} = \frac{\partial}{\partial t}\boldsymbol{D} + \boldsymbol{J}_e.$$

Prove that in free space these can be written

$$\nabla F = J_e + J_m I,$$

where $J_m = (\rho_m + \boldsymbol{J}_m)\gamma_0$. A duality transformation of the $\boldsymbol{E}$ and $\boldsymbol{B}$ fields is defined by

$$\boldsymbol{E}' = \boldsymbol{E}\cos(\alpha) + \boldsymbol{B}\sin(\alpha), \qquad \boldsymbol{B}' = \boldsymbol{B}\cos(\alpha) - \boldsymbol{B}\sin(\alpha).$$

Prove that this can be written compactly as $F' = F\mathrm{e}^{-I\alpha}$. Hence find the equivalent transformation law for the source terms such that the equations remain invariant, and prove that the electromagnetic energy-momentum tensor is also invariant under a duality transformation.

7.3 A particle follows the trajectory $x_0(\tau)$, with velocity $v = \dot{x}$ and acceleration $\dot{v}$. If X is the retarded null vector connecting the point x to the worldline, show that the electromagnetic field at x is given by

$$F = \frac{q}{4\pi}\frac{X\wedge v + \frac{1}{2}X\Omega_v X}{(X\cdot v)^3},$$

where $\Omega_v = \dot{v}\wedge v$. Prove directly that F satisfies $\nabla F = 0$ off the particle worldline.

7.4 Prove the following formulae relating the retarded A and F fields for a point charge to the null vector X:

$$A = -\frac{q}{8\pi\epsilon_0}\nabla^2 X, \qquad F = -\frac{q}{8\pi\epsilon_0}\nabla^3 X.$$

These expressions are of interest in the 'action at a distance' formulation of electrodynamics, as discussed by Wheeler and Feynman (1949).

7.5 Confirm that, at large distances for the source, the radiation fields due to both linearly and circularly accelerating charges go as

$$F_{rad} \approx \frac{1}{r}(1+\boldsymbol{\sigma}_r)\boldsymbol{a},$$

where $\boldsymbol{\sigma}_r \cdot \boldsymbol{a} = 0$.

7.6 From the solution for the fields due to a point charge in a circular orbit (section 7.3.4), explain why synchrotron radiation arrives in pulses.

7.7 For the κ defined in equation (7.149), verify that $\kappa\boldsymbol{\sigma}_3\kappa^\dagger = s_i\boldsymbol{\sigma}_i$, where s_i are Stokes parameters.

7.8 A rotor R relates two frames by $\mathsf{e}'_\mu = R\mathsf{e}_\mu\tilde{R}$. In both frames the vector e_3 vector is defined by

$$\mathsf{e}_3 = \mathsf{e}'_3 = \frac{k - k\cdot\mathsf{e}_0\,\mathsf{e}_0}{k\cdot\mathsf{e}_0},$$

where k is a fixed null vector. Prove that for this relation to be valid for both frames we must have

$$Rk\tilde{R} = \lambda k.$$

How many degrees of freedom are left in the rotor R if this equation holds?

7.9 In optical problems we are regularly interested in the effects of a planar aperture on incident plane waves. Suppose that the aperture lies in the $z=0$ plane, and we are interested in the fields in the region $z>0$. By introducing the Green's function

$$G'(\boldsymbol{r};\boldsymbol{r}') = G(\boldsymbol{r}-\boldsymbol{r}') - G(\boldsymbol{r}-\bar{\boldsymbol{r}}'),$$

where $\bar{\boldsymbol{r}} = -\boldsymbol{\sigma}_3\boldsymbol{r}\boldsymbol{\sigma}_3$, prove that the field in the region $z>0$ is given by

$$F_s(\boldsymbol{r}') = \int dx\,dy\,\frac{z'\mathrm{e}^{I\omega d}}{2\pi d^3}(1-i\omega d)F_s(x,y,0), \tag{E7.1}$$

where $d = |\boldsymbol{r}-\boldsymbol{r}'|$. In the Kirchoff approximation we assume that F_s over the aperture can be taken as the incident plane wave. By working in the large r and small angle limit, prove the Fraunhofer result that the transmitted amplitude is controlled by the Fourier transform of the aperture function.

7.10 Repeat the analysis of the previous question for a two-dimensional arrangement. You will need to understand some of the properties of Hankel functions.

8

Quantum theory and spinors

In this chapter we study the application of geometric algebra to both non-relativistic and relativistic quantum mechanics. We concentrate on the quantum theory of spin-1/2 particles, whose dynamics is described by the Pauli and Dirac equations. For interactions where spin and relativity are not important the dynamics reduces to that of the Schrödinger equation. There are many good textbooks describing this topic and we will make no attempt to cover it here. We assume, furthermore, that most readers have a basic understanding of quantum mechanics, and are familiar with the concepts of states and operators.

Both the Pauli and Dirac matrices arise naturally as representations of the geometric algebras of space and spacetime. It is no surprise, then, that much of quantum theory finds a natural expression within geometric algebra. To achieve this, however, one must reconsider the standard interpretation of the quantum spin operators. Like much discussion of the interpretation of quantum theory, certain issues raised here are controversial. There is no question about the validity of our algebraic approach, however, and little doubt about its advantages. Whether the algebraic simplifications obtained here are indicative of a deeper structure embedded in quantum mechanics is an open question.

In this chapter we only consider the quantum theory of single particles in background fields. Multiparticle systems are considered in the following chapter. Amongst the results discussed in this section are the angular separation of the Dirac equation, and a method of calculating cross sections that avoids the need for spin sums. Both of these results are used in chapter 14 for studying the behaviour of fermions in gravitational backgrounds.

8.1 Non-relativistic quantum spin

The Stern–Gerlach experiment was the first to demonstrate the quantum nature of the magnetic moment. In this experiment a beam of particles passes through

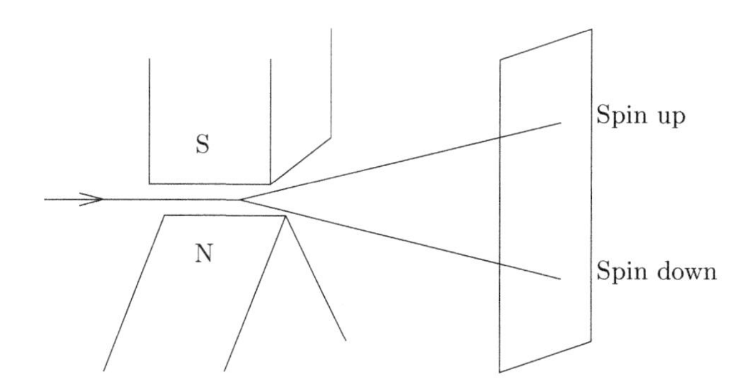

Figure 8.1 *The Stern–Gerlach experiment.* A particle beam is sent through a highly non-uniform $\boldsymbol{B}$ field. What emerges is a set of discrete, evenly-spaced beams.

a non-uniform magnetic field $\boldsymbol{B}$. Classically, one would expect the force on each particle to be governed by the equation

$$\boldsymbol{f} = \boldsymbol{\mu} \cdot \boldsymbol{\nabla} \boldsymbol{B}, \tag{8.1}$$

where $\boldsymbol{\mu}$ is the magnetic moment. This would give rise to a continuous distribution after passing through the field. Instead, what is observed is a number of evenly-spaced discrete bands (figure 8.1). The magnetic moment is quantised in the same manner as angular momentum.

When silver atoms are used to make up the beam there is a further surprise: only two beams emerge on the far side. Silver atoms contain a single electron in their outermost shell, so it looks as if electrons have an intrinsic angular momentum which can take only two values. This is known as its *spin*, though no classical picture should be inferred from this name. The double-valued nature of the spin suggests that the electron's wavefunction should contain two terms, representing a superposition of the possible spin states,

$$|\psi\rangle = \alpha|\uparrow\rangle + \beta|\downarrow\rangle, \tag{8.2}$$

where α and β are complex numbers. Such a state can be represented in matrix form as the *spinor*

$$|\psi\rangle = \begin{pmatrix} \alpha \\ \beta \end{pmatrix}. \tag{8.3}$$

If we align the z axis with the spin-up direction, then the operator returning the spin along the z axis must be

$$\hat{s}_3 = \lambda \begin{pmatrix} 1 & 0 \\ 0 & -1 \end{pmatrix}, \tag{8.4}$$

where λ is to be determined. The spin is added to the orbital angular momentum to give a conserved total angular momentum operator $\hat{\boldsymbol{j}} = \hat{\boldsymbol{l}} + \hat{\boldsymbol{s}}$. For this to make sense the spin operators should have the same commutation relations as the angular momentum operators $\hat{l}_i$,

$$\hat{l}_i = -i\hbar\epsilon_{ijk}x_j\partial_k, \quad [\hat{l}_i, \hat{l}_j] = i\hbar\epsilon_{ijk}\hat{l}_k. \tag{8.5}$$

This is sufficient to specify the remaining operators, up to an arbitrary phase (see exercise 8.1). The result is that the spin operators are given by

$$\hat{s}_k = \tfrac{1}{2}\hbar\hat{\sigma}_k, \tag{8.6}$$

where the $\hat{\sigma}_k$ are the familiar *Pauli matrices*

$$\hat{\sigma}_1 = \begin{pmatrix} 0 & 1 \\ 1 & 0 \end{pmatrix}, \quad \hat{\sigma}_2 = \begin{pmatrix} 0 & -i \\ i & 0 \end{pmatrix}, \quad \hat{\sigma}_3 = \begin{pmatrix} 1 & 0 \\ 0 & -1 \end{pmatrix}. \tag{8.7}$$

The 'hat' notation is used to record the fact that these are viewed explicitly as matrix operators, rather than as elements of a geometric algebra. The Pauli matrices satisfy the commutation relations,

$$[\hat{\sigma}_i, \hat{\sigma}_j] = 2i\epsilon_{ijk}\hat{\sigma}_k. \tag{8.8}$$

They also have the property that two different matrices *anticommute*,

$$\hat{\sigma}_1\hat{\sigma}_2 + \hat{\sigma}_2\hat{\sigma}_1 = 0, \quad \text{etc.} \tag{8.9}$$

and all of the matrices square to the identity matrix,

$$\hat{\sigma}_1^2 = \hat{\sigma}_2^2 = \hat{\sigma}_3^2 = \mathsf{I}. \tag{8.10}$$

These are precisely the relations obeyed by a set of orthonormal vectors in space. We denote such a set by $\{\boldsymbol{\sigma}_k\}$. The crucial distinction is that the Pauli matrices are operators in quantum isospace, whereas the $\{\boldsymbol{\sigma}_k\}$ are vectors in real space.

The $\hat{\sigma}_k$ operators act on two-component complex spinors as described in equation (8.3). Spinors belong to two-dimensional complex vector space, so have four real degrees of freedom. A natural question to ask is whether an equivalent representation can be found in terms of real multivectors, such that the matrix action is replaced by multiplication by the $\{\boldsymbol{\sigma}_k\}$ vectors. To find a natural way to do this we consider the observables of a spinor. These are the eigenvalues of Hermitian operators and, for two-state systems, the relevant operators are the Pauli matrices. We therefore form the three observables

$$s_k = \tfrac{1}{2}\hbar n_k = \langle\psi|\hat{s}_k|\psi\rangle. \tag{8.11}$$

The n_k are the components of a single vector in the quantum theory of spin. Focusing attention on the components of this vector, we have

$$\begin{aligned} n_1 &= \langle\psi|\hat{\sigma}_1|\psi\rangle = \alpha\beta^* + \alpha^*\beta, \\ n_2 &= \langle\psi|\hat{\sigma}_2|\psi\rangle = i(\alpha\beta^* - \alpha^*\beta), \\ n_3 &= \langle\psi|\hat{\sigma}_3|\psi\rangle = \alpha\alpha^* - \beta\beta^*. \end{aligned} \tag{8.12}$$

The magnitude of the vector with components n_k is

$$\begin{aligned}|\boldsymbol{n}|^2 &= (\alpha\beta^* + \alpha^*\beta)^2 - (\alpha\beta^* - \alpha^*\beta)^2 + (\alpha\alpha^* - \beta\beta^*)^2 \\ &= (|\alpha|^2 + |\beta|^2)^2 = \langle\psi|\psi\rangle^2.\end{aligned} \tag{8.13}$$

So, provided the state is normalised to 1, the vector $\boldsymbol{n}$ must have unit length. We can therefore introduce polar coordinates and write

$$\begin{aligned}n_1 &= \sin(\theta)\cos(\phi), \\ n_2 &= \sin(\theta)\sin(\phi), \\ n_3 &= \cos(\theta).\end{aligned} \tag{8.14}$$

Comparing equation (8.14) with equation (8.12) we see that we must have

$$\alpha = \cos(\theta/2)\mathrm{e}^{i\gamma}, \qquad \beta = \sin(\theta/2)\mathrm{e}^{i\delta} \tag{8.15}$$

where $\delta - \gamma = \phi$. It follows that the spinor can be written in terms of the polar coordinates of the vector observable as

$$|\psi\rangle = \begin{pmatrix}\cos(\theta/2)\mathrm{e}^{-i\phi/2} \\ \sin(\theta/2)\mathrm{e}^{i\phi/2}\end{pmatrix} \mathrm{e}^{i(\gamma+\delta)/2}. \tag{8.16}$$

The overall phase factor can be ignored, and what remains is a description in terms of half-angles. This suggests a strong analogy with *rotors*. To investigate this analogy, we use the idea that polar coordinates can be viewed as part of an instruction to rotate the 3 axis onto the chosen vector. To expose this we write the vector $\boldsymbol{n}$ as

$$\boldsymbol{n} = \sin(\theta)\big(\cos(\phi)\boldsymbol{\sigma}_1 + \sin(\phi)\boldsymbol{\sigma}_2\big) + \cos(\theta)\boldsymbol{\sigma}_3. \tag{8.17}$$

This can be written

$$\boldsymbol{n} = R\boldsymbol{\sigma}_3 R^\dagger, \tag{8.18}$$

where

$$R = \mathrm{e}^{-\phi I\boldsymbol{\sigma}_3/2}\mathrm{e}^{-\theta I\boldsymbol{\sigma}_2/2}. \tag{8.19}$$

This suggests that there should be a natural map between the normalised spinor of equation (8.16) and the rotor R. Both belong to linear spaces of real dimension four and both are normalised. Expanding out the rotor R the following one-to-one map is found:

$$|\psi\rangle = \begin{pmatrix}a^0 + ia^3 \\ -a^2 + ia^1\end{pmatrix} \leftrightarrow \psi = a^0 + a^k I\boldsymbol{\sigma}_k. \tag{8.20}$$

This map will enable us to perform all operations involving spinors without leaving the geometric algebra of space. Throughout this chapter we use the $\leftrightarrow$ symbol to denote a one-to-one map between conventional quantum mechanics and the multivector equivalent. We will continue to refer to the multivector ψ

as a spinor. On this scheme the spin-up and spin-down basis states $|\uparrow\rangle$ and $|\downarrow\rangle$ become

$$|\uparrow\rangle \leftrightarrow 1 \qquad |\downarrow\rangle \leftrightarrow -I\boldsymbol{\sigma}_2. \tag{8.21}$$

One can immediately see for these that the vectors of observables have components $(0,0,\pm 1)$, as required.

8.1.1 Pauli operators

Now that a suitable one-to-one map has been found, we need to find a representation for Pauli operators acting on the multivector version of a spinor. It turns out that the action of the quantum $\hat{\sigma}_k$ operators on a state $|\psi\rangle$ is equivalent to the following operation on ψ:

$$\hat{\sigma}_k|\psi\rangle \leftrightarrow \boldsymbol{\sigma}_k\psi\boldsymbol{\sigma}_3 \quad (k=1,2,3). \tag{8.22}$$

The $\boldsymbol{\sigma}_3$ on the right-hand side ensures that the multivector remains in the even subalgebra. The choice of vector does not break rotational covariance, in the same way that choosing the $\hat{\sigma}_3$ matrix to be diagonal does not alter the rotational covariance of the Pauli theory. One can explicitly verify that the translation procedure of equation (8.20) and equation (8.22) is consistent by routine computation; for example

$$\hat{\sigma}_1|\psi\rangle = \begin{pmatrix} -a^2+ia^1 \\ a^0+ia^3 \end{pmatrix} \leftrightarrow -a^2+a^1I\boldsymbol{\sigma}_3 - a^0I\boldsymbol{\sigma}_2 + a^3I\boldsymbol{\sigma}_1 = \boldsymbol{\sigma}_1\psi\boldsymbol{\sigma}_3. \tag{8.23}$$

The remaining cases, for $\hat{\sigma}_2$ and $\hat{\sigma}_3$ can be checked equally easily.

Now that we have a translation for the action of the Pauli matrices, we can find the equivalent of multiplying by the unit imaginary i. To find this we note that

$$\hat{\sigma}_1\hat{\sigma}_2\hat{\sigma}_3 = \begin{pmatrix} i & 0 \\ 0 & i \end{pmatrix}, \tag{8.24}$$

so multiplication of both components of $|\psi\rangle$ by i can be achieved by multiplying by the product of the three matrix operators. We therefore arrive at the translation

$$i|\psi\rangle \leftrightarrow \boldsymbol{\sigma}_1\boldsymbol{\sigma}_2\boldsymbol{\sigma}_3\psi(\boldsymbol{\sigma}_3)^3 = \psi I\boldsymbol{\sigma}_3. \tag{8.25}$$

So, on this scheme, the unit imaginary of quantum theory is replaced by right multiplication by the *bivector* $I\boldsymbol{\sigma}_3$. This is certainly suggestive, though it should be borne in mind that this conclusion is a feature of our chosen representation. The appearance of the bivector $I\boldsymbol{\sigma}_3$ is to be expected, since the vector of observables $\boldsymbol{s} = s_k\boldsymbol{\sigma}_k$ was formed by rotating the $\boldsymbol{\sigma}_3$ vector. This vector is unchanged by rotations in the $I\boldsymbol{\sigma}_3$ plane, which provides a geometric picture of phase invariance.

8.1.2 Observables in the Pauli theory

We next need to establish the quantum inner product for our multivector form of a spinor. We first note that the Hermitian adjoint operation has $\hat{\sigma}_k^\dagger = \hat{\sigma}_k$, and reverses the order of all products. This is precisely the same as the reversion operation for multivectors in three dimensions, so the dagger symbol can be used consistently for both operations. The quantum inner product is

$$\langle\psi|\phi\rangle = (\psi_1^*, \psi_2^*) \begin{pmatrix} \phi_1 \\ \phi_2 \end{pmatrix} = \psi_1^*\phi_1 + \psi_2^*\phi_2, \tag{8.26}$$

where we ignore spatial integrals. For a wide range of problems the spatial and spin components of the wave function can be separated. If this is not the case then the quantum inner product should also contain an integral over all space. The result of the real part of the inner product is reproduced by

$$\mathrm{Re}\langle\psi|\phi\rangle \leftrightarrow \langle\psi^\dagger\phi\rangle, \tag{8.27}$$

so that, for example,

$$\langle\psi|\psi\rangle \leftrightarrow \langle\psi^\dagger\psi\rangle = \langle(a^0 - a^j I\boldsymbol{\sigma}_j)(a^0 + a^k I\boldsymbol{\sigma}_k)\rangle = \sum_{\alpha=0}^{3} a^\alpha a^\alpha. \tag{8.28}$$

Since

$$\langle\psi|\phi\rangle = \mathrm{Re}\langle\psi|\phi\rangle - i\mathrm{Re}\langle\psi|i\phi\rangle, \tag{8.29}$$

the full inner product can be written

$$\langle\psi|\phi\rangle \leftrightarrow \langle\psi^\dagger\phi\rangle - \langle\psi^\dagger\phi I\boldsymbol{\sigma}_3\rangle I\boldsymbol{\sigma}_3. \tag{8.30}$$

The right-hand side projects out the 1 and $I\boldsymbol{\sigma}_3$ components from the geometric product $\psi^\dagger\phi$. The result of this projection on a multivector A is written $\langle A\rangle_q$. For even-grade multivectors in three dimensions this projection has the simple form

$$\langle A\rangle_q = \tfrac{1}{2}(A + \boldsymbol{\sigma}_3 A\boldsymbol{\sigma}_3). \tag{8.31}$$

If the result of an inner product is used to multiply a second multivector, one has to remember to keep the terms in $I\boldsymbol{\sigma}_3$ to the right of the multivector. This might appear a slightly clumsy procedure at first, but it is easy to establish conventions so that manipulations are just as efficient as in the standard treatment. Furthermore, the fact that all manipulations are now performed within the geometric algebra framework offers a number of new ways to simplify the analysis of a range of problems.

8.1.3 The spin vector

As a check on the consistency of our scheme, we return to the expectation value of the spin in the k-direction, $\langle\psi|\hat{s}_k|\psi\rangle$. For this we require

$$\langle\psi|\hat{\sigma}_k|\psi\rangle \leftrightarrow \langle\psi^\dagger\boldsymbol{\sigma}_k\psi\boldsymbol{\sigma}_3\rangle - \langle\psi^\dagger\boldsymbol{\sigma}_k\psi I\rangle I\boldsymbol{\sigma}_3. \tag{8.32}$$

Since $\psi^\dagger I\boldsymbol{\sigma}_k\psi$ reverses to give minus itself it has zero scalar part, so the final term on the right-hand side vanishes. This is to be expected, as the $\hat{\sigma}_k$ are Hermitian operators. For the remaining term we note that in three dimensions $\psi\boldsymbol{\sigma}_3\psi^\dagger$ is both odd-grade and reverses to itself, so is a pure vector. We therefore define the spin vector

$$\boldsymbol{s} = \tfrac{1}{2}\hbar\psi\boldsymbol{\sigma}_3\psi^\dagger. \tag{8.33}$$

The quantum expectation now reduces to

$$\langle\psi|\hat{s}_k|\psi\rangle = \tfrac{1}{2}\hbar\langle\boldsymbol{\sigma}_k\psi\boldsymbol{\sigma}_3\psi^\dagger\rangle = \boldsymbol{\sigma}_k\cdot\boldsymbol{s}. \tag{8.34}$$

This new expression has a rather different interpretation to that usually encountered in quantum theory. Rather than forming the expectation value of a quantum operator, we are simply projecting out the kth component of the vector $\boldsymbol{s}$. Working with the vector $\boldsymbol{s}$ may appear to raise questions about whether we are free to talk about all three components of the spin vector. This is in fact consistent with the results of spin measurements, if we view the spin measurement apparatus as acting more as a spin polariser. This is discussed in Doran *et al.* (1996b).

The rotor description introduced at the start of this section is recovered by first defining the scalar

$$\rho = \psi\psi^\dagger. \tag{8.35}$$

The spinor ψ then decomposes into

$$\psi = \rho^{1/2}R, \tag{8.36}$$

where $R = \rho^{-1/2}\psi$. The multivector R satisfies $RR^\dagger = 1$, so is a rotor. In this approach, Pauli spinors are nothing but unnormalised rotors. The spin vector $\boldsymbol{s}$ can now be written as

$$\boldsymbol{s} = \tfrac{1}{2}\hbar\rho R\boldsymbol{\sigma}_3R^\dagger, \tag{8.37}$$

which recovers the form of equation (8.18).

The double-sided construction of the expectation value of equation (8.32) contains an instruction to rotate the fixed $\boldsymbol{\sigma}_3$ axis into the spin direction and dilate it. It might appear here that we are singling out some preferred direction in space. But in fact all we are doing is utilising an idea from rigid-body dynamics, as discussed in section 3.4.3. The $\boldsymbol{\sigma}_3$ on the right of ψ represents a vector in a 'reference' frame. All physical vectors, like $\boldsymbol{s}$, are obtained by rotating this frame

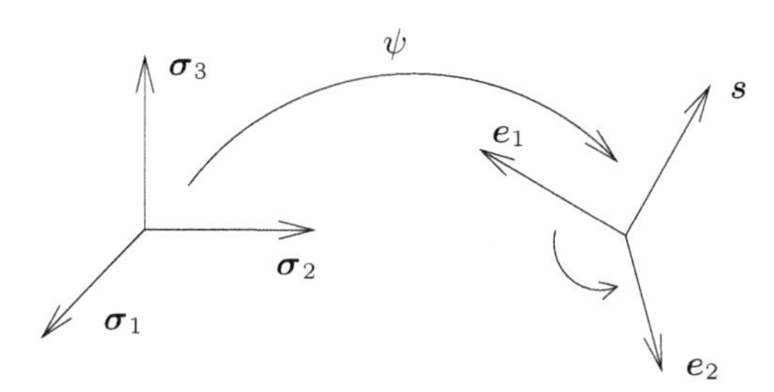

Figure 8.2 *The spin vector.* The normalised spinor ψ transforms the initial reference frame onto the frame $\{e_k\}$. The vector e_3 is the spin vector. A phase transformation of ψ generates a rotation in the e_1e_2 plane. Such a transformation is unobservable, so the e_1 and e_2 vectors are also unobservable.

onto the physical values (see figure 8.2). There is nothing special about $\boldsymbol{\sigma}_3$ — one can choose any (constant) reference frame and use the appropriate rotation onto $\boldsymbol{s}$, in the same way that there is nothing special about the orientation of the reference configuration of a rigid body. In rigid-body mechanics this freedom is usually employed to align the reference configuration with the initial state of the body. In quantum theory the convention is to work with the z axis as the reference vector.

8.1.4 Rotating spinors

Suppose that the vector $\boldsymbol{s}$ is to be rotated to a new vector $R_0\boldsymbol{s}R_0^\dagger$. To achieve this the spinor ψ must transform according to

$$\psi \mapsto R_0\psi. \tag{8.38}$$

Now suppose that for R_0 we use the rotor R_θ,

$$R_\theta = \exp(-\hat{B}\theta/2), \tag{8.39}$$

where $\hat{B}^2 = -1$ is a constant bivector. The resulting spinor is

$$\psi' = R_\theta\psi = \mathrm{e}^{-\hat{B}\theta/2}\psi. \tag{8.40}$$

We now start to increase θ from 0 through to 2π, so that $\theta = 2\pi$ corresponds to a 2π rotation, bringing all observables back to their original values. But under this we see that ψ transforms to

$$\psi' = \mathrm{e}^{-\hat{B}\pi}\psi = \big(\cos(\pi) - \hat{B}\sin(\pi)\big)\psi = -\psi. \tag{8.41}$$

The spinor changes sign! If a spin vector is rotated through 2π, the wavefunction does not come back to itself, but instead transforms to minus its original value.

This change of sign of a state vector under 2π rotations is the distinguishing property of spin-1/2 fermions in quantum theory. Once one sees the rotor derivation of this result, however, it is rather less mysterious. Indeed, there are classical phenomena involving systems of linked rotations that show precisely the same property. One example is the 4π symmetry observed when rotating an arm holding a tray. For a more detailed discussion if this point, see chapter 41 of *Gravitation* by Misner, Thorne & Wheeler (1973). A linear space which is acted on in a single-sided manner by rotors forms a carrier space for a spin representation of the rotation group. Elements of such a space are generally called spinors, which is why that name is adopted for our representation in terms of even multivectors.

8.1.5 Quantum particles in a magnetic field

Particles with non-zero spin also have a magnetic moment which is proportional to the spin. This is expressed as the operator relation

$$\hat{\mu}_k = \gamma \hat{s}_k, \tag{8.42}$$

where $\hat{\mu}_k$ is the magnetic moment operator, γ is the gyromagnetic ratio and $\hat{s}_k$ is the spin operator. The gyromagnetic ratio is usually written in the form

$$\gamma = g\frac{q}{2m}, \tag{8.43}$$

where m is the particle mass, q is the charge and g is the reduced gyromagnetic ratio. The reduced gyromagnetic ratios are determined experimentally to be

$$\begin{array}{lll} \text{electron} & g_e = 2 & (\text{actually } 2(1 + \alpha/2\pi + \cdots)), \\ \text{proton} & g_p = 5.587, & \\ \text{neutron} & g_n = -3.826 & (\text{using proton charge}). \end{array}$$

The value for the neutron is negative because its spin and magnetic moment are antiparallel. All of the above are spin-1/2 particles for which we have $\hat{s}_k = (\hbar/2)\hat{\sigma}_k$.

Now suppose that the particle is placed in a magnetic field, and that all of the spatial dynamics has been separated out. We introduce the Hamiltonian operator

$$\hat{H} = -\tfrac{1}{2}\gamma\hbar B_k \hat{\sigma}_k = -\hat{\mu}_k B_k. \tag{8.44}$$

The spin state at time t is then written as

$$|\psi(t)\rangle = \alpha(t)|\uparrow\rangle + \beta(t)|\downarrow\rangle, \tag{8.45}$$

with α and β general complex coefficients. The dynamical equation for these coefficients is given by the time-dependent Schrödinger equation

$$\hat{H}|\psi\rangle = i\hbar \frac{d|\psi\rangle}{dt}. \tag{8.46}$$

This equation can be hard to analyse, conventionally, because it involves a pair of coupled differential equations for α and β. Instead, let us see what the Schrödinger equation looks like in the geometric algebra formulation. We first write the equation in the form

$$\frac{d|\psi\rangle}{dt} = \tfrac{1}{2}\gamma i B_k \hat{\sigma}_k |\psi\rangle. \tag{8.47}$$

Now replacing $|\psi\rangle$ by the multivector ψ we see that the left-hand side is simply $\dot{\psi}$, where the dot denotes the time derivative. The right-hand side involves multiplication of the spinor $|\psi\rangle$ by $i\hat{\sigma}_k$, which we replace by

$$i\hat{\sigma}_k|\psi\rangle \leftrightarrow \boldsymbol{\sigma}_k \psi \boldsymbol{\sigma}_3 (I\boldsymbol{\sigma}_3) = I\boldsymbol{\sigma}_k \psi. \tag{8.48}$$

The Schrödinger equation (8.46) is therefore simply

$$\dot{\psi} = \tfrac{1}{2}\gamma B_k I\boldsymbol{\sigma}_k \psi = \tfrac{1}{2}\gamma I\boldsymbol{B}\psi, \tag{8.49}$$

where $\boldsymbol{B} = B_k \boldsymbol{\sigma}_k$. If we now decompose ψ into $\rho^{1/2}R$ we see that

$$\dot{\psi}\psi^\dagger = \tfrac{1}{2}\dot{\rho} + \rho \dot{R} R^\dagger = \tfrac{1}{2}\rho\gamma I\boldsymbol{B}. \tag{8.50}$$

The right-hand side is a bivector, so ρ must be constant. This is to be expected, as the evolution should be unitary. The dynamics now reduces to

$$\dot{R} = \tfrac{1}{2}\gamma I\boldsymbol{B}R, \tag{8.51}$$

so the quantum theory of a spin-1/2 particle in a magnetic field reduces to a simple rotor equation. This is very natural, if one thinks about the behaviour of particles in magnetic fields, and is an important justification for our approach.

Recovering a rotor equation explains the difficulty of the traditional analysis based on a pair of coupled equations for the components of $|\psi\rangle$. This approach fails to capture the fact that there is a rotor underlying the dynamics, and so carries along a redundant degree of freedom in the normalisation. In addition, the separation of a rotor into a pair of components is far from natural. For example, suppose that $\boldsymbol{B}$ is a constant field. The rotor equation integrates immediately to give

$$\psi(t) = \mathrm{e}^{\gamma I\boldsymbol{B}t/2}\psi_0. \tag{8.52}$$

The spin vector $\boldsymbol{s}$ therefore just precesses in the $I\boldsymbol{B}$ plane at a rate $\omega_0 = \gamma|\boldsymbol{B}|$. Even this simple result is rather more difficult to establish when working with the components of $|\psi\rangle$.

8.1.6 NMR and magnetic resonance imaging

A more interesting example of a particle in a magnetic field is provided by nuclear magnetic resonance, or NMR. Suppose that the $\boldsymbol{B}$ field includes an oscillatory field $(B_1\cos(\omega t), B_1\sin(\omega t), 0)$ together with a constant field along the z axis. This oscillatory field induces transitions (spin-flips) between the up and down states, which differ in energy because of the constant component of the field. This is a very interesting system of great practical importance. It is the basis of magnetic resonance imaging and Rabi molecular beam spectroscopy.

To study this system we first write the $\boldsymbol{B}$ field as

$$B_1\big(\cos(\omega t)\boldsymbol{\sigma}_1 + \sin(\omega t)\boldsymbol{\sigma}_2\big) + B_0\boldsymbol{\sigma}_3 = S(B_1\boldsymbol{\sigma}_1 + B_0\boldsymbol{\sigma}_3)S^\dagger, \tag{8.53}$$

where

$$S = \mathrm{e}^{-\omega t I\boldsymbol{\sigma}_3/2}. \tag{8.54}$$

We now define

$$\boldsymbol{B}_c = B_1\boldsymbol{\sigma}_1 + B_0\boldsymbol{\sigma}_3 \tag{8.55}$$

so that we can write $\boldsymbol{B} = S\boldsymbol{B}_c S^\dagger$. The rotor equation now simplifies to

$$S^\dagger\dot{\psi} = \tfrac{1}{2}\gamma I\boldsymbol{B}_c S^\dagger\psi, \tag{8.56}$$

where we have pre-multiplied by $S^\dagger$, and we continue to use ψ for the normalised rotor. Now noting that

$$\dot{S}^\dagger = \tfrac{1}{2}\omega I\boldsymbol{\sigma}_3 S^\dagger \tag{8.57}$$

we see that

$$\frac{d}{dt}(S^\dagger\psi) = \tfrac{1}{2}(\gamma I\boldsymbol{B}_c + \omega I\boldsymbol{\sigma}_3)S^\dagger\psi. \tag{8.58}$$

It is now $S^\dagger\psi$ that satisfies a rotor equation with a constant field. The solution is straightforward:

$$S^\dagger\psi(t) = \exp\Big(\tfrac{1}{2}\gamma t\, I\boldsymbol{B}_c + \tfrac{1}{2}\omega t\, I\boldsymbol{\sigma}_3\Big)\psi_0, \tag{8.59}$$

and we arrive at

$$\psi(t) = \exp\Big(-\tfrac{1}{2}\omega t\, I\boldsymbol{\sigma}_3\Big)\exp\Big(\tfrac{1}{2}(\omega_0+\omega)t\, I\boldsymbol{\sigma}_3 + \tfrac{1}{2}\omega_1 t\, I\boldsymbol{\sigma}_1\Big)\psi_0, \tag{8.60}$$

where $\omega_1 = \gamma B_1$. There are three separate frequencies in this solution, which contains a wealth of interesting physics.

To complete our analysis we must relate our solution to the results of experiments. Suppose that at time $t = 0$ we switch on the oscillating field. The particle is initially in a spin-up state, so $\psi_0 = 1$, which also ensures that the state is normalised. The probability that at time t the particle is in the spin-down state is

$$P_\downarrow = |\langle\downarrow|\psi(t)\rangle|^2. \tag{8.61}$$

We therefore need to form the inner product

$$\langle\downarrow|\psi(t)\rangle \leftrightarrow \langle I\boldsymbol{\sigma}_2\psi\rangle_q = \langle I\boldsymbol{\sigma}_2\psi\rangle - I\boldsymbol{\sigma}_3\langle I\boldsymbol{\sigma}_1\psi\rangle. \tag{8.62}$$

To find this inner product we write

$$\psi(t) = \mathrm{e}^{-\omega t I\boldsymbol{\sigma}_3/2}\big(\cos(\alpha t/2) + I\hat{\boldsymbol{B}}\sin(\alpha t/2)\big), \tag{8.63}$$

where

$$\hat{\boldsymbol{B}} = \frac{(\omega_0+\omega)\boldsymbol{\sigma}_3 + \omega_1\boldsymbol{\sigma}_1}{\alpha} \quad \text{and} \quad \alpha = \sqrt{(\omega+\omega_0)^2+\omega_1^2}. \tag{8.64}$$

The only term giving a contribution in the $I\boldsymbol{\sigma}_1$ and $I\boldsymbol{\sigma}_2$ planes is that in $\omega_1 I\boldsymbol{\sigma}_1/\alpha$. We therefore have

$$\langle I\boldsymbol{\sigma}_2\psi\rangle_q = \frac{\omega_1\sin(\alpha t/2)}{\alpha}\,\mathrm{e}^{-\omega t I\boldsymbol{\sigma}_3/2} I\boldsymbol{\sigma}_3 \tag{8.65}$$

and the probability is immediately

$$P_\downarrow = \left(\frac{\omega_1\sin(\alpha t/2)}{\alpha}\right)^2. \tag{8.66}$$

The maximum value is at $\alpha t = \pi$, and the probability at this time is maximised by choosing α as small as possible. This is achieved by setting $\omega = -\omega_0 = -\gamma B_0$. This is the *spin resonance condition* which is the basis of NMR spectroscopy.

8.2 Relativistic quantum states

The relativistic quantum dynamics of a spin-1/2 particle is described by the Dirac theory. The Dirac matrix operators are

$$\hat{\gamma}_0 = \begin{pmatrix} \mathsf{I} & 0 \\ 0 & -\mathsf{I} \end{pmatrix}, \quad \hat{\gamma}_k = \begin{pmatrix} 0 & -\hat{\sigma}_k \\ \hat{\sigma}_k & 0 \end{pmatrix}, \quad \hat{\gamma}_5 = \begin{pmatrix} 0 & \mathsf{I} \\ \mathsf{I} & 0 \end{pmatrix}, \tag{8.67}$$

where $\hat{\gamma}_5 = -i\hat{\gamma}_0\hat{\gamma}_1\hat{\gamma}_2\hat{\gamma}_3$ and I is the 2×2 identity matrix. These matrices act on Dirac spinors, which have four complex components (eight real degrees of freedom). We follow an analogous procedure to the Pauli case and map these spinors onto elements of the eight-dimensional even subalgebra of the spacetime algebra. Dirac spinors can be visualised as decomposing into 'upper' and 'lower' components,

$$|\psi\rangle = \begin{pmatrix} |\phi\rangle \\ |\eta\rangle \end{pmatrix}, \tag{8.68}$$

where $|\phi\rangle$ and $|\eta\rangle$ are a pair of two-component spinors. We already know how to represent these as multivectors ϕ and η, which lie in the space of scalars +

relative bivectors. Our map from the Dirac spinor onto an element of the full eight-dimensional subalgebra is simply

$$|\psi\rangle = \begin{pmatrix} |\phi\rangle \\ |\eta\rangle \end{pmatrix} \leftrightarrow \psi = \phi + \eta\boldsymbol{\sigma}_3. \tag{8.69}$$

The action of the Dirac matrix operators now becomes,

$$\begin{aligned} \hat{\gamma}_\mu|\psi\rangle &\leftrightarrow \gamma_\mu\psi\gamma_0 \quad (\mu = 0, \dots, 3), \\ i|\psi\rangle &\leftrightarrow \psi\, I\boldsymbol{\sigma}_3, \\ \hat{\gamma}_5|\psi\rangle &\leftrightarrow \psi\boldsymbol{\sigma}_3. \end{aligned} \tag{8.70}$$

Again, verifying the details of this map is a matter of routine computation. One feature is that we now have two 'reference' vectors that can appear on the right-hand side of ψ: γ_0 and γ_3. That is, the relative vector $\boldsymbol{\sigma}_3$ used in the Pauli theory has been decomposed into a spacelike and a timelike direction. As in the Pauli theory, these reference vectors multiplying ψ from the right do not break Lorentz covariance, as all observables are formed by rotating these reference vectors onto the frame of observables. Since $I\boldsymbol{\sigma}_3$ and γ_0 commute, our use of right-multiplication by $I\boldsymbol{\sigma}_3$ for the complex structure remains consistent.

The goal of our approach is to perform all calculations without ever having to introduce an explicit matrix representation. The explicit map of equation (8.69) is for column spinors written in the Dirac–Pauli representation, but it is a simple matter to establish similar maps for other representations. All one needs to do is find the unitary matrix which transforms the second representation into the Dirac–Pauli one, and then apply the map of equation (8.69). All of the matrix operators are then guaranteed to have the equivalence defined in equation (8.70). Certain other operations, such as complex conjugation, depend on the particular representation. But rather than think of these as the same operation in different representations, it is simpler to view them as different operations which can be applied to the multivector ψ.

In order to discuss the observables of the Dirac theory, we must first distinguish between the Hermitian and Dirac adjoints. The Hermitian adjoint is written as usual as $\langle\psi|$. The Dirac adjoint is written as $\langle\bar{\psi}|$ and is defined by

$$\langle\bar{\psi}| = (\langle\psi_u|, -\langle\psi_l|), \tag{8.71}$$

where the subscripts u and l refer to the upper and lower components. It is the Dirac adjoint which gives Lorentz-covariant observables. The Dirac inner product decomposes into

$$\langle\bar{\psi}|\phi\rangle = \langle\psi_u|\phi_u\rangle - \langle\psi_l|\phi_l\rangle. \tag{8.72}$$

This has the equivalent form

$$\langle\psi_u^\dagger\phi_u\rangle_q - \langle\psi_l^\dagger\phi_l\rangle_q = \langle(\psi_u^\dagger - \boldsymbol{\sigma}_3\psi_l^\dagger)(\phi_u + \phi_l\boldsymbol{\sigma}_3)\rangle_q = \langle\tilde{\psi}\phi\rangle_q. \tag{8.73}$$

So the Dirac adjoint is replaced by the manifestly covariant operation of spacetime reversion in the spacetime algebra formulation. The Hermitian adjoint now becomes

$$\langle\psi| \leftrightarrow \psi^\dagger = \gamma_0\tilde{\psi}\gamma_0, \tag{8.74}$$

which defines the meaning of the dagger symbol in the full spacetime algebra. Clearly, this operation requires singling out a preferred timelike vector, so is not covariant. In the relative space defined by γ_0, the Hermitian adjoint reduces to the non-relativistic reverse operation, so our notation is consistent with the use of the dagger for the reverse in three-dimensional space.

We can now look at the main observables formed from a Dirac spinor. The first is the current

$$J_\mu = \langle\bar{\psi}|\hat{\gamma}_\mu|\psi\rangle \leftrightarrow \langle\tilde{\psi}\gamma_\mu\psi\gamma_0\rangle - \langle\tilde{\psi}\gamma_\mu\psi I\gamma_3\rangle I\boldsymbol{\sigma}_3. \tag{8.75}$$

The final term contains $\langle\gamma_\mu\psi I\gamma_3\tilde{\psi}\rangle$. This vanishes because $\psi I\gamma_3\tilde{\psi}$ is odd-grade and reverses to minus itself, so is a pure trivector. Similarly, $\psi\gamma_0\tilde{\psi}$ is a pure vector, and we are left with

$$J_\mu = \langle\bar{\psi}|\hat{\gamma}_\mu|\psi\rangle \leftrightarrow \gamma_\mu\cdot(\psi\gamma_0\tilde{\psi}). \tag{8.76}$$

As with the Pauli theory, the operation of taking the expectation value of a matrix operator is replaced by that of picking out a component of a vector. We can therefore reconstitute the full vector J and write

$$J = \psi\gamma_0\tilde{\psi} \tag{8.77}$$

for the first of our observables.

To gain some further insight into the form of J, and its formation from ψ, we introduce the scalar + pseudoscalar quantity $\psi\tilde{\psi}$ as

$$\psi\tilde{\psi} = \rho e^{I\beta}. \tag{8.78}$$

Factoring this out from ψ, we define the spacetime rotor R:

$$R = \psi\rho^{-1/2}e^{-I\beta/2}, \qquad R\tilde{R} = 1. \tag{8.79}$$

(If $\rho = 0$ a slightly different procedure can be used.) We have now decomposed the spinor ψ into

$$\psi = \rho^{1/2}e^{I\beta/2}R, \tag{8.80}$$

which separates out a density ρ and the rotor R. The remaining factor of β is curious. It turns out that plane-wave particle states have $\beta = 0$, whereas antiparticle states have $\beta = \pi$. The picture for bound state wavefunctions is more complicated, however, and β appears to act as a remnant of multiparticle

Bilinear covariant	Standard form	STA equivalent	Frame-free form
Scalar	$\langle\bar{\psi}\|\psi\rangle$	$\langle\psi\tilde{\psi}\rangle$	$\rho\cos(\beta)$
Vector	$\langle\bar{\psi}\|\hat{\gamma}_\mu\|\psi\rangle$	$\gamma_\mu\cdot(\psi\gamma_0\tilde{\psi})$	$\psi\gamma_0\tilde{\psi} = J$
Bivector	$\langle\bar{\psi}\|i\hat{\gamma}_{\mu\nu}\|\psi\rangle$	$(\gamma_\mu\wedge\gamma_\nu)\cdot(\psi I\boldsymbol{\sigma}_3\tilde{\psi})$	$\psi I\boldsymbol{\sigma}_3\tilde{\psi} = S$
Pseudovector	$\langle\bar{\psi}\|\hat{\gamma}_\mu\hat{\gamma}_5\|\psi\rangle$	$\gamma_\mu\cdot(\psi\gamma_3\tilde{\psi})$	$\psi\gamma_3\tilde{\psi} = s$
Pseudoscalar	$\langle\bar{\psi}\|i\hat{\gamma}_5\|\psi\rangle$	$\langle\psi\tilde{\psi}I\rangle$	$-\rho\sin(\beta)$

Table 8.1 *Observables in the Dirac theory.* The standard expressions for the bilinear covariants are shown, together with their spacetime algebra (STA) equivalents.

effects from the full quantum field theory. With this decomposition of ψ, the current becomes

$$J = \psi\gamma_0\tilde{\psi} = \rho e^{I\beta/2}R\gamma_0\tilde{R}e^{I\beta/2} = \rho R\gamma_0\tilde{R}. \tag{8.81}$$

So the rotor is now an instruction to rotate γ_0 onto the direction of the current. This is precisely the picture we adopted in section 5.5 for studying the dynamics of a relativistic point particle.

A similar picture emerges for the spin. In relativistic mechanics angular momentum is a bivector quantity. Accordingly, the spin observables form a rank-2 antisymmetric tensor, with components given by

$$\langle\bar{\psi}|i\tfrac{1}{2}(\hat{\gamma}_\mu\hat{\gamma}_\nu - \hat{\gamma}_\nu\hat{\gamma}_\mu)|\psi\rangle \leftrightarrow \langle\tilde{\psi}\gamma_\mu\wedge\gamma_\nu\psi I\boldsymbol{\sigma}_3\rangle_q = \langle\gamma_\mu\wedge\gamma_\nu\psi I\boldsymbol{\sigma}_3\tilde{\psi}\rangle, \tag{8.82}$$

where again there is no imaginary component. This time we are picking out the components of the spin bivector S, given by

$$S = \psi I\boldsymbol{\sigma}_3\tilde{\psi}. \tag{8.83}$$

This is the natural spacetime generalisation of the Pauli result of equation (8.18). (Factors of $\hbar/2$ can always be inserted when required.) There are five such observables in all, which are summarised in Table 8.1. Of particular interest is the spin vector $s = \rho R\gamma_3\tilde{R}$. This justifies the classical model of spin introduced in section 5.5.6, where it was shown that the rotor form of the Lorentz force law naturally gives rise to a reduced gyromagnetic ratio of $g = 2$.

8.3 The Dirac equation

While much of the preceding discussion is both suggestive about the role of spinors in quantum theory, and algebraically very useful, one has to remember that quantum mechanics deals with *wave equations*. We therefore need to construct a relativistic wave equation for our Dirac spinor ψ, where ψ is an element of the eight-dimensional even subalgebra of the spacetime algebra. The relativistic wave equation for a spin-1/2 particle is the *Dirac equation*. This is a first-order wave equation, which is both Lorentz-invariant and has a future-pointing conserved current.

Like Pauli spinors, ψ is also subject to a single-sided rotor transformation law, $\psi \mapsto R\psi$, where R is a Lorentz rotor. To write down a covariant equation, we can therefore only place other covariant objects on the left of ψ. The available objects are any scalar or pseudoscalar, the vector derivative ∇ and any gauge fields describing interactions. On the right of ψ we can place combinations of γ_0, γ_3 and $I\boldsymbol{\sigma}_3$. The first equation we could write down is simply

$$\nabla\psi = 0. \tag{8.84}$$

This is the spacetime generalisation of the Cauchy–Riemann equations, as described in section 6.3. Remarkably, this equation does describe the behaviour of fermions — it is the wave equation for a (massless) *neutrino*. Any solution to this decomposes into two separate solutions by writing

$$\psi = \psi\tfrac{1}{2}(1+\boldsymbol{\sigma}_3) + \psi\tfrac{1}{2}(1-\boldsymbol{\sigma}_3) = \psi_+ + \psi_-. \tag{8.85}$$

The separate solutions ψ_+ and ψ_- are the right-handed and left-handed helicity eigenstates. For neutrinos, nature only appears to make use of the left-handed solutions. A more complete treatment of this subject involves the *electroweak* theory. (In fact, recent experiments point towards neutrinos carrying a small mass, whose origin can be explained by an interaction with the Higgs field.)

The formal operator identification of $i\partial_\mu$ with p_μ tells us that any wavefunction for a free massive particle should satisfy the Klein–Gordon equation $\nabla^2\psi = -m^2\psi$. We therefore need to add to the right-hand side of equation (8.84) a term that is linear in the particle mass m and that generates $-m^2\psi$ on squaring the operator. The natural covariant vector to form on the left of ψ is the momentum $\gamma^\mu p_\mu$. In terms of this operator we are led to an equation of the form

$$p\psi = m\psi a_0, \tag{8.86}$$

where a_0 is some multivector to be determined. It is immediately clear that a_0 must have odd grade, and must square to $+1$. The obvious candidate is γ_0, so that ψ contains a rotor to transform γ_0 to the velocity p/m. We therefore arrive at the equation

$$\nabla\psi I\boldsymbol{\sigma}_3 = m\psi\gamma_0. \tag{8.87}$$

This is the *Dirac equation* in its spacetime algebra form. This is easily seen to be equivalent to the matrix form of the equation

$$\hat{\gamma}^u\mu(\partial_\mu - ieA_\mu)|\psi\rangle = m|\psi\rangle, \tag{8.88}$$

where the electromagnetic vector potential has been included. The full Dirac equation is now

$$\nabla\psi I\boldsymbol{\sigma}_3 - eA\psi = m\psi\gamma_0. \tag{8.89}$$

A remarkable feature of this formulation is that the equation and all of its observables have been captured in the *real* algebra of spacetime, with no need for a unit imaginary. This suggests that interpretations of quantum mechanics that place great significance in the need for complex numbers are wide off the mark.

8.3.1 Symmetries and currents

The subject of the symmetries of the Dirac equation, and their conjugate currents, is discussed more fully in chapter 12. Here we highlight the main results. There are three important discrete symmetry operations: charge conjugation, parity and time reversal, denoted C, P and T respectively. Following the conventions of Bjorken & Drell (1964) we find that

$$\begin{aligned} \hat{P}|\psi\rangle &\leftrightarrow \gamma_0\psi(\bar{x})\gamma_0, \\ \hat{C}|\psi\rangle &\leftrightarrow \psi\boldsymbol{\sigma}_1, \\ \hat{T}|\psi\rangle &\leftrightarrow I\gamma_0\psi(-\bar{x})\gamma_1, \end{aligned} \tag{8.90}$$

where $\bar{x} = \gamma_0 x\gamma_0$ is (minus) the reflection of x in the timelike γ_0 axis. The combined CPT symmetry corresponds to

$$\psi \mapsto -I\psi(-x) \tag{8.91}$$

so that CPT symmetry does not require singling out a preferred timelike vector.

Amongst the continuous symmetries of the Dirac equation, the most significant is local electromagnetic gauge invariance. The equation is unchanged in physical content if we make the simultaneous replacements

$$\psi \mapsto \psi e^{\alpha I\boldsymbol{\sigma}_3}, \qquad eA \mapsto eA - \nabla\alpha. \tag{8.92}$$

The conserved current conjugate to this symmetry is the Dirac current $J = \psi\gamma_0\tilde{\psi}$. This satisfies

$$\begin{aligned} \nabla\cdot J &= \langle\nabla\psi\gamma_0\tilde{\psi}\rangle + \langle\psi\gamma_0\dot{\tilde{\psi}}\dot{\nabla}\rangle \\ &= -2\langle(eA\psi\gamma_0 + m\psi)I\boldsymbol{\sigma}_3\tilde{\psi}\rangle \\ &= 0 \end{aligned} \tag{8.93}$$

and so is conserved even in the presence of a background field. This is important. It means that single fermions cannot be created or destroyed. This feature was initially viewed as a great strength of the Dirac equation, though ultimately it is its biggest weakness. Fermion pairs, such as an electron and a positron, can be created and destroyed — a process which cannot be explained by the Dirac equation alone. These are many-body problems and are described by *quantum field theory.*

The timelike component of J in the γ_0 frame, say, is

$$J_0 = \gamma_0 \cdot J = \langle \gamma_0 \tilde{\psi} \gamma_0 \psi \rangle = \langle \psi^\dagger \psi \rangle > 0, \tag{8.94}$$

which is *positive definite.* This is interpreted as a probability density, and localised wave functions are usually normalised such that

$$\int d^3x \, J_0 = 1. \tag{8.95}$$

Arriving at a relativistic theory with a consistent probabilistic interpretation was Dirac's original goal.

8.3.2 Plane-wave states

A positive energy plane-wave state is defined by

$$\psi = \psi_0 e^{-I\boldsymbol{\sigma}_3 p \cdot x}, \tag{8.96}$$

where ψ_0 is a constant spinor. The Dirac equation (8.87) tells us that ψ_0 satisfies

$$p\psi_0 = m\psi_0\gamma_0, \tag{8.97}$$

and post-multiplying by $\tilde{\psi}_0$ we see that

$$p\psi_0\tilde{\psi}_0 = mJ. \tag{8.98}$$

Recalling that we have $\psi\tilde{\psi} = \rho e^{i\beta}$, and noting that both p and J are vectors, we see that we must have $\exp(i\beta) = \pm 1$. For positive energy states the timelike component of p is positive, as is the timelike component of J, so we take the positive solution $\beta = 0$. It follows that ψ_0 is then simply a rotor with a normalisation constant. The proper boost L taking $m\gamma_0$ onto the momentum has

$$p = mL\gamma_0\tilde{L} = mL^2\gamma_0, \tag{8.99}$$

and from section 5.4.4 the solution is

$$L = \frac{m + p\gamma_0}{[2m(m + p\cdot\gamma_0)]^{1/2}} = \frac{E + m + \boldsymbol{p}}{[2m(E+m)]^{1/2}}, \tag{8.100}$$

where $p\gamma_0 = E + \boldsymbol{p}$. The full spinor ψ_0 is LU, where U is a spatial rotor in the γ_0 frame, so is a Pauli spinor.

Negative-energy solutions have a phase factor of $\exp(+I\boldsymbol{\sigma}_3 p\cdot x)$, with $E = \gamma_0 \cdot p > 0$. For these we have $-p\psi\tilde{\psi} = mJ$ so it is clear that we now need $\beta = \pi$. Positive and negative energy plane wave states can therefore be summarised by

$$\begin{aligned} \text{positive energy:} \quad & \psi^{(+)}(x) = L(p)U_r e^{-I\boldsymbol{\sigma}_3 p\cdot x}, \\ \text{negative energy:} \quad & \psi^{(-)}(x) = L(p)U_r I e^{I\boldsymbol{\sigma}_3 p\cdot x}, \end{aligned} \tag{8.101}$$

with $L(p)$ given by equation (8.100). The subscript r on the spatial rotors labels the spin state, with $U_0 = 1$, $U_1 = -I\boldsymbol{\sigma}_2$. These plane wave solutions are the fundamental components of *scattering theory*.

8.3.3 Hamiltonian form and the Pauli equation

The problem of how to best formulate operator techniques within spacetime algebra is little more than a question of finding a good notation. We could of course borrow the traditional Dirac 'bra-ket' notation, but we have already seen that the bilinear covariants are better handled without it. It is easier instead to just juxtapose the operator and the wavefunction on which it acts. But we saw in section 8.2 that the operators often act double-sidedly on the spinor ψ. This is not a problem, as the only permitted right-sided operations are multiplication by γ_0 or $I\boldsymbol{\sigma}_3$, and these operations commute. Our notation can therefore safely suppress these right-sided multiplications and gather all operations on the left. The overhat notation is useful to achieve this and we define

$$\hat{\gamma}_\mu \psi = \gamma_\mu \psi \gamma_0. \tag{8.102}$$

It should be borne in mind that all operations are now defined in the spacetime algebra, so the $\hat{\gamma}_\mu$ are not to be read as matrix operators, as they were in section 8.2. Of course, the action of the operators in either system is identical.

It is also useful to have a symbol for the operation of right-sided multiplication by $I\boldsymbol{\sigma}_3$. The symbol j carries the correct connotations of an operator that commutes with all others and squares to -1, and we define

$$j\psi = \psi I\boldsymbol{\sigma}_3. \tag{8.103}$$

The Dirac equation can now be written in the 'operator' form

$$j\hat{\nabla}\psi - e\hat{A}\psi = m\psi, \tag{8.104}$$

where

$$\hat{\nabla}\psi = \nabla\psi\gamma_0 \quad \text{and} \quad \hat{A}\psi = A\psi\gamma_0. \tag{8.105}$$

Writing the Dirac equation in the form (8.104) does not add anything new, but does confirm that we have an efficient notation for handling operators. One might ask why we have preferred the j symbol over the more obvious i. One reason is historical. In much of the spacetime algebra literature it has been common practice to denote the spacetime pseudoscalar with a small i. We now feel that this is a misleading notation, but it is commonplace. In addition, there are occasions when we may wish to formally complexify the spacetime algebra, as was the case for electromagnetic scattering, covered in section 7.5. To avoid confusion with either of these cases we have chosen to denote right-multiplication of ψ by $I\boldsymbol{\sigma}_3$ as $j\psi$ in both this and the following chapter.

To express the Dirac equation in Hamiltonian form we simply multiply from the left by γ_0. The resulting equation, with the dimensional constants temporarily put back in, is

$$j\hbar\partial_t\psi = c\hat{\boldsymbol{p}}\psi + eV\psi - ce\boldsymbol{A}\psi + mc^2\bar{\psi}, \tag{8.106}$$

where

$$\begin{aligned} \hat{\boldsymbol{p}}\psi &= -j\hbar\boldsymbol{\nabla}\psi, \\ \bar{\psi} &= \gamma_0\psi\gamma_0, \\ \gamma_0 A &= V - c\boldsymbol{A}. \end{aligned} \tag{8.107}$$

Choosing a Hamiltonian is a non-covariant operation, since it picks out a preferred timelike direction. The Hamiltonian relative to the γ_0 direction is the operator on the right-hand side of equation (8.106).

As an application of the Hamiltonian formulation, consider the non-relativistic reduction of the Dirac equation. This can be achieved formally via the *Foldy–Wouthuysen* transformation. For details we refer the reader to Itzykson & Zuber (1980). While the theoretical motivation for this transformation is clear, it can be hard to compute in all but the simplest cases. A simpler approach, dating back to Feynman, is to separate out the fast-oscillating component of the waves and then split into separate equations for the Pauli-even and Pauli-odd components of ψ. We write (with $\hbar = 1$ and the factors of c kept in)

$$\psi = (\phi + \eta)e^{-I\sigma_3 mc^2 t}, \tag{8.108}$$

where $\bar{\phi} = \phi$ (Pauli-even) and $\bar{\eta} = -\eta$ (Pauli-odd). The Dirac equation (8.106) now splits into the two equations

$$\begin{aligned} \mathcal{E}\phi - c\mathcal{O}\eta &= 0, \\ (\mathcal{E} + 2mc^2)\eta - c\mathcal{O}\phi &= 0, \end{aligned} \tag{8.109}$$

where

$$\begin{aligned} \mathcal{E}\phi &= (j\partial_t - eV)\phi, \\ \mathcal{O}\phi &= (\hat{\boldsymbol{p}} - e\boldsymbol{A})\phi. \end{aligned} \tag{8.110}$$

The formal solution to the second of equations (8.109) is

$$\eta = \frac{1}{2mc}\left(1 + \frac{\mathcal{E}}{2mc^2}\right)^{-1}\mathcal{O}\phi, \tag{8.111}$$

where the inverse on the right-hand side denotes a power series. Provided the expectation value of $\mathcal{E}$ is smaller than $2mc^2$ (which it is in the non-relativistic limit) the series should converge. The remaining equation for ϕ is

$$\mathcal{E}\phi - \frac{\mathcal{O}}{2m}\left(1 - \frac{\mathcal{E}}{2mc^2} + \cdots\right)\mathcal{O}\phi = 0, \tag{8.112}$$

which can be expanded out to the desired order of magnitude. There is little point in going beyond the first relativistic correction, so we approximate equation (8.112) by

$$\mathcal{E}\phi + \frac{\mathcal{O}\mathcal{E}\mathcal{O}}{4m^2c^2}\phi = \frac{\mathcal{O}^2}{2m}\phi. \tag{8.113}$$

We seek an equation of the form $\mathcal{E}\phi = \mathcal{H}\phi$, where $\mathcal{H}$ is the non-relativistic Hamiltonian. We therefore need to replace the $\mathcal{O}\mathcal{E}\mathcal{O}$ term in equation (8.113) by a term that does not involve $\mathcal{E}$. To achieve this we write

$$2\mathcal{O}\mathcal{E}\mathcal{O} = [\mathcal{O},[\mathcal{E},\mathcal{O}]] + \mathcal{E}\mathcal{O}^2 + \mathcal{O}^2\mathcal{E} \tag{8.114}$$

so that equation (8.113) becomes

$$\mathcal{E}\phi = \frac{\mathcal{O}^2}{2m}\phi - \frac{\mathcal{E}\mathcal{O}^2 + \mathcal{O}^2\mathcal{E}}{8m^2c^2}\phi - \frac{1}{8m^2c^2}[\mathcal{O},[\mathcal{E},\mathcal{O}]]\phi. \tag{8.115}$$

We can now make the approximation

$$\mathcal{E}\phi \approx \frac{\mathcal{O}^2}{2m}\phi, \tag{8.116}$$

so that equation (8.113) can be approximated by

$$\mathcal{E}\phi = \frac{\mathcal{O}^2}{2m}\phi - \frac{1}{8m^2c^2}[\mathcal{O},[\mathcal{E},\mathcal{O}]]\phi - \frac{\mathcal{O}^4}{8m^3c^2}\phi, \tag{8.117}$$

which is valid to order c^{-2}.

To evaluate the commutators we first need

$$[\mathcal{E},\mathcal{O}] = -je(\partial_t \boldsymbol{A} + \boldsymbol{\nabla} V) = je\boldsymbol{E}. \tag{8.118}$$

There are no time derivatives left in this commutator, so we do achieve a sensible non-relativistic Hamiltonian. The full commutator required in equation (8.117) is

$$\begin{aligned}[\mathcal{O},[\mathcal{E},\mathcal{O}]] &= [-j\boldsymbol{\nabla} - e\boldsymbol{A}, je\boldsymbol{E}] \\ &= (e\boldsymbol{\nabla}\boldsymbol{E}) - 2e\boldsymbol{E}\wedge\boldsymbol{\nabla} - 2je^2\boldsymbol{A}\wedge\boldsymbol{E}.\end{aligned} \tag{8.119}$$

The various operators (8.110) and (8.119) can now be substituted into equation (8.117) to yield the Pauli equation

$$\begin{aligned}\frac{\partial\phi}{\partial t}I\boldsymbol{\sigma}_3 = {}& \frac{1}{2m}(\hat{\boldsymbol{p}} - e\boldsymbol{A})^2\phi + eV\phi - \frac{\hat{\boldsymbol{p}}^4}{8m^3c^2}\phi \\ &- \frac{1}{8m^2c^2}\big(e(\boldsymbol{\nabla}\boldsymbol{E} - 2\boldsymbol{E}\wedge\boldsymbol{\nabla})\phi - 2e^2\boldsymbol{A}\wedge\boldsymbol{E}\phi I\boldsymbol{\sigma}_3\big),\end{aligned} \tag{8.120}$$

which is written entirely in the geometric algebra of three-dimensional space. In the standard approach, the geometric product in the $\boldsymbol{\nabla}\boldsymbol{E}$ term of equation (8.120) is split into a 'spin-orbit' term $\boldsymbol{\nabla}\wedge\boldsymbol{E}$ and the 'Darwin' term $\boldsymbol{\nabla}\cdot\boldsymbol{E}$.

The spacetime algebra approach reveals that these terms arise from a single source.

A similar approximation scheme can be adopted for the observables of the Dirac theory. For example the current, $\psi\gamma_0\tilde{\psi}$, has a three-vector part:

$$\boldsymbol{J} = (\psi\gamma_0\tilde{\psi})\wedge\gamma_0 = \phi\eta^\dagger + \eta\phi^\dagger. \tag{8.121}$$

This is approximated to leading order by

$$\boldsymbol{J} \approx -\frac{1}{m}(\langle\boldsymbol{\nabla}\phi I\boldsymbol{\sigma}_3\phi^\dagger\rangle_1 - \boldsymbol{A}\phi\phi^\dagger), \tag{8.122}$$

where the $\langle\,\rangle_1$ projects onto the grade-1 components of the Pauli algebra. Not all applications of the Pauli theory correctly identify (8.122) as the conserved current in the Pauli theory — an inconsistency first pointed out by Hestenes & Gurtler (1971).

8.4 Central potentials

Suppose now that we restrict our discussion to problems described by a central potential $V = V(r)$, $\boldsymbol{A} = 0$, where $r = |\boldsymbol{x}|$. The full Hamiltonian, denoted $\mathcal{H}$, reduces to

$$j\hbar\partial_t\psi = \mathcal{H}\psi = -j\boldsymbol{\nabla}\psi + eV(r)\psi + m\bar{\psi}. \tag{8.123}$$

Quantum states are classified in terms of eigenstates of operators that commute with the Hamiltonian $\mathcal{H}$, because the accompanying quantum numbers are conserved in time. Of particular importance are the angular-momentum operators $\hat{L}_i$, defined by

$$\hat{L}_i = -i\epsilon_{ijk}x_j\partial_k. \tag{8.124}$$

These are the components of the bivector operator $i\boldsymbol{x}\wedge\boldsymbol{\nabla}$. We therefore define the operators

$$L_B = jB\cdot(\boldsymbol{x}\wedge\boldsymbol{\nabla}), \tag{8.125}$$

where B is a relative bivector. Throughout this section interior and exterior products refer to the (Pauli) algebra of space. Writing $B = I\boldsymbol{\sigma}_i$ recovers the component form. The L_B operators satisfy the commutation relations

$$[L_{B_1}, L_{B_2}] = -jL_{B_1\times B_2}, \tag{8.126}$$

where $B_1\times B_2$ denotes the commutator product. The angular-momentum commutation relations directly encode the bivector commutation relations, which are those of the Lie algebra of the rotation group (see chapter 11). One naturally expects this group to arise as it represents a symmetry of the potential.

If we now form the commutator of L_B with the Hamiltonian $\mathcal{H}$ we obtain a result that is, initially, disconcerting. The scalar operator L_B commutes with the bar operator $\psi \mapsto \bar{\psi}$, but for the momentum term we find that

$$[B\cdot(\boldsymbol{x}\wedge\boldsymbol{\nabla}), \boldsymbol{\nabla}] = -\dot{\boldsymbol{\nabla}} B\cdot(\dot{\boldsymbol{x}}\wedge\boldsymbol{\nabla}) = B\times\boldsymbol{\nabla}. \tag{8.127}$$

The commutator does not vanish, so orbital angular momentum does not yield a conserved quantum number in relativistic physics. But, since $B\times\boldsymbol{\nabla} = \frac{1}{2}(B\boldsymbol{\nabla} - \boldsymbol{\nabla}B)$, we can write equation (8.127) as

$$[B\cdot(\boldsymbol{x}\wedge\boldsymbol{\nabla}) - \tfrac{1}{2}B, \mathcal{H}] = 0. \tag{8.128}$$

We therefore recover a conserved angular momentum operator by defining

$$J_B = L_B - \tfrac{1}{2}jB. \tag{8.129}$$

In conventional notation this is

$$\hat{J}_i = \hat{L}_i + \tfrac{1}{2}\hat{\Sigma}_i, \tag{8.130}$$

where $\hat{\Sigma}_i = (i/2)\epsilon_{ijk}\hat{\gamma}_j\hat{\gamma}_k$. The extra term of $B/2$ accounts for the spin-1/2 nature of Dirac particles. If we look for eigenstates of the J_3 operator, we see that the spin contribution to this is

$$-\tfrac{1}{2}jI\boldsymbol{\sigma}_3\psi = \tfrac{1}{2}\boldsymbol{\sigma}_3\psi\boldsymbol{\sigma}_3. \tag{8.131}$$

In the non-relativistic Pauli theory the eigenstates of this operator are simply 1 and $-I\boldsymbol{\sigma}_2$, with eigenvalues $\pm 1/2$. In the relativistic theory the separate spin and orbital operators are not conserved, and it is only the combined J_B operators that commute with the Hamiltonian.

The geometric algebra derivation employed here highlights some interesting features. Stripping away all of the extraneous terms, the result rests solely on the commutation properties of the $B\cdot(\boldsymbol{x}\wedge\boldsymbol{\nabla})$ and $\boldsymbol{\nabla}$ operators. The factor of 1/2 would therefore be present in any dimension, and so has no special relation to the three-dimensional rotation group. Furthermore, in writing $J_B = L_B - \frac{1}{2}jB$ we are forming an explicit sum of a scalar and a bivector. The standard notation of equation (8.130) encourages us to view these as the sum of two vector operators!

8.4.1 Spherical monogenics

The spherical monogenics play a key role in the solution of the Dirac equation for problems with radial symmetry. These are Pauli spinors (even elements of the Pauli algebra) that satisfy the eigenvalue equation

$$-\boldsymbol{x}\wedge\boldsymbol{\nabla}\psi = l\psi. \tag{8.132}$$

These functions arise naturally as solutions of the three-dimensional generalisation of the Cauchy–Riemann equations

$$\boldsymbol{\nabla}\Psi = 0. \tag{8.133}$$

Solutions of this equation are known in the Clifford analysis literature as monogenics. Looking for solutions which separate into $\Psi = r^l \psi(\theta,\phi)$ yields equation (8.132), where (r,θ,ϕ) is a standard set of polar coordinates. The solutions of equation (8.132) are called spherical monogenics, or *spin-weighted spherical harmonics* (with weight 1/2).

To analyse the properties of equation (8.132) we first note that

$$[J_B, \boldsymbol{x}\wedge\boldsymbol{\nabla}] = 0, \tag{8.134}$$

which is proved in the same manner as equation (8.128). It follows that ψ can simultaneously be an eigenstate of the $\boldsymbol{x}\wedge\boldsymbol{\nabla}$ operator and one of the J_B operators. To simplify the notation we now define

$$J_k\psi = J_{I\sigma_k}\psi = \big((I\sigma_k)\cdot(\boldsymbol{x}\wedge\boldsymbol{\nabla}) - \tfrac{1}{2}I\sigma_k\big)\psi \boldsymbol{I\sigma}_3. \tag{8.135}$$

We choose ψ to be an eigenstate of J_3. We label this state as $\psi(l,\mu)$, so

$$-\boldsymbol{x}\wedge\boldsymbol{\nabla}\psi(l,\mu) = l\psi(l,\mu), \qquad J_3\psi(l,\mu) = \mu\psi(l,\mu). \tag{8.136}$$

The J_i operators satisfy

$$\begin{aligned} J_iJ_i\psi(l,\mu) &= 3/4\psi - 2\boldsymbol{x}\wedge\boldsymbol{\nabla}\psi + \boldsymbol{x}\wedge\boldsymbol{\nabla}(\boldsymbol{x}\wedge\boldsymbol{\nabla}\psi) \\ &= (l+1/2)(l+3/2)\psi(l,\mu), \end{aligned} \tag{8.137}$$

so the $\psi(l,\mu)$ are also eigenstates of J_iJ_i.

We next introduce the ladder operators J_+ and J_-, defined by

$$\begin{aligned} J_+ &= J_1 + jJ_2, \\ J_- &= J_1 - jJ_2. \end{aligned} \tag{8.138}$$

It is a simple matter to prove the following results:

$$\begin{aligned} [J_+, J_-] &= 2J_3, & J_iJ_i &= J_-J_+ + J_3 + {J_3}^2, \\ [J_\pm, J_3] &= \mp J_\pm, & J_iJ_i &= J_+J_- - J_3 + {J_3}^2. \end{aligned} \tag{8.139}$$

The raising operator J_+ increases the eigenvalue of J_3 by an integer. But, for fixed l, μ must ultimately attain some maximum value. Denoting this value as μ_+, we must reach a state for which

$$J_+\psi(l,\mu_+) = 0. \tag{8.140}$$

Acting on this state with J_iJ_i and using one of the results in equation (8.139) we find that

$$(l+1/2)(l+3/2) = \mu_+(\mu_+ + 1). \tag{8.141}$$

Since l is positive and μ_+ represents an upper bound, it follows that

$$\mu_+ = l + 1/2. \tag{8.142}$$

There must similarly be a lowest eigenvalue of J_3 and a corresponding state with

$$J_-\psi(l,\mu_-) = 0. \tag{8.143}$$

In this case we find that

$$(l+1/2)(l+3/2) = \mu_-(\mu_- - 1), \tag{8.144}$$

hence $\mu_- = -(l+1/2)$. The spectrum of eigenvalues of J_3 therefore ranges from $(l+1/2)$ to $-(l+1/2)$, a total of $2(l+1)$ states. Since the J_3 eigenvalues are always of the form (integer $+1/2$), it is simpler to label the spherical monogenics with a pair of integers. We therefore write the spherical monogenics as ψ_l^m, where

$$-\boldsymbol{x}\wedge\boldsymbol{\nabla}\psi_l^m = l\psi_l^m \qquad l \geq 0 \tag{8.145}$$

and

$$J_3\psi_l^m = (m+\tfrac{1}{2})\psi_l^m \qquad -1-l \leq m \leq l. \tag{8.146}$$

To find an explicit form for the ψ_l^m we first construct the highest m case. This satisfies

$$J_+\psi_l^l = 0 \tag{8.147}$$

and it is not hard to see that this equation is solved by

$$\psi_l^l \propto \sin^l(\theta)\,\mathrm{e}^{-l\phi I\boldsymbol{\sigma}_3}. \tag{8.148}$$

This is the angular part of the monogenic function $(x+yI\boldsymbol{\sigma}_3)^l$. Introducing a convenient factor, we write

$$\psi_l^l = (2l+1)P_l^l(\cos(\theta))\,\mathrm{e}^{l\phi I\boldsymbol{\sigma}_3}. \tag{8.149}$$

Our convention for the associated Legendre polynomials follows Gradshteyn & Ryzhik (1994), so we have

$$P_l^m(x) = \frac{(-1)^m}{2^l l!}(1-x^2)^{m/2}\frac{d^{l+m}}{dx^{l+m}}(x^2-1)^l. \tag{8.150}$$

(Some useful recursion relations for the associated Legendre polynomials are discussed in the exercises.) The lowering operator J_- has the following effect on ψ:

$$J_-\psi = \big(-\partial_\theta\psi + \cot(\theta)\,\partial_\phi\psi I\boldsymbol{\sigma}_3\big)\mathrm{e}^{-\phi I\boldsymbol{\sigma}_3} - I\boldsymbol{\sigma}_2\,\tfrac{1}{2}(\psi + \boldsymbol{\sigma}_3\psi\boldsymbol{\sigma}_3). \tag{8.151}$$

The final term just projects out the $\{1, I\boldsymbol{\sigma}_3\}$ terms and multiplies them by $-I\boldsymbol{\sigma}_2$. This is the analog of the lowering matrix in the standard formalism. The derivatives acting on ψ_l^l form

$$\big(-\partial_\theta\psi_l^l + \cot(\theta)\,\partial_\phi\psi_l^l I\boldsymbol{\sigma}_3\big)\mathrm{e}^{-\phi I\boldsymbol{\sigma}_3} = (2l+1)2lP_l^{l-1}(\cos(\theta))\mathrm{e}^{(l-1)\phi I\boldsymbol{\sigma}_3}, \tag{8.152}$$

and, if we use the result that

$$\sigma_\phi = \sigma_2 e^{\phi I \sigma_3}, \tag{8.153}$$

we find that

$$\psi_l^{l-1} \propto \big(2l P_l^{l-1}(\cos(\theta)) - P_l^l(\cos(\theta)) I \sigma_\phi\big) e^{(l-1)\phi I \sigma_3}. \tag{8.154}$$

Proceeding in this manner, we are led to the following formula for the spherical monogenics:

$$\psi_l^m = \big((l+m+1)P_l^m(\cos(\theta)) - P_l^{m+1}(\cos(\theta)) I \sigma_\phi\big) e^{m\phi I \sigma_3}, \tag{8.155}$$

in which l is a positive integer or zero, m ranges from $-(l+1)$ to l and the P_l^m are taken to be zero if $|m| > l$. The positive- and negative-m states are related by

$$P_l^{-m}(x) = (-1)^m \frac{(l-m)!}{(l+m)!} P_l^m(x), \tag{8.156}$$

from which it can be shown that

$$\psi_l^m(-I\sigma_2) = (-1)^m \frac{(l+m+1)!}{(l-m)!} \psi_l^{-(m+1)}. \tag{8.157}$$

The spherical monogenics presented here are unnormalised. Normalisation factors are not hard to compute, and we find that

$$\int_0^\pi d\theta \int_0^{2\pi} d\phi \, \sin(\theta) \, \psi_l^m \psi_l^{m\dagger} = 4\pi \frac{(l+m+1)!}{(l-m)!}. \tag{8.158}$$

If σ_r denotes the unit radial vector, $\sigma_r = x/r$ we find that

$$x \wedge \nabla \sigma_r = 2\sigma_r. \tag{8.159}$$

It follows that

$$-x \wedge \nabla(\sigma_r \psi \sigma_3) = -(l+2)\sigma_r \psi \sigma_3, \tag{8.160}$$

which provides an equation for the negative-l eigenstates. The possible eigenvalues and degeneracies are summarised in Table 8.2. One curious feature of this table is that we appear to be missing a line for the eigenvalue $l = -1$. In fact solutions for this case do exist, but they contain singularities which render them unnormalisable. For example, the functions

$$\frac{I\sigma_\phi}{\sin(\theta)}, \quad \text{and} \quad \frac{e^{-I\sigma_3\phi}}{\sin(\theta)} \tag{8.161}$$

have $l = -1$ and J_3 eigenvalues $+1/2$ and $-1/2$ respectively. Both solutions are singular along the z axis, however, which limits their physical relevance.

l	Eigenvalues of J_3	Degeneracy
$\vdots$	$\vdots$	$\vdots$
2	$5/2,\dots,-5/2$	6
1	$3/2,\dots,-3/2$	4
0	$1/2,\dots,-1/2$	2
(-1)	?	?
-2	$1/2,\dots,-1/2$	2
$\vdots$	$\vdots$	$\vdots$

Table 8.2 *Eigenvalues and degeneracies for the* ψ_l^m *monogenics.*

8.4.2 The radial equations

We can use the angular monogenics to construct eigenfunctions of the Dirac Hamiltonian of equation (8.123). Since the J_B operators commute with $\mathcal{H}$, ψ can be placed in an eigenstate of J_3. The operator J_iJ_i must also commute with $\mathcal{H}$, so $(l+1/2)(l+3/2)$ is a good quantum number. The operator $\boldsymbol{x}\wedge\boldsymbol{\nabla}$ does not commute with $\mathcal{H}$, however, so both the ψ_l^m and $\boldsymbol{\sigma}_r\psi_l^m\boldsymbol{\sigma}_3$ monogenics are needed in the solution. While $\boldsymbol{x}\wedge\boldsymbol{\nabla}$ does not commute with $\mathcal{H}$, the operator

$$\hat{K} = \hat{\gamma}_0(1-\boldsymbol{x}\wedge\boldsymbol{\nabla}) \tag{8.162}$$

does, as follows from

$$[\hat{\gamma}_0(1-\boldsymbol{x}\wedge\boldsymbol{\nabla}),\boldsymbol{\nabla}] = 2\hat{\gamma}_0\boldsymbol{\nabla} - \hat{\gamma}_0\dot{\boldsymbol{\nabla}}\dot{\boldsymbol{x}}\wedge\boldsymbol{\nabla} = 0. \tag{8.163}$$

We should therefore work with eigenstates of the $\hat{K}$ operator. This implies that $\psi(\boldsymbol{x})$ can be written for positive l as either

$$\psi(\boldsymbol{x},l+1) = \psi_l^m u(r) + \boldsymbol{\sigma}_r\psi_l^m v(r) I\boldsymbol{\sigma}_3 \tag{8.164}$$

or

$$\psi(\boldsymbol{x},-(l+1)) = \boldsymbol{\sigma}_r\psi_l^m\boldsymbol{\sigma}_3 u(r) + \psi_l^m I v(r). \tag{8.165}$$

In both cases the second label in $\psi(\boldsymbol{x},l+1)$ specifies the eigenvalue of $\hat{K}$. It is useful to denote this by κ, so we have

$$\hat{K}\psi = \kappa\psi, \qquad \kappa = \dots,-2,-1,1,2,\dots \tag{8.166}$$

and κ is a non-zero positive or negative integer.

In equations (8.164) and (8.165) the radial functions $u(r)$ and $v(r)$ are 'complex' combinations of 1 and $I\boldsymbol{\sigma}_3$. In the case of the Hamiltonian of (8.123), with $V(r)$ real, it turns out that the real and imaginary equations decouple, and it is sufficient to treat $u(r)$ and $v(r)$ as real, scalar quantities. On substituting our trial functions into the Hamiltonian, we find that the radial equations reduce to

$$\begin{pmatrix} u' \\ v' \end{pmatrix} = \begin{pmatrix} (\kappa - 1)/r & -(E - eV(r) + m) \\ E - eV(r) - m & (-\kappa - 1)/r \end{pmatrix} \begin{pmatrix} u \\ v \end{pmatrix}. \tag{8.167}$$

The same equation holds for all values of κ. This successfully separates the Dirac equation in any radially-symmetric potential. As one might expect, we arrive at a pair of coupled first-order equations, as opposed to the single second-order equation familiar from Schrödinger theory.

8.4.3 The hydrogen atom

The radial equations describing the relativistic quantum theory of the hydrogen atom are obtained simply by setting $eV = -Z\alpha/r$, where $\alpha = e^2/4\pi$ is the fine structure constant and Z is the atomic charge. The solution of the radial equations is described in most textbooks on relativistic quantum mechanics. The conclusion is that the radial dependence is governed by a pair of *hypergeometric functions,* which generalise the Laguerre polynomials of the non-relativistic theory. Rather than reproduce the analysis here, we instead present a more direct method of solving the equations, first given by Eddington (1936) in his unconventional *Relativity Theory of Protons and Electrons.*

We start with the equation

$$-j\boldsymbol{\nabla}\psi - \frac{Z\alpha}{r}\psi + m\hat{\gamma}_0\psi = E\psi. \tag{8.168}$$

We assume that ψ is in an eigenstate of $\hat{K}$, so we can write

$$\boldsymbol{x}\wedge\boldsymbol{\nabla}\psi = \psi - \kappa\hat{\gamma}_0\psi. \tag{8.169}$$

We now pre-multiply the Dirac equation by $j\boldsymbol{x}$ and rearrange to find

$$r\partial_r\psi + \psi - \kappa\hat{\gamma}_0\psi = j\boldsymbol{x}\left(E + \frac{Z\alpha}{r}\right)\psi - jm\boldsymbol{x}\hat{\gamma}_0\psi. \tag{8.170}$$

On introducing the reduced function $\Psi = r\psi$ the equation simplifies to

$$\partial_r\Psi = j\boldsymbol{\sigma}_r(E - m\hat{\gamma}_0)\Psi + \frac{1}{r}(jZ\alpha\boldsymbol{\sigma}_r + \kappa\hat{\gamma}_0)\Psi. \tag{8.171}$$

We accordingly define the two operators

$$\hat{F} = -j\boldsymbol{\sigma}_r(E - m\hat{\gamma}_0), \qquad \hat{G} = -(jZ\alpha\boldsymbol{\sigma}_r + \kappa\hat{\gamma}_0), \tag{8.172}$$

so that the Dirac equation reduces to

$$\partial_r\Psi + \left(\hat{F} + \frac{\hat{G}}{r}\right)\Psi = 0. \tag{8.173}$$

The $\hat{F}$ and $\hat{G}$ operators satisfy

$$\begin{aligned} \hat{F}^2 &= m^2 - E^2 = f^2, \\ \hat{G}^2 &= \kappa^2 - (Z\alpha)^2 = \nu^2, \end{aligned} \tag{8.174}$$

which define f and ν. The operators also satisfy the anticommutation relation

$$\hat{F}\hat{G} + \hat{G}\hat{F} = -2Z\alpha E. \tag{8.175}$$

The next step is to transform to the dimensionless variable $x = fr$ and remove the large-x behaviour by setting

$$\Psi = \Phi \mathrm{e}^{-x}. \tag{8.176}$$

The function Φ now satisfies

$$\partial_x \Phi + \frac{\hat{G}}{x}\Phi + \Big(\frac{\hat{F}}{f} - 1\Big)\Phi = 0. \tag{8.177}$$

We are now in a position to consider a power series solution, so we set

$$\Phi = x^s \sum_{n=0} C_n x^n, \tag{8.178}$$

where the C_n are all multivectors. (In Eddington's original notation these are his 'e-numbers'.) The recursion relation is first-order and is given simply by

$$(n + s + \hat{G})C_n = -\left(\frac{\hat{F}}{f} - 1\right) C_{n-1}. \tag{8.179}$$

Setting $n = 0$ we see that

$$\big(s + \hat{G}\big)C_0 = 0. \tag{8.180}$$

Acting on this equation with the operator $\big(s - \hat{G}\big)$ we see that we must have $s^2 = \hat{G}^2 = \nu^2$. We set $s = \nu$ in order that the wavefunction is well behaved at the origin.

With the small and large x behaviour now separated out, all that remains is the power series. One can show that, in order for ψ to fall to zero at large distances, the series must terminate. We therefore set $C_{n+1} = 0$, and it follows that

$$\left(\frac{\hat{F}}{f} - 1\right) C_n = 0, \quad \text{or} \quad \hat{F}C_n = fC_n. \tag{8.181}$$

But we also have

$$\left(\frac{\hat{F}}{f} + 1\right)(n + \nu + \hat{G})C_n = -\left(\frac{\hat{F}}{f} + 1\right)\left(\frac{\hat{F}}{f} - 1\right) C_{n-1} = 0, \tag{8.182}$$

so

$$\left(2(n+\nu)+\hat{G}+\frac{\hat{F}}{f}\hat{G}\right)C_n = 0. \tag{8.183}$$

If we write this as

$$\left(2(n+\nu)+\frac{1}{f}(\hat{G}\hat{F}+\hat{F}\hat{G})\right)C_n = 0, \tag{8.184}$$

we find that we must have

$$n+\nu-\frac{Z\alpha E}{f} = 0. \tag{8.185}$$

This is precisely our energy quantisation condition. The equation is equivalent to

$$\frac{E}{(m^2-E^2)^{1/2}} = \frac{n+\nu}{Z\alpha}, \tag{8.186}$$

which rearranges to the standard formula

$$E^2 = m^2\left(1-\frac{(Z\alpha)^2}{n^2+2n\nu+\kappa^2}\right), \tag{8.187}$$

where n is a non-negative integer.

The non-relativistic formula for the energy levels is recovered by first recalling that $\alpha \approx 1/137$ is small. We can therefore approximate to

$$\nu \approx |\kappa| = l+1, \tag{8.188}$$

where $l \geq 0$ and

$$E \approx m\left(1-\frac{(Z\alpha)^2}{2}\frac{1}{n^2+2n(l+1)+(l+1)^2}\right). \tag{8.189}$$

Subtracting off the rest mass energy we are left with the non-relativistic expression

$$E_{NR} = -m\frac{(Z\alpha)^2}{2}\frac{1}{(n+l+1)^2} = -\frac{mZ^2e^4}{32\pi^2\epsilon_0^2\hbar^2}\frac{1}{n'^2}, \tag{8.190}$$

where $n' = n+l+1$ and the dimensional constants have been reinserted. We have recovered the familiar Bohr formula for the energy levels. This derivation shows that the relativistic quantum number n differs from the Bohr quantum number n'.

Expanding to next order we find that

$$E_{NR} = -m\frac{(Z\alpha)^2}{2n'^2} - m\frac{(Z\alpha)^4}{2n'^4}\left(\frac{n'}{l+1}-\frac{3}{4}\right). \tag{8.191}$$

The first relativistic correction shows that the binding energy is increased slightly from the non-relativistic value, and also introduces some dependence on the

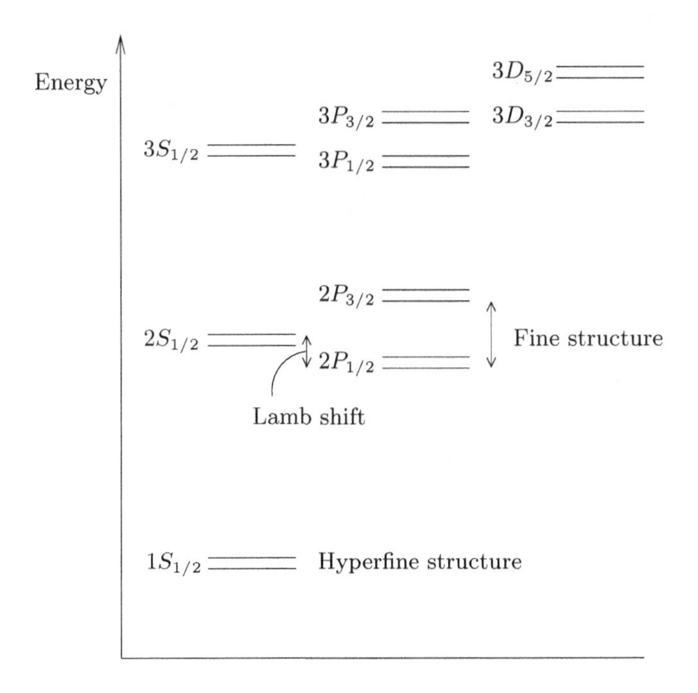

Figure 8.3 *Hydrogen atom energy levels.* The diagram illustrates how various degeneracies are broken by relativistic and spin effects. The Dirac equation accounts for the fine structure. The hyperfine structure is due to interaction with the magnetic moment of the nucleus. The Lamb shift is explained by quantum field theory. It lifts the degeneracy between the $S_{1/2}$ and $P_{1/2}$ states.

angular quantum number l. This lifts some degeneracies present in the non-relativistic solution. The various corrections contributing to the energy levels are shown in figure 8.3. A more complete analysis also requires replacing the electron mass m by the reduced mass of the two-body system. This introduces corrections of the same order of the relativistic corrections, but only affects the overall scale.

8.5 Scattering theory

Many of the experimental tests of Dirac theory, and quantum electrodynamics in general, are based on the results of scattering. Here we see how our new formulation can help to simplify these calculations through its handling of spin.

To aid this analysis it is useful to introduce the energy projection operators

$$\Lambda_{\pm}\psi = \frac{1}{2m}(m\psi \pm p\psi\gamma_0), \tag{8.192}$$

which project onto particle and antiparticle states.

A key role in relativistic quantum theory is played by Feynman propagators, which provide a means of imposing causal boundary conditions. We start by replacing the Dirac equation with the integral equation

$$\psi(x) = \psi_i(x) + e\int d^4x'\, S_F(x-x')A(x')\psi(x')\gamma_0, \tag{8.193}$$

where $\psi_i(x)$ is the asymptotic in-state and solves the free-particle equation, and $S_F(x-x')$ is the propagator. Substituting (8.193) into the Dirac equation, we find that $S_F(x-x')$ must satisfy

$$j\nabla_x S_F(x-x')\psi(x')\gamma_0 - mS_F(x-x')\psi(x') = \delta^4(x-x')\psi(x'). \tag{8.194}$$

The solution to this equation is

$$S_F(x-x')\psi(x') = \int \frac{d^4p}{(2\pi)^4}\,\frac{p\psi(x')\gamma_0 + m\psi(x')}{p^2-m^2+j\epsilon}\mathrm{e}^{-jp\cdot(x-x')}. \tag{8.195}$$

The factor of $j\epsilon$ is a mnemonic device to tell us how to negotiate the poles in the complex energy integral, which is performed first. The factor ensures positive-frequency waves propagate into the future ($t > t'$) and negative-frequency waves propagate into the past ($t' > t$). The result of performing the energy integration is summarised in the expression

$$S_F(x) = -2mj\int \frac{d^3\boldsymbol{p}}{(2\pi)^3}\,\frac{1}{2E}\Big(\theta(t)\Lambda_+\mathrm{e}^{-jp\cdot x} + \theta(-t)\Lambda_-\mathrm{e}^{jp\cdot x}\Big), \tag{8.196}$$

where $E = +\sqrt{\boldsymbol{p}^2+m^2}$.

There are other choices of relativistic propagator, which may be appropriate in other settings. For classical electromagnetism, for example, it is necessary to work with retarded propagators. If one constructs a closed spacetime surface integral, with boundary conditions consistent with the field equations, then the choice of propagator is irrelevant, since they all differ by a spacetime monogenic function. In most applications, however, we do not work like this. Instead we work with initial data, which we seek to propagate to a later time in such a way that the final result is consistent with imposing causal boundary conditions. In this case one has to use the Feynman propagator for quantum fields.

8.5.1 Electron scattering

In scattering calculations we write the wavefunction as the sum of an incoming plane wave and a scattered beam,

$$\psi(x) = \psi_i(x) + \psi_{diff}(x). \tag{8.197}$$

At asymptotically large times ψ_{diff} is given by

$$\psi_{diff}(x) = -2mje \int d^4x' \int \frac{d^3\boldsymbol{p}}{(2\pi)^3} \frac{1}{2E} \Lambda_+ \big(A(x')\psi(x')\gamma_0\big) \mathrm{e}^{-jp\cdot(x-x')}. \tag{8.198}$$

This can be written as a sum over final states

$$\psi_{diff}(x) = \int \frac{d^3\boldsymbol{p}_f}{(2\pi)^3} \frac{1}{2E_f} \psi_f \mathrm{e}^{-jp_f\cdot x}, \tag{8.199}$$

where the final states are plane waves with

$$\psi_f = -je \int d^4x' \big(p_f A(x')\psi(x') + mA(x')\psi(x')\gamma_0\big) \mathrm{e}^{jp_f\cdot x'}. \tag{8.200}$$

The number of scattered particles is given by (recalling that $J = \psi\gamma_0\tilde{\psi}$)

$$\int d^3\boldsymbol{x}\, \gamma_0 \cdot J_{diff} = \int \frac{d^3\boldsymbol{p}_f}{(2\pi)^3} \frac{1}{2E_f} \left(\frac{\gamma_0 \cdot J_f}{2E_f} \right) = \int \frac{d^3\boldsymbol{p}_f}{(2\pi)^3} \frac{1}{2E_f} N_f, \tag{8.201}$$

where N_f is the number density per Lorentz-invariant phase space interval:

$$N_f = \frac{\gamma_0 \cdot J_f}{2E_f} = \frac{\gamma_0 \cdot (\psi_f \gamma_0 \tilde{\psi}_f)}{2E_f} = \frac{\rho_f}{2m}. \tag{8.202}$$

The integral equation (8.193) is the basis for a perturbative approach to solving the Dirac equation in an external field. We seek the full propagator S_A which satisfies

$$\big(j\nabla_2 - eA(x_2)\big) S_A(x_2, x_1)\gamma_0 - mS_A(x_2, x_1) = \delta^4(x_2 - x_1). \tag{8.203}$$

The iterative solution to this is provided by

$$\begin{aligned} S_A(x_f, x_i) &= S_F(x_f - x_i) + \int d^4x_1\, S_F(x_f - x_1) e\hat{A}(x_1) S_F(x_1 - x_i) \\ &+ \iint d^4x_1\, d^4x_2\, S_F(x_f - x_1) e\hat{A}(x_1) S_F(x_1 - x_2) e\hat{A}(x_2) S_F(x_2 - x_i) + \cdots, \end{aligned} \tag{8.204}$$

which is the basis for a diagrammatic representation of a scattering calculation. In the Born approximation we work to first order and truncate the series for S_A

after the first interaction term. Assuming incident plane waves of momentum p_i, so that $\psi_i(x) = \psi_i \exp(-jp_i \cdot x)$, we find that the final states become

$$\begin{aligned}\psi_f &= -je \int d^4x' \, (p_f A(x') + A(x')p_i) \psi_i e^{jq\cdot x'} \\ &= -je\big(p_f A(q) + A(q)p_i\big)\psi_i,\end{aligned} \tag{8.205}$$

where $q = p_f - p_i$ is the change in momentum, and $A(q)$ is the Fourier transform of the electromagnetic potential. The form of the result here is quite typical, and in general we can write

$$\psi_f = S_{fi}\psi_i, \tag{8.206}$$

where S_{fi} is the *scattering operator.* This is a multivector that takes initial states into final states. Since both ψ_i and ψ_f are plane-wave particle states, we must have

$$S_{fi}\tilde{S}_{fi} = \rho_{fi}, \tag{8.207}$$

where ρ_{fi} is a scalar quantity (which determines the cross section). We can therefore decompose S_{fi} as

$$S_{fi} = \rho_{fi}^{1/2} R_{fi}, \tag{8.208}$$

where R_{fi} is a rotor. This rotor takes the initial momentum to the final momentum,

$$R_{fi} p_i \tilde{R}_{fi} = p_f. \tag{8.209}$$

8.5.2 Spin effects in scattering

The multivector S_{fi} depends on the initial and final momenta and, in some cases, the initial spin. The final spin is determined from the initial spin by the rotation encoded in S_{fi}. If s_i and s_f denote the initial and final (unit) spin vectors, we have

$$s_f = R_{fi} s_i \tilde{R}_{fi}. \tag{8.210}$$

Sometimes it is of greater interest to separate out the boost terms in R_{fi} to isolate a pure rotation in the γ_0 frame. This tells us directly what happens to the spin vector in the electron's rest frame. With L_i and L_f the appropriate pure boosts, we define the rest spin scattering operator

$$U_{fi} = \tilde{L}_f R_{fi} L_i. \tag{8.211}$$

This satisfies

$$U_{fi}\gamma_0 \tilde{U}_{fi} = \frac{1}{m}\tilde{L}_f R_{fi} p_i \tilde{R}_{fi} L_f = \gamma_0, \tag{8.212}$$

so is a pure rotation in the γ_0 frame.

The fact that $p_f S_{fi} = S_{fi} p_i$ ensures that S_{fi} is always of the form

$$S_{fi} = -j(p_f M + M p_i), \tag{8.213}$$

where M is an odd-grade multivector. In the Born approximation of equation (8.205), for example, we have $M = eA(q)$. In general, M can contain both real and imaginary terms, so we must write

$$S_{fi}\psi_i = -j\big(p_f(M_r + jM_j) + (M_r + jM_j)p_i\big)\psi_i, \tag{8.214}$$

where M_j and M_r are independent of j. We can now use

$$j\psi_i = \psi_i I\boldsymbol{\sigma}_3 = \hat{S}_i\psi_i, \tag{8.215}$$

where $\hat{S}_i$ is the initial unit spin bivector. Since $\hat{S}_i$ and p_i commute, S_{fi} can still be written in the form of equation (8.213), with

$$M = M_r + M_j\hat{S}_i. \tag{8.216}$$

So M remains a real multivector, which now depends on the initial spin. This scheme is helpful if we are interested in any spin-dependent features of the scattering process.

8.5.3 Positron scattering and pair annihilation

Adapting the preceding results to positron scattering is straightforward. In this case a negative-energy plane wave arrives from the future and scatters into the past, so we set

$$\psi_i(x) = \psi_2 e^{jp_i\cdot x}, \quad \psi_f(x) = \psi_f e^{jp_f\cdot x}. \tag{8.217}$$

In this case repeating the analysis gives

$$S_{fi}\psi_i = -j(-p_f M\psi_i + M\psi_i\gamma_0), \tag{8.218}$$

which we can write as

$$S_{fi} = j(p_f M + M p_i). \tag{8.219}$$

This amounts to simply swapping the sign of S_{fi}. In the Born approximation, q is replaced by $-q$ in the Fourier transform of $A(x)$, which will alter the factor M if $A(x)$ is complex.

The other case to consider is when the incoming electron is scattered into the past, corresponding to pair annihilation. In this case we have

$$S_{fi} = -j(-p_2 M + M p_1), \tag{8.220}$$

where p_1 and p_2 are the incoming momenta of the electron and positron respectively. We decompose S_{fi} as

$$S_{fi} = \rho_{fi}^{1/2} I R_{fi}, \tag{8.221}$$

since S_{fi} must now contain a factor of I to map electrons into positrons. This form for S_{fi} implies that

$$S_{fi}\tilde{S}_{fi} = -\rho_{fi}. \tag{8.222}$$

The minus sign reflects the fact that the transformation between initial and final momenta is not proper orthochronous.

8.5.4 Cross sections

We must now relate our results to the cross sections measured in experiments. The scattering rate into the final states, per unit volume, per unit time, is given by

$$W_{fi} = \frac{1}{VT}N_f = \frac{1}{VT}\frac{\gamma_0 \cdot J_f}{2E_f} = \frac{\rho_f}{2mVT}, \tag{8.223}$$

where V and T denote the total volume and time respectively. The density ρ_f is given by

$$\rho_f = |S_{fi}\tilde{S}_{fi}|\rho_i = \rho_{fi}\rho_i. \tag{8.224}$$

Here $S_{fi}\tilde{S}_{fi} = \pm\rho_{fi}$, where the plus sign corresponds to electron to electron and positron to positron scattering, and the minus sign to electron–positron annihilation.

The differential cross section is defined as

$$d\sigma = \frac{W_{fi}}{\text{target density} \times \text{incident flux}}. \tag{8.225}$$

When S_{fi} is of the form

$$S_{fi} = -j(2\pi)^4\delta^4(P_f - P_i)T_{fi}, \tag{8.226}$$

where the δ-function ensures conservation of total momentum, we have

$$|S_{fi}|^2 = VT(2\pi)^4\delta^4(P_f - P_i)|T_{fi}|^2. \tag{8.227}$$

Working in the J_i frame the target density is just ρ_i so, writing the incident flux as χ, we have

$$d\sigma = \frac{1}{2m\chi}(2\pi)^4\delta^4(P_f - P_i)|T_{fi}|^2. \tag{8.228}$$

Alternatively we may be interested in an elastic scattering with just energy conservation ($E_f = E_i$) and

$$S_{fi} = -j2\pi\delta(E_f - E_i)T_{fi}. \tag{8.229}$$

In this case

$$|S_{fi}|^2 = 2\pi T\delta(E_f - E_i)|T_{fi}|^2. \tag{8.230}$$

A target density of $1/V$ and an incident flux of $|\boldsymbol{J}_i| = \rho_i|\boldsymbol{p}_i|/m$ then gives

$$d\sigma = \frac{\pi}{|\boldsymbol{p}_i|}\delta(E_f - E_i)|T_{fi}|^2. \tag{8.231}$$

The total cross section is obtained by integrating over the available phase space. For the case of a single particle scattering elastically we find that

$$\sigma = \int \frac{d^3\boldsymbol{p}_f}{(2\pi)^3}\frac{1}{2E_f}\frac{\pi}{|\boldsymbol{p}_i|}\delta(E_f - E_i)|T_{fi}|^2 = \int d\Omega\, \frac{|T_{fi}|^2}{16\pi^2}. \tag{8.232}$$

This is usually expressed in terms of the differential cross section per solid angle:

$$\frac{d\sigma}{d\Omega_f} = \frac{|T_{fi}|^2}{16\pi^2}. \tag{8.233}$$

8.5.5 Coulomb scattering

As an application of our formalism consider Coulomb scattering from a nucleus, with the external field defined by

$$A(x) = \frac{-Ze}{4\pi|\boldsymbol{x}|}\gamma_0. \tag{8.234}$$

Working with the first Born approximation, M is given by $M = eA(q)$, where $A(q)$ is the Fourier transform of $A(x)$ given by

$$A(q) = -\frac{2\pi Ze}{\boldsymbol{q}^2}\delta(E_f - E_i)\gamma_0 \tag{8.235}$$

and $q{\cdot}\gamma_0 = E_f - E_i$. Writing

$$S_{fi} = -j2\pi\delta(E_f - E_i)T_{fi} \tag{8.236}$$

and using energy conservation we find that

$$T_{fi} = -\frac{Ze^2}{\boldsymbol{q}^2}(2E + \boldsymbol{q}). \tag{8.237}$$

The cross section is therefore given by the Mott scattering formula:

$$\frac{d\sigma}{d\Omega_f} = \frac{Z^2\alpha^2}{\boldsymbol{q}^4}(4E^2 - \boldsymbol{q}^2) = \frac{Z^2\alpha^2}{4\boldsymbol{p}^2\beta^2\sin^4(\theta/2)}\left(1 - \beta^2\sin^2(\theta/2)\right), \tag{8.238}$$

where

$$\boldsymbol{q}^2 = (\boldsymbol{p}_f - \boldsymbol{p}_i)^2 = 2\boldsymbol{p}^2\big(1 - \cos(\theta)\big) \quad \text{and} \quad \beta = |\boldsymbol{p}|/E. \tag{8.239}$$

The angle θ measures the deviation between the incoming and scattered beams. In the low velocity limit the Mott result reduces to the Rutherford formula. The result is independent of the sign of the nuclear charge and, to this order, is obtained for both electron and positron scattering.

A significant feature of this derivation is that no spin sums are required. Instead, all the spin dependence is contained in the directional information in T_{fi}. As well as being computationally more efficient, this method for organising cross section calculations offers deeper insights into the structure of the theory. For Coulomb scattering the spin information is contained in the rotor

$$R_{fi} = \frac{p_f\gamma_0 + \gamma_0 p_i}{4E^2 - \boldsymbol{q}^2} \propto L_f^2 + \tilde{L}_i^2, \tag{8.240}$$

where L_f and L_i are the pure boosts from γ_0 to p_f and p_i respectively. The behaviour of the rest spin is governed by the unnormalised rotor

$$U_{fi} = \tilde{L}_f(L_f^2 + \tilde{L}_i^2)L_i = L_f L_i + \tilde{L}_f\tilde{L}_i, = 2\big((E+m)^2 + \boldsymbol{p}_f\boldsymbol{p}_i\big). \tag{8.241}$$

It follows that the rest-spin vector precesses in the $\boldsymbol{p}_f \wedge \boldsymbol{p}_i$ plane through an angle δ, where

$$\tan(\delta/2) = \frac{\sin(\theta)}{(E+m)/(E-m) + \cos(\theta)}. \tag{8.242}$$

This method of calculating the spin precession for Coulomb scattering was first described by Hestenes (1982a).

8.5.6 Compton scattering

Compton scattering is the process in which an electron scatters off a photon. To lowest order there are two Feynman diagrams to consider, shown in figure 8.4. The preceding analysis follows through with little modification, and gives rise two terms of the form

$$M_1 = e^2 \iint d^4x_1\, d^4x_2\, \frac{d^4p}{(2\pi)^4}\, A_1(x_1) \frac{pA_2(x_2) + A_2(x_2)p_i}{p^2 - m^2 + j\epsilon} \times \mathrm{e}^{jx_1\cdot(p_f - p)} \mathrm{e}^{jx_2\cdot(p - p_i)}, \tag{8.243}$$

where

$$A(x) = \epsilon \mathrm{e}^{\mp jk\cdot x} \tag{8.244}$$

is the (complex) vector potential. The vector ϵ denotes the polarisation state, so $k\cdot\epsilon = 0$ and $\epsilon^2 = -1$. In relativistic quantum theory there appears to be no alternative but to work with a fully complex vector potential.

Performing the integrations and summing the two contributions we arrive at

$$M = e^2(2\pi)^4\delta^4(P)\left(\epsilon_f \frac{(p_i + k_i)\epsilon_i + \epsilon_i p_i}{2k_i\cdot p_i} - \epsilon_i \frac{(p_i - k_f)\epsilon_f + \epsilon_f p_i}{2p_i\cdot k_f}\right), \tag{8.245}$$

where $P = p_f + k_f - p_i - k_i$, so that the δ-function enforces momentum conservation. Gauge invariance means that we can set $p_i\cdot\epsilon_i = p_i\cdot\epsilon_f = 0$, in which case

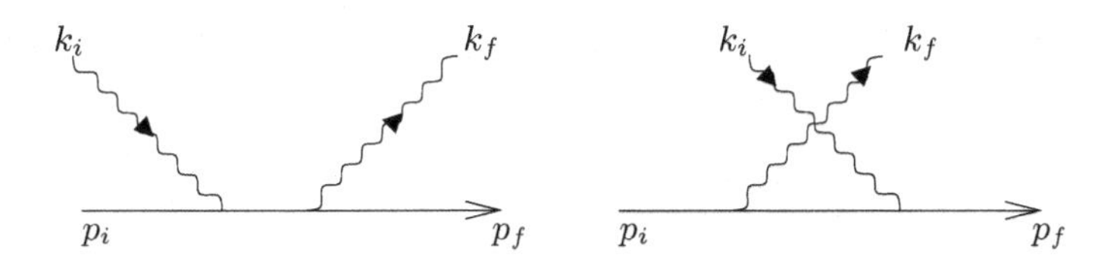

Figure 8.4 *Compton scattering.* Two diagrams contribute to the amplitude, to lowest order.

M simplifies to

$$M = e^2(2\pi)^4\delta^4(p_f + k_f - p_i - k_i)\left(\frac{\epsilon_f k_i \epsilon_i}{2k_i\cdot p_i} + \frac{\epsilon_i k_f \epsilon_f}{2p_i\cdot k_f}\right). \tag{8.246}$$

We now set

$$S_{fi} = -j(2\pi)^4\delta^4(p_f + k_f - p_i - k_i)T_{fi}, \tag{8.247}$$

so that the cross section is given by equation (8.228). After a little work, and making use of momentum conservation, we find that

$$|T_{fi}|^2 = e^4\left(4(\epsilon_i\cdot\epsilon_f)^2 - 2 + \frac{p_i\cdot k_f}{p_i\cdot k_i} + \frac{p_i\cdot k_i}{p_i\cdot k_f}\right). \tag{8.248}$$

This is all that is required to calculate the cross section in any desired frame. Again, this derivation applies regardless of the initial electron spin.

The same scheme can be applied to a wide range of relativistic scattering problems. In all cases the spacetime algebra formulation provides a simpler and clearer method for handling the spin, as it does not force us to work with a preferred basis set. In section 14.4.1 the same formalism is applied to scattering from a black hole. At some point, however, it is necessary to face questions of second quantisation and the construction of a relativistic multiparticle quantum theory. This is discussed in the following chapter.

8.6 Notes

A significant amount of new notation was introduced in this chapter, relating to how spinors are handled in spacetime algebra. Much of this is important in later chapters, and the most useful results of this approach are summarised in table 8.3.

Quantum mechanics has probably been the most widely researched application of geometric algebra to date. Many authors have carried out investigations into whether the spacetime algebra formulation of Dirac theory offers any deeper insights into the nature of quantum theory. Among the most interesting of these are Hestenes' work on zitterbewegung (1990), and his comments on the nature

Pauli spinors	$\vert\psi\rangle = \begin{pmatrix} a^0 + ia^3 \\ -a^2 + ia^1 \end{pmatrix} \leftrightarrow \psi = a^0 + a^k I\boldsymbol{\sigma}_k$
Pauli operators	$\hat{\sigma}_k\vert\psi\rangle \leftrightarrow \boldsymbol{\sigma}_k\psi\boldsymbol{\sigma}_3$ $i\vert\psi\rangle \leftrightarrow \psi I\boldsymbol{\sigma}_3 = j\psi$ $\langle\psi\vert\psi'\rangle \leftrightarrow \langle\psi^\dagger\psi'\rangle_q = \frac{1}{2}(\psi^\dagger\psi' + \boldsymbol{\sigma}_3\psi^\dagger\psi'\boldsymbol{\sigma}_3)$
Pauli observables	$\rho = \psi\psi^\dagger$ $\boldsymbol{s} = \frac{1}{2}\psi\boldsymbol{\sigma}_3\psi^\dagger$
Dirac spinors	$\begin{pmatrix} \vert\phi\rangle \\ \vert\eta\rangle \end{pmatrix} \leftrightarrow \psi = \phi + \eta\boldsymbol{\sigma}_3$
Dirac operators	$\hat{\gamma}_\mu\vert\psi\rangle \leftrightarrow \gamma_\mu\psi\gamma_0$ $j\vert\psi\rangle \leftrightarrow \psi\boldsymbol{i\sigma}_3$ $\hat{\gamma}_5\vert\psi\rangle \leftrightarrow \psi\boldsymbol{\sigma}_3$ $\langle\bar{\psi}\vert\psi'\rangle \leftrightarrow \langle\tilde{\psi}\psi'\rangle_q$
Dirac equation	$\nabla\psi I\boldsymbol{\sigma}_3 - eA\psi = m\psi\gamma_0$
Dirac observables	$\rho e^{i\beta} = \psi\tilde{\psi} \quad J = \psi\gamma_0\tilde{\psi}$ $S = \psi\boldsymbol{i\sigma}_3\tilde{\psi} \quad s = \psi\gamma_3\tilde{\psi}$
Plane-wave states	$\psi^{(+)}(x) = L(p)\Phi e^{-\boldsymbol{i\sigma}_3 p\cdot x}$ $\psi^{(-)}(x) = L(p)\Phi\boldsymbol{\sigma}_3 e^{\boldsymbol{i\sigma}_3 p\cdot x}$ $L(p) = (p\gamma_0 + m)/\sqrt{2m(E+m)}$

Table 8.3 *Quantum states and operators.* This table summarises the main features of the spacetime algebra representation of Pauli and Dirac spinors and operators.

of the electroweak group (1982b). Many authors have advocated spacetime algebra as a better computational tool for Dirac theory than the explicit matrix formulation (augmented with various spin sum rules). A summary of these ideas is contained in the paper 'Electron scattering without spin sums' by Lewis *et al.* (2001). Elsewhere, a similar approach has been applied to modelling a spin measurement (Challinor *et al.* 1996) and to the results of tunnelling experiments (Gull *et al.* 1993b). Much of this work is summarised in the review 'Spacetime algebra and electron physics' by Doran *et al.* (1996b).

There is no shortage of good textbooks describing standard formulations of Dirac theory and quantum electrodynamics. We particularly made use of the classic texts by Itzykson & Zuber (1980), and Bjorken & Drell (1964). For a detailed exposition of the solution of the Dirac equation in various backgrounds one can do little better than Greiner's *Relativistic Quantum Mechanics* (1990).

Also recommended is Grandy's *Relativistic Quantum Mechanics of Leptons and Fields* (1991) which, unusually, does not shy away from the more problematic areas of the conceptual foundations of quantum field theory.

8.7 Exercises

8.1 The spin matrix operators $\hat{s}_k$ are defined as a set of 2×2 Hermitian matrices satisfying the commutation relations $[\hat{s}_i, \hat{s}_j] = i\hbar\epsilon_{ijk}\hat{s}_k$. Given that $\hat{s}_3$ is defined by

$$\hat{s}_3 = \lambda \begin{pmatrix} 1 & 0 \\ 0 & -1 \end{pmatrix},$$

show that the remaining matrices are unique, up to an overall choice of phase. Find λ and show that we can choose the phase such that $\hat{s}_k = \hbar/2\,\hat{\sigma}_k$.

8.2 Verify that the equivalence between Pauli spinors and even multivectors defined in equation (8.20) is consistent with the operator equivalences

$$\hat{\sigma}_k|\psi\rangle \leftrightarrow \boldsymbol{\sigma}_k\psi\boldsymbol{\sigma}_3 \quad (k = 1, 2, 3).$$

8.3 Suppose that two spin-1/2 states are represented by the even multivectors ϕ and ψ, and the accompanying spin vectors are

$$\boldsymbol{s}_1 = \phi\boldsymbol{\sigma}_3\tilde{\phi} \quad \text{and} \quad \boldsymbol{s}_2 = \psi\boldsymbol{\sigma}_3\tilde{\psi}.$$

Prove that the quantum mechanical formula for the probability of measuring state ϕ in state ψ satisfies

$$P = \frac{|\langle\phi|\psi\rangle|^2}{\langle\phi|\phi\rangle\langle\psi|\psi\rangle} = \tfrac{1}{2}(1 + \cos(\theta))$$

where θ is the angle between $\boldsymbol{s}_1$ and $\boldsymbol{s}_2$.

8.4 Verify that the Pauli inner product is invariant under both spatial rotations and gauge transformations (i.e. rotations in the $I\boldsymbol{\sigma}_3$ plane applied to the right of the spinor ψ). Repeat the analysis for Dirac spinors.

8.5 Prove that the angular momentum operators $L_B = jB\cdot(\boldsymbol{x}\wedge\boldsymbol{\nabla})$ satisfy

$$[L_{B_1}, L_{B_2}] = -jL_{B_1 \times B_2}.$$

8.6 Prove that, in any dimension,

$$[B\cdot(x\wedge\nabla) - \tfrac{1}{2}B, \nabla] = 0,$$

where B is a bivector.

8.7 The Majorana representation is defined in terms of a set of real matrices. Prove that the complex conjugation operation in this representation has the spacetime algebra equivalent

$$|\psi\rangle^*_{Maj} \leftrightarrow \psi \boldsymbol{\sigma}_2.$$

Confirm that this anticommutes with the operation of multiplying by the imaginary.

8.8 Prove that the associated Legendre polynomials satisfy the following recursion relations:

$$(1-x^2)\frac{dP_l^m(x)}{dx} + mxP_l^m(x) = -(1-x^2)^{1/2}P_l^{m+1}(x),$$

$$(1-x^2)\frac{dP_l^m(x)}{dx} - mxP_l^m(x) = (1-x^2)^{1/2}(l+m)(l-m+1)P_l^{m-1}(x).$$

8.9 Prove that the spherical monogenics satisfy

$$\int d\Omega\, \langle \psi_l^{m\dagger} \psi_{l'}^{m'} \rangle_q = \delta^{mm'} \delta_{ll'} 4\pi \frac{(l+m+1)!}{(l-m)!}.$$

8.10 From the result of equation (8.248), show that the cross section for scattering of a photon of a free electron (initially at rest) is determined by the Klein–Nishina formula

$$\frac{d\sigma}{d\Omega} = \frac{\alpha^2}{4m^2}\left(\frac{\omega_f}{\omega_i}\right)^2 \left(\frac{\omega_f}{\omega_i} + \frac{\omega_i}{\omega_f} + 4(\epsilon_f \cdot \epsilon_i)^2 - 2\right).$$

9

Multiparticle states and quantum entanglement

The previous chapter dealt with the quantum theory of single particles in a background field. In this chapter we turn to the study of multiparticle quantum theory. In many ways, this subject is even more strange than the single-particle theory, as it forces us to face up to the phenomenon of quantum entanglement. The basic idea is simple enough to state. The joint state of a two-particle system is described by a *tensor product* state of the form $|\psi\rangle \otimes |\phi\rangle$. This is usually abbreviated to $|\psi\rangle|\phi\rangle$. Quantum theory allows for linear complex superpositions of multiparticle states, which allows us to consider states which have no classical counterpart. An example is the spin singlet state

$$|\varepsilon\rangle = \frac{1}{\sqrt{2}}\big(|0\rangle|1\rangle - |1\rangle|0\rangle\big). \tag{9.1}$$

States such as these are referred to as being *entangled*. The name reflects the fact that observables for the two particles remain correlated, even if measurements are performed in such a way that communication between the particles is impossible. The rapidly evolving subject of *quantum information processing* is largely concerned with the properties of entangled states, and the prospects they offer for quantum computation.

Quantum entanglement is all around us, though rarely in a form we can exploit. Typically, a state may entangle with its environment to form a new *pure* state. (A pure state is one that can be described by a single wavefunction, which may or may not be entangled.) The problem is that our knowledge of the state of the environment is highly limited. All we can measure are the observables of our initial state. In this case the wavefunction formulation is of little practical value, and instead we have to consider equations for the evolution of the observables themselves. This is usually handled by employing a representation in terms of *density matrices*. These lead naturally to concepts of quantum statistical physics and quantum definitions of entropy.

In this chapter we explore how these concepts can be formulated in the language of geometric algebra. One of the essential mysteries of quantum theory is the origin of this tensor product construction. The tensor product is used in constructing both multiparticle states and many of the operators acting on these states. So the first challenge is to find a representation of the tensor product in terms of the geometric product. This is surprisingly simple to do, though only once we have introduced the idea of a relativistic configuration space. The geometric algebra of such a space is called the *multiparticle spacetime algebra* and it provides the ideal algebraic structure for studying multiparticle states and operators. This has applications in a wealth of subjects, from NMR spectroscopy to quantum information processing, some of which are discussed below. Most of these applications concern non-relativistic multiparticle quantum mechanics. Later in this chapter we turn to a discussion of the insights that this new approach can bring to relativistic multiparticle quantum theory. There we find a simple, geometric encoding of the Pauli principle, which opens up a route through to the full quantum field theory.

9.1 Many-body quantum theory

In order to set the context for this chapter, we start with a review of the basics of multiparticle quantum theory. We concentrate in particular on two-particle systems, which illustrate many of the necessary properties. The key concept is that the quantum theory of n-particles is *not* described by a set of n single wavefunctions. Instead, it is described by *one* wavefunction that encodes the entire state of the system of n particles. Unsurprisingly, the equations governing the evolution of such a wavefunction can be extraordinarily complex.

For a wide range of problems one can separate position degrees of freedom from internal (spin) degrees of freedom. This is typically the case in non-relativistic physics, particularly if the electromagnetic field can be treated as constant. In this case the position degrees of freedom are handled by the many-body Schrödinger equation. The spin degrees of freedom in many ways represent a cleaner system to study, as they describe the quantum theory of n two-state systems. This illustrates the two most important features of multiparticle quantum theory: the exponential increase in the size of state space, and the existence of entangled states.

9.1.1 The two-body Schrödinger equation

Two-particle states are described by a single wavefunction $\psi(\boldsymbol{r}_1, \boldsymbol{r}_2)$. The joint vectors $(\boldsymbol{r}_1, \boldsymbol{r}_2)$ define an abstract six-dimensional configuration space over which ψ defines a complex-valued function. This sort of configuration space is a useful tool in classical mechanics, and in quantum theory it is indispensable. The kinetic energy operator is given by the sum of the individual operators:

$$\hat{K} = -\frac{\hbar^2 \boldsymbol{\nabla}_1^2}{2m_1} - \frac{\hbar^2 \boldsymbol{\nabla}_2^2}{2m_2}. \tag{9.2}$$

The subscripts refer to the individual particles, and m_i is the mass of particle i. The two-particle Schrödinger equation is now

$$i\hbar \frac{\partial \psi}{\partial t} = -\frac{\hbar^2 \boldsymbol{\nabla}_1^2}{2m_1}\psi - \frac{\hbar^2 \boldsymbol{\nabla}_2^2}{2m_2}\psi + V(\boldsymbol{r}_1, \boldsymbol{r}_2)\psi. \tag{9.3}$$

As a simple example, consider the bound state Coulomb problem

$$-\frac{\hbar^2 \boldsymbol{\nabla}_1^2}{2m_1}\psi - \frac{\hbar^2 \boldsymbol{\nabla}_2^2}{2m_2}\psi - \frac{q_1 q_2}{4\pi\epsilon_0 r}\psi = E\psi, \tag{9.4}$$

where r is the Euclidean distance between the points $\boldsymbol{r}_1$ and $\boldsymbol{r}_2$. This problem is separated in a similar manner to the classical Kepler problem (see section 3.2). We introduce the vectors

$$\boldsymbol{r} = \boldsymbol{r}_1 - \boldsymbol{r}_2, \qquad \frac{\boldsymbol{R}}{\mu} = \frac{\boldsymbol{r}_1}{m_1} + \frac{\boldsymbol{r}_2}{m_2}, \tag{9.5}$$

where μ is the reduced mass. In terms of these new variables the Schrödinger equation becomes

$$-\frac{\hbar^2 \boldsymbol{\nabla}_r^2}{2\mu}\psi - \frac{\hbar^2 \boldsymbol{\nabla}_R^2}{2M}\psi - \frac{q_1 q_2}{4\pi\epsilon_0 r}\psi = E\psi. \tag{9.6}$$

We can now find separable solutions to this equation by setting

$$\psi(\boldsymbol{r}_1, \boldsymbol{r}_2) = \phi(\boldsymbol{r})\Psi(\boldsymbol{R}). \tag{9.7}$$

The wavefunction Ψ satisfies a free-particle equation, which corresponds classically to the motion of the centre of mass. The remaining term, $\phi(\boldsymbol{r})$, satisfies the equivalent single-particle equation, with the mass given by the reduced mass of the two particles.

This basic example illustrates how quantum mechanics accounts for multiparticle interactions. There is a single wavefunction, which simultaneously accounts for the properties of all of the particles. In many cases this wavefunction decomposes into the product of a number of simpler wavefunctions, but this is not always the case. One can construct states that cannot be decomposed into a single direct product state. An important example of this arises when the two particles in question are identical. In this case one can see immediately that if $\psi(\boldsymbol{r}_1, \boldsymbol{r}_2)$ is an eigenstate of a two-particle Hamiltonian, then so to is $\psi(\boldsymbol{r}_2, \boldsymbol{r}_1)$. The operator that switches particle labels like this is called the particle interchange operator $\hat{P}$, and it commutes with all physically-acceptable Hamiltonians. Since it commutes with the Hamiltonian, and squares to the identity operation, there are two possible eigenstates of $\hat{P}$. These are

$$\psi_{\pm} = \psi(\boldsymbol{r}_1, \boldsymbol{r}_2) \pm \psi(\boldsymbol{r}_2, \boldsymbol{r}_1). \tag{9.8}$$

These two possibilities are the only ones that arise physically, and give rise to the distinction between fermions (minus sign) and bosons (plus sign). Here we see the first indications of some new physical possibilities entering in multiparticle interactions. Quantum theory remains linear, so one can form complex superpositions of the n-particle wavefunctions. These superpositions can have new properties not present in the single-particle theory.

9.1.2 Spin states

Ignoring the spatial dependence and concentrating instead on the internal spin degrees of freedom, a spin-1/2 state can be written as a complex superposition of 'up' and 'down' states, which we will denote as $|0\rangle$ and $|1\rangle$. Now suppose that a second particle is introduced, so that system 1 is in the state $|\psi\rangle$ and system 2 is in the state $|\phi\rangle$. The joint state of the system is described by the *tensor product* state

$$|\Psi\rangle = |\psi\rangle \otimes |\phi\rangle, \tag{9.9}$$

which is abbreviated to $|\psi\rangle|\phi\rangle$. The total set of possible states is described by the basis

$$\begin{aligned} |00\rangle &= |0\rangle|0\rangle, \qquad & |01\rangle &= |0\rangle|1\rangle, \\ |10\rangle &= |1\rangle|0\rangle, \qquad & |11\rangle &= |1\rangle|1\rangle. \end{aligned} \tag{9.10}$$

This illustrates an important phenomenon of multiparticle quantum theory. The number of available states grows as 2^n, so large systems have an enormously larger state space than their classical counterparts. Superpositions of these basis states will, in general, produce states which cannot be written as a single tensor product of the form $|\psi\rangle|\phi\rangle$. Such states are entangled. A standard example is the singlet state of equation (9.1). One feature of these entangled states is that they provide 'short-cuts' through Hilbert space between classical states. The speed-up this can offer is often at the core of algorithms designed to exploit the possibilities offered by quantum computation.

A challenge faced by theorists looking for ways to exploit these ideas is how best to classify multiparticle entanglement. The problem is to describe concisely the properties of a state that are unchanged under local unitary operations. Local operations consist of unitary transformations applied entirely to one particle. They correspond to operations applied to a single particle in the laboratory. Features of the state that are unchanged by these operations relate to joint properties of the particles, in particular how entangled they are.

To date, only two-particle (or 'bipartite') systems have been fully understood. A general state of two particles can be written

$$\Psi = \sum_{i,j} \alpha_{ij} |i\rangle \otimes |j\rangle, \tag{9.11}$$

where the $|i\rangle$ denote some orthonormal basis. The *Schmidt decomposition* (which is little more than a singular-value decomposition of α_{ij}) tells us that one can always construct a basis such that

$$\Psi = \sum_i \beta_i |i'\rangle \otimes |i'\rangle. \tag{9.12}$$

The β_i are all real parameters that tell us directly how much entanglement is present. These parameters are unchanged under local transformations of the state Ψ. An important example of the Schmidt decomposition, which we shall revisit frequently, is for systems of two entangled spinors. For these we find that a general state can be written explicitly as

$$\begin{aligned}|\psi\rangle =& \rho^{1/2}\mathrm{e}^{i\chi}\left(\cos(\alpha/2)\mathrm{e}^{i\tau/2}\begin{pmatrix}\cos(\theta_1/2)\mathrm{e}^{-i\phi_1/2}\\ \sin(\theta_1/2)\mathrm{e}^{i\phi_1/2}\end{pmatrix}\otimes\begin{pmatrix}\cos(\theta_2/2)\mathrm{e}^{-i\phi_2/2}\\ \sin(\theta_2/2)\mathrm{e}^{i\phi_2/2}\end{pmatrix}\right.\\ &\left.+\sin(\alpha/2)\mathrm{e}^{-i\tau/2}\begin{pmatrix}\sin(\theta_1/2)\mathrm{e}^{-i\phi_1/2}\\ -\cos(\theta_1/2)\mathrm{e}^{i\phi_1/2}\end{pmatrix}\otimes\begin{pmatrix}\sin(\theta_2/2)\mathrm{e}^{-i\phi_2/2}\\ -\cos(\theta_2/2)\mathrm{e}^{i\phi_2/2}\end{pmatrix}\right). \end{aligned} \tag{9.13}$$

In this decomposition we arrange that $0 \leq \alpha \leq \pi/4$, so that the decomposition is unique (save for certain special cases).

9.1.3 Pure and mixed states

So far the discussion has focused entirely on *pure* states, which can be described in terms of a single wavefunction. For many applications, however, such a description is inappropriate. Suppose, for example, that we are studying spin states in an NMR experiment. The spin states are only partially coherent, and one works in terms of ensemble averages. For example, the average spin vector (or polarisation) is given by

$$\boldsymbol{p} = \frac{1}{n}\sum_{i=1}^{n} \hat{\boldsymbol{s}}_i. \tag{9.14}$$

Unless all of the spin vectors are precisely aligned (a coherent state), the polarisation vector will not have unit length and so cannot be generated by a single wavefunction. Instead, we turn to a formulation in terms of density matrices. The density matrix for a normalised pure state is

$$\hat{\rho} = |\psi\rangle\langle\psi|, \tag{9.15}$$

which is necessarily a Hermitian matrix. All of the observables associated with the state $|\psi\rangle$ can be obtained from the density matrix by writing

$$\langle\psi|\hat{Q}|\psi\rangle = \mathrm{tr}(\hat{\rho}\hat{Q}). \tag{9.16}$$

For an incoherent mixture (a mixed state) the density matrix is the weighted sum of the matrices for the pure states:

$$\hat{\rho} = \sum_{i=1}^{n} p_i |\psi_i\rangle\langle\psi_i|. \tag{9.17}$$

The real coefficients satisfy

$$\sum_{i=1}^{n} p_i = 1, \tag{9.18}$$

which ensures that the density matrix has unit trace. The definition of $\hat{\rho}$ ensures that all observables are constructed from the appropriate averages of the pure states. In principle, the state of any system is described by a Hermitian density matrix, which is constrained to be positive-semidefinite and to have unit trace. All observables are then formed according to equation (9.16).

The need for a density matrix can be seen in a second way, as a consequence of entanglement. Suppose that we are interested in the state of particle 1, but that this particle has been allowed to entangle with a second particle 2, forming the pure state $|\psi\rangle$. The density matrix for the two-particle system is again described by equation (9.15). But we can only perform measurements of particle 1. The effective density matrix for particle 1 is obtained by performing a *partial trace* of $\hat{\rho}$ to trace out the degrees of freedom associated with particle 2. We therefore define

$$\hat{\rho}_1 = \mathrm{tr}_2 \hat{\rho}, \tag{9.19}$$

where the sum runs over the space of particle 2. One can easily check that, in the case where the particles are entangled, $\hat{\rho}_1$ is no longer the density matrix for a pure state. The most extreme example of this is the singlet state (9.1) mentioned in the introduction. In the obvious basis, the singlet state can be written as

$$|\varepsilon\rangle = \frac{1}{\sqrt{2}}(0,\, 1,\, -1,\, 0)^{\dagger}. \tag{9.20}$$

The density matrix for this state is

$$\hat{\rho} = |\varepsilon\rangle\langle\varepsilon| = \frac{1}{2}\begin{pmatrix} 0 & 0 & 0 & 0 \\ 0 & 1 & -1 & 0 \\ 0 & -1 & 1 & 0 \\ 0 & 0 & 0 & 0 \end{pmatrix}. \tag{9.21}$$

This is appropriate for a pure state, as the matrix satisfies $\hat{\rho}^2 = \hat{\rho}$. But if we now form the partial trace over the second particle we are left with

$$\hat{\rho}_1 = \frac{1}{2}\begin{pmatrix} 1 & 0 \\ 0 & 1 \end{pmatrix}. \tag{9.22}$$

This is the density matrix for a totally unpolarised state, which is to be expected, since there can be no directional information in the singlet state. Clearly, $\hat{\rho}_1$ cannot be generated by a single-particle pure state.

9.2 Multiparticle spacetime algebra

The key to constructing a suitable geometric framework for multiparticle quantum theory involves the full, relativistic spacetime algebra. This is because it is only the relativistic treatment which exposes the nature of the $\boldsymbol{\sigma}_i$ as spacetime bivectors. This is crucial for determining their algebraic properties as further particles are added. The n-particle spacetime algebra is the geometric algebra of $4n$-dimensional relativistic configuration space. We call this the multiparticle spacetime algebra. A basis is for this is constructed by taking n sets of basis vectors $\{\gamma_\mu^a\}$, where the superscript labels the particle space. These satisfy the orthogonality conditions

$$\gamma_\mu^a \gamma_\nu^b + \gamma_\nu^b \gamma_\mu^a = \begin{cases} 0 & a \neq b \\ 2\eta_{\mu\nu} & a = b \end{cases}, \tag{9.23}$$

which are summarised in the single formula

$$\gamma_\mu^a \cdot \gamma_\nu^b = \delta^{ab} \eta_{\mu\nu}. \tag{9.24}$$

There is nothing uniquely quantum-mechanical in this construction. A system of three classical particles could be described by a set of three trajectories in a single space, or by one path in a nine-dimensional space. The extra dimensions label the properties of each individual particle, and are not to be thought of as existing in anything other than a mathematical sense. One unusual feature concerning relativistic configuration space is that it requires a separate copy of the time dimension for each particle, as well as the three spatial dimensions. This is required in order that the algebra is fully Lorentz-covariant. The presence of multiple time coordinates can complicate the evolution equations in the relativistic theory. Fortunately, the non-relativistic reduction does not suffer from this problem as all of the individual time coordinates are identified with a single absolute time.

As in the single-particle case, the even subalgebra of each copy of the spacetime algebra defines an algebra for relative space. We perform all spacetime splits with the vector γ_0, using a separate copy of this vector in each particle's space. A basis set of relative vectors is then defined by

$$\boldsymbol{\sigma}_i^a = \gamma_i^a \gamma_0^a. \tag{9.25}$$

Again, superscripts label the particle space in which the object appears, and subscripts are retained for the coordinate frame. We do not enforce the summation convention for superscripted indices in this chapter. If we now consider bivectors from spaces 1 and 2, we find that the basis elements satisfy

$$\sigma_i^1 \sigma_j^2 = \gamma_i^1 \gamma_0^1 \gamma_j^2 \gamma_0^2 = \gamma_i^1 \gamma_j^2 \gamma_0^2 \gamma_0^1 = \gamma_j^2 \gamma_0^2 \gamma_i^1 \gamma_0^1 = \sigma_j^2 \sigma_i^1. \tag{9.26}$$

The basis elements *commute*, rather than anticommute. This solves the problem of how to represent the tensor product in geometric algebra. The geometric product $\sigma_i^a \sigma_j^b$ *is* the tensor product. Since single particle states are constructed out of geometric algebra elements, this gives a natural origin for tensor product states in the multiparticle case. This property only holds because the relative vectors σ_i^a are constructed as spacetime bivectors.

The pseudoscalar for each particle space is defined in the obvious way, so that

$$I^a = \gamma_0^a \gamma_1^a \gamma_2^a \gamma_3^a. \tag{9.27}$$

Relative bivectors in each space take the form $I^a \sigma_k^a$. Wherever possible we abbreviate these by dropping the first particle label, so that

$$I\sigma_k^a = I^a \sigma_k^a. \tag{9.28}$$

The reverse operation in the multiparticle spacetime algebra is denoted with a tilde, and reverses the order of products of all relativistic vectors. Wherever possible we use this operation when forming observables. The Hermitian adjoint in each space can be constructed by inserting appropriate factors of γ_0^a.

9.2.1 Non-relativistic states and the correlator

In the single-particle theory, non-relativistic states are constructed from the even subalgebra of the Pauli algebra. A basis for these is provided by the set $\{1, I\sigma_k\}$. When forming multiparticle states we take tensor products of the individual particle states. Since the tensor product and geometric product are equivalent in the multiparticle spacetime algebra, a complete basis is provided by the set

$$\{1,\, I\sigma_k^1,\, I\sigma_k^2,\, I\sigma_j^1\, I\sigma_k^2\}. \tag{9.29}$$

But these basis elements span a 16-dimensional real space, whereas the state space for two spin-1/2 particles is a four-dimensional complex space — only eight real degrees of freedom. What has gone wrong? The answer lies in our treatment of the complex structure. Quantum theory works with a single unit imaginary i, but in our two-particle algebra we now have two bivectors playing the role of i: $I\sigma_3^1$ and $I\sigma_3^2$. Right-multiplication of a state by either of these has to result in the same state in order for the geometric algebra treatment to faithfully mirror standard quantum mechanics. That is, we must have

$$\psi I\sigma_3^1 = \psi I\sigma_3^2. \tag{9.30}$$

Rearranging this, we find that

$$\psi = -\psi I\sigma_3^1\, I\sigma_3^2 = \psi \tfrac{1}{2}(1 - I\sigma_3^1\, I\sigma_3^2). \tag{9.31}$$

This tells us what we must do. If we define

$$E = \tfrac{1}{2}(1 - I\boldsymbol{\sigma}_3^1 I\boldsymbol{\sigma}_3^2), \tag{9.32}$$

we find that

$$E^2 = E. \tag{9.33}$$

So right-multiplication by E is a *projection operation*. If we include this factor on the right of all states we halve the number of (real) degrees of freedom from 16 to the expected 8.

The spacetime algebra representation of a direct-product two-particle Pauli spinor is now given by $\psi^1\phi^2 E$, where ψ^1 and ϕ^2 are spinors (even multivectors) in their own spaces. A complete basis for two-particle spin states is provided by

$$\begin{aligned} |0\rangle|0\rangle &\leftrightarrow E, \\ |0\rangle|1\rangle &\leftrightarrow -I\boldsymbol{\sigma}_2^2 E, \\ |1\rangle|0\rangle &\leftrightarrow -I\boldsymbol{\sigma}_2^1 E, \\ |1\rangle|1\rangle &\leftrightarrow I\boldsymbol{\sigma}_2^1 I\boldsymbol{\sigma}_2^2 E. \end{aligned} \tag{9.34}$$

We further define

$$J = EI\boldsymbol{\sigma}_3^1 = EI\boldsymbol{\sigma}_3^2 = \tfrac{1}{2}(I\boldsymbol{\sigma}_3^1 + I\boldsymbol{\sigma}_3^2), \tag{9.35}$$

so that

$$J^2 = -E. \tag{9.36}$$

Right-sided multiplication by J takes on the role of multiplication by the quantum imaginary i for multiparticle states.

This procedure extends simply to higher multiplicities. All that is required is to find the 'quantum correlator' E_n satisfying

$$E_n I\boldsymbol{\sigma}_3^a = E_n I\boldsymbol{\sigma}_3^b = J_n \quad \text{for all } a,\, b. \tag{9.37}$$

E_n can be constructed by picking out the $a = 1$ space, say, and correlating all the other spaces to this, so that

$$E_n = \prod_{b=2}^{n} \tfrac{1}{2}(1 - I\boldsymbol{\sigma}_3^1 I\boldsymbol{\sigma}_3^b). \tag{9.38}$$

The value of E_n is independent of which of the n spaces is singled out and correlated to. The complex structure is defined by

$$J_n = E_n I\boldsymbol{\sigma}_3^a, \tag{9.39}$$

where $I\boldsymbol{\sigma}_3^a$ can be chosen from any of the n spaces. To illustrate this consider the case of $n = 3$, where

$$\begin{aligned} E_3 &= \tfrac{1}{4}(1 - I\boldsymbol{\sigma}_3^1 I\boldsymbol{\sigma}_3^2)(1 - I\boldsymbol{\sigma}_3^1 I\boldsymbol{\sigma}_3^3) \\ &= \tfrac{1}{4}(1 - I\boldsymbol{\sigma}_3^1 I\boldsymbol{\sigma}_3^2 - I\boldsymbol{\sigma}_3^1 I\boldsymbol{\sigma}_3^3 - I\boldsymbol{\sigma}_3^2 I\boldsymbol{\sigma}_3^3) \end{aligned} \tag{9.40}$$

and

$$J_3 = \tfrac{1}{4}(I\boldsymbol{\sigma}_3^1 + I\boldsymbol{\sigma}_3^2 + I\boldsymbol{\sigma}_3^3 - I\boldsymbol{\sigma}_3^1 I\boldsymbol{\sigma}_3^2 I\boldsymbol{\sigma}_3^3). \tag{9.41}$$

Both E_3 and J_3 are symmetric under permutations of their indices.

9.2.2 Operators and observables

All of the operators defined for the single-particle spacetime algebra extend naturally to the multiparticle algebra. In the two-particle case, for example, we have

$$i\hat{\sigma}_k \otimes \hat{\mathsf{I}}|\psi\rangle \leftrightarrow I\boldsymbol{\sigma}_k^1\psi, \tag{9.42}$$

$$\hat{\mathsf{I}} \otimes i\hat{\sigma}_k|\psi\rangle \leftrightarrow I\boldsymbol{\sigma}_k^2\psi, \tag{9.43}$$

where $\hat{\mathsf{I}}$ is the 2×2 identity matrix and a factor of E is implicit in the spinor ψ. For the Hermitian operators we form, for example,

$$\hat{\sigma}_k \otimes \hat{\mathsf{I}}|\psi\rangle \leftrightarrow -I\boldsymbol{\sigma}_k^1\psi J = \boldsymbol{\sigma}_k^1\psi\boldsymbol{\sigma}_3^1. \tag{9.44}$$

This generalises in the obvious way, so that

$$\hat{\mathsf{I}} \otimes \cdots \otimes \hat{\sigma}_k^a \otimes \cdots \otimes \hat{\mathsf{I}}|\psi\rangle \leftrightarrow \boldsymbol{\sigma}_k^a\psi\boldsymbol{\sigma}_3^a. \tag{9.45}$$

We continue to adopt the j symbol as a convenient shorthand notation for the complex structure, so

$$i|\psi\rangle \leftrightarrow j\psi = \psi J = \psi I\boldsymbol{\sigma}_3^a. \tag{9.46}$$

The quantum inner product is now

$$\langle\psi|\phi\rangle \leftrightarrow 2^{n-1}\big(\langle\phi E\tilde{\psi}\rangle - \langle\phi J\tilde{\psi}\rangle j\big). \tag{9.47}$$

The factor of E in the real part is not strictly necessary as it is always present in the spinors, but including it does provide a neat symmetry between the real and imaginary parts. The factor of 2^{n-1} guarantees complete consistency with the standard quantum inner product, as it ensures that the state E has unit norm.

Suppose that we now form the observables in the two-particle case. We find that

$$\langle\psi|\,\hat{\sigma}_k \otimes \hat{\mathsf{I}}\,|\psi\rangle \leftrightarrow -2I\boldsymbol{\sigma}_k^1\cdot(\psi J\tilde{\psi}) \tag{9.48}$$

and

$$\langle\psi|\,\hat{\sigma}_j \otimes \hat{\sigma}_k\,|\psi\rangle \leftrightarrow -2(I\boldsymbol{\sigma}_j^1 I\boldsymbol{\sigma}_k^2)\cdot(\psi E\tilde{\psi}). \tag{9.49}$$

All of the observables one can construct are therefore contained in the multivectors $\psi E\tilde{\psi}$ and $\psi J\tilde{\psi}$. This generalises to arbitrary particle numbers. To see why, we use the fact that any density matrix can be expanded in terms of products of Hermitian operators, as in the two-particle expansion

$$\hat{\rho} = |\psi\rangle\langle\psi| = \frac{1}{4}(\hat{1}\otimes\hat{1} + a_k\,\hat{\sigma}_k\otimes\hat{1} + b_k\,\hat{1}\otimes\hat{\sigma}_k + c_{jk}\,\hat{\sigma}_j\otimes\hat{\sigma}_k). \tag{9.50}$$

The various coefficients are found by taking inner products with the appropriate combinations of operators. Each of these corresponds to picking out a term in $\psi E\tilde{\psi}$ or $\psi J\tilde{\psi}$. If an even number of Pauli matrices is involved we pick out a term in $\psi E\tilde{\psi}$, and an odd number picks out a term in $\psi J\tilde{\psi}$. In general, $\psi E\tilde{\psi}$ contains terms of grades 0, 4, ..., and $\psi J\tilde{\psi}$ contains terms of grade 2, 6, ... These account for all the coefficients in the density matrix, and hence for all the observables that can be formed from ψ.

An advantage of working directly with the observables $\psi E\tilde{\psi}$ and $\psi J\tilde{\psi}$ is that the partial trace operation has a simple interpretation. If we want to form the partial trace over the ath particle, we simply remove all terms from the observables with a contribution in the ath particle space. No actual trace operation is required. Furthermore, this operation of discarding information is precisely the correct physical picture for the partial trace operation — we are discarding the (often unknown) information associated with a particle in one or more spaces. A minor complication in this approach is that $\psi J\tilde{\psi}$ gives rise to anti-Hermitian terms, whereas the density matrix is Hermitian. One way round this is to correlate all of the pseudoscalars together and then dualise all bivectors back to vectors. This is the approach favoured by Havel and coworkers in their work on NMR spectroscopy. Alternatively, one can simply ignore this feature and work directly with the observables $\psi E\tilde{\psi}$ and $\psi J\tilde{\psi}$. When presented with a general density matrix one often needs to pull it apart into sums of terms like this anyway (the product operator expansion), so it makes sense to work directly with the multivector observables when they are available.

9.3 Systems of two particles

Many of the preceding ideas are most simply illustrated for the case of a system of two particles. For these, the Schmidt decomposition of equation (9.13) provides a useful formulation for a general state. The geometric algebra version of this is rather more compact, however, as we now establish. First, we define the spinor

$$\psi(\theta,\phi) = \mathrm{e}^{-\phi I\boldsymbol{\sigma}_3/2}\,\mathrm{e}^{-\theta I\boldsymbol{\sigma}_2/2}. \tag{9.51}$$

We also need a representation of the state orthogonal to this, which is

$$\begin{pmatrix} \sin(\theta/2)\mathrm{e}^{-i\phi/2} \\ -\cos(\theta/2)\mathrm{e}^{i\phi/2} \end{pmatrix} \leftrightarrow \psi(\theta,\phi)I\boldsymbol{\sigma}_2. \tag{9.52}$$

Now we are in a position to construct the multiparticle spacetime algebra version of the Schmidt decomposition. We replace equation (9.13) with

$$\begin{aligned}\psi =&\rho^{1/2}\Big(\cos(\alpha/2)\psi^1(\theta_1,\phi_1)\psi^2(\theta_2,\phi_2)\mathrm{e}^{J\tau/2}\\ &+\sin(\alpha/2)\psi^1(\theta_1,\phi_1)\psi^2(\theta_2,\phi_2)I\sigma_2^1 I\sigma_2^2\mathrm{e}^{-J\tau/2}\Big)\mathrm{e}^{J\chi}E\\ =&\rho^{1/2}\psi^1(\theta_1,\phi_1)\psi^2(\theta_2,\phi_2)\mathrm{e}^{J\tau/2}\left(\cos(\alpha/2)+\sin(\alpha/2)I\sigma_2^1 I\sigma_2^2\right)\mathrm{e}^{J\chi}E. \end{aligned} \tag{9.53}$$

We now define the individual rotors

$$R=\psi(\theta_1,\phi_1)\mathrm{e}^{I\sigma_3\tau/4}, \quad S=\psi(\theta_2,\phi_2)\mathrm{e}^{I\sigma_3\tau/4}, \tag{9.54}$$

so that the wavefunction ψ simplifies to

$$\psi=\rho^{1/2}R^1S^2\big(\cos(\alpha/2)+\sin(\alpha/2)I\sigma_2^1I\sigma_2^2\big)\mathrm{e}^{J\chi}E. \tag{9.55}$$

This gives a compact, general form for an arbitrary two-particle state. The degrees of freedom are held in an overall magnitude and phase, two separate rotors in the individual particle spaces, and a single entanglement angle α. In total this gives nine degrees of freedom, so one must be redundant. This redundancy lies in the single-particle rotors. If we take

$$R\mapsto R\mathrm{e}^{I\sigma_3\beta}, \quad S\mapsto S\mathrm{e}^{-I\sigma_3\beta} \tag{9.56}$$

then the overall wavefunction ψ is unchanged. In practice this redundancy is not a problem, and the form of equation (9.55) turns out to be extremely useful.

9.3.1 Observables for two-particle states

The individual rotors R^1 and S^2 generate rotations in their own spaces. These are equivalent to local unitary transformations. The novel features associated with the observables for a two-particle system arise from the entanglement angle α. To study this we first form the bivector observable $\psi J\tilde{\psi}$:

$$\begin{aligned}\psi J\tilde{\psi} =&R^1S^2\big(\cos(\alpha/2)+\sin(\alpha/2)I\sigma_2^1 I\sigma_2^2\big)J\big(\cos(\alpha/2)+\sin(\alpha/2)I\sigma_2^1 I\sigma_2^2\big)\tilde{R}^1\tilde{S}^2\\ =&\tfrac{1}{2}R^1S^2\big(\cos^2(\alpha/2)-\sin^2(\alpha/2)\big)(I\sigma_3^1+I\sigma_3^2)\tilde{R}^1\tilde{S}^2\\ =&\tfrac{1}{2}\cos(\alpha)\big((RI\sigma_3\tilde{R})^1+(SI\sigma_3\tilde{S})^2\big),\end{aligned} \tag{9.57}$$

where we have assumed that $\rho=1$. This result extends the definition of the spin bivector to multiparticle systems. One can immediately see that the lengths of the bivectors are no longer fixed, but instead depend on the entanglement. Only in the case of zero entanglement are the spin bivectors unit length.

The remaining observables are contained in

$$\psi E\tilde{\psi}=\tfrac{1}{2}R^1S^2\Big(1-I\sigma_3^1I\sigma_3^2+\sin(\alpha)(I\sigma_2^1I\sigma_2^2-I\sigma_1^1I\sigma_1^2)\Big)\tilde{R}^1\tilde{S}^2. \tag{9.58}$$

To make this result clearer we introduce the notation

$$A_k = RI\boldsymbol{\sigma}_k\tilde{R}, \quad B_k = SI\boldsymbol{\sigma}_k\tilde{S}, \tag{9.59}$$

so that

$$2\psi E\tilde{\psi} = 1 - A_3^1 B_3^2 + \sin(\alpha)(A_2^1 B_2^2 - A_1^1 B_1^2). \tag{9.60}$$

The scalar part confirms that the state is normalised correctly. The 4-vector part contains an interesting new term, which goes as $A_2^1 B_2^2 - A_1^1 B_1^2$. None of the individual A_1, A_2, B_1, or B_2 bivectors is accessible to measurement in the single-particle case as they are not phase-invariant. But in the two-particle case these terms do start to influence the observables. This is one of essential differences between classical and quantum models of spin.

9.3.2 *Density matrices and probabilities*

Now that we have all of the observables, we have also found all of the terms in the density matrix. Of particular interest are the results of partial traces, where we discard the information associated with one of the particles. If we throw out all of the information about the second particle, for example, what remains is the single-particle density matrix

$$\hat{\rho} = \tfrac{1}{2}(1 + \boldsymbol{p}), \tag{9.61}$$

where the polarisation vector is given by

$$\boldsymbol{p} = \cos(\alpha) R\boldsymbol{\sigma}_3\tilde{R}. \tag{9.62}$$

This vector no longer has unit length, so the density matrix is that of a mixed state. Entanglement with a second particle has led to a loss of coherence of the first particle. This process, by which entanglement produces decoherence, is central to attempts to explain the emergence of classical physics from quantum theory.

For two particles we see that there is a symmetry between the degree of entanglement. If we perform a partial trace over particle 1, the polarisation vector for the second particle also has its length reduced by a factor of $\cos(\alpha)$. More generally the picture is less simple, and much work remains in understanding entanglement beyond the bipartite case.

A further application of the preceding is to calculate the overlap probability for the inner product of two states. Given two normalised states we have

$$P(\psi, \phi) = |\langle\psi|\phi\rangle|^2 = \mathrm{tr}(\hat{\rho}_\psi \hat{\rho}_\phi). \tag{9.63}$$

The degrees of freedom in the density matrices are contained in $\psi E\tilde{\psi}$ and $\psi J\tilde{\psi}$, with equivalent expressions for ϕ. When forming the inner product between two density matrices, the only terms that can arise are inner products between these observables. A little work confirms that we can write, in the n-particle case,

$$P(\psi,\phi) = 2^{n-2}\langle(\psi E\tilde{\psi})(\phi E\tilde{\phi})\rangle - 2^{n-2}\langle(\psi J\tilde{\psi})(\phi J\tilde{\phi})\rangle. \tag{9.64}$$

Expressions like this are unique to the geometric algebra approach. The expression confirms that once one has found the two multivector observables for a state, one has all of the available information to hand.

As an example, suppose that we are presented with two *separable* states, ψ and ϕ. For separable states we know that the observables take the forms

$$2\psi J\tilde{\psi} = A^1 + B^2, \qquad 2\psi E\tilde{\psi} = 1 - A^1 B^2 \tag{9.65}$$

and

$$2\phi J\tilde{\phi} = C^1 + D^2, \qquad 2\phi E\tilde{\phi} = 1 - C^1 D^2, \tag{9.66}$$

where each of the A^1, B^2, C^1 and D^2 are unit bivectors. We can now write

$$\begin{aligned} P(\psi,\phi) &= \tfrac{1}{4}\langle(1 - A^1B^2)(1 - C^1D^2) - (A^1 + B^2)(C^1 + D^2)\rangle \\ &= \tfrac{1}{4}(1 + A\cdot C\, B\cdot D - A\cdot C - B\cdot D) \\ &= \tfrac{1}{2}(1 - A\cdot C)\, \tfrac{1}{2}(1 - B\cdot D). \end{aligned} \tag{9.67}$$

This confirms the probability is the product of the separate single-particle probabilities. If one of the states is entangled this result no longer holds, as we see in the following section.

9.3.3 The singlet state

As a further example of entanglement we now study some of the properties of the non-relativistic spin singlet state. This is

$$|\varepsilon\rangle = \frac{1}{\sqrt{2}}\big(|0\rangle|1\rangle - |1\rangle|0\rangle\big). \tag{9.68}$$

This is represented in the two-particle spacetime algebra by the multivector

$$\varepsilon = \frac{1}{\sqrt{2}}(I\boldsymbol{\sigma}_2^1 - I\boldsymbol{\sigma}_2^2)E. \tag{9.69}$$

The properties of ε are more easily seen by writing

$$\varepsilon = \tfrac{1}{2}(1 + I\boldsymbol{\sigma}_2^1 I\boldsymbol{\sigma}_2^2)\tfrac{1}{2}(1 + I\boldsymbol{\sigma}_3^1 I\boldsymbol{\sigma}_3^2)\sqrt{2}\, I\boldsymbol{\sigma}_2^1, \tag{9.70}$$

which shows how ε contains the commuting idempotents $(1 + I\boldsymbol{\sigma}_2^1 I\boldsymbol{\sigma}_2^2)/2$ and $(1 + I\boldsymbol{\sigma}_3^1 I\boldsymbol{\sigma}_3^2)/2$. Identifying these idempotents tells us immediately that

$$I\boldsymbol{\sigma}_2^1\varepsilon = \tfrac{1}{2}(I\boldsymbol{\sigma}_2^1 - I\boldsymbol{\sigma}_2^2)\tfrac{1}{2}(1 + I\boldsymbol{\sigma}_3^1 I\boldsymbol{\sigma}_3^2)\sqrt{2} I\boldsymbol{\sigma}_2^1 = -I\boldsymbol{\sigma}_2^2\varepsilon \tag{9.71}$$

and

$$I\boldsymbol{\sigma}_3^1\varepsilon = -I\boldsymbol{\sigma}_3^2\varepsilon. \tag{9.72}$$

If follows that

$$I\sigma_1^1\varepsilon = I\sigma_3^1 I\sigma_2^1\varepsilon = -I\sigma_2^2 I\sigma_3^1\varepsilon = I\sigma_2^2 I\sigma_3^2\varepsilon = -I\sigma_1^2\varepsilon. \tag{9.73}$$

Combining these results, if M^1 is an arbitrary even element in the Pauli algebra ($M^1 = M_0 + M_k I\sigma_k^1$), ε satisfies

$$M^1\varepsilon = \tilde{M}^2\varepsilon. \tag{9.74}$$

Here M^1 and M^2 denote the same multivector, but expressed in space 1 or space 2.

Equation (9.74) provides a novel demonstration of the rotational invariance of ε. Under a joint rotation in two-particle space, a spinor ψ transforms to $R^1R^2\psi$, where R^1 and R^2 are copies of the same rotor but acting in the two different spaces. From equation (9.74) it follows that, under such a rotation, ε transforms as

$$\varepsilon \mapsto R^1R^2\varepsilon = R^1\tilde{R}^1\varepsilon = \varepsilon, \tag{9.75}$$

so that ε is a genuine two-particle rotational scalar.

If we now form the observables from ε we find that

$$2\varepsilon E\tilde{\varepsilon} = 1 + \sum_{k=1}^{3} I\sigma_k^1 I\sigma_k^2 \tag{9.76}$$

and

$$\varepsilon J\tilde{\varepsilon} = 0. \tag{9.77}$$

The latter has to hold, as there are no rotationally-invariant bivector observables. Equation (9.76) identifies a new two-particle invariant, which we can write as

$$\sum_{k=1}^{3} I\sigma_k^1 I\sigma_k^2 = 2\varepsilon\tilde{\varepsilon} - 1. \tag{9.78}$$

This is invariant under joint rotations in the two particles spaces. This multivector equation contains the essence of the matrix result

$$\sum_{k=1}^{3} \hat{\sigma}^{a}_{k\,a'}\,\hat{\sigma}^{b}_{k\,b'} = 2\delta^a_{b'}\,\delta^b_{a'} - \delta^a_{a'}\,\delta^b_{b'}, \tag{9.79}$$

where a, b, a', b' label the matrix components. In standard quantum mechanics this invariant would be thought of as arising from the 'inner product' of the spin vectors $\hat{\sigma}_i^1$ and $\hat{\sigma}_i^2$. Here, we have seen that the invariant arises in a completely different way, as a component of the multivector $\varepsilon\tilde{\varepsilon}$.

The fact that $\varepsilon J\tilde{\varepsilon} = 0$ confirms that the reduced density matrix for either particle space is simply one-half of the identity matrix, as established in equation (9.22). It follows that all directions are equally likely. If we align our measuring apparatus along some given axis and measure the state of particle 1, then both up and down have equal probabilities of 1/2. Suppose now that we construct a joint measurement on the singlet state. We can model this as the overlap probability between ψ and the separable state

$$\phi = R^1 S^2 E. \tag{9.80}$$

Denoting the spin directions by

$$R I\boldsymbol{\sigma}_3 \tilde{R} = P, \quad S I\boldsymbol{\sigma}_3 \tilde{S} = Q, \tag{9.81}$$

we find that, from equation (9.64),

$$\begin{aligned} P(\psi, \phi) &= \langle \tfrac{1}{2}(1 - P^1 Q^2) \tfrac{1}{2}(1 + I\sigma_k^1\, I\sigma_k^2) \rangle \\ &= \tfrac{1}{4}(1 - P\cdot(I\boldsymbol{\sigma}_k)\, Q\cdot(I\boldsymbol{\sigma}_k)) \\ &= \tfrac{1}{4}\big(1 - \cos(\theta)\big) \end{aligned} \tag{9.82}$$

where θ is the angle between the spin bivectors P and Q. So, for example, the probability that both measurements result in the particles having the same spin ($\theta = 0$) is zero, as expected. Similarly, if the measuring devices are aligned, the probability that particle 1 is up and particle 2 is down is 1/2, whereas if there was no entanglement present the probability would be the product of the separate single-particle measurements (resulting in 1/4).

Some consequences of equation (9.82) run counter to our intuitions about locality and causality. In particular, it is impossible to reproduce the statistics of equation (9.82) if we assume that the individual particles both know which spin state they are in prior to measurement. These contradictions are embodied in the famous *Bell inequalities*. The behaviour of entangled states has now been tested experimentally, and the results confirm all of the predictions of quantum mechanics. The results are unchanged even if the measurements are performed in such a way that the particles cannot be in causal contact. This does not provide any conflict with special relativity, as entangled states cannot be used to exchange classical information at faster than the speed of light. The reason is that the presence of entanglement can only be inferred when the separate measurements on the two subsystems are compared. Without knowing which measurements observer 1 is performing, observer 2 cannot extract any useful classical information from an entangled state.

For many years the properties of entangled states were explored largely as a theoretical investigation into the nature of quantum theory. Now, however, physicists are starting to view quantum entanglement as a resource that can be controlled in the laboratory. To date our control of entangled states is limited, but it is improving rapidly, and many predict that before long we will see the first viable quantum computers able to exploit this new resource.

9.4 Relativistic states and operators

The ideas developed for the multiparticle Pauli algebra extend immediately to the relativistic domain. A single-particle relativistic state is described by an arbitrary even element of the full spacetime algebra. Accordingly, a two-particle state is constructed from the tensor product of two such states. This results is a space of of $8 \times 8 = 64$ real dimensions. Post-multiplying the direct-product space by the quantum correlator E reduces to 32 real dimensions, which are equivalent to the 16 complex dimensions employed in standard two-particle relativistic quantum theory. All the single-particle operators and observables discussed in section 8.2 extend in fairly obvious ways.

To begin, the individual matrix operators have the equivalent action

$$\begin{aligned} \hat{\gamma}_\mu \otimes \hat{\mathsf{I}}|\psi\rangle &\leftrightarrow \gamma_\mu^1 \psi \gamma_0^1, \\ \hat{\mathsf{I}} \otimes \hat{\gamma}_\mu|\psi\rangle &\leftrightarrow \gamma_\mu^2 \psi \gamma_0^2, \end{aligned} \tag{9.83}$$

where $\hat{\mathsf{I}}$ denotes the 4×4 identity matrix. The multiparticle spacetime algebra operators *commute*, as they must in order to represent the tensor product. The result of the action of $\gamma_\mu^1 \psi \gamma_0^1$, for example, does not take us outside the two-particle state space, since the factor of γ_0^1 on the right-hand side commutes with the correlator E. The remaining matrix operators are easily constructed now, for example

$$\hat{\gamma}_\mu \hat{\gamma}_\nu \otimes \hat{\mathsf{I}}|\psi\rangle \leftrightarrow \gamma_\mu^1 \gamma_\nu^1 \psi. \tag{9.84}$$

The role of multiplication by the unit imaginary i is still played by right-multiplication by J, and the individual helicity projection operators become

$$\hat{\gamma}_5 \otimes \hat{\mathsf{I}}|\psi\rangle \leftrightarrow -I^1 \psi J = \psi \boldsymbol{\sigma}_3^1. \tag{9.85}$$

Relativistic observables are also constructed in a similar manner to the single-particle case. We form geometric products $\psi \Sigma \tilde{\psi}$, where Σ is any combination of γ_0 and γ_3 from either space. The result is then guaranteed to be Lorentz-covariant and phase-invariant. The first observable to consider is the multivector

$$\psi\tilde{\psi} = \psi E \tilde{\psi} = \langle \psi E \tilde{\psi} \rangle_{0,8} + \langle \psi E \tilde{\psi} \rangle_4. \tag{9.86}$$

The grade-0 and grade-8 terms are the two-particle generalisation of the scalar + pseudoscalar combination $\psi\tilde{\psi} = \rho \exp(i\beta)$ found at the single-particle level. The 4-vector part generalises the entanglement terms found in the non-relativistic case. This allows for a relativistic definition of entanglement, which is important for a detailed study of the relationship between locality and entanglement.

Next, we form two-particle current and spin vectors:

$$\mathcal{J} = \langle \psi(\gamma_0^1 + \gamma_0^2)\tilde{\psi} \rangle_1, \tag{9.87}$$

$$s = \langle \psi(\gamma_3^1 + \gamma_3^2)\tilde{\psi} \rangle_1. \tag{9.88}$$

(The calligraphic symbol $\mathcal{J}$ is used to avoid confusion with the correlated bivector J.) The full observables will contain grade-1 and grade-5 terms. For direct-product states the latter are seen to arise from the presence of a β factor in either of the single-particle states. Finally, we can also define the spin bivector S by

$$S = \langle \psi J \tilde{\psi} \rangle_2. \tag{9.89}$$

These expressions show how easy it is to generalise the single-particle formulae to the multiparticle case.

9.4.1 The relativistic singlet state

In the non-relativistic theory the spin singlet state has a special significance, both in being maximally entangled, and in its invariance under joint rotations in the two-particle space. An interesting question is whether we can construct a relativistic analogue that plays the role of a Lorentz singlet. Recalling the definition of ε (9.69), the property that ensured ε was a singlet state was that

$$I\boldsymbol{\sigma}_k^1 \varepsilon = -I\boldsymbol{\sigma}_k^2 \varepsilon, \qquad k = 1, \ldots, 3. \tag{9.90}$$

In addition to (9.90) a relativistic singlet state, which we will denote as η, must satisfy

$$\boldsymbol{\sigma}_k^1 \eta = -\boldsymbol{\sigma}_k^2 \eta, \qquad k = 1, \ldots, 3. \tag{9.91}$$

It follows that η satisfies

$$I^1 \eta = \boldsymbol{\sigma}_1^1 \boldsymbol{\sigma}_2^1 \boldsymbol{\sigma}_3^1 \eta = -\boldsymbol{\sigma}_3^2 \boldsymbol{\sigma}_2^2 \boldsymbol{\sigma}_1^2 \eta = I^2 \eta. \tag{9.92}$$

For this to hold, η must contain a factor of $(1 - I^1 I^2)$. We can therefore construct a Lorentz single state by multiplying ε by $(1 - I^1 I^2)$, and we define

$$\eta = (I\boldsymbol{\sigma}_2^1 - I\boldsymbol{\sigma}_2^2) \tfrac{1}{2} (1 - I\boldsymbol{\sigma}_3^1 I\boldsymbol{\sigma}_3^2) \tfrac{1}{2} (1 - I^1 I^2). \tag{9.93}$$

This is normalised so that $2\langle \eta E \tilde{\eta} \rangle = 1$. The properties of η can be summarised as

$$M^1 \eta = \tilde{M}^2 \eta, \tag{9.94}$$

where M is an even multivector in either the particle-1 or particle-2 spacetime algebra. The proof that η is a relativistic invariant now reduces to the simple identity

$$R^1 R^2 \eta = R^1 \tilde{R}^1 \eta = \eta, \tag{9.95}$$

where R is a single-particle relativistic rotor.

Equation (9.94) can be seen as originating from a more primitive relation between vectors in the separate spaces. Using the result that $\gamma_0^1\gamma_0^2$ commutes with η, we can derive

$$\begin{aligned}\gamma_\mu^1\eta\gamma_0^1 &= \gamma_\mu^1\gamma_0^1\gamma_0^2\eta\gamma_0^2\gamma_0^1\gamma_0^1\\ &= \gamma_0^2(\gamma_\mu\gamma_0)^1\eta\gamma_0^2\\ &= \gamma_\mu^2\eta\gamma_0^2.\end{aligned} \tag{9.96}$$

For an arbitrary vector a we can now write

$$a^1\eta\gamma_0^1 = a^2\eta\gamma_0^2. \tag{9.97}$$

Equation (9.94) follows immediately from equation (9.97) by writing

$$\begin{aligned}a^1b^1\eta &= a^1b^2\eta\gamma_0^2\gamma_0^1\\ &= b^2a^2\eta\gamma_0^2\gamma_0^2\\ &= b^2a^2\eta.\end{aligned} \tag{9.98}$$

Equation (9.97) can therefore be viewed as the fundamental property of the relativistic invariant η.

The invariant η can be used to construct a series of observables that are also invariant under coupled rotations in the two spaces. The first is

$$2\eta E\tilde{\eta} = (1 - I^1I^2) - (\boldsymbol{\sigma}_k^1\boldsymbol{\sigma}_k^2 - I\boldsymbol{\sigma}_k^1 I\boldsymbol{\sigma}_k^2). \tag{9.99}$$

The scalar and pseudoscalar (grade-8) terms are clearly invariants, and the 4-vector term, $(\boldsymbol{\sigma}_k^1\boldsymbol{\sigma}_k^2 - I\boldsymbol{\sigma}_k^1 I\boldsymbol{\sigma}_k^2)$, is a Lorentz invariant because it is a contraction over a complete bivector basis in the two spaces. Next we consider the multivector

$$\begin{aligned}2\eta\gamma_0^1\gamma_0^2\tilde{\eta} &= \gamma_0^1\gamma_0^2 - I^1I^2\gamma_k^1\gamma_k^2 - I^1I^2\gamma_0^1\gamma_0^2 - \gamma_k^1\gamma_k^2)\\ &= (\gamma_0^1\gamma_0^2 - \gamma_k^1\gamma_k^2)(1 - I^1I^2).\end{aligned} \tag{9.100}$$

The essential invariant here is the bivector

$$K = \gamma_\mu^1\wedge\gamma^{\mu 2}, \tag{9.101}$$

and the invariants from (9.100) are simply K and KI^1I^2. The bivector K takes the form of a 'doubling' bivector, which will be encountered again in section 11.4.

From the definition of K in equation (9.101), we find that

$$\begin{aligned}K\wedge K &= -2\gamma_0^1\gamma_0^2\gamma_k^1\gamma_k^2 + (\gamma_k^1\gamma_k^2)\wedge(\gamma_j^1\gamma_j^2)\\ &= 2(\boldsymbol{\sigma}_k^1\boldsymbol{\sigma}_k^2 - I\boldsymbol{\sigma}_k^1 I\boldsymbol{\sigma}_k^2),\end{aligned} \tag{9.102}$$

which recovers the grade-4 invariant found in equation (9.99). The full set of two-particle invariants constructed from K are summarised in table 9.1. These invariants are regularly employed in constructing interaction terms in multiparticle wave equations.

Invariant	Type of interaction	Grade
1	Scalar	0
K	Vector	2
$K\wedge K$	Bivector	4
I^1I^2K	Pseudovector	6
I^1I^2	Pseudoscalar	8

Table 9.1 *Relativistic invariants in the two-particle algebra.*

9.4.2 Multiparticle wave equations

The question of how to construct a valid, relativistic, multiparticle wave equation has troubled physicists almost from the moment Dirac proposed his equation. The question is far from settled, and the current preferred option is to ignore the question where possible and instead work within the framework of perturbative quantum field theory. This approach runs into difficulties when analysing bound states, however, and for these problems the need for a suitable wave equation is particularly acute. The main candidate for a relativistic two-particle system is the Bethe–Salpeter equation. Written in the multiparticle spacetime algebra, this equation is

$$(j\hat{\nabla}_r^1 - m_1)(j\hat{\nabla}_s^2 - m_2)\psi(r,s) = \mathcal{I}(r,s)\psi(r,s) \tag{9.103}$$

where $\mathcal{I}(r,s)$ is an integral operator representing the interparticle interaction, and ∇_r^1 and ∇_s^2 denote vector derivatives with respect to r^1 and s^2 respectively. The combined vector

$$x = r^1 + s^2 = r^\mu\gamma_\mu^1 + s^\mu\gamma_\mu^2 \tag{9.104}$$

is the full position vector in eight-dimensional configuration space.

One slightly unsatisfactory feature of equation (9.103) is that it is not first-order. This has led researchers to propose a number of alternative equations, typically with the aim of providing a more detailed analysis of two-body bound state systems such as the hydrogen atom, or positronium. One such equation is

$$(\nabla_r^1\psi\gamma_0^1 + \nabla_s^2\psi\gamma_0^2)J = (m_1+m_2)\psi. \tag{9.105}$$

As well as being first order, this equation also has the required property that it is satisfied by direct products of single-particle solutions. But a problem is that any distinction between the particle masses has been lost, since only the total mass enters. A second candidate equation, which does keep the masses distinct, is

$$\left(\frac{\nabla_r^1}{m_1} + \frac{\nabla_s^2}{m_2}\right)\psi(x)J = \psi(x)(\gamma_0^1+\gamma_0^2). \tag{9.106}$$

This equation has a number of attractive features, not least of which is that the mass enters in a manner that is highly suggestive of gravitational interactions. A potential weakness of this equation is that the state space can no longer be restricted to sums of direct products of individual states. Instead we have to widen the state space to include the entire (correlated) even subalgebra of the two-particle spacetime algebra. This doubles the number of degrees of freedom, and it is not clear that this doubling can be physical.

Practically all candidate two-particle wave equations have difficulties in performing a separation into centre-of-mass and relative coordinates. This is symptomatic of the fact that the centre of mass cannot be defined sensibly even in classical relativistic dynamics. Usually some approximation scheme has to be employed to avoid this problem, even when looking for bound state solutions. While the question of finding a suitable wave equation remains an interesting challenge, one should be wary of the fact that the mass term in the Dirac equation is essentially a remainder from a more complicated interaction with the Higgs boson. The electroweak theory immediately forces us to consider particle doublets, and it could be that one has to consider multiparticle extensions of these in order to arrive at a satisfactory theory.

9.4.3 The Pauli principle

In quantum theory, indistinguishable particles must obey either Fermi–Dirac or Bose–Einstein statistics. For fermions this requirement results in the Pauli exclusion principle that no two particles can occupy a state in which their properties are identical. The Pauli principle is usually enforced in one of two ways in relativistic quantum theory. At the level of multiparticle wave mechanics, antisymmetrisation is enforced by using a Slater determinant representation of a state. At the level of quantum field theory, however, antisymmetrisation is a consequence of the anticommutation of the creation and annihilation operators for fermions. Here we are interested in the former approach, and look to achieve the antisymmetrisation in a simple geometrical manner.

We start by introducing the grade-4 multivector

$$I_P = \Gamma_0\Gamma_1\Gamma_2\Gamma_3, \tag{9.107}$$

where

$$\Gamma_\mu = \frac{1}{\sqrt{2}}\left(\gamma_\mu^1 + \gamma_\mu^2\right). \tag{9.108}$$

It is a simple matter to verify that I_P has the properties

$$I_P^2 = -1 \tag{9.109}$$

and

$$I_P\gamma_\mu^1 I_P = \gamma_\mu^2, \quad I_P\gamma_\mu^2 I_P = \gamma_\mu^1. \tag{9.110}$$

It follows that I_P functions as a geometrical version of the particle exchange operator. In particular, acting on the eight-dimensional position vector $x = r^1 + s^2$ we find that

$$I_P x I_P = r^2 + s^1 \tag{9.111}$$

where

$$r^2 = \gamma^2_\mu r^\mu, \quad s^1 = \gamma^1_\mu s^\mu. \tag{9.112}$$

So I_P can be used to interchange the coordinates of particles 1 and 2. Next we must confirm that I_P is independent of the choice of initial frame. Suppose that instead we had started with the rotated frame $\{R\gamma_\mu \tilde{R}\}$, with

$$\Gamma'_\mu = \frac{1}{\sqrt{2}}(R^1\gamma^1_\mu \tilde{R}^1 + R^2\gamma^2_\mu \tilde{R}^2) = R^1R^2\Gamma_\mu \tilde{R}^2\tilde{R}^1. \tag{9.113}$$

The new Γ'_μ vectors give rise to the rotated 4-vector

$$I'_P = R^1R^2 I_P \tilde{R}^2\tilde{R}^1. \tag{9.114}$$

But, acting on a bivector in particle space 1, we find that

$$I_P a^1 \wedge b^1 I_P = -(I_P a^1 I_P)\wedge(I_P b^1 I_P) = -a^2\wedge b^2, \tag{9.115}$$

and the same is true of an arbitrary even element in either space. More generally, the operation $M \mapsto I_P M I_P$ applied to an even element in one of the particle spaces flips it to the other particle space and changes sign, while applied to an odd element it just flips the particle space. It follows that

$$I_P \tilde{R}^2\tilde{R}^1 = \tilde{R}^1 I_P \tilde{R}^1 = \tilde{R}^1\tilde{R}^2 I_P, \tag{9.116}$$

and substituting this into (9.114) we find that $I'_P = I_P$. It follows that I_P is independent of the chosen orthonormal frame, as required.

We can now use the 4-vector I_P to encode the Pauli exchange principle geometrically. Let $\psi(x)$ be a wavefunction for two electrons. The state

$$\psi(x)' = -I_P \psi(I_P x I_P) I_P, \tag{9.117}$$

then swaps the position dependence, and interchanges the space of the multivector components of ψ. The antisymmetrised state is therefore

$$\psi_-(x) = \psi(x) + I_P \psi(I_P x I_P) I_P. \tag{9.118}$$

For n-particle systems the extension is straightforward, as we require that the wavefunction is invariant under the interchange enforced by the I_Ps constructed from each pair of particles.

For a single Dirac particle the probability current $J = \psi\gamma_0\tilde{\psi}$ has zero divergence, and can therefore be used to define streamlines. These are valuable for understanding a range of phenomena, such as wavepacket tunnelling and spin measurement. We now illustrate how these ideas extend to the multiparticle domain. The two-particle current is

$$\mathcal{J} = \langle \psi(\gamma_0^1 + \gamma_0^2)\tilde{\psi} \rangle_1, \tag{9.119}$$

as defined in equation (9.87). The vector $\mathcal{J}$ has components in both particle-1 and particle-2 spaces, which we write as

$$\mathcal{J} = \mathcal{J}_1^1 + \mathcal{J}_2^2. \tag{9.120}$$

For sums of separable solutions to the single-particle equations, the individual currents are both conserved:

$$\nabla^1 \cdot \mathcal{J}_1^1 = \nabla^2 \cdot \mathcal{J}_2^2 = 0. \tag{9.121}$$

It follows that the full current $\mathcal{J}$ is conserved in 8-dimensional space, so its streamlines never cross there. The streamlines of the individual particles, however, are obtained by integrating $\mathcal{J}_1$ and $\mathcal{J}_2$ in a single spacetime, and these can cross if plotted in the same space. For example, suppose that the wavefunction is just

$$\psi = \phi^1(r^1)\chi^2(s^2)E, \tag{9.122}$$

where ϕ and χ are Gaussian wavepackets moving in opposite directions. Since the distinguishable case is assumed, no Pauli antisymmetrisation is used. One can easily confirm that for this case the streamlines and the wavepackets simply pass straight through each other.

But suppose now that we assume indistinguishability, and apply the Pauli symmetrisation procedure to the wavefunction of equation (9.122). We arrive at the state

$$\psi = \left(\phi^1(r^1)\chi^2(s^2) - \chi^1(r^2)\phi^2(s^1)\right) E, \tag{9.123}$$

from which we form $\mathcal{J}_1$ and $\mathcal{J}_2$, as before. Figure 9.1 shows the streamlines that result from these currents. In the left-hand plot both particles are in the same spin state. The corrugated appearance of the lines near the origin is the result of the streamlines having to pass through a region of highly oscillatory destructive interference, since the probability of both particles occupying the same position (the origin) with the same spin state is zero. The right-hand plot is for two particles in different spin states. Again, the streamlines are seen to repel. The reason for this can be found in the symmetry properties of the two-particle current. Given that the wavefunction ψ has been antisymmetrised according to equation (9.118), the current must satisfy

$$I_P \mathcal{J}(I_P x I_P) I_P = \mathcal{J}(x). \tag{9.124}$$

It follows that at the same spacetime position, encoded by $I_P x I_P = x$ in the two-particle algebra, the two currents $\mathcal{J}_1$ and $\mathcal{J}_2$ are equal. Hence, if two streamlines

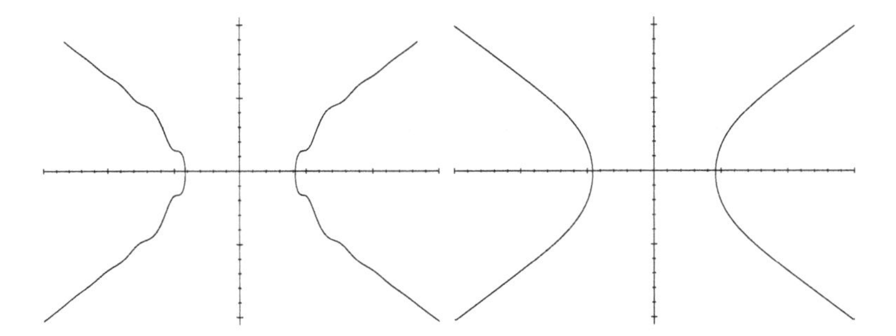

Figure 9.1 *Streamlines for an antisymmetrised two-particle wavefunction.* The wavefunction is $\psi = \big(\phi^1(r^1)\chi^2(s^2) - \chi^1(r^2)\phi^2(s^1)\big)E$. The individual wavepackets pass through each other, but the streamlines from separate particles do not cross. The left-hand figure has both particles with spins aligned in the $+z$ direction. The right-hand figure shows particles with opposite spins, with ϕ in the $+z$ direction, and χ in the $-z$ direction.

ever met, they could never separate again. For the simulations presented here, the symmetry of the set-up implies that the spatial currents at the origin are both zero. As the particles approach the origin, they are forced to slow up. The delay means that they are then swept back in the direction they have just come from by the wavepacket travelling through from the other side. This repulsion has its origin in indistinguishability, and the spin of the states exerts only a marginal effect.

9.5 Two-spinor calculus

The ideas introduced in this chapter can be employed to construct a geometric algebra version of the two-spinor calculus developed by Penrose & Rindler (1984). The building blocks of their approach are two-component complex spinors, denoted κ^A and $\bar{\omega}^{A'}$. Indices are raised and lowered with the antisymmetric tensor ϵ_{AB}. In the spacetime algebra version both κ^A and κ_A have the same multivector equivalent, which we write as

$$\kappa^A \leftrightarrow \kappa\tfrac{1}{2}(1+\boldsymbol{\sigma}_3). \tag{9.125}$$

The presence of the idempotent $(1+\boldsymbol{\sigma}_3)/2$ allows us to restrict κ to the Pauli-even algebra, as any Pauli-odd terms can be multiplied on the right by $\boldsymbol{\sigma}_3$ to convert them back to the even subspace. This ensures that κ has four real degrees of freedom, as required. Under a Lorentz transformation the full spinor transforms to

$$R\kappa\tfrac{1}{2}(1+\boldsymbol{\sigma}_3) = \kappa'\tfrac{1}{2}(1+\boldsymbol{\sigma}_3), \tag{9.126}$$

where R is a Lorentz rotor. If we decompose the rotor R into Pauli-even and Pauli-odd terms, $R = R_+ + R_-$, then κ' is given by

$$\kappa' = R_+\kappa + R_-\kappa\boldsymbol{\sigma}_3. \tag{9.127}$$

The decomposition into Pauli-even and Pauli-odd terms is frame-dependent, as it depends on the choice of the γ_0 direction. But by augmenting κ with the $(1+\boldsymbol{\sigma}_3)/2$ idempotent we ensure that the full object is a proper Lorentz-covariant spinor.

The opposite idempotent, $(1-\boldsymbol{\sigma}_3)/2$, also generates a valid two-spinor which belongs to a second linear space (or module). This is the $\bar{\omega}^{A'}$ spinor in the notation of Penrose & Rindler, which we translate to

$$\bar{\omega}^{A'} \leftrightarrow = -\omega I\boldsymbol{\sigma}_2\tfrac{1}{2}(1-\boldsymbol{\sigma}_3). \tag{9.128}$$

The factor of $-I\boldsymbol{\sigma}_2$ is a matter of convention, and is inserted to simplify some of the later expressions. Under a Lorentz transformation we see that the Pauli-even element ω transforms as

$$\omega \mapsto \omega' = R_+\omega - R_-\omega\boldsymbol{\sigma}_3. \tag{9.129}$$

So κ and ω have different transformation laws: they belong to distinct carrier spaces of representations of the Lorentz group.

The power of the two-spinor calculus is the ease with which vector and tensor objects are generated from the basic two-spinors. As emphasised by Penrose & Rindler, this makes the calculus equally useful for both classical and quantum applications. It is instructive to see how this looks from the geometric algebra point of view. Unsurprisingly, what we discover is that the two-spinor calculus is a highly abstract and sophisticated means of introducing the geometric product to tensor manipulations. Once this is understood, much of the apparatus of the two-spinor calculus can be stripped away, and one is left with the now familiar spacetime algebra approach to relativistic physics.

9.5.1 Two-spinor observables

In two-spinor calculus one forms tensor objects from pairs of two-spinors, for example $\kappa^A\bar{\kappa}^{A'}$. To formulate this in the multiparticle spacetime algebra we simply multiply together the appropriate spinors, putting each spinor in its own copy of the spacetime algebra. In this way we replicate the tensor product implicit in writing $\kappa^A\bar{\kappa}^{A'}$. The result is that we form the object

$$\kappa^A\bar{\kappa}^{A'} \leftrightarrow -\kappa^1\tfrac{1}{2}(1+\boldsymbol{\sigma}_3^1)\kappa^2 I\boldsymbol{\sigma}_2^2\tfrac{1}{2}(1-\boldsymbol{\sigma}_3^2)\tfrac{1}{2}(1-I\boldsymbol{\sigma}_3^1 I\boldsymbol{\sigma}_3^2). \tag{9.130}$$

$\frac{1}{2}(1+\boldsymbol{\sigma}_3^1)\frac{1}{2}(1-\boldsymbol{\sigma}_3^2)E = -\frac{1}{2}(\gamma_0^1+\gamma_3^1)I\boldsymbol{\sigma}_2^1\bar{\epsilon}\gamma_0^1$
$\frac{1}{2}(1-\boldsymbol{\sigma}_3^1)\frac{1}{2}(1+\boldsymbol{\sigma}_3^2)E = -\frac{1}{2}(\gamma_0^1-\gamma_3^1)I\boldsymbol{\sigma}_2^1\epsilon\gamma_0^1$
$\frac{1}{2}(1+\boldsymbol{\sigma}_3^1)\frac{1}{2}(1+\boldsymbol{\sigma}_3^2)E = -\frac{1}{2}(\boldsymbol{\sigma}_1^1+I\boldsymbol{\sigma}_2^1)\epsilon$
$\frac{1}{2}(1-\boldsymbol{\sigma}_3^1)\frac{1}{2}(1-\boldsymbol{\sigma}_3^2)E = -\frac{1}{2}(-\boldsymbol{\sigma}_1^1+I\boldsymbol{\sigma}_2^1)\bar{\epsilon}$

Table 9.2 *Two-spinor identities.* The identities listed here can be used to convert any expression involving a pair of two-spinors into an equivalent multivector.

As it stands this looks rather clumsy, but the various idempotents hide what is really going on. The key is to expose the Lorentz singlet structure hidden in the combination of idempotents. To achieve this we define two new Lorentz singlet states

$$\epsilon = \eta\tfrac{1}{2}(1+\boldsymbol{\sigma}_3^1), \quad \bar{\epsilon} = \eta\tfrac{1}{2}(1-\boldsymbol{\sigma}_3^2), \tag{9.131}$$

where η is the Lorentz singlet defined in equation (9.93). These new states both satisfy the essential equation

$$M^1\epsilon = \tilde{M}^2\epsilon, \quad M^1\bar{\epsilon} = \tilde{M}^2\bar{\epsilon}, \tag{9.132}$$

where M is an even-grade multivector. The reason is that any idempotents applied on the right of η cannot affect the result of equation (9.94). Expanding out in full, and rearranging the idempotents, we find that

$$\begin{aligned} \epsilon &= (I\boldsymbol{\sigma}_2^1 - I\boldsymbol{\sigma}_2^2)\tfrac{1}{2}(1+\boldsymbol{\sigma}_3^1)\tfrac{1}{2}(1+\boldsymbol{\sigma}_3^2)E, \\ \bar{\epsilon} &= (I\boldsymbol{\sigma}_2^1 - I\boldsymbol{\sigma}_2^2)\tfrac{1}{2}(1-\boldsymbol{\sigma}_3^1)\tfrac{1}{2}(1-\boldsymbol{\sigma}_3^2)E. \end{aligned} \tag{9.133}$$

These relations can manipulated to give, for example,

$$\begin{aligned} I\boldsymbol{\sigma}_2^1\epsilon &= -(1+I\boldsymbol{\sigma}_2^1 I\boldsymbol{\sigma}_2^2)\tfrac{1}{2}(1+\boldsymbol{\sigma}_3^1)\tfrac{1}{2}(1+\boldsymbol{\sigma}_3^2)E, \\ \boldsymbol{\sigma}_1^1\epsilon &= -(1-I\boldsymbol{\sigma}_2^1 I\boldsymbol{\sigma}_2^2)\tfrac{1}{2}(1+\boldsymbol{\sigma}_3^1)\tfrac{1}{2}(1+\boldsymbol{\sigma}_3^2)E. \end{aligned} \tag{9.134}$$

It follows that

$$\tfrac{1}{2}(1+\boldsymbol{\sigma}_3^1)\tfrac{1}{2}(1+\boldsymbol{\sigma}_3^2)E = -\tfrac{1}{2}(\boldsymbol{\sigma}_1^1 + I\boldsymbol{\sigma}_2^2)\epsilon. \tag{9.135}$$

There are four such identities in total, which are listed in table 9.2.

The results given in table 9.2 enable us to immediately convert any two-spinor expression into an equivalent multivector in the spacetime algebra. For example, returning to equation (9.130), we form

$$\begin{aligned} -\kappa^1\kappa^2 I\boldsymbol{\sigma}_2^2\tfrac{1}{2}(1+\boldsymbol{\sigma}_3^1)\tfrac{1}{2}(1-\boldsymbol{\sigma}_3^2)E &= \kappa^1\kappa^2\tfrac{1}{2}(\gamma_0^1+\gamma_3^1)\bar{\epsilon}\gamma_0^1 \\ &= \tfrac{1}{2}\big(\kappa(\gamma_0+\gamma_3)\tilde{\kappa}\big)^1\bar{\epsilon}\gamma_0^1. \end{aligned} \tag{9.136}$$

The key term in this expression is the null vector $\kappa(\gamma_0+\gamma_3)\tilde{\kappa}$, which is constructed in the familiar manner for relativistic observables. A feature of the two-spinor calculus is that it lends itself to formulating most quantities in terms of null vectors. The origin of these can be traced back to the original $(1\pm\boldsymbol{\sigma}_3)/2$ idempotents, which contain the null vector $\gamma_0\pm\gamma_3$. These are rotated and dilated onto spacetime null vectors through the application of a spinor.

9.5.2 The two-spinor inner product

A Lorentz-invariant inner product for a pair of two-spinors is constructed from the antisymmetric combination

$$\kappa^A\omega_A = -\kappa_0\omega_1 + \kappa_1\omega_0, \tag{9.137}$$

where the subscripts here denote complex components of a two-spinor. The result of the inner product is a Lorentz-invariant complex scalar. The antisymmetry of the inner product tells us that we should form the equivalent expression

$$\begin{aligned}(\kappa^1\omega^2-\kappa^2\omega^1)\tfrac{1}{2}(1+\boldsymbol{\sigma}_3^1)\tfrac{1}{2}(1+\boldsymbol{\sigma}_3^2)E &= -\tfrac{1}{2}\big(\kappa(\boldsymbol{\sigma}_1+I\boldsymbol{\sigma}_2)\tilde{\omega}-\omega(\boldsymbol{\sigma}_1+I\boldsymbol{\sigma}_2)\tilde{\kappa}\big)^1\epsilon \\ &= -\langle\kappa(\boldsymbol{\sigma}_1+I\boldsymbol{\sigma}_2)\tilde{\omega}\rangle_{0,4}^1\epsilon. \end{aligned} \tag{9.138}$$

The antisymmetric product picks out the scalar and pseudoscalar parts of the quantity $\kappa(\boldsymbol{\sigma}_1+I\boldsymbol{\sigma}_2)\tilde{\omega}$. This is sensible, as these are the two terms that are invariant under Lorentz transformations.

The fact that we form a scalar + pseudoscalar combination reveals a second important feature of the two-spinor calculus, which is that the unit imaginary is a representation of the spacetime pseudoscalar. The complex structure therefore has a concrete, geometric significance, which is one reason why two-spinor techniques have proved popular in general relativity, for example. Further insight into the form of the two-spinor inner product is gained by assembling the full even multivector

$$\psi = \kappa\tfrac{1}{2}(1+\boldsymbol{\sigma}_3) + \omega I\boldsymbol{\sigma}_2\tfrac{1}{2}(1-\boldsymbol{\sigma}_3). \tag{9.139}$$

The essential term in the two-spinor inner product is now reproduced by

$$\begin{aligned}\psi\tilde{\psi} &= -\kappa\tfrac{1}{2}(1+\boldsymbol{\sigma}_3)I\boldsymbol{\sigma}_2\tilde{\omega} + \omega I\boldsymbol{\sigma}_2\tfrac{1}{2}(1-\boldsymbol{\sigma}_3)\tilde{\kappa} \\ &= -\langle\kappa(\boldsymbol{\sigma}_1+I\boldsymbol{\sigma}_2)\tilde{\omega}\rangle_{0,4}, \end{aligned} \tag{9.140}$$

so the inner products pick up both the scalar and pseudoscalar parts of a full Dirac spinor product $\psi\tilde{\psi}$. This form makes the Lorentz invariance of the product quite transparent. Interchanging κ and ω in ψ of equation (9.139) is achieved by right-multiplication by $\boldsymbol{\sigma}_1$, which immediately reverses the sign of $\psi\tilde{\psi}$.

9.5.3 Spin-frames and the null tetrad

An important concept in the two-spinor calculus is that of a *spin-frame*. This consists of a pair of two-spinors, κ^A and ω^A say, normalised such that $\kappa^A\omega_A = 1$. In terms of the spinor ψ of equation (9.139), this normalisation condition amounts to saying that ψ satisfies $\psi\tilde{\psi} = 1$. A normalised spin-frame is therefore the two-spinor encoding of a spacetime rotor. This realisation also sheds light on the associated concept of a *null tetrad*. In terms of the spin frame $\{\kappa^A, \omega^A\}$, the associated null tetrad is defined as follows:

$$\begin{aligned} l^a &= \kappa^A\bar{\kappa}^{A'} \leftrightarrow \big(\kappa(\gamma_0+\gamma_3)\tilde{\kappa}\big)^1\bar{\epsilon}\gamma_0^1, \\ n^a &= \omega^A\bar{\omega}^{A'} \leftrightarrow \big(\omega(\gamma_0+\gamma_3)\tilde{\omega}\big)^1\bar{\epsilon}\gamma_0^1, \\ m^a &= \kappa^A\bar{\omega}^{A'} \leftrightarrow \big(\kappa(\gamma_0+\gamma_3)\tilde{\omega}\big)^1\bar{\epsilon}\gamma_0^1, \\ \bar{m}^a &= \omega^A\bar{\kappa}^{A'} \leftrightarrow \big(\omega(\gamma_0+\gamma_3)\tilde{\kappa}\big)^1\bar{\epsilon}\gamma_0^1. \end{aligned} \tag{9.141}$$

In each case we have projected into a single copy of the spacetime algebra to form a geometric multivector. To simplify these expressions we introduce the rotor R defined by

$$R = \kappa\tfrac{1}{2}(1+\boldsymbol{\sigma}_3) + \omega I\boldsymbol{\sigma}_2\tfrac{1}{2}(1-\boldsymbol{\sigma}_3). \tag{9.142}$$

It follows that

$$\begin{aligned} R(\gamma_1+I\gamma_2)\tilde{R} &= -\kappa\gamma_1(1+\boldsymbol{\sigma}_3)I\boldsymbol{\sigma}_2\tilde{\omega} \\ &= \kappa(\gamma_0+\gamma_3)\tilde{\omega}. \end{aligned} \tag{9.143}$$

The null tetrad induced by a normalised spin-frame can now be written in the spacetime algebra as

$$\begin{aligned} l &= R(\gamma_0+\gamma_3)\tilde{R}, \qquad & m &= R(\gamma_1+I\gamma_2)\tilde{R}, \\ n &= R(\gamma_0-\gamma_3)\tilde{R}, \qquad & \bar{m} &= R(\gamma_1-I\gamma_2)\tilde{R}. \end{aligned} \tag{9.144}$$

(One can chose alternative normalisations, if required). The complex vectors m^a and $\bar{m}^a$ of the two-spinor calculus have now been replaced by vector + trivector combinations. This agrees with the earlier observation that the imaginary scalar in the two-spinor calculus plays the role of the spacetime pseudoscalar. The multivectors in a null tetrad satisfy the anticommutation relations

$$\{l,n\} = 4, \quad \{m,\bar{m}\} = 4, \quad \text{all others} = 0. \tag{9.145}$$

These relations provide a framework for the formulation of supersymmetric quantum theory within the multiparticle spacetime algebra.

9.6 Notes

The multiparticle spacetime algebra was introduced in the paper 'States and operators in the spacetime algebra' by Doran, Lasenby & Gull (1993a). Since its introduction the multiparticle spacetime algebra has been developed by a range of researchers. For introductions see the papers by Parker & Doran (2002) and Havel & Doran (2000a,2002b). Of particular interest are the papers by Somaroo et al. (1998,1999) and Havel et al. (2001), which show how the multiparticle spacetime algebra can be applied to great effect in the theory of quantum information processing. These researchers were primarily motivated by the desire to create quantum gates in an NMR environment, though their observations can be applied to quantum computation in general. For a good introduction into the subject of quantum information, we recommend the course notes made available by Preskill (1998).

The subject of relativistic multiparticle quantum theory has been tackled by many authors. The most authoritative discussions are contained in the papers by Salpeter & Bethe (1951), Salpeter (1952), Breit (1929) and Feynman (1961). A more modern perspective is contained in the discussions in Itzykson & Zuber (1980) and Grandy (1991). For more recent attempts at constructing a two-particle version of the Dirac equation, see the papers by Galeao & Ferreira (1992), Cook (1988) and Koide (1982). A summary of the multiparticle spacetime algebra approach to this problem is contained in Doran et al.(1996b).

The two-spinor calculus is described in the pair of books '*Spinors and Spacetime*' volumes I and II by Penrose & Rindler (1984,1986). The spacetime algebra version of two-spinor calculus is described in more detail in 'Geometric algebra and its application to mathematical physics' by Doran (1994), with additional material contained in the paper '2-spinors, twistors and supersymmetry in the spacetime algebra' by Lasenby et al. (1993b). The conventions adopted in this book differ slightly from those adopted in many of the earlier papers.

9.7 Exercises

9.1 Explain how the two-particle Schrödinger equation for the Coulomb problem is reduced to the effective single-particle equation

$$-\frac{\hbar^2\boldsymbol{\nabla}^2}{2\mu}\psi - \frac{q_1 q_2}{4\pi\epsilon_0 r}\psi = E\psi,$$

where μ is the reduced mass.

9.2 Given that $\psi(\theta,\phi) = \exp(-\phi I\boldsymbol{\sigma}_3/2)\exp(-\theta I\boldsymbol{\sigma}_2/2)$, prove that

$$\begin{pmatrix} \sin(\theta/2)\mathrm{e}^{-i\phi/2} \\ -\cos(\theta/2)\mathrm{e}^{i\phi/2} \end{pmatrix} \leftrightarrow \psi(\theta,\phi)I\boldsymbol{\sigma}_2.$$

Confirm that this state is orthogonal to $\psi(\theta,\phi)$.

9.3 The interaction energy of two dipoles is given classically by

$$E = \frac{\mu_0}{4\pi}\left(\frac{\boldsymbol{\mu}_1\cdot\boldsymbol{\mu}_2}{r^3} - 3\frac{\boldsymbol{\mu}_1\cdot\boldsymbol{r}\,\boldsymbol{\mu}_2\cdot\boldsymbol{r}}{r^5}\right),$$

where $\boldsymbol{\mu}_i$ denotes the magnetic moment of particle i. For a quantum system of spin 1/2 particles we replace the magnetic moment vectors with the operators $\hat{\mu}_k = (\gamma\hbar/2)\hat{\sigma}_k$. Given that $\boldsymbol{n} = \boldsymbol{r}/r$, show that the Hamiltonian operator takes the form of the 4-vector

$$H = -\frac{d}{4}\left(\sum_{k-1}^{3} I\boldsymbol{\sigma}_k^1\, I\boldsymbol{\sigma}_k^2 - 3\,I\boldsymbol{n}^1\, I\boldsymbol{n}^2\right)$$

and find an expression for d. Can you solve the two-particle Schrödinger equation with this Hamiltonian?

9.4 ψ and ϕ are a pair of non-relativistic multiparticle states. Prove that the overlap probability between the two states can be written

$$P(\psi,\phi) = \frac{\langle(\psi E\tilde{\psi})(\phi E\tilde{\phi})\rangle - \langle(\psi J\tilde{\psi})(\phi J\tilde{\phi})\rangle}{2\langle\psi E\tilde{\psi}\rangle\langle\phi E\tilde{\phi}\rangle}.$$

9.5 Investigate the properties of the $l = 1$, $m = 0$ state

$$|\psi\rangle = |0\rangle|1\rangle + |1\rangle|0\rangle.$$

Is this state maximally entangled?

9.6 The β_μ operators that act on states in the two-particle relativistic algebra are defined by:

$$\beta_\mu(\psi) = \tfrac{1}{2}\left(\gamma_\mu^1\psi\gamma_0^1 + \gamma_\mu^2\psi\gamma_0^2\right).$$

Verify that these operators generate the *Duffin–Kemmer* ring

$$\beta_\mu\beta_\nu\beta_\rho + \beta_\rho\beta_\nu\beta_\mu = \eta_{\nu\rho}\beta_\mu + \eta_{\nu\mu}\beta_\rho.$$

9.7 The multiparticle wavefunction ψ is constructed from superpositions of states of the form $\phi^1(r^1)\chi^2(s^2)$, where ϕ and χ satisfy the single-particle Dirac equation. Prove that the individual currents $\mathcal{J}_1^1$ and $\mathcal{J}_2^2$ are conserved, where

$$\mathcal{J}_1^1 + \mathcal{J}_2^2 = \langle\psi(\gamma_0^1 + \gamma_0^2)\tilde{\psi}\rangle_1.$$

9.8 In the two-spinor calculus the two-component complex vector κ^A is acted on by a 2×2 complex matrix R . Prove that R is a representation of the Lorentz rotor group if $\det \mathsf{R} = 1$. (This defines the Lie group Sl(2,C).) Hence establish that the antisymmetric combination $\kappa^0\omega^1 - \kappa^1\omega^0$ is a Lorentz scalar.

9.9 The two-spinor calculus version of the Dirac equation is

$$\nabla^{A'A}\kappa_A = \mu\bar{\omega}^{A'},$$
$$\nabla^{AA'}\bar{\omega}_{A'} = \mu\kappa^A,$$

where $\mu = m/\sqrt{2}$. Prove that these equations are equivalent to the single equation $\nabla\psi I\boldsymbol{\sigma}_3 = m\psi\gamma_0$ and give an expression for ψ in terms of κ^A and $\bar{\omega}_{A'}$.

9.10 A null tetrad is defined by the set

$$l = R(\gamma_0 + \gamma_3)\tilde{R}, \qquad m = R(\gamma_1 + I\gamma_2)\tilde{R},$$
$$n = R(\gamma_0 - \gamma_3)\tilde{R}, \qquad \bar{m} = R(\gamma_1 - I\gamma_2)\tilde{R}.$$

Prove that these satisfy the anticommutation relations

$$\{l, n\} = 4, \quad \{m, \bar{m}\} = 4, \quad \text{all others} = 0.$$

10

Geometry

In the preceding chapters of this book we have dealt entirely with a single geometric interpretation of the elements of a geometric algebra. But the relationship between algebra and geometry is seldom unique. Geometric problems can be studied using a variety of algebraic techniques, and the same algebraic result can typically be pictured in a variety of different ways. In this chapter, we explore a range of alternative geometric systems, and discover how geometric algebra can be applied to each of them. We will find that there is no unique interpretation forced on the multivectors of a given grade. For example, to date we have viewed bivectors solely as directed plane segments. But in projective geometry a bivector represents a line, and in conformal geometry a bivector can represent a pair of points.

Ideas from geometry have always been a prime motivating factor in the development of mathematics. By the nineteenth century mathematicians were familiar with affine, Euclidean, spherical, hyperbolic, projective and inversive geometries. The unifying framework for studying these geometries was provided by the *Kleinian viewpoint*. Under this view a geometry consists of a space of points, together with a group of transformations mapping the points onto themselves. Any property of a particular geometry must be invariant under the action of the associated symmetry group. Klein was thus able to unite various geometries by describing how some symmetry groups are subgroups of larger groups. For example, Euclidean geometry is a subgeometry of affine geometry, because the group of Euclidean transformations is a subgroup of the group of affine transformations.

In this chapter we will see how the various classical geometries, and their associated groups, are handled in geometric algebra. But we will also go further by addressing the question of how to represent various geometric primitives in the most compact and efficient way. The Kleinian viewpoint achieves a united approach to classical geometry, but it does not help much when it comes to addressing problems of how to perform calculations efficiently. For example, circles are as much geometric primitives in Euclidean geometry as points, lines a planes. But how should circles be represented as algebraic entities? Storing a point and a radius is unsatisfactory, as this representation involves objects of different grades. In this chapter we answer this question by showing that both lines and circles are represented as *trivectors* in the conformal model of Euclidean geometry.

We begin with the study of projective geometry. The addition of an extra dimension allows us to create an algebra of incidence relations between points, lines and planes in space. We then return to Euclidean geometry, but rather than viewing this as a subgeometry of projective geometry (the Kleinian viewpoint), we will instead increase the dimension once more to establish a conformal representation of Euclidean geometry. The beauty of this construction is that the group of Euclidean transformations can now be formulated as a rotor group. Euclidean invariants are then constructed as inner products between multivectors. This framework allows us to extend the projective treatment of incidence relations to include circles and spheres.

A further attractive feature of the conformal model is that Euclidean, spherical and hyperbolic geometries are all handled in the same framework. This allows the Poincaré disc model of non-Euclidean geometry in the plane to be extended seamlessly to higher dimensions. Of particular importance is the clarification of the role of complex coordinates in planar non-Euclidean geometry. Much of their utility rests on features of the conformal group of the plane that do not extend naturally. Instead, we work within the framework of *real* geometric algebra to obtain results which are independent of dimension. Finally in this chapter we turn to spacetime geometry. The conformal model for spacetime is of considerable importance in formulations of supersymmetric theories of gravity, and also lies at the heart of the twistor program. We display some surprising links between these ideas and the multiparticle spacetime algebra described in chapter 9. Throughout this chapter we denote the vector space with signature p, q by $\mathcal{V}(p, q)$, and the geometric algebra of this space by $\mathcal{G}(p, q)$.

10.1 Projective geometry

There was a time when projective geometry formed a large part of undergraduate mathematics courses. For various reasons the subject fell out of fashion in the twentieth century, making way for the more relevant subject of differential geometry. But in recent years projective geometry has enjoyed a resurgence due to its importance in the computer graphics industry. For example, the routines at the core of the OpenGL graphics language are built on a projective representation of three-dimensional space.

The key idea in projective geometry is that points in space are represented as

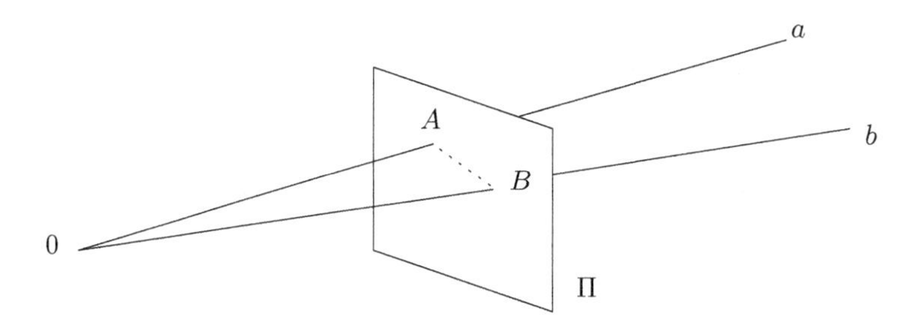

Figure 10.1 *Projective geometry.* Points in the projective plane are represented by vectors in a space one dimension higher. The plane Π does not intersect the origin 0.

vectors in a space of one dimension higher. For example, points in the projective plane are represented as vectors in three-dimensional space (see figure 10.1). The magnitude of the vector is unimportant, as both a and λa represent the same point. This representation of points is said to be *homogeneous.* The two key operations in projective geometry are the *join* and *meet.* The join of two points, for example, is the line between them. Forming the join raises the grade, and the join can usually be encoded algebraically via the exterior product (this was Grassmann's original motivation for introducing his exterior algebra). The meet is used for forming intersections, such as two lines in a plane meeting at a point. The meet is traditionally encoded via the notion of duality, and in geometric algebra the role of the meet is played by the inner product. Operations such as the meet and join do not depend on the metric, so in projective geometry we have a non-metric interpretation of the inner product. This is an important point. Some authors have argued that, because geometric algebra is built on a quadratic form, it is intimately tied to metric geometry. This view is incorrect, as we demonstrate below.

10.1.1 The projective line

The simplest place to start is with a one-dimensional line. The 'Euclidean' model of the line consists of labelling each point with a real number. But there are drawbacks with this representation of a line. Geometrically, all points on the line are equal. But algebraically there are two exceptional points on the line. The first is the origin, which is represented by the algebraically special number zero. The second is the point at infinity, which becomes important when we start to consider projective transformations. The resolution of both of these problems is to represent points in the line as vectors in two-dimensional space. In this way the point x is replaced by a pair of homogeneous coordinates (x_1, x_2), with

$$x = \frac{x_1}{x_2}. \tag{10.1}$$

One can immediately see that the origin is represented by the non-zero vector $(0, 1)$, and that the point at infinity is $(1, 0)$.

If the vectors $\{\mathsf{e}_1, \mathsf{e}_2\}$ denote an orthonormal frame for two-dimensional space, we can set

$$\boldsymbol{x} = x_1 \mathsf{e}_1 + x_2 \mathsf{e}_2. \tag{10.2}$$

The set of all non-zero vectors $\boldsymbol{x}$ constitute the projective line, RP^1. The fact that the origin is excluded implies that in projective spaces one loses linearity. This is obvious from the fact that $\boldsymbol{x}$ and $\lambda\boldsymbol{x}$ represent the same point, so linear combinations do not make geometric sense. Indeed, no geometric significance can be attached to the addition of two points in projective geometry. One cannot form midpoints, for example, as distances and angles are not projective invariants.

The projective group consists of the group of general linear transformations applied to vectors in projective space. For the case of the projective line this group is defined by transformations of the form

$$\begin{pmatrix} x_1 \\ x_2 \end{pmatrix} \mapsto \begin{pmatrix} a & b \\ c & d \end{pmatrix} \begin{pmatrix} x_1 \\ x_2 \end{pmatrix} = \begin{pmatrix} ax_1 + bx_2 \\ cx_1 + dx_2 \end{pmatrix}, \qquad ab - bc \neq 0. \tag{10.3}$$

In terms of points on the line, this transformation corresponds to

$$x \mapsto x' = \frac{ax + b}{cx + d}. \tag{10.4}$$

The group action includes dilations, inversions and translations. The last are obtained for the case $c = 0$, $a/d = 1$. The fact that translations become *linear* transformations in projective geometry is of considerable importance. In three-dimensional geometry, for example, both rotations and translations can be encoded as 4×4 matrices. While this may appear to be an overly-complicated representation, it makes stringing together a series of translations and rotations a straightforward exercise. This is important in computer graphics, and is the representation employed in all OpenGL routines.

In geometric algebra notation we write a general linear transformation as the map $\boldsymbol{x} \mapsto \mathsf{f}(\boldsymbol{x})$, where $\det(\mathsf{f}) \neq 0$. Valid geometric statements in projective geometry must be invariant under such transformations, which is a strong restriction. Inner products between projective vectors (points) are clearly not invariant under projective transformations. The outer product does transform sensibly, however, due to the properties of the outermorphism. For example, suppose that the points α and β are represented projectively by

$$\boldsymbol{a} = \alpha \mathsf{e}_1 + \mathsf{e}_2, \quad \boldsymbol{b} = \beta \mathsf{e}_1 + \mathsf{e}_2. \tag{10.5}$$

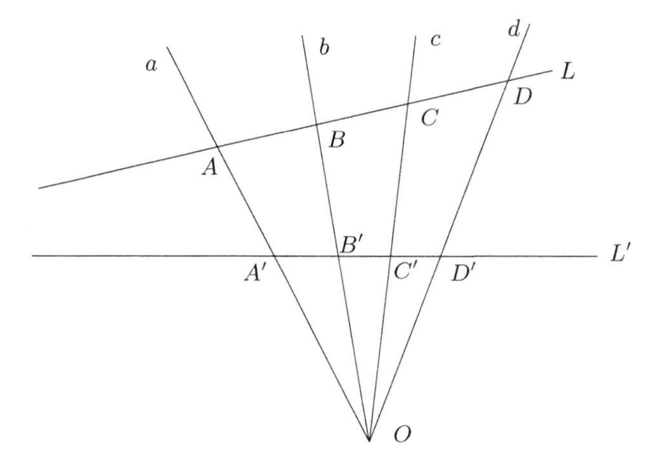

Figure 10.2 *The cross ratio.* Points on the lines L and L' represent two different projective views of the same vectors in space. The cross ratio of the four points is the same on both lines.

The outer product of these is

$$\boldsymbol{a}\wedge\boldsymbol{b} = (\alpha - \beta)\mathsf{e}_1\wedge\mathsf{e}_2, \tag{10.6}$$

which is controlled by the distance between the points on the line. Under a projective transformation in two dimensions

$$\mathsf{e}_1\wedge\mathsf{e}_2 \mapsto \mathsf{f}(\mathsf{e}_1\wedge\mathsf{e}_2) = \det\,(\mathsf{f})\,\mathsf{e}_1\wedge\mathsf{e}_2, \tag{10.7}$$

which is just an overall scaling.

The fact that distances between points are scaled under a projective transformation provides us with an important projective invariant for four points on a line. This is formed from ratios of lengths along a line. We must further ensure that the ratio is invariant under individual rescaling of individual vectors to be a true projective invariant. We therefore define the *cross ratio* of four points, A, B, C, D, by

$$(ABCD) = \frac{AC}{BC}\frac{BD}{AD} = \frac{\boldsymbol{a}\wedge\boldsymbol{c}}{\boldsymbol{b}\wedge\boldsymbol{c}}\frac{\boldsymbol{b}\wedge\boldsymbol{d}}{\boldsymbol{a}\wedge\boldsymbol{d}}, \tag{10.8}$$

where AB denotes the distance between A and B. Given any four points on a line, their cross ratio is a projective invariant (see figure 10.2). The figure illustrates one possible geometric interpretation of a projective transformation, which is that the line onto which points are projected is transformed to a new line. Invariants such as the cross ratio are important in computer vision where, for example, we seek to extract three-dimensional information from a series of two-dimensional scenes. Knowledge of invariants can help establish point matches between the scenes.

10.1.2 The projective plane

Rather more interesting than the case of a line is that of the projective plane. Points in the plane are now represented by vectors in the three-dimensional algebra $\mathcal{G}(3,0)$. Figure 10.1 shows that the line between the points a and b is the result of projecting the plane defined by a and b onto the projective plane. We therefore define the *join* of the points a and b by

$$\text{join}(a,\, b) = a \wedge b. \tag{10.9}$$

Bivectors thus define lines in projective geometry. The line itself is recovered by solving the equation

$$a \wedge b \wedge x = 0. \tag{10.10}$$

This equation is solved by

$$x = \lambda a + \mu b, \tag{10.11}$$

which defines the set of projective points on the line joining A and B.

By taking exterior products of vectors we define (projectively) higher dimensional objects. For example, the join of a point a and a line $b \wedge c$ is the plane defined by the trivector $a \wedge b \wedge c$. Three points on a line cannot define a projected area, so for these we must have

$$a \wedge b \wedge c = 0 \quad \Rightarrow \quad a,\, b,\, c \text{ collinear.} \tag{10.12}$$

This was the condition used to recover the points x on the line $a \wedge b$. The join itself can be slightly more problematic. Given three points one cannot just write that their join is $a \wedge b \wedge c$, as the result may be zero. Instead the join is defined as the smallest subspace containing a, b and c. If they are collinear, then the join is the common line. This is well defined mathematically, but is hard to encode computationally. The problem is that the finite precision used on computers means that testing for zero is unreliable. Wherever possible it is safer to avoid defining the join and instead work with the exterior product.

Projective geometry deals with relationships that are invariant under projective transformations. The join is one such concept — as two points are transformed the line joining them transforms in the obvious way:

$$a \wedge b \mapsto \mathsf{f}(a) \wedge \mathsf{f}(b) = \mathsf{f}(a \wedge b). \tag{10.13}$$

So, for example, the statement that three points lie on a line ($a \wedge b \wedge c = 0$) is unchanged by a projective transformation. Similarly, the statement that three lines intersect at a point must also be a projective invariant. We therefore seek an algebraic encoding of the intersection of two lines. This is the called the *meet*, usually denoted with the $\vee$ symbol. Before we can encode this, however, we need to define the dual. In the projective plane, points and lines are represented as vectors and bivectors in $\mathcal{G}(3,0)$. We know that these can be interchanged via a duality transformation, which amounts to multiplying by the pseudoscalar I. In this way every point has a dual line, and vice versa. The geometric picture associated with duality depends on the embedding plane.

If we denote the dual of A by A^*, the meet $A \vee B$ is defined by the 'de Morgan' rule

$$(A \vee B)^* = A^* \wedge B^*. \tag{10.14}$$

For a pair of lines in a plane, this amounts to

$$A \vee B = -I(IA)\wedge(IB) = I\, A \times B = A \cdot (IB) = (IA) \cdot B. \tag{10.15}$$

These formulae show how the inner product can be used to encode the meet, without imposing a metric on projective space. The expression

$$A \vee B = I\, A \times B \tag{10.16}$$

shows how the construction works. In three dimensions, $A \times B$ is the plane perpendicular to A and B, and $I\, A \times B$ is the line perpendicular to this plane, through the origin. This is therefore the line common to both planes, so projectively gives the point of intersection of two lines.

The meet of two distinct lines in a plane always results in a non-zero point. If the lines are parallel then their meet returns the point at infinity. Parallelism is not a projective invariant, however, so under a projective transformation two parallel lines can transform to lines intersecting at a finite point. This illustrates the fact that the point at infinity does not necessarily stay at infinity under projective transformations. It is instructive to see how the meet itself transforms under a projective transformation. Using the results of section 4.4, we find that

$$\begin{aligned} A \vee B \mapsto \mathsf{f}(A) \vee \mathsf{f}(B) &= I\, \big(I\mathsf{f}(A)\big) \wedge \big(I\mathsf{f}(B)\big) \\ &= \det\,(\mathsf{f})^2\, I\, \bar{\mathsf{f}}^{-1}(IA) \wedge \bar{\mathsf{f}}^{-1}(IB) \\ &= \det\,(\mathsf{f})^2\, I\, \bar{\mathsf{f}}^{-1}\big((IA)\wedge(IB)\big) \\ &= \det\,(\mathsf{f})\, \mathsf{f}\big(I\,(IA)\wedge(IB)\big). \end{aligned} \tag{10.17}$$

We can summarise this result as

$$\mathsf{f}(A) \vee \mathsf{f}(B) = \det\,(\mathsf{f})\, \mathsf{f}(A \vee B). \tag{10.18}$$

But in projective geometry, a and λa represent the same point, so the factor of $\det\,(\mathsf{f})$ does not affect the resulting point. This confirms that under a projective transformation the meet transforms as required.

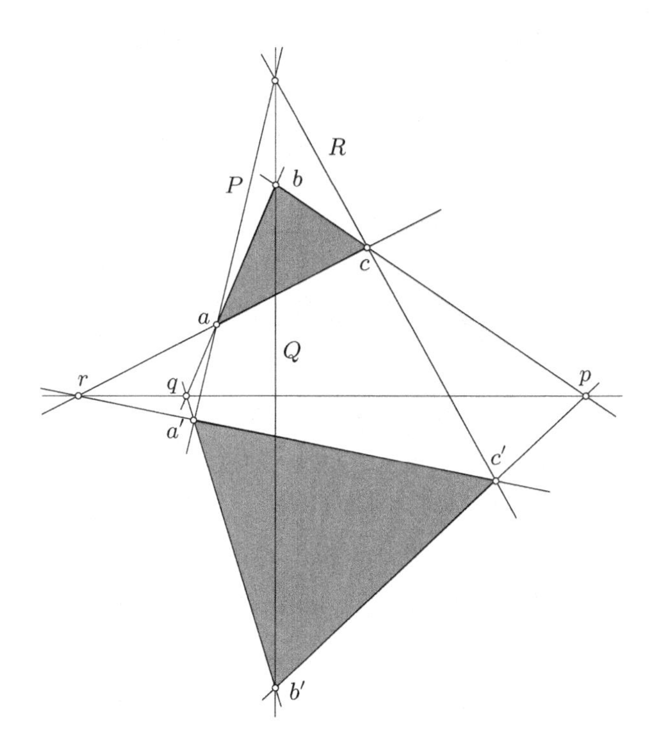

Figure 10.3 *Desargues' theorem.* The lines P, Q, R meet at a point if and only if the points p, q, r lie on a line. The two triangles are then projectively related.

The condition that three lines meet at a common point requires that the meet of two lines lies on a third line, which goes as

$$(A \vee B)\wedge C = (I\, A\times B)\wedge C = 0. \tag{10.19}$$

Dualising this result we obtain the condition

$$\langle (A\times B)C\rangle = \langle ABC\rangle = 0, \quad \Rightarrow \quad A,\ B,\ C \text{ coincident.} \tag{10.20}$$

This is an extremely simple algebraic encoding of the statement that three lines (represented by bivectors) all meet at a common point. Equations like this demonstrate how powerful geometric algebra can be when applied in a projective setting.

As an application consider Desargues' theorem, which is illustrated in figure 10.3. The points a, b, c and a', b', c' define two triangles. The associated lines are defined by

$$A = b\wedge c, \quad B = c\wedge a, \quad C = a\wedge b, \tag{10.21}$$

with the same definitions holding for A', B', C' in terms of a', b', c'. The two sets of vertices determine the lines

$$P = a\wedge a', \quad Q = b\wedge b', \quad R = c\wedge c', \tag{10.22}$$

and the two sets of lines determine the points

$$p = A\times A'\, I, \quad q = B\times B'\, I, \quad r = C\times C'\, I. \tag{10.23}$$

Desargues' theorem states that, if p, q, r lie on a common line, then P, Q and R all meet at a common point. The latter condition requires

$$\langle PQR\rangle = \langle a\wedge a'\, b\wedge b'\, c\wedge c'\rangle = 0. \tag{10.24}$$

Similarly, for p, q, r to fall on a line we form

$$\begin{aligned} p\wedge q\wedge r &= \langle A\times A'\, I\, B\times B'\, I\, C\times C'\, I\rangle_3 \\ &= -I\langle A\times A'\, B\times B'\, C\times C'\rangle. \end{aligned} \tag{10.25}$$

Desargues' theorem is then proved by the algebraic identity

$$\langle a\wedge b\wedge c\, a'\wedge b'\wedge c'\rangle\langle a\wedge a'\, b\wedge b'\, c\wedge c'\rangle = \langle A\times A'\, B\times B'\, C\times C'\rangle, \tag{10.26}$$

the proof of which is left as an exercise. The left-hand side vanishes if and only if the lines P, Q, R meet at a point. The right-hand side vanishes if and only if the points p, q, r lie on a line. This proves the theorem. The complex geometry illustrated in figure 10.3 has therefore been reduced to a straightforward algebraic identity.

We can find a simple generalisation of the cross ratio for the case of the projective plane. From the derivation of the cross ratio, it is clear that any analogous object for the plane must involve ratios of trivectors. These represent areas in the projective plane. For example, suppose we have six points in space with position vectors $a_1, \ldots, a_6$. These produce the six projected points $A_1, \ldots, A_6$. An invariant is formed by

$$\frac{a_5\wedge a_4\wedge a_3}{a_5\wedge a_1\wedge a_3}\frac{a_6\wedge a_2\wedge a_1}{a_6\wedge a_2\wedge a_4} = \frac{A_{543}}{A_{513}}\frac{A_{621}}{A_{624}}, \tag{10.27}$$

where A_{ijk} is the projected area of the triangle with vertices A_i, A_j, A_k. Again, elementary algebraic reasoning quickly yields a geometrically significant result.

10.1.3 Homogeneous coordinates and projective splits

In typical applications of projective geometry we are interested in the relationship between coordinates in an image plane (for example in terms of pixels relative to some origin) and the three-dimensional position vector. Suppose that the origin in the image plane is defined by the vector n, which is perpendicular to the plane. The line on the image plane from the origin to the image point is represented by the bivector $a\wedge n$ (see figure 10.4) . The vector OA belongs to a two-dimensional

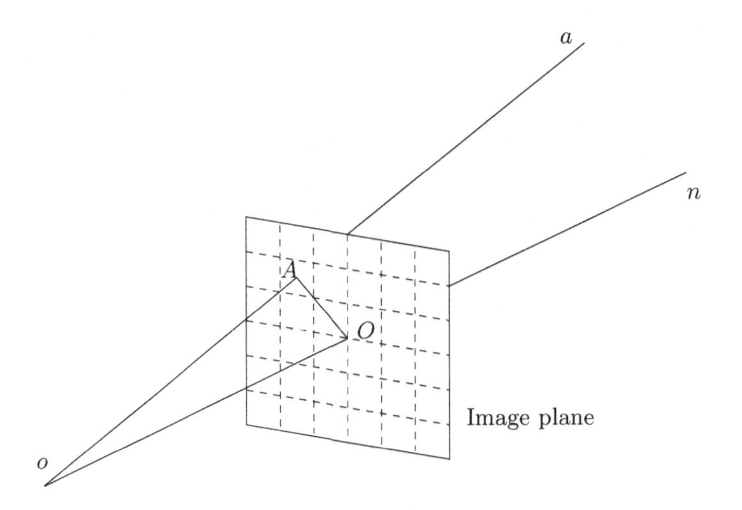

Figure 10.4 *The image plane.* Vectors in the image plane, OA, are described by bivectors in $\mathcal{G}(3,0)$. The point A can be expressed in terms of *homogeneous* coordinates in the image plane.

geometric algebra. We can relate this directly to the three-dimensional algebra by first writing

$$n + OA = \lambda a. \tag{10.28}$$

Contracting with n, we find that $\lambda = n^2(a\cdot n)^{-1}$. It follows that

$$OA = \frac{a\,n^2 - a\cdot n\,n}{a\cdot n} = \frac{a\wedge n}{a\cdot n}n. \tag{10.29}$$

If we now drop the final factor of n, we obtain a bivector that is homogeneous in both a and n. In this way we can directly represent the line OA in two dimensions with the bivector

$$A = \frac{a\wedge n}{a\cdot n}. \tag{10.30}$$

This is the *projective split*, first introduced in chapter 5 as a means of relating physics as seen by observers with different velocities.

The map of equation (10.30) relates bivectors in a higher dimensional space to vectors in a space of dimension one lower. If we introduce a coordinate frame $\{\mathsf{e}_i\}$, with e_3 in the n direction, we see that the coordinates of the image of $a = a_i\mathsf{e}_i$ are

$$A = \frac{a_1}{a_3}\mathsf{e}_1\mathsf{e}_3 + \frac{a_2}{a_3}\mathsf{e}_2\mathsf{e}_3 = A_1\boldsymbol{E}_1 + A_2\boldsymbol{E}_2. \tag{10.31}$$

This equation defines the *homogeneous coordinates* A_i:

$$A_i = \frac{a_i}{a_3}. \tag{10.32}$$

Homogeneous coordinates are independent of scale and it is these that are usually measured in a camera projection of a scene. The bivectors $(\boldsymbol{E}_1, \boldsymbol{E}_2)$ act as generators for a two-dimensional geometric algebra. If the vectors in the projective space are all Euclidean, the $\boldsymbol{E}_i$ bivectors will have negative square. If necessary, this can be avoided by letting e_3 be an anti-Euclidean vector. The projective split is an elegant scheme for relating results in projective space to Euclidean space one dimension lower. Algebraically, the projective split rests on the isomorphism

$$\mathcal{G}^+(p+1, q) \simeq \mathcal{G}(q, p). \tag{10.33}$$

This states that the even subalgebra of the geometric algebra with signature $(p+1, q)$ is isomorphic to the algebra with signature (q, p). The projective split is not always the best way to map from projective space back to Euclidean space, however, as constructing a set of bivectors can be an unnecessary complication. Often it is simpler to choose an orthonormal frame, with n one of the frame vectors, and then scale all vectors x such that $n \cdot x = 1$.

10.1.4 Projective geometry in three dimensions

To handle complicated three-dimensional problems in a projective framework we require a four-dimensional geometric algebra. The basic elements of four-dimensional geometric algebra will be familiar from relativity and the spacetime algebra, though now the elements are given a projective interpretation. The algebra of a four-dimensional space contains six bivectors, which represent lines in three dimensions. As in the planar case, the important feature of the projective framework is that we are free from the restriction that all lines pass through the origin. The line through the points a and b is again represented by the bivector $a \wedge b$. This is a *blade*, as must be the case for any bivector representing a line. Any bivector blade $B = a \wedge b$ must satisfy the algebraic condition

$$B \wedge B = a \wedge b \wedge a \wedge b = 0, \tag{10.34}$$

which removes one degree of freedom from the six components needed to specify an arbitrary bivector. This is known at the Plücker condition. If the vector e_4 defines the projection into Euclidean space, the line $a \wedge b$ has coordinates

$$a \wedge b = (\boldsymbol{a} + \mathsf{e}_4) \wedge (\boldsymbol{b} + \mathsf{e}_4) = \boldsymbol{a} \wedge \boldsymbol{b} + (\boldsymbol{a} - \boldsymbol{b}) \wedge \mathsf{e}_4, \tag{10.35}$$

where $\boldsymbol{a}$ and $\boldsymbol{b}$ denote vectors in the three-dimensional space. The bivector B therefore encodes a line as a combination of a tangent $(\boldsymbol{b} - \boldsymbol{a})$ and a moment $\boldsymbol{a} \wedge \boldsymbol{b}$. These are the Plücker coordinates for a line.

Given two lines as bivectors B and B', the test that they intersect in three dimensions is that their join does not span all of projective space, which implies that

$$B \wedge B' = 0. \tag{10.36}$$

This provides a projective interpretation for commuting bivectors in four dimensions. Commuting (orthogonal) bivectors have BB' equalling a multiple of the pseudoscalar. Projectively, these can be interpreted as two lines in three dimensions that do not share a common point. As mentioned earlier, the problem with a test such as equation (10.36) is that one can never guarantee to obtain zero when working to finite numerical precision. In practice, then, one tends to avoid trying to find the intersection of two lines in the three dimensions, unless there is good reason to believe that they intersect at a point.

The exterior product of three vectors in projective space results in the trivector encoding the plane containing the three points. One of the most frequently encountered problems is finding the point of intersection of a line L and a plane P. This is given by

$$x = P\cdot(IL), \tag{10.37}$$

where I is the four-dimensional pseudoscalar. This will always return a point, provided the line does not lie entirely in the plane. Similarly, the intersection of two planes in three dimensions must result in a line. Algebraically, this line is encoded by the bivector

$$L = (IP_1)\cdot P_2 = I\, P_1 \times P_2, \tag{10.38}$$

where P_1 and P_2 are the two planes. Such projective formulae are important in computer vision and graphics applications.

10.2 Conformal geometry

Projective geometry does provide an efficient framework for handling Euclidean geometry. Euclidean geometry is a subgeometry of projective geometry, so any valid result in the latter must hold in the former. But there are some limitations to the projective viewpoint. Euclidean concepts, like lengths and angles, are not straightforwardly encoded, and the related concepts of circles and spheres are equally awkward. Conformal geometry provides an elegant solution to this problem. The key is to introduce a further dimension of opposite signature, so that points in a space of signature (p, q) are modelled as null vectors in a space of signature $(p+1, q+1)$. That is, points in $\mathcal{V}(p, q)$ are represented by null vectors in $\mathcal{V}(p+1, q+1)$. Projective geometry is retained as a subset of conformal geometry, but the range of geometric primitives is extended to include circles and spheres.

We denote a point in $\mathcal{V}(p, q)$ by x, and its conformal representation by X. We continue to employ the spacetime notation of using the tilde symbol to denote

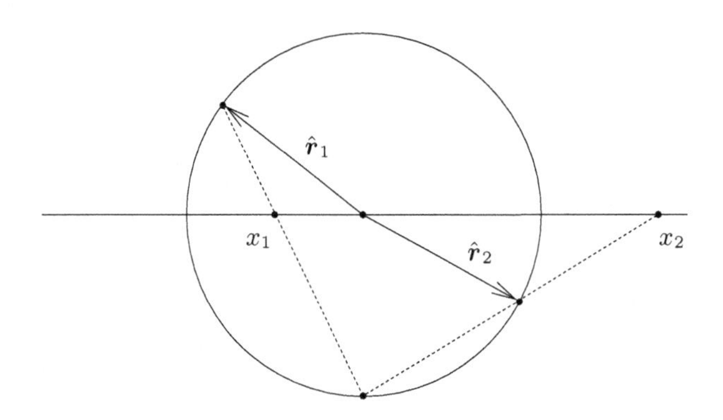

Figure 10.5 *A stereographic projection.* The line is mapped into the unit circle, so the points on the line x_1 and x_2 are mapped to the unit vectors $\hat{r}_1$ and $\hat{r}_2$. The origin and infinity are mapped to opposite points on the circle.

the reverse operation for a general multivector in any geometric algebra. A basis set of vectors for $\mathcal{G}(p,q)$ is denoted by $\{e_i\}$, and the two additional vectors $\{e, \bar{e}\}$ complete this to an orthonormal basis for $\mathcal{G}(p+1, q+1)$.

10.2.1 Stereographic projection of a line

We illustrate the general construction by starting with the simple case of a line. In projective geometry points on a line are modeled as two-dimensional vectors. The conformal model is established from a slightly different starting point, using the *stereographic projection.* Under a stereographic projection, points on a line are mapped to the unit circle in a plane (see figure 10.5). Points on the unit circle in two dimensions are represented by

$$\hat{r} = \cos(\theta)\, e_1 + \sin(\theta)\, e_2. \tag{10.39}$$

The corresponding point on the line is given by

$$x = \frac{\cos(\theta)}{1 + \sin(\theta)}. \tag{10.40}$$

This relation inverts simply to give

$$\cos(\theta) = \frac{2x}{1+x^2}, \quad \sin(\theta) = \frac{1-x^2}{1+x^2}. \tag{10.41}$$

So far we have achieved a representation of the line in terms of a circle in two dimensions. But the constraint that the vector has unit magnitude means that we have lost homogeneity. To get round this we introduce a third vector, $\bar{e}$, which has negative signature,

$$\bar{e}^2 = -1, \tag{10.42}$$

and we assume that $\bar{e}$ is orthogonal to e_1 and e_2. We can now replace the unit vector $\hat{\boldsymbol{r}}$ with the null vector X, where

$$X = \cos(\theta)\, e_1 + \sin(\theta)\, e_2 + \bar{e} = \frac{2x}{1+x^2} e_1 + \frac{1-x^2}{1+x^2} e_2 + \bar{e}. \tag{10.43}$$

The vector X satisfies $X^2 = 0$, so is null.

The equation $X^2 = 0$ is homogeneous. If it is satisfied for X, it is satisfied for λX. We can therefore move to a homogeneous representation and let both X and λX represent the same point. Multiplying by $(1 + x^2)$ we establish the conformal representation

$$X = 2xe_1 + (1 - x^2)e_2 + (1 + x^2)\bar{e}. \tag{10.44}$$

This is the basic representation we use throughout. To establish a more general notation we first replace the vector e_2 by $-e$. We therefore have

$$e^2 = 1, \qquad \bar{e}^2 = -1, \qquad e\cdot\bar{e} = 0. \tag{10.45}$$

The vectors e and $\bar{e}$ are then the two extra vectors that extend the space $\mathcal{V}(p, q)$ to $\mathcal{V}(p+1, q+1)$. Frequently, it is more convenient to work with a null basis for the extra dimensions. We define

$$n = e + \bar{e}, \qquad \bar{n} = e - \bar{e}. \tag{10.46}$$

These vectors satisfy

$$n^2 = \bar{n}^2 = 0, \qquad n\cdot\bar{n} = 2. \tag{10.47}$$

The vector X is now

$$X = 2xe_1 + x^2 n - \bar{n}. \tag{10.48}$$

It is straightforward to confirm that this is a null vector. The set of all null vectors in this space form a cone, and the real number line is modelled by the intersection of this cone and a plane. The construction is illustrated in figure 10.6.

10.2.2 Conformal model of Euclidean space

The form of equation (10.48) generalises easily. If x is an element of $\mathcal{V}(p, q)$, we set

$$F(x) = X = x^2 n + 2x - \bar{n}, \tag{10.49}$$

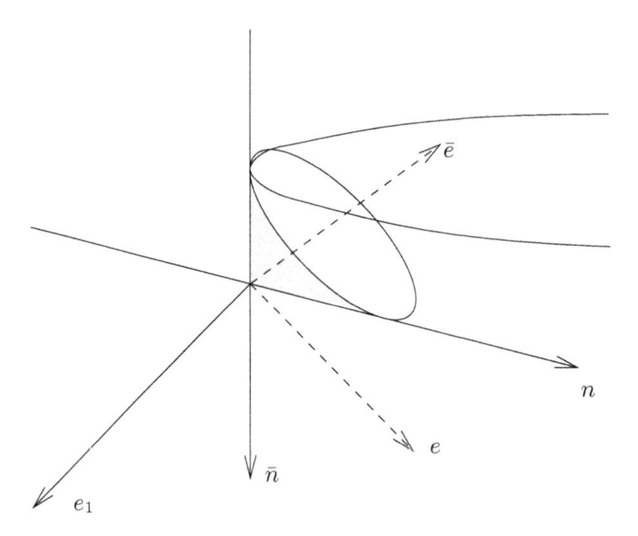

Figure 10.6 *The conformal model of a line.* Points on the line are represented by null vectors in three dimensions. These lie on a cone, and the intersection of the cone with a plane recovers the point.

which is a null vector in $\mathcal{V}(p+1, q+1)$. This vector can be obtained simply via the map,

$$F(x) = -(x-e)n(x-e), \tag{10.50}$$

which is a reflection of the null vector n in the plane perpendicular to $(x-e)$. The result must therefore be a new null vector. The presence of the vector e removes any ambiguity in handling the origin $x = 0$. The map $F(x)$ is non-linear so, as with projective geometry, we move to a non-linear representation of points in conformal geometry.

More generally, any null vector in $\mathcal{V}(p+1, q+1)$ can be written as

$$X = \lambda(x^2 n + 2x - \bar{n}), \tag{10.51}$$

with λ a scalar. This provides a projective map between $\mathcal{V}(p+1, q+1)$ and $\mathcal{V}(p, q)$. The family of null vectors, $\lambda(x^2 n + 2x - \bar{n})$, in $\mathcal{V}(p+1, q+1)$ correspond to the single point $x \in \mathcal{V}(p, q)$. Given an arbitrary null vector X, it is frequently useful to convert it to the standard form of equation (10.49). This is achieved by setting

$$X \mapsto -2\frac{X}{X \cdot n}. \tag{10.52}$$

This map is similar to that employed in constructing a standard embedding in projective geometry. The status of the vector n is clear here — it represents the point at infinity.

Given two null vectors X and Y, in standard form, their inner product is

$$X \cdot Y = (x^2 n + 2x - \bar{n}) \cdot (y^2 n + 2y - \bar{n})$$
$$= -2x^2 - 2y^2 + 4x \cdot y$$
$$= -2(x-y)^2. \tag{10.53}$$

This result is of fundamental importance to the conformal model of Euclidean geometry. The inner product in conformal space encodes the *distance* between points in Euclidean space. It follows that any transformation of null vectors in $\mathcal{V}(p+1, q+1)$ which leaves inner products invariant can correspond to a transformation in $\mathcal{V}(p,q)$ which leaves angles and distances invariant. In the next section we discuss these transformations in detail.

10.3 Conformal transformations

The study of the main geometric primitives in conformal geometry is simplified by first understanding the nature of the conformal group. For points x, y in $\mathcal{V}(p,q)$ the definition of a conformal transformation is that it leaves angles invariant. So, if f is a map from $\mathcal{V}(p,q)$ to itself, then f is a conformal transformation if

$$\mathsf{f}(a) \cdot \mathsf{f}(b) = \lambda a \cdot b, \quad \forall a, b \in \mathcal{V}(p,q), \tag{10.54}$$

where

$$\mathsf{f}(a) = a \cdot \nabla f(x). \tag{10.55}$$

While $\mathsf{f}(a)$ is a linear map at each point x, the conformal transformation $f(x)$ is not restricted to being linear. Conformal transformations form a group, the conformal group, the main elements of which are translations, rotations, dilations and inversions. We now study each of these in turn.

10.3.1 Translations

To begin, consider the fundamental operation of translation in the space $\mathcal{V}(p,q)$. This is *not* a linear operation in $\mathcal{V}(p,q)$, but does become linear in the projective framework. In the conformal model we achieve a further refinement, as translations can now be handled by rotors. Consider the rotor

$$R = T_a = \mathrm{e}^{na/2}, \tag{10.56}$$

where $a \in \mathcal{V}(p,q)$, so that $a \cdot n = 0$. The generator for the rotor is a null bivector, so the Taylor series for T_a terminates after two terms:

$$T_a = 1 + \frac{na}{2}. \tag{10.57}$$

The rotor T_a transforms the null vectors n and $\bar{n}$ into

$$T_a n \tilde{T}_a = n + \tfrac{1}{2}nan + \tfrac{1}{2}nan + \tfrac{1}{4}nanan = n \tag{10.58}$$

and

$$T_a \bar{n} \tilde{T}_a = \bar{n} - 2a - a^2 n. \tag{10.59}$$

Acting on a vector $x \in \mathcal{V}(p,q)$ we similarly obtain

$$T_a x \tilde{T}_a = x + n(a\cdot x). \tag{10.60}$$

Combining these we find that

$$\begin{aligned} T_a F(x) \tilde{T}_a &= x^2 n + 2(x + a\cdot x\, n) - (\bar{n} - 2a - a^2 n) \\ &= (x+a)^2 n + 2(x+a) - \bar{n} \\ &= F(x+a), \end{aligned} \tag{10.61}$$

which performs the conformal version of the translation $x \mapsto x + a$. Translations are handled as rotations in conformal space, and the rotor group provides a double-cover representation of a translation. The identity

$$\tilde{T}_a = T_{-a} \tag{10.62}$$

ensures that the inverse transformation in conformal space corresponds to a translation in the opposite direction, as required.

10.3.2 Rotations

Next, suppose that we rotate the vector x about the origin in $\mathcal{V}(p,q)$. This is achieved with the rotor $R \in \mathcal{G}(p,q)$ via the familiar transformation $x \mapsto x' = Rx\tilde{R}$. The image of the transformed point is

$$\begin{aligned} F(x') &= x'^2 n + 2Rx\tilde{R} - \bar{n} \\ &= R(x^2 n + 2x - \bar{n})\tilde{R} = RF(x)\tilde{R}. \end{aligned} \tag{10.63}$$

This holds because R is an even element in $\mathcal{G}(p,q)$, so must commute with both n and $\bar{n}$. Rotations about the origin therefore take the same form in either space.

Suppose instead that we wish to rotate about the point $a \in \mathcal{V}(p,q)$. This can be achieved by translating a to the origin, rotating and then translating forward again. In terms of $X = F(x)$ the result is

$$X \mapsto T_a R T_{-a} X \tilde{T}_{-a} \tilde{R} \tilde{T}_a = R' X \tilde{R}'. \tag{10.64}$$

The rotation is now controlled by the rotor

$$R' = T_a R \tilde{T}_a = \left(1 + \frac{na}{2}\right) R \left(1 + \frac{an}{2}\right). \tag{10.65}$$

So, as expected, the conformal model has freed us from treating the origin as a special point. Rotations about any point are handled in the same manner, and are still generated by a bivector blade. Similar observations hold for reflections, but we delay a full treatment of these until we have described how lines and surfaces are handled in the conformal model. The preceding formulae for translations and rotations form the basis of the subject of *screw theory*, which has its origins in the nineteenth century.

10.3.3 Inversions

Rotations and translations are elements of the Euclidean group, as they leave distances between points invariant. This is a subgroup of the larger conformal group, which only leaves angles invariant. The conformal group essentially contains two further transformations: inversions and dilations. An inversion in the origin consists of the map

$$x \mapsto \frac{x}{x^2}. \tag{10.66}$$

The conformal vector corresponding to the inverted point is

$$F(x^{-1}) = x^{-2}n + 2x^{-1} - \bar{n} = \frac{1}{x^2}(n + 2x - x^2\bar{n}). \tag{10.67}$$

But in conformal space points are represented homogeneously, so the pre-factor of x^{-2} can be ignored. In conformal space an inversion in the origin consists solely of the map

$$n \mapsto -\bar{n}, \qquad \bar{n} \mapsto -n. \tag{10.68}$$

This is generated by a reflection in e, since

$$-ene = -ee\bar{n} = -\bar{n}. \tag{10.69}$$

We can therefore write

$$-eF(x)e = x^2 F(x^{-1}), \tag{10.70}$$

which shows that inversions in $\mathcal{V}(p,q)$ are represented as reflections in the conformal space $\mathcal{V}(p+1,q+1)$. As both X and $-X$ are homogeneous representations of the same point, it is irrelevant whether we take $-e(\ldots)e$ or $e(\ldots)e$ as the reflection. In the following we will use $e(...)e$ for convenience.

A reflection in e corresponds to an inversion in the origin in Euclidean space. To find the generator of an inversion in an arbitrary point a, we translate to the origin, invert and translate forward again. The resulting generator is then

$$T_a e T_{-a} = \left(1 + \frac{na}{2}\right) e \left(1 + \frac{an}{2}\right) = e - a - \frac{a^2}{2}n. \tag{10.71}$$

Now, recalling that $e = (n + \bar{n})/2$, the generating vector can also be written as

$$T_a e T_{-a} = \tfrac{1}{2}\big(n - F(a)\big) = \tfrac{1}{2}(n - A). \tag{10.72}$$

A reflection in $(n - F(a))$ therefore achieves an inversion about the point a in Euclidean space. As with translations, a nonlinear transformation in Euclidean space has been linearised by moving to a conformal representation of points. The generator of an inversion is a vector with positive square. In section 10.5.1 we see how these vectors are related to circles and spheres.

10.3.4 Dilations

A dilation in the origin is given by

$$x \mapsto x' = \mathrm{e}^{-\alpha} x, \tag{10.73}$$

where α is a scalar. Clearly, this transformation does not alter angles, so is a conformal transformation. The null vector corresponding to the transformed point is

$$F(x') = \mathrm{e}^{-\alpha}(x^2 \mathrm{e}^{-\alpha} n + 2x + \mathrm{e}^{\alpha}\bar{n}). \tag{10.74}$$

Clearly the map we need to achieve is

$$n \mapsto \mathrm{e}^{-\alpha} n, \qquad \bar{n} \mapsto \mathrm{e}^{\alpha}\bar{n}. \tag{10.75}$$

This transformation does not alter the inner product of n and $\bar{n}$, so can be represented with a rotor. As the vector x is unchanged, the rotor can only be generated by the timelike bivector $e\bar{e}$. If we set

$$N = e\bar{e} = \tfrac{1}{2}\bar{n}\wedge n \tag{10.76}$$

then N satisfies

$$Nn = -n = -nN, \qquad N\bar{n} = \bar{n} = -\bar{n}N, \qquad N^2 = 1. \tag{10.77}$$

We now introduce the rotor

$$D_\alpha = \mathrm{e}^{\alpha N/2} = \cosh(\alpha/2) + \sinh(\alpha/2)\, N. \tag{10.78}$$

This rotor satisfies

$$\begin{aligned} D_\alpha n \tilde{D}_\alpha &= \mathrm{e}^{-\alpha} n, \\ D_\alpha \bar{n} \tilde{D}_\alpha &= \mathrm{e}^{\alpha} \bar{n} \end{aligned} \tag{10.79}$$

and so carries out the required transformation. We can therefore write

$$F(\mathrm{e}^{-\alpha} x) = \mathrm{e}^{-\alpha} D_\alpha F(x) \tilde{D}_\alpha, \tag{10.80}$$

which confirms that a dilation in the origin is represented by a simple rotor in conformal space. To achieve a dilation about an arbitrary point a we form

$$D'_\alpha = T_a D_\alpha \tilde{T}_a = \mathrm{e}^{\alpha N'/2}, \tag{10.81}$$

where the generator is now

$$N' = T_a N\tilde{T}_a = \tfrac{1}{2}T_a\bar{n}\wedge n\tilde{T}_a = -\tfrac{1}{2}A\wedge n, \tag{10.82}$$

with $A = F(a)$. A dilation about a is therefore generated by

$$D'_\alpha = \exp(-\alpha A\wedge n/4) = \exp\left(\frac{\alpha}{2}\frac{A\wedge n}{A\cdot n}\right). \tag{10.83}$$

The generator is governed by two null vectors, one for the point about which the dilation is performed and one for the point at infinity.

10.3.5 Special conformal transformations

A special conformal transformation consists of an inversion in the origin, a translation and a further inversion in the origin. We can therefore handle these in terms of the representations we have already established. In Euclidean space the effect of a conformal transformation can be written as

$$x \mapsto \frac{x + ax^2}{1 + 2a\cdot x + a^2x^2} = x\frac{1}{1 + ax} = \frac{1}{1 + xa}x. \tag{10.84}$$

The final expressions confirm that a special conformal transformation corresponds to a position-dependent rotation and dilation in Euclidean space, so does leave angles unchanged. To construct the equivalent rotor in $\mathcal{G}(p+1, q+1)$ we form

$$K_a = eT_ae = 1 - \frac{\bar{n}a}{2}, \tag{10.85}$$

which ensures that $K_aF(x)\tilde{K}_a$ is a special conformal transformation. Explicitly, we have

$$F\left(x\frac{1}{1 + ax}\right) = (1 + 2a\cdot x + a^2x^2)^{-1}K_aF(x)\tilde{K}_a \tag{10.86}$$

and again we can ignore the pre-factor and use $K_aF(x)\tilde{K}_a$ as the homogeneous representation of the result of a special conformal transformation.

10.3.6 Euclidean transformations

The group of Euclidean transformations is a subgroup of the full conformal group. The additional restriction is that lengths as well as angles are invariant. Equation (10.53) showed that the inner product of two null vectors is related to the Euclidean distance between the corresponding points. To establish a homogeneous formula, we must write

$$|a - b|^2 = -2\frac{A\cdot B}{A\cdot n\, B\cdot n}, \tag{10.87}$$

which is homogeneous on A and B. The Euclidean group can now be seen to be the subgroup of the conformal group which leaves n invariant. This is sensible, as the point at infinity should stay there under a Euclidean transformation. The Euclidean group is thus the *stability group* of a null vector in conformal space. The group of generators of reflections and rotations in conformal space which leave n invariant then provide a double cover of the Euclidean group. Equation (10.87) returns the Euclidean distance between points. If the vector n is replaced by e or $\bar{e}$ we can transform to distance measures in hyperbolic or spherical geometry. This makes it a simple exercise to attach different geometric pictures to algebraic results in conformal space.

10.4 Geometric primitives in conformal space

Now that we have seen how points are encoded in conformal space, we can begin to build up more complex geometric objects. As in projective geometry, we expect that a multivector blade L will encode a geometric object via the equation

$$L\wedge X = 0, \qquad X^2 = 0. \tag{10.88}$$

The question, then, is what type of object does each grade of multivector return. One important result we can exploit is that $X^2 = 0$ is unchanged if $X \mapsto RX\tilde{R}$. So, if a geometric object is specified by L via equation (10.88), it follows that

$$R(L\wedge X)\tilde{R} = (RL\tilde{R})\wedge(RX\tilde{R}) = 0. \tag{10.89}$$

We can therefore transform the object L with a general element of the conformal group to obtain a new object. Similar considerations hold for incidence relations. Since conformal transformations only preserve angles, and do not necessarily map straight lines to straight lines, the range of objects we can describe by simple blades is clearly going to be larger than in projective geometry.

10.4.1 Bivectors and points

A pair of points in Euclidean space are represented by two null vectors in a space of two dimensions higher. We know that the inner product in this space returns information about distances. The next question to ask is what is the significance of the outer product of two vectors. If A and B are null vectors, we form the bivector

$$G = A\wedge B. \tag{10.90}$$

The bivector G has magnitude

$$G^2 = (AB - A\cdot B)(-BA + A\cdot B) = (A\cdot B)^2, \tag{10.91}$$

which shows that G is *timelike*, borrowing the terminology of special relativity. It follows that G contains a pair of null vectors. If we look for solutions to the equation

$$G\wedge X = 0, \qquad X^2 = 0, \tag{10.92}$$

the only solutions are the two null vectors contained in G. These are precisely A and B, so the bivector encodes the two points directly. In the conformal model, no information is lost in forming the exterior product of two null vectors. Spacelike bivectors, with $B^2 < 0$, do not contain any null vectors, so in this case there are no solutions to $B\wedge X = 0$ with $X^2 = 0$. The critical case of $B^2 = 0$ implies that B contains a single null vector.

Given a timelike bivector, $B^2 > 0$, we require an efficient means of finding the two null vectors in the plane. This can be achieved without solving any quadratic equations as follows. Pick an arbitrary vector a, with a partial projection in the plane, $a\cdot B \neq 0$. If the underlying space is Euclidean, one can use the vector $\bar{e}$, since all timelike bivectors contain a factor of this. Now remove the component of a outside the plane by defining

$$a' = a - a\wedge\hat{B}\,\hat{B}, \tag{10.93}$$

where $\hat{B} = B/|B|$ is normalised so that $\hat{B}^2 = 1$. If a' is already null then it defines one of the required vectors. If not, then one can form two null vectors in the B plane by writing

$$A_\pm = a' \pm a'\hat{B}. \tag{10.94}$$

One can easily confirm that $A_\pm$ are both null vectors, and so return the desired points.

10.4.2 Trivectors, lines and circles

If a bivector now only represents a pair of points, the obvious question is how do we describe a line? Suppose we construct the line through the points a and b in $\mathcal{V}(p,q)$. A point on the line is given by

$$x = \lambda a + (1-\lambda)b. \tag{10.95}$$

The conformal version of this line is

$$\begin{aligned} F(x) &= \big(\lambda^2 a^2 + 2\lambda(1-\lambda)a\cdot b + (1-\lambda)^2 b\big)n + 2\lambda a + 2(1-\lambda)b - \bar{n} \\ &= \lambda A + (1-\lambda)B + \tfrac{1}{2}\lambda(1-\lambda)A\cdot B\, n, \end{aligned} \tag{10.96}$$

and any multiple of this encodes the same point on the line. It is clear, then, that a conformal point X is a linear combination of A, B and n, subject to the constraint that $X^2 = 0$. This is summarised by

$$(A\wedge B\wedge n)\wedge X = 0, \qquad X^2 = 0. \tag{10.97}$$

So it is *trivectors* that represent lines in conformal geometry. This illustrates a general feature of the conformal model — geometric objects are represented by multivectors of one grade higher than their projective counterpart. The extra degree of freedom is absorbed by the constraint that $X^2 = 0$.

As stated above, if we apply a conformal transformation to a trivector representing a line, we must obtain a new line. But there is no reason to expect this to be straight. To see what else can result, consider a simple inversion in the origin. Suppose that (x_1, x_2) denote a pair of Cartesian coordinates for the Euclidean plane, and consider the line $x_1 = 1$. Points on the line have components $(1, x_2)$, with $-\infty \leq x_2 \leq +\infty$. The image of this line under an inversion in the origin has coordinates (x_1', x_2'), where

$$x_1' = \frac{1}{1+x_2^2}, \qquad x_2' = \frac{x_2}{1+x_2^2}. \tag{10.98}$$

It is now straightforward to show that

$$(x_1' - \tfrac{1}{2})^2 + (x_2')^2 = \left(\tfrac{1}{2}\right)^2. \tag{10.99}$$

Hence inversion of a line produces a *circle*, centred on $(1/2, 0)$ and with radius $1/2$.

It follows that a general trivector in conformal space can encode a circle, with a line representing the special case of infinite radius. This is entirely sensible, as three distinct points are required to specify a circle. The points define a plane, and any three non-collinear points in a plane specify a unique circle. So, given three points A_1, A_2, A_3, the circle through all three is defined by

$$A_1 \wedge A_2 \wedge A_3 \wedge X = 0, \tag{10.100}$$

together with the restriction (often unstated) that $X^2 = 0$. The trivector

$$L = A_1 \wedge A_2 \wedge A_3 \tag{10.101}$$

therefore encodes a unique circle in conformal geometry. The test that the points lie on a straight line is that the circle passes through the point at infinity,

$$L \wedge n = 0 \quad \Rightarrow \quad \text{straight line}. \tag{10.102}$$

This explains why our earlier derivation of the line through A_1 and A_2 led to the trivector $A_1 \wedge A_2 \wedge n$, which explicitly includes the point at infinity. Unlike tests for linear dependence, testing for zero in equation (10.102) is numerically acceptable. The reason is that the magnitude of $L \wedge n$ controls the deviation from straightness. If precision is limited, one can then define how close $L \wedge n$ should be to zero in order for the line to be treated as straight. This is quite different to linear independence, where the concept of 'nearly independent' makes no sense.

Given that a trivector L encodes a circle, we should expect to be able to extract the key geometric properties of the circle directly from L. In particular, we seek

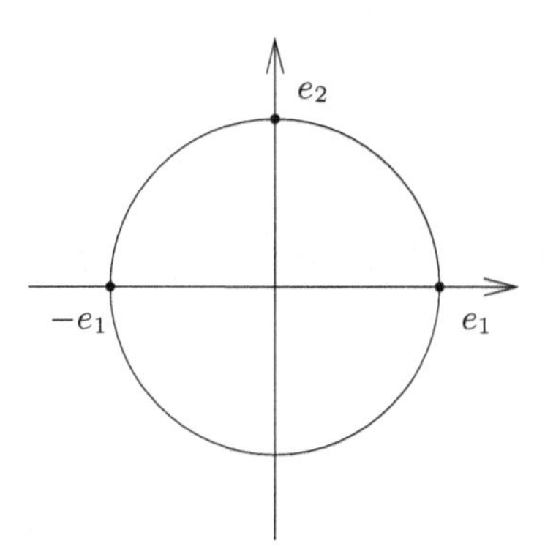

Figure 10.7 *The unit circle.* Three reference points are marked on the circle.

expressions for the centre and radius of the circle. (The plane containing the circle is specified by the 4-vector $L \wedge n$, as we explain in the following section.) Any circle in a plane can be mapped onto any other by a translation and a dilation. Under that latter we find that

$$L \wedge n \mapsto (D_\alpha L \tilde{D}_\alpha) \wedge n = \mathrm{e}^{\alpha} D_\alpha (L \wedge n) \tilde{D}_\alpha. \tag{10.103}$$

It follows that $(L \wedge n)^2$ scales as the inverse square of the radius. Next, consider the unit circle in the circle in the xy plane, and take as three points on the circle those shown in figure 10.7. The trivector for this circle is

$$L_0 = F(e_1) \wedge F(e_2) \wedge F(-e_1) = 16 e_1 e_2 \bar{e}. \tag{10.104}$$

It follows that

$$\frac{L_0^2}{(L_0 \wedge n)^2} = -1, \tag{10.105}$$

which is (minus) the square of the radius of the unit circle. We can translate and dilate this into any circle we choose, so the radius ρ of the circle encoded by the trivector L is given by

$$\rho^2 = -\frac{L^2}{(L \wedge n)^2}. \tag{10.106}$$

This is a further illustration of how metric information is carried around in the homogeneous framework of the conformal model. If L represents a straight line we know that $L \wedge n = 0$, so the radius we obtain is infinite.

Similar reasoning produces a formula for the centre of a circle. Essentially the only objects we have to work with are L and n. If we form LnL for the case of the unit circle we obtain

$$L_0 n L_0 \propto e_1 e_2 \bar{e} n \bar{e} e_1 e_2 = -\bar{n}. \tag{10.107}$$

But $\bar{n}$ is the null vector for the origin, so this expression has returned the desired point. Again, we can translate and dilate this result to obtain an arbitrary circle, and we find in general that the centre C of the circle L is obtained by

$$C = LnL. \tag{10.108}$$

We will see in section 10.5.5 that the operation $L \dots L$ generates a reflection in a circle. Equation (10.108) then says that the centre of a circle is the image of the point at infinity under a reflection in the circle.

10.4.3 4-vectors, spheres and planes

We can apply the same reasoning for lines and circles to the case of planes and spheres and, for mixed signature spaces, hyperboloids. Suppose initially that the points a, b, c define a plane in $\mathcal{V}(p, q)$, so that an arbitrary point in the plane is given by

$$x = \alpha a + \beta b + \gamma c, \qquad \alpha + \beta + \gamma = 1. \tag{10.109}$$

The conformal representation of x is

$$X = \alpha A + \beta B + \gamma C + \delta n, \tag{10.110}$$

where $A = f(a)$ etc., and

$$\delta = \tfrac{1}{2}(\alpha\beta A \cdot B + \alpha\gamma A \cdot C + \beta\gamma B \cdot C). \tag{10.111}$$

Varying α and β, together with the freedom to scale $F(x)$, now produces general null combinations of the vectors A, B, C and n. The equation for the plane can then be written

$$A \wedge B \wedge C \wedge n \wedge X = 0. \tag{10.112}$$

The plane passes through the points defined by A, B, C and the point at infinity n. We can therefore see that a general plane in conformal space is defined by four points.

If the four points in question do not lie on a (flat) plane, then the 4-vector formed from their outer product defines a *sphere*. To see this we again consider inversion in the origin, this time applied to the $x_1 = 1$ plane. A point on the plane has coordinates $(1, x_2, x_3)$, and under an inversion this maps to the point with coordinates

$$x_1' = \frac{1}{1 + x_2^2 + x_3^2}, \quad x_2' = \frac{y}{1 + x_2^2 + x_3^2}, \quad x_3' = \frac{z}{1 + x_2^2 + x_3^2}. \tag{10.113}$$

The new coordinates satisfy

$$(x_1' - \tfrac{1}{2})^2 + (x_2')^2 + (x_3')^2 = (\tfrac{1}{2})^2, \tag{10.114}$$

which is the equation of a sphere. Inversion thus interchanges planes and spheres. In particular, the point at infinity n is transformed to the origin $\bar{n}$ under inversion, which is now one of the points on the sphere.

Given any four distinct points $A_1, \ldots, A_4$, not all on a line or circle, the equation of the unique sphere through all four points is

$$A_1 \wedge A_2 \wedge A_3 \wedge A_4 \wedge X = P \wedge X = 0, \tag{10.115}$$

so the sphere is defined by the 4-vector $P = A_1 \wedge A_2 \wedge A_3 \wedge A_4$. The sphere is flat (a plane) if it passes through the point at infinity, the test for which is

$$A_1 \wedge A_2 \wedge A_3 \wedge A_4 \wedge n = P \wedge n = 0. \tag{10.116}$$

The 4-vector P contains all of the relevant geometric information for a sphere. The radius of the sphere ρ is given by

$$\rho^2 = \frac{P^2}{(P \wedge n)^2}, \tag{10.117}$$

as is easily confirmed for the case of the unit sphere, $P = e_1 e_2 e_3 \bar{e}$. Similarly, the centre of the sphere $C = F(c)$ is given by

$$C = PnP. \tag{10.118}$$

These formulae are the obvious generalisations of the results derived for circles.

10.5 Intersection and reflection in conformal space

One of the most significant advantages of the conformal approach to Euclidean geometry is the ease with which it solves complicated intersection problems. So, for example, finding the circle of intersection of two spheres is now no more complicated than finding the line of intersection of two planes. In addition, the concept of reflection is generalised in conformal space to include reflection in a sphere. This provides a very compact means of encoding the key concepts of inversive geometry.

10.5.1 Duality in conformal space

The concept of duality is key to intersecting objects in projective space, and the same is true in conformal space. Suppose that we start with the Euclidean plane, modelled in $\mathcal{G}(3,1)$. Duality in this algebra interchanges spacelike and timelike bivectors. It also maps trivectors to vectors, and vice versa. A trivector encodes a line, or circle, so the dual of the circle C is a vector c, where

$$c = C^* = IC \tag{10.119}$$

and I is the pseudoscalar for $\mathcal{G}(3,1)$. The equation for the circle, $X \wedge C = 0$, can now be written in dual form and reduces to

$$X \cdot c = -I(X \wedge C) = 0. \tag{10.120}$$

The radius of the circle is now given by

$$\rho^2 = \frac{c^2}{(c \cdot n)^2}, \tag{10.121}$$

as the vector dual to a circle has positive signature. This picture provides us with an alternative view of the concept of a point as being a circle of zero radius.

Similar considerations hold for spheres in three-dimensional space. These are represented as 4-vectors in $\mathcal{G}(4,1)$, so their dual is a vector. We write

$$s = S^* = IS, \tag{10.122}$$

where I is the pseudoscalar, so that the equation of a sphere becomes

$$X \cdot s = I(X \wedge S) = 0. \tag{10.123}$$

The radius of the sphere is again given by

$$\rho^2 = \frac{s^2}{(s \cdot n)^2}, \tag{10.124}$$

so that points are spheres of zero radius. One can see that this is sensible by considering an alternative equation for a sphere. Suppose we are interested in the sphere with centre C and radius ρ^2. The equation for this can be written

$$-2\frac{X \cdot C}{X \cdot n \, C \cdot n} = \rho^2. \tag{10.125}$$

Rearranging, this equation becomes

$$X \cdot (2C + \rho^2 C \cdot n \, n) = 0, \tag{10.126}$$

and if C is in standard form, $C = F(c)$, we obtain

$$X \cdot (F(c) - \rho^2 n) = 0. \tag{10.127}$$

We can therefore identify $s = S^*$ with the vector $F(c) - \rho^2 n$, which neatly encodes the centre and radius of the sphere in a single vector. Whether the 4-vector S or its dual vector s is most useful depends on whether the sphere is specified by four points lying on it, or by its centre and radius. For a given sphere s we can now write

$$s = \lambda(2C + \rho^2 C \cdot n \, n). \tag{10.128}$$

It is then straightforward to confirm that the radius is given by equation (10.124). The centre of the circle can be recovered from

$$\frac{C}{C\cdot n} = \frac{s}{s\cdot n} - \frac{\rho^2}{2}n = \frac{sns}{2(s\cdot n)^2}. \tag{10.129}$$

The sns form for the centre of a sphere is dual to the SnS expression found in equation (10.118).

10.5.2 Intersection of two lines in a plane

As a simple example of intersection in the conformal model, consider the intersection of two lines in a Euclidean plane. The lines are described by trivectors L_1 and L_2 in $\mathcal{G}(3,1)$. The intersection is described by the bivector

$$B = (L_1^* \wedge L_2^*)^* = I(L_1 \times L_2), \tag{10.130}$$

where I is the conformal pseudoscalar. The bivector B can contain zero, one or two points, depending on the sign of its square, as described in section 10.4.1. This is to be expected, as distinct circles can intersect at a maximum of two points. If the lines are both straight, then one of the points of intersection will be at infinity, and $B \wedge n = 0$.

To verify this result, consider the case of two straight lines, both passing through the origin, and with the first line in the a direction and the second in the b direction. With suitable normalisation we can write

$$L_1 = aN, \qquad L_2 = bN, \tag{10.131}$$

where $N = e\bar{e}$. The intersection of L_1 and L_2 is controlled by

$$B = I\, a\wedge b \propto N \tag{10.132}$$

and the bivector N contains the null vectors n and $\bar{n}$. This confirms that the lines intersect at the origin and infinity. Applying conformal transformations to this result ensures that it holds for all lines in a plane, whether the lines are straight or circular. The formulae for L_1 and L_2 also show that their inner product is related to the angle between the lines,

$$\langle L_1 L_2 \rangle = a\cdot b. \tag{10.133}$$

We can therefore write

$$\cos(\theta) = \frac{\langle L_1 L_2 \rangle}{|L_1|\,|L_2|}, \tag{10.134}$$

where $|L| = \sqrt{(L^2)}$. This equation returns the angle between two lines. The quantity is invariant under the full conformal group, and not just the Euclidean group, because angles are conformal invariants. It follows that the same formula must hold even if L_1 and L_2 describe circles. The angle between two circles is the angle made by their tangent vectors at the point of intersection. Two circles intersect at a right angle, therefore, if

$$\langle L_1 L_2 \rangle = 0. \tag{10.135}$$

This result can equally be expressed in terms of the dual vectors l_1 and l_2.

10.5.3 Intersection of a line and a surface

Now suppose that the 4-vector P defines a plane or sphere in three-dimensional Euclidean space, and we wish to find the point of intersection with a line described by the trivector L. The algebra proceeds entirely as expected and we arrive at the bivector

$$B = (P^* \wedge L^*)^* = (IP) \cdot L = I \langle PL \rangle_3. \tag{10.136}$$

This bivector can again describe zero, one or two points, depending on the sign of its square. This setup describes all possible intersections between lines or circles, and planes or spheres — an extremely wide range of applications. Precisely the same algebra enables us to answer whether a ring in space intersects a given plane, or whether a straight line passes through a sphere.

10.5.4 Surface intersections

Next, suppose we wish to intersect two surfaces in three dimensions. Suppose that these are spheres defined by the 4-vectors S_1 and S_2. Their intersection is described by the trivector

$$L = I(S_1 \times S_2). \tag{10.137}$$

This trivector directly encodes the circle formed from the intersection of two spheres. As with the bivector case, the sign of L^2 defines whether or not two surfaces intersect. If $L^2 > 0$ then the surfaces do intersect. If $L^2 = 0$ then the surfaces intersect at a point. Tests such as this are extremely helpful in graphics applications.

We can similarly express the intersection in terms of the dual vectors s_1 and s_2 as

$$L = I\, s_1 \wedge s_2. \tag{10.138}$$

As a check, the point X lies on both spheres if

$$X \cdot s_1 = X \cdot s_2 = 0. \tag{10.139}$$

It follows that

$$X \cdot (s_1 \wedge s_2) = X \cdot s_1\, s_2 - X \cdot s_2\, s_1 = 0. \tag{10.140}$$

The dual result is that $X\wedge(I\, s_1\wedge s_2) = 0$, which confirms that X lies in the space defined by the trivector L.

10.5.5 Reflections in conformal space

At various points in previous sections we have obtained formulae which generate reflections. We now discuss these more systematically. In section 2.6 we established that the vector obtained by reflecting a in the hyperplane perpendicular to l, $l^2 = 1$, is $-lal$. But this formula assumes that the line and plane intersect at the origin. We seek a more general expression, valid for an arbitrary line and plane. Let P denote the plane and L the line we wish to reflect in the plane, then the obvious candidate for the reflected line L' is

$$L' = PLP. \tag{10.141}$$

(The sign of this is irrelevant in conformal space.) To verify that this is correct, suppose that L passes through the origin in the a direction,

$$L = aN_3 \tag{10.142}$$

and the plane P is defined by the origin and the directions b and c,

$$P = b\wedge cN. \tag{10.143}$$

In this case

$$L' = b\wedge c\, a\, b\wedge c\, N = \big(-(I_3\, b\wedge c)a(I_3\, b\wedge c)\big)N, \tag{10.144}$$

where I_3 is the three-dimensional pseudoscalar. This result achieves the required result. The vector a is reflected in the $b\wedge c$ plane to obtain the desired direction. The outer product with N then defines the line through the origin with the required direction. Equation (10.141) is correct at the origin, so therefore holds for all lines and planes, by conformal invariance.

There are a number of significant consequences of equation (10.141). The first is that it recovers the correct line in three dimensions without having to to find the point of reflection. The second is that it is straightforward to chain together multiple reflections by forming successive products with planes. In this way complicated reflections can be easily composed, all the time keeping track of the direction and position of the resultant line. A further consequence is that the same reflection formula must hold for higher dimensional objects. Suppose, for example, we wish to reflect the sphere S in the plane P. The result is

$$S' = PSP. \tag{10.145}$$

This type of equation is extremely useful in dealing with wave propagation, where a wavefront is modelled as a series of expanding spheres.

Conformal invariance of the reflection formula (10.141) ensures that the same formula holds for reflection in a circle, or in a sphere. For example, suppose we wish to carry out a reflection in the unit circle in two-dimensional Euclidean space. The circle is defined by $L_0 = e_1 e_2 \bar{e}$, and the dual vector is

$$IL_0 = e. \tag{10.146}$$

Reflection in the unit circle is therefore performed by the operation

$$M \mapsto eMe. \tag{10.147}$$

This is an inversion, as discussed in section 10.3.3. In this manner, the main results of inversive geometry are easily formulated in terms of reflections in conformal space.

10.6 Non-Euclidean geometry

The sudden growth in the subject of geometry in the nineteenth century was stimulated in part by the discovery of geometries with very different properties to Euclidean space. These were obtained by a simple modification of Euclid's *parallel postulate*. For Euclidean geometry this states that, given any line l and a point P not on the line, there exists a unique line through P in the plane of l and P which does not meet l. This is then a line parallel to l. For many centuries this postulate was viewed as problematic, as it cannot be easily experimentally verified. As a result, mathematicians attempted to remove the parallel postulate by proving it from the remaining, uncontroversial, postulates of Euclidean geometry. This enterprise proved fruitless, and the reason why was discovered by Lobachevskii and Bolyai in the 1820s. One can replace the parallel postulate with a different postulate, and obtain a new, mathematically acceptable geometry.

There are in fact two alternative geometries one can obtain, by replacing the statement that there is a *single* line through P which does not intersect l with either an infinite number or zero. The case of an infinite number produces *hyperbolic* geometry, which is the non-Euclidean geometry constructed by Lobachevskii and Bolyai. (In this section 'non-Euclidean' usually refers to the hyperbolic case.) The case of zero lines produces spherical geometry. Intuitively, the spherical case corresponds to space curling up, so that all (straight) lines meet somewhere, and the hyperbolic case corresponds to space curving outwards, so that lines do not meet. From the more modern perspective of Riemannian geometry, we are talking about homogeneous, isotropic spaces, which have no preferred points or directions. These can have positive, zero or negative curvature, corresponding to spherical, Euclidean and hyperbolic geometries. Today, the question of which of these correctly describes the universe on the largest scales remains an outstanding problem in cosmology.

An extremely attractive feature of the conformal model of Euclidean geometry

Figure 10.8 *Circle limit III* by Maurits Escher. ©2002 Cordon Art B.V., Baarn, Holland.

is that, with little modification, it can be applied to both hyperbolic and spherical geometries as well. In essence, the geometry reduces to a choice of the point at infinity, which in turn fixes the distance measure. This idea replaces the concept of the *absolute conic*, adopted in classical projective geometry as a means of imposing a distance measure. In this section we illustrate these ideas with a discussion of the conformal approach to planar hyperbolic geometry. As a concrete model of this we concentrate on the Poincaré disc. This version of hyperbolic geometry is mathematically very appealing, and also gives rise to some beautiful graphic designs, as popularised in the prints of Maurits Escher (see figure 10.8).

10.6.1 The Poincaré disc

The Poincaré disc $\mathcal{D}$ consists of the set of points in the plane a distance $r < 1$ from the origin. At first sight this may not appear to be homogeneous, but in

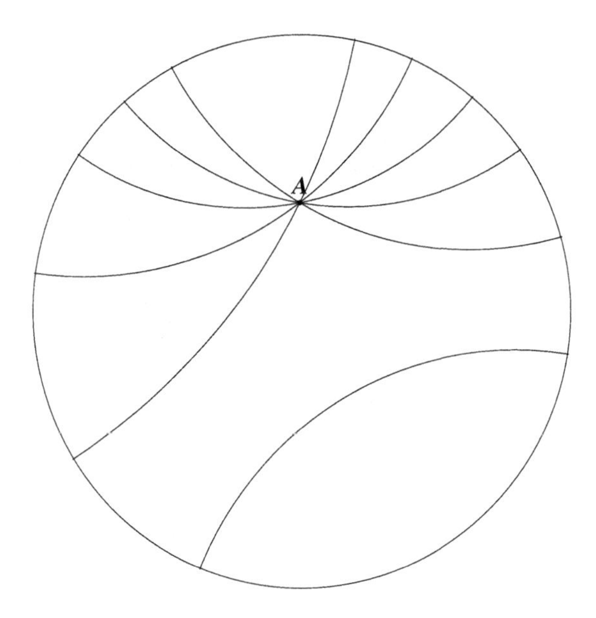

Figure 10.9 *The Poincaré disc.* Points inside the disc represent points in a hyperbolic space. A set of d-lines are also shown. These are (Euclidean) circles that intersect the unit circle at right angles. The d-lines through A illustrate the parallel postulate for hyperbolic geometry.

fact the nature of the geometry will ensure that there is nothing special about the origin. Note that points on the unit circle $r = 1$ are *not* included in this model of hyperbolic geometry. The key to this geometry is the concept of a non-Euclidean straight line. These are called d-lines, and represent geodesics in hyperbolic geometry. A d-line consists of a section of a Euclidean circle which intersects the unit circle at a right angle. Examples of d-lines are illustrated in figure 10.9. Given any two points in the Poincaré disc there is a unique d-line through them, which represents the 'straight' line between the points. It is now clear that for any point not on a given d-line l, there are an infinite number of d-lines through the point which do not intersect l.

We can now begin to encode these concepts in the conformal setting. We continue to denote points in the plane with homogeneous null vectors in precisely the same manner as the Euclidean case. Suppose, then, that X and Y are the conformal vectors representing two points in the disc. The set of all circles through these two points consists of trivectors of the form $X \wedge Y \wedge A$, where A is an additional point. But we require that the d-line intersects the unit circle at right angles. The unit circle is described by the trivector Ie, where I is the pseudoscalar in $\mathcal{G}(3,1)$. If a line L is perpendicular to the unit circle it satisfies

$$(Ie)\cdot L = I(e\wedge L) = 0. \tag{10.148}$$

It follows that all d-lines contain a factor of e. The d-line through X and Y must therefore be described by the trivector

$$L = X\wedge Y\wedge e. \tag{10.149}$$

One can see now that a general scheme is beginning to emerge. Everywhere in the Euclidean treatment that the vector n appears it is replaced in hyperbolic geometry by the vector e. This vector represents the circle at infinity.

Given a pair of d-lines, they can either miss each other, or intersect at a point in the disc $\mathcal{D}$. If they intersect, the angle between the lines is given by the Euclidean formula

$$\cos(\theta) = \frac{L_1\cdot L_2}{|L_1|\,|L_2|}. \tag{10.150}$$

It follows that angles are preserved by a general conformal transformation in hyperbolic geometry. A non-Euclidean transformation takes d-lines to d-lines. The transformation must therefore map (Euclidean) circles to circles, while preserving orthogonality with e. The group of non-Euclidean transformations must therefore be the subgroup of the conformal group which leaves e invariant. This is confirmed in the following section, where we find the appropriate distance measure for non-Euclidean geometry.

The fact that the point at infinity is represented by e, as opposed to n in the Euclidean counterpart, provides an additional operation in non-Euclidean geometry. This is inversion in e:

$$X \mapsto eXe. \tag{10.151}$$

As all non-Euclidean transformations leave e invariant, all geometric relations remain unchanged under this inversion. Geometrically, the interpretation of the inversion is quite clear. It maps everything inside the Poincaré disc to a 'dual' version outside the disc. In this dual space incidence relations and distances are unchanged from their counterparts inside the disc.

10.6.2 Non-Euclidean translations and distance

The key to finding the correct distance measure in non-Euclidean geometry is to first generalise the concept of a translation. Given points X and Y we know that the d-line connecting them is defined by $X\wedge Y\wedge e$. This is the non-Euclidean concept of a straight line. A non-Euclidean translation must therefore move points along this line. Such a transformation must take X to Y, but must also leave e invariant. The generator for such a transformation is the bivector

$$B = (X\wedge Y\wedge e)e = Le, \tag{10.152}$$

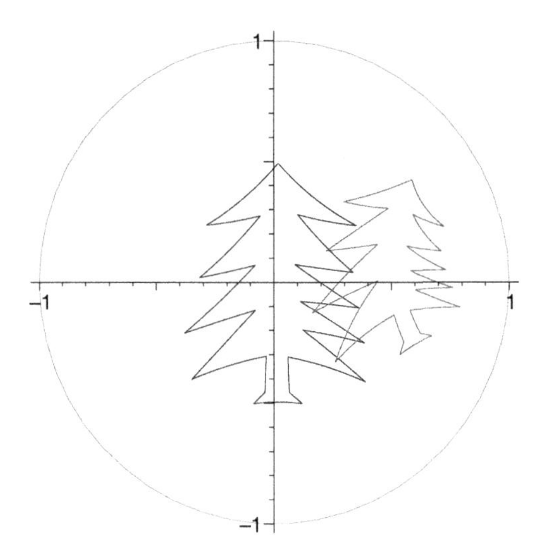

Figure 10.10 *A non-Euclidean translation.* The figure near the origin is translated via a boost to give the distorted figure on the right. This distortion in the Poincaré disc is one way of visualising the effect of a Lorentz boost in spacetime.

where $L = X \wedge Y \wedge e$. We find immediately that

$$B^2 = L^2 > 0, \tag{10.153}$$

so non-Euclidean translations are *hyperbolic* transformations, as one might expect. An example of such a translation is shown in figure 10.10.

We next define

$$\hat{B} = \frac{B}{|B|}, \qquad \hat{B}^2 = 1, \tag{10.154}$$

so that we can write

$$Y = \mathrm{e}^{\alpha \hat{B}/2} X \mathrm{e}^{-\alpha \hat{B}/2}. \tag{10.155}$$

By varying α we obtain the set of points along the d-line through X and Y. To obtain a distance measure, we first require a formula for α. If we decompose X into

$$X = X\hat{B}^2 = X \cdot \hat{B}\, \hat{B} + X \wedge \hat{B}\, \hat{B} \tag{10.156}$$

we obtain

$$Y = X \wedge \hat{B}\, \hat{B} + \cosh(\alpha)\, X \cdot \hat{B}\, \hat{B} - \sinh(\alpha)\, X \cdot \hat{B}. \tag{10.157}$$

The right-hand side must give zero when contracted with Y, so

$$\langle X \wedge \hat{B}\, \hat{B} \wedge Y \rangle + \cosh(\alpha) \langle X \cdot \hat{B}\, \hat{B} \cdot Y \rangle + \sinh(\alpha)\, (X \wedge Y) \cdot \hat{B} = 0. \tag{10.158}$$

To simplify this equation we first find

$$X\wedge\hat{B} = \frac{X\wedge(X\wedge Y\wedge e\, e)}{|B|} = \frac{e\cdot X\, L}{|L|} \tag{10.159}$$

and

$$(X\wedge Y)\cdot\hat{B} = \frac{L^2}{|B|} = |L|. \tag{10.160}$$

It follows that

$$e\cdot X\, e\cdot Y + \cosh(\alpha)(X\cdot Y - e\cdot X\, e\cdot Y) + \sinh(\alpha)\,|L| = 0, \tag{10.161}$$

the solution to which is

$$\cosh(\alpha) = 1 - \frac{X\cdot Y}{X\cdot e\, Y\cdot e}. \tag{10.162}$$

The half-angle formula is more relevant for the distance measure, and we find that

$$\sinh^2(\alpha/2) = -\frac{X\cdot Y}{2X\cdot e\, Y\cdot e}. \tag{10.163}$$

This closely mirrors the Euclidean expression, with n replaced by e.

There are a number of obvious properties that a distance measure must satisfy. Among these is the additive property that

$$d(X_1, X_2) + d(X_2, X_3) = d(X_1, X_3) \tag{10.164}$$

for any three points X_1, X_2, X_3 in this order along a d-line. Returning to the translation formula of equation (10.155), suppose that Z is a third point along the line, beyond Y. We can write

$$Z = \mathrm{e}^{\beta\hat{B}/2} Y \mathrm{e}^{-\beta\hat{B}/2} = \mathrm{e}^{(\alpha+\beta)\hat{B}} X \mathrm{e}^{-(\alpha+\beta)\hat{B}/2}. \tag{10.165}$$

Clearly it is hyperbolic angles that must form the appropriate distance measure. No other function satisfies the additive property. We therefore define the non-Euclidean distance by

$$d(x, y) = 2\sinh^{-1}\left(-\frac{X\cdot Y}{2X\cdot e\, Y\cdot e}\right)^{1/2}. \tag{10.166}$$

In terms of the position vectors x and y in the Poincaré disc we can write

$$d(x, y) = 2\sinh^{-1}\left(\frac{|x-y|^2}{(1-x^2)(1-y^2)}\right)^{1/2}, \tag{10.167}$$

where the modulus refers to the Euclidean distance. The presence of the arcsinh function in the definition of distance reflects the fact that, in hyperbolic geometry, generators of translations have positive square and the appropriate distance measure is the hyperbolic angle. Similarly, in spherical geometry translations correspond to rotations, and it is the trigonometric angle which plays the role of distance. Euclidean geometry is therefore unique in that the generators of translations are *null* bivectors. For these, combining translations reduces to the addition of bivectors, and hence we recover the standard definition of Euclidean distance.

10.6.3 Metrics and physical units

The derivation of the non-Euclidean distance formula of equation (10.166) forces us to face an issue that has been ignored to date. Physical distances are dimensional quantities, whereas our formulae for distances in both Euclidean and non-Euclidean geometries are manifestly dimensionless, as they are homogeneous in X. To resolve this we cannot just demand that the vector x has dimensions, as this would imply that the conformal vector X contained terms of mixed dimensions. Neither can this problem be circumvented by assigning dimensions of distance to $\bar{n}$ and $(\text{distance})^{-1}$ to n, as then e has mixed dimensions, and the non-Euclidean formula of (10.166) is non-sensical.

The resolution is to introduce a fundamental length scale, λ, which is a positive scalar with the dimensions of length. If the vector x has dimensions of length, the conformal representation is then given by

$$X = \frac{1}{2\lambda^2}\left(x^2 n + 2\lambda x - \lambda^2 \bar{n}\right). \tag{10.168}$$

This representation ensures that X remains dimensionless, and is nothing more than the conformal representation of x/λ. Physical distances can then be converted into a dimensionally meaningful form by including appropriate factors of λ. Curiously, the introduction of λ into the spacetime conformal model has many similarities to the introduction of a cosmological constant $\Lambda = \lambda^2$.

We can make contact with the metric encoding of distance by finding the infinitesimal distance between the points x and $x + dx$. This defines the line element

$$ds^2 = 4\lambda^4 \frac{dx^2}{(\lambda^2 - x^2)^2}, \tag{10.169}$$

where the factors of λ have been included and x is assumed to have dimensions of distance. This line element is more often seen in polar coordinates, where it takes the form

$$ds^2 = \frac{4\lambda^4}{(\lambda^2 - r^2)^2}(dr^2 + r^2 d\theta^2). \tag{10.170}$$

This is the line element for a space of constant negative curvature, expressed in terms of conformal coordinates. The coordinates are conformal because the line element is that of a flat space multiplied by a scaling function. The geodesics in this geometry are precisely the d-lines in the Poincaré disc. The Riemann curvature for this metric shows that the space has uniform negative curvature, so the space is indeed homogeneous and isotropic — there are no preferred points or directions. The centre of the disc is not a special point, and indeed it can be translated to any other point by 'boosting' along a d-line.

10.6.4 Midpoints and circles in non-Euclidean geometry

Now that we have a conformal encoding of a straight line and of distance in non-Euclidean geometry, we can proceed to discuss concepts such as the midpoint of two points, and of the set of points a constant distance from a given point (a non-Euclidean circle). Suppose that A and B are the conformal vectors of two points in the Poincaré disc. Their midpoint C lies on the line $L = A \wedge B \wedge e$ and is equidistant from both A and B. The latter condition implies that

$$\frac{C \cdot A}{C \cdot e \, A \cdot e} = \frac{C \cdot B}{C \cdot e \, B \cdot e}. \tag{10.171}$$

Both of the conditions for C are easily satisfied by setting

$$C = \frac{A}{2A \cdot e} + \frac{B}{2B \cdot e} + \alpha e, \tag{10.172}$$

where α must be chosen such that $C^2 = 0$. Normalising to $C \cdot e = -1$ we find that the midpoint is

$$C = -\frac{1}{\sqrt{1+\delta}} \left(\frac{A}{2A \cdot e} + \frac{B}{2B \cdot e} + (\sqrt{1+\delta} - 1)e \right), \tag{10.173}$$

where

$$\delta = -\frac{A \cdot B}{2A \cdot e \, B \cdot e}. \tag{10.174}$$

An equation such as this is rather harder to achieve without access to the conformal model.

Next suppose we wish to find the set of points a constant (non-Euclidean) distance from the point C. This defines a non-Euclidean circle with centre C. From equation (10.166), any point X on the circle must satisfy

$$-\frac{X \cdot C}{2X \cdot e \, C \cdot e} = \text{constant} = \alpha^2, \tag{10.175}$$

so that the radius is $\sinh^{-1}(\alpha)$. It follows that

$$X \cdot (C + 2\alpha^2 C \cdot e \, e) = 0. \tag{10.176}$$

If we define s by

$$s = C + 2\alpha^2 C \cdot e \, e \tag{10.177}$$

we see that $s^2 > 0$, and the circle is defined by $X \cdot s = 0$. But this is precisely the formula for a circle in Euclidean geometry, so non-Euclidean circles still appear as ordinary circles when plotted in the Poincaré disc. The only difference is the

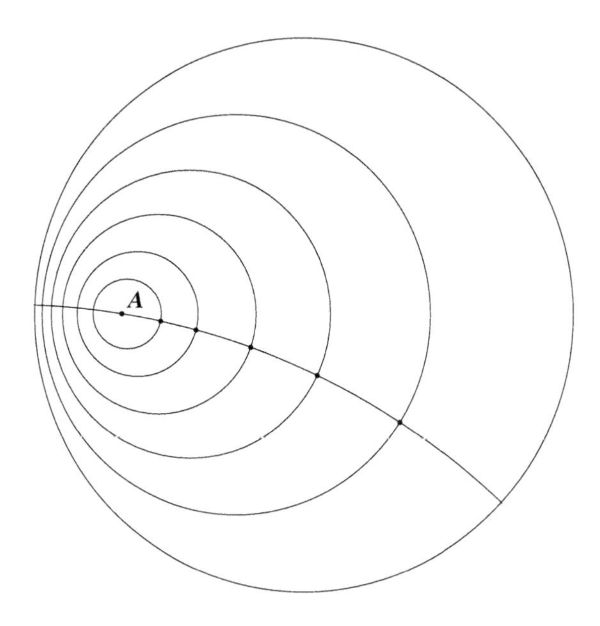

Figure 10.11 *Non-Euclidean circles.* A series of non-Euclidean circles with differing radii are shown, all about the common centre A. A d-line through A is also shown. This intersects each circle at a right angle.

interpretation of their centre. The Euclidean centre of the circle s, defined by sns, does not coincide with the non-Euclidean centre C. This is illustrated in figure 10.11.

Suppose that A, B and C are three points in the Poincaré disc. We can still define the line L through these points by

$$L = A \wedge B \wedge C, \tag{10.178}$$

and this defines the circle through the three points regardless of the geometry we are working in. All that is different in the two geometries is the position of the midpoint and the size of the radius. The test that the three points lie on a d-line is simply that $L \wedge e = 0$. Again, the Euclidean formula holds, but with n replaced by e. Similar comments apply to other operations in conformal space, such as reflection. Given a line L, points are reflected in this line by the map $X \mapsto LXL$. This formula is appropriate in both Euclidean and non-Euclidean geometry. In the non-Euclidean case it is not hard to verify that LXL corresponds to first finding the d-line through X intersecting L at right angles, and then finding the point on this line an equal non-Euclidean distance on the other side. This is as one would expect for the definition of reflection in a line.

10.6.5 A unified framework for geometry

We have so far seen how Euclidean and hyperbolic geometries can both be handled in terms of null vectors in conformal space. The key concept is the vector representing the point at infinity, which remains invariant under the appropriate symmetry group. The full conformal group of a space with signature (p, q) is the orthogonal group $O(p+1, q+1)$. The group of Euclidean transformations is the subgroup of $O(p+1, q+1)$ that leaves the vector n invariant. The hyperbolic group is the subgroup of $O(p+1, q+1)$ which leaves e invariant. For the case of planar geometry, with signature $(2,0)$, the hyperbolic group is $O(2,1)$. The Killing form for this group is non-degenerate (see chapter 11), which makes hyperbolic geometry a useful way of compactifying a flat space.

The remaining planar geometry to consider is spherical geometry. By now, it should come as little surprise that spherical geometry is handled in the conformal framework in terms of transformations which leave the vector $\bar{e}$ invariant. For the case of the plane, the conformal algebra has signature $(3,1)$, with $\bar{e}$ the basis vector with negative signature. The subgroup of the conformal group which leaves $\bar{e}$ invariant is therefore the orthogonal group $O(3,0)$, which is the group one expects for a 2-sphere. The distance measure for spherical geometry is

$$d(x, y) = 2\lambda \sin^{-1}\left(-\frac{X\cdot Y}{2X\cdot\bar{e}\,Y\cdot\bar{e}}\right)^{1/2}, \tag{10.179}$$

with $\bar{e}$ replacing n in the obvious manner. To see that this expression is correct, suppose that we write

$$\frac{X}{X\cdot\bar{e}} = \hat{x} - \bar{e}, \tag{10.180}$$

where $\hat{x}$ is a unit vector built in the three-dimensional space spanned by the vectors e_1, e_2 and e. With $Y/Y\cdot\bar{e}$ written in the same way we find that

$$-\frac{X\cdot Y}{2X\cdot\bar{e}\,Y\cdot\bar{e}} = \frac{1-\hat{x}\cdot\hat{y}}{2} = \sin^2(\theta/2), \tag{10.181}$$

where θ is the angle between the unit vectors on the 2-sphere. The distance measure is then precisely the angle θ multiplied by the dimensional quantity λ, which represents the radius of the sphere.

Conformal geometry provides a unified framework for the three types of planar geometry because in all cases the conformal groups are the same. That is, the group of transformations of sphere that leave angles in the sphere unchanged is the same as for the plane and the hyperboloid. In all cases the group is $O(3,1)$. The geometries are then recovered by a choice of distance measure. In classical projective geometry the distance measure is defined by the introduction of the *absolute conic*. All lines intersect this conic in a pair of points. The distance between two points A and B is then found from the four-point ratio between A, B, and the two points of intersection of the line through A and B and the absolute conic. In this way all geometries are united in the framework of projective geometry. But there is a price to pay for this scheme — all coordinates have to be complex, to ensure that all lines intersect the conic in two points. Recovering a real geometry is then rather clumsy. In addition, the conformal group is not a subgroup of the projective group, so much of the elegant unity exhibited by the three geometries is lost. Conformal geometry is a more powerful framework for a unified treatment of these geometries. Furthermore, the conformal approach can be applied to spaces of any dimension with little modification. Trivectors represent lines and circles, 4-vectors represent planes and spheres, and so on.

So far we have restricted ourselves to a single view of the various geometries, but the discussion of the sphere illustrates that there are many different ways of representing the underlying geometry. To begin with, we have plotted points on the Euclidean plane according the the formula

$$x = -\frac{X \wedge N}{X \cdot n} N, \tag{10.182}$$

where $N = e\bar{e}$. This is the natural scheme for plotting on a Euclidean piece of paper, as it ensures that the angle between lines on the paper is the correct angle in each of the three geometries. Euclidean geometry plotted in this way recovers the obvious standard picture of Euclidean geometry. Hyperbolic geometry led to the Poincaré disc model, in which hyperbolic lines appear as circles. For spherical geometry the 'straight lines' are great circles on a sphere. On the plane these also plot as circles. This time the condition is that all circles intersect the unit circle at antipodal points. This then defines the spherical line between two points (see figure 10.12). This view of spherical geometry is precisely that obtained from a stereographic projection of the sphere onto the plane. This is not a surprise, as the conformal model was initially constructed in terms of a stereographic projection, with the $\bar{e}$ vector then enabling us to move to a homogeneous framework. In this representation of spherical geometry the map

$$X \mapsto \bar{e} X \bar{e} \tag{10.183}$$

is a symmetry operation. This maps points to their antipodal opposites on the sphere. In the planar view this transformation is an inversion in the unit circle, followed by a reflection in the origin.

We now have three separate geometries, all with conformal representations in the plane such that the true angle between lines is the same as that measured on the plane. The price for such a representation is that straight lines in spherical and hyperbolic geometries do not appear straight in the plane. But we could equally choose to replace the map of equation (10.182) with an alternative rule of how to plot the null vector X on a planar piece of paper. The natural alternatives to consider are replacing the vector n with e and $\bar{e}$. In total we then have three different planar realisations of each of the two-dimensional geometries. First,

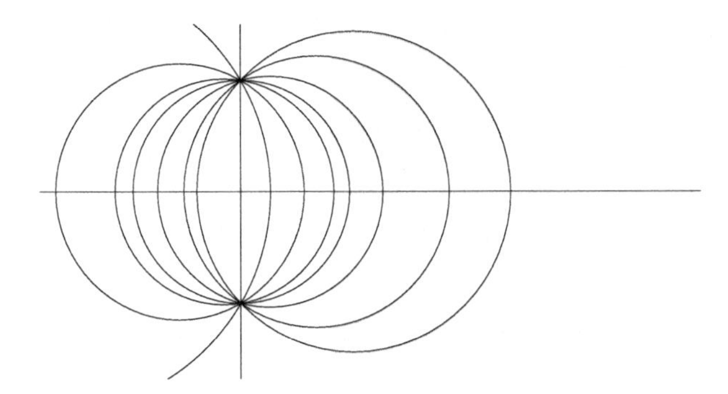

Figure 10.12 *Stereographic view of spherical geometry.* All great circles on the 2-sphere project onto circles in the plane which intersect the unit circle (shown in bold) at antipodal points. A series of such lines are shown.

suppose we define

$$y = \frac{X \wedge N}{X \cdot e} N. \tag{10.184}$$

In terms of the vector x we have

$$y = \frac{2x}{1 - x^2}, \tag{10.185}$$

which represents a radial rescaling. Euclidean straight lines now appear as hyperbolae or ellipses, depending on whether or not the original line intersected the disc. If the line intersected the disc then the map of equation (10.185) has two branches and defines a hyperbola. If the line misses the disc then an ellipse is obtained. In all cases the image lines pass through the origin, as this is the image of the point at infinity.

The fact that the map of equation (10.185) is two-to-one means it has little use as a version of Euclidean geometry. It is better suited to hyperbolic geometry, as one might expect, as the Poincaré disc is now mapped onto the entire plane. Hyperbolic straight lines now appear as (single-branch) hyperbolae on the Euclidean page, all with their asymptotes crossing at the origin. If the dual space outside the disc is included in the map, then this generates the second branch of each hyperbola. Points then occur in pairs, with each point paired with its image under reflection in the origin. Finally, we can consider spherical geometry as viewed on a plane through the map of equation (10.185). This defines a standard projective map between a sphere and the plane. Antipodal points on the sphere define the same point on the plane and spherical straight lines appear as straight lines.

Similarly, we can consider plotting vectors in the plane according to

$$y = -\frac{X \wedge N}{X \cdot \bar{e}} N = -\frac{F(x) \wedge N}{F(x) \cdot \bar{e}} N \tag{10.186}$$

or in terms of the vector x

$$y = \frac{2x}{1+x^2}. \tag{10.187}$$

This defines a one-to-one map of the unit disc onto itself, and a two-to-one map of the entire plane onto the disc. Euclidean straight lines now appear plotted as ellipses inside the unit disc. This construction involves forming a stereographic projection of the plane onto the 2-sphere, so that lines map to circles on the sphere. The sphere is then mapped onto the plane by viewing from above, so that circles on the sphere map to ellipses. All ellipses pass through the origin, as this is the image of the point at infinity.

Similar comments apply to spherical geometry. Spherical lines are great circles on the sphere, and viewed in the plane according to equation (10.187) great circles appear as ellipses centred on the origin and touching the unit circle at their endpoints. The two-to-one form of the projection means that circle intersections are not faithfully represented in the disc as some of the apparent intersections are actually caused by points on opposite sides of the plane. Finally, we consider plotting hyperbolic geometry in the view of equation (10.187). The disc maps onto itself, so we do have a faithful representation of hyperbolic geometry. This is a representation in which hyperbolic lines appear straight on the page, though angles are not rendered correctly, and non-Euclidean circles appear as ellipses.

As well as viewing each geometry on the Euclidean plane, we can also picture the geometries on a sphere or a hyperboloid. The spherical picture is obtained in equation (10.180), and the hyperboloid view is similarly obtained by setting

$$\frac{X}{X \cdot e} = \hat{x} + e, \tag{10.188}$$

where $\hat{x}^2 = -1$. The set of $\hat{x}$ defines a pair of hyperbolic sheets in the space defined by the vectors $\{e_1, e_2, \bar{e}\}$. The fact that two sheets are obtained explains why some views of hyperbolic geometry end up with points represented twice. So, as well as three geometries (defined by a transformation group) and a variety of plotting schemes, we also have a choice of space to draw on, providing a large number of alternative schemes for studying the three geometries. At the back of all of this is a single algebraic scheme, based on the geometric algebra of conformal space. Any algebraic result involving products of null vectors immediately produces a geometric theorem in each geometry, which can be viewed in a variety of different ways.

10.7 Spacetime conformal geometry

As a final application of the conformal approach to geometry we turn to spacetime. The conformal geometric algebra for a spacetime with signature $(1,3)$ is the six-dimensional algebra with signature $(2,4)$. The algebra $\mathcal{G}(2,4)$ contains 64 terms, which decompose into graded subspaces of dimensions 1, 6, 15, 20, 15, 6 and 1. As a basis for this space we use the standard spacetime algebra basis $\{\gamma_\mu\}$, together with the additional vectors $\{e, \bar{e}\}$. The pseudoscalar I is defined by

$$I = \gamma_0\gamma_1\gamma_2\gamma_3 e\bar{e}. \tag{10.189}$$

This has negative norm, $I^2 = -1$. The conformal algebra allows us to simply encode ideas such as closed circles in spacetime, or light-spheres centred on an arbitrary point.

The conformal algebra of spacetime also arises classically in a slightly different setting. In conformal geometry, circles and spheres are represented homogeneously as trivectors and 4-vectors. These are unoriented because L and $-L$ are used to encode the same object. A method of dealing with oriented spheres was developed by Sophus Lie and is called Lie sphere geometry. A sphere in three dimensions can be represented by a vector s in the conformal algebra $\mathcal{G}(4,1)$, with $s^2 > 0$. Lie sphere geometry is obtained by introducing a further basis vector of negative signature, f, and replacing s by the null vector

$$\bar{s} = s + |s|f, \qquad \bar{s}^2 = 0. \tag{10.190}$$

Now the spheres encoded by s and $-s$ have different representations as null vectors in a space of signature $(4,2)$. This algebra is ideally suited to handling the contact geometry of spheres. The signature shows that this space is isomorphic to the conformal algebra of spacetime, so in a sense the introduction of the vector f can be thought of as introducing a time direction. A sphere can then be viewed as a light-sphere allowed to grow for a certain time. Orientation for spheres is then handled by distinguishing between incoming and outgoing light-spheres.

The conformal geometry of spacetime is a rich and important subject. The Poincaré group of spacetime translations and rotations is a subgroup of the full conformal group, but in a number of subjects in theoretical physics, including supersymmetry and supergravity, it is the full conformal group that is relevant. One reason is that conformal symmetry is present in most massless theories. This symmetry then has consequences that can carry over to the massive regime. We will not develop the classical approach to spacetime conformal geometry further here. Instead, we concentrate on an alternative route through to conformal geometry, which unites the multiparticle spacetime algebra of chapter 9 with the concept of a *twistor*.

10.7.1 The spacetime conformal group

For most of this chapter we have avoided detailed descriptions of the relationships between the groups involved in the geometric algebra formulation of conformal geometry. For the following, however, it is helpful to have a clearer picture of precisely how the various groups fit together. The subject of Lie groups in general is discussed in chapter 11. The spacetime conformal group $C(1,3)$ consists of spacetime maps $x \mapsto f(x)$ that preserve angles. This is the definition first encountered in section 10.3. The group of orthogonal transformations $O(2,4)$ is a double-cover representation of the conformal group, because in conformal space both X and $-X$ represent the same spacetime point. As with Lorentz transformations, we are typically interested in the restricted conformal group. This consists of transformations that preserve orientation and time sense, and contains translations, proper orthochronous rotations, dilations and special conformal transformations. The restricted orthogonal group, $SO^+(2,4)$, is a double-cover representation of the restricted conformal group.

We can form a double-cover representation of $SO^+(2,4)$ by writing all restricted orthogonal transformations as rotor transformations $a \mapsto Ra\tilde{R}$. The group of conformal rotors, denoted $\text{spin}^+(2,4)$, is therefore a four-fold covering of the restricted conformal group. The rotor group in $\mathcal{G}(2,4)$ is isomorphic to the Lie group $\text{SU}(2,2)$. It follows that the action of the restricted conformal group can be represented in terms of complex linear transformations of four-dimensional vectors, in a complex space of signature $(2,2)$. This is the basis of the *twistor* program, initiated by Roger Penrose. Twistors were introduced as objects describing the geometry of spacetime at a 'pre-metric' level, one of the aims being to provide a route to a quantum theory of gravity. Instead of points and a metric, twistors represent incidence relations between null rays. Spacetime points and their metric relations then emerge as a secondary concept, corresponding to the points of intersection of null lines.

As a first step in understanding the twistor program, we establish a concrete representation of the conformal group within the spacetime algebra. The key to this is the observation that the spinor inner product

$$\langle \tilde{\psi}\phi \rangle_q = \langle \tilde{\psi}\phi \rangle - \langle \tilde{\psi}\phi I\boldsymbol{\sigma}_3 \rangle I\boldsymbol{\sigma}_3 \tag{10.191}$$

defines a complex space with precisely the required metric. The complex structure is represented by right-multiplication by combinations of 1 and $I\boldsymbol{\sigma}_3$, as discussed in chapter 8. We continue to refer to ψ and ϕ as spinors, as they are acted on by a spin representation of the restricted conformal group. To establish a representation in terms of operators on ψ, we first form a representation of the bivectors in $\mathcal{G}(2,4)$ as

$$\begin{aligned} e\gamma_\mu &\leftrightarrow \gamma_\mu \psi \gamma_0 I\boldsymbol{\sigma}_3 = \gamma_\mu \psi I \gamma_3, \\ \bar{e}\gamma_\mu &\leftrightarrow I\gamma_\mu \psi \gamma_0. \end{aligned} \tag{10.192}$$

A representation of the even subalgebra of $\mathcal{G}(2,4)$, and hence an arbitrary rotor, can be constructed from these bivectors. The representation of each of the operations in the restricted conformal group can now be constructed from the rotors found in section 10.3. We use the same symbol for the spinor representation of the transformations as the vector case. A translation by the vector a has the spin representation

$$T_a(\psi) = \psi + a\psi I\gamma_3 \tfrac{1}{2}(1+\boldsymbol{\sigma}_3). \tag{10.193}$$

The spinor inner product of equation (10.191) is invariant under this transformation. To confirm this, suppose that we set

$$\psi' = T_a(\psi) \quad \text{and} \quad \phi' = T_a(\phi). \tag{10.194}$$

The quantum inner product contains the terms

$$\begin{aligned}\langle\tilde{\psi}'\phi'\rangle &= \langle(\phi + a\phi I\gamma_3 \tfrac{1}{2}(1+\boldsymbol{\sigma}_3))(\tilde{\psi} - \tfrac{1}{2}(1-\boldsymbol{\sigma}_3) I\gamma_3\tilde{\psi}a)\rangle \\ &= \langle\tilde{\psi}\phi\rangle\end{aligned} \tag{10.195}$$

and

$$\begin{aligned}\langle\tilde{\psi}'\phi' I\boldsymbol{\sigma}_3\rangle &= \langle(\phi + a\phi I\gamma_3 \tfrac{1}{2}(1+\boldsymbol{\sigma}_3))I\boldsymbol{\sigma}_3(\tilde{\psi} - \tfrac{1}{2}(1-\boldsymbol{\sigma}_3) I\gamma_3\tilde{\psi}a)\rangle \\ &= \langle\tilde{\psi}\phi I\boldsymbol{\sigma}_3\rangle.\end{aligned} \tag{10.196}$$

It follows that

$$\langle\tilde{\psi}'\phi'\rangle_q = \langle\tilde{\psi}\phi\rangle_q, \tag{10.197}$$

as expected.

The spinor representation of a rotation about the origin is precisely the spacetime algebra rotor, so we can write

$$R_0(\psi) = R\psi, \tag{10.198}$$

where R_0 denotes a rotation in the origin, and R is a spacetime rotor. Rotations about arbitrary points are constructed from combinations of translations and rotations. The dilation $x \mapsto \exp(\alpha)x$ has the spinor representation

$$D_\alpha(\psi) = \psi e^{\alpha\boldsymbol{\sigma}_3/2}. \tag{10.199}$$

This represents a dilation in the origin. Dilations about a general point are also obtained from a combination of translations and a dilation in the origin. The representation of the restricted conformal group is completed by the special conformal transformations, which are represented by

$$K_a(\psi) = \psi - a\psi I\gamma_3 \tfrac{1}{2}(1-\boldsymbol{\sigma}_3). \tag{10.200}$$

It is a routine exercise to confirm that the preceding operations do form a spin representation of the restricted conformal group.

The full conformal group includes inversions. These can be represented as *antiunitary* operators. An inversion in the origin is represented by

$$\psi \mapsto \psi' = \psi I \boldsymbol{\sigma}_2. \tag{10.201}$$

The effect of this on the inner product of equation (10.191) is that we form

$$\langle \tilde{\psi}' \phi' \rangle_q = \langle \tilde{\phi} \psi \rangle_q = \left(\langle \tilde{\psi} \phi \rangle_q \right)^{\sim}. \tag{10.202}$$

This representation of an inversion in the origin satisfies

$$D_\alpha(\psi I \boldsymbol{\sigma}_2) = D_{-\alpha}(\psi) I \boldsymbol{\sigma}_2, \tag{10.203}$$

as required.

10.7.2 Multiparticle representation of conformal vectors

We have defined a carrier space for a spin-1/2 representation of the spacetime conformal group. A vector representation of the conformal groups can therefore be constructed from quadratic combinations of spinors. Spinors can be thought of as belonging to a complex four-dimensional space. The tensor product space therefor contains 16 complex degrees of freedom. This decomposes into a ten-dimensional symmetric space and six-dimensional antisymmetric space. The six complex degrees of freedom in the antisymmetric representation are precisely the dimensions required to construct a conformal vector. The ten-dimensional symmetric space has 20 real degrees of freedom, and forms a representation of trivectors in conformal spacetime.

In principle, then, we will form complex vectors in conformal spacetime. But for a special class of spinor the conformal vector is real. If we translate a constant spinor by the position vector $r = x^\mu \gamma_\mu$ we form the object

$$T_r(\psi) = \psi + r\psi I \gamma_3 \tfrac{1}{2}\left(1 + \boldsymbol{\sigma}_3\right), \tag{10.204}$$

which is the spacetime algebra version of a *twistor*. A twistor is essentially a spacetime algebra spinor with a particular position dependence. The key to constructing a real conformal vector from an antisymmetric pair of twistors is to impose the conditions that they are both null, and orthogonal. Suppose that we set

$$\mathsf{X} = T_r(\psi), \quad \mathsf{Z} = T_r(\phi). \tag{10.205}$$

The conditions that these generate a real conformal vector are then

$$\langle \tilde{\mathsf{X}} \mathsf{X} \rangle_q = \langle \tilde{\mathsf{Z}} \mathsf{Z} \rangle_q = \langle \tilde{\mathsf{X}} \mathsf{Z} \rangle_q = 0. \tag{10.206}$$

The position dependence in X and Z does not affect the inner product, so the same conditions must also be satisfied by ψ and ϕ. Choosing appropriate spinors satisfying these relationships essentially amounts to a choice of origin. The most straightforward way to satisfy the requirements is to set

$$\mathsf{X} = \omega\tfrac{1}{2}(1-\boldsymbol{\sigma}_3) + r\omega I\gamma_3\tfrac{1}{2}(1+\boldsymbol{\sigma}_3) \tag{10.207}$$

and

$$\mathsf{Z} = \kappa\tfrac{1}{2}(1-\boldsymbol{\sigma}_3) + r\kappa I\gamma_3\tfrac{1}{2}(1+\boldsymbol{\sigma}_3), \tag{10.208}$$

where ω and κ are Pauli spinors (spinors in the spacetime algebra that commute with γ_0).

To construct a vector from the two twistors X and Z we form their antisymmetrised tensor product in the multiparticle spacetime algebra. We therefore construct the multivector

$$\psi_r = (\mathsf{X}^1\mathsf{Z}^2 - \mathsf{Z}^1\mathsf{X}^2)E, \tag{10.209}$$

where the notation follows section 9.2. If we now make use of the results in table 9.2 we find that

$$\psi_r = (r\cdot r\,\epsilon - r^1\eta\gamma_0^1 J - \bar{\epsilon})\langle I\boldsymbol{\sigma}_2\tilde{\kappa}\omega\rangle_q, \tag{10.210}$$

where η is the Lorentz singlet state defined in equation (9.93), and ϵ and $\bar{\epsilon}$ are defined by

$$\epsilon = \eta\tfrac{1}{2}(1+\boldsymbol{\sigma}_3^1), \quad \bar{\epsilon} = \eta\tfrac{1}{2}(1-\boldsymbol{\sigma}_3^1). \tag{10.211}$$

The two-particle state ψ closely resembles our standard encoding of a point as a null vector in conformal space. The singlet state ϵ represents the point at infinity, and is the spacetime algebra version of the infinity twistor. The opposite ideal, $\bar{\epsilon}$, represents the origin ($r = 0$).

More generally, given arbitrary single-particle spinors, we arrive at a complex six-dimensional vector. Restricting to the real subspace, a general point in this space can be written as the state

$$\psi_P = (V-W)\epsilon + P^1\eta\gamma_0^1 + (V+W)\bar{\epsilon}, \tag{10.212}$$

where

$$P = T\gamma_0 + X\gamma_1 + Y\gamma_2 + Z\gamma_3. \tag{10.213}$$

To form the inner product of such states we require the results that

$$\langle\tilde{\epsilon}\epsilon\rangle_q = \langle\tilde{\bar{\epsilon}}\bar{\epsilon}\rangle_q = 0, \qquad 4\langle\tilde{\epsilon}\bar{\epsilon}\rangle_q = 1. \tag{10.214}$$

Now forming the quantum norm for the state ψ_P we find that

$$2\langle\tilde{\psi}_P\psi_P\rangle_q = T^2 + V^2 - W^2 - X^2 - Y^2 - Z^2. \tag{10.215}$$

So (V, W, T, X, Y, Z) are the coordinates of a six-dimensional vector in a space with signature $(2, 4)$. This establishes the map between a two-particle antisymmetrised spinor and a conformal vector.

Our 'real' state ψ_r can be cast into standard form by removing the complex factor on the right-hand side and setting

$$\psi_r \mapsto \frac{\psi_r}{4\langle\tilde{\psi}_r \epsilon\rangle_q}. \tag{10.216}$$

Once this is done, all reference to the original ω and κ spinors is removed. The inner product between two two-particle states ψ_r and ϕ_s, where ϕ_s represents the point s, returns

$$-\frac{\langle\tilde{\psi}_r \phi_s\rangle_q}{4\langle\tilde{\psi}_r \epsilon\rangle_q \langle\tilde{\phi}_s \epsilon\rangle_q} = (r-s)\cdot(r-s). \tag{10.217}$$

The multiparticle inner product therefore recovers the square of the spacetime distance between points. This result is one reason why points are encoded through pairs of *null* twistors.

We have now established a complete representation of conformal vectors for spacetime in terms of antisymmetrised products of a class of spinors, each evaluated in a single copy of the spacetime algebra. We should now check that our representation of the conformal group through its action on spinors induces the correct vector representation in the two-particle algebra. We start with our standard multiparticle representation of a conformal vector as

$$\psi_r = r\cdot r\, \epsilon - r^1 \eta \gamma_0^1 J - \bar{\epsilon}. \tag{10.218}$$

The first operation to consider is a translation. The spinor representation of a translation by a induces the map

$$\psi_r \mapsto \psi_r' = T_{a^1} T_{a^2} \psi_r. \tag{10.219}$$

After some algebra we establish that

$$\psi_r' = (r+a)\cdot(r+a)\, \epsilon - (r+a)^1 \eta \gamma_0^1 J - \bar{\epsilon}, \tag{10.220}$$

as required.

Next consider a Lorentz rotation centred on the origin. These are easily accomplished as they correspond to multiplying the single-particle spinor by the appropriate rotor. This induces the map

$$\begin{aligned}\psi_r \mapsto R^1 R^2 \psi_r &= r\cdot r\, R^1 R^2 \epsilon - R^1 r^1 R^2 \eta \gamma_0^1 J - R^1 R^2 \bar{\epsilon} \\ &= r\cdot r\, \epsilon - (R r \tilde{R})^1 \eta \gamma_0^1 J - \bar{\epsilon},\end{aligned} \tag{10.221}$$

which achieves the desired rotation. Reflections in planes through the origin are equally easily achieved through the single-particle antiunitary operation

$$\psi \mapsto I a \psi \gamma_2, \tag{10.222}$$

where a is the normal vector to the plane of reflection. Applied to the two-particle state we obtain

$$\psi_r \mapsto a\cdot a\big(r\cdot r\,\epsilon + (ara^{-1})^1\eta\gamma_0^1 J - \bar{\epsilon}\big), \tag{10.223}$$

which is the conformal representation of the reflected vector $-ara^{-1}$. As we also have a representation of translations, we can rotate and reflect about an arbitrary point.

Inversions in the origin are handled in conformal space by an operation that swaps the vectors representing the origin and infinity. In the multiparticle setting we must therefore interchange ϵ and $\bar{\epsilon}$, which is achieved by right-multiplication by $I\boldsymbol{\sigma}_2^1 I\boldsymbol{\sigma}_2^2$,

$$\begin{aligned}\psi_r \mapsto \psi_r I\boldsymbol{\sigma}_2^1\, I\boldsymbol{\sigma}_2^2 &= -r\cdot r\,\bar{\epsilon} + r^1\eta\gamma_0^1 J + \epsilon \\ &= -r\cdot r(r'\cdot r'\,\epsilon - (r')^1\eta\gamma_0^1 J - \bar{\epsilon}),\end{aligned} \tag{10.224}$$

where $r' = r/(r\cdot r)$. Dilations in the origin are performed in a similar manner, this time by scaling ϵ and $\bar{\epsilon}$ through opposite amounts. This is successfully achieved by the two-particle map induced by equation (10.199),

$$\psi_r \mapsto \psi_r' = \psi_r \mathrm{e}^{\alpha/2(\boldsymbol{\sigma}_3^1 + \boldsymbol{\sigma}_3^2)}. \tag{10.225}$$

Special conformal transformations are also handled in the obvious way as the two-particle extension of the K_a operator of equation (10.200). This completes the description of the conformal group in the two-particle spacetime algebra setting.

Conformal spacetime geometry can be formulated in an entirely 'quantum' language in terms of multiparticle states built from spinor representations of the conformal group. This link between multiparticle quantum theory and conformal geometry is quite remarkable, and is the basis for the twistor programme. But one obvious question remains — is this abstract quantum-mechanical formulation necessary, if all one is interested is the conformal geometric algebra of spacetime? If the twistor programme is simply a highly convoluted way of discussing conformal geometric algebra, then the answer is no. The question is whether there is anything more fundamental about the quantum framework of the twistor approach.

Advocates of the twistor program would argue that the route we have followed here, which embeds a twistor within the spacetime algebra, reverses the logic which initially motivates twistors. The idea is that they exist at a pre-metric level, so that the spacetime interval between points emerges from a particular two-particle quantum inner product. This hints at a route to a quantum theory of gravity, where distance becomes a quantum observable. But much of the initial promise of this work remains unfulfilled, and twistors are no longer the most popular candidate for a quantum theory of gravity. For classical applications to real spacetime geometry it does appear that all twistor methods have direct counterparts in the geometric algebra $\mathcal{G}(2,4)$, and the latter approach avoids much of the additional formal baggage required when employing twistors.

10.8 Notes

The authors would like to thank Joan Lasenby for her help in writing this chapter.
The subjects discussed in this chapter range from the foundations of algebraic geometry, dating back to the nineteenth century and before, through to some very modern applications. An excellent introduction to geometry is the book *Geometry* by Brannan, Esplen & Gray (1999). Projective geometry is described in the classic text by Semple & Kneebone (1998), and Lie sphere geometry is described by Cecil (1992). A valuable tool for studying two-dimensional geometry is the software package *Cinderella*, written by Richter-Gebert and Kortenkamp. This package was used to produce a number of the illustrations in this chapter.

The geometric algebra formulation of projective geometry is described in the pair of important papers 'The design of linear algebra and geometry' by Hestenes and 'Projective geometry with Clifford algebra' by Hestenes & Ziegler (both 1991). These papers also include preliminary discussions of conformal geometry, though the approach is different to that taken here. Projective geometry is particularly relevant to the field of computer graphics, and some applications of geometric algebra in this area are discussed in the papers by Stevenson & Lasenby (1998) and Perwass & Lasenby (1998).

The systematic study of conformal geometry with geometric algebra was only initiated in the 1990s and is one of the fastest developing areas of current research. Some of the earliest developments are contained in *Clifford Algebra to Geometric Calculus* by Hestenes & Sobczyk (1984), and in the paper 'Distance geometry and geometric algebra' by Dress & Havel (1993), which emphasises the role of the conformal metric. Uncovering the roles of the various geometric primitives in conformal space was initiated by Hestenes (2001) in the paper 'Old wine in new bottles: a new algebraic framework for computational geometry' and is described in detail in the papers by Hestenes, Li & Rockwood (1999a,b). Applications to the study of surfaces are described in the paper 'Surface evolution and representation using geometric algebra' by Lasenby & Lasenby (2000b), and a range of further applications are discussed in the proceedings of the 2001 conference *Applications of Geometric Algebra in Computer Science and Engineering* (Dorst, Doran & Lasenby, 2002). The rapid development of the subject has meant that a consistent notation is yet to be established by all authors.

The unification of Euclidean and non-Euclidean geometry in the conformal framework is also described in the series of papers by Hestenes, Li & Rockwood (1999a,b) and in a separate paper by Li (2001). The development in this chapter goes further than these papers in giving a concrete realisation of traditional methods within the geometric algebra framework. Twistor techniques are described in volume II of *Spinors and Space-time* by Penrose & Rindler (1986). A preliminary discussion of how twistors are incorporated into spacetime algebra is contained in the paper '2-spinors, twistors and supersymmetry in the spacetime algebra' by Lasenby, Doran & Gull (1993b). The multiparticle description of conformal vectors is discussed in the paper 'Applications of geometric algebra in physics and links with engineering' by Lasenby & Lasenby (2000a). Due to a printing error all dot products in this paper appear as deltas, though once one knows this the paper is readable!

10.9 Exercises

10.1 Let A, B, C, D denote four points on a line, and write their cross ratio as $(ABCD)$. Given that $(ABCD) = k$, prove that

$$(BACD) = (ABDC) = 1/k$$

and

$$(ACBD) = (DBCA) = 1 - k.$$

10.2 Prove that the cross ratio of four collinear points is a projective invariant, regardless of the size of the space containing the line.

10.3 Given four points in a plane, no three of which are collinear, prove that there exists a projective transformation that maps these to any second set of four points, where again no three are collinear.

10.4 The vectors a, b, c, a', b', c' all belong to $\mathcal{G}(3,0)$. From these we define the bivectors

$$A = b \wedge c, \quad B = c \wedge a, \quad C = a \wedge b,$$

with the same definitions holding for A', B', C'. Prove that

$$\langle A \times A' \, B \times B' \, C \times C' \rangle = \langle a \wedge b \wedge c \, a' \wedge b' \wedge c' \rangle \langle a \wedge a' \, b \wedge b' \, c \wedge c' \rangle.$$

This proves Desargues' theorem for two triangles in a common plane. Does the theorem still hold in three dimensions when the triangles lie on different planes?

10.5 Given six vectors $a_1, \ldots, a_6$ representing points in the projective plane, prove that

$$\frac{a_5 \wedge a_4 \wedge a_3}{a_5 \wedge a_1 \wedge a_3} \frac{a_6 \wedge a_2 \wedge a_1}{a_6 \wedge a_2 \wedge a_4} = \frac{A_{543}}{A_{513}} \frac{A_{621}}{A_{624}},$$

where A_{ijk} is the area of the triangle whose vertices are described projectively by the vectors a_i, a_j, a_k. How does this ratio of areas transform under a projective transformation?

10.6 A Möbius transformation in the complex plane is defined by

$$z \mapsto z' = \frac{az+b}{cz+d},$$

where a, b, c, d are complex numbers. Prove that, viewed as a map of the complex plane onto itself, a Möbius transformation is a conformal transformation. Can all conformal transformations in the plane be represented as Möbius transformations? If not, which operation is missing?

10.7 Find the general form of the rotor, in conformal space, for a rotation through θ in the $a \wedge b$ plane, about the point with position vector a.

10.8 A special conformal transformation in Euclidean space corresponds to a combination of an inversion in the origin, a translation by b and a further inversion in the origin. Prove that the result of this can be written

$$x \mapsto = x \frac{1}{1+bx}.$$

Hence show that the linear function $\mathsf{f}(a) = a \cdot \nabla x$ is given by

$$\mathsf{f}(a) = \frac{(1+bx)a(1+xb)}{(1+2b \cdot x + b^2x^2)^2}.$$

Why does this transformation leave angles unchanged?

10.9 Given a conformal bivector B, with $B^2 > 0$, why does this encode a pair of Euclidean points? Prove that the midpoint of these two points is described by

$$C = BnB.$$

10.10 Two circles in a Euclidean plane are described by conformal trivectors L_1 and L_2. By expressing the dual vectors l_1 and l_2 in terms of the centre and radius of the circles, confirm directly that the circles intersect at right angles if

$$l_1 \cdot l_2 = 0.$$

10.11 The conformal vector X denotes a point lying on the circle L, $L \wedge X = 0$, where L is a trivector. Prove that the tangent vector T to the circle at X can be written

$$T = (X \cdot L) \wedge n.$$

10.12 A non-Euclidean translation along the line through X and Y is generated by the bivector $B = Le$, where

$$L = X \wedge Y \wedge e.$$

Prove that the hyperbolic angle α which takes us from X to Y is given by

$$\cosh(\alpha) = 1 - \frac{X \cdot Y}{X \cdot e \, Y \cdot e}.$$

10.13 The line element over the Poincaré disc is defined by

$$ds^2 = \frac{1}{1-r^2}(dr^2 + r^2 d\theta^2),$$

where r and θ are polar coordinates and $r < 1$. Prove that geodesics in this geometry all intersect the circle $r = 1$ at right angles.

10.14 Suppose that ψ is an even element of the spacetime algebra. This is acted on by the following linear transformations:

$$\begin{aligned}
R_0(\psi) &= R\psi, \\
T_a(\psi) &= \psi + a\psi I\gamma_3 \tfrac{1}{2}\left(1 + \boldsymbol{\sigma}_3\right), \\
D_\alpha(\psi) &= \psi e^{\alpha\boldsymbol{\sigma}_3/2}, \\
K_a(\psi) &= \psi - a\psi I\gamma_3 \tfrac{1}{2}\left(1 - \boldsymbol{\sigma}_3\right),
\end{aligned}$$

where R is a spacetime rotor. Prove that this set of linear transformations generate a representation of the restricted conformal group of spacetime.

11

Further topics in calculus and group theory

In this chapter we collect together a number of diverse algebraic ideas and techniques. The first part of the chapter deals with some advanced topics in calculus. We introduce the *multivector derivative*, which is a valuable tool in Lagrangian analysis. We also show how the vector derivative can be adapted to provide a compact notation for studying linear functions. We then extend the multivector derivative to the case where we differentiate with respect to a linear function. Finally in this part we look briefly at Grassmann calculus, which is a major ingredient in modern quantum field theory.

The second major topic covered in this chapter is the theory of Lie groups. We provide a detailed analysis of spin groups over a real geometric algebra. By introducing invariant bivectors we show how both the unitary and general linear groups can be represented in terms of spin groups. It then follows that all Lie algebras can be represented as bivector algebras under the commutator product. Working in this way we construct the main Lie groups as subgroups of rotation groups. This is a valuable alternative procedure to the more common method of describing Lie groups in terms of matrices. Throughout this chapter we use the tilde symbol for the reverse, $\tilde{R}$. This avoids confusion with the Hermitian conjugate, which is required in section 11.4 on complex structures.

11.1 Multivector calculus

Before extending our analysis of linear functions in geometric algebra, we first discuss differentiation with respect to a multivector. Suppose that the multivector F is an arbitrary function of some multivector argument X, $F = F(X)$. The derivative of F with respect to X in the A direction is defined by

$$A*\partial_X F(X) = \lim_{\tau \mapsto 0} \frac{F(X + \tau A) - F(X)}{\tau}, \tag{11.1}$$

where $A*B = \langle AB \rangle$. The multivector derivative ∂_X is defined in terms of its directional derivatives by

$$\frac{\partial}{\partial X} = \partial_X = \sum_{i<\cdots<j} \mathsf{e}^i \wedge \cdots \wedge \mathsf{e}^j (\mathsf{e}_j \wedge \cdots \wedge \mathsf{e}_i) * \partial_X, \tag{11.2}$$

where the $\{\mathsf{e}^i\}$ are a set of frame vectors for the space of interest. The definition shows how the multivector derivative ∂_X inherits the multivector properties of its argument X, as well as a calculus from equation (11.1). This is the natural generalisation of the vector derivative ∇ to a general multivector.

Most of the properties of the multivector derivative follow from the result that

$$\partial_X \langle XA \rangle = P_X(A), \tag{11.3}$$

where $P_X(A)$ is the projection of A onto the grades contained in X. Leibniz's rule is then used to build up results for more complicated functions. We employ the same rules for the multivector derivative as for the vector derivative. The derivative acts on objects to its immediate right unless brackets are present. If the ∂_X is intended to only act on B then this is written as $\dot{\partial}_X A \dot{B}$, where the overdot denotes the multivector on which the derivative acts. For example, Leibniz's rule can be written as

$$\partial_X (AB) = \dot{\partial}_X \dot{A} B + \dot{\partial}_X A \dot{B}. \tag{11.4}$$

As an example, suppose that ψ is a general even element. The derivative of the scalar product $\langle \psi \tilde{\psi} \rangle$ is

$$\partial_\psi \langle \psi \tilde{\psi} \rangle = \dot{\partial}_\psi \langle \dot{\psi} \tilde{\psi} \rangle + \dot{\partial}_\psi \langle \psi \dot{\tilde{\psi}} \rangle = 2\tilde{\psi}. \tag{11.5}$$

For the second term we used the result that

$$\dot{\partial}_\psi \langle \psi \dot{\tilde{\psi}} \rangle = \dot{\partial}_\psi \langle \dot{\psi} \tilde{\psi} \rangle = \tilde{\psi}, \tag{11.6}$$

which follows from the fact that any scalar term reverses to give itself. This result for the derivative of $\langle \psi \tilde{\psi} \rangle$ can be verified rather more laboriously by expanding out in a basis.

11.1.1 The vector derivative and multilinear algebra

The derivative with respect to a vector was first introduced in chapter 6 as an essential component of field theory. Here we exploit the properties of the vector derivative in a rather different setting. Suppose that a denotes an arbitrary vector. We write the derivative with respect a as ∂_a. Algebraically, this derivative has the properties of a vector. It is essentially the same object as the vector derivative, except that we are not differentiating with respect to the position dependence of a function. Instead we will use ∂_a to differentiate a variety of expressions that are linear in a. Introducing the tools of calculus may appear unnecessary for the analysis of linear algebra, but the notation does have some practical advantages. Combinations of a and ∂_a can be used to perform contractions and protractions without having to introduce a basis frame. For example, the results of section 4.3.2 can be summarised in the compact formulae

$$\begin{aligned}\partial_a a\cdot A_r &= rA_r,\\ \partial_a a\wedge A_r &= (n-r)A_r,\\ \partial_a A_r a &= (-1)^r(n-2r)A_r.\end{aligned} \tag{11.7}$$

Similarly, the vector derivative allows the trace of a linear function to be written simply as

$$\mathrm{tr}(\mathsf{f}) = \partial_a\cdot\mathsf{f}(a). \tag{11.8}$$

The trace is the first of a series of scalar invariants that can be defined from f. These are compactly handled using the vector derivative. Suppose that $\{a_1, a_2, \ldots, a_n\}$ denote a set of n independent vectors. We define the multivector variable

$$a_{(r)} = a_1\wedge a_2\wedge\cdots\wedge a_r \tag{11.9}$$

with the associated derivative

$$\partial_{(r)} = \frac{1}{r!}\partial_{a_r}\wedge\partial_{a_{r-1}}\wedge\cdots\wedge\partial_{a_1}. \tag{11.10}$$

Since

$$\langle A_r\wedge\partial_a\, a\wedge B_r\rangle = (n-r)\langle A_r B_r\rangle, \tag{11.11}$$

it follows that

$$\partial_{(r)}a_{(r)} = \frac{n!}{(n-r)!\,r!} = \binom{n}{r}. \tag{11.12}$$

We also make the further abbreviation

$$\mathsf{f}(a_{(r)}) = \mathsf{f}(a_1)\wedge\mathsf{f}(a_2)\wedge\cdots\wedge\mathsf{f}(a_r) = \mathsf{f}_{(r)}. \tag{11.13}$$

This notation allows us to write

$$\partial_{(1)}\cdot\mathsf{f}_{(1)} = \partial_{a_1}\cdot\mathsf{f}(a_1) = \mathrm{tr}(\mathsf{f}) \tag{11.14}$$

and

$$\partial_{(n)}\mathsf{f}_{(n)} = \partial_{(n)}a_{(n)}\mathrm{det}\,(\mathsf{f}) = \mathrm{det}\,(\mathsf{f}). \tag{11.15}$$

These two invariants are clearly special cases of the range of invariants $\partial_{(r)}\cdot\mathsf{f}_{(r)}$.

To understand the importance of the $\partial_{(r)}\cdot\mathsf{f}_{(r)}$ invariants, consider the characteristic polynomial for f. This is formed by constructing the determinant of the function $\mathsf{G}(a) = \mathsf{f}(a) - \lambda a$, which yields

$$\begin{aligned}\det(\mathsf{G}) &= \partial_{(n)}\mathsf{G}_{(n)} \\ &= \partial_{(n)}\big(\mathsf{f}(a_1) - \lambda a_1\big)\wedge\big(\mathsf{f}(a_2) - \lambda a_2\big)\wedge\cdots\wedge\big(\mathsf{f}(a_n) - \lambda a_n\big) \\ &= \partial_{(n)}\big(\mathsf{f}_{(n)} - n\lambda\mathsf{f}_{(n-1)}\wedge a_n + \cdots + (-\lambda)^n a_{(n)}\big). \end{aligned} \tag{11.16}$$

A general term in this expression goes as

$$(-\lambda)^s\binom{n}{s}\partial_{(n)}\cdot\big(\mathsf{f}_{(n-s)}\wedge a_{n-s+1}\wedge\cdots\wedge a_n\big) = (-\lambda)^s\partial_{(n-s)}\cdot\mathsf{f}_{(n-s)}. \tag{11.17}$$

It follows that the characteristic polynomial is simply

$$C(\lambda) = \sum_{s=0}^{n}(-\lambda)^{n-s}\partial_{(s)}\cdot\mathsf{f}_{(s)}, \tag{11.18}$$

where $\partial_{(0)}\cdot\mathsf{f}_{(0)} = 1$. This expression clearly demonstrates the significance of the invariant quantities $\partial_{(r)}\cdot\mathsf{f}_{(r)}$.

The *Cayley–Hamilton* theorem states that

$$\sum_{s=0}^{n}(-1)^{n-s}\partial_{(s)}\cdot\mathsf{f}_{(s)}\,\mathsf{f}^{n-s}(a) = 0, \tag{11.19}$$

where $\mathsf{f}^r(a)$ denotes the r-fold application of f on a. This says that a linear function satisfies its own characteristic equation. The theorem can be proved quite generally without any assumptions about the form of f — it applies for any linear function, in any linear space of any dimension and signature. An immediate consequence is that, if e is an eigenvector of f,

$$\mathsf{f}(e) = \lambda e, \tag{11.20}$$

then λ automatically satisfies the characteristic equation.

11.1.2 Calculus for linear functions

As well as the ability to differentiate with respect to a multivector, it is also very useful to build up results for the derivative with respect to a linear function. We start by introducing a fixed frame $\{\mathsf{e}_i\}$, and define the scalar coefficients

$$\mathsf{f}_{ij} = \mathsf{e}_i\cdot\mathsf{f}(\mathsf{e}_j). \tag{11.21}$$

Now consider the derivative with respect to f_{ij} of the scalar $\mathsf{f}(b)\cdot c$. This is

$$\begin{aligned}\partial_{\mathsf{f}_{ij}}\mathsf{f}(b)\cdot c &= \partial_{\mathsf{f}_{ij}}(\mathsf{f}_{lk}b^k c^l) \\ &= c^i b^j. \end{aligned} \tag{11.22}$$

Multiplying both sides of this equation by $a\cdot\mathsf{e}_j\,\mathsf{e}_i$ we obtain

$$a\cdot\mathsf{e}_j\,\mathsf{e}_i\partial_{\mathsf{f}_{ij}}\mathsf{f}(b)\cdot c = a\cdot b\, c, \tag{11.23}$$

which assembles a frame-independent vector on the right-hand side. It follows that the operator $a \cdot \mathsf{e}_j\, \mathsf{e}_i \partial_{f_{ij}}$ must also be frame-independent. We therefore define the vector-valued differential operator $\partial_{\mathsf{f}(a)}$ by

$$\partial_{\mathsf{f}(a)} = a\cdot\mathsf{e}_j\, \mathsf{e}_i \partial_{f_{ij}}. \tag{11.24}$$

The essential property of $\partial_{\mathsf{f}(a)}$ is

$$\partial_{\mathsf{f}(a)} \mathsf{f}(b)\cdot c = a\cdot b\, c, \tag{11.25}$$

which simply restates equation (11.23). As with the vector derivative, $\partial_{\mathsf{f}(a)}$ has the algebraic properties of a vector, which can be exploited in analysing a range of expressions.

Equation (11.25), together with Leibniz's rule, is sufficient to derive the main results for the $\partial_{\mathsf{f}(a)}$ operator. For example, suppose that B is a bivector, and we construct

$$\begin{aligned}\partial_{\mathsf{f}(a)} \langle \mathsf{f}(b\wedge c) B\rangle &= \dot{\partial}_{\mathsf{f}(a)} \langle \dot{\mathsf{f}}(b)\mathsf{f}(c)B\rangle - \dot{\partial}_{\mathsf{f}(a)} \langle \dot{\mathsf{f}}(c)\mathsf{f}(b)B\rangle \\ &= a\cdot b\, \mathsf{f}(c)\cdot B - a\cdot c\, \mathsf{f}(b)\cdot B \\ &= \mathsf{f}\big(a\cdot(b\wedge c)\big)\cdot B.\end{aligned} \tag{11.26}$$

This extends by linearity to give

$$\partial_{\mathsf{f}(a)} \langle \mathsf{f}(A) B\rangle = \mathsf{f}(a\cdot A)\cdot B, \tag{11.27}$$

where A and B are both bivectors. Proceeding in this manner, we obtain the general formula

$$\partial_{\mathsf{f}(a)} \langle \mathsf{f}(A) B\rangle = \sum_r \langle \mathsf{f}(a\cdot A_r) B_r\rangle_1. \tag{11.28}$$

For a fixed grade-r multivector A_r, we can now write

$$\begin{aligned}\partial_{\mathsf{f}(a)} \mathsf{f}(A_r) &= \partial_{\mathsf{f}(a)} \langle \mathsf{f}(A_r)\dot{X}_r\rangle \dot{\partial}_{X_r} \\ &= \mathsf{f}(a\cdot A_r)\cdot \dot{X}_r\, \dot{\partial}_{X_r} \\ &= (n-r+1)\mathsf{f}(a\cdot A_r).\end{aligned} \tag{11.29}$$

This is a very powerful result. For example, suppose that for A_r we take the pseudoscalar I. We obtain

$$\partial_{\mathsf{f}(a)} \mathsf{f}(I) = \partial_{\mathsf{f}(a)} \det\,(\mathsf{f}) I = \mathsf{f}(a\cdot I). \tag{11.30}$$

It follows that

$$\partial_{\mathsf{f}(a)} \det\,(\mathsf{f}) = \det\,(\mathsf{f}) \bar{\mathsf{f}}^{-1}(a), \tag{11.31}$$

where we have used equation (4.152) This derivation is considerably more compact than any available to conventional matrix/tensor methods.

Equation (11.28) can be used to derive formulae for the functional derivative of the adjoint. The general result is

$$\begin{aligned}\partial_{\mathsf{f}(a)}\bar{\mathsf{f}}(A_r) &= \partial_{\mathsf{f}(a)}\langle \mathsf{f}(\dot{X}_r)A_r\rangle \dot{\partial}_{X_r} \\ &= \mathsf{f}(a\cdot\dot{X}_r)\cdot A_r\,\dot{\partial}_{X_r}.\end{aligned} \tag{11.32}$$

When A is a vector, this admits the simpler form

$$\partial_{\mathsf{f}(a)}\bar{\mathsf{f}}(b) = ba. \tag{11.33}$$

If f is a symmetric function then $\mathsf{f} = \bar{\mathsf{f}}$. But this fact cannot be exploited when differentiating with respect to f, since f_{ij} and f_{ji} must be treated as independent variables for the purposes of calculus.

11.2 Grassmann calculus

For most of his lifetime, Grassmann's work on algebra and geometry was largely ignored by the wider mathematical community. Today, however, Grassmann algebra is a fundamental ingredient in theoretical physics. Fermionic creation operators generate a Grassmann algebra, and Grassmann (anticommuting) variables are important components of path-integral quantisation, supersymmetry and string theory. In this section we describe how the main algebraic results of Grassmann calculus can be formulated in a straightforward manner within geometric algebra. This reverses the standard approach, by which one progresses from Grassmann to Clifford algebra via quantization.

Suppose that $\{\zeta_i\}$ are a set of n Grassmann variables, satisfying the anticommutation relations

$$\{\zeta_i, \zeta_j\} = 0. \tag{11.34}$$

The Grassmann variables $\{\zeta_i\}$ are mapped into geometric algebra by introducing a set of n linearly independent vectors $\{e_i\}$. We do not need to specify any properties for their inner products, though some calculations are performed more easily if we assume that the $\{e_i\}$ belong to a Euclidean algebra. The role of the product of Grassmann variables is taken over by the exterior product in geometric algebra, so we write

$$\zeta_i\zeta_j \;\leftrightarrow\; e_i \wedge e_j. \tag{11.35}$$

Equation (11.34) is satisfied by virtue of the antisymmetry of the exterior product. Any combination of Grassmann variables can now be replaced in the obvious manner by a multivector.

In order for the above scheme to have computational power, we need a translation for the Grassmann calculus introduced by Berezin. In this calculus, differentiation is defined by the rules

$$\frac{\partial \zeta_j}{\partial \zeta_i} = \delta_{ij}, \qquad \zeta_j \frac{\overleftarrow{\partial}}{\partial \zeta_i} = \delta_{ij}, \tag{11.36}$$

together with the graded Leibniz rule,

$$\frac{\partial}{\partial \zeta_i}(f_1 f_2) = \frac{\partial f_1}{\partial \zeta_i} f_2 + (-1)^{[f_1]} f_1 \frac{\partial f_2}{\partial \zeta_i}, \tag{11.37}$$

where $[f_1]$ is the parity of f_1. The parity of a Grassmann variable is determined by whether it contains an even or odd number of vectors. Berezin differentiation is handled within the algebra generated by the $\{e_i\}$ frame by introducing the reciprocal frame $\{e^i\}$, and replacing

$$\frac{\partial}{\partial \zeta_i} f \leftrightarrow e^i \cdot f \tag{11.38}$$

so that

$$\frac{\partial \zeta_j}{\partial \zeta_i} \leftrightarrow e^i \cdot e_j = \delta^i_j. \tag{11.39}$$

The graded Leibniz rule follows from the basic identities of geometric algebra. For example, if f_1 and f_2 are grade-1 and so are treated as vectors in geometric algebra, then the rule (11.37) simply restates the familiar result

$$e^i \cdot (f_1 \wedge f_2) = e^i \cdot f_1 \, f_2 - f_1 \, e^i \cdot f_2. \tag{11.40}$$

Right action by a Grassmann derivative operator translates in a similar manner:

$$(f) \frac{\overleftarrow{\partial}}{\partial \zeta_i} \leftrightarrow f \cdot e^i. \tag{11.41}$$

The standard results for Grassmann calculus follow simply from this basic translation scheme.

Grassmann integration is defined to be essentially the same operation as right differentiation:

$$\int f(\zeta) d\zeta_n d\zeta_{n-1} \cdots d\zeta_1 = f(\zeta) \frac{\overleftarrow{\partial}}{\partial \zeta_n} \frac{\overleftarrow{\partial}}{\partial \zeta_{n-1}} \cdots \frac{\overleftarrow{\partial}}{\partial \zeta_1}. \tag{11.42}$$

The equivalent operation in geometric algebra is therefore a right-sided contraction, as given in equation (11.38). The most important formula is that for the total integral

$$\int f(\zeta) d\zeta_n d\zeta_{n-1} \cdots d\zeta_1 \leftrightarrow (\cdots((F \cdot e^n) \cdot e^{n-1}) \cdots) \cdot e^1 = \langle F E^n \rangle, \tag{11.43}$$

where F is the multivector equivalent of $f(\zeta)$ and E^n is the pseudoscalar for the $\{e^i\}$ vectors,

$$E^n = e^n \wedge e^{n-1} \wedge \cdots \wedge e^1. \tag{11.44}$$

Equation (11.43) does nothing more than pick out the coefficient of the pseudoscalar part of F.

A 'change of variables' is performed by a linear transformation f, with

$$e_i' = \mathsf{f}(e_i), \qquad e^{i\prime} = \bar{\mathsf{f}}^{-1}(e^i). \tag{11.45}$$

It follows that

$$E_n' = \det\,(\mathsf{f})\,E_n, \qquad E^{n\prime} = \det\,(\mathsf{f})^{-1}\,E^n, \tag{11.46}$$

so that a change of variables in a Grassmann multiple integral picks up a Jacobian factor of $\det\,(\mathsf{f})^{-1}$. This contrasts with the factor of $\det\,(\mathsf{f})$ for a Riemannian integral. In a similar manner all of the main results of Grassmann calculus can be derived in geometric algebra. Often these derivations are simpler, as access to the geometric product offers a quick route through the algebra.

11.3 Lie groups

In earlier chapters we saw that rotors form a continuous group, in the same way that rotations do. Continuous groups of this type are called *Lie groups*, after the mathematician Sophus Lie, and they play an important role in a wide range of subjects in physics. Lie groups contain an infinite number of elements but, like vector spaces, the elements can usually be written in terms of a finite number of parameters. For example, three-dimensional rotations can be parameterised in terms of the three Euler angles. The reason is that the elements of the group belong to a topological space — the *group manifold*. In two-dimensional Euclidean space all rotors correspond to phase factors, so the rotor group manifold is the unit circle. Every point on the circle corresponds to a distinct rotor.

Similarly, in three dimensions rotors are built from the space of scalars and bivectors. The only condition they have to satisfy is that $R\tilde{R} = 1$. Suppose that we write

$$R = x_0 + x_1 I e_1 + x_2 I e_2 + x_3 I e_3. \tag{11.47}$$

Then

$$R\tilde{R} = {x_0}^2 + {x_1}^2 + {x_2}^2 + {x_3}^2 = 1. \tag{11.48}$$

This defines a unit vector in the four-dimensional space spanned by $\{x_0, x_i\}$. The group manifold is therefore the set of unit vectors in four-dimensional space. This is called a 3-sphere S^3 — it is the four-dimensional analogue of the surface of a ball. In higher dimensions the rotor group manifolds become increasingly more complicated.

Since all rotations are generated by the double-sided formula $Ra\tilde{R}$, both R and $-R$ correspond to the same rotation. The group manifold for three-dimensional rotations, rather than for the rotors themselves, is therefore more complicated that S^3. It involves taking a 3-sphere and projectively identifying opposite points. The fact that the group manifold for rotors is somewhat simpler than that for rotations has many applications. If the orientation of a rigid body is described by a rotor, the configuration space for the dynamics of the rigid body is a 3-sphere. This is important when looking for best-fit rotations, or extrapolating between two rotations to find their midpoint. The group manifold is also the appropriate setting for a Lagrangian treatment. This has implications for constructing conjugate momenta, which are essential for the transition to a quantum theory. Applications of this include the rotational energy levels of molecules, many of which can be viewed as rigid bodies.

11.3.1 Formal definitions

The fact that the elements of a Lie group belong to a manifold is sufficient to provide an abstract definition of a general Lie group. A Lie group is defined as a manifold, $\mathcal{M}$, together with a product $\phi(x,y)$. Points on the manifold can be labelled with vectors $\{x,y\}$, which can be viewed as lying in a higher dimensional embedding space (as with the 3-sphere). The product $\phi(x,y)$ takes as its argument two points in the manifold, and returns a third. This encodes the group product. The final set of conditions apply to $\phi(x,y)$ and ensure that the product has the correct group properties. These are

(i) *Closure.* $\phi(x,y) \in \mathcal{M} \quad \forall x,y \in \mathcal{M}$.
(ii) *Identity.* There exists an element $e \in \mathcal{M}$ such that $\phi(e,x) = \phi(x,e) = x$, $\forall x \in \mathcal{M}$.
(iii) *Inverse.* For every element $x \in \mathcal{M}$ there exists a unique element $\bar{x}$ such that $\phi(x,\bar{x}) = \phi(\bar{x},x) = e$.
(iv) *Associativity.* $\phi(\phi(x,y),z) = \phi(x,\phi(y,z)), \quad \forall x,y,z \in \mathcal{M}$.

Any manifold with a product defined on it with the preceding properties is called a Lie group manifold. Many of the group properties of the group can be uncovered by examining the properties near the identity element. The product then induces a *Lie bracket* structure on elements of the tangent space at the identity. The tangent space is a linear space and the vectors in this space, together with their bracket, form a Lie algebra.

11.3.2 Spin groups and the bivector algebra

The general theory of Lie groups is rather too abstract for our purposes. Instead, we will adopt a different approach to the subject by concentrating on the properties of rotors, and their associated spin groups. The Lie algebra of a spin group is defined by a set of bivectors. We will establish that every Lie algebra can be represented as a bivector algebra, and that every matrix Lie group can be represented in terms of a spin group.

Before proceeding, we need to clarify some of the terminology for the various groups discussed in this chapter. We let $\mathcal{G}(p,q)$ denote the geometric algebra of a space of signature p,q, and write $\mathcal{V}$ for the space of grade-1 vectors. The orthogonal group $\mathrm{O}(p,q)$ is the set of all linear transformations f mapping $\mathcal{V} \mapsto \mathcal{V}$ that preserve the inner product. That is,

$$\bar{\mathsf{f}}\mathsf{f}(a) = a \quad \forall a \in \mathcal{V}. \tag{11.49}$$

Orthogonal transformations can have determinant 1 or -1. The special orthogonal group $\mathrm{SO}(p,q)$ is the subgroup of $\mathrm{O}(p,q)$ of linear transformations with determinant 1. Orthogonal transformations can be constructed from series of reflections, each of which can be written as

$$a \mapsto -mam^{-1}, \tag{11.50}$$

where m is a non-null vector. Reflections have determinant -1, so do not belong to $\mathrm{SO}(p,q)$. If we restrict m to be a unit vector, $m^2 = \pm 1$, then the set of all unit vectors form a group under the geometric product. This is called the *pin* group, $\mathrm{Pin}(p,q)$. The pin group is a double-cover representation of the orthogonal group. The elements of the pin group all satisfy

$$M\tilde{M} = \pm 1 \quad \forall M \in \mathrm{Pin}(p,q). \tag{11.51}$$

The elements of the pin group split into those of even grade, and those of odd grade. The even-grade elements form a subgroup called the *spin* group, $\mathrm{Spin}(p,q)$. The spin group consists of even-grade multivectors $S \in \mathcal{G}(p,q)$ satisfying

$$SaS^{-1} \in \mathcal{V} \quad \forall a \in \mathcal{V}, \qquad S\tilde{S} = \pm 1. \tag{11.52}$$

The transformations defined by S all have determinant $+1$, so the spin group is a double-cover representation of the special orthogonal group $\mathrm{SO}(p,q)$.

Rotors are elements of the spin group satisfying the further constraint that $R\tilde{R} = 1$. These define the *rotor* group, sometimes denoted $\mathrm{Spin}^+(p,q)$. For rotors we have $R^{-1} = \tilde{R}$, and their action on multivectors is defined by the familiar double-sided formula

$$M \mapsto RM\tilde{R}. \tag{11.53}$$

With the exception of rotors in $\mathcal{G}(1,1)$, the rotor group is a subgroup of the spin group consisting of elements that are *connected* to the identity. That is, all elements of the rotor group can be connected to the identity by a single unbroken path in the group manifold. It follows that rotors form a double-cover representation of the connected subgroup of $\mathrm{SO}(p,q)$. For Euclidean spaces the special orthogonal group is connected, and for these spaces there is no distinction between the spin group and rotor group. In mixed signature spaces the spin group differs from the rotor group by the direct product with a discrete group. For example, the rotor group in spacetime is a representation of the group of proper orthochronous transformations (see section 5.4).

In Euclidean spaces we know that all rotations can be written as the exponential of a bivector. The natural question now is can any rotor be written as the exponential of a bivector? To answer this question, consider a family of rotors $R(\lambda)$, which specifies a path on the rotor group manifold. Differentiating the normalisation condition $R\tilde{R} = 1$ we find that

$$\frac{d}{d\lambda}(R\tilde{R}) = 0 = R'\tilde{R} + R\tilde{R}', \tag{11.54}$$

where the primes denote differentiation with respect to λ. Now define the set vectors

$$a(\lambda) = R(\lambda)a_0\tilde{R}(\lambda), \tag{11.55}$$

where a_0 is some fixed initial vector. Differentiating this expression we find that

$$\frac{d}{d\lambda}a(\lambda) = R'a_0\tilde{R} + Ra_0\tilde{R}' = (R'\tilde{R})a(\lambda) - a(\lambda)(R'\tilde{R}). \tag{11.56}$$

The quantity $R'\tilde{R}$ reverses to minus itself, so can only contain terms of grade 2, 6, 10 etc. But the commutator of $R'\tilde{R}$ with any vector must return another vector, otherwise the derivative of $a(\lambda)$ would grow non-vector terms. It follows that $R'\tilde{R}$ can only contain a bivector component. We can therefore write

$$\frac{d}{d\lambda}R(\lambda) = -\tfrac{1}{2}B(\lambda)R(\lambda). \tag{11.57}$$

Locally, around any rotor, we can write

$$R(\lambda + \delta\lambda) = (1 - \tfrac{1}{2}\delta\lambda\, B)R(\lambda) = \exp(-\delta\lambda\, B/2)R(\lambda). \tag{11.58}$$

In this way, bivectors capture all of the local information about the rotor group. All 'nearby' rotors differ by a term that is the exponential of a bivector.

Now suppose we look for paths satisfying

$$R(0) = 1, \qquad R(\lambda + \mu) = R(\lambda)R(\mu). \tag{11.59}$$

The set $R(\lambda)$ form a one-parameter subgroup of the rotor group. For the case of three-dimensional rotations the interpretation of this subgroup is clear — it is the group of all rotations in a fixed plane. For this path we find that

$$\begin{aligned}\frac{d}{d\lambda}R(\lambda + \mu) &= -\tfrac{1}{2}B(\lambda + \mu)R(\lambda + \mu)\\ &= \frac{d}{d\lambda}\big(R(\lambda)R(\mu)\big)\\ &= -\tfrac{1}{2}B(\lambda)R(\lambda)R(\mu).\end{aligned} \tag{11.60}$$

It follows that B is constant along this curve. We can therefore integrate equation (11.57) to get

$$R(\lambda) = e^{-\lambda B/2}. \tag{11.61}$$

This confirms that all rotors near the origin can be written as the exponential of a bivector. For Euclidean space it turns out that all rotors lie on a path described by equation (11.59) and so can be written as the exponential of a bivector. This is not the case in mixed signature spaces, though it does turn out that in Lorentzian spaces every rotor can be written as

$$R(\lambda) = \pm e^{-\lambda B/2}. \tag{11.62}$$

It is instructive to establish the inverse result that the exponential of a bivector always returns a rotor. To see this, return to the one-parameter family of vectors

$$a(\lambda) = e^{-\lambda B/2} a_0 e^{\lambda B/2}. \tag{11.63}$$

To establish that these are the result of rotations we need only establish that a is a vector, as the remaining properties follow automatically. Differentiating with respect to λ, we find that

$$\begin{aligned} \frac{da}{d\lambda} &= e^{-\lambda B/2} a_0 \cdot B \, e^{\lambda B/2}, \\ \frac{d^2 a}{d\lambda^2} &= e^{-\lambda B/2} (a_0 \cdot B) \cdot B \, e^{-\lambda B/2} \quad \text{etc.} \end{aligned} \tag{11.64}$$

For every extra derivative we pick up a further inner product with the bivector B. It follows that every term in the Taylor series of $a(\lambda)$ is a vector, and the overall operation is grade-preserving, as it must be. We have also proved the following useful Taylor expansion:

$$e^{-B/2} a e^{B/2} = a + a \cdot B + \frac{1}{2!}(a \cdot B) \cdot B + \cdots . \tag{11.65}$$

This series is convergent for all bivectors B.

11.3.3 Examples of rotor groups

The preceding definitions are illustrated neatly by the algebras $\mathcal{G}(1,1)$ and $\mathcal{G}(1,2)$. First suppose that γ_0 and γ_1 are basis vectors for $\mathcal{G}(1,1)$, with $\gamma_0^2 = 1$ and $\gamma_1^2 = -1$. The spin group consists of even-grade elements, which take the form $\alpha + \beta \gamma_1 \gamma_0$. The restriction that $\psi \tilde{\psi} = \pm 1$ becomes

$$\alpha^2 - \beta^2 = \pm 1, \tag{11.66}$$

which defines four unconnected hyperbolic curves. The rotor group consists of the subgroup for which $\alpha^2 - \beta^2 = 1$. This defines two unconnected branches of a hyperbola, so the rotor group in $\mathcal{G}(1,1)$ is not connected. For the case of Euclidean spaces the scalar product $\langle\psi\tilde{\psi}\rangle$ is positive definite, so there is no difference between the spin and rotor groups, which are always connected.

Now suppose we add a further vector γ_2 of negative signature, and write a general even element as

$$R = R_0 + R_1\gamma_1\gamma_0 + R_2\gamma_2\gamma_0 + R_3\gamma_1\gamma_2. \tag{11.67}$$

The rotor group is specified by the single extra condition that $R\tilde{R} = 1$, which becomes

$$(R_0)^2 - (R_1)^2 - (R_2)^2 + (R_3)^2 = 1. \tag{11.68}$$

It follows that we can write

$$R = \cosh(\alpha)\big(\cos(\theta) + \sin(\theta)\gamma_1\gamma_2\big) + \sinh(\alpha)\big(\cos(\phi) + \sin(\phi)\gamma_1\gamma_2\big)\gamma_1\gamma_0. \tag{11.69}$$

This parameterisation confirms that the group must now be connected. Given an arbitrary rotor we simply find the values of the parameters (α, θ, ϕ), then smoothly run them down to zero to establish a path in the group manifold that connects the rotor to the identity. The reason we can do this in $\mathcal{G}(1,2)$ but could not in $\mathcal{G}(1,1)$ is that the former contains a bivector generator of negative signature. This ensures that -1 is connected to the identity. Among all algebras $\mathcal{G}(p,q)$, with $p+q > 1$, the algebra $\mathcal{G}(1,1)$ is unique in containing no bivector with negative square.

While the rotor group in $\mathcal{G}(1,2)$ is connected, it is straightforward to construct examples of rotors that cannot be written as the exponential of a bivector. For example, consider the rotor

$$R = \exp\big((\gamma_0 + \gamma_1)\gamma_2\big) = 1 + (\gamma_0 + \gamma_1)\gamma_2. \tag{11.70}$$

While this rotor clearly is the exponential of a bivector, it is impossible to write the rotor $-R$ in this way. This is why the strongest statement that can be made about rotors in a mixed signature space is that they can be written as $\pm\exp(-B/2)$.

11.3.4 The bivector algebra

The operation of commuting a multivector with a bivector is always grade-preserving. In particular, the commutator of a bivector with a second bivector produces a third bivector. That is, the space of bivectors is closed under the commutator product. This closed algebra defines the *Lie algebra* of the associated rotor group. The group is formed from the algebra by the act of exponentiation. The commutator of two bivectors expresses the fact that rotations do not commute. If we apply a pair of rotations, and then perform the back rotations in the incorrect order, the result is the new rotation

$$Ra\tilde{R} = \tilde{R}_2\tilde{R}_1(R_2R_1a\tilde{R}_1\tilde{R}_2)R_1R_2. \tag{11.71}$$

Now suppose that we are working close to the identity, so that we can write

$$R = \mathrm{e}^{-B/2} = \mathrm{e}^{B_2/2}\mathrm{e}^{B_1/2}\mathrm{e}^{-B_2/2}\mathrm{e}^{-B_1/2}. \tag{11.72}$$

Expanding the exponentials we find that

$$B = B_1 \times B_2 + \text{higher order terms}. \tag{11.73}$$

This is an example of a more general result known as the *Baker–Campbell–Hausdorff* formula. This states that if

$$\mathrm{e}^{C} = \mathrm{e}^{A}\mathrm{e}^{B}, \tag{11.74}$$

then we have

$$C = A + B + A \times B + \frac{1}{3}\big(A\times(A\times B) + B\times(B\times A)\big) + \cdots. \tag{11.75}$$

The series converges for generators of rotors sufficiently close to the identity. (The precise definition of 'sufficiently close' was clarified by Hausdorff.)

Now suppose that we write

$$R_1 = \exp(-\lambda B_1/2), \qquad R_2 = \exp(-\lambda B_2/2), \tag{11.76}$$

so that $R(\lambda)$ is a path in the group manifold. Equation (11.73) ensures that

$$R(\lambda) = 1 - \lambda^2 B_1 \times B_2/2 + \cdots. \tag{11.77}$$

In the tangent space at the identity the new generator is the commutator of the two original bivectors. The bivector algebra must therefore be closed under the commutator product. This is the way in which the local structure of a rotor group around the identity is passed to the bivector algebra. In the abstract theory of Lie groups, the Lie algebra elements are acted on by the Lie bracket, which is antisymmetric and satisfies the *Jacobi identity*. For a rotor group the Lie bracket is simply the commutator product for bivectors. The Jacobi identity for the Lie algebra then reduces to the identity

$$(A\times B)\times C + (C\times A)\times B + (B\times C)\times A = 0, \tag{11.78}$$

which holds for any three bivectors A, B and C.

11.3.5 Structure constants and the Killing form

Suppose now that we introduce a basis set of bivectors $\{B_i\}$. The commutator of any pair of these returns a third bivector, which can also be expanded in terms of the basis set. We can therefore write

$$B_j \times B_k = C^i_{jk} B_i. \tag{11.79}$$

The C^i_{jk} are called the *structure constants* of the Lie algebra. They provide one of the most compact encodings of the group properties, since knowledge of the bracket structure is sufficient to recover most of the properties of the group. The structure constants also provide a route to solving the problem of classifying all possible Lie algebras over the real and complex fields. The solution of this problem was a significant achievement, completed by the mathematician Élie Cartan.

The *adjoint* representation of a Lie group is defined in terms of functions mapping the Lie algebra onto itself. Every element of a Lie group induces an adjoint representation through its action on the Lie algebra. For the case of rotor groups the Lie algebra is the bivector algebra, and the adjoint representation consists of a map of the form

$$B \mapsto RB\tilde{R} = \mathrm{Ad}_R(B). \tag{11.80}$$

It is immediately clear that this representation satisfies

$$\mathrm{Ad}_{R_1}\big(\mathrm{Ad}_{R_2}(B)\big) = \mathrm{Ad}_{R_1R_2}(B). \tag{11.81}$$

The adjoint representation of the group induces an adjoint representation $\mathrm{ad}_{A/2}$ of the Lie algebra as

$$\mathrm{ad}_{A/2}(B) = A\times B. \tag{11.82}$$

The adjoint representation of an element of the Lie algebra can be considered as a linear map on the space of bivectors. The matrix corresponding to the adjoint representation of the basis bivector B_j is defined by the structure coefficients

$$(\mathrm{ad}_{B_j})^i_k = 2C^i_{jk}. \tag{11.83}$$

The Killing form for a Lie algebra is defined through the adjoint representation as

$$K(A,B) = \mathrm{tr}(\mathrm{ad}_A \mathrm{ad}_B). \tag{11.84}$$

Up to an irrelevant normalisation, the Killing form for a bivector algebra is simply the inner product

$$K(A,B) = A\cdot B, \tag{11.85}$$

which is the definition we shall adopt. It is immediately clear that rotor groups in Euclidean space have a negative-definite Killing form. An algebra with a negative-definite Killing form is said to be of compact type, and the associated Lie group is compact.

11.4 Complex structures and unitary groups

So far we have only dealt with the properties of real rotation groups, but it turns out that this is sufficient for us to uncover the properties of all Lie algebras. We can start to see how this works by studying how complex groups fit into our *real* geometric algebra. The ideas developed in this section are useful in a number of areas, particularly Hamiltonian dynamics and geometric quantum mechanics.

11.4.1 Complex spaces

The simplest algebraic way to define a complex structure is to introduce a commuting scalar quantity j with the property $j^2 = -1$, and to add the assumption that all linear superpositions are now taken over the complex field. A more attractive, geometric alternative is to work in a real space of dimension $2n$ and introduce a bivector in this space to play the role of the complex structure. We saw in section 6.3 that complex analysis can be performed in the geometric algebra of the real two-dimensional plane with the role of the unit imaginary played by the unit pseudoscalar. Here we generalise this idea to an n-dimensional complex space.

Our starting point is a real n-dimensional vector space. Suppose that this has some arbitrary basis $\{e_k\}$, which need not be orthonormal. Now introduce a further set of n-vectors $\{f_k\}$ perpendicular to the $\{e_k\}$, with the properties

$$f_i \cdot f_j = e_i \cdot e_j, \qquad f_i \cdot e_j = 0, \tag{11.86}$$

which hold for all $i, j = 1, \dots, n$. From these vectors we construct the bivector

$$J = \sum_{i=1}^{n} e_i \wedge f^i = e_i \wedge f^i, \tag{11.87}$$

where the $\{f^k\}$ are the reciprocal vectors to the $\{f_k\}$ frame. For this and the following section we assume that repeated indices are summed from $1, \dots, n$. The bivector J is independent of the initial choice of frame $\{e_i\}$. To see this, introduce a second pair of frames $\{e'_i\}$ and $\{f'_i\}$ related in the same manner as the $\{e_k\}$, $\{f_k\}$ pair. For these we find that

$$J' = e'_i \wedge f'^i = e'_i \cdot e^j \, e_j \wedge f'^i = f'_i \cdot f^j e_j \wedge f'^i = e_j \wedge f^j = J. \tag{11.88}$$

In particular, if the $\{e_k\}$ frame is chosen to be orthonormal, we find that

$$J = e_1 f_1 + e_2 f_2 + \cdots + e_n f_n = J_1 + J_2 + \cdots + J_n. \tag{11.89}$$

Each bivector blade J_i then provides the complex structure for the ith plane.

To understand the properties of the bivector J we first form the products

$$e_i \cdot J = e_i \cdot e_j \, f^j = f_i \cdot f_j \, f^j = f_i \tag{11.90}$$

and

$$f_i \cdot J = -e_j \, f_i \cdot f^j = -e_i. \tag{11.91}$$

It follows that

$$\begin{aligned}(e_i\cdot J)\cdot J &= f_i\cdot J = -e_i,\\ (f_i\cdot J)\cdot J &= -e_i\cdot J = -f_i,\end{aligned} \tag{11.92}$$

and hence that

$$(a\cdot J)\cdot J = -a, \tag{11.93}$$

for any vector a. We can now see how J will take over the role of the unit imaginary. For example, the analogue of phase rotations is generated by the bivector J, which describes a series of coupled rotations in each of the J_i planes. A Taylor expansion then yields

$$\begin{aligned}\mathrm{e}^{-J\phi/2}a\mathrm{e}^{J\phi/2} &= a + \phi\, a\cdot J + \frac{\phi^2}{2!}(a\cdot J)\cdot J\cdots\\ &= \cos(\phi)a + \sin(\phi)a\cdot J.\end{aligned} \tag{11.94}$$

The map $a \mapsto a\cdot J$ is therefore a $\pi/2$ rotation. Setting $\phi = \pi$ we also see that

$$a\mathrm{e}^{J\pi/2} = -\mathrm{e}^{J\pi/2}a, \tag{11.95}$$

so $\exp(J\pi/2)$ anticommutes with every vector in the algebra. The only multivector with this property is the pseudoscalar, so we have

$$\mathrm{e}^{J\pi/2} = I_{2n}, \tag{11.96}$$

where I_{2n} is the pseudoscalar of the $2n$-dimensional algebra.

Next we need a means of distinguishing the real and imaginary parts of a vector. As with the two-dimensional case, this requires picking out a preferred set of directions to represent the real axes. As a matter of convention we choose to identify these with the original $\{e_k\}$ vectors. A real vector a in the $2n$-dimensional algebra can now be mapped to a set of complex coefficients $\{a_i\}$ as follows:

$$a_i = a\cdot e_i + j\, a\cdot f_i. \tag{11.97}$$

The complex inner product therefore becomes

$$\begin{aligned}\langle a|b\rangle = a^i b_i^* &= (a\cdot e^i + j\, a\cdot f^i)(b\cdot e_i - j\, b\cdot f_i)\\ &= a\cdot e^i\, b\cdot e_i + a\cdot f^i\, b\cdot f_i + j(a\cdot f^i\, b\cdot e_i - a\cdot e^i\, b\cdot f_i)\\ &= a\cdot b + j(a\wedge b)\cdot J.\end{aligned} \tag{11.98}$$

This shows that the complex inner product combines two geometrically distinct terms. The real part is the usual vector inner product, and it follows immediately that $a^i a_i^* = a^2$. The imaginary part is an antisymmetric product formed by projecting the bivector $a\wedge b$ onto J. Antisymmetric products such as these play an important role in symplectic geometry and Hamiltonian mechanics.

11.4.2 Unitary transformations

We are free to consider any linear function defined over our $2n$-dimensional vector space. However, only a subset of these can be represented by complex matrices — those that observe the complex structure. These transformations are linear over the complex field, so must satisfy

$$\mathsf{f}(\alpha a + \beta a\cdot J) = \alpha \mathsf{f}(a) + \beta \mathsf{f}(a)\cdot J. \tag{11.99}$$

It follows that complex linear transformations satisfy

$$\mathsf{f}(a\cdot J) = \mathsf{f}(a)\cdot J \tag{11.100}$$

for any vector a in the $2n$-dimensional vector space.

The study of complex linear functions now reduces to the study of functions satisfying the condition (11.100). For example, the matrix operation of Hermitian conjugation has

$$\langle a|\mathsf{f}(b)\rangle = \langle \mathsf{f}^\dagger(a)|b\rangle. \tag{11.101}$$

By considering the various terms in this identity we see immediately that the Hermitian adjoint is the same as the familiar adjoint function $\bar{\mathsf{f}}$. That is, $\mathsf{f}^\dagger = \bar{\mathsf{f}}$. This explains why it is Hermitian conjugation that is so important in analysing complex matrices. Similarly, suppose that a is a complex eigenvector of the complex function f. This implies that

$$\mathsf{f}(a) = \alpha a + \beta a\cdot J. \tag{11.102}$$

Clearly, if a satisfies this equation, then $a\cdot J$ satisfies

$$\mathsf{f}(a\cdot J) = \alpha a\cdot J - \beta a. \tag{11.103}$$

It follows that $a\wedge(a\cdot J)$ is an eigenbivector, with

$$\mathsf{f}\big(a\wedge(a\cdot J)\big) = (\alpha^2+\beta^2)a\wedge(a\cdot J). \tag{11.104}$$

Next we need to establish the invariance group of the Hermitian inner product. This group must leave invariant both terms in equation (11.98). This includes the inner product $a\cdot b$, which tells us that the invariance group is built from reflections and rotations. The fact that the linear transformations preserve the complex structure then ensures that the antisymmetric term is also invariant. To see this, suppose that f satisfies $\bar{\mathsf{f}} = \mathsf{f}^{-1}$, together with equation (11.100). It follows that

$$\big(\mathsf{f}(a)\wedge\mathsf{f}(b)\big)\cdot J = \mathsf{f}(a)\cdot\big(\mathsf{f}(b)\cdot J\big) = \mathsf{f}(a)\cdot\mathsf{f}(b\cdot J) = (a\wedge b)\cdot J. \tag{11.105}$$

This result can be summarised concisely as

$$\mathsf{f}(J) = J. \tag{11.106}$$

Unitary groups are therefore constructed from reflections and rotations which leave J invariant. For a reflection to satisfy this constraint would require that the vector generator m satisfies

$$mJm^{-1} = J. \tag{11.107}$$

But this implies that $m \cdot J = 0$, and hence that $(m \cdot J) \cdot J = -m = 0$. There are therefore no vector generators of reflections, and hence all unitary transformations are generated by elements of the spin group. So far we have not specified the underlying signature, so our description applies equally to the unitary groups $\mathrm{U}(n)$ and $\mathrm{U}(p,q)$. These groups can be represented in terms of even multivectors in $\mathcal{G}(2n,0)$ and $\mathcal{G}(2p,2q)$ respectively.

To simplify matters, we now restrict to the Euclidean case, so we seek a rotor description of the unitary group $\mathrm{U}(n)$. The spin group and rotor group in $\mathcal{G}(2n,0)$ are the same, so the unitary group has a double-cover representation in terms of rotors satisfying

$$RJ\tilde{R} = J. \tag{11.108}$$

Writing $R = \exp(-B/2)$, we see that the bivector generators of the unitary group must satisfy

$$B \times J = 0. \tag{11.109}$$

This defines a bivector representation of the Lie algebra $\mathrm{u}(n)$ of the unitary group $\mathrm{U}(n)$. We can construct bivectors satisfying equation (11.109) by first using the Jacobi identity to prove that

$$\begin{aligned} \big((a \cdot J) \wedge (b \cdot J)\big) \times J &= -(a \cdot J) \wedge b + (b \cdot J) \wedge a \\ &= -(a \wedge b) \times J. \end{aligned} \tag{11.110}$$

It follows that

$$\big(a \wedge b + (a \cdot J) \wedge (b \cdot J)\big) \times J = 0. \tag{11.111}$$

Any bivector of the form on the left-hand side will therefore commute with J. Suppose now that the $\{e_i\}$ and $\{f_i\}$ are orthonormal vectors. We can work through all combinations of these to arrive at the bivector algebra in table 11.1. Establishing the closure of this algebra under the commutator product is straightforward. The bivector algebra contains J, which commutes with all other elements and is responsible for a global phase term. Removing this term defines the Lie algebra $\mathrm{su}(n)$ of the special unitary group $\mathrm{SU}(n)$. The analysis can be repeated with a different signature base space to construct a bivector representation of the Lie algebra $\mathrm{u}(p,q)$.

11.5 The general linear group

We have seen how to represent both rotation groups and unitary groups in terms of spin groups. We will now see how all matrix groups can be represented by spin

E_{ij}	$=$	$e_ie_j + f_if_j$	$(i < j = 1, \dots, n)$
F_{ij}	$=$	$e_if_j - f_ie_j$	$(i < j = 1, \dots, n)$
J_i	$=$	e_if_i	$(i = 1, \dots, n)$

Table 11.1 *The Lie algebra* $\mathrm{u}(n)$. The bivectors all belong to the geometric algebra $\mathcal{G}(2n, 0)$, and the vectors $\{e_i\}$ and $\{f_i\}$ form an orthonormal basis for this algebra. The complex structure is generated by the bivector $J = J_1 + \cdots + J_n$.

groups, and hence that all possible Lie algebras can be represented as bivector algebras. This is a significant motivation for the treatment adopted in this chapter. Formulating general linear functions as rotors is achieved by working in a balanced algebra, generated by equal numbers of vectors with positive and negative square. Some of the algebraic considerations for these types of algebra were encountered in the discussions of spacetime and conformal geometry.

11.5.1 The balanced algebra $\mathcal{G}(n, n)$

Suppose that the vectors $\{e_i\}$ span a non-degenerate space of unspecified signature. We introduce a second frame $\{f_k\}$, orthogonal to the first and with *opposite* signature, with the properties

$$f_i \cdot f_j = -e_i \cdot e_j, \qquad e_i \cdot f_j = 0. \tag{11.112}$$

The vectors $\{e_i, f_i\}$ therefore generate the algebra $\mathcal{G}(n, n)$, regardless of the signature of the original $\{e_i\}$ space. We next introduce the balanced analogue of the complex bivector J by defining

$$K = e_i \wedge f^i. \tag{11.113}$$

This has the properties that

$$e_i \cdot K = e_i \cdot e_j \, f^j = -f_i \cdot f_j \, f^j = -f_i \tag{11.114}$$

and

$$f_i \cdot K = -f_i \cdot f^j \, e_j = -e_i. \tag{11.115}$$

It follows that

$$(a \cdot K) \cdot K = K \cdot (K \cdot a) = a \quad \forall a \in \mathcal{V}. \tag{11.116}$$

There is therefore a crucial sign difference compared with the complex bivector J. This means that K does not generate a complex structure, but instead generates a *null* structure. To see this, we first form

$$(a \cdot K)^2 = -\big((a \cdot K) \cdot K\big) \cdot a = -a^2, \tag{11.117}$$

so the vector $a{\cdot}K$ has opposite signature to a. Given a general vector $a \in \mathcal{G}(n,n)$ we can define two separate null vectors by writing

$$a = \tfrac{1}{2}(a + a{\cdot}K) + \tfrac{1}{2}(a - a{\cdot}K). \tag{11.118}$$

In this way the vector space $\mathcal{V}$ of $\mathcal{G}(n,n)$ splits into two null spaces, $\mathcal{V}_+$ and $\mathcal{V}_-$. Vectors in $\mathcal{V}_+$ satisfy

$$a_+{\cdot}K = a_+ \quad \forall a_+ \in \mathcal{V}_+, \tag{11.119}$$

with a similar expression (with a minus sign) holding for $\mathcal{V}_-$. Both of the spaces $\mathcal{V}_+$ and $\mathcal{V}_-$ are entirely null, and they are dual spaces to one another. Working entirely with vectors in $\mathcal{V}_+$ is a further way of formulating a Grassmann algebra within geometric algebra.

11.5.2 Linear transformations

We will shortly demonstrate that every linear function acting on an n-dimensional vector space, $a \mapsto \mathsf{f}(a)$, can be represented in $\mathcal{V}_+$ by a transformation of the form

$$a_+ \mapsto M a_+ M^{-1}. \tag{11.120}$$

Here M belongs to a subgroup of the spin group for $\mathcal{G}(n,n)$, and a_+ is the image of a in $\mathcal{V}_+$ defined by

$$a_+ = a + a{\cdot}K. \tag{11.121}$$

In this sense we form a double-cover representation of the general linear group. The relevant subgroup consists of transformations that map the subspaces $\mathcal{V}_+$ and $\mathcal{V}_-$ entirely within themselves. For this to hold we require that

$$(M a_+ M^{-1}){\cdot}K = M a_+ M^{-1}, \tag{11.122}$$

so we must have

$$\begin{aligned} a_+ &= M^{-1}\,(M a_+ M^{-1}){\cdot}K\, M \\ &= M^{-1}\tfrac{1}{2}(M a_+ M^{-1} K - K M a_+ M^{-1}) M \\ &= a_+{\cdot}(M^{-1} K M). \end{aligned} \tag{11.123}$$

It follows that we require $M^{-1}KM = K$, or

$$MK = KM. \tag{11.124}$$

As with the unitary case, M must belong to the spin group. The bivector generators of this group must commute with K. The Jacobi identity ensures that the commutator product of two bivectors that commute with K results in

E_{ij}	$=$	$e_ie_j - f_if_j$	$(i < j = 1, \dots, n)$
F_{ij}	$=$	$e_if_j - f_ie_j$	$(i < j = 1, \dots, n)$
K_i	$=$	e_if_i	$(i = 1, \dots, n)$

Table 11.2 *The Lie algebra* gl(n). The bivectors all belong to the geometric algebra $\mathcal{G}(n,n)$. The $\{e_i\}$ vectors are orthonormal with positive signature, and the $\{f_i\}$ are orthonormal with negative signature. The algebra contains the bivector $K = K_1 + \cdots + K_n$, which generates the Abelian subgroup of global dilations. Factoring out this bivector produces the algebra sl(n).

a third that also commutes with K. We proceed as with the unitary group and construct

$$\big((a\cdot K)\wedge(b\cdot K)\big)\times K = a\wedge(b\cdot K) + (a\cdot K)\wedge b = (a\wedge b)\times K, \tag{11.125}$$

so that

$$\big(a\wedge b - (a\cdot K)\wedge(b\cdot K)\big)\times K = 0. \tag{11.126}$$

We can again run through all combinations of the basis bivectors to obtain the basis for the Lie algebra of the general linear group listed in table 11.2. The difference in structure between the Lie algebras of the linear group and the unitary group is due solely to the different signatures of their underlying spaces.

The remaining step is to give an explicit construction of a representation of a linear transformation as an element of the spin group. The key to this is the singular value decomposition of section 4.4.8. This decomposition shows that any $n \times n$ matrix (with non-zero determinant) can be decomposed into a positive-definite diagonal matrix sandwiched between two orthogonal matrices. To find a suitable encoding in terms of rotors, all we have to do is find representations of orthogonal transformations and positive dilations.

Rotations are clearly present as they are generated by the E_{ij} bivectors in the Lie algebra of table 11.2. These bivectors jointly rotate the $\{e_i\}$ and $\{f_i\}$ vectors by the same amount. But the orthogonal group also includes reflections, so we need to represent these as well. Suppose the reflection in $\mathcal{G}(p,q)$ is generated by the unit vector n, $n^2 = 1$. We define

$$\bar{n} = n\cdot K, \qquad \bar{n}^2 = -1, \tag{11.127}$$

and consider the multivector $n\bar{n}$. This satisfies

$$n\bar{n}K = 2n\,\bar{n}\cdot K + nK\bar{n} = 2(n^2 + \bar{n}^2) + Kn\bar{n} = Kn\bar{n}, \tag{11.128}$$

so the bivector does commute with K. But since

$$n\bar{n}(n\bar{n})^{\sim} = -1 \tag{11.129}$$

this bivector is not a rotor. It belongs to the spin group, but not the rotor group. The action of $n\bar{n}$ on vectors $a_+ \in \mathcal{V}_+$ results in the vector

$$-n\bar{n}a_+\bar{n}n = -n\bar{n}a\bar{n}n - (n\bar{n}a\bar{n}n)\cdot K = -nan - (nan)\cdot K, \tag{11.130}$$

where a is the original vector, in the same space as n. Since $\bar{n}$ is in the orthogonal space generated by the $\{f_i\}$ vectors, $\bar{n}$ anticommutes with a. Equation (11.130) is the required result for a reflection. The need to include reflections forces us to work with elements of the full spin group in $\mathcal{G}(n,n)$.

The final step is to see how dilations are formulated with rotors. Suppose that we now require a positive dilation in the n direction. We again form the bivector $n\bar{n}$, which is constructed from the F_{ij} and K_i Lie algebra generators. With $n_+ = n + \bar{n}$ the equivalent of the vector n in $\mathcal{V}_+$, we find that

$$\begin{aligned} \mathrm{e}^{-\lambda n\bar{n}/2} n_+ \mathrm{e}^{\lambda n\bar{n}/2} &= \big(\cosh(\lambda) - n\bar{n}\sinh(\lambda)\big)(n+\bar{n}) \\ &= \mathrm{e}^{\lambda} n_+, \end{aligned} \tag{11.131}$$

which is a pure dilation. Furthermore, any vector perpendicular to n has an image in $\mathcal{V}_+$ that commutes with $n\bar{n}$ and so is unaffected by the action of the rotor. These are precisely the required properties of the positive dilation, which completes the construction.

We now have an alternative means of representing every matrix group within geometric algebra. Since *all* Lie algebras can be represented by matrices, we have proved that all Lie algebras can be realised as bivector algebras. The accompanying Lie group elements can then all be written as even products of unit vectors. This is potentially a very powerful idea. One immediate construct one can form this way is the *tensor product* of two linear functions. All one requires for this is a separate copy of the algebra $\mathcal{G}(n,n)$ for each linear operator. As with the multiparticle spacetime algebra construction of chapter 9, the generators of each space are orthogonal, so anticommute. It follows that even elements from either space commute. So rotors from either space can be multiplied commutatively, forming a spinor representation of the tensor product. The combined rotor generates the correct tensor product action on vectors in the combined space. The tensor product can therefore be constructed from the geometric product.

11.6 Notes

The multivector derivative and the use of the vector derivative in analysing linear functions are described in detail in the book *Clifford Algebra to Geometric Calculus* by Hestenes & Sobczyk (1984). This book also contains an elegant proof of the Cayley–Hamilton theorem, and details of the geometric algebra approach to Lie group theory. Some further material is contained in the 'Lectures in geometric algebra' by Doran *et al.* (1996a).

The basis of Grassmann calculus is described in *The Method of Second Quantisation* by Berezin (1966). A summary of the main results from this is contained in the appendices to the paper 'Particle spin dynamics as the Grassmann variant of classical mechanics' by Berezin and Marinov (1977). More recently, Grassmann calculus has been extended to the field of superanalysis, as described in the books by Berezin (1987) and de Witt (1984). Similar themes also reappear in the subject of non-commutative geometry, as discussed by Connes & Lott (1990) and Coquereaux, Jadczyk & Kastler (1991). The geometric algebra treatment of Grassmann calculus was introduced in the papers 'Grassmann calculus, pseudoclassical mechanics and geometric algebra' by Lasenby, Doran & Gull (1993c) and 'Grassmann mechanics, multivector derivatives and geometric algebra' by Doran, Lasenby & Gull (1993b). Some additional material is contained in the thesis by Doran (1994). These works also show how the super-Lie bracket, and super-Lie algebras, can be formulated within geometric algebra.

The subject of Lie groups is covered in an enormous range of textbooks. The series entitled *Group Theory in Physics* by Cornwell (1984a,1984b,1989) are particularly recommended, as are the books by Georgi (1982) and Gilmore (1974). The subject of pin and spin groups has also been discussed widely. Thorough treatments can be found in the books *An Introduction to Spinors and Geometry* by Benn & Tucker (1988) and *Clifford Algebras and Spinors* by Lounesto (1997). The construction of the general linear group in terms of rotors was first described in the paper 'Lie groups as spin groups' by Doran *et al.* (1993). The thesis by Doran (1994) contains explicit constructions of a number of further Lie algebras, including symplectic and quaternionic algebras.

11.7 Exercises

11.1 The function f maps vectors to vectors in the spacetime algebra according to

$$\mathsf{f}(a) = a + \alpha a\cdot\gamma_+ \, \gamma_+,$$

where γ_+ is the null vector $\gamma_0 + \gamma_3$. Find the characteristic equation satisfied by f. What are the roots of the characteristic polynomial and how many independent eigenvectors are there? Verify that f satisfies its own characteristic equation.

11.2 Suppose that the vectors γ_0, γ_1 form an orthogonal basis for a space of signature $(1,1)$. Show that the linear function f_1,

$$\mathsf{f}_1(a) = -12 a\cdot\gamma_0 \, \gamma_0 + 2a\cdot\gamma_0 \, \gamma_1 + 2a\cdot\gamma_1 \, \gamma_0 + a\cdot\gamma_1 \, \gamma_1,$$

has no symmetric square root. Similarly, show that the function f_2,

$$\mathsf{f}_2(a) = 8a\cdot\gamma_0 \, \gamma_0 + a\cdot\gamma_0 \, \gamma_1 + a\cdot\gamma_1 \, \gamma_0 - a\cdot\gamma_1 \, \gamma_1,$$

has two symmetric square roots, and find them both.

11.3 The function $\phi(\lambda)$ is defined by

$$\phi(\lambda) = \det\big(\exp(\lambda \mathsf{f})\big)$$

where f is a linear function. The exponential function is defined by the power series

$$\exp(\lambda \mathsf{f})(a) = \sum_{r=0}^{\infty} \frac{\lambda^r}{r!} \mathsf{f}^r(a)$$

where $\mathsf{f}^r(a)$ denotes the r-fold application of f and $\mathsf{f}^0(a) = a$. Prove that $\phi(\lambda)$ satisfies

$$\frac{d\phi}{d\lambda} = \partial_a \cdot \mathsf{f}(a)\, \phi(\lambda),$$

and hence prove that

$$\det\big(\exp(\mathsf{f})\big) = \exp\big(\partial_a \cdot \mathsf{f}(a)\big).$$

11.4 Prove the following results for the functional derivative:

$$\partial_{\mathsf{f}(a)} \partial_b \cdot \mathsf{f}^r(b) = r \mathsf{f}^{r-1}(a), \quad r \geq 1,$$
$$\partial_{\mathsf{f}(a)} \langle \bar{\mathsf{f}}^{-1}(A_r) B_r \rangle = -\langle \bar{\mathsf{f}}^{-1}(a) \cdot B_r\, \bar{\mathsf{f}}^{-1}(A_r) \rangle_1.$$

11.5 Given a non-singular function f in Euclidean space, the function ε is defined by

$$\varepsilon = \tfrac{1}{2} \ln\big(\bar{\mathsf{f}}\, \mathsf{f}\big). \tag{E11.1}$$

The logarithm can be defined either by a power series, or by diagonalising $\bar{\mathsf{f}}\mathsf{f}$ and taking the logarithm of the eigenvalues. Prove that

$$\partial_{\mathsf{f}(a)} \partial_b \cdot \varepsilon(b) = \bar{\mathsf{f}}^{-1}(a),$$
$$\partial_{\mathsf{f}(a)} \partial_b \cdot \varepsilon^2(b) = \bar{\mathsf{f}}^{-1} \varepsilon(a).$$

11.6 Prove that left and right-sided Grassmann derivatives commute.

11.7 Suppose that x, y and e are unit vectors in $\mathcal{G}(4,0)$, with the pseudoscalar denoted by I. Prove that the product $\phi(x,y)$, where

$$\phi(x,y) = \langle x e y (1 + I) \rangle_1,$$

satisfies all the axioms of a Lie group product, with e the identity element. Which group does this product define?

11.8 The multivector R is defined by

$$R = -1 - (\gamma_0 + \gamma_1)\gamma_2,$$

where $\{\gamma_0, \gamma_1, \gamma_2\}$ are an orthonormal basis for $\mathcal{G}(1,2)$. Prove that R is a rotor, and that it is impossible to find a bivector B such that $R = \exp(-B/2)$.

11.9 The vectors $\{e_i, f_i\}, i = 1, \ldots, n$ form an orthonormal basis for $\mathcal{G}(2n, 0)$. The Lie algebra $\mathrm{u}(n)$ is defined by the following bivectors:

$$
\begin{aligned}
E_{ij} &= e_i e_j + f_i f_j & (i < j = 1, \ldots, n), \\
F_{ij} &= e_i f_j - f_i e_j & (i < j = 1, \ldots, n), \\
J_i &= e_i f_i.
\end{aligned}
$$

Prove that this algebra is closed under the commutator product. Hence find the structure constants of the unitary group.

11.10 Prove that the Lie algebras $\mathrm{su}(4)$ and $\mathrm{so}(6)$ are isomorphic. Repeat the analysis for the case of $\mathrm{su}(2, 2)$ and $\mathrm{so}(2, 4)$. This latter isomorphism is important in the theory of twistors.

12

Lagrangian and Hamiltonian techniques

The Lagrangian formulation of mechanics is popular in practically all modern treatments of the subject. The ideas date back to the pioneering work of Euler, Lagrange and Hamilton, who showed how the equations of Newtonian dynamics could be derived from variational principles. In these, the evolution of a system is viewed as a path in some parameter space. The path the system follows is one which extremises a quantity called the *action*, which is the integral of the Lagrangian with respect to the evolution parameter (usually time). The mathematics behind this approach was clear from the outset, but a thorough physical understanding had to wait until the arrival of quantum theory. In the path-integral formulation of quantum mechanics a particle is viewed as simultaneously following all possible paths. By assigning a phase factor to the action for each path and summing these, one obtains the amplitude for a quantum process. The classical limit can then be understood as resulting from trajectories that reinforce the amplitude. In this manner classical trajectories emerge as those which make the action stationary.

A closely related idea is the Hamiltonian formulation of dynamics. The advantage of this approach is that it produces a set of first-order equations, making it well suited to numerical methods. The Hamiltonian approach also exposes the appropriate geometry for classical dynamical systems, which is a symplectic manifold. The Lagrangian and Hamiltonian formulations are well suited to studying the role of symmetry in physics. Any symmetry present in the Lagrangian will remain present in the equations of motion, and will produce a set of possible paths all related by the appropriate symmetry group. In this chapter we will touch on many of these ideas, and provide a number of Lagrangians for systems of physical interest. We also show how the method can be extended to the case of a multivector Lagrangian, which establishes contact with the systems studied in pseudoclassical mechanics.

12.1 The Euler–Lagrange equations

Suppose that a system is described by the multivector variables X_i, $i = 1, \ldots, n$. (The use of multivector variables makes this derivation slightly more general than usually seen.) The Lagrangian L is a scalar-valued function of X_i and $\dot{X}_i$, and possibly time, where the dot denotes the derivative with respect to time. The action for the system is

$$S = \int_{t_1}^{t_2} dt\, L(X_i, \dot{X}_i, t), \tag{12.1}$$

and we seek the equations for a path for which the action is stationary. The solution to this problem is standard application of variational calculus. We write

$$X_i(t) = X_i^0(t) + \epsilon Y_i(t), \tag{12.2}$$

where Y_i is a multivector containing the same grades as X_i and which vanishes at the endpoints, ϵ is a scalar, and X_i^0 represents the extremal path. It follows that the action must satisfy

$$\left.\frac{dS}{d\epsilon}\right|_{\epsilon=0} = 0, \tag{12.3}$$

in order to ensure that X_i^0 is a stationary solution. The chain rule now gives

$$\begin{aligned}\left.\frac{dS}{d\epsilon}\right|_{\epsilon=0} &= \int_{t_1}^{t_2} dt \sum_{i=1}^{n} \bigl(Y_i {*} \partial_{X_i} L + \dot{Y}_i {*} \partial_{\dot{X}_i} L\bigr) \\ &= \int_{t_1}^{t_2} dt \sum_{i=1}^{n} Y_i {*} \Bigl(\partial_{X_i} L - \frac{d}{dt}(\partial_{\dot{X}_i} L)\Bigr),\end{aligned} \tag{12.4}$$

where $A{*}B = \langle AB \rangle$. This integral must equal zero for all paths Y_i, from which we can read off the Euler–Lagrange equations in the form

$$\frac{\partial L}{\partial X_i} - \frac{d}{dt}\left(\frac{\partial L}{\partial \dot{X}_i}\right) = 0, \qquad \forall i = 1, \ldots, n. \tag{12.5}$$

The multivector derivative ensures that there are as many equations as there are grades present in the X_i, which implies we have precisely the same number of equations as there are degrees of freedom in the system.

12.1.1 Symmetries and conservation laws

Suppose now that we consider a scalar-parameterised transformation of the dynamical variables, so that we have

$$X_i' = X_i'(X_i, \alpha). \tag{12.6}$$

We further assume that $\alpha = 0$ corresponds to the identity transformation (this restriction can be removed if necessary). The first-order change in X_i is denoted by δX_i, where

$$\delta X_i = \left.\frac{\partial X_i'}{\partial \alpha}\right|_{\alpha=0}. \tag{12.7}$$

We define the new Lagrangian

$$L'(X_i, \dot{X}_i) = L(X_i', \dot{X}_i'), \tag{12.8}$$

which is obtained from L simply by replacing each of the dynamical variables by their transformed equivalent. The chain rule now gives

$$\left.\frac{dL'}{d\alpha}\right|_{\alpha=0} = \sum_{i=1}^{n} \Big((\delta X_i)*\partial_{X_i} L + (\delta \dot{X}_i)*\partial_{\dot{X}_i'} L \Big). \tag{12.9}$$

If we now suppose that the X_i satisfy the Euler–Lagrange equations, we can rewrite the right-hand side as a total derivative to obtain

$$\left.\frac{dL'}{d\alpha}\right|_{\alpha=0} = \frac{d}{dt} \sum_{i=1}^{n} \Big((\delta X_i)*\partial_{\dot{X}_i} L \Big). \tag{12.10}$$

This result applies for any transformation, and can be used in a number of ways.

If the transformation is a symmetry of the Lagrangian, then L' is independent of α. In this case we immediately establish that a conjugate quantity is conserved. That is, symmetries of the Lagrangian produce conjugate conserved quantities. This is Noether's theorem, and it is valuable for extracting conserved quantities from dynamical systems. The fact that the derivation of equation (12.10) assumed the equations of motion were satisfied means that the quantity is conserved 'on-shell'. Some symmetries can also be extended 'off-shell', which becomes an important issue in quantum and supersymmetric systems.

An important application of equation (12.10) is to the case of time translation,

$$X_i'(t, \alpha) = X_i(t + \alpha), \tag{12.11}$$

so that

$$\left.\frac{\partial X_i'}{\partial \alpha}\right|_{\alpha=0} = \dot{X}_i. \tag{12.12}$$

If there is no explicit time dependence in the Lagrangian, then equation (12.10) gives

$$\frac{dL}{dt} = \frac{d}{dt} \sum_{i=1}^{n} \Big(\dot{X}_i*\partial_{\dot{X}_i} L \Big). \tag{12.13}$$

We therefore define the conserved *Hamiltonian* by

$$H = \sum_{i=1}^{n} \dot{X}_i*\partial_{\dot{X}_i} L - L. \tag{12.14}$$

This is more often written in terms of the *generalised momenta*

$$P_i = \partial_{\dot{X}_i} L, \tag{12.15}$$

so that

$$H = \sum_{i=1}^{n} \dot{X}_i * P_i - L. \tag{12.16}$$

The Hamiltonian gives the total energy in the system, and is conserved for systems with no explicit time dependence.

12.1.2 Point particle actions

The simplest application of the Lagrangian framework is for a particle moving in three dimensions in an external potential $V(\boldsymbol{x})$. The Lagrangian is the difference between the kinetic and potential energies,

$$L = \frac{m\boldsymbol{v}^2}{2} - V(\boldsymbol{x}), \tag{12.17}$$

where $\boldsymbol{v} = \dot{\boldsymbol{x}}$. The Euler–Lagrange equations give

$$m\dot{\boldsymbol{v}} = -\boldsymbol{\nabla} V, \tag{12.18}$$

which identifies $-\boldsymbol{\nabla} V$ with the force on a particle. The Hamiltonian is

$$H = \frac{\boldsymbol{p}^2}{2m} + V, \tag{12.19}$$

where $\boldsymbol{p} = m\boldsymbol{v}$. The Hamiltonian is conserved if V is independent of time.

The relativistic action for a free point particle raises some new issues. We begin with the simplest form of the action, which is

$$S = -m \int dt \, (1 - \dot{\boldsymbol{x}}^2)^{1/2}, \tag{12.20}$$

where the overdot denotes the derivative with respect to time t, and we work in units with $c = 1$. The momentum is

$$\boldsymbol{p} = \frac{\partial L}{\partial \dot{\boldsymbol{x}}} = \frac{m\dot{\boldsymbol{x}}}{(1 - \dot{\boldsymbol{x}}^2)^{1/2}}, \tag{12.21}$$

and the equations of motion state that $\boldsymbol{p}$ is constant. The Hamiltonian is

$$H = \boldsymbol{p} \cdot \dot{\boldsymbol{x}} - L = (\boldsymbol{p}^2 + m^2)^{1/2}, \tag{12.22}$$

and is also conserved.

The fact that the energy and momentum are dealt with differently is unsatisfactory from the point of view of Lorentz invariance, so we seek an alternative formulation which is manifestly covariant. This can be achieved from the observation that the action is equivalent to

$$S = -m \int d\lambda \, (x' \cdot x')^{1/2}, \tag{12.23}$$

where $x' = \partial_\lambda x(\lambda)$. This integral is unchanged under a reparameterisation of the trajectory. By identifying λ with t we recover equation (12.20), and setting λ equal to the proper time τ we see that the action is $-m$ times the proper time along the path. Variation with respect to the relativistic position x now produces

$$\frac{d}{d\lambda}\left(\frac{mx'}{(x' \cdot x')^{1/2}}\right) = 0. \tag{12.24}$$

If we now set λ equal to the proper time the left-hand side becomes m times the relativistic acceleration $\dot{v}$, where overdots now denote the derivative with respect to proper time.

Interaction with an electromagnetic field is included through a term in $-qx' \cdot A$, producing the action

$$S = \int d\lambda \, \big(-m(x' \cdot x')^{1/2} - qx' \cdot A(x)\big). \tag{12.25}$$

Variation with respect to x now produces

$$-q\nabla A(x) \cdot x' + \frac{d}{d\lambda}\left(m\frac{x'}{(x' \cdot x')^{1/2}} + qA(x)\right) = 0. \tag{12.26}$$

Setting λ equal to the proper time, we find that

$$\begin{aligned} m\dot{v} &= q\big(\nabla A(x) \cdot v - v \cdot \nabla A(x)\big) \\ &= qF \cdot v, \end{aligned} \tag{12.27}$$

where $F = \nabla \wedge A$. We therefore recover the Lorentz force law, as discussed in section 5.5.3.

The square root in the free-particle action of equation (12.23) is often inconvenient, and can be removed by the inclusion of an *einbein*. This is a scalar function $e(\lambda)$, which has the transformation property under reparameterisations that

$$e(\nu) = \frac{d\lambda}{d\nu} e(\lambda), \tag{12.28}$$

where $\nu(\lambda)$ denotes a new parameterisation for the trajectory. The action can now be written in the equivalent form

$$S = -\tfrac{1}{2} \int d\lambda \, \big(e^{-1} x' \cdot x' + m^2 e\big). \tag{12.29}$$

Variation of e produces

$$e = \frac{(x'\cdot x')^{1/2}}{m}, \tag{12.30}$$

and substitution of this back into the action recovers equation (12.23). A first-order form of the action can also be developed by introducing the momentum p and writing

$$S = \int d\lambda \left(-p\cdot x' + \frac{e}{2}(p^2 - m^2)\right). \tag{12.31}$$

Variation of e produces the constraint equation $p^2 = m^2$, and variation of p produces $x' = ep$. This ensures that e is again given by equation (12.30). Finally, the x variation determines

$$p' = \frac{d}{d\lambda}\left(\frac{mx'}{(x'\cdot x')^{1/2}}\right) = 0, \tag{12.32}$$

recovering the desired equation. In each of these cases interaction with an electromagnetic field is included through a term in $-qx' \cdot A$. Moving to a reparameterisation-invariant formulation ensures that Lorentz covariance is manifest, but it limits the use of Hamiltonian techniques. Hamiltonians deal with energy, so picking out a Hamiltonian almost always implies breaking manifest Lorentz covariance.

12.1.3 Rigid-body dynamics

As a further application, consider a rigid body as discussed in section 3.4.3. The configuration of the body is described by the variables $\boldsymbol{x}_0(t)$ and $R(t)$, where $\boldsymbol{x}_0$ is the position of the centre of mass, and R is a three-dimensional rotor. We will ignore the motion of the centre of mass and concentrate on the rotational degrees of freedom. We also assume for simplicity that the object is freely rotating, so the Lagrangian is given by the rotational energy,

$$L = -\tfrac{1}{2}\Omega_B\cdot\mathcal{I}(\Omega_B). \tag{12.33}$$

Here $\mathcal{I}(B)$ is the inertia tensor, and

$$\Omega_B = -2R^\dagger \dot{R}, \tag{12.34}$$

where the dagger denotes the reverse operation in three dimensions.

The fact that the degrees of freedom are described by a rotor presents a slight problem. Rotors belong to a Lie group, and so form a group manifold. The Lagrangian is then a function defined for paths on the group manifold, which makes the Euler–Lagrange equations slightly more difficult to write down. There are two main methods of proceeding. The first is to introduce an explicit parameterisation of R, such as the Euler angles, and to compute the Lagrangian in terms of these. This has the disadvantage of introducing a fixed coordinate system, making it difficult to assemble the final equations into a coordinate-free form. The structure of the rotor group provides a more elegant alternative. We replace the rotor R by an arbitrary even element (a spinor) ψ. The constraint $\psi\psi^\dagger = 1$ is enforced through the inclusion of a Lagrange multiplier. This method allows us to use the coordinate-free apparatus of multivector calculus in the variational principle and leads quickly to the full set of Euler equations.

Our Lagrangian is now

$$L(\psi, \dot{\psi}) = -\tfrac{1}{2}\Omega_B \cdot \mathcal{I}(\Omega_B) - \lambda(\psi\psi^\dagger - 1), \tag{12.35}$$

where the dynamical variable is the spinor ψ, and λ is a Lagrange multiplier. The bivector Ω_B is determined from ψ by

$$\Omega_B = -\psi^\dagger\dot{\psi} + \dot{\psi}^\dagger\psi, \tag{12.36}$$

which is a bivector, as required. The Euler–Lagrange equations reduce to the single multivector equation

$$\partial_\psi L - \frac{d}{dt}(\partial_{\dot{\psi}} L) = 0. \tag{12.37}$$

The symmetry of the inertia tensor simplifies the derivatives, and we obtain

$$\begin{aligned} \partial_\psi\big(-\tfrac{1}{2}\Omega_B \cdot \mathcal{I}(\Omega_B)\big) &= -2\mathcal{I}(\Omega_B)\dot{\psi}^\dagger, \\ \partial_{\dot{\psi}}\big(-\tfrac{1}{2}\Omega_B \cdot \mathcal{I}(\Omega_B)\big) &= 2\mathcal{I}(\Omega_B)\psi^\dagger, \end{aligned} \tag{12.38}$$

where we have used the results of section 11.1. After reversing, the Euler–Lagrange equation for ψ is simply

$$\frac{d}{dt}\big(\psi\mathcal{I}(\Omega_B)\big) + \dot{\psi}\mathcal{I}(\Omega_B) = \lambda\psi. \tag{12.39}$$

Variation with respect to the Lagrange multiplier λ enforces the constraint that $\psi\psi^\dagger = 1$, which means we can now replace ψ with the rotor R. We therefore arrive at the equation

$$\mathcal{I}(\dot{\Omega}_B) - \Omega_B\mathcal{I}(\Omega_B) = \lambda. \tag{12.40}$$

The scalar part of this equation determines λ and shows that, in the absence of any applied couple, the rotational energy is a constant of the motion. The bivector part of equation (12.40) recovers the familiar equation

$$\mathcal{I}(\dot{\Omega}_B) - \Omega_B \times \mathcal{I}(\Omega_B) = 0, \tag{12.41}$$

as found in section 3.4.3. The Lagrange multiplier has avoided any need for handling the rotor group manifold.

12.2 Classical models for spin-1/2 particles

The use of non-relativistic spinors in describing the dynamics of a rigid body demonstrates that spinors are not necessarily restricted to applications in quantum mechanics. This is significant in addressing the question: what is the classical analogue of the Dirac equation? That is, what classical dynamical system produces the Dirac equation on quantisation? There have been many attempts to answer this question, and in the following sections we investigate two of them.

12.2.1 Rotor dynamics

For our first classical model of a fermion, we start with the Lagrangian for the Dirac field. Following the notation of section 8.2 this is

$$L_{Dirac} = \langle \nabla\psi I\gamma_3\tilde{\psi} - m\psi\tilde{\psi}\rangle. \tag{12.42}$$

The properties of this Lagrangian are studied in detail in chapter 13. Focusing on the first (kinetic) term, we can write this as

$$\langle \nabla\psi I\gamma_3\tilde{\psi}\rangle = \langle \nabla\psi \boldsymbol{I\sigma}_3\psi^{-1}\psi\gamma_0\tilde{\psi}\rangle = \langle J\nabla\psi\boldsymbol{I\sigma}_3\psi^{-1}\rangle, \tag{12.43}$$

where $J = \psi\gamma_0\tilde{\psi}$ is the Dirac current. The streamlines of J describe how the probability density flows through spacetime. To reduce to a point-particle model, we assume that only the derivatives along a streamline are important and that the density is concentrated entirely on one streamline. This streamline is then identified with the particle worldline, and the kinetic term becomes

$$\langle J\cdot\nabla\psi\boldsymbol{I\sigma}_3\psi^{-1}\rangle = \langle \psi'\boldsymbol{I\sigma}_3\psi^{-1}\rangle, \tag{12.44}$$

where the prime denotes the derivative with respect to some parameter along the worldline. Now recall from section 8.2 that a Dirac spinor decomposes into

$$\psi = \rho^{1/2}\mathrm{e}^{I\beta/2}R, \tag{12.45}$$

where ρ and β are scalars, and R is a Lorentz rotor (a member of the connected subgroup of the spin group). The inverse, ψ^{-1}, is therefore

$$\psi^{-1} = \rho^{-1/2}\mathrm{e}^{-I\beta/2}\tilde{R}. \tag{12.46}$$

Substituting this parameterisation into equation (12.44), we find that

$$\langle \psi'\boldsymbol{I\sigma}_3\psi^{-1}\rangle = \langle R'\boldsymbol{I\sigma}_3\tilde{R}\rangle. \tag{12.47}$$

The dynamics are now parameterised by a Lorentz rotor, as opposed to a full spinor. Given that the magnitude of a spinor is related to the quantum concept of probability density, it is sensible that the classical model should only depend on the rotor component.

To complete the model we need to impose the condition that the current $\psi\gamma_0\tilde{\psi}$

defines the tangent to the worldline. This is achieved by including a Lagrange multiplier to enforce the constraint that

$$x' = eR\gamma_0\tilde{R}, \tag{12.48}$$

where e is an einbein. Finally, the mass term $m\psi\tilde{\psi}$ becomes simply em, where again the einbein ensures reparameterisation invariance. The full Lagrangian is now

$$L(x, x', R, R', p, e) = \langle R' I\boldsymbol{\sigma}_3\tilde{R} - p(x' - eR\gamma_0\tilde{R}) - em\rangle, \tag{12.49}$$

and the action is formed by integrating this with respect to the evolution parameter λ. The p equation returns the constraint of equation (12.48), and the einbein e returns

$$p\cdot(R\gamma_0\tilde{R}) = m. \tag{12.50}$$

After variation we can choose the parameterisation such that $e = 1$, and x' is replaced by $\dot{x}$, with dots denoting the derivative with respect to proper time along the worldline $x(\tau)$. It follows that $p \cdot \dot{x} = m$. Clearly, then, we can identify p with the momentum. The x variation then says that the momentum is constant.

The final equation requires varying R, which lies on the group manifold of the rotor group $\mathrm{Spin}^+(1,3)$. This variation can be performed in a number of ways. We could extend the technique employed for rigid-body mechanics, and relax the normalisation constraint so that R becomes a full spinor. The normalisation is then enforced by a pair of Lagrange multipliers (one each for the scalar and pseudoscalar terms). However, we can avoid this by returning to the original form of the Lagrangian in terms of ψ and replacing the relevant terms by

$$\langle R' I\boldsymbol{\sigma}_3\tilde{R} + epR\gamma_0\tilde{R}\rangle = \langle \psi' I\boldsymbol{\sigma}_3\psi^{-1} + ep\psi\gamma_0\tilde{\psi}/\rho\rangle, \tag{12.51}$$

where $\rho = |\psi\gamma_0\tilde{\psi}|$. This form ensures that L is only dependent on the rotor component of ψ, but still allows us to vary L with respect to ψ. This is easier than constructing the derivative on the group manifold. To proceed we need a pair of additional results. The first is that

$$\partial_\psi\langle M\psi^{-1}\rangle = -\psi^{-1}M\psi^{-1}, \tag{12.52}$$

which holds for any even multivector M. The second is that

$$2\rho\partial_\psi\rho = \partial_\psi\langle\psi\gamma_0\tilde{\psi}\psi\gamma_0\tilde{\psi}\rangle = 4\gamma_0\tilde{\psi}\psi\gamma_0\tilde{\psi}, \tag{12.53}$$

which implies that

$$\partial_\psi\rho = 2\rho\psi^{-1}. \tag{12.54}$$

The ψ variation now produces (after setting λ equal to proper time τ)

$$-\psi^{-1}\dot{\psi}I\boldsymbol{\sigma}_3\psi^{-1} + \frac{e}{\rho}(2\gamma_0\tilde{\psi}p - 2\psi^{-1}\langle p\psi\gamma_0\tilde{\psi}\rangle) - \frac{d}{d\tau}(I\boldsymbol{\sigma}_3\psi^{-1}) = 0. \tag{12.55}$$

On multiplying through by ψ we obtain

$$\dot{S} + 2p\wedge\dot{x} = 0, \tag{12.56}$$

where

$$S = \psi I\boldsymbol{\sigma}_3\psi^{-1} = RI\boldsymbol{\sigma}_3\tilde{R}. \tag{12.57}$$

The rotor variation therefore produces an equation which states that the total angular momentum is conserved. This shows that the classical model has many of the desired features. Linear momentum is conserved, and the spin-1/2 nature of the particle is captured in the total angular momentum.

The simplest solution to the equations of motion has $m\dot{x} = p$, so that the particle is at rest in the p frame. The spin bivector is also constant, as one would expect in the absence of interaction. There are a range of further solutions, however, which are of interest. Suppose that we align γ_0 with momentum, and write

$$p = \frac{m}{\cosh(\alpha)}\gamma_0 = m^*\gamma_0, \tag{12.58}$$

which defines the 'effective mass' m^*. The equations of motion are then solved by

$$\begin{aligned} R &= \mathrm{e}^{I\boldsymbol{\sigma}_3 m^*\tau}\mathrm{e}^{\alpha\boldsymbol{\sigma}_2/2}, \\ x &= \tau\cosh(\alpha)\,\gamma_0 - \frac{\sinh(\alpha)}{2m^*}\gamma_1\mathrm{e}^{-2I\boldsymbol{\sigma}_3 m^*\tau}. \end{aligned} \tag{12.59}$$

The total angular momentum is

$$\tfrac{1}{2}S + p\wedge x = \tfrac{1}{2}\cosh(\alpha)\,I\boldsymbol{\sigma}_3, \tag{12.60}$$

which is constant. This solution describes a particle rotating at angular frequency $2m/\cosh(\alpha)$ (as measured by the proper time), and with a radius of

$$r_0 = \frac{1}{2m}\sinh(\alpha)\cosh(\alpha). \tag{12.61}$$

As α increases, the momentum goes 'off-shell', and the particle can 'borrow' energy to execute a circular motion and feel out its surroundings. This model therefore captures some aspects of fermionic quantum mechanics, exhibiting a form of zitterbewegung, while still describing a point-particle trajectory.

For many applications the model constructed here is unnecessarily complicated, and we instead choose to work with the somewhat simpler Lagrangian

$$L(x, x', \psi, \psi', p, e) = \langle\psi' I\boldsymbol{\sigma}_3\tilde{\psi} - p(x' - e\psi\gamma_0\tilde{\psi}) - em\psi\tilde{\psi}\rangle. \tag{12.62}$$

Global phase invariance of L ensures that $\langle\psi\tilde{\psi}\rangle$ is constant and can be set to 1. If the initial conditions are chosen suitably, one can also show that the 4-vector part of $\psi\tilde{\psi}$ remains zero, and the motion reduces to that of the previous model. An open question is whether either of these models produces the Dirac equation on quantisation. The problem is that a path-integral quantisation involves the group manifold of $\text{Spin}^+(1,3)$, which is non-compact. In addition, the Lagrangian is first order, which can give rise to complications in the path integral.

A deficiency of the classical model is exposed when we couple the particle to the electromagnetic field. If we consider the phase transformation

$$R \mapsto Re^{I\boldsymbol{\sigma}_3\phi}, \tag{12.63}$$

then this introduces a term going as $-\partial_\lambda \phi = -x' \cdot (\nabla \phi)$ into the Lagrangian. Local phase invariance is therefore restored by modifying the Lagrangian to

$$L(x, x', R, R', p, e) = \langle R' I\boldsymbol{\sigma}_3 \tilde{R} - p(x' - eR\gamma_0\tilde{R}) - qx' \cdot A - em \rangle, \tag{12.64}$$

where A is the electromagnetic vector potential. The $qx' \cdot A$ term is the natural point-particle equivalent of the interaction term $qJ \cdot A$ in the Dirac Lagrangian. Variation now modifies the p equation in the expected manner to read

$$\dot{p} = qF \cdot \dot{x}. \tag{12.65}$$

But the spin equation is not affected — we do not naturally pick up the $g = 2$ behaviour for the gyromagnetic ratio of a spin-1/2 particle. This is disappointing, given that the A term is all that is required to guarantee that $g = 2$ in Dirac theory. The problem can be rectified by introducing a further term into the Lagrangian, going as

$$L_g = \left\langle -\frac{q}{2m} F R I\boldsymbol{\sigma}_3 \tilde{R} \right\rangle = \left\langle -\frac{q}{2m} F \psi I\boldsymbol{\sigma}_3 \psi^{-1} \right\rangle. \tag{12.66}$$

This modifies both the R and p equations to give

$$\begin{aligned} \dot{S} &= 2\dot{x} \wedge p + \frac{q}{m} F \times S, \\ \dot{p} &= qF \cdot \dot{x} + \frac{q}{2m} \dot{\nabla} \dot{F}(x) \cdot S. \end{aligned} \tag{12.67}$$

These equations have the expected form for a particle with $g = 2$, but the value of the gyromagnetic ratio has been put in by hand.

12.2.2 Pseudoclassical mechanics

A quite different approach to the classical mechanics of a spin-1/2 particle is provided by pseudoclassical mechanics, which introduces the interesting new concept of a multivector-valued Lagrangian. We only consider the simplest case of a non-relativistic model. The model is motivated by the idea that the spin operators satisfy

$$\hat{s}_i \hat{s}_j + \hat{s}_j \hat{s}_i = \frac{\hbar^2}{2} \delta_{ij}. \tag{12.68}$$

The classical analogue of these relations should have zero on the right-hand side, so the particle is described by a set of anticommuting Grassmann variables.

This argument runs contrary to the viewpoint of this book, which is that there is nothing at all quantum-mechanical about a Clifford algebra, but the model itself is interesting. We introduce a set of three Grassmann variables $\{\zeta_i\}$ and define the Lagrangian

$$L = \tfrac{1}{2}\zeta_i\dot{\zeta}_i - \tfrac{1}{2}\epsilon_{ijk}\omega_i\zeta_j\zeta_k, \tag{12.69}$$

where the ω_i are constants. Following the prescription of section 11.2 we replace the Grassmann variables with a set of three vectors $\{e_i\}$ under the exterior product. The Lagrangian then becomes

$$L = \tfrac{1}{2}e_i\wedge\dot{e}_i - \omega, \tag{12.70}$$

where

$$\omega = \tfrac{1}{2}\epsilon_{ijk}\omega_i e_j e_k = \omega_1(e_2\wedge e_3) + \omega_2(e_3\wedge e_1) + \omega_3(e_1\wedge e_2). \tag{12.71}$$

The Lagrangian is now a *bivector*, and not simply a scalar. This raises an immediate question — how can the variational principle be applied to a multivector? The answer is that all components of the Lagrangian must remain stationary under variation. Suppose that we contract L with an arbitrary bivector B to form the scalar $\langle LB\rangle$. Variation of this produces the Euler–Lagrange equation

$$\partial_{e_i}\langle LB\rangle - \frac{d}{dt}(\partial_{\dot{e}_i}\langle LB\rangle) = 0. \tag{12.72}$$

Treating the $\{e_i\}$ as vector variables, we arrive at the equation

$$(\dot{e}_i + \epsilon_{ijk}\omega_j e_k)\cdot B = 0. \tag{12.73}$$

But we must demand that this vanishes for all possible B, from which we extract the equation

$$\dot{e}_i + \epsilon_{ijk}\omega_j e_k = 0. \tag{12.74}$$

This is the general method for handling multivector Lagrangians. The contraction with any constant multivector must result in a scalar Lagrangian which is stationary when the equations of motion are satisfied. Equation (12.73) illustrates a further feature. For a fixed B, equation (12.73) is not sufficient to extract the full set of equations. It is only by allowing B to vary, and hence treat the Lagrangian as a bivector, that the full equations are extracted.

To solve equation (12.74) we first establish that ω is constant,

$$\dot{\omega} = 0, \tag{12.75}$$

which follows immediately from the equation of motion. Next we introduce the reciprocal frame $\{e^i\}$ and write the equation of motion in the form

$$\dot{e}_i = e^i\cdot\omega. \tag{12.76}$$

Now suppose that we define the symmetric function g by

$$\mathsf{g}(a) = \sum_{i=1}^{3} a\cdot e_i \, e_i, \tag{12.77}$$

so that $\mathsf{g}(e^i) = e_i$. The function g is a form of metric for the non-orthonormal frame e_i. On differentiating $\mathsf{g}(a)$, holding a constant, we find that

$$\frac{d}{dt}\mathsf{g}(a) = \sum_{i=1}^{3}\big(a\cdot(e^i\cdot\omega)e_i + a\cdot e_i \, e^i\cdot\omega\big) = \omega\cdot a + a\cdot\omega = 0. \tag{12.78}$$

It follows that the function g is constant, even though the e_i vectors vary in time. The motion is found by introducing the square root of g, which satisfies

$$\bar{\mathsf{h}}\mathsf{h}(a) = \mathsf{g}(a), \qquad \bar{\mathsf{h}} = \mathsf{h}. \tag{12.79}$$

This function is found by diagonalising g and taking the square root of the eigenvalues. It follows that

$$\delta_i^j = e_i\cdot e^j = \mathsf{g}(e^i)\cdot e^j = \mathsf{h}(e^i)\cdot\mathsf{h}(e^j). \tag{12.80}$$

The vectors $\mathsf{h}(e^i)$ are therefore orthonormal, so we write

$$f_i = \mathsf{h}(e^i), \qquad f_i\cdot f_j = \delta_{ij}. \tag{12.81}$$

These vectors satisfy

$$\dot{f}_i = f_i\cdot\Omega, \tag{12.82}$$

where

$$\Omega = \omega_1 f_2 f_3 + \omega_2 f_3 f_1 + \omega_3 f_1 f_2. \tag{12.83}$$

Since $\mathsf{h}(\Omega) = \omega$, we see that Ω is a constant bivector. It follows that the $\{f_i\}$ frame simply rotates at a constant frequency in the Ω plane. The solution for the e_i vectors is therefore

$$e_i(t) = \mathsf{h}^{-1}\big(\mathrm{e}^{-\Omega t/2} f_i(0)\mathrm{e}^{\Omega t/2}\big). \tag{12.84}$$

The only motion taking place in this system is that a fixed set of orthonormal vectors is rotating in a constant plane, and the resulting frame is then distorted by a constant symmetric function. A simple picture of this type is fairly typical of pseudoclassical systems when analysed in this manner.

12.3 Hamiltonian techniques

The Hamiltonian formulation of mechanics is important in a range of applications, not least because of its superior handling of numerical issues. We start by forming Hamilton's equations in local coordinates, before placing Hamiltonian dynamics in a more geometric setting. Suppose that a dynamical system is described in terms of a Lagrangian $L(q_i, \dot{q}_i, t)$, where the $\{q_i\}$ are a set of n coordinates for configuration space. The Euler–Lagrange equations are

$$\frac{d}{dt}\left(\frac{\partial L}{\partial \dot{q}_i}\right) = \frac{\partial L}{\partial q_i}. \tag{12.85}$$

These equations typically result in a set of n second-order equations that relate the generalised momenta to the forces in the system. The Euler–Lagrange equations are equivalent to the set of $2n$ first-order equations

$$\dot{q}_i = \frac{\partial H}{\partial p_i}, \qquad \dot{p}_i = -\frac{\partial H}{\partial q_i}. \tag{12.86}$$

These are Hamilton's equations. The Hamiltonian $H(q_i, p_i, t)$ is given by

$$H(q_i, p_i, t) = \sum_{i=1}^{n} p_i \dot{q}_i - L(q_i, \dot{q}_i, t) \tag{12.87}$$

in which the $\dot{q}_i$ are expressed in terms of the p_i by inverting the equations

$$p_i = \frac{\partial L}{\partial \dot{q}_i}. \tag{12.88}$$

The transformation from a Lagrangian to a Hamiltonian framework is called a *Legendre transformation*. We move from considering dynamics in n-dimensional configuration space to a $2n$-dimensional *phase space*.

If the Hamiltonian is independent of time we can immediately see that it is conserved. That is, H gives the conserved energy in the system. The proof is straightforward:

$$\frac{dH}{dt} = \sum_{i=1}^{n}\left(\dot{q}_i \frac{\partial H}{\partial q_i} + \dot{p}_i \frac{\partial H}{\partial p_i}\right) = 0. \tag{12.89}$$

Phase space provides a very useful way of analysing the motion and stability of complicated systems. As a simple example, consider a pendulum consisting of a mass m attached to a rigid rod of length a. The configuration of the system is described by a single angle θ, and the Lagrangian is

$$L = \frac{ma^2\dot{\theta}^2}{2} + mga\cos(\theta). \tag{12.90}$$

The Hamiltonian is therefore

$$H = \frac{p_\theta^2}{2ma^2} - mga\cos(\theta), \tag{12.91}$$

and this is conserved. The trajectories of the system can be visualised in terms of a phase-space portrait, which plots surfaces of constant H in phase space. Sample trajectories are shown in figure 12.1. The figure illustrates how the phase portrait can capture global aspects of the system, such as the behaviour of the system as the energy gets close to value for which the pendulum can complete a full loop.

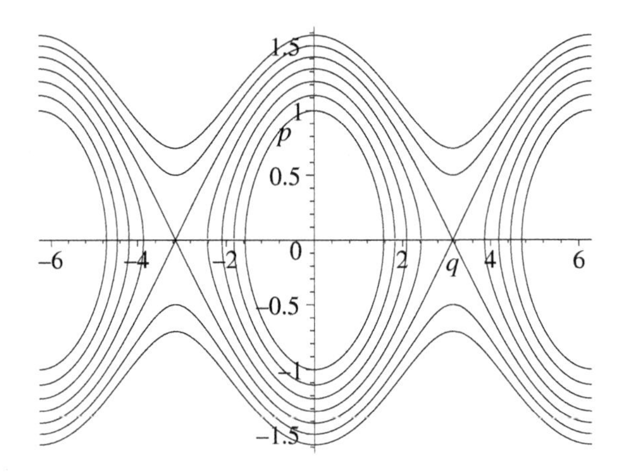

Figure 12.1 *A phase portrait.* The q coordinate represents the angle and p is the canonical momentum. The Hamiltonian is $p^2 - \cos(q)$. As H approaches 1 a bifurcation appears, corresponding to the energy for which the pendulum can complete a loop. The system is periodic, so the phase portrait can be thought of as wrapping up into a cylinder

12.3.1 Symplectic geometry

The natural setting for Hamilton's equations is provided by *symplectic geometry*. A symplectic manifold (M, Ω) consists of a $2n$-dimensional manifold M together with a closed, non-degenerate 2-form Ω. We will assume that n is finite so as to avoid discussion of the technicalities of infinite-dimensional spaces. We can analyse this structure using the apparatus of vector manifolds, described in section 6.5. A symplectic manifold does not have a metric structure, so we must take care not to employ the metric induced in the vector manifold by its embedding. This means we must distinguish tangent and cotangent spaces, as we can only apply the inner product between tangent and cotangent vectors. We denote the tangent space at x by T_xM, and the cotangent space T_x^*M.

The covariant vector derivative is denoted by ∇, and always results in a multivector that is intrinsic to the manifold (in section 6.5.3 this derivative was denoted D). The 2-form Ω is a bivector field evaluated in the cotangent space. The statement that Ω is closed is simply

$$\nabla \wedge \Omega = 0. \tag{12.92}$$

This is required in order that the Poisson bracket satisfies the Jacobi identity.

The condition that Ω is non-degenerate is simply that

$$\Omega(x)\cdot a \neq 0, \qquad \forall a \neq 0,\ a \in T_xM. \tag{12.93}$$

If we view $\Omega(x)\cdot a$ as a linear map from T_xM to T_x^*M, then Ω being non-degenerate implies that the map has non-zero determinant, so is invertible. The inverse map is generated by a second bivector, which we label J. This second bivector lies in the tangent space, and can be viewed as the inverse of Ω. The two bivectors are related by the pair of equations

$$\begin{aligned} J\cdot(\Omega\cdot a) &= a, \quad \forall a \in T_xM, \\ \Omega\cdot(J\cdot a^*) &= a^*, \quad \forall a^* \in T_x^*M. \end{aligned} \tag{12.94}$$

The properties of Ω and J can be understood simply by introducing a set of local coordinates (p^i, q^i) over M. In terms of these we define the tangent vectors

$$e_i = \frac{\partial x}{\partial p^i}, \qquad f_i = \frac{\partial x}{\partial q^i}, \tag{12.95}$$

and the cotangent vectors

$$e^i = \nabla p^i, \qquad f^i = \nabla q^i. \tag{12.96}$$

We then set

$$\Omega = \sum_{i=1}^{n} f^i \wedge e^i, \qquad J = \sum_{i=1}^{n} e_i \wedge f_i. \tag{12.97}$$

So J and Ω both have a similar structure to the complex bivector introduced in section 11.4. By construction, Ω is clearly closed. It is also straightforward to verify the relations

$$\begin{aligned} \Omega\cdot e_i &= f^i, \qquad & \Omega\cdot f_i &= -e^i, \\ J\cdot e^i &= -f_i, \qquad & J\cdot f^i &= e_i, \end{aligned} \tag{12.98}$$

which confirm that equations (12.94) are satisfied.

The Hamiltonian $H(x,t)$ is a scalar field defined over M, and the dynamics of the system are governed by the equation

$$\dot{x} = (\nabla H)\cdot J. \tag{12.99}$$

This is an identity between tangent vectors. In terms of local coordinates this equation becomes

$$\dot{p}^i e_i + \dot{q}^i f_i = \left(e^i \frac{\partial H}{\partial p^i} + f^i \frac{\partial H}{\partial q^i}\right)\cdot J = f_i \frac{\partial H}{\partial p^i} - e_i \frac{\partial H}{\partial q^i}, \tag{12.100}$$

where repeated indices are summed over $1, \ldots, n$. We therefore recover Hamilton's equations in local coordinates.

12.3.2 Conservation theorems and the Poisson bracket

We now restrict to the case where H is independent of time t. Suppose that a scalar function $f(x)$ is defined over phase space. The evolution of this along a phase space trajectory $x(t)$ is determined by

$$\dot{f} = \dot{x}\cdot\nabla f = (\nabla f\wedge\nabla H)\cdot J. \tag{12.101}$$

It follows immediately that $\dot{H} = 0$. A further consequence follows if H is invariant along some direction a in phase space. If we form the directional derivative of H we obtain

$$a\cdot\nabla H = \big(J\cdot(\Omega\cdot a)\big)\cdot\nabla H = (\Omega\cdot a)\cdot\dot{x}. \tag{12.102}$$

So if H is unchanged in the a direction we have

$$(\Omega\cdot a)\cdot\dot{x} = 0, \tag{12.103}$$

so all flows are perpendicular to the cotangent vector $\Omega\cdot a$.

The equation for the evolution of f leads naturally to the definition of the Poisson bracket of a Hamiltonian system. Given two scalar fields f and g the Poisson bracket is defined by

$$\{f(x), g(x)\} = (\nabla f\wedge\nabla g)\cdot J. \tag{12.104}$$

In terms of local coordinates this takes the more familiar form

$$\{f(x), g(x)\} = \sum_i \left(\frac{\partial f}{\partial q_i}\frac{\partial g}{\partial p_i} - \frac{\partial f}{\partial p_i}\frac{\partial g}{\partial q_i}\right). \tag{12.105}$$

The geometric form neatly brings out the antisymmetry of the Poisson bracket. It follows, for example, that the Poisson bracket with the Hamiltonian returns the time development of a scalar field:

$$\{f, H\} = (\nabla f\wedge\nabla H)\cdot J = \dot{f}. \tag{12.106}$$

Poisson brackets and the Hamiltonian formulation of dynamics provide a natural route through to quantum mechanics, where Poisson brackets are replaced by operator commutation relations.

An important property satisfied by the Poisson bracket is the Jacobi identity

$$\{\{f, g\}, h\} + \{\{g, h\}, f\} + \{\{h, f\}, g\} = 0, \tag{12.107}$$

which is easily confirmed in terms of local coordinates. This identity links the Poisson bracket structure to a Lie algebra structure. The identity is satisfied by any symplectic manifold, as we now establish. We first write

$$\{f, g\} = \big(\nabla\wedge(f\nabla g)\big)\cdot J = (J\cdot\nabla)\cdot(f\nabla g), \tag{12.108}$$

which follows from the identity that $\nabla\wedge\nabla = 0$ (every exact form is closed). The Jacobi identity now reduces to

$$(J\cdot\nabla)\cdot\big((\nabla f\wedge\nabla g)\cdot J\,\nabla h + \text{cyclic permutations}\big) = 0. \tag{12.109}$$

If we define

$$T = \nabla f\wedge\nabla g\wedge\nabla h, \tag{12.110}$$

then equation (12.109) simplifies to

$$(J\cdot\nabla)\cdot(J\cdot T) = 0. \tag{12.111}$$

To simplify this equation further we employ the identity

$$(B\wedge B)\cdot a = 2B\wedge(B\cdot a), \tag{12.112}$$

which holds for any bivector B and vector a. We can now write

$$(J\cdot\nabla)\cdot(J\cdot T) = (J\cdot\dot{\nabla})\cdot(\dot{J}\cdot T) + \tfrac{1}{2}(J\wedge J)\cdot(\nabla\wedge T) \tag{12.113}$$

and the final term here vanishes as T is exact. It follows that the Jacobi identity reduces to the condition

$$(J\cdot\nabla)\wedge J = 0. \tag{12.114}$$

The final task is to demonstrate that this identity for J is equivalent to the statement that Ω is closed. Equation (12.114) is evaluated entirely in tangent space. If we use Ω to map each term into the cotangent space we arrive at the equivalent expression

$$e^{\alpha}\wedge e^{\beta}\wedge\dot{\nabla}\langle\dot{J}(\Omega\cdot e_{\alpha})\wedge(\Omega\cdot e_{\beta})\rangle = 0, \tag{12.115}$$

where Greek indices are summed from 1 to $2n$, with the first n covering the e_i frame, and the second n covering the f_i frame. This identity is equivalent to

$$\nabla\wedge\big(e^{\alpha}\wedge e^{\beta}\langle J(\Omega\cdot e_{\alpha})\wedge(\Omega\cdot e_{\beta})\rangle\big) - e^{\alpha}\wedge e^{\beta}\wedge\nabla\langle\check{J}(\Omega\cdot e_{\alpha})\wedge(\Omega\cdot e_{\beta})\rangle = 0, \tag{12.116}$$

where the check denotes that J is not differentiated in the second expression. The frame derivatives in this expression can all be shown to vanish, which leaves

$$2\nabla\wedge\Omega - e^{\alpha}\wedge e^{\beta}\wedge\dot{\nabla}\langle J(\dot{\Omega}\cdot e_{\alpha})\wedge(\Omega\cdot e_{\beta}) + J(\Omega\cdot e_{\alpha})\wedge(\dot{\Omega}\cdot e_{\beta})\rangle = 0. \tag{12.117}$$

This is equivalent to

$$-2\nabla\wedge\Omega = 0, \tag{12.118}$$

which proves the main result. Any symplectic manifold admits a Poisson bracket structure that satisfies the Jacobi identity. As such, any symplectic manifold can form the basis for a Hamiltonian system.

12.3.3 The cotangent bundle

In practice, the phase space for a Hamiltonian system is often the cotangent bundle of configuration space. This works as follows. Suppose that Q denotes configuration space. It is a manifold with potentially non-trivial topology. At each point q in the manifold we define the cotangent space T_q^*Q. If q^i form a set of local coordinates in configuration space, then ∇q^i define a set of basis vectors for T_q^*Q. An arbitrary cotangent vector can be written as $p_i \nabla q^i$, so the p_i can be used as coordinates for T_q^*Q. Now consider the bundle of all tangent spaces, T^*Q. This is a manifold, and a general point in T^*Q is specified by the set of $2n$ coordinates (q^i, p_i). The first n of these locate the position over Q, and the second n locate a point within the cotangent space. The cotangent bundle T^*Q is a symplectic manifold, with the symplectic structure defined by

$$\Omega = \sum_{i=1}^{n} \nabla q^i \wedge \nabla p_i. \tag{12.119}$$

The reason this structure often arises is that, while there may be constraints placed on configuration space, there are usually no restrictions in momentum space. Returning to the case of the simple pendulum, Q is a circle since θ is a periodic coordinate. But there are no such constraints on $\dot{\theta}$, so the cotangent space is a line. The manifold T^*Q can therefore be visualised as a cylinder, which is the phase space for the pendulum.

12.3.4 Canonical transformations

Suppose that (M_1, Ω_1) and (M_2, Ω_2) are two symplectic manifolds. We let f denote a map from M_1 to M_2, which we write as

$$x' = f(x), \quad x \in M_1, \quad x' \in M_2. \tag{12.120}$$

This map is canonical if it respects the symplectic structure. That is, we must have

$$\bar{\mathsf{f}}(\Omega_2) = \Omega_1, \qquad \mathsf{f}(J_1) = J_2. \tag{12.121}$$

Here

$$\mathsf{f}(a) = a \cdot \nabla f(x) \tag{12.122}$$

is a map from $T_x M_1$ to $T_{x'} M_2$. The fact that Ω is non-degenerate means that we can define a volume form on either manifold by

$$V = \frac{1}{n!} \langle (\Omega)^n \rangle_{2n} = \frac{1}{n!} \langle \Omega \wedge \Omega \wedge \cdots \wedge \Omega \rangle_{2n}. \tag{12.123}$$

The map $\bar{\mathsf{f}}$ must preserve this volume form, so has non-zero determinant. It follows that f is invertible for a canonical transformation, and hence so too is f.

As a check, the Poisson bracket structure remains intact as

$$J_1 \cdot (\nabla f \wedge \nabla g) = J_2 \cdot \bar{\mathsf{f}}^{-1}(\nabla f \wedge \nabla g) = J_2 \cdot (\nabla_2 f \wedge \nabla_2 g), \tag{12.124}$$

where $\nabla_2 = \bar{\mathsf{f}}^{-1}(\nabla)$ is the vector derivative on M_2. It follows that the dynamics can be formulated on either M_1 or M_2, and the physical results will remain unchanged. This is potentially a very powerful result. The set of all possible symplectic transformations is large, and there may well by a suitable transformation which can dramatically simplify the dynamics. This is particularly evident when one notices that symplectic transformations can mix up the position and momentum coordinates in one space. These transformations are richer than simply converting between configuration spaces.

In some applications, phase space is simply R^{2n}, and the bivector J is constant. In this case we can consider canonical transformations which map phase space onto itself. For these the map f is canonical if and only if

$$\mathsf{f}(J) = J. \tag{12.125}$$

Linear transformations satisfying this identity define the *symplectic group*. This can be analysed using the spin group approach developed in section 11.4.

12.4 Lagrangian field theory

The Lagrangian approach to classical dynamics extends to field theory, which can be viewed as the dynamics of systems with an infinite number of degrees of freedom. There are some technical issues connected with the infinite size of the configuration space, but we will not discuss these here. Suppose that the system of interest depends on a field $\psi(x)$, where for simplicity we will assume that x is a spacetime vector. This does not restrict us to relativistic theories, as there is no need to restrict the Lagrangian to be Lorentz-invariant. The action is now defined as an integral over a region of spacetime by

$$S = \int d^4x \, \mathcal{L}(\psi, \partial_\mu \psi, x), \tag{12.126}$$

where $\mathcal{L}$ is the *Lagrangian density* and x^μ are a set of fixed orthonormal coordinates for spacetime. More general coordinate systems are easily accommodated with the inclusion of suitable factors of the Jacobian.

The derivation of the Euler–Lagrange equations proceeds precisely as in section 12.1. We assume that $\psi_0(x)$ represents the extremal path, satisfying the desired boundary conditions, and look for variations of the form

$$\psi(x) = \psi_0(x) + \epsilon\phi(x). \tag{12.127}$$

Here $\phi(x)$ is a field of the same form as $\psi(x)$, which vanishes over the boundary.

The first-order variation in the action is (summation convention in force)

$$\left.\frac{dS}{d\epsilon}\right|_{\epsilon=0} = \int d^4x \left(\phi(x) * \frac{\partial \mathcal{L}}{\partial \psi} + \frac{\partial \phi}{\partial x^\mu} * \frac{\partial \mathcal{L}}{\partial(\partial_\mu \psi)} \right). \tag{12.128}$$

The final term is integrated by parts, and the boundary term vanishes. We therefore find that

$$\left.\frac{dS}{d\epsilon}\right|_{\epsilon=0} = \int d^4x \, \phi(x) * \left(\frac{\partial \mathcal{L}}{\partial \psi} - \frac{\partial}{\partial x^\mu} \left(\frac{\partial \mathcal{L}}{\partial(\partial_\mu \psi)} \right) \right), \tag{12.129}$$

from which we can read off the variational equations as

$$\frac{\partial \mathcal{L}}{\partial \psi} - \frac{\partial}{\partial x^\mu} \left(\frac{\partial \mathcal{L}}{\partial(\partial_\mu \psi)} \right) = 0. \tag{12.130}$$

If more fields are present we obtain an equation of this form for each field. Our main applications of these equations are in the following chapters, where we discuss gauge theories and gravitation. Here we illustrate the equations with a pair of examples concerned with elastic and fluid materials.

12.4.1 Hyperelastic materials

The equations of continuum mechanics, which govern an elastic body, were derived in section 6.6. For certain elastic materials it is possible to obtain these equations from a variational principle. We follow the notation of section 6.6, so f is the displacement field, $\mathsf{f}(\boldsymbol{a})$ is the directional derivative of f, and $\mathsf{C} = \bar{\mathsf{f}}\mathsf{f}$ is the Cauchy–Green tensor. The materials of interest here are called *hyperelastic*. These are defined by the property that, in the absence of external fields, their internal energy U is a function of C only. A suitable action for this system is

$$S = \int dt \, d^3x \left(\frac{\rho(x)}{2} \dot{f}^2 - U(\partial_i f) \right), \tag{12.131}$$

from which we can read off $\mathcal{L}$. Overdots denote the derivative with respect to time, and the integral runs over the space of the reference copy of the body.

The Euler–Lagrange equations are found entirely from the variation of the action with respect to the displacement field f. Since the Lagrangian depends on f through only its time and space derivatives, the Euler–Lagrange equations are

$$\frac{\partial}{\partial t} \left(\frac{\partial \mathcal{L}}{\partial \dot{f}} \right) + \frac{\partial}{\partial x^i} \left(\frac{\partial \mathcal{L}}{\partial(\partial_i f)} \right) = 0. \tag{12.132}$$

These simplify to

$$\rho \dot{\boldsymbol{v}} = \frac{\partial}{\partial x^i} \left(\frac{\partial U}{\partial(\partial_i f)} \right), \tag{12.133}$$

where $\boldsymbol{v} = \dot{y}$ is the local velocity of the body. Comparison with equation (6.319) tells us that we must have

$$\mathsf{T}(\mathsf{e}^i) = \frac{\partial U}{\partial(\partial_i f)}, \tag{12.134}$$

where $\{\mathsf{e}_i\}$ is the (fixed) coordinate frame defined by the x^i coordinates. To simplify the right-hand side we employ the derivative with respect to a linear function, as defined in section 11.1.2. From the definition of equation (11.24) we have

$$\frac{\partial}{\partial \mathsf{f}(\mathsf{e}^i)} = \mathsf{e}_j \frac{\partial}{\partial \mathsf{f}_{ji}}. \tag{12.135}$$

The scalars f_{ji} are defined by

$$\mathsf{f}_{ji} = \mathsf{e}_j \cdot \mathsf{f}(\mathsf{e}_i) = \mathsf{e}_j \cdot (\partial_i f). \tag{12.136}$$

These are the components of the vector $\partial_i f$, so we can write

$$\frac{\partial}{\partial(\partial_i f)} = \mathsf{e}_j \frac{\partial}{\partial \mathsf{f}_{ji}} = \frac{\partial}{\partial \mathsf{f}(\mathsf{e}^i)}. \tag{12.137}$$

The notation for the derivative with respect to $\mathsf{f}(\mathsf{e}^i)$ is slightly misleading, as f is never evaluated on e^i. Instead, we have

$$\mathsf{T}(\boldsymbol{a}) = \partial_{\mathsf{f}(\boldsymbol{a})} U(\mathsf{C}). \tag{12.138}$$

The fact that U is a function of $\mathsf{C} = \bar{\mathsf{f}}\mathsf{f}$ ensures that the second Piola–Kirchoff stress $\mathcal{T} = \mathsf{f}^{-1}\mathsf{T}$ is a symmetric function.

To make further progress we must specify the precise form of U, which amounts to specifying the constitutive properties of the system. The simplest hyperelastic materials to consider are isotropic and homogeneous. For these the internal energy can only depend on the principal stretches:

$$W = W(\lambda_1, \lambda_2, \lambda_3). \tag{12.139}$$

Even within this class there are a large variety of models one can consider. To obtain a linear model the energy should be quadratic in the strains, where linear in this context refers to the relationship between the stress and strain tensors, and not to the underlying dynamics. A natural model to consider is to define the strain by

$$\mathcal{E}(\boldsymbol{a}) = \mathsf{C}^{1/2}(\boldsymbol{a}) - \boldsymbol{a}, \tag{12.140}$$

and set

$$\begin{aligned} U(\mathcal{E}) &= G\,\mathrm{tr}(\mathcal{E}^2) + (B/2 - G/3)\,\mathrm{tr}(\mathcal{E})^2 \\ &= G\big(\mathrm{tr}(\mathsf{C}) - 2\mathrm{tr}(\mathsf{C}^{1/2}) + 3\big) + (B/2 - G/3)\big(\mathrm{tr}(\mathsf{C}^{1/2}) - 3\big)^2, \end{aligned} \tag{12.141}$$

where B and G are respectively the (constant) bulk and the shear moduli. To find the stress tensor we need the derivative of $\mathrm{tr}((\bar{\mathsf{f}}\mathsf{f})^{1/2})$. To evaluate this we first write

$$(\bar{\mathsf{f}}\mathsf{f})^{1/2} = \exp\big(\tfrac{1}{2}\ln(\mathsf{C})\big) = \exp(\mathcal{E}_{\ln}), \tag{12.142}$$

where

$$\mathcal{E}_{\ln} = \tfrac{1}{2}\ln(\mathsf{C}). \tag{12.143}$$

We can now make use of the result

$$\partial_{\mathsf{f}(\boldsymbol{a})}\mathrm{tr}(\mathcal{E}_{\ln}^{n}) = n\bar{\mathsf{f}}^{-1}\mathcal{E}_{\ln}^{n-1}(\boldsymbol{a}) \tag{12.144}$$

to prove that

$$\partial_{\mathsf{f}(\boldsymbol{a})}\mathrm{tr}\big((\bar{\mathsf{f}}\mathsf{f})^{1/2}\big) = \bar{\mathsf{f}}^{-1}(\bar{\mathsf{f}}\mathsf{f})^{1/2}(\boldsymbol{a}). \tag{12.145}$$

The stress tensor T therefore evaluates to

$$\begin{aligned}\mathsf{T} &= 2G(\mathsf{f} - \bar{\mathsf{f}}^{-1}(\bar{\mathsf{f}}\mathsf{f})^{1/2}) + (B - 2G/3)\mathrm{tr}(\mathcal{E})\bar{\mathsf{f}}^{-1}(\bar{\mathsf{f}}\mathsf{f})^{1/2} \\ &= \bar{\mathsf{f}}^{-1}(\bar{\mathsf{f}}\mathsf{f})^{1/2}\big(2G\mathcal{E} + (B - 2G/3)\mathrm{tr}(\mathcal{E})\mathsf{I}\big),\end{aligned} \tag{12.146}$$

where I is the identity transformation. The bracketed term is the expression we would expect to see in a linear theory. The extra pre-factor can be understood in terms of a singular-value decomposition of f. We write

$$\mathsf{f} = \mathsf{R}\mathsf{C}^{1/2}, \tag{12.147}$$

where R is a rotation. We then find that

$$\bar{\mathsf{f}}^{-1}\mathsf{C}^{1/2} = \mathsf{R}\mathsf{C}^{-1/2}\mathsf{C}^{1/2} = \mathsf{R}, \tag{12.148}$$

which recovers the rotation. We can now write

$$\mathsf{T}(\boldsymbol{a}) = \mathsf{R}\big((2G\mathcal{E}(\boldsymbol{a}) + (B - 2G/3)\mathrm{tr}(\mathcal{E})\boldsymbol{a}\big). \tag{12.149}$$

This can be understood as a linear function of $\mathcal{E}$, followed by a rotation to align the principal axes in the reference configuration with those in the body.

The definition of $\mathcal{E}_{\ln}$ raises the interesting prospect that this could be used as an alternative definition of the strain. For an isotropic, homogeneous media this amounts to choosing an energy density of

$$U_{\ln} = G\big((\ln\lambda_1)^2 + (\ln\lambda_2)^2 + (\ln\lambda_3)^2\big) + (B/2 - G/3)\big(\ln(\lambda_1\lambda_2\lambda_3)\big)^2. \tag{12.150}$$

This definition has the same behaviour under small deflection as the potential energy of equation (12.141), but differences emerge as the stresses build up. In essence, the logarithmic definition of energy defines a material which retains its elastic properties no matter what shape it is stretched into. This limits the application of $U_{\ln}$ for modelling physical objects, though it may well be of use in computer graphics simulations, as routines built on $U_{\ln}$ will not break down when forces become large.

12.4.2 Relativistic fluid dynamics

The field equations for a relativistic fluid can be formulated in a number of different ways. Here we give a fairly direct derivation, albeit from a slightly surprising starting point. We start with the action integral

$$S = \int d^4x \, (-\varepsilon + J\cdot(\nabla\lambda) - \mu J\cdot\nabla\eta), \tag{12.151}$$

where $J(x)$ is a spacetime current, ε is the total energy density and η is the entropy. The current can be written as

$$J = \rho v, \qquad v^2 = 1, \tag{12.152}$$

and we assume that ε is a function of ρ and η only, which we write as

$$\varepsilon = \rho(1 + e(\rho, \eta)). \tag{12.153}$$

The remaining terms λ and μ are Lagrange multipliers enforcing the two constraints

$$\nabla\cdot J = 0, \qquad v\cdot\nabla\eta = 0. \tag{12.154}$$

These are two of the four equations of motion. The first constraint is that the current is conserved, so the total number of particles in the system is constant. The second constraint says that entropy is constant along the field lines of J. The various constraints and assumptions ensure that we are describing a relativistic ideal fluid.

Variation with respect to η yields the equation

$$\frac{\partial e}{\partial \eta} = v\cdot\nabla\mu, \tag{12.155}$$

and variation with respect to J produces

$$v(1+e) + v\rho\frac{\partial e}{\partial \rho} = \nabla\lambda - \mu\nabla\eta. \tag{12.156}$$

In the derivation of this equation we have employed the result that

$$\partial_J f(\rho) = v\frac{\partial f}{\partial \rho}. \tag{12.157}$$

Next, we define the pressure P by

$$P = \rho^2\frac{\partial e}{\partial \rho}, \tag{12.158}$$

so that equation (12.156) becomes

$$v(\varepsilon + P) = \rho(\nabla\lambda - \mu\nabla\eta). \tag{12.159}$$

The final step is to remove the Lagrange multipliers by employing the constraint equations. First we contract equation (12.159) with v to obtain

$$\varepsilon + P = \rho v \cdot \nabla \lambda = J \cdot (\nabla \lambda). \tag{12.160}$$

Next we differentiate equation (12.159) to obtain

$$\begin{aligned} J \cdot \nabla(\nabla\lambda - \mu\nabla\eta) &= v \cdot \nabla\big(v(\varepsilon + P)\big) + v(\varepsilon + P) J \cdot \nabla(\rho^{-1}) \\ &= v \cdot \nabla\big(v(\varepsilon + P)\big) + v(\varepsilon + P)\nabla \cdot v. \end{aligned} \tag{12.161}$$

The left-hand side is manipulated as follows:

$$\begin{aligned} J \cdot \nabla(\nabla\lambda - \mu\nabla\eta) &= \nabla(J \cdot \nabla\ \lambda) - \dot{\nabla}\dot{J} \cdot \nabla\lambda - \rho\frac{\partial e}{\partial \eta}\nabla\eta + \mu\dot{\nabla}\dot{J} \cdot \nabla\eta \\ &= \nabla(\varepsilon + P) - \rho\frac{\partial e}{\partial \eta}\nabla\eta - \dot{\nabla}\dot{J} \cdot v\frac{(\varepsilon + P)}{\rho} \\ &= \nabla P + \nabla\rho\frac{\partial \varepsilon}{\partial \rho} - \frac{(\varepsilon + P)}{\rho}\nabla\rho \\ &= \nabla P. \end{aligned} \tag{12.162}$$

We therefore arrive at the equation

$$v \cdot \nabla\big(v(\varepsilon + P)\big) + v(\varepsilon + P)\nabla \cdot v = \nabla P, \tag{12.163}$$

which describes a relativistic ideal fluid. This is more clearly seen if we introduce the relativistic stress-energy tensor $\mathsf{T}(a)$, which is defined by

$$\mathsf{T}(a) = (\varepsilon + P)a \cdot v\, v - Pa. \tag{12.164}$$

The rest frame of the fluid is defined locally by v. We find that $\mathsf{T}(v) = \varepsilon v$, so that ε is the local energy density, as required. In any spacelike direction n perpendicular to v we have $\mathsf{T}(n) = -Pn$, which shows that the local stress is governed by an isotropic pressure P. These are the relativistic definitions of the stress-energy tensor for an ideal fluid. The field equations reduce to the single conservation equation

$$\dot{\mathsf{T}}(\dot{\nabla}) = 0, \tag{12.165}$$

which expresses relativistic conservation of the stress-energy tensor. Electromagnetic coupling is included simply with the addition of the term $-qJ \cdot A$ to the Lagrangian density.

12.5 Notes

The Lagrangian formulation of mechanics is described in a wide range of books. *Analytical Mechanics* by Hand & Finch (1998) contains a detailed introduction and, despite its name, *Introduction to Mechanics and Symmetry* by Marsden & Ratiu (1994) contains a more detailed description of Lagrangian and Hamiltonian methods and symplectic geometry. Further applications, including relativistic fluid dynamics, are contained in *The Variational Principles of Dynamics* by Kupershmidt (1992).

Pseudoclassical mechanics was introduced by Berezin & Marinov (1977). Further references are contained in the notes to chapter 11. Similar ideas to those developed in this chapter have been applied in the supersymmetric setting by Heumann & Manton (2000). The section on spinor models of relativistic spin-1/2 point particles was motivated by the initial work of Barut & Zanghi (1984). The description given here contains a number of refinements, many of which are also discussed in Doran (1994). A detailed discussion of the complexities involved in performing a path-integral quantisation of such systems is given by Barut & Duru (1989).

12.6 Exercises

12.1 A relativistic action for a point particle is defined by

$$S = \int d\lambda \left(-p\cdot\dot{x} + \frac{e}{2}(p^2 - m^2) - qx'\cdot A(x)\right),$$

where A is an external field representing the electromagnetic vector potential. Vary S with respect to x, p and e to obtain the Lorentz force law.

12.2 Prove that

$$\partial_\psi \langle M\psi^{-1}\rangle = -\psi^{-1}M\psi^{-1},$$

where ψ and M are even multivectors.

12.3 The configuration of a rigid body is described by a rotor R. If we relax the normalisation of R and replace it by ψ, explain why we can write Ω_B as

$$\Omega_B = -2\tilde{R}\dot{R} = -\psi^{-1}\dot{\psi} + \dot{\psi}^\dagger {\psi^\dagger}^{-1}.$$

Now define the Lagrangian

$$L(\psi, \dot{\psi}) = -\tfrac{1}{2}\Omega_B \cdot \mathcal{I}(\Omega_B),$$

where $\mathcal{I}$ is the inertia tensor. Find the Euler–Lagrange equation for variation with respect to ψ. Prove that this produces the equation of motion $\dot{J} = 0$, where J is the angular momentum. Why does this method work?

12.4 One classical model for a spin-1/2 particle describes the motion in terms of a rotor R and a momentum p. The rotor determines the quantities

$$\dot{x} = R\gamma_0\tilde{R}, \qquad S = RI\boldsymbol{\sigma}_3\tilde{R}$$

and the equations of motion are

$$\dot{p} = 0, \quad p\cdot\dot{x} = m, \quad \dot{S} + 2\dot{x}\wedge p = 0.$$

Verify that these are solved by

$$p = \frac{m}{\cosh(\alpha)}\gamma_0 = m^*\gamma_0, \qquad R = \mathrm{e}^{I\boldsymbol{\sigma}_3 m^* \tau}\mathrm{e}^{\alpha\boldsymbol{\sigma}_2/2}.$$

Integrate $\dot{x}$ to find the trajectory of the particle and comment on its properties.

12.5 Find the equations of motion for the Lagrangian

$$L = \langle \psi' I\boldsymbol{\sigma}_3\tilde{\psi} - p(x' - e\psi\gamma_0\tilde{\psi}) - em\psi\tilde{\psi} - qx'\cdot A\rangle,$$

where ψ is a spinor and $A(x)$ is an external electromagnetic vector potential. Comment on the form of the solutions.

12.6 A set of vectors satisfy the equations

$$\dot{e}_i + \epsilon_{ijk}\omega_j e_k = 0,$$

where the ω_i are constant. Prove that the volume element E is constant, where

$$E = e_1\wedge e_2\wedge e_3.$$

12.7 The relativistic Hamiltonian for a charged particle in three dimensions is defined by

$$H(\boldsymbol{p}, \boldsymbol{x}, t) = ((\boldsymbol{p} - q\boldsymbol{A})^2 + m^2)^{1/2} + q\phi,$$

where $\phi + \boldsymbol{A} = A\gamma_0$ and the vector potential A is a function of $\boldsymbol{x}$ and t. Find Hamilton's equations and prove that these recover the Lorentz force law.

12.8 Fill in the missing steps in the proof that a closed non-degenerate 2-form in a symplectic manifold guarantees that the Poisson bracket satisfies the Jacobi identity.

12.9 A system is described by the Hamiltonian

$$H(p, q) = \frac{1}{2}\left(\frac{1}{q^2} + p^2 q^4\right).$$

Find a canonical transformation which maps this onto the Hamiltonian for a simple harmonic oscillator.

12.10 The total energy in a hyperelastic medium is given by

$$E = \int d^3x \left(\frac{\rho(x)}{2}\dot{f}^2 + U\right).$$

Prove that the energy flow per unit area perpendicular to $\boldsymbol{n}$ is given by $\dot{f}\cdot\mathsf{T}(\boldsymbol{n})$, where $\boldsymbol{n}$ is a vector in the reference configuration.

12.11 An incompressible elastic material is one for which det $\mathsf{f} = 1$. The Mooney–Rivlin model for rubber is as an incompressible material with internal energy

$$U = \alpha(\lambda_1^2 + \lambda_2^2 + \lambda_3^2 - 3) + \beta((\lambda_2\lambda_3)^2 + (\lambda_3\lambda_1)^2 + (\lambda_1\lambda_2)^2 - 3),$$

where the λ_i denote the principal strains. Analyse the properties of this material under uniform pressure. What happens when two of the λ_i pass through $4^{1/3}$?

12.12 A hyperelastic material is defined with an energy density

$$U_{\ln} = G\big((\ln \lambda_1)^2 + (\ln \lambda_2)^2 + (\ln \lambda_3)^2\big) + (B/2 - G/3)\big(\ln(\lambda_1\lambda_2\lambda_3)\big)^2.$$

Prove that when this system is placed under isotropic pressure P we have

$$-3B \ln \lambda = P.$$

13

Symmetry and gauge theory

The fundamental forces of nature can all be described in terms of *gauge theories*. Not long after the advent of quantum theory, physicists realised that electromagnetic interactions arise from demanding invariance of quantum wave equations under local changes of phase. This idea was later extended by Yang and Mills, who showed how to construct theories based on more complicated, non-commutative Lie groups. This is the basis for the *standard model* of the electroweak and strong interactions. Around this time physicists also turned their attention to gravitation, and discovered that general relativity could also be formulated as a gauge theory. But this time there was a price to pay. The existence of spinor fields means that the simple geometric structure of general relativity has to be modified by the inclusion of a torsion field, leading to an Einstein–Cartan theory. For clarity, we use the term *general relativity* to refer to the theory defined by Einstein, with zero torsion and the connection given by the Christoffel symbol. The extended theory, with torsion present, is referred to as Einstein–Cartan theory.

While gauge theory is the dominant method in particle physics, it is less popular as a means of analysing gravitational interactions. This is, in part, due to the perception that the gauge theory equations are more complicated than their geometric counterparts. In this and the following chapter we argue that this apparent complexity is a reflection of the inappropriate mathematical techniques typically employed when analysing the gauge theory equations. The spacetime algebra provides the appropriate setting for a gauge formulation of gravity and, applied carefully, this approach is often *easier* to compute with than the metric formulation. We demonstrate that, in the absence of torsion and highly esoteric topology, the gauge and metric approaches produce the same physical predictions.

We begin with a discussion of symmetry in the Maxwell and Dirac theories. Our starting point is the field Lagrangian, which we analyse using Noether's theorem. In particular, we use this to extract the canonical energy-momentum tensor, which is conserved in the absence of external fields. We then turn to the wider subject of gauge theories, before deriving the properties of the gauge fields for gravitation. This chapter concludes with a derivation of the gravitational field equations, and a discussion of the observable quantities in the theory. For the source matter, observables are contained in the functional energy-momentum tensor, which is closely related to the canonical tensor. Applications of the field equations are contained in chapter 14. Throughout the present chapter various results and notation from chapter 11 are assumed without comment.

13.1 Conservation laws in field theory

In section 12.4 we derived the Euler–Lagrange equations for field theory, and demonstrated how to apply these to the cases of elasticity and relativistic fluid dynamics. In this section we concentrate on conservation theorems for Lagrangian field theory. As all of the applications that will concern us are to relativistic field theory, we assume from the outset that the we are describing field theory in a (flat) spacetime. Given a Lagrangian density $\mathcal{L}(\psi_i, \partial_\mu \psi_i)$, where ψ_i, $i = 1, \ldots, n$ are a set of multivector fields, the Euler–Lagrange equations governing the evolution of the system are

$$\frac{\partial \mathcal{L}}{\partial \psi_i} - \frac{\partial}{\partial x^\mu}\left(\frac{\partial \mathcal{L}}{\partial(\partial_\mu \psi_i)}\right) = 0, \tag{13.1}$$

where $x^\mu = \gamma^\mu \cdot x$ are a set of fixed orthonormal coordinates. For the applications of interest here the final equations can always be assembled into a frame-free form. Curvilinear coordinates can then be introduced to analyse these equations, if desired.

To obtain a version of Noether's theorem appropriate for field theory we follow the derivation of section 12.1.1. For simplicity we assume that only one field is present. The results are easily extended to the case of more fields by summing over all of the fields present. Suppose that $\psi'(x)$ is a new field obtained from $\psi(x)$ by a scalar-parameterised transformation of the form

$$\psi'(x) = f(\psi(x), \alpha), \tag{13.2}$$

with $\alpha = 0$ corresponding to the identity. We again define

$$\delta\psi = \left.\frac{\partial \psi'}{\partial \alpha}\right|_{\alpha=0}. \tag{13.3}$$

With $\mathcal{L}'$ denoting the original Lagrangian evaluated on the transformed fields we find that

$$\left.\frac{\partial \mathcal{L}'}{\partial \alpha}\right|_{\alpha=0} = (\delta\psi)*\frac{\partial \mathcal{L}}{\partial \psi} + \partial_\mu(\delta\psi)*\frac{\partial \mathcal{L}}{\partial(\partial_\mu \psi)}$$
$$= \frac{\partial}{\partial x^\mu}\left((\delta\psi)*\frac{\partial \mathcal{L}}{\partial(\partial_\mu\psi)}\right). \tag{13.4}$$

This equation relates the change in the Lagrangian to the divergence of the current J, where

$$J = \gamma_\mu\,(\delta\psi)*\frac{\partial \mathcal{L}}{\partial(\partial_\mu\psi)}. \tag{13.5}$$

If the transformation is a symmetry of the system then $\mathcal{L}'$ is independent of α. In this case we immediately establish that the conjugate current is conserved, that is,

$$\nabla\cdot J = 0. \tag{13.6}$$

Symmetries of a field Lagrangian therefore give rise to conserved currents. These in turn define Lorentz-invariant constants via

$$Q = \int d^3x\, J^0, \tag{13.7}$$

where $J^0 = J\cdot\gamma^0$ is the density measured in the γ_0 frame. The fact that this is constant follows from

$$\frac{dQ}{dt} = \int d^3x\, \frac{\partial J^0}{\partial t} = \int d^3x\, \boldsymbol{\nabla}\cdot\boldsymbol{J} = 0, \tag{13.8}$$

where we assume that the current $\boldsymbol{J}$ falls off sufficiently fast at infinity. The value of Q is constant, and independent of the spatial hypersurface used to define the integral.

If the transformation involves a change in the spacetime dependence, Noether's theorem does apply, but we have to be careful in defining the transformation law for $\mathcal{L}$. Suppose that we define

$$\psi'(x) = \psi(x'), \tag{13.9}$$

where

$$x' = f(x). \tag{13.10}$$

The differential is defined in the usual way as

$$\mathsf{f}(a) = a\cdot\nabla f(x). \tag{13.11}$$

The transformed action is

$$S = \int d^4x\, \mathcal{L}(\psi(x'))$$
$$= \int d^4x' \det\,(\mathsf{f})^{-1}\mathcal{L}(\psi(x')) \tag{13.12}$$

from which we see that the correct definition of the transformed Lagrangian is

$$\mathcal{L}'(\psi'(x)) = \det\,(\mathsf{f})^{-1}\mathcal{L}(\psi(x')). \tag{13.13}$$

This transformation law demonstrates that $\mathcal{L}$ is indeed a Lagrangian *density*.

13.1.1 Spacetime symmetries

One of the most important spacetime symmetries is translational invariance. All fundamental theories are assumed to give rise to the same physical predictions, independent of the position of the fields in (flat) spacetime. That is, the background space is assumed to be homogeneous. A more careful discussion of this principle, and its relation to gravitation, is contained in section 13.4.1. In terms of the Lagrangian, this principle is encoded in the statement that all x dependence enters $\mathcal{L}$ through the fields. In this case we can apply Noether's theorem to extract a conserved quantity, though we could proceed equally simply by differentiating $\mathcal{L}$ directly to obtain

$$\begin{aligned} a\cdot\nabla\mathcal{L} &= (a\cdot\nabla\psi)*\frac{\partial\mathcal{L}}{\partial\psi} + \big(a\cdot\nabla(\partial_\mu\psi)\big)*\frac{\partial\mathcal{L}}{\partial(\partial_\mu\psi)} \\ &= \frac{\partial}{\partial x^\mu}\left((a\cdot\nabla\psi)*\frac{\partial\mathcal{L}}{\partial(\partial_\mu\psi)}\right), \end{aligned} \tag{13.14}$$

where the field equations have been assumed. We can therefore define the conserved current conjugate to translations by

$$\mathsf{T}(a) = \gamma_\mu(a\cdot\nabla\psi)*\frac{\partial\mathcal{L}}{\partial(\partial_\mu\psi)} - a\mathcal{L}. \tag{13.15}$$

This defines a linear function of a, called the *canonical energy-momentum tensor*. This is a conserved tensor if the system is invariant under translations, so

$$\nabla\cdot\mathsf{T}(a) = 0, \quad \forall\,\text{constant } a. \tag{13.16}$$

The canonical energy-momentum tensor need not be symmetric, and its adjoint is found to be

$$\bar{\mathsf{T}}(a) = \partial_b\langle\mathsf{T}(b)a\rangle = a\cdot\gamma_\mu\,\dot{\nabla}\Big\langle\dot{\psi}*\frac{\partial\mathcal{L}}{\partial(\partial_\mu\psi)}\Big\rangle - a\mathcal{L}. \tag{13.17}$$

The conservation equation for the adjoint tensor is

$$\dot{\bar{\mathsf{T}}}(\dot{\nabla}) = 0. \tag{13.18}$$

If more than one field is present, the energy-momentum tensor is the sum of the individual contributions from each field.

One can similarly define a conserved tensor conjugate to rotations. This time the assumption is that spacetime is isotropic, so does not contain any preferred directions except those defined by the fields themselves. The derivation is slightly more complicated now, as the fields transform in different ways depending on their spins. For all cases we have

$$x' = \tilde{R}xR, \qquad R = e^{\alpha B/2}, \tag{13.19}$$

and in general we can write

$$\delta\psi = -B\cdot(x\wedge\nabla)\psi + \delta_B\psi, \tag{13.20}$$

where ψ is a general field, and the precise form of $\delta_B\psi$ depends on the spin. The transformation $x' = \tilde{R}xR$ has unit Jacobian, so Noether's theorem gives

$$-B\cdot(x\wedge\nabla)\mathcal{L} = \frac{\partial}{\partial x^\mu}\left(\left(-B\cdot(x\wedge\nabla)\psi + \delta_B\psi\right)*\frac{\partial\mathcal{L}}{\partial(\partial_\mu\psi)}\right). \tag{13.21}$$

We can therefore read off the canonical angular momentum tensor $\mathsf{J}(B)$, where

$$\begin{aligned}\mathsf{J}(B) &= \gamma_\mu\left(-B\cdot(x\wedge\nabla)\psi + \delta_B\psi\right)*\frac{\partial\mathcal{L}}{\partial(\partial_\mu\psi)} + B\cdot x\,\mathcal{L} \\ &= \mathsf{T}(x\cdot B) + (\delta_B\psi)*\frac{\partial\mathcal{L}}{\partial(\partial_\mu\psi)}.\end{aligned} \tag{13.22}$$

This is a vector-valued linear function of the bivector B, which is conserved for all constant B.

The adjoint function $\bar{\mathsf{J}}(a)$ is often easier to work with. This evaluates to

$$\bar{\mathsf{J}}(a) = \partial_B\langle\mathsf{J}(B)a\rangle = \bar{\mathsf{T}}(a)\wedge x + \mathsf{S}(a), \tag{13.23}$$

which is a bivector-valued linear function of the vector a. The form of $\bar{\mathsf{J}}(a)$ generalises the point-particle definition of angular momentum to the field theory setting. The term $\mathsf{S}(a)$ is the canonical spin tensor,

$$\mathsf{S}(a) = a\cdot\gamma_\mu\,\partial_B\left\langle(\delta_B\psi)\frac{\partial\mathcal{L}}{\partial(\partial_\mu\psi)}\right\rangle. \tag{13.24}$$

The conservation equation for $\bar{\mathsf{J}}$ states that

$$\dot{\bar{\mathsf{J}}}(\dot{\nabla}) = 0 = \dot{\bar{\mathsf{T}}}(\dot{\nabla})\wedge x + \bar{\mathsf{T}}(\dot{\nabla})\wedge\dot{x} + \dot{\mathsf{S}}(\dot{\nabla}). \tag{13.25}$$

Since the energy-momentum tensor is also conserved, conservation of angular momentum reduces to the equation

$$\bar{\mathsf{T}}(\partial_a)\wedge a + \dot{\mathsf{S}}(\dot{\nabla}) = 0. \tag{13.26}$$

So, in any homogeneous, isotropic, relativistic field theory, the antisymmetric part of the canonical energy-momentum tensor is a total divergence.

13.2 Electromagnetism

As a first application of the preceding results we consider electromagnetism. The dynamical variable in electromagnetism is the vector potential A, and the electromagnetic Lagrangian density is

$$\mathcal{L} = \tfrac{1}{2}F \cdot F - A \cdot J, \tag{13.27}$$

where $F = \nabla \wedge A$, and A couples to an external current J. An electromagnetic gauge transformation is defined by

$$A \mapsto A + \nabla\phi(x), \tag{13.28}$$

where $\phi(x)$ is a scalar field. Gauge invariance of the Lagrangian is ensured by requiring that the current J is conserved. The field equation is

$$-J - \frac{\partial}{\partial x^\mu}\left(\frac{1}{2}\frac{\partial}{\partial(\partial_\mu A)}\langle FF\rangle\right) = -J - \frac{\partial}{\partial x^\mu}(\nabla \wedge A)\cdot\gamma^\mu = 0, \tag{13.29}$$

which simplifies to the familiar equation

$$\nabla \cdot F = J. \tag{13.30}$$

The remaining Maxwell equation, $\nabla \wedge F = 0$, follows from the definition of F in terms of A.

13.2.1 The electromagnetic energy-momentum tensor

To calculate the free-field energy-momentum tensor, we set $J = 0$ and work with the Lagrangian density

$$\mathcal{L}_0 = \tfrac{1}{2}\langle F^2\rangle. \tag{13.31}$$

Equation (13.15) yields the energy-momentum tensor

$$\mathsf{T}(a) = (a \cdot \nabla A) \cdot F - \tfrac{1}{2}a\langle F^2\rangle. \tag{13.32}$$

This expression is somewhat unsatisfactory as it stands, as it is not gauge-invariant. In order to find a gauge-invariant form of the energy-momentum tensor we write

$$a \cdot \nabla A = a \cdot F + \dot{\nabla}(\dot{A} \cdot a). \tag{13.33}$$

If we now employ the field equations we can write

$$\mathsf{T}(a) = F \cdot (F \cdot a) - \tfrac{1}{2}a\, F \cdot F + \nabla \cdot (A \cdot a\, F). \tag{13.34}$$

The first two terms are gauge-invariant, and the final term is a total divergence. In most classical applications the total divergence can be ignored, as its integral over any finite volume results in a boundary term which can be set to zero. In quantum field theory the issue of how to handle gauge invariance is more complicated. Typically, manifest gauge invariance is lost at the level of the quantum field equations, and only recovered in the physical predictions of the theory. With the boundary term removed, the remaining terms recover the familiar classical free-field electromagnetic energy-momentum tensor,

$$\begin{aligned}\mathsf{T}_{em}(a) &= F\cdot(F\cdot a) - \tfrac{1}{2}a\,F\cdot F \\ &= \tfrac{1}{2}Fa\tilde{F},\end{aligned} \tag{13.35}$$

as found in section 7.2.3. This tensor is gauge-invariant, traceless and symmetric. It is also equal to the functional energy-momentum tensor, defined in section 13.5.4.

13.2.2 Angular momentum in electromagnetism

The canonical angular momentum is found by considering the symmetry transformation

$$A'(x) = RA(x')\tilde{R}, \tag{13.36}$$

with R and x' as defined in equation (13.19). The transformation law for x implies that

$$\nabla_{x'} = \tilde{R}\nabla R, \tag{13.37}$$

so that the new field satisfies

$$\nabla\wedge A' = R\,\nabla_{x'}\wedge A(x')\,\tilde{R} = RF(x')\tilde{R}. \tag{13.38}$$

It follows that the transformed free-field Lagrangian only depends on α through the transformed position dependence, as required for isotropy. We also find that

$$\delta A = B\cdot A - (B\cdot x)\cdot\nabla A, \tag{13.39}$$

so equation (13.22) gives

$$\mathsf{J}(B) = \big(B\cdot A - (B\cdot x)\cdot\nabla A\big)\cdot F + \tfrac{1}{2}B\cdot x\langle F^2\rangle. \tag{13.40}$$

As with the canonical energy-momentum tensor, the angular momentum tensor is not manifestly gauge-invariant. This time we write

$$\begin{aligned}(B\cdot x)\cdot\nabla A &= (B\cdot x)\cdot(\nabla\wedge A) + \dot{\nabla}(B\cdot x)\cdot\dot{A} \\ &= (B\cdot x)\cdot F + \nabla\big((B\cdot x)\cdot A\big) + B\cdot A,\end{aligned} \tag{13.41}$$

so that

$$\mathsf{J}(B) = -\big((B\cdot x)\cdot F\big)\cdot F + \tfrac{1}{2}B\cdot x\langle F^2\rangle - \nabla\cdot\big((B\cdot x)\cdot A\,F\big). \tag{13.42}$$

The final term is again a total divergence which can be ignored. We therefore define

$$\mathsf{J}_{em}(B) = -\big((B\cdot x)\cdot F\big)\cdot F + \tfrac{1}{2}B\cdot x\langle F^2\rangle = \mathsf{T}_{em}(x\cdot B), \tag{13.43}$$

which is now manifestly gauge-invariant. The adjoint is simply

$$\bar{\mathsf{J}}_{em}(a) = \mathsf{T}_{em}(a)\wedge x. \qquad (13.44)$$

Conservation of angular momentum implies that

$$\nabla\cdot\mathsf{T}_{em}(x\cdot B) = \partial_a\cdot\mathsf{T}_{em}(a\cdot B) = (\bar{\mathsf{T}}_{em}(\partial_a)\wedge a)\cdot B = 0. \qquad (13.45)$$

This holds because $\mathsf{T}_{em}(a)$ is symmetric.

The redefinition of the energy-momentum and angular momentum tensors for electromagnetism removes the spin term and absorbs it directly into $\mathsf{T}_{em}(a)\wedge x$. This guarantees that the fields are gauge-invariant, but suppresses the spin-1 nature of the electromagnetic field. For gravitational interactions the canonical energy-momentum and spin tensors are not as important as their functional equivalents. In the case of electromagnetism, the latter are guaranteed to be (electromagnetic) gauge-invariant, and the spin contribution does turn out to vanish.

13.2.3 Conformal invariance of free-field electromagnetism

In addition to invariance under Poincaré transformations, free-field electromagnetism is invariant under the full conformal group of spacetime. Conformal geometry is discussed in detail in chapter 10. Here we are interested in the field theory manifestation of conformal invariance. We start by considering an arbitrary displacement, $x' = f(x)$. Gauge invariance tells us that A must transform in the same manner as ∇ (it is a 1-form), so we define

$$A'(x) = \bar{\mathsf{f}}(A(x')). \qquad (13.46)$$

The electromagnetic field strength therefore transforms to

$$\nabla\wedge A'(x) = \bar{\mathsf{f}}\big(\bar{\mathsf{f}}^{-1}(\nabla)\wedge A(x')\big) = \bar{\mathsf{f}}\big(F(x')\big), \qquad (13.47)$$

where we have made use of the results

$$\dot{\nabla}\wedge\dot{\bar{\mathsf{f}}}(a) = 0 \qquad (13.48)$$

and

$$\nabla_{x'} = \bar{\mathsf{f}}^{-1}(\nabla). \qquad (13.49)$$

These formulae are derived in section 6.5.6. The transformed Lagrangian density is now

$$\mathcal{L}' = \tfrac{1}{2}\det\,(\mathsf{f})^{-1}\,\langle\bar{\mathsf{f}}(F(x'))\bar{\mathsf{f}}(F(x'))\rangle. \qquad (13.50)$$

We therefore define a symmetry of the action integral if f satisfies

$$\bar{\mathsf{f}}(A)\cdot\bar{\mathsf{f}}(B) = \det\,(\mathsf{f})\,A\cdot B \qquad (13.51)$$

for any pair of bivectors A and B. This is clearly satisfied by any orthogonal transformation, but it is also satisfied by dilations. The Lagrangian for the free electromagnetic field is therefore symmetric under any displacement whose derivative is a local orthogonal transformation coupled with a dilation. This defines the conformal group.

As a simple example, consider the dilation $x' = \exp(\alpha)x$. For this transformation Noether's theorem gives

$$x\cdot\nabla\mathcal{L} = -4\mathcal{L} + \nabla\cdot\left(\gamma_\mu\,(A + x\cdot\nabla\,A)*\frac{\partial\mathcal{L}}{\partial(\partial_\mu A)}\right), \tag{13.52}$$

from which we extract the conserved current

$$J = \mathsf{T}(x) + A\cdot F = \mathsf{T}_{em}(x) + \nabla\cdot(A\cdot x\,F). \tag{13.53}$$

The final term is the divergence of a bivector so is automatically conserved. Dilation invariance therefore tells us that

$$\nabla\cdot\mathsf{T}_{em}(x) = 0, \tag{13.54}$$

which holds because T_{em} is conserved and traceless. The latter property is typical of scale-invariant theories.

Similarly, a special conformal transformation maps the position vector x to x', where

$$x' = f(x) = (x^{-1} + \alpha a)^{-1} = x(1 + \alpha a x)^{-1}. \tag{13.55}$$

The derivative transformation is

$$\mathsf{f}(b) = b\cdot\nabla f(x) = (1 + \alpha x a)^{-1} b (1 + \alpha a x)^{-1}, \tag{13.56}$$

which is a local rotation and dilation. The determinant is

$$\det\,(\mathsf{f}) = (1 + 2\alpha a\cdot x + \alpha^2 a^2 x^2)^{-4}. \tag{13.57}$$

We also find that

$$\left.\frac{\partial x'}{\partial\alpha}\right|_{\alpha=0} = -xax \tag{13.58}$$

and

$$\left.\frac{\partial}{\partial\alpha}\det\,(\mathsf{f})^{-1}\right|_{\alpha=0} = 8x\cdot a. \tag{13.59}$$

Noether's theorem for special conformal transformations can then be shown to produce the conserved tensor $\mathsf{T}_{em}(xax)$. Conservation again follows from the properties of T_{em}.

13.3 Dirac theory

The free-field Dirac Lagrangian is

$$\mathcal{L} = \langle \nabla \psi I \gamma_3 \tilde{\psi} - m\psi\tilde{\psi} \rangle, \tag{13.60}$$

where ψ is a spinor field. Variation with respect to ψ produces the Euler–Lagrange equation

$$(\nabla \psi I \gamma_3)^\sim - 2m\tilde{\psi} + \frac{\partial}{\partial x^\mu}(I\gamma_3\tilde{\psi}\gamma^\mu) = 0, \tag{13.61}$$

which reverses to recover the Dirac equation in the form

$$\nabla \psi I \gamma_3 = m\psi. \tag{13.62}$$

This derivation departs from that given in many textbooks, as we do not consider ψ and $\tilde{\psi}$ as independent variables. Instead we view $\mathcal{L}$ as a real scalar function of a single field ψ. An immediate consequence of the field equations is that $\mathcal{L} = 0$ when the Dirac equation is satisfied. This behaviour is typical of first-order systems.

13.3.1 Spacetime transformations

The canonical energy-momentum tensor for the Dirac field is easily found,

$$\begin{aligned} \mathsf{T}_D(a) &= \gamma_\mu \langle a\cdot\nabla\psi I\gamma_3\tilde{\psi}\gamma^\mu\rangle - a\mathcal{L} \\ &= \langle a\cdot\nabla\psi I\gamma_3\tilde{\psi}\rangle_1. \end{aligned} \tag{13.63}$$

This energy-momentum tensor is not symmetric. Its adjoint is

$$\bar{\mathsf{T}}_D(a) = \dot{\nabla}\langle \dot{\psi} I\gamma_3\tilde{\psi} a\rangle, \tag{13.64}$$

and the antisymmetric term is governed by the bivector

$$\partial_a \wedge \mathsf{T}_D(a) = \dot{\nabla}\wedge\langle \dot{\psi} I\gamma_3\tilde{\psi}\rangle_1. \tag{13.65}$$

This bivector can be written as

$$\begin{aligned} \dot{\nabla}\wedge\langle \dot{\psi} I\gamma_3\tilde{\psi}\rangle_1 &= \big\langle \langle \nabla\psi I\gamma_3\tilde{\psi} - \dot{\nabla}\langle\dot{\psi}I\gamma_3\tilde{\psi}\rangle_3\big\rangle_2 \\ &= -\tfrac{1}{2}\nabla\cdot(\psi I\gamma_3\tilde{\psi}). \end{aligned} \tag{13.66}$$

So, as stated in section 13.1.1, the antisymmetric component of the energy-momentum tensor is a total divergence. In this case we can write

$$\partial_a \wedge \mathsf{T}_D(a) = -\tfrac{1}{2}\nabla\cdot S, \tag{13.67}$$

where S is the spin trivector

$$S = \psi I \gamma_3 \tilde{\psi}. \tag{13.68}$$

Rotational invariance follows from the transformation

$$\psi'(x) = R\psi(x') \tag{13.69}$$

with R and x' as defined in equation (13.19). The wavefunction ψ is subject to the single-sided transformation law appropriate for spinors. One can easily show that the rotors cancel out of the transformed Lagrangian, and the conjugate angular momentum is

$$\mathsf{J}(B) = \langle(-B\cdot x)\cdot\nabla\psi I\gamma_3\tilde{\psi}\rangle_1 + \tfrac{1}{2}B\cdot(\psi I\gamma_3\tilde{\psi}). \tag{13.70}$$

The adjoint gives

$$\bar{\mathsf{J}}(a) = \bar{\mathsf{T}}(a)\wedge x + \tfrac{1}{2}a\cdot S, \tag{13.71}$$

which neatly exposes the spin contribution to the angular momentum. Comparison with equation (12.56) confirms that the point-particle models discussed in section 12.2.1 do correctly capture the properties of the field angular momentum.

The mass term in the free-field Dirac Lagrangian is the sole term breaking conformal invariance. Spacetime spinors have a conformal weight of 3/2, so dilations are defined by

$$\psi'(x) = \mathrm{e}^{3\alpha/2}\psi(\mathrm{e}^{\alpha}x). \tag{13.72}$$

For this transformation, Noether's theorem gives rise to the canonical vector $\mathsf{T}_D(x)$, which satisfies the partial conservation law

$$\nabla\cdot\mathsf{T}_D(x) = \langle m\psi\tilde{\psi}\rangle. \tag{13.73}$$

Special conformal transformations are also interesting to consider. With the transformation as defined in equation (13.55), we write the derivative transformation as

$$a\cdot\nabla x' = \mathsf{f}(a) = \frac{1}{\rho}Ra\tilde{R}, \tag{13.74}$$

where

$$\rho = 1 + 2\alpha a\cdot x + \alpha^2 a^2 x^2, \qquad R = \frac{1+\alpha a x}{\rho^{1/2}}. \tag{13.75}$$

We define the transformed spinor by

$$\psi'(x) = \frac{1}{\rho^{3/2}}\tilde{R}\psi(x') = (1+\alpha a x)^{-2}(1+\alpha x a)^{-1}\psi(x'). \tag{13.76}$$

This transformation of ψ defines a symmetry of the action because of the remarkable result that

$$\nabla\big((1+\alpha a x)^{-2}(1+\alpha x a)^{-1}\big) = 0. \tag{13.77}$$

It follows that

$$\nabla\psi'(x) = \frac{1}{\rho^{5/2}}\tilde{R}\nabla_{x'}\psi(x'), \tag{13.78}$$

which is precisely the transformation required in the Dirac action. More generally, a special conformal transformation can be applied to any spacetime monogenic to obtain a new monogenic function. Equation (13.77) is an example of the general result that

$$\nabla \left(\frac{1+\alpha x a}{(1+2\alpha x\cdot a+\alpha^2 a^2 x^2)^{n/2}} \right) = 0, \tag{13.79}$$

which holds in an n-dimensional space of arbitrary signature.

The conserved tensor conjugate to special conformal transformations, T_c, is found from Noether's theorem to be

$$\mathsf{T}_c(a) = \mathsf{T}_D(xax) + (a\wedge x)\cdot S. \tag{13.80}$$

The partial conservation law for this is

$$\nabla\cdot\mathsf{T}_c(a) = 2m\, a\cdot x\langle\psi\tilde{\psi}\rangle. \tag{13.81}$$

For both dilations and special conformal transformations we recover a genuine conservation law if the mass m is set to zero. This is the basis for an important technique in quantum field theory. In high-energy experiments it is often a reasonable approximation to treat the particles as massless. One can then take advantage of the conformal symmetry to compute a range of consequences for the outcome of experiment. Typically, these predictions will be valid up to order m/E, where E is the energy.

13.3.2 Internal symmetries and phase invariance

As well as spacetime symmetries there are a number of internal symmetries of the Dirac action we can consider. The first of these is the duality transformation

$$\psi' = \psi e^{I\alpha}. \tag{13.82}$$

Equation (13.4) produces the relation

$$\nabla\cdot(\psi\gamma_3\tilde{\psi}) = 2\langle mI\psi\tilde{\psi}\rangle. \tag{13.83}$$

So the spin vector defines a conserved current in the massless limit. This is the partially-conserved axial current, which is important in scattering calculations.

Further transformations to consider are internal rotations of the form

$$\psi' = \psi e^{\alpha B}, \tag{13.84}$$

where B is a bivector. In this case equation (13.4) reduces to

$$\nabla\cdot\big(\psi\, B\cdot(I\gamma_3)\,\tilde{\psi}\big) = 0, \tag{13.85}$$

where we have applied the Dirac equation. This yields conserved currents for any component of B which commutes with γ_3. This space is spanned by $\boldsymbol{\sigma}_1$, $\boldsymbol{\sigma}_2$

and $I\boldsymbol{\sigma}_3$. Of these, only $I\boldsymbol{\sigma}_3$ has the additional property of leaving invariant the observable current $\psi\gamma_0\tilde{\psi}$. This is the case of a phase transformation, and the conjugate conserved quantity is precisely the current J, so

$$\nabla\cdot J = 0, \qquad J = \psi\gamma_0\tilde{\psi}. \tag{13.86}$$

This is an example of the general result in quantum theory that phase invariance ensures that probability density is conserved, and wavefunction evolution is unitary.

The phase transformation law

$$\psi \mapsto \psi' = \psi e^{\phi I\boldsymbol{\sigma}_3} \tag{13.87}$$

is a *global* symmetry of the Lagrangian, because ϕ is a constant. If ψ satisfies the Dirac equation, then so to does ψ'. We arrive at a *gauge theory* if we convert this global symmetry to a local one. There are a number of reasons for believing that this is a sensible way to construct interactions in field theory. One motivation is from the structure of the physical statements that can be extracted from Dirac theory. Quantum theory makes predictions about the values of *observables*, which are formed from inner products between spinors, $\langle\psi|\phi\rangle$. These inner products are invariant under local changes of phase. Similarly, quantum theory can make statements about the equality of two spinor expressions, for example

$$\psi = \psi_1 + \psi_2. \tag{13.88}$$

This might decompose ψ into two orthogonal eigenstates of some operator. Again, if all spinors pick up the same locally-varying phase factor then the physical predictions are unchanged. In addition, a global change of phase corresponds to simultaneously changing the phase of the wavefunction everywhere in the universe. While this can be conceived of mathematically, it does not make a great deal of physical sense. The ultimate motivation, however, comes from the fact that gauge theories are spectacularly successful. All of the known fundamental forces can be described by the procedure of turning a global symmetry into a local symmetry.

13.3.3 Covariant derivatives and minimal coupling

Now that we are clear on the motivation, we must find how to modify the Dirac equation in order that phase changes become a local symmetry. This is the prototype gauge theory. We start by writing

$$\psi' = \psi R, \tag{13.89}$$

where R is a position-dependent rotor. We will later set $R = \exp(I\boldsymbol{\sigma}_3\phi(x))$. This slightly more general formulation eases the transition to the more complicated cases of electroweak and gravitational interactions. The equation for ψ' now includes the term

$$\nabla\psi' = \gamma^\mu\big(\partial_\mu\psi R + \psi\partial_\mu R\big). \tag{13.90}$$

We need to modify the ∇ operator to be able to cancel out the term in the derivative of R. We therefore define a new, *covariant* derivative operator D, where

$$D\psi = \gamma^\mu D_\mu\psi. \tag{13.91}$$

The directional covariant derivatives D_μ contain an extra term going as

$$D_\mu\psi = \partial_\mu\psi + \tfrac{1}{2}\psi\Omega_\mu, \tag{13.92}$$

where Ω_μ is a multivector field whose nature and transformation properties we have to determine. (The factor of $1/2$ is inserted for later convenience.) The index indicates that Ω_μ is a linear function. We can therefore write

$$\Omega_\mu = \Omega(\gamma_\mu) = \Omega(\gamma_\mu; x), \tag{13.93}$$

which defines the linear function $\Omega(a) = \Omega(a; x)$. The x dependence records the fact that the field will in general be a function of position. This label is usually suppressed. In later applications we will make strong use of the index-free form $\Omega(a)$.

The behaviour we require is that under a local rotation, D should transform in such a way that ψR is still a solution of the modified equation. So, with D transforming to D', we require that

$$D'(\psi R) = (D\psi)R \tag{13.94}$$

for any R. We expect that D' should have the same functional form as D, so we also have

$$D'\psi = \gamma^\mu\big(\partial_\mu\psi + \tfrac{1}{2}\psi\Omega'_\mu\big). \tag{13.95}$$

Equation (13.94) therefore gives

$$\begin{aligned} D'(\psi R) &= \gamma^\mu\big(\partial_\mu\psi R + \psi\partial_\mu R + \tfrac{1}{2}\psi R\Omega'_\mu\big) \\ &= \gamma^\mu\big(\partial_\mu\psi + \tfrac{1}{2}\psi\Omega_\mu\big)R. \end{aligned} \tag{13.96}$$

From this we can read off that

$$\partial_\mu R + \tfrac{1}{2}R\Omega'_\mu = \tfrac{1}{2}\Omega_\mu R, \tag{13.97}$$

which establishes the transformation law

$$\Omega'_\mu = \tilde{R}\Omega_\mu R - 2\tilde{R}\partial_\mu R. \tag{13.98}$$

Now R is a rotor, so $2\tilde{R}\partial_\mu R$ is a member of the Lie algebra of the rotor group. It follows that this term is a pure bivector, so Ω_μ must also contain a bivector term if it is to cancel a term in $2\tilde{R}\partial_\mu R$. We assume that this is the only term present the Ω_μ field. This is the minimal assumption, and is referred to as defining *minimal coupling*.

The important point in this derivation is that we have used the form of the term $-2\tilde{R}\partial_\mu R$ to say what type of object Ω_μ is. We are *not* asserting that Ω_μ is equal to $-2\tilde{R}\partial_\mu R$. On the contrary, as will become apparent later, if Ω_μ was given by the gradient of a rotor in this manner it would give rise to a vanishing field strength and therefore be of no physical interest. This step, of taking a term arising from a derivative (like $-2\tilde{R}\partial_\mu R$ here), and generalizing it to a field *not* in general derivable from a derivative, is the essence of the gauging process. The Ω_μ term in the covariant derivative is called a *connection*. In general, connections take their values in the Lie algebra of the associated symmetry group. Many of the symmetry groups we consider are rotor groups, so for these the connections are bivector fields.

13.3.4 The minimally coupled Dirac equation

Returning to electromagnetism, we are concerned with the restricted class of rotations that take place entirely in the $\gamma_2\gamma_1$ plane. In this case, writing $R = \exp(I\boldsymbol{\sigma}_3\phi)$, we have

$$-2\tilde{R}\partial_\mu R = -2\mathrm{e}^{-I\boldsymbol{\sigma}_3\phi}\partial_\mu\phi\mathrm{e}^{I\boldsymbol{\sigma}_3\phi}I\boldsymbol{\sigma}_3 = -2\gamma_\mu\cdot(\nabla\phi)I\boldsymbol{\sigma}_3. \tag{13.99}$$

In generalizing to Ω_μ, we see that this must take the form

$$\Omega_\mu = -\lambda\gamma_\mu\cdot A\, I\boldsymbol{\sigma}_3 \tag{13.100}$$

or, in frame-free notation,

$$\Omega(a) = -\lambda a\cdot A\, I\boldsymbol{\sigma}_3. \tag{13.101}$$

Here A is a spacetime vector field, and λ is some coupling constant. We now reassemble our full, covariant Dirac equation to obtain

$$D\psi I\gamma_3 = \gamma^\mu\big(\partial_\mu\psi - \tfrac{1}{2}\lambda\psi\,\gamma_\mu\cdot A\, I\boldsymbol{\sigma}_3\big)I\gamma_3 = m\psi. \tag{13.102}$$

This simplifies to give

$$\nabla\psi I\gamma_3 - \tfrac{1}{2}\lambda A\psi\gamma_0 = m\psi, \tag{13.103}$$

and we see that the contraction between the γ^μ frame and the connection in equation (13.102) assembles to give a vector multiplying ψ from the left. It is clear that for an electron we require $\lambda = 2e$, so the *minimally coupled* Dirac equation is

$$\nabla\psi I\boldsymbol{\sigma}_3 - eA\psi = m\psi\gamma_0, \tag{13.104}$$

as studied in section 8.3. A local phase transformation of ψ now induces the transformation

$$eA \mapsto eA - \nabla\phi, \tag{13.105}$$

which we recognise as an electromagnetic change of gauge. By adding an interaction term solely in A we are making the simplest possible modification to the original equation, which is the essence of minimal coupling. We could, for example, add further terms in F, or F^2 multiplying ψ, and the equation would still be gauge-invariant. It appears, however, that this possibility is not required for describing the fundamental forces. Why this should be so is unknown.

13.3.5 The gauge field strength

Now that we have introduced the gauge fields the next step is to construct the observable (gauge-invariant) quantities associated with them. For electromagnetism we know that these are the $\boldsymbol{E}$ and $\boldsymbol{B}$ fields, which form part of the *field strength tensor*. This is found in general by commuting covariant derivatives. We form

$$\begin{aligned}[D_\mu, D_\nu]\psi &= D_\mu\big(\partial_\nu\psi + \tfrac{1}{2}\psi\Omega_\nu\big) - D_\nu\big(\partial_\mu\psi + \tfrac{1}{2}\psi\Omega_\mu\big) \\ &= \tfrac{1}{2}\psi\big(\partial_\mu\Omega_\nu - \partial_\nu\Omega_\mu - \Omega_\mu\times\Omega_\nu\big).\end{aligned} \tag{13.106}$$

Despite the fact that we formed commutators of derivatives on ψ, all of the derivatives of ψ have cancelled, and we are left with a single object

$$\mathsf{F}_{\mu\nu} = \mathsf{F}(\gamma_\mu\wedge\gamma_\nu) = \partial_\mu\Omega_\nu - \partial_\nu\Omega_\mu - \Omega_\mu\times\Omega_\nu. \tag{13.107}$$

This is a bivector-valued linear function of the bivector argument $\gamma_\mu\wedge\gamma_\nu$. The construction of this object guarantees that under a change of gauge

$$\mathsf{F}_{\mu\nu} \mapsto \mathsf{F}'_{\mu\nu} = \tilde{R}\mathsf{F}_{\mu\nu}R. \tag{13.108}$$

This transformation tells us that the field strength transforms *covariantly* under changes of gauge.

Specialising to the case of electromagnetism, where $\Omega_\mu = -2e\gamma_\mu\cdot A\, I\boldsymbol{\sigma}_3$, we find that the term multiplying ψ contains

$$\begin{aligned}(-2e)^{-1}\mathsf{F}_{\mu\nu} &= \partial_\mu(\gamma_\nu\cdot A\, I\boldsymbol{\sigma}_3) - \partial_\nu(\gamma_\mu\cdot A\, I\boldsymbol{\sigma}_3) - \gamma_\mu\cdot A\,\gamma_\nu\cdot A\, I\boldsymbol{\sigma}_3\times I\boldsymbol{\sigma}_3 \\ &= (\gamma_\nu\wedge\gamma_\mu)\cdot(\nabla\wedge A)I\boldsymbol{\sigma}_3 \\ &= (\gamma_\nu\wedge\gamma_\mu)\cdot F\, I\boldsymbol{\sigma}_3.\end{aligned} \tag{13.109}$$

This is a function that maps the bivector $\gamma_\nu\wedge\gamma_\mu$ linearly onto a pure phase term. For most applications of electromagnetism it is sensible to lose the mapping nature of the field strength and instead work directly with the bivector F. For more complicated gauge fields this is not appropriate. In forming the commutator of covariant derivatives we have extracted the correct field strength, $F = \nabla \wedge A$, which encodes the physically measurable content of the electromagnetic field. The electromagnetic field strength is *invariant* under a change of gauge, as opposed to covariant. This is because the underlying gauge group, U(1), is a commutative group, so the rotors cancel out in equation (13.108). The picture is less simple for non-commutative Lie groups.

13.3.6 Electroweak symmetry

A full treatment of electroweak gauge theory requires the apparatus of quantum field theory, which is beyond the scope of this book. Here we give a simplified treatment, concentrating entirely on the fermionic sector for an electron and a neutrino. The left-handed particles in this sector are assembled into a doublet

$$L_e = \begin{pmatrix} |\nu_e\rangle \\ |e_l\rangle \end{pmatrix} \tag{13.110}$$

and the right-handed particles consist of a singlet state $|e_r\rangle$. The kets denote Dirac spinors, projected into their left-handed or right-handed states. The left-hand doublet is acted on by SU(2) matrices, which transform the upper and lower components into linear superpositions of $|\nu_e\rangle$ and $|e_l\rangle$. To construct an equivalent group action in spacetime algebra, we introduce the spinor ψ_l, where

$$\begin{pmatrix} |\nu_e\rangle \\ |e_l\rangle \end{pmatrix} \leftrightarrow \psi_l = \psi_e \tfrac{1}{2}(1 - \boldsymbol{\sigma}_3) - \psi_\nu I \boldsymbol{\sigma}_2 \tfrac{1}{2}(1 + \boldsymbol{\sigma}_3). \tag{13.111}$$

Here ψ_e and ψ_ν are the spacetime algebra equivalents of the $|e_l\rangle$ and $|\nu_e\rangle$ spinors, as defined by the map of equation (8.69). This map ensures that the action of the generators of the SU(2) group become

$$\hat{\sigma}_k L_e \leftrightarrow \psi_l \boldsymbol{\sigma}_k, \tag{13.112}$$

and hence

$$i L_e \leftrightarrow -\psi_l I. \tag{13.113}$$

So all transformations are now carried out on the right-hand side of ψ_l, and are of the class discussed in section 13.3.2.

The kinetic term in the Lagrangian for the left-handed doublet is usually written as

$$\bar{L}_e i \not{D} L_e = \langle \bar{\nu}_e | i \not{D} | \nu_e \rangle + \langle \bar{e}_l | i \not{D} | e_l \rangle, \tag{13.114}$$

which has the multivector equivalent

$$\mathcal{L}_l = \langle \nabla \psi_\nu \tfrac{1}{2}(1 - \boldsymbol{\sigma}_3) I \gamma_3 \tilde{\psi}_\nu + \nabla \psi_e \tfrac{1}{2}(1 - \boldsymbol{\sigma}_3) I \gamma_3 \tilde{\psi}_e \rangle. \tag{13.115}$$

Now

$$\begin{aligned}\tfrac{1}{2}(1-\boldsymbol{\sigma}_3)I\gamma_3\tilde{\psi}_e &= -\tfrac{1}{2}(1-\boldsymbol{\sigma}_3)I\gamma_0\tilde{\psi}_l,\\ \tfrac{1}{2}(1-\boldsymbol{\sigma}_3)I\gamma_3\tilde{\psi}_\nu &= I\boldsymbol{\sigma}_2\tfrac{1}{2}(1+\boldsymbol{\sigma}_3)I\gamma_0\tilde{\psi}_l,\end{aligned} \tag{13.116}$$

so

$$\begin{aligned}\mathcal{L}_l &= -\langle\nabla\big(\psi_e\tfrac{1}{2}(1-\boldsymbol{\sigma}_3) - \psi_\nu I\boldsymbol{\sigma}_2\tfrac{1}{2}(1+\boldsymbol{\sigma}_3)\big)I\gamma_0\tilde{\psi}_l\rangle\\ &= -\langle\nabla\psi_l I\gamma_0\tilde{\psi}_l\rangle.\end{aligned} \tag{13.117}$$

The left-handed fermionic sector of the electroweak Lagrangian is similar to the Dirac Lagrangian, but with γ_3 replaced by γ_0. The internal symmetry group is therefore defined by transformations of the form

$$\psi \mapsto \psi e^M, \tag{13.118}$$

where M is any even multivector that satisfies

$$\exp(M)\gamma_0\exp(\tilde{M}) = \gamma_0. \tag{13.119}$$

This picks out the set of bivectors that commute with γ_0, and the pseudoscalar. The former define an SU(2) group, and the latter is a U(1) phase term. The Lagrangian therefore has the expected SU(2)×U(1) symmetry of electroweak theory, encoded in a very natural way in the spacetime algebra.

The right-handed sector of the electroweak theory involves a singlet state

$$\psi_r = \psi_e\tfrac{1}{2}(1+\boldsymbol{\sigma}_3). \tag{13.120}$$

The kinetic term for this is

$$\langle\nabla\psi_r I\gamma_3\tilde{\psi}_r\rangle = -\langle\nabla\psi_e\tfrac{1}{2}(\gamma_0+\gamma_3)I\tilde{\psi}_e\rangle. \tag{13.121}$$

Mass terms are introduced via interaction with the Higgs field, which can be modelled straightforwardly as an interaction between left-handed and right-handed particles. A global SU(2) transformation is described by

$$\psi_l \mapsto \psi_l R, \tag{13.122}$$

where R is a rotor satisfying $R\gamma_0\tilde{R} = \gamma_0$. This is converted to a local symmetry following the procedure of section 13.3.2, which tells us that the connection consists of bivectors which commute with γ_0. The U(1) connection is a multiple of the pseudoscalar. The field strength is defined similarly, and one can proceed to model spontaneous symmetry breaking using this scheme. At some point, however, it is necessary to adopt a quantum field theory perspective, and replace the wavefunctions described here by operators acting on the quantum vacuum.

13.4 Gauge principles for gravitation

We have so far described electromagnetism and electroweak forces in terms of gauge theories. We now turn our attention to gravity. Our aim is to model gravitational interactions in terms of gauge fields defined in the spacetime algebra. This initially appears to be a radical departure from general relativity, but in fact the two approaches converge in a manner that sheds light on the physical structure of the theory. Spacetime algebra is the geometric algebra of *flat* spacetime, and the introduction of fields cannot alter this basic property. What then are we to make of the standard arguments that spacetime is curved? The answer is that all of these arguments involve light paths, or measuring rods, or similar devices, and all of these processes are also modelled by fields. Since all physical quantities correspond to fields, the *absolute* position and orientation of particles or fields in our background spacetime is not measurable. It drops out of all physical calculations. The only predictions that can be extracted are relative relations between fields. Ensuring that this property is true locally means there is no conflict with any of the principles by which one is traditionally led to general relativity, and naturally guides us in the direction of a gauge theory.

To illustrate these considerations, consider possible relations between quantum fields. Suppose that $\psi_1(x)$ and $\psi_2(x)$ are spinor fields. A physical statement could be a simple relation of equality:

$$\psi_1(x) = \psi_2(x). \tag{13.123}$$

But all this statement says is that at a point where one field has a particular value, then the second field has the same value. This statement is completely independent of where we choose to place the fields in the spacetime algebra. And, more importantly, it is totally independent of where we choose to locate other values of the fields. We could equally well introduce two new fields

$$\psi_1'(x) = \psi_1(x'), \quad \psi_2'(x) = \psi_2(x'), \tag{13.124}$$

where x' is an arbitrary function of position x. The statement $\psi_1'(x) = \psi_2'(x)$ contains precisely the same physical content as the original equation.

The same picture emerges if both fields are acted on by a spacetime rotor, giving rise to new fields

$$\psi_1' = R\psi_1, \quad \psi_2' = R\psi_2. \tag{13.125}$$

Again, the statement $\psi_1' = \psi_2'$ has the same physical content as the original equation. Similar considerations apply to the observables formed from ψ, such as the vector $J = \psi\gamma_0\tilde{\psi}$. Replacing ψ by ψ' produces the new vector $J' = RJ\tilde{R}$. Invariance of the equations under this transformation ensures that the absolute direction of vectors in the spacetime algebra is not measurable, only the relative orientation of two physical vectors is measurable. We now have a clear mathematical statement of the invariance properties we want to establish. The next task is to study the form of the gauge fields needed to enforce this invariance.

13.4.1 Displacements

We write $x' = f(x)$ for an arbitrary (differentiable) map between spacetime position vectors. The transformation we are interested in is where the field $\psi(x)$ is transformed to the new field

$$\psi'(x) = \psi(x'). \tag{13.126}$$

The map $f(x)$ should not be thought of as a map between manifolds, or as moving points around. The function $f(x)$ is just a rule for relating one position vector to another within a single vector space. It is the fields that are transformed in this space. We need a good name for this operation of moving fields around. One possibility is *translation*, but this suggests a rigid map where all fields are translated by the same amount. Mathematicians favour the term *diffeomorphism*, but this usually refers to a map between distinct manifolds. We prefer to use the term *displacement*, which does suggest the concept of moving a field around from one point to another in an arbitrary manner.

The next step is to consider the behaviour of the derivative of ψ. With the displacement denoted by $x' = f(x)$, and the derivative defined by

$$\mathsf{f}(a) = a\cdot\nabla f(x), \tag{13.127}$$

we know that the vector derivative satisfies

$$\nabla_x = \bar{\mathsf{f}}(\nabla_{x'}). \tag{13.128}$$

So, for example, if $\psi(x)$ is a spinor, and $\psi'(x) = \psi(x')$, we have

$$\nabla\psi'(x) = \bar{\mathsf{f}}(\nabla_{x'})\psi(x'). \tag{13.129}$$

To formulate a version of the Dirac action that is invariant under arbitrary displacements, we must introduce a gauge field that removes the effect of the $\bar{\mathsf{f}}$ function. This field will then assemble with the vector derivative to form an object which, under displacements, simply reevaluates to the derivative with respect to the new position vector. We construct such an object by replacing ∇ with a new derivative $\bar{\mathsf{h}}(\nabla)$, where

$$\bar{\mathsf{h}}(a) = \bar{\mathsf{h}}(a;x) \tag{13.130}$$

is a position-dependent linear function of a. We again suppress this position dependence where clarity permits.

Under displacements the gauge field $\bar{\mathsf{h}}$ must transform such that

$$\bar{\mathsf{h}}'(\nabla_{x'}) = \bar{\mathsf{h}}(\nabla_x) = \bar{\mathsf{h}}\bar{\mathsf{f}}(\nabla_{x'}). \tag{13.131}$$

Explicitly, the transformation law for $\bar{\mathsf{h}}$ under displacements must be

$$\bar{\mathsf{h}}'(a;x) = \bar{\mathsf{h}}\big(\bar{\mathsf{f}}^{-1}(a);x'\big), \tag{13.132}$$

or, suppressing the position dependence,

$$\bar{\mathsf{h}}'(a) = \bar{\mathsf{h}}\bar{\mathsf{f}}^{-1}(a). \tag{13.133}$$

This must hold for any arbitrary vector a. This transformation law is different to that encountered in the gauge theories discussed previously, as the gauge field acts directly on ∇. The $\bar{\mathsf{h}}$ field is therefore not a connection in the conventional Yang–Mills sense. It is clear, however, that the $\bar{\mathsf{h}}$ field embodies the idea of ensuring that a symmetry is local, so can sensibly be called a gauge field. Since $\bar{\mathsf{h}}(a)$ is an arbitrary, position-dependent linear function of a, it has $4 \times 4 = 16$ degrees of freedom.

We can now systematically replace every occurrence of ∇ with $\bar{\mathsf{h}}(\nabla)$, and all our equations will be invariant under arbitrary displacements. In particular, the Dirac Lagrangian density is now modified to read

$$\mathcal{L} = \det(\mathsf{h})^{-1}\left\langle \bar{\mathsf{h}}(\nabla)\psi I\gamma_3\tilde{\psi} - m\psi\tilde{\psi}\right\rangle. \tag{13.134}$$

This now transforms covariantly under arbitrary displacements of the fields. Similarly, we can consider the proper time or distance along a trajectory $x(\lambda)$. In the absence of gravitational fields this is

$$S = \int d\lambda \left|\frac{\partial x}{\partial \lambda}\cdot\frac{\partial x}{\partial \lambda}\right|^{1/2}. \tag{13.135}$$

Under a displacement the path transforms to $f(x(\lambda))$, so the tangent vector transforms to

$$\partial_\lambda f\big(x(\lambda)\big) = \mathsf{f}(\partial_\lambda x). \tag{13.136}$$

We can therefore construct a gauge-invariant interval by setting

$$S = \int d\lambda \left|\mathsf{h}^{-1}(x')\cdot\mathsf{h}^{-1}(x')\right|^{1/2}, \tag{13.137}$$

where

$$x' = \frac{\partial x(\lambda)}{\partial \lambda}. \tag{13.138}$$

This distance is now invariant under displacements, so is a physically-observable quantity.

We now see that tangent vectors pick up a factor of h^{-1} and cotangent vectors a factor of $\bar{\mathsf{h}}$. Spinors are not acted on by the h function. Next we establish contact with more familiar constructions of general relativity. Suppose that x^μ denote an arbitrary coordinate system, with frame vectors denoted by

$$e_\mu = \frac{\partial x}{\partial x^\mu}, \qquad e^\mu = \nabla x^\mu. \tag{13.139}$$

In terms of this coordinate system, equation (13.137) involves the term

$$\mathsf{h}^{-1}(x')\cdot\mathsf{h}^{-1}(x') = \frac{\partial x^\mu}{\partial\lambda}\frac{\partial x^\mu}{\partial\lambda}\mathsf{h}^{-1}(e_\mu)\cdot\mathsf{h}^{-1}(e_\nu). \tag{13.140}$$

If we define the vectors

$$g_\mu = \mathsf{h}^{-1}(e_\mu), \qquad g^\mu = \bar{\mathsf{h}}(e^\mu). \tag{13.141}$$

then we can write the preceding term as

$$\mathsf{h}^{-1}(x')\cdot\mathsf{h}^{-1}(x') = \frac{\partial x^\mu}{\partial\lambda}\frac{\partial x^\mu}{\partial\lambda}g_\mu\cdot g_\nu. \tag{13.142}$$

Equation (13.137) is therefore equivalent to the line interval in general relativity if we set the metric equal to

$$g_{\mu\nu} = g_\mu\cdot g_\nu = \mathsf{h}^{-1}(e_\mu)\cdot\mathsf{h}^{-1}(e_\nu). \tag{13.143}$$

The gauge field h is therefore a form of square root of the metric, which allows us to replace the metric inner product with the inner product in the spacetime algebra. In this sense, h is closely related to the concept of a spacetime orthonormal tetrad or *vierbein*. A vierbein is obtained from the h field by defining

$$\begin{aligned} e_\mu{}^i &= g_\mu\cdot\gamma^i, \\ e^\mu{}_i &= g^\mu\cdot\gamma_i, \end{aligned} \tag{13.144}$$

where both i and μ run from 0 to 4. The advantage of working directly with the h field is that it frees us from any coordinate frame. Coordinate frames are best introduced at a later date, when the geometry of a given problem usually dictates the appropriate coordinate system.

Now that we have recovered the metric, the obvious question is what has happened to the original flat space? It has not gone away, as all fields take their values over this space. In fact, there are now three distinct spaces of objects we can discuss. We refer to these as the tangent, cotangent and covariant spaces. Tangent vectors are of the form e_μ. Inner products between these are not gauge-invariant, and hence not physically meaningful. Similarly, cotangent vectors are of the form of e^μ, and the inner product of cotangent vectors is also an unphysical quantity. The inner product between tangent and cotangent vectors does produce a gauge-invariant quantity, so can correspond to a physical observable. Tangent and cotangent vectors can be interchanged via the metric, which maps one space into the other. In frame-free form, we can write

$$a^* = \bar{\mathsf{h}}^{-1}\mathsf{h}^{-1}(a) = \mathsf{g}(a). \tag{13.145}$$

The tangent and cotangent spaces, and the metric map between them, are the traditional elements of general relativity. Our third space, of covariant objects,

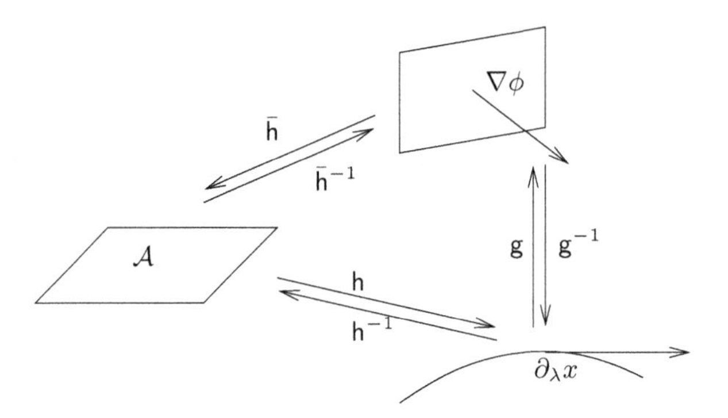

Figure 13.1 *Gauge fields for gravitation.* There are three vector spaces involved, consisting of tangent vectors $\partial_\lambda x$, cotangent vectors $\nabla\phi$ and covariant fields $\mathcal{A}$. The h field maps between these. The metric tensor maps between tangent and cotangent vectors, so is given by $\mathsf{g} = \bar{\mathsf{h}}^{-1}\mathsf{h}^{-1}$. Gauge-invariant quantities are formed from the scalar product of a tangent and cotangent vector, or from a pair of covariant vectors.

is unique to the gauge theory formulation. This space consists of objects whose transformation law under displacements is

$$\phi'(x) = \phi(x'). \tag{13.146}$$

This defines what it means to transform covariantly under displacements. These include velocity vectors of the form $\mathsf{h}^{-1}(\partial_\lambda x)$, gradients of the form $\bar{\mathsf{h}}(\nabla)\phi$, and spinor fields. Inner products between covariant vectors produce covariant scalars, which can be physically observable.

The various fields and spaces involved are depicted in figure 13.1. The advantage of the gauge theory viewpoint, coupled with the application of spacetime algebra, is that we can now take full advantage of the space of covariant objects when analysing the gravitational field equations. This turns out to have many advantages, both conceptually and computationally. The possibilities afforded by this space have been overlooked in most treatments of gauge theory gravity. One immediate question posed by figure 13.1 is whether the insistence on the existence of a map from a curved spacetime onto a flat one has any topological consequences. The answer is yes, though the restrictions are not as severe as one might expect. Many apparently topological constructions, such as cosmic strings and closed universe models, are easily handled in the gauge theory framework. Others, such as wormholes connecting multiple universes, do not fit so easily because they require a modification of the initial assumption that the background space is topologically flat. Models incorporating these effects can be constructed, though their motivation is less clear from the gauge theory perspective, as aspects of the theory have to be put in by hand initially.

13.4.2 Rotations

Now that we have discovered the metric tensor within the gauge approach we could immediately write down the familiar equations of general relativity. But we seek a theory formulated entirely in terms of covariant vectors, and this requires the existence of a second gauge field. As well as invariance under displacements, we require that our wave equation be invariant under the transformation

$$\psi \mapsto \psi' = R\psi, \tag{13.147}$$

where R is an arbitrary, position-dependent spacetime rotor. We are now back in the territory of section 13.3.3, with the difference that the rotor multiplies ψ from the left, instead of the right. To convert ∂_μ into a covariant derivative, we add a bivector connection Ω_μ and define

$$D_\mu\psi = \partial_\mu\psi + \tfrac{1}{2}\Omega_\mu\psi. \tag{13.148}$$

The connection Ω_μ is a position-dependent bivector, subject to the transformation law

$$\Omega_\mu \mapsto \Omega'(a) = R\Omega_\mu\tilde{R} - 2\partial_\mu R\tilde{R}. \tag{13.149}$$

Since R is an arbitrary rotor there is no constraint on the blades that Ω_μ can contain, so Ω_μ has $6 \times 4 = 24$ degrees of freedom.

With the rotation gauge field included, the fully covariant Dirac action now reads, with the electromagnetic term included,

$$S = \int d^4x \det(\mathsf{h})^{-1}\left\langle \bar{\mathsf{h}}(\gamma^\mu)(\partial_\mu\psi + \tfrac{1}{2}\Omega_\mu\psi)I\gamma_3\tilde{\psi} - e\bar{\mathsf{h}}(A)\psi\gamma_0\tilde{\psi} - m\psi\tilde{\psi}\right\rangle. \tag{13.150}$$

The value of this action should be unchanged under local displacements and rotations. To establish this we need to complete the set of transformation properties for the gravitational gauge fields. First, we need to define how Ω_μ transforms under displacements. For this it is easier to use the notation $\Omega(a;x)$ for the linear argument and position dependence of the connection. Since $\Omega(a)$ picks up a term in $a\cdot\nabla R\tilde{R}$ under local rotations, we see that the appropriate transformation law under displacements is

$$\Omega'(a;x) = \Omega(\mathsf{f}(a);x'). \tag{13.151}$$

The connection in the action of equation (13.150) is contracted to form the object

$$\bar{\mathsf{h}}(\gamma^\mu)\Omega_\mu = \bar{\mathsf{h}}(\partial_a)\Omega(a). \tag{13.152}$$

So under a displacement this transforms to

$$\bar{\mathsf{h}}'(\partial_a)\Omega'(a) = \bar{\mathsf{h}}(\bar{\mathsf{f}}^{-1}(\partial_a); x')\,\Omega(\mathsf{f}(a); x') = \bar{\mathsf{h}}(\partial_a; x')\Omega(a; x'), \tag{13.153}$$

which is precisely the behaviour we require.

Similarly, we can establish the behaviour of the h field under rotations from the kinetic term in the covariant Dirac action. Under a local rotation this transforms to

$$\left\langle \bar{\mathsf{h}}'(\gamma^\mu)(\partial_\mu\psi' + \tfrac{1}{2}\Omega'_\mu\psi')I\gamma_3\tilde{\psi}'\right\rangle = \left\langle \tilde{R}\bar{\mathsf{h}}'(\gamma^\mu)R(\partial_\mu\psi + \tfrac{1}{2}\Omega_\mu\psi)I\gamma_3\tilde{\psi}\right\rangle. \tag{13.154}$$

So under rotations we must have

$$\bar{\mathsf{h}}(a) \mapsto \bar{\mathsf{h}}'(a) = R\bar{\mathsf{h}}(a)\tilde{R}. \tag{13.155}$$

The same transformation law is obeyed by vectors of the form $\mathsf{h}^{-1}(a)$, where a is a tangent vector. This guarantees that inner products between tangent and cotangent vectors are gauge-invariant, as required. The action of equation (13.150) now contains all of the local symmetries we require. The coupling of the electromagnetic vector potential A follows from the fact that A generalises the gradient of a scalar, so is a cotangent vector. This is acted on by $\bar{\mathsf{h}}$ to establish a covariant vector.

13.4.3 The Dirac equation in a gravitational background

We have so far established invariance at the level of the Dirac action, which led us to the action of equation (13.150). We now vary this action with respect to ψ, treating all other fields as external, to obtain the full, minimally-coupled Dirac equation. After reversing, variation with respect to ψ produces the equation

$$\begin{aligned}\bar{\mathsf{h}}(\nabla)\psi I\gamma_3 &+ \tfrac{1}{2}\bar{\mathsf{h}}(\gamma^\mu)\Omega_\mu\psi I\gamma_3 + \tfrac{1}{2}\Omega_\mu\bar{\mathsf{h}}(\gamma^\mu)\psi I\gamma_3 \\ &- 2e\bar{\mathsf{h}}(A)\psi\gamma_0 - 2m\psi = -\frac{\partial}{\partial x^\mu}\big(\det(\mathsf{h})^{-1}\bar{\mathsf{h}}(\gamma^\mu)\psi I\gamma_3\big)\det(\mathsf{h}).\end{aligned} \tag{13.156}$$

This simplifies to

$$\bar{\mathsf{h}}(\gamma^\mu)(\partial_\mu\psi + \tfrac{1}{2}\Omega_\mu\psi)I\gamma_3 - e\bar{\mathsf{h}}(A)\psi\gamma_0 = m\psi + \tfrac{1}{2}t\psi I\gamma_3, \tag{13.157}$$

where the vector t is defined by

$$t = \det(\mathsf{h})\partial_\mu\big(\det(\mathsf{h})^{-1}\bar{\mathsf{h}}(\gamma^\mu)\big) + \Omega_\mu{\cdot}\bar{\mathsf{h}}(\gamma^\mu). \tag{13.158}$$

Here we encounter an initial surprise. The minimally-coupled Dirac action only produces the expected Dirac equation if the vector t is zero. We will establish the circumstances when this holds once we have discovered the full gravitational field equations. With t assumed to equal zero, we obtain the expected equation, which we write as

$$D\psi I\gamma_3 - e\mathcal{A}\psi\gamma_0 = m\psi. \tag{13.159}$$

Here we introduce the notation

$$D\psi = \bar{\mathsf{h}}(\gamma^\mu) D_\mu \psi = \bar{\mathsf{h}}(\gamma^\mu)(\partial_\mu \psi + \tfrac{1}{2}\Omega_\mu \psi) \qquad (13.160)$$

and

$$\bar{\mathsf{h}}(A) = \mathcal{A}. \qquad (13.161)$$

In this latter definition we begin to introduce the useful notation of writing fully covariant multivectors in calligraphic font.

13.4.4 Covariant derivatives for observables

Having established the form of the gravitational covariant derivative for a spinor, it is a simple matter to establish the form of the derivatives of the observables formed from a spinor. In general, these observables have the form

$$\mathcal{M} = \psi \Gamma \tilde{\psi}, \qquad (13.162)$$

where Γ is a constant multivector formed from combinations of γ_0, γ_3 and $I\boldsymbol{\sigma}_3$. The observable $\mathcal{M}$ inherits its transformation properties from the spinor ψ, so under displacements $\mathcal{M}$ transforms as

$$\mathcal{M}(x) \mapsto \mathcal{M}'(x) = \mathcal{M}(x') \qquad (13.163)$$

and under rotations $\mathcal{M}$ transforms as

$$\mathcal{M} \mapsto \mathcal{M}' = R\mathcal{M}\tilde{R}. \qquad (13.164)$$

Multivectors with these transformation properties are said to be (fully) *covariant*. Scalars formed from inner products of these quantities account for the physical observables in the theory.

If we now form the partial derivative of $\mathcal{M}$ we obtain

$$\partial_\mu \mathcal{M} = (\partial_\mu \psi)\Gamma\tilde{\psi} + \psi\Gamma(\partial_\mu\psi)^\sim. \qquad (13.165)$$

There is no need to restrict to orthonormal coordinates, so we can take ∂_μ as the derivative with respect to an arbitrary coordinate system, with coordinate frame $\{e_\mu\}$. We immediately see how to construct a covariant derivative for $\mathcal{M}$. We simply replace spinor directional derivatives with their covariant versions and form

$$\begin{aligned}(D_\mu\psi)\Gamma\tilde{\psi} + \psi\Gamma(D_\mu\psi)^\sim &= \partial_\mu(\psi\Gamma\tilde{\psi}) + \tfrac{1}{2}\Omega_\mu\psi\Gamma\tilde{\psi} - \tfrac{1}{2}\psi\Gamma\tilde{\psi}\Omega_\mu \\ &= \partial_\mu(\psi\Gamma\tilde{\psi}) + \Omega_\mu \times (\psi\Gamma\tilde{\psi}),\end{aligned} \qquad (13.166)$$

where

$$\Omega_\mu = \Omega(e_\mu). \qquad (13.167)$$

We therefore define the covariant derivative $\mathcal{D}_\mu$ by

$$\mathcal{D}_\mu \mathcal{M} = \partial_\mu \mathcal{M} + \Omega_\mu \times \mathcal{M}. \tag{13.168}$$

This is the form appropriate for acting on covariant multivectors, including observables formed from spinors. The commutator with the bivector Ω_μ has two important properties. The first is that it is grade-preserving, so the full $\mathcal{D}_\mu$ operator preserves grade. The second is that

$$\Omega_\mu \times (AB) = (\Omega_\mu \times A)B + A(\Omega_\mu \times B), \tag{13.169}$$

which holds for any multivectors A and B. This ensures that $\mathcal{D}_a$ is a *derivation.* That is, it satisfies Leibniz's rule

$$\mathcal{D}_\mu(AB) = (\mathcal{D}_\mu A)B + A(\mathcal{D}_\mu B). \tag{13.170}$$

These properties of preserving grade and satisfying Leibniz's rule are necessary for $\mathcal{D}_\mu$ to be a suitable generalisation of a directional derivative.

We can assemble a full, covariant version of the vector derivative by writing

$$\mathcal{D} = \bar{\mathsf{h}}(e^\mu)\mathcal{D}_\mu = g^\mu \mathcal{D}_\mu, \tag{13.171}$$

where $g^\mu = \bar{\mathsf{h}}(e^\mu)$. This acts on covariant multivectors to raise and lower the grade by one. We can also write

$$\mathcal{D}\mathcal{M} = \mathcal{D}\cdot\mathcal{M} + \mathcal{D}\wedge\mathcal{M}, \tag{13.172}$$

where $\mathcal{M}$ is a homogeneous-grade multivector, and

$$\begin{aligned} \mathcal{D}\cdot\mathcal{M} &= g^\mu \cdot(\mathcal{D}_\mu \mathcal{M}), \\ \mathcal{D}\wedge\mathcal{M} &= g^\mu \wedge(\mathcal{D}_\mu \mathcal{M}). \end{aligned} \tag{13.173}$$

It is also sometimes convenient to write the directional covariant derivative as $a\cdot\mathcal{D}$, where

$$a\cdot\mathcal{D}\,\mathcal{M} = a\cdot g^\mu \mathcal{D}_\mu \mathcal{M}. \tag{13.174}$$

We are now beginning to assemble a very powerful, compact notation for the main operators in gauge theory gravitation.

13.5 The gravitational field equations

The price we pay for ensuring that the Dirac action is invariant under local rotations is the introduction of two gauge fields: the vector-valued function $\mathsf{h}(a)$ and the bivector-valued $\Omega(a)$. These in total have 40 degrees of freedom. Our next task is to construct suitable equations for these gauge fields. As with the Dirac equation, our ultimate goal is to formulate the equations in terms of covariant objects, where the physical content of the theory is clearest. The alternative approach is to work entirely in terms of the metric $g_{\mu\nu}$. This is invariant under rotations, so all reference to the rotation gauge is removed. The end result is a set of second-order equations that are notoriously difficult to solve. The gauge theory approach, with its focus on gauge-covariant objects, provides a number of new solution strategies, both for analytical and numerical work.

Our method for constructing covariant field equations is to find a covariant Lagrangian and vary this. The resulting equations are then guaranteed to be covariant. Our first task, then, is to find covariant forms of the field strengths for the gravitational gauge fields. From these we can construct covariant scalar quantities, which can act as a Lagrangian density.

13.5.1 The rotation-gauge field strength

The field strength for the $\Omega(a)$ connection is found in the standard way by considering commutators of covariant derivatives. We define

$$[D_\mu, D_\nu]\psi = \tfrac{1}{2}\mathsf{R}_{\mu\nu}\psi, \tag{13.175}$$

so that

$$\mathsf{R}_{\mu\nu} = \partial_\mu\Omega_\nu - \partial_\nu\Omega_\mu + \Omega_\mu\times\Omega_\nu. \tag{13.176}$$

A frame-free notation is introduced by first writing

$$\mathsf{R}_{\mu\nu} = \mathsf{R}(e_\mu\wedge e_\nu), \tag{13.177}$$

where the $\{e_\mu\}$ vectors are the coordinate frame defined by the x^μ. We can therefore write

$$\mathsf{R}(a\wedge b) = a\cdot\nabla\Omega(b) - b\cdot\nabla\Omega(a) + \Omega(a)\times\Omega(b). \tag{13.178}$$

Whenever we adopt this notation we assume that the vector arguments a and b are constant. Since the right-hand side is antisymmetric on a and b, the field strength depends only on the bivector $a\wedge b$. This linear action on bivector blades is extended to general bivectors by defining

$$\mathsf{R}(a\wedge b + c\wedge d) = \mathsf{R}(a\wedge b) + \mathsf{R}(c\wedge d). \tag{13.179}$$

This means that we can write the field strength as

$$\mathsf{R}(B) = \mathsf{R}(B; x), \tag{13.180}$$

which is a position-dependent, linear function of the bivector B. The field strength is a general bivector, as there are no restrictions on the form of $\Omega(a)$. This means that $\mathsf{R}(a\wedge b)$ has 36 degrees of freedom, as opposed to the rather simpler six of electromagnetism.

Unlike the electromagnetic case of equation (13.109), the commutator term $\Omega(a)\times\Omega(b)$ has not cancelled out. This has an important consequence for the field equations — they are no longer *linear*. If we add together two configurations of $\Omega(a)$, the field strength of the resultant $\Omega(a)$ is not the same as that from the superposition of the original field strengths. This makes the gravitational field equations much more difficult to solve than those of electromagnetism.

The definition of $\mathsf{R}(B)$ in terms of commutators makes it easy to establish its transformation properties under rotation gauge transformations. We see that

$$[D'_\mu, D'_\nu]\psi' = \tfrac{1}{2}\mathsf{R}'(e_\mu\wedge e_\nu)R\psi = R[D_\mu, D_\nu]\psi = \tfrac{1}{2}R\mathsf{R}(e_\mu\wedge e_\nu)\psi, \tag{13.181}$$

from which we can read off that

$$\mathsf{R}'(B) = R\mathsf{R}(B)\tilde{R}. \tag{13.182}$$

Unlike electromagnetism, the field strength now transforms under gauge transformations, albeit in a straightforward way.

Under displacements, $\Omega(a)$ transforms as defined in equation (13.153). It follows that the field strength transforms to

$$\begin{aligned}\mathsf{R}'(e_\mu\wedge e_\nu) &= \partial_\mu\Omega'(e_\nu) - \partial_\nu\Omega'(e_\mu) + \Omega'(e_\nu)\times\Omega'(e_\mu)\\ &= \mathsf{f}(e_\mu)\cdot\dot{\nabla}_{x'}\dot{\Omega}\big(\mathsf{f}(e_\nu);x'\big) - \mathsf{f}(e_\nu)\cdot\dot{\nabla}_{x'}\dot{\Omega}\big(\mathsf{f}(e_\mu);x'\big) + \Omega'(e_\mu)\times\Omega'(e_\nu)\\ &\quad + \Omega\big(\partial_\mu\mathsf{f}(e_\nu) - \partial_\nu\mathsf{f}(e_\mu);x'\big)\\ &= \mathsf{R}\big(\mathsf{f}(e_\mu\wedge e_\nu);x'\big) + \Omega\big(\partial_\mu\mathsf{f}(e_\nu) - \partial_\nu\mathsf{f}(e_\mu);x'\big).\end{aligned} \tag{13.183}$$

But we know that

$$\partial_\mu\mathsf{f}(e_\nu) - \partial_\nu\mathsf{f}(e_\mu) = \partial_\mu\partial_\nu f(x) - \partial_\nu\partial_\mu f(x) = 0, \tag{13.184}$$

so the field strength has the simple displacement transformation law

$$\mathsf{R}(B) \mapsto \mathsf{R}'(B) = \mathsf{R}\big(\mathsf{f}(B);x'\big). \tag{13.185}$$

We see that $\mathsf{R}'(B)$ picks up a term in $\mathsf{f}(B)$ under displacements, so is not fully covariant. To form a covariant tensor we insert a term in $\mathsf{h}(a)$ into $\mathsf{R}(B)$ and define the covariant field strength

$$\mathcal{R}(B) = \mathsf{R}\big(\mathsf{h}(B)\big). \tag{13.186}$$

The factor of $\mathsf{h}(B)$ in this definition alters the transformation properties under rotations. Since $\bar{\mathsf{h}}$ transforms according to equation (13.155), the adjoint transforms as

$$\mathsf{h}(a) \mapsto \mathsf{h}'(a) = \partial_b\langle aR\bar{\mathsf{h}}(b)\tilde{R}\rangle = \mathsf{h}(\tilde{R}aR). \tag{13.187}$$

The transformation properties of $\mathcal{R}(B)$ are therefore summarised by:

$$\begin{aligned}\text{displacements:}\quad & \mathcal{R}'(B,x) = \mathcal{R}(B,x'),\\ \text{rotations:}\quad & \mathcal{R}'(B) = R\mathcal{R}(\tilde{R}BR)\tilde{R}.\end{aligned} \tag{13.188}$$

These are precisely the properties we require, and they define a *covariant tensor*. The rotation law may look complicated, but it is quite natural. For example, suppose that $\mathcal{R}(B)$ simply amounts to the instruction 'dilate all fields by the factor α'. This is a physical statement, so ought to be true in all gauges. The original statement corresponds to

$$\mathcal{R}(B) = \alpha B. \tag{13.189}$$

The transformed field is then

$$\mathcal{R}'(B) = R\mathcal{R}(\tilde{R}BR)\tilde{R} = R(\alpha\tilde{R}BR)\tilde{R} = \alpha B, \tag{13.190}$$

so does contain to the same physical information. The function $\mathcal{R}(B)$ plays the same role in the gauge theory approach as the curvature tensor in general relativity, so we refer to $\mathcal{R}(B)$ as the *Riemann tensor*. We continue to employ the notational device of writing covariant tensors in calligraphic symbols to help keep track of which objects are gauge-invariant.

13.5.2 The displacement-gauge field strength

The displacement gauge field couples to the vector derivative to form the object $\bar{\mathsf{h}}(\nabla)$. This coupling is different to that of the connection for the rotation gauge field, and we cannot use the commutator of covariant derivatives to obtain the field strength. Indeed, the precise definition and meaning of the field strength for the displacement gauge are unclear. Here we motivate a definition that has the desired properties and is physically plausible.

The main property we require of a field strength is that it should vanish if the field is obtained by a pure gauge transformation. If we start with the identity and apply a displacement, the induced h field is given by

$$\bar{\mathsf{h}}(a) = \bar{\mathsf{f}}^{-1}(a). \tag{13.191}$$

One of the properties satisfied by a pure displacement is that

$$\nabla\wedge\bar{\mathsf{f}}(a) = 0. \tag{13.192}$$

So h will define a pure gauge transformation if it satisfies

$$\nabla\wedge\bar{\mathsf{h}}^{-1}(a) = 0, \tag{13.193}$$

where we temporarily ignore the rotation gauge. The left-hand side is our candidate object for the field strength. The task now is to make it covariant.

We know that the vector derivative ∇ picks up a factor of $\bar{\mathsf{h}}$ to convert it to covariant form. Since $\bar{\mathsf{h}}^{-1}$ transforms in the same way as ∇, we can define a displacement-gauge covariant object $H(a)$ as

$$H(a) = -\bar{\mathsf{h}}\big(\nabla\wedge\bar{\mathsf{h}}^{-1}(a)\big) = \bar{\mathsf{h}}(\dot{\nabla})\wedge\dot{\bar{\mathsf{h}}}\bar{\mathsf{h}}^{-1}(a). \tag{13.194}$$

This is a bivector-valued function of its vector argument. The final step is to convert the derivative to one that is covariant under rotations. This is straightforward since $\bar{\mathsf{h}}$ transforms as a vector under rotations. We therefore define

$$\mathcal{H}(a) = \bar{\mathsf{h}}(\partial_b)\wedge\Big(b\cdot\dot{\nabla}\dot{\bar{\mathsf{h}}}\bar{\mathsf{h}}^{-1}(a) + \Omega(b)\cdot a\Big), \tag{13.195}$$

or, in terms of a coordinate frame,

$$\mathcal{H}(g^\mu) = g^\alpha\wedge(\mathcal{D}_\alpha g_\mu) = \mathcal{D}\wedge g^\mu, \tag{13.196}$$

where we have applied that result that $\nabla\wedge e^\mu = 0$.

The tensor $\mathcal{H}(a)$ is covariant under displacements and rotations, so transforms covariantly as

$$\begin{aligned} \text{displacements:} \quad & \mathcal{H}'(a,x) = \mathcal{H}(a,x'), \\ \text{rotations:} \quad & \mathcal{H}'(a) = R\mathcal{H}(\tilde{R}aR)\tilde{R}. \end{aligned} \tag{13.197}$$

As we will soon see, the object we have defined is in fact the *torsion* tensor, a bivector-valued function of a vector with $6\times 4 = 24$ degrees of freedom. This is the appropriate number for the field strength of the displacement gauge, as a displacement is specified by four degrees of freedom. In the simplest formulation of the field equations, the torsion is equated with the spin of the matter. It is therefore a pure contact term, and usually extremely small. One can justify this on dimensional grounds. The two field strengths we have defined, $\mathcal{H}(a)$ and $\mathcal{R}(B)$, differ in dimensions by a factor of length. This is because $\Omega(a)$ has dimensions of $(\text{length})^{-1}$, whereas $\bar{\mathsf{h}}(a)$ is dimensionless. The only fundamental length scale that could relate these is the Planck length, l_P, which is tiny. The natural scale for $\mathsf{S}(a)$ is therefore l_P times $\mathcal{R}(B)$, making it negligible compared to the Riemann tensor.

13.5.3 The gravitational action

We have now defined two covariant tensors from the gravitational gauge fields — the Riemann and torsion tensors. We next require a scalar term to act as the Lagrangian density for gravitation. There are a number of quadratic scalars we can derive from the gauge fields, but only one scalar is linear in the field strength. This is important, as one can again argue on dimensional grounds that higher order terms should be reduced by factors of the Planck length.

We first define the contractions of the Riemann tensor. The first is the *Ricci tensor*:

$$\mathcal{R}(b) = \partial_a\cdot\mathcal{R}(a\wedge b). \tag{13.198}$$

By construction, this is a tensor. The Ricci tensor can be contracted further to defined the *Ricci scalar*

$$\mathcal{R} = \partial_a\cdot\mathcal{R}(a). \tag{13.199}$$

We use the same symbol to denote the Riemann tensor, Ricci tensor and Ricci scalar, and distinguish between these by their argument. The Ricci scalar is a covariant scalar field, so is invariant under rotations and transforms covariantly under displacements. The Ricci scalar is the first scalar observable we have constructed from the gravitational fields, and is the simplest candidate for the Lagrangian density. We therefore suppose that the overall action integral is of the form

$$S = \int |d^4x| \det(\mathsf{h})^{-1}(\tfrac{1}{2}\mathcal{R} + \Lambda - \kappa\mathcal{L}_m), \tag{13.200}$$

where $\mathcal{L}_m$ describes the matter content and $\kappa = 8\pi G$. We have also included the cosmological constant Λ, though for most applications we set this to zero. The independent dynamical variables are $\bar{\mathsf{h}}(a)$ and $\Omega(a)$, and we assume that $\mathcal{L}_m$ contains no second-order derivatives, so that $\bar{\mathsf{h}}(a)$ and $\Omega(a)$ appear undifferentiated in the matter Lagrangian.

The $\bar{\mathsf{h}}$ field is undifferentiated in the entire action, as we have not included any terms in $\mathcal{H}(a)$. The Euler–Lagrange equation for $\bar{\mathsf{h}}$ is simply

$$\partial_{\bar{\mathsf{h}}(a)}\big(\det(\mathsf{h})^{-1}(\mathcal{R}/2 + \Lambda - \kappa\mathcal{L}_m)\big) = 0. \tag{13.201}$$

Employing the results of section 11.1.2 we find that

$$\partial_{\bar{\mathsf{h}}(a)}\det(\mathsf{h})^{-1} = -\det(\mathsf{h})^{-1}\mathsf{h}^{-1}(a) \tag{13.202}$$

and

$$\begin{aligned}\partial_{\bar{\mathsf{h}}(a)}\mathcal{R} &= \partial_{\bar{\mathsf{h}}(a)}\left\langle \bar{\mathsf{h}}(\partial_c\wedge\partial_b)\mathsf{R}(b\wedge c)\right\rangle \\ &= 2\bar{\mathsf{h}}(\partial_b)\cdot\mathsf{R}(b\wedge a).\end{aligned} \tag{13.203}$$

It follows that

$$\partial_{\bar{\mathsf{h}}(a)}\Big(\mathcal{R}\det(\mathsf{h})^{-1}\Big) = 2\mathcal{G}\Big(\mathsf{h}^{-1}(a)\Big)\det(\mathsf{h})^{-1}, \tag{13.204}$$

where $\mathcal{G}$ is the Einstein tensor,

$$\mathcal{G}(a) = \mathcal{R}(a) - \tfrac{1}{2}a\mathcal{R}. \tag{13.205}$$

We now define the *functional* matter energy-momentum tensor $\mathcal{T}(a)$ by

$$\det(\mathsf{h})\partial_{\bar{\mathsf{h}}(a)}(\mathcal{L}_m\det(\mathsf{h})^{-1}) = \mathcal{T}\Big(\mathsf{h}^{-1}(a)\Big). \tag{13.206}$$

We therefore arrive at the first of our field equations,

$$\mathcal{G}(a) - \Lambda a = \kappa\mathcal{T}(a). \tag{13.207}$$

This is the gauge theory statement of Einstein's equation. The source term in the Einstein equations in the functional energy-momentum tensor, not the canonical one. The form of this is discussed once we have found the remaining field equations for the rotation gauge field.

The Euler–Lagrange field equation from $\Omega(a)$ is, after multiplying through by $\det(\mathsf{h})$,

$$\frac{\partial \mathcal{R}}{\partial \Omega(a)} - \det(\mathsf{h})\frac{\partial}{\partial x^\mu}\left(\frac{\partial \mathcal{R}}{\partial(\partial_\mu \Omega(a))}\det(\mathsf{h})^{-1}\right) = 2\kappa\frac{\partial \mathcal{L}_m}{\partial \Omega(a)}, \tag{13.208}$$

where we have employed the assumption that $\Omega(a)$ does not contain any coupling to matter through its derivatives, and have temporarily reverted to an orthonormal coordinate system. The right-hand side defines the matter *spin tensor*

$$\mathsf{S}(a) = \frac{\partial \mathcal{L}_m}{\partial \Omega(a)}. \tag{13.209}$$

This has the covariant form

$$\mathcal{S}(a) = \mathsf{S}\big(\bar{\mathsf{h}}^{-1}(a)\big), \tag{13.210}$$

which is a covariant tensor. For the left-hand side we use the results

$$\partial_{\Omega(a)}\big\langle \bar{\mathsf{h}}(\partial_d\wedge\partial_c)\Omega(c)\times\Omega(d)\big\rangle = 2\Omega(b)\times\bar{\mathsf{h}}(\partial_b\wedge a)$$

and

$$\frac{\partial}{\partial(\partial_\mu\Omega(a))}\big\langle \bar{\mathsf{h}}(\partial_d\wedge\partial_c)\big(c\cdot\nabla\Omega(d) - d\cdot\nabla\Omega(c)\big)\big\rangle = 2\bar{\mathsf{h}}(a\wedge\gamma^\mu). \tag{13.211}$$

Combining these results, equation (13.208) becomes

$$\begin{aligned}\bar{\mathsf{h}}(\dot{\nabla})\wedge\dot{\bar{\mathsf{h}}}(a) + \det(\mathsf{h})\partial_\mu\Big(\bar{\mathsf{h}}(\gamma^\mu)\det(\mathsf{h})^{-1}\Big)\wedge\bar{\mathsf{h}}(a)&\\ + \Omega(b)\times\bar{\mathsf{h}}(\partial_b\wedge a) &= \kappa\mathsf{S}(a).\end{aligned} \tag{13.212}$$

Recalling the definitions of $\mathcal{H}(a)$ and t, from equations (13.195) and (13.158) respectively, the second field equation has the covariant form

$$\mathcal{H}(a) + t\wedge a = \kappa\mathcal{S}(a). \tag{13.213}$$

So, as stated, $\mathcal{H}$ is governed by the matter spin density.

The second field equation (13.213) simplifies further once we form the contraction of the torsion tensor $\mathcal{H}(a)$. This is

$$\partial_a\cdot\mathcal{H}(a) = \mathcal{D}_\mu\bar{\mathsf{h}}(\gamma^\mu) - \bar{\mathsf{h}}(\dot{\nabla})\,\mathsf{h}^{-1}(\gamma_\mu)\cdot\dot{\bar{\mathsf{h}}}(\gamma^\mu). \tag{13.214}$$

But we can now use

$$\begin{aligned}\mathsf{h}^{-1}(\gamma_\mu)\cdot\big(\partial_\nu\bar{\mathsf{h}}(\gamma^\mu)\big) &= \big\langle\big(\partial_\nu\bar{\mathsf{h}}(\gamma_0)\big)\wedge\bar{\mathsf{h}}(\gamma_1\wedge\gamma_2\wedge\gamma_3)I^{-1}\det(\mathsf{h})^{-1}\big\rangle + \cdots\\ &= \det(\mathsf{h})^{-1}\partial_\nu\det(\mathsf{h}),\end{aligned} \tag{13.215}$$

to write

$$\partial_a\cdot\mathcal{H}(a) = \det(\mathsf{h})\mathcal{D}_\mu\big(\det(\mathsf{h})^{-1}\bar{\mathsf{h}}(\gamma^\mu)\big) = t. \tag{13.216}$$

So the vector t which appeared in the Dirac equation is the contraction of the torsion tensor. On contracting equation (13.213) we find that

$$-2t = \kappa\partial_a\cdot\mathcal{S}(a), \tag{13.217}$$

which directly relates t to the matter spin density. The second field equation can now be written as

$$\mathcal{H}(a) = \kappa\mathcal{S}(a) + \tfrac{1}{2}\kappa\big(\partial_b\cdot\mathcal{S}(b)\big)\wedge a. \tag{13.218}$$

This equation directly relates the torsion to the matter spin density.

13.5.4 The matter content

To illustrate the structure of the source terms we return to the covariant Maxwell and Dirac Lagrangian densities. First consider free-field electromagnetism. Under displacements, the vector potential A transforms as a cotangent vector (1-form):

$$A(x) \mapsto A'(x) = \bar{\mathsf{f}}\big(A(x')\big), \tag{13.219}$$

and the field strength F transforms as a 2-form:

$$F \mapsto F'(x) = \nabla\wedge A'(x) = \bar{\mathsf{f}}\big(F(x')\big). \tag{13.220}$$

The covariant field strength is therefore defined by

$$\mathcal{F} = \bar{\mathsf{h}}(F) = \bar{\mathsf{h}}(\nabla\wedge A), \tag{13.221}$$

and the covariant Lagrangian density for the electromagnetic field is

$$\mathcal{L}_{em} = \tfrac{1}{2}\mathcal{F}\cdot\mathcal{F}. \tag{13.222}$$

The functional energy-momentum tensor is defined by

$$\begin{aligned}\mathcal{T}_{em}\big(\mathsf{h}^{-1}(a)\big) &= \det\,(\mathsf{h})\partial_{\bar{\mathsf{h}}(a)}\Big(\tfrac{1}{2}\mathcal{F}\cdot\mathcal{F}\det\,(\mathsf{h})^{-1}\Big) \\ &= \bar{\mathsf{h}}(a\cdot F)\cdot\mathcal{F} - \mathsf{h}^{-1}(a).\end{aligned} \tag{13.223}$$

So we obtain

$$\mathcal{T}_{em}(a) = (a\cdot\mathcal{F})\cdot\mathcal{F} - a = -\tfrac{1}{2}\mathcal{F}a\mathcal{F}. \tag{13.224}$$

This is precisely the form we would expect for the covariant generalisation of the electromagnetic field strength. Unlike the canonical definition, there is no issue about the tensor being electromagnetic gauge-invariant, and the tensor is automatically symmetric. Furthermore, there is no coupling to $\Omega(a)$, so the electromagnetic spin density is zero. We will discover in section 13.6 that, if the spin tensor is zero, the functional energy-momentum tensor must also be symmetric.

As an example of a field with non-vanishing spin density we next consider the

Dirac theory. With the electromagnetic coupling included, the covariant action is defined by equation (13.150). The functional energy-momentum tensor is simply

$$\mathcal{T}_D(a) = \left\langle a\cdot g^\mu D_\mu \psi I\gamma_3\tilde{\psi}\right\rangle_1 - ea\cdot\mathcal{A}\,\psi\gamma_0\tilde{\psi}. \tag{13.225}$$

This is manifestly a covariant tensor, though it is not necessarily symmetric. The spin density is

$$\mathsf{S}_D(a) = \tfrac{1}{2}\bar{\mathsf{h}}(a)\cdot\left(\psi I\gamma_3\tilde{\psi}\right) \tag{13.226}$$

or, covariantly,

$$\mathcal{S}_D(a) = \tfrac{1}{2}a\cdot\left(\psi I\gamma_3\tilde{\psi}\right) = \tfrac{1}{2}a\cdot S, \tag{13.227}$$

where S is the spin trivector. In the limit where gravitational interactions are turned off, the functional definitions agree with the canonical energy-momentum and angular momentum tensors.

The form of the Dirac spin has an important consequence. If we form the contraction we find that

$$2\partial_a\cdot\mathcal{S}(a) = \partial_a\cdot(a\cdot S) = 0, \tag{13.228}$$

so the torsion vector t vanishes. This is reassuring, as it implies that the minimally-coupled Dirac action produces the minimally-coupled Dirac equation on variation. Equation (13.228) is satisfied by scalar, Dirac and Yang–Mills fields. An exception is provided by a vector field that is often introduced to ensure local dilation invariance. There are good reasons for introducing such a field, though any interactions it might generate are likely to be on the scale of quantum gravity and are not discussed here.

As a further example of a source field for gravitation, we consider the case of an ideal fluid. This is the simplest form of matter energy-momentum tensor one can consider, and generates an important class of models. The action for an ideal fluid was introduced in section 12.4.2, and the only modification required to convert to a covariant action is multiplication of the energy density by $\det(\mathsf{h})^{-1}$:

$$S = \int d^4x\left(-\det(\mathsf{h})^{-1}\varepsilon + J\cdot(\nabla\lambda) - \mu J\cdot\nabla\eta\right). \tag{13.229}$$

The Lagrange multiplier terms are both unaffected by the presence of a gravitational field. The covariant current density is

$$\mathcal{J} = \det(\mathsf{h})\mathsf{h}^{-1}(J) = \rho v, \tag{13.230}$$

where $v^2 = 1$ (see section 13.5.6). The energy density ε therefore depends on the h field through its dependence on ρ. We find that

$$\partial_{\bar{\mathsf{h}}(a)}\rho^2 = 2\rho^2\left(\mathsf{h}^{-1}(a) - \mathsf{h}^{-1}(a)\cdot v\, v\right), \tag{13.231}$$

so the functional stress-energy tensor is

$$\mathcal{T}(a) = -\rho(a - a\cdot v\, v)\frac{\partial \varepsilon}{\partial \rho} + a\varepsilon. \tag{13.232}$$

Recalling the definition of the pressure from equation (12.158), we are left with

$$\begin{aligned}\mathcal{T}(a) &= -(a - a\cdot v\, v)(\varepsilon + P) + a\varepsilon \\ &= (\varepsilon + P)a\cdot v\, v - Pa.\end{aligned} \tag{13.233}$$

This is precisely the form we expect, with v now a covariant vector satisfying the constraint $v^2 = 1$. The actual form of v is gauge-dependent, a fact we can exploit to our advantage in applications by choosing a gauge where v has a simple form.

13.5.5 The torsion-free equations and general relativity

For many applications the matter spin density is negligible. It is a quantum effect, and the macroscopic spin of an object is usually extremely small as all of the individual constituents cancel out. In the case where the spin can be ignored the second field equation becomes

$$\mathcal{H}(a) = 0. \tag{13.234}$$

If we replace a by a general cotangent vector A, this equation can be written

$$\mathcal{D}\wedge\bar{\mathsf{h}}(A) = \bar{\mathsf{h}}(\nabla\wedge A), \tag{13.235}$$

which is extremely useful in practice. This equation says that antisymmetrised partial and covariant derivatives produce the same result. We will now establish that the spinless gauge field equations are (locally) equivalent to those of general relativity. Many of the relevant equations for Riemannian geometry were derived in section 6.5.5.

To begin, we define the connection by

$$\mathcal{D}_\mu g_\nu = \Gamma^\alpha_{\mu\nu} g_\alpha, \tag{13.236}$$

so that

$$\Gamma^\lambda_{\mu\nu} = g^\lambda\cdot(\mathcal{D}_\mu g_\nu). \tag{13.237}$$

It follows that the directional covariant derivative of a vector $\mathcal{A} = A^\mu g_\mu$ has components

$$\begin{aligned}\mathcal{D}_\mu\mathcal{A} &= \mathcal{D}_\mu(A^\alpha g_\alpha) \\ &= (\partial_\mu A^\alpha)g_\alpha + A^\alpha\Gamma^\beta_{\mu\alpha}g_\beta \\ &= (\partial_\mu A^\alpha + \Gamma^\alpha_{\mu\beta}A^\beta)g_\alpha,\end{aligned} \tag{13.238}$$

which recovers the general relativistic expression.

If we recall from equation (13.143) that the metric is given by $g_{\mu\nu} = g_\mu \cdot g_\nu$, we can now write

$$\partial_\mu g_{\nu\lambda} = (\mathcal{D}_\mu g_\nu)\cdot g_\lambda + g_\nu \cdot (\mathcal{D}_\mu g_\lambda), \tag{13.239}$$

so that

$$\partial_\mu g_{\nu\lambda} = \Gamma^\alpha_{\mu\nu} g_{\alpha\lambda} + \Gamma^\alpha_{\mu\lambda} g_{\alpha\nu}. \tag{13.240}$$

This is the metric compatibility condition for the connection. The second important condition on the connection, for pure general relativity, is antisymmetry. This follows from the torsion-free condition, since

$$\begin{aligned} 0 = (g_\mu \wedge g_\nu)\cdot(\mathcal{D}\wedge g^\alpha) &= g_\mu(\mathcal{D}_\nu g^\alpha) - g_\nu(\mathcal{D}_\mu g^\alpha) \\ &= g^\alpha \cdot (\mathcal{D}_\mu g_\nu - \mathcal{D}_\nu g_\mu). \end{aligned} \tag{13.241}$$

We can therefore read off that

$$\mathcal{D}_\mu g_\nu - \mathcal{D}_\nu g_\mu = 0. \tag{13.242}$$

It follows that, in the absence of torsion,

$$\Gamma^\alpha_{\mu\nu} - \Gamma^\alpha_{\nu\mu} = 0. \tag{13.243}$$

This equation and equation (13.240) together define the Christoffel connection. The equations can be inverted to recover the connection in terms of derivatives of the metric. Rather than reproduce the standard derivation at this point, we will instead demonstrate how to invert equation (13.234) to find $\Omega(a)$ in terms of the h field.

Returning to the definition of the $H(a)$ and $\mathcal{H}(a)$ tensors of equations (13.194) and (13.195), the absence of torsion tells us that

$$-H(a) = \bar{\mathsf{h}}(\partial_b)\wedge\big(\Omega(b)\cdot a\big). \tag{13.244}$$

At this point it is useful to introduce the displacement-gauge-covariant connection

$$\omega(a) = \Omega\big(\mathsf{h}(a)\big). \tag{13.245}$$

Under displacements this transforms covariantly,

$$\omega'(a;x) = \omega(a;x'). \tag{13.246}$$

Under rotations the transformation law for $\omega(a)$ is somewhat more complicated than that for $\Omega(a)$, so it is usually preferable to deal with the latter when discussing rotation-gauge transformations. Equation (13.244) now becomes

$$\partial_b \wedge \big(\omega(b)\cdot a\big) = -H(a), \tag{13.247}$$

which gives $\omega(a)$ in terms of h and its derivatives. To solve this we first compute

$$\partial_a \wedge \partial_b \wedge \big(\omega(b)\cdot a\big) = 2\partial_b \wedge \omega(b) = -\partial_b \wedge H(b). \tag{13.248}$$

Now, taking the inner product with a again, we obtain

$$\omega(a) - \partial_b \wedge (a \cdot \omega(b)) = -\tfrac{1}{2} a \cdot (\partial_b \wedge H(b)). \tag{13.249}$$

We can therefore write

$$\omega(a) = H(a) - \tfrac{1}{2} a \cdot (\partial_b \wedge H(b)), \tag{13.250}$$

which enables us to compute $\omega(a)$ directly. In the presence of spin an additional term built from the spin tensor is added to the right-hand side. One can now convert the solution for $\omega(a)$ into a set of Christoffel coefficients, if desired. One disadvantage of the latter is that they mix up gauge terms with terms induced by a choice of curvilinear coordinates. From the manifold viewpoint this is sensible, but it is less natural in the gauge theory context.

Next we turn to the form of the Riemann tensor in general relativity. In terms of the connection, this is

$$\begin{aligned} R_{\mu\nu\rho}{}^{\sigma} &= \partial_\mu \Gamma^{\sigma}_{\nu\rho} - \partial_\nu \Gamma^{\sigma}_{\mu\rho} + \Gamma^{\sigma}_{\mu\alpha}\Gamma^{\alpha}_{\nu\rho} - \Gamma^{\sigma}_{\nu\alpha}\Gamma^{\alpha}_{\mu\rho} \\ &= \partial_\mu \big(g^\sigma \cdot (\mathcal{D}_\nu g_\rho)\big) - \partial_\nu \big(g^\sigma \cdot (\mathcal{D}_\mu g_\rho)\big) - (\mathcal{D}_\mu g^\sigma) \cdot (\mathcal{D}_\nu g_\rho) + (\mathcal{D}_\nu g^\sigma) \cdot (\mathcal{D}_\mu g_\rho) \\ &= g^\sigma \cdot (\mathcal{D}_\mu \mathcal{D}_\nu g_\rho - \mathcal{D}_\nu \mathcal{D}_\mu g_\rho), \end{aligned} \tag{13.251}$$

from which we can read off that

$$R_{\mu\nu\rho}{}^{\sigma} = \mathcal{R}(g_\mu \wedge g_\nu) \cdot (g_\rho \wedge g^\sigma). \tag{13.252}$$

This converts directly between the gauge theory and tensor formulations of gravity. One can also check that the contractions defined earlier are all equivalent to their general relativistic counterparts, so the gauge theory equation (13.207), in the torsion-free case, has the same content as the Einstein equations. The main differences between the two theories are topological in nature, and one can argue that such considerations are beyond the scope of the (local) theory of general relativity anyway.

13.5.6 Currents and Killing vectors

The gauge theory we have constructed is founded on an action principle in a flat spacetime. It follows that Noether's theorem still holds, and that symmetries of the action result in a conserved vector current J. Every such vector has a corresponding covariant equivalent. To find this we first write

$$\nabla \cdot J = I \nabla \wedge (IJ) = 0, \tag{13.253}$$

so, assuming no torsion is present, we have

$$\bar{\mathsf{h}}\big((\nabla \wedge (IJ)\big) = \mathcal{D} \wedge \bar{\mathsf{h}}(IJ) = 0. \tag{13.254}$$

We can therefore write

$$I\mathcal{J} = \bar{\mathsf{h}}(IJ) = I\mathsf{h}^{-1}(J)\det(\mathsf{h}), \tag{13.255}$$

which defines the covariant current $\mathcal{J}$ in terms of J. The covariant vector $\mathcal{J}$ then satisfies

$$\mathcal{D}\cdot\mathcal{J} = 0. \tag{13.256}$$

There is a vector $\mathcal{J}$ conjugate to each continuous symmetry of the action. If we attempt to find conserved vectors conjugate to translations and rotations, however, we do not discover any new information. In both cases the conjugate tensor turns out to be zero once the field equations are employed. This is due to the manner of the coupling of the h field. Variation with respect to h can be viewed as defining the total energy-momentum tensor, and this is zero because there is no derivative term for the h field in the action. It is traditional, of course, to single out (minus) the gravitational contribution to the total energy-momentum tensor (the Einstein tensor), and then equate this to the matter energy-momentum tensor.

A covariantly-conserved vector $\mathcal{J}$ gives rise to a conserved scalar because it can always be converted back to a non-covariant vector J satisfying $\nabla\cdot J = 0$. The same is not true of covariant conservation of a tensor, such as $\mathcal{G}(a)$. Tensors only give rise to useful conserved quantities in the presence of additional symmetries of the Lagrangian. This is the case when the h field is independent of the derivative along a global vector field. In this case one can construct a coordinate system such that the metric $g_{\mu\nu}$ is independent of one of the coordinates. If we call this x^0, we have

$$\frac{\partial}{\partial x^0} g_{\mu\nu} = g_\mu\cdot(g_0\cdot\mathcal{D}g_\nu) + g_\nu\cdot(g_0\cdot\mathcal{D}g_\mu) = 0. \tag{13.257}$$

But, for a coordinate frame in the absence of torsion, equation (13.242) holds and we have

$$g_\mu\cdot(g_\nu\cdot\mathcal{D}\mathcal{K}) + g_\nu\cdot(g_\mu\cdot\mathcal{D}\mathcal{K}) = 0, \tag{13.258}$$

where $\mathcal{K} = g_0$ is the covariant Killing vector. In coordinate-free form we can write

$$a\cdot(b\cdot\mathcal{D}\mathcal{K}) + b\cdot(a\cdot\mathcal{D}\mathcal{K}) = 0 \tag{13.259}$$

for any two vector fields a and b. This can be used as an alternative definition for a Killing vector. Contracting with $\partial_a\cdot\partial_b$ immediately tells us that $\mathcal{K}$ is divergenceless.

13.5.7 Point particle motion

General relativity typically models observers as point particles following geodesic paths, as defined by the geodesic equation. But the gauge approach has dealt solely with the properties of classical and quantum fields. To complete the proof of the equivalence of the gauge approach and general relativity, we must recover the geodesic equation from the minimally-coupled Dirac equation. In coordinate form, the geodesic equation is

$$\dot{v}^\mu + v^\alpha v^\beta \Gamma^\mu_{\alpha\beta} = 0, \tag{13.260}$$

where $v^\mu = \dot{x}^\mu$ and the overdots denote the derivative with respect to proper time. This is defined such that

$$g_{\mu\nu} v^\mu v^\nu = 1. \tag{13.261}$$

To convert to covariant form we introduce the vector

$$v = v_\mu g^\mu = \mathsf{h}^{-1}(\dot{x}), \qquad v^2 = 1. \tag{13.262}$$

This is a covariant vector, though for aesthetic reasons we do not write this in a calligraphic font. The derivative with respect to proper time is

$$\partial_\tau = \dot{x}^\mu \partial_\mu = v \cdot \bar{\mathsf{h}}(\nabla). \tag{13.263}$$

The geodesic equation (13.260) can be now be written

$$\partial_\tau v - v^\mu \partial_\tau g_\mu + v^\alpha v^\beta (\mathcal{D}_\alpha g_\beta) = \dot{v} + \omega(v) \cdot v = 0. \tag{13.264}$$

The gauge theory form of the the geodesic equation is therefore

$$v \cdot \mathcal{D} v = \dot{v} + \omega(v) \cdot v = 0. \tag{13.265}$$

This equation is also recovered by finding the paths that minimise the proper time interval

$$S = \int d\lambda \, |\mathsf{h}^{-1}(x') \cdot \mathsf{h}^{-1}(x')|^{1/2}. \tag{13.266}$$

Geodesics are classified into timelike, lightlike or spacelike according to the value of v^2, which can be $+1$, 0 or -1 respectively. Point particles with mass follow timelike geodesics.

The process by which classical paths are recovered from Dirac theory is discussed in section 12.2.1. The essential term in the action is the kinetic one, which we manipulate in the same way to write

$$\det(\mathsf{h})^{-1} \langle D\psi I \gamma_3 \tilde{\psi} \rangle = \det(\mathsf{h})^{-1} \langle \mathcal{J} D\psi I \boldsymbol{\sigma}_3 \psi^{-1} \rangle, \tag{13.267}$$

where $\mathcal{J} = \psi \gamma_0 \tilde{\psi}$. Equation (13.255) relates the covariant current $\mathcal{J}$ to the divergenceless current J. The classical limit is formed by concentrating the density onto a single streamline of J and ignoring terms in the action perpendicular to the flow. The action therefore contains the term

$$\det(\mathsf{h})^{-1} \left\langle \mathcal{J} \cdot g^\mu \, D_\mu \psi I \boldsymbol{\sigma}_3 \psi^{-1} \right\rangle = \left\langle (\psi' + \tfrac{1}{2} \Omega(x') \psi) I \boldsymbol{\sigma}_3 \psi^{-1} \right\rangle. \tag{13.268}$$

Separating out the rotor dependence, as before, and converting to proper time derivatives, the equations of motion are

$$v\cdot\mathcal{D}S + 2p\wedge v = 0 \tag{13.269}$$

and

$$v\cdot\mathcal{D}p = 0. \tag{13.270}$$

Here $v = \mathsf{h}^{-1}(x') = R\gamma_0\tilde{R}$ and $S = RI\boldsymbol{\sigma}_3\tilde{R}$. Classical point-particle motion is recovered by setting the spin to zero, so that p and v are aligned, and fixing $p\cdot v = m$. In this case we recover precisely the geodesic equation.

This derivation is unusual, but it is important for two reasons. The geodesic equation tells us that point particles follow the same paths regardless of their mass and so implies the equivalence of gravitational and inertial mass. This is the weak equivalence principle, a fundamental ingredient in general relativity. From the gauge theory perspective, the weak equivalence principle is derived from the classical limit of the Dirac equation. The only principle invoked in constructing the covariant Dirac equation was minimal coupling, so at one level this has the consequence of enforcing the weak equivalence principle. One can also argue that minimal coupling is the essence of the full equivalence principle, which tells us how physics should appear locally to a freely-falling observer. The second important feature of this derivation is that it points out the limitations of the weak equivalence principle. Both the wave nature of matter and the existence of quantum spin ensure that the geodesic equation is an approximation, and there are many quantum effects in gravitational backgrounds (such as black hole absorption) where the particle mass is important.

If a Killing vector is present, equation (13.259) tells us that

$$v\cdot(v\cdot\mathcal{D}\mathcal{K}) = 0. \tag{13.271}$$

So, for a particle satisfying the geodesic equation, we find that

$$\partial_\tau(v\cdot\mathcal{K}) = v\cdot\mathcal{D}(v\cdot\mathcal{K}) = \mathcal{K}\cdot(v\cdot\mathcal{D}v) + v\cdot(v\cdot\mathcal{D}\mathcal{K}) = 0. \tag{13.272}$$

It follows that the quantity $v\cdot\mathcal{K}$ is conserved along the worldline of a freely-falling particle. For stationary matter configurations, this can be used to define the conserved energy of the particle.

13.5.8 Electromagnetism in a gravitational background

The electromagnetic vector potential A ensures that the Dirac equation is covariant under local phase transformations. In equation (13.222) we found that the covariant action integral for the electromagnetic field in a gravitational background is given by

$$S = \int |d^4x| (\det \mathsf{h})^{-1}\, \tfrac{1}{2}\mathcal{F}\cdot\mathcal{F}, \tag{13.273}$$

where

$$\mathcal{F} = \bar{\mathsf{h}}(F). \tag{13.274}$$

The field strength $\mathcal{F}$ is covariant under local translations and rotations, as well as being phase-invariant.

We can include a source term by adding an $\mathcal{A}\cdot\mathcal{J}$ term, where $\mathcal{J}$ is a covariant vector. For example, when coupling to a fermion $\mathcal{J}$ is given by the Dirac current $\psi\gamma_0\tilde{\psi}$. The full action integral is therefore

$$S = \int |d^4x| (\det \mathsf{h})^{-1} \left(\tfrac{1}{2}\mathcal{F}\cdot\mathcal{F} + \mathcal{A}\cdot\mathcal{J}\right). \tag{13.275}$$

To find the field equations for electromagnetism we vary this integral with respect to the underlying dynamical variable A, with $\bar{\mathsf{h}}$ and $\mathcal{J}$ treated as external fields. The result is the equation

$$\nabla\cdot\left(\mathsf{h}\bar{\mathsf{h}}(\nabla\wedge A)\det(\mathsf{h})^{-1}\right) = J, \tag{13.276}$$

where

$$J = \det(\mathsf{h})^{-1}\mathsf{h}(\mathcal{J}). \tag{13.277}$$

Equation (13.276) combines with the identity $\nabla\wedge F = 0$ to form the full set of Maxwell equations in a gravitational background. Some insight into these equations is provided by performing a spacetime split and writing

$$\begin{aligned} \boldsymbol{E} + cI\boldsymbol{B} &= F, \\ \boldsymbol{D} + I\boldsymbol{H}/c &= \epsilon_0 \mathsf{h}\bar{\mathsf{h}}(F)\det(\mathsf{h})^{-1}, \end{aligned} \tag{13.278}$$

where we have temporarily included the factors of c and ϵ_0. In terms of these variables Maxwell's equations can be written in the familiar forms

$$\begin{aligned} \boldsymbol{\nabla}\cdot\boldsymbol{B} &= 0, & \boldsymbol{\nabla}\cdot\boldsymbol{D} &= \rho, \\ \boldsymbol{\nabla}\wedge\boldsymbol{E} + I\frac{\partial\boldsymbol{B}}{\partial t} &= 0, & \frac{\partial\boldsymbol{D}}{\partial t} + \boldsymbol{\nabla}\cdot(I\boldsymbol{H}) &= -\boldsymbol{J}, \end{aligned} \tag{13.279}$$

where $J\gamma_0 = \rho + \boldsymbol{J}$. These forms of the equations illustrate how the $\det(\mathsf{h})^{-1}\mathsf{h}\bar{\mathsf{h}}$ is a generalized permittivity/permeability tensor, defining the properties of the space through which the electromagnetic field propagates. For example, the bending of light by the sun can be easily understood in terms of the properties of the dielectric defined by the $\bar{\mathsf{h}}$ field exterior to it.

So far, however, we have failed to achieve a covariant form of the Maxwell equations. We have, furthermore, failed to unite the separate equations into a single equation. To find a covariant equation, we simplify matters by ignoring torsion effects, so that we can write

$$\mathcal{D}\wedge\mathcal{F} = \bar{\mathsf{h}}(\nabla\wedge F) = 0. \tag{13.280}$$

Next, we use a double-duality transformation to write the left-hand side of equation (13.276) as

$$\begin{aligned}\nabla\cdot\big(\mathsf{h}(\mathcal{F})\det(\mathsf{h})^{-1}\big) &= I\nabla\wedge(I\mathsf{h}(\mathcal{F})\det(\mathsf{h})^{-1}) \\ &= I\nabla\wedge(\bar{\mathsf{h}}^{-1}(I\mathcal{F})) \\ &= I\bar{\mathsf{h}}^{-1}\big(\mathcal{D}\wedge(I\mathcal{F})\big).\end{aligned} \tag{13.281}$$

Equation (13.276) now becomes

$$\mathcal{D}\cdot\mathcal{F} = \mathcal{J}, \tag{13.282}$$

and equations (13.280) and (13.282) combine into the single covariant equation

$$\mathcal{D}\mathcal{F} = \mathcal{J}. \tag{13.283}$$

This achieves our objective. Equation (13.283) is manifestly covariant and generalises the free-field Maxwell equations to a gravitational background in an obvious and natural manner. In the presence of torsion an additional term appears in the covariant expression of the Maxwell equations. But in such circumstances the spin fields generating the torsion are likely to interact strongly with the electromagnetic field and swamp most interesting gravitational effects.

13.6 The structure of the Riemann tensor

The Riemann tensor $\mathcal{R}(B)$ contains a remarkable amount of algebraic structure, much of which is hidden in the tensor calculus approach. Again, we assume that there is no torsion present, so that the second field equation reduces to (13.234). Writing $\mathcal{A} = \bar{\mathsf{h}}(A)$ we have

$$\mathcal{D}\wedge\mathcal{A} = \bar{\mathsf{h}}(\nabla\wedge A), \tag{13.284}$$

so

$$\mathcal{D}\wedge(\mathcal{D}\wedge\mathcal{A}) = \bar{\mathsf{h}}(\nabla\wedge\nabla\wedge A) = 0. \tag{13.285}$$

It follows that

$$\begin{aligned}g^\mu\wedge\big(\mathcal{D}_\mu(g^\nu\wedge(\mathcal{D}_\nu\mathcal{A}))\big) &= g^\mu\wedge g^\nu\wedge\big(\mathcal{D}_\mu\mathcal{D}_\nu\mathcal{A}\big) \\ &= \tfrac{1}{2}g^\mu\wedge g^\nu\wedge\big([\mathcal{D}_\mu,\mathcal{D}_\nu]\mathcal{A}\big) \\ &= g^\mu\wedge g^\nu\wedge(\mathsf{R}_{\mu\nu}\times\mathcal{A}).\end{aligned} \tag{13.286}$$

So, for any multivector $\mathcal{M}$,

$$\partial_a\wedge\partial_b\wedge(\mathcal{R}(a\wedge b)\times\mathcal{M}) = 0, \tag{13.287}$$

which is a covariant equation.

To analyse equation (13.287) further we set $\mathcal{M}$ equal to the vector c, and protract with ∂_c to form

$$\partial_c\wedge\partial_a\wedge\partial_b\wedge\big(\mathcal{R}(a\wedge b)\times c\big) = -2\partial_a\wedge\partial_b\wedge\mathcal{R}(a\wedge b) = 0. \tag{13.288}$$

Now forming the inner product with c we obtain

$$2\partial_a\wedge\mathcal{R}(a\wedge c) + \partial_a\wedge\partial_b\wedge\big(\mathcal{R}(a\wedge b)\times c\big) = 0, \tag{13.289}$$

so that we are are left with the compact identity

$$\partial_a\wedge\mathcal{R}(a\wedge b) = 0. \tag{13.290}$$

This summarises *all* of the symmetries of $\mathcal{R}(B)$ in the case of zero torsion. Equation (13.290) says that the trivector $\partial_a\wedge\mathcal{R}(a\wedge b)$ vanishes for all values of the vector b, so gives a set of $4\times 4 = 16$ equations. These reduce the number of independent degrees of freedom in $\mathcal{R}(B)$ from 36 to 20, the expected number for general relativity. Contracting equation (13.290) we obtain

$$\partial_b\cdot\big(\partial_a\wedge\mathcal{R}(a\wedge b)\big) = \partial_a\wedge\mathcal{R}(a) = 0, \tag{13.291}$$

which shows that the Ricci tensor $\mathcal{R}(a)$ is symmetric. The same is therefore true of the Einstein tensor. In the absence of any spin-torsion interactions, the matter energy-momentum tensor must also be symmetric, as is the case for electromagnetism and the relativistic fluid. The covariant Riemann tensor satisfies the further useful identities,

$$\begin{aligned}\partial_c\wedge\big(a\cdot\mathcal{R}(c\wedge b)\big) &= \mathcal{R}(a\wedge b),\\ (B\cdot\partial_a)\cdot\mathcal{R}(a\wedge b) &= -\partial_a\, B\cdot\mathcal{R}(a\wedge b).\end{aligned} \tag{13.292}$$

It follows that

$$\partial_b\wedge\big((B\cdot\partial_a)\cdot\mathcal{R}(a\wedge b)\big) = -2\mathcal{R}(B) = -\partial_b\wedge\partial_a\langle B\mathcal{R}(a\wedge b)\rangle. \tag{13.293}$$

The Riemann tensor is therefore also symmetric,

$$B_1\cdot\mathcal{R}(B_2) = B_2\cdot\mathcal{R}(B_1). \tag{13.294}$$

That is, $\mathcal{R}(B) = \bar{\mathcal{R}}(B)$.

13.6.1 The Weyl tensor

The structure of the Riemann tensor is more clearly seen by separating out the matter content, as contained in the Ricci tensor. Since the contraction of $\mathcal{R}(a\wedge b)$ results in the Ricci tensor $\mathcal{R}(a)$, we expect that $\mathcal{R}(a\wedge b)$ will contain a term in $\mathcal{R}(a)\wedge b$. This must be matched with a term in $a\wedge\mathcal{R}(b)$, since it is only the sum of these that is a function of $a\wedge b$. Contracting this sum we obtain

$$\begin{aligned}\partial_a\cdot\big(\mathcal{R}(a)\wedge b + a\wedge\mathcal{R}(b)\big) &= b\mathcal{R} - \mathcal{R}(b) + 4\mathcal{R}(b) - \mathcal{R}(b)\\ &= 2\mathcal{R}(b) + b\mathcal{R},\end{aligned} \tag{13.295}$$

and it follows that

$$\partial_a \cdot \left(\tfrac{1}{2}\big(\mathcal{R}(a)\wedge b + a\wedge\mathcal{R}(b)\big) - \tfrac{1}{6}a\wedge b\mathcal{R}\right) = \mathcal{R}(b). \tag{13.296}$$

We can therefore write

$$\mathcal{R}(a\wedge b) = \mathcal{W}(a\wedge b) + \tfrac{1}{2}\big(\mathcal{R}(a)\wedge b + a\wedge\mathcal{R}(b)\big) - \tfrac{1}{6}a\wedge b\mathcal{R}, \tag{13.297}$$

where $\mathcal{W}(B)$ is the *Weyl tensor.*

From its definition the Weyl tensor must satisfy

$$\partial_a \cdot \mathcal{W}(a\wedge b) = 0. \tag{13.298}$$

As the Ricci tensor is symmetric, we also have

$$\partial_a \wedge \big(\tfrac{1}{2}\left(\mathcal{R}(a)\wedge b + a\wedge\mathcal{R}(b)\right) - \tfrac{1}{6}a\wedge b\mathcal{R}\big) = 0, \tag{13.299}$$

so the Weyl tensor also satisfies

$$\partial_a \wedge \mathcal{W}(a) = 0. \tag{13.300}$$

Equations (13.298) and (13.300) combine into the single equation

$$\partial_a \mathcal{W}(a\wedge b) = 0. \tag{13.301}$$

This compact equation is unique to the geometric algebra formulation, as it involves the geometric product. To study the consequences of equation (13.301) it is useful to introduce the $\{\gamma_\mu\}$ frame and write the four equations for b equalling each of the γ_μ vectors as

$$\begin{aligned}\boldsymbol{\sigma}_1\mathcal{W}(\boldsymbol{\sigma}_1) + \boldsymbol{\sigma}_2\mathcal{W}(\boldsymbol{\sigma}_2) + \boldsymbol{\sigma}_3\mathcal{W}(\boldsymbol{\sigma}_3) &= 0,\\ \boldsymbol{\sigma}_1\mathcal{W}(\boldsymbol{\sigma}_1) - I\boldsymbol{\sigma}_2\mathcal{W}(I\boldsymbol{\sigma}_2) - I\boldsymbol{\sigma}_3\mathcal{W}(I\boldsymbol{\sigma}_3) &= 0,\\ -I\boldsymbol{\sigma}_1\mathcal{W}(I\boldsymbol{\sigma}_1) + \boldsymbol{\sigma}_2\mathcal{W}(\boldsymbol{\sigma}_2) - I\boldsymbol{\sigma}_3\mathcal{W}(I\boldsymbol{\sigma}_3) &= 0,\\ -I\boldsymbol{\sigma}_1\mathcal{W}(I\boldsymbol{\sigma}_1) - I\boldsymbol{\sigma}_2\mathcal{W}(I\boldsymbol{\sigma}_2) + \boldsymbol{\sigma}_3\mathcal{W}(\boldsymbol{\sigma}_3) &= 0.\end{aligned} \tag{13.302}$$

Summing the final three equations, and employing the first, produces

$$I\boldsymbol{\sigma}_k\mathcal{W}(I\boldsymbol{\sigma}_k) = 0. \tag{13.303}$$

Substituting this into each of the final three equations produces

$$\mathcal{W}(I\boldsymbol{\sigma}_k) = I\mathcal{W}(\boldsymbol{\sigma}_k), \tag{13.304}$$

and it follows that the Weyl tensor satisfies

$$\mathcal{W}(IB) = I\mathcal{W}(B). \tag{13.305}$$

This says that the Weyl tensor is *self-dual.* In the two-spinor formalism of Penrose and Rindler the duality of the Weyl tensor is expressed in terms of a complex formulation. The spacetime algebra shows that this complex structure arises geometrically through the properties of the pseudoscalar.

Given the self-duality of the Weyl tensor, the remaining content of equation (13.302) is summarised by

$$\boldsymbol{\sigma}_k \mathcal{W}(\boldsymbol{\sigma}_k) = 0. \tag{13.306}$$

This equation says that, viewed as a three-dimensional complex linear function, $\mathcal{W}(B)$ is symmetric and traceless. This gives $\mathcal{W}(B)$ five complex, or ten real degrees of freedom. The gauge-invariant information is held in the complex eigenvalues of $\mathcal{W}(B)$, since these are invariant under rotations. As these must sum to zero, only two are independent. This leaves a set of four real intrinsic scalar quantities.

Overall, $\mathcal{R}(B)$ has 20 degrees of freedom, six of which are contained in the freedom to perform arbitrary local rotations. Of the remaining 14 physical degrees of freedom, four are contained in the two complex eigenvalues of $\mathcal{W}(B)$, and a further four in the real eigenvalues of the matter stress-energy tensor. The six remaining physical degrees of freedom determine the rotation between the frame that diagonalises $\mathcal{G}(a)$ and the frame that diagonalises $\mathcal{W}(B)$. This identification of the physical degrees of freedom contained in $\mathcal{R}(B)$ is physically very revealing and extremely useful in guiding solution strategies.

13.6.2 The Bianchi identities

Further information about the Riemann tensor is contained in the Bianchi identities. These follow immediately from the Jacobi identity in the form

$$[\mathcal{D}_\alpha, [\mathcal{D}_\beta, \mathcal{D}_\gamma]]\mathcal{A} + \text{cyclic permutations} = 0. \tag{13.307}$$

It follows that

$$\mathcal{D}_\alpha \mathsf{R}_{\beta\gamma} + \text{cyclic permutations} = 0, \tag{13.308}$$

which we need to express as a fully covariant relation. We start by forming the adjoint relation,

$$\partial_a \wedge \partial_b \wedge \partial_c \langle \big(a \cdot \nabla R(b \wedge c) + \Omega(a) \times R(b \wedge c)\big) B \rangle = 0, \tag{13.309}$$

which simplifies to

$$\nabla \wedge \bar{\mathsf{R}}(B) - \partial_a \wedge \bar{\mathsf{R}}\big(\Omega(a) \times B\big) = 0, \tag{13.310}$$

where B is a constant bivector. To make further progress we again assume that the torsion vanishes. The Riemann tensor is then symmetric, so

$$\bar{\mathsf{R}}(B) = \bar{\mathsf{h}}^{-1}\mathsf{R}\mathsf{h}(B) = \bar{\mathsf{h}}^{-1}\mathcal{R}(B). \tag{13.311}$$

We can therefore write

$$\nabla \wedge \big(\bar{\mathsf{h}}^{-1}\mathcal{R}(B)\big) - \partial_a \wedge \bar{\mathsf{h}}^{-1}\mathcal{R}\big(\Omega(a) \times B\big) = 0. \tag{13.312}$$

Now acting on this equation with $\bar{\mathsf{h}}$ and using equation (13.235), we establish the covariant result

$$\mathcal{D}\wedge\mathcal{R}(B) - \partial_a\wedge\mathcal{R}(\omega(a)\times B) = 0. \tag{13.313}$$

This result takes a more natural form when B becomes an arbitrary function of position, and we write the Bianchi identity as

$$\partial_a\wedge\big(a\cdot\mathcal{D}\mathcal{R}(B) - \mathcal{R}(a\cdot\mathcal{D}B)\big) = 0. \tag{13.314}$$

We can extend the overdot notation of section 11.1 in the natural manner to write equation (13.314) as

$$\dot{\mathcal{D}}\wedge\dot{\mathcal{R}}(B) = 0. \tag{13.315}$$

This is a highly compact, elegant expression of the Bianchi identity, though it is often easier to use the more explicit form of equation (13.314).

The contracted Bianchi identity is obtained from

$$\begin{aligned}(\partial_a\wedge\partial_b)\cdot\big(\dot{\mathcal{D}}\wedge\dot{\mathcal{R}}(a\wedge b)\big) &= \partial_a\cdot\big(\dot{\mathcal{R}}(a\wedge\dot{\mathcal{D}}) + \dot{\mathcal{D}}\dot{\mathcal{R}}(a)\big)\\ &= 2\dot{\mathcal{R}}(\dot{\mathcal{D}}) - \mathcal{D}\mathcal{R},\end{aligned} \tag{13.316}$$

from which we find

$$\dot{\mathcal{G}}(\dot{\mathcal{D}}) = 0. \tag{13.317}$$

The adjoint form of this equation is sometimes more useful:

$$\dot{\mathcal{D}}\cdot\dot{\mathcal{G}}(a) = \mathcal{D}\cdot\mathcal{G}(a) - \partial_b\cdot\mathcal{G}(b\cdot\mathcal{D}\,a) = 0. \tag{13.318}$$

This is the covariant expression of conservation of the Einstein tensor. It follows that the total matter energy-momentum tensor must satisfy the same relation. With the gravitational interaction turned off, the free-field (or flat-space) energy-momentum tensor must be symmetric and divergence-free. This is the case for the functional electromagnetic and fluid energy-momentum tensors. This is not true of the Dirac theory, where the presence of spin alters many of the preceding results and distorts much of the elegant structure of pure general relativity.

The covariant conservation equation (13.318) does not give rise to conserved vector currents, and hence conserved scalars, unless a further symmetry is present in the gravitational fields. In this case one can construct a Killing vector $\mathcal{K}$ satisfying equation (13.259). This is sufficient to prove that

$$\mathcal{G}(\partial_a)\cdot(a\mathcal{D}\mathcal{K}) = 0, \tag{13.319}$$

which holds because $\mathcal{G}(a)$ is a symmetric tensor. It follows that

$$\mathcal{D}\cdot\big(\mathcal{G}(\mathcal{K})\big) = \dot{\mathcal{D}}\cdot\dot{\mathcal{G}}(\mathcal{K}) - \partial_a\cdot\mathcal{G}(a\cdot\mathcal{D}\mathcal{K}) = 0, \tag{13.320}$$

which yields a covariantly conserved vector. This can be converted to a spacetime current and hence to a conserved scalar quantity.

13.7 Notes

Lagrangian field theory is discussed in many textbooks, particularly those that go on to treat quantum field theory. The texts by Itzykson & Zuber (1980) and Bjorken & Drell (1964) are again recommended, as are the book by Cheng & Li (1984) and the set of lecture notes by Coleman (1985). The history of gauge theories in the twentieth century is described in the set of collected papers edited by Taylor (2001). The use of the multivector derivative in analysing field Lagrangians was introduced in the paper by Lasenby, Doran & Gull (1993a), and further refinements are contained in the thesis by Doran (1994).

The discovery that gravity could be treated as a gauge theory was made initially by Utiyama (1956) and Kibble (1961). An attempt at a quantum treatment along the lines suggested by Kibble was made by Feynman and is contained in the *Feynman Lectures on Gravitation* (Feynman, Morningo & Wagner, 1995). The application of spacetime algebra in the context of classical general relativity was promoted by Hestenes in the book *Space-Time Algebra* (1966) and the paper 'Curvature calculations with spacetime algebra' (1986). Many other authors have followed this route and a considerable literature now exists on applications of Clifford algebra in general relativity. Rather than attempt to list all of these, and run the risk of offending anyone we miss out, we recommend searching the main pre-print archives on the keyword 'Clifford'.

The particular combination of the gauge treatment of gravity and the spacetime algebra developed here was first presented in full in the paper 'Gravity, gauge theories and geometric algebra', by Lasenby, Doran & Gull (1998). This contains an extensive list of references and we refer the reader there for further material. The form of the field equations in the presence of torsion is discussed in Doran et al.(1998). Readers of these papers, and the preceding chapter, will notice that the notation and conventions for this subject have not yet settled down. We believe that this chapter represents an advance over previous work, but doubtless there is still room for improvement. While it has not been employed in this chapter, we do recommend the underbar/overbar notation for linear functions in hand-written work. This helps keep track of the form of various objects, and avoids the problem of using different fonts to distinguish objects. Unfortunately, this notation tends to look too cluttered when typeset, which is why underbars are not employed in this book.

13.8 Exercises

13.1 The physical energy-momentum tensor for free-field electromagnetism is defined by

$$\mathsf{T}_{em}(a) = -\tfrac{1}{2}FaF.$$

Prove that each of $\mathsf{T}_{em}(x)$, $\mathsf{T}_{em}(a)$, $\mathsf{T}_{em}(xax)$ and $\mathsf{T}_{em}(B\cdot x)$ is conserved. How many independent conserved constants can one construct from these? How does this relate to the dimension of the spacetime conformal group?

13.2 Prove that, in a space of dimension n,

$$\nabla\left(\frac{1+xa}{(1+2x\cdot a+a^2x^2)^{n/2}}\right)=0,$$

where a is an arbitrary vector.

13.3 The field ψ satisfies the minimally-coupled Dirac equation. Prove that

$$\nabla\cdot(\psi\gamma_1\tilde{\psi}) = 2eA\cdot(\psi\gamma_2\tilde{\psi}),$$
$$\nabla\cdot(\psi\gamma_2\tilde{\psi}) = -2eA\cdot(\psi\gamma_1\tilde{\psi}).$$

Can you derive these relations from a transformation applied to the Dirac Lagrangian?

13.4 The coupled Maxwell–Dirac Lagrangian is defined by

$$\mathcal{L} = \langle\nabla\psi I\gamma_3\tilde{\psi} - eA\psi\gamma_0\tilde{\psi} - m\psi\tilde{\psi}\rangle.$$

Find the canonical energy-momentum tensor. Prove that $\mathcal{L}$ is unchanged in form by the transformations

$$\psi(x)\mapsto R\psi(x'), \qquad A(x)\mapsto RA(x')\tilde{R},$$

where $x'=\tilde{R}xR$ and R is a constant rotor. Find the conserved tensor conjugate to this transformation.

13.5 The gravitational field strength is defined in terms of the bivector connection Ω_μ by

$$\mathsf{R}_{\mu\nu} = \partial_\mu\Omega_\nu - \partial_\nu\Omega_\mu + \Omega_\mu\times\Omega_\nu.$$

Verify that this vanishes if

$$\Omega_\mu = -2\partial_\mu R\tilde{R},$$

where R is a spacetime rotor.

13.6 Prove that, for non-vanishing spin, the $\omega(a)$ field is given by

$$\omega(a) = H(a) - \frac{1}{2}a\cdot(\partial_b\wedge H(b)) + \kappa\mathcal{S}(a) - \frac{3}{2}\kappa a\cdot\big(\partial_b\wedge\mathcal{S}(b)\big).$$

13.7 Prove that, in the case of zero torsion, timelike paths which minimise the proper time

$$S=\int d\lambda\,\big(\mathsf{h}^{-1}(x')\cdot\mathsf{h}^{-1}(x')\big)^{1/2}$$

satisfy the geodesic equation $v\cdot\mathcal{D}v=0$, where $v=\mathsf{h}^{-1}(\dot{x})$ and $v^2=1$.

14

Gravitation

In this chapter we explore the content of the gravitational field equations derived in section 13.5. In covariant notation these equations are

$$\begin{aligned} \mathcal{G}(a) - \Lambda a &= \kappa \mathcal{T}(a), \\ \mathcal{H}(a) &= \kappa \mathcal{S}(a) + \tfrac{1}{2}\kappa\big(\partial_b \cdot \mathcal{S}(b)\big) \wedge a, \end{aligned} \tag{14.1}$$

where $\kappa = 8\pi G$, Λ is the cosmological constant, $\mathcal{G}(a)$ and $\mathcal{H}(a)$ denote the Einstein and torsion tensors, and the matter sources are determined by the total energy-momentum tensor $\mathcal{T}(a)$ and the spin tensor $\mathcal{S}(a)$. Locally, the field equations define an Einstein–Cartan theory of gravitation.

We start this chapter with a discussion of the various strategies we can adopt for solving the field equations. In particular, we focus on a new technique that is unique to the gauge theory approach. Of course, the physical content of the equations does not depend on the method of solution. But the field equations have proved so resistant to analysis that it is important to have a wide range of analytical approaches at our disposal. Most of the applications of interest do not involve macroscopic spin, so the torsion is set to zero. The only exception is when we consider self-consistent cosmological models for a single spinor field in a gravitational background.

As a first application of our solution method we study spherically-symmetric, time-dependent systems. This setup is sufficiently general to use for studying non-rotating stars and black holes, and also cosmology. We study the properties of both classical and quantum matter in these backgrounds, looking in detail at scattering and absorption processes around a black hole. We then turn to static cylindrical systems. These are of limited astrophysical interest, but they do demonstrate some important features of our solution method. In particular we find that, for certain matter distributions, the gravitational fields admit closed timelike curves. These matter distributions can give rise to violations of causality, which are therefore not ruled out by the theory without further assumptions. We end this chapter with a discussion of axially-symmetric fields and the Kerr solution. We give a novel derivation of the Kerr solution, which exposes a remarkable algebraic structure hidden in other approaches. We also describe a version of the Kerr solution that illustrates many of its physical features in a straightforward manner.

14.1 Solving the field equations

The traditional approach to solving the gravitational field equations in general relativity is to start with the metric $g_{\mu\nu}$. In equation (13.143) we showed that the metric is recovered from the $\mathsf{h}(a)$ gauge field by setting

$$g_{\mu\nu} = g_\mu \cdot g_\nu = \mathsf{h}^{-1}(e_\mu) \cdot \mathsf{h}^{-1}(e_\nu), \tag{14.2}$$

where the $\{e_\mu\}$ comprise a coordinate frame. The metric $g_{\mu\nu}$ is invariant under rotation-gauge transformations, so working in terms of the metric removes this gauge freedom from the outset. The result is that the field equations become a set of non-linear, second-order differential equations for the terms in $g_{\mu\nu}$. Any metric is potentially a solution of the field equations — one where the matter energy-momentum tensor is determined by the corresponding Einstein tensor. But this is seldom useful, as what is required is a solution for a given matter distribution. This is an extremely difficult problem.

A related shortcoming of the metric approach is that it is extremely difficult to set up a consistent perturbative scheme. The problem is that the metric is gauge-dependent, so it is not apparent which quantities can be treated as small. This can only be defined consistently in terms of covariant scalars, as these are the only gauge-invariant quantities. Clearly, then, we should aim to solve the equations directly in terms of these quantities. Such a method is described here, and applied to a range of problems in this chapter.

We start by focusing on objects that transform covariantly under displacements. For ease of reference we call these *intrinsic* objects. Unlike the metric formulation, the class of intrinsic objects in the gauge treatment extends beyond scalars to include general multivectors and functions. For example, each of $\bar{\mathsf{h}}(\nabla)$, $\omega(a)$ and $\mathcal{R}(B)$ are intrinsic objects. The task is to formulate the field equations directly in terms of these objects. We assume that the spin is negligible, so that the second field equation in (14.1) states that the torsion is zero. The method we describe here can therefore be directly applied to problems in general relativity.

The torsion equation relates derivatives of $\bar{\mathsf{h}}(a)$ to the $\omega(a)$ field, where $\omega(a)$ is defined in equation (13.245). The torsion equation can be written as

$$\bar{\mathsf{h}}(\dot{\nabla}) \wedge \dot{\bar{\mathsf{h}}}(c) = -\partial_d \wedge \big(\omega(d) \cdot \bar{\mathsf{h}}(c)\big), \tag{14.3}$$

which we contract with $a\wedge b$ to form

$$\begin{aligned}\langle b\wedge a\,\bar{\mathsf{h}}(\dot{\nabla})\wedge\dot{\bar{\mathsf{h}}}(c)\rangle &= -\langle b\wedge a\,\partial_d\wedge\big(\omega(d)\cdot\bar{\mathsf{h}}(c)\big)\rangle \\ &= \big(a\cdot\omega(b)-b\cdot\omega(a)\big)\cdot\bar{\mathsf{h}}(c).\end{aligned} \tag{14.4}$$

The essential operator on the left-hand side is the directional derivative $a\cdot\bar{\mathsf{h}}(\nabla)$. This turns out to be the key operator in our approach, and we write this as

$$L_a = a\cdot\bar{\mathsf{h}}(\nabla). \tag{14.5}$$

The use of L_a for this operator should not be confused with the quantum-mechanical angular momentum operators, though their properties are analysed in a similar way. In terms of L_a the torsion equation becomes

$$\big(\dot{L}_a\dot{\mathsf{h}}(b)-\dot{L}_b\dot{\mathsf{h}}(a)\big)\cdot c = \big(a\cdot\omega(b)-b\cdot\omega(a)\big)\cdot\bar{\mathsf{h}}(c), \tag{14.6}$$

where, as usual, the overdots determine the scope of a differential operator.

The information contained in the torsion equation is summarised neatly in terms of the commutator bracket of the L_a operators. We find that the commutator of L_a and L_b is

$$\begin{aligned}[L_a,L_b] &= \big(L_a\mathsf{h}(b)-L_b\mathsf{h}(a)\big)\cdot\nabla \\ &= \big(\dot{L}_a\dot{\mathsf{h}}(b)-\dot{L}_b\dot{\mathsf{h}}(a)\big)\cdot\nabla + (L_ab-L_ba)\cdot\bar{\mathsf{h}}(\nabla) \\ &= \big(a\cdot\omega(b)-b\cdot\omega(a)+L_ab-L_ba\big)\cdot\bar{\mathsf{h}}(\nabla).\end{aligned} \tag{14.7}$$

We can therefore write

$$[L_a,L_b] = L_c, \tag{14.8}$$

where

$$c = a\cdot\omega(b)-b\cdot\omega(a)+L_ab-L_ba = a\cdot\mathcal{D}b - b\cdot\mathcal{D}a. \tag{14.9}$$

This bracket structure summarises the intrinsic content of the torsion equation in a very convenient manner. If spin is present, the right-hand side of equation (14.9) is modified in a straightforward way to include spin-dependent terms.

The key to our strategy is that we delay any explicit solution for $\omega(a)$ until after further gauge fixing has been performed. Instead, we let $\omega(a)$ take on a suitably general form, consistent with the form of the $\bar{\mathsf{h}}$ function. This is often best achieved with the aid of a symbolic algebra package, though it is possible, if tedious, to perform the calculations by hand. Once a general form for $\omega(a)$ has been found, the relationship between $\bar{\mathsf{h}}(a)$ and $\omega(a)$ is then encoded intrinsically in the commutation relations of the L_a.

The next object to form is the Riemann tensor $\mathcal{R}(B)$. This is constructed in terms of abstract first-order derivatives of the $\omega(a)$ and additional quadratic terms. We see this by writing

$$\begin{aligned}\mathcal{R}(a\wedge b) &= \dot{L}_a\dot{\Omega}\big(\mathsf{h}(a)\big) - \dot{L}_b\dot{\Omega}\big(\mathsf{h}(a)\big) + \omega(a)\times\omega(b) \\ &= L_a\omega(b) - L_b\omega(a) + \omega(a)\times\omega(b) - \Omega\big(L_a\mathsf{h}(b) - L_b\mathsf{h}(a)\big),\end{aligned} \tag{14.10}$$

so that we have

$$\mathcal{R}(a\wedge b) = L_a\omega(b) - L_b\omega(a) + \omega(a)\times\omega(b) - \omega(c), \tag{14.11}$$

where c is given by equation (14.9). Equation (14.11) enables $\mathcal{R}(B)$ to be calculated entirely in terms of intrinsic quantities. Once the general form of the Riemann tensor is found, we can start to employ the rotation-gauge freedom to convert $\mathcal{R}(B)$ to a suitably simple expression. This gauge fixing is crucial in order to arrive at a set of equations that are not underconstrained. The gauge fixing is now performed directly at the level of the covariant variables. This gives the method great power, as one can motivate gauge choices on sensible physical grounds, rather than blind guesswork at the level of the metric.

With $\mathcal{R}(B)$ suitably fixed, we arrive at a set of relations between first-order abstract derivatives of the $\omega(a)$, quadratic terms in $\omega(a)$ and matter terms. The next step is to impose the Bianchi identities, which ensure overall consistency of the equations with the bracket structure. Once all this is achieved, one arrives at a fully intrinsic set of equations. Solving these equations usually involves searching for natural integrating factors. The final step is to make an explicit position gauge choice of the h function. The natural way to do this is often to ensure that the form of $\bar{\mathsf{h}}(a)$ is such that the integrating factors are expressed simply in terms of the chosen coordinates. This description is quite abstract, but in the following sections we apply this scheme to a range of physical problems. These should illustrate how the scheme is applied in practice. We start with the simplest case of spherically-symmetric, torsion-free systems.

14.2 Spherically-symmetric systems

To solve the field equations for spherically-symmetric systems, we first introduce the standard polar coordinates (t, r, θ, ϕ). In terms of the fixed $\{\gamma_\mu\}$ frame we write

$$\begin{aligned} t &= x\cdot\gamma_0, & \cos(\theta) &= \frac{x\cdot\gamma^3}{r}, \\ r &= \sqrt{(x\wedge\gamma_0)^2}, & \tan(\phi) &= \frac{x\cdot\gamma^2}{x\cdot\gamma^1}. \end{aligned} \tag{14.12}$$

The associated coordinate frame is

$$\begin{aligned} e_t &= \gamma_0, \\ e_r &= \sin(\theta)\big(\cos(\phi)\,\gamma_1 + \sin(\phi)\,\gamma_2\big) + \cos(\theta)\,\gamma_3, \\ e_\theta &= r\cos(\theta)\big(\cos(\phi)\,\gamma_1 + \sin(\phi)\,\gamma_2\big) - r\sin(\theta)\,\gamma_3, \\ e_\phi &= r\sin(\theta)\big(-\sin(\phi)\,\gamma_1 + \cos(\phi)\,\gamma_2\big), \end{aligned} \tag{14.13}$$

and we will also make use of the unit vectors $\hat{\theta}$ and $\hat{\phi}$ defined by

$$\hat{\theta} = \frac{1}{r}e_\theta, \qquad \hat{\phi} = \frac{r}{\sin(\theta)}e_\phi. \tag{14.14}$$

From these we define the unit bivectors

$$\sigma_r = e_r e_t, \qquad \sigma_\theta = \hat{\theta} e_t, \qquad \sigma_\phi = \hat{\phi} e_t. \tag{14.15}$$

For applications in gravity there is little reason to write these spatial bivectors in bold face, so we break the convention adopted earlier in this book and leave the unit bivectors in ordinary face. We work throughout in natural units $c = \hbar = G = 1$, so that $\kappa = 8\pi$, and in the first instance we set the cosmological constant to zero.

14.2.1 The spherical equations

Our first step towards a solution is to decide a suitable form for the $\bar{\mathsf{h}}$ function consistent with spherical symmetry. The form we use is

$$\begin{aligned} \bar{\mathsf{h}}(e^t) &= f_1 e^t + f_2 e^r, \\ \bar{\mathsf{h}}(e^r) &= g_1 e^r + g_2 e^t, \\ \bar{\mathsf{h}}(e^\theta) &= \alpha e^\theta, \\ \bar{\mathsf{h}}(e^\phi) &= \alpha e^\phi, \end{aligned} \tag{14.16}$$

where f_1, f_2, g_1, g_2 and α are all functions of t and r only. The only rotation-gauge freedom in this system is the freedom to perform a boost in the σ_r direction. This freedom will be employed later to simplify the equations. Our remaining position-gauge freedom lies in the freedom to reparameterise t and r, which does not affect the general form of $\bar{\mathsf{h}}(a)$. A natural parameterisation will emerge once the physical variables have been identified.

To find a general form $\omega(a)$ consistent with the $\bar{\mathsf{h}}$ function of equation (14.16), we substitute the latter into equation (13.250) and look at the general algebraic form of $\omega(a)$. Where the coefficients in $\omega(a)$ contain derivatives of terms from $\bar{\mathsf{h}}(a)$ new symbols are introduced. Undifferentiated terms from $\bar{\mathsf{h}}(a)$ appearing in $\omega(a)$ arise from frame derivatives and are left in explicitly. The result is that

	σ_r	σ_θ	σ_ϕ
$e_t\cdot\mathcal{D}$	0	$GI\sigma_\phi$	$-GI\sigma_\theta$
$e_r\cdot\mathcal{D}$	0	$FI\sigma_\phi$	$-FI\sigma_\theta$
$\hat{\theta}\cdot\mathcal{D}$	$T\sigma_\theta - SI\sigma_\phi$	$-T\sigma_r$	$SI\sigma_r$
$\hat{\phi}\cdot\mathcal{D}$	$T\sigma_\phi + SI\sigma_\theta$	$-SI\sigma_r$	$-T\sigma_r$

Table 14.1 *Covariant derivatives of the polar-frame unit timelike bivectors.*

we can write

$$\begin{aligned} \omega(e_t) &= Ge_re_t, \\ \omega(e_r) &= Fe_re_t, \\ \omega(\hat{\theta}) &= S\hat{\theta}e_t + (T-\alpha/r)e_r\hat{\theta}, \\ \omega(\hat{\phi}) &= S\hat{\phi}e_t + (T-\alpha/r)e_r\hat{\phi}, \end{aligned} \tag{14.17}$$

where G, F, S and T are functions of t and r only. The important feature of these functions is that they transform covariantly under displacements of r and t.

To define a suitable bracket structure we first introduce the operators

$$\begin{aligned} L_t &= e_t\cdot\bar{\mathsf{h}}(\nabla), \qquad & L_{\hat{\theta}} &= \hat{\theta}\cdot\bar{\mathsf{h}}(\nabla), \\ L_r &= e_r\cdot\bar{\mathsf{h}}(\nabla), \qquad & L_{\hat{\phi}} &= \hat{\phi}\cdot\bar{\mathsf{h}}(\nabla). \end{aligned} \tag{14.18}$$

Equation (14.8), together with our form for $\omega(a)$, yields the relations

$$\begin{aligned} [L_t, L_r] &= GL_t - FL_r, \qquad & [L_r, L_{\hat{\theta}}] &= -TL_{\hat{\theta}}, \\ [L_t, L_{\hat{\theta}}] &= -SL_{\hat{\theta}}, \qquad & [L_r, L_{\hat{\phi}}] &= -TL_{\hat{\phi}}, \\ [L_t, L_{\hat{\phi}}] &= -SL_{\hat{\phi}}, \qquad & [L_{\hat{\theta}}, L_{\hat{\phi}}] &= 0. \end{aligned} \tag{14.19}$$

A set of bracket relations such as these is the first step in writing the field equations in an entirely intrinsic form. The use of orthonormal vectors in expressing these relations brings out the structure most clearly.

Next we seek an intrinsic form of the Riemann tensor. This calculation is simplified by making use of the results in table 14.1. The bracket relations enable us to calculate the derivatives of α/r by writing

$$\begin{aligned} L_t(\alpha/r) &= L_tL_{\hat{\theta}}\theta = [L_t, L_{\hat{\theta}}]\theta = -S\alpha/r, \\ L_r(\alpha/r) &= L_rL_{\hat{\theta}}\theta = [L_r, L_{\hat{\theta}}]\theta = -T\alpha/r. \end{aligned} \tag{14.20}$$

Application of equation (14.11) is now straightforward, and leads to the Riemann tensor

$$
\begin{aligned}
\mathcal{R}(\sigma_r) &= (L_r G - L_t F + G^2 - F^2)\sigma_r, \\
\mathcal{R}(\sigma_\theta) &= (-L_t S + GT - S^2)\sigma_\theta + (L_t T + ST - SG) I\sigma_\phi, \\
\mathcal{R}(\sigma_\phi) &= (-L_t S + GT - S^2)\sigma_\phi - (L_t T + ST - SG) I\sigma_\theta, \\
\mathcal{R}(I\sigma_\phi) &= (L_r T + T^2 - FS) I\sigma_\phi - (L_r S + ST - FT)\sigma_\theta, \\
\mathcal{R}(I\sigma_\theta) &= (L_r T + T^2 - FS) I\sigma_\theta + (L_r S + ST - FT)\sigma_\phi, \\
\mathcal{R}(I\sigma_r) &= (-S^2 + T^2 - (\alpha/r)^2) I\sigma_r.
\end{aligned}
\tag{14.21}
$$

We must next decide on the form of matter energy-momentum tensor that the gravitational fields couple to. We assume that the matter is modelled by an ideal fluid, as discussed in section 13.5.4, so we can write

$$
\mathcal{T}(a) = (\rho + p) a \cdot v\, v - pa, \tag{14.22}
$$

where ρ is the energy density, p is the pressure and v is the covariant fluid velocity ($v^2 = 1$). Radial symmetry means that v can only lie in the e_t and e_r directions, so v must take the form

$$
v = \cosh(\chi)\, e_t + \sinh(\chi)\, e_r. \tag{14.23}
$$

But, in restricting the $\bar{\mathsf{h}}$ function to the form of equation (14.16), we retained the gauge freedom to perform arbitrary radial boosts. This freedom can now be employed to set $v = e_t$, so that the matter energy-momentum tensor becomes

$$
\mathcal{T}(a) = (\rho + p) a \cdot e_t\, e_t - pa. \tag{14.24}
$$

There is no physical content in the choice $v = e_t$ as all physical relations must be independent of gauge choices. The choice simply fixes the rotation gauge in such a way that the energy-momentum tensor takes on the simplest form. This removes all rotation-gauge freedom — an essential step in the solution method, since all non-physical degrees of freedom must be removed before one can achieve a complete set of physical equations.

In section 13.6.1 we saw how to decompose the Riemann tensor into a source term and the Weyl tensor. The source term can be written

$$
\mathcal{R}(a\wedge b) - \mathcal{W}(a\wedge b) = \frac{4\pi}{3}\big(3a\wedge\mathcal{T}(b) + 3\mathcal{T}(a)\wedge b - 2\mathcal{T}\, a\wedge b\big), \tag{14.25}
$$

where $\mathcal{T} = \partial_a \cdot \mathcal{T}(a)$ is the trace of the matter energy-momentum tensor. With $\mathcal{T}(a)$ given by equation (14.24), $\mathcal{R}(B)$ is restricted to the form

$$
\mathcal{R}(B) = \mathcal{W}(B) + \frac{4\pi}{3}\big(3(\rho + p) B \cdot e_t\, e_t - 2\rho B\big). \tag{14.26}
$$

Comparing this with equation (14.21) we see that the Weyl tensor must have the general form

$$\begin{aligned}
\mathcal{W}(\sigma_r) &= \alpha_1\sigma_r, & \mathcal{W}(I\sigma_r) &= \alpha_4 I\sigma_r, \\
\mathcal{W}(\sigma_\theta) &= \alpha_2\sigma_\theta + \beta_1 I\sigma_\phi, & \mathcal{W}(I\sigma_\theta) &= \alpha_3 I\sigma_\theta + \beta_2\sigma_\phi, \\
\mathcal{W}(\sigma_\phi) &= \alpha_2\sigma_\phi - \beta_1 I\sigma_\theta, & \mathcal{W}(I\sigma_\phi) &= \alpha_3 I\sigma_\phi - \beta_2\sigma_\theta.
\end{aligned} \tag{14.27}$$

Here each of the α_i represents a combination of intrinsic objects.

The torsionless gravitational field equations ensure that the Weyl tensor is self-dual and symmetric. The former implies that $\alpha_1 = \alpha_4$, $\alpha_2 = \alpha_3$ and $\beta_1 = -\beta_2$, and the latter implies that $\beta_1 = \beta_2$. It follows that $\beta_1 = \beta_2 = 0$. Finally, $\mathcal{W}(B)$ must be traceless, which requires that $\alpha_1 + 2\alpha_2 = 0$. Taken together, these conditions reduce $\mathcal{W}(B)$ to the form

$$\mathcal{W}(B) = \frac{\alpha_1}{4}(B + 3\sigma_r B\sigma_r). \tag{14.28}$$

This is of Petrov type D. From the form of $\mathcal{R}(I\sigma_r)$ we can see that

$$\alpha_1 = \frac{8\pi\rho}{3} - S^2 + T^2 - \frac{\alpha^2}{r^2}. \tag{14.29}$$

If we now define β by

$$4\beta = -S^2 + T^2 - \frac{\alpha^2}{r^2}, \tag{14.30}$$

then the full Riemann tensor can be written as

$$\mathcal{R}(B) = \left(\beta + \frac{2\pi}{3}\rho\right)(B + 3\sigma_r B\sigma_r) + \frac{4\pi}{3}(3(\rho+p)B\cdot e_t e_t - 2\rho B). \tag{14.31}$$

We compare this with equation (14.21) to obtain the following set of equations:

$$\begin{aligned}
L_t S &= 2\beta + GT - S^2 - 4\pi p, \\
L_t T &= S(G - T), \\
L_r S &= T(F - S), \\
L_r T &= -2\beta + FS - T^2 - 4\pi\rho, \\
L_r G - L_t F &= F^2 - G^2 + 4\beta + 4\pi(\rho + p).
\end{aligned} \tag{14.32}$$

We are now close to our goal of a complete set of intrinsic equations. The remaining step is to enforce the Bianchi identities. The only identity that contains new information in our present setup is the contracted Bianchi identity defined in section 13.6.2, which guarantees covariant conservation of the energy-momentum tensor. For an ideal fluid this results in the pair of equations

$$\begin{aligned}
\mathcal{D}\cdot(\rho v) + p\mathcal{D}\cdot v &= 0, \\
(\rho + p)(v\cdot\mathcal{D}v)\wedge v - (\mathcal{D}p)\wedge v &= 0.
\end{aligned} \tag{14.33}$$

The quantity $(v\cdot\mathcal{D}v)\wedge v$ is the covariant acceleration bivector, so the second equation relates the acceleration to the pressure gradient. For the case of spherically-symmetric fields, these equations reduce to

$$\begin{aligned} L_t\rho &= -(F+2S)(\rho+p), \\ L_r p &= -G(\rho+p). \end{aligned} \tag{14.34}$$

The latter of these identifies G as the radial acceleration. The full Bianchi identities now turn out to be satisfied as a consequence of the contracted identities and the bracket relation

$$[L_t, L_r] = GL_t - FL_r. \tag{14.35}$$

This completes our derivation of the intrinsic equations. The full set is defined by equations (14.20), (14.32), the contracted identities (14.34) and the bracket structure of equation (14.35). The equation structure is closed, as the bracket relation (14.35) is consistent with the known derivatives. The derivation of such a set of equations is the basic aim of our method. The equations deal solely with objects that transform covariantly under displacements, and many of these quantities have direct physical significance.

14.2.2 Solving the spherical equations

To solve the intrinsic equation structure we first form the derivatives of β to obtain

$$\begin{aligned} L_t\beta + 3S\beta &= 2\pi Sp, \\ L_r\beta + 3T\beta &= -2\pi T\rho. \end{aligned} \tag{14.36}$$

These results suggest that we should look for an integrating factor for the L_t+S and L_r+T operators. Such a function, X say, should have the properties that

$$L_tX = SX, \qquad L_rX = TX. \tag{14.37}$$

A function with these properties can exist only if the derivatives are consistent with the bracket relation of equation (14.35). This is checked by forming

$$\begin{aligned} [L_t, L_r]X &= L_t(TX) - L_r(SX) \\ &= X(L_tT - L_rS) \\ &= X(SG - FT) \\ &= GL_tX - FL_rX, \end{aligned} \tag{14.38}$$

which confirms that the properties of X are consistent with the bracket structure. In fact, we can see from equation (14.20) that r/α has the desired properties. Integrating factors of this type often arise as natural, intrinsically-defined coordinates, and the form of the solution is usually simplest when expressed directly in terms of these. Since the position-gauge freedom in the r direction has not yet been fixed, it is natural to set $\alpha = 1$, so that r plays the role of the integrating factor directly. We will confirm shortly that this gauge choice ensures that r is a physically meaningful quantity.

With the radial scale fixed by setting $\alpha = 1$, we can now make some further simplifications. From the form of the $\bar{\mathsf{h}}$ function in equation (14.16), together with equation (14.37), we see that

$$\begin{aligned} g_1 &= L_r r = Tr, \\ g_2 &= L_t r = Sr. \end{aligned} \tag{14.39}$$

This replaces two functions in the bivector connection in favour of terms in $\bar{\mathsf{h}}(a)$. We also define

$$M = -2r^3\beta = \frac{r}{2}(g_2{}^2 - g_1{}^2 + 1), \tag{14.40}$$

which satisfies

$$\begin{aligned} L_t M &= -4\pi r^2 g_2 p, \\ L_r M &= 4\pi r^2 g_1 \rho. \end{aligned} \tag{14.41}$$

The latter suggests that M plays the role of an intrinsic mass.

So far we have defined the natural distance scale, but have not yet found a natural time coordinate. Such a coordinate is required to complete the solution, so we now look for additional criteria to motivate this choice. We are currently free to perform an arbitrary r and t-dependent displacement along the e_t direction. This gives us complete freedom in the choice of f_2 function. If we now invert equation (14.41) to find the coordinate derivatives of M we obtain

$$\begin{aligned} \frac{\partial M}{\partial t} &= \frac{-4\pi g_1 g_2 r^2(\rho + p)}{f_1 g_1 - f_2 g_2}, \\ \frac{\partial M}{\partial r} &= \frac{4\pi r^2(f_1 g_1 \rho + f_2 g_2 p)}{f_1 g_1 - f_2 g_2}. \end{aligned} \tag{14.42}$$

The second equation reduces to a simple classical relation if we choose $f_2 = 0$, as we then obtain

$$\partial_r M = 4\pi r^2 \rho. \tag{14.43}$$

This says that, at constant time t, $M(r,t)$ is determined by the amount of mass-energy in a sphere of radius r.

With f_2 set to zero we can now use the bracket structure to solve for f_1. We have

$$L_t = f_1\partial_t + g_2\partial_r, \qquad L_r = g_1\partial_r, \tag{14.44}$$

so the bracket relation of equation (14.35) implies that

$$L_r f_1 = -G f_1. \tag{14.45}$$

It follows that

$$f_1 = \epsilon(t) \exp\left(-\int^r \frac{G(s)}{g_1(s)}\, ds\right). \tag{14.46}$$

The function $\epsilon(t)$ can be absorbed by a further t-dependent rescaling along e_t, which will not reintroduce a term in f_2. In the $f_2 = 0$ gauge we can therefore reduce to a system in which

$$f_1 = \exp\left(-\int^r \frac{G(s)}{g_1(s)}\, ds\right). \tag{14.47}$$

The physical explanation for why the $f_2 = 0$ gauge is a very natural one to work in emerges when we set the pressure to zero. In this case equation (14.34) forces G to be zero, and equation (14.47) then sets $f_1 = 1$. A (free-falling) particle comoving with the fluid has covariant velocity $v = e_t$, so the trajectory of this particle is defined by

$$\dot{t} e_t + \dot{r} e_r = \mathsf{h}(e_t) = e_t + g_2\, e_r, \tag{14.48}$$

where the dots denote differentiation with respect to the proper time. Since $\dot{t} = 1$ the time coordinate t matches the proper time of all observers comoving with the fluid. In this sense, the time coordinate that has emerged behaves like a global Newtonian time on which all observers can agree (provided all clocks are correlated initially). By employing the various gauge choices outlined above, and casting the dynamics in terms of the t coordinate, we are ensuring that (when $p = 0$) the physics is formulated from the viewpoint of freely-falling observers. We then expect that the gravitational equations should take on a clear, physical form, which is indeed the case.

As a further illustration of this point, it is clear from (14.48) that g_2 represents a radial velocity for the particle. In the absence of pressure, the rate of change of mass is given by

$$\partial_t M = -4\pi r^2 g_2 \rho. \tag{14.49}$$

This equation equates the work with the rate of flow of energy density. Similarly, equation (14.40), written in the form

$$\frac{(g_2)^2}{2} - \frac{M}{r} = \frac{1}{2}\Big((g_1)^2 - 1\Big), \tag{14.50}$$

is also now familiar from Newtonian physics — it is a Bernoulli equation for zero pressure and total (non-relativistic) energy $({g_1}^2 - 1)/2$. When pressure is included, the purely Newtonian interpretation starts to break down, due mainly to the fact that pressure can act as a source of gravitation. But it remains the case that the gauge choices described here pick out what appears to be the most natural set of equations for studying spherically-symmetric systems.

The system of equations we have now derived is summarised in table 14.2. We

The $\bar{\mathsf{h}}$ field	$\bar{\mathsf{h}}(e^t) = f_1 e^t$ $\bar{\mathsf{h}}(e^r) = g_1 e^r + g_2 e^t$ $\bar{\mathsf{h}}(e^\theta) = e^\theta$ $\bar{\mathsf{h}}(e^\phi) = e^\phi$
The ω field	$\omega(e_t) = G e_r e_t$ $\omega(e_r) = F e_r e_t$ $\omega(\hat{\theta}) = g_2/r\, \hat{\theta} e_t + (g_1 - 1)/r\, e_r \hat{\theta}$ $\omega(\hat{\phi}) = g_2/r\, \hat{\phi} e_t + (g_1 - 1)/r\, e_r \hat{\phi}$
Directional derivatives	$L_t = f_1 \partial_t + g_2 \partial_r$ $L_r = g_1 \partial_r$
Equations for G and F	$L_t g_1 = G g_2$ $L_r g_2 = F g_1$ $f_1 = \exp\{\int^r -G/g_1\, ds\}$
Definition of M	$M = \frac{1}{2} r({g_2}^2 - {g_1}^2 + 1)$
Remaining derivatives	$L_t g_2 = G g_1 - M/r^2 - 4\pi r p$ $L_r g_1 = F g_2 + M/r^2 - 4\pi r \rho$
Matter derivatives	$L_t M = -4\pi r^2 g_2 p$ $L_r M = 4\pi r^2 g_1 \rho$ $L_t \rho = -(2g_2/r + F)(\rho + p)$ $L_r p = -G(\rho + p)$
Riemann tensor	$\mathcal{R}(B) = 4\pi\big((\rho + p) B \cdot e_t e_t - 2\rho/3\, B\big)$ $\quad -\frac{1}{2}(M/r^3 - 4\pi\rho/3)(B + 3\sigma_r B \sigma_r)$
Energy-momentum tensor	$\mathcal{T}(a) = (\rho + p) a \cdot e_t e_t - pa$

Table 14.2 *Gravitational equations governing a radially-symmetric perfect fluid.* An equation of state and initial data $\rho(r, t_0)$ and $g_2(r, t_0)$ determine the future evolution of the system.

refer to this system as defining the *Newtonian* gauge, since so many equations take on an almost Newtonian form. Of course, this should not distract from the fact that we have solved the full, relativistic gravitational field equations. The system of equations in table 14.2 underlies a wide range of phenomena in relativistic astrophysics and cosmology. One aspect of these equations is immediately apparent. Given an equation of state $p = p(\rho)$, and initial data in the form of the density $\rho(r, t_0)$ and the velocity $g_2(r, t_0)$, the future evolution of the system is fully determined. This is because ρ determines p and M on a time slice, and the definition of M determines g_1. The equations for $L_r p$, $L_r g_1$

and $L_r g_2$ then determine the remaining information on the time slice. Finally, the $L_t M$ and $L_t g_2$ equations can be used to update the information to the next time slice, and the process can start again. The equations can therefore be implemented numerically as a simple set of first-order update equations. This is important for a wide range of applications.

14.2.3 Static matter distributions

As a first application of the equations governing a spherically-symmetric system, we consider a static matter distribution. This solution is appropriate for a non-rotating spherical source. The density and pressure are now functions of r only. The mass is given by

$$M(r) = \int_0^r 4\pi s^2 \rho(s)\, ds \tag{14.51}$$

and it follows that

$$L_t M = 4\pi r^2 g_2 \rho = -4\pi r^2 g_2 p. \tag{14.52}$$

For any physical matter distribution ρ and p must both be positive, in which case equation (14.52) can only be satisfied if g_2 vanishes. It follows that $F = 0$ as well, so for static, extended objects we have

$$g_2 = F = 0. \tag{14.53}$$

Since g_2 is zero, g_1 is given simply in terms of $M(r)$ by

$${g_1}^2 = 1 - \frac{2M(r)}{r}. \tag{14.54}$$

For this to hold we require that $2M(r) < r$. This condition says that a horizon has not formed anywhere in the object.

The remaining equation of use is that for $L_t g_2$, which now gives

$$G g_1 = \frac{M(r)}{r^2} + 4\pi r p. \tag{14.55}$$

Equations (14.54) and (14.55) combine with that for $L_r p$ to produce the *Oppenheimer–Volkov* equation

$$\frac{\partial p}{\partial r} = -\frac{(\rho + p)(M(r) + 4\pi r^3 p)}{r(r - 2M(r))}. \tag{14.56}$$

This is the force balance equation appropriate for a relativistic matter distribution. The line element generated by our solution is

$$ds^2 = \frac{1}{(f_1)^2} dt^2 - \frac{r}{r - 2M(r)} dr^2 - r^2\, d\theta^2 - r^2 \sin^2(\theta)\, d\phi^2, \tag{14.57}$$

where f_1 is given by equation (14.47). The solution extends straightforwardly to the region outside the star. In this region M is constant, and

$$f_1 = 1/g_1 = (1 - 2M/r)^{-1/2}. \tag{14.58}$$

We therefore recover the *Schwarzschild* line element. This is the solution used for some of the most famous tests of general relativity, including those for the bending of light and the perihelion precession of Mercury. Clearly, the gauge theory framework does not alter any of these results.

14.3 Schwarzschild black holes

Perhaps the most famous solution of the Einstein equations (apart from Lorentzian spacetime) is the Schwarzschild solution for a black hole. This solution describes the gravitational fields surrounding a point source of matter, of total gravitational mass M. One form of this solution is described by the line element of equation (14.57) for the case of constant M. But this is ill defined at $r = 2M$ which, as we shall soon discover, defines an *event horizon*. This tells us that our gauge choice has not yielded a satisfactory global solution, so we must return to the field equations to discover what went wrong.

For a point source located at the origin we have $\rho = p = 0$ everywhere away from the source. The matter equations therefore reduce to

$$L_t M = L_r M = 0, \tag{14.59}$$

which tells us that the mass M is constant. The remaining equations simplify to

$$\begin{aligned} L_t g_1 &= G g_2, \\ L_r g_2 &= F g_1, \\ {g_1}^2 - {g_2}^2 &= 1 - 2M/r. \end{aligned} \tag{14.60}$$

No further equations yield new information, so we have an underdetermined system of equations. Despite all of the gauge-fixing steps taken to arrive at the set of equations summarised in table 14.2, for vacuum fields some additional gauge fixing is still required. The reason for this is that, in the vacuum region, the Riemann tensor reduces to

$$\mathcal{R}(B) = -\frac{M}{2r^3}(B + 3\sigma_r B \sigma_r). \tag{14.61}$$

This tensor is now invariant under boosts in the σ_r plane, whereas previously the presence of the fluid velocity in the Riemann tensor vector broke this symmetry. The appearance of this new symmetry in the matter-free case manifests itself as a new freedom in the choice of the $\bar{\mathsf{h}}$ function.

Given this new freedom, we can look for a choice of g_1 and g_2 which simplifies the equations. If we attempt to reproduce the Schwarzschild line element we have to set $g_2 = 0$, but then we immediately run into difficulties with g_1, which is not defined for $r < 2M$. We must therefore look for an alternative gauge choice. A suitable candidate, motivated by the pressure-free equations, is provided by the simple choice

$$g_1 = 1. \tag{14.62}$$

It follows that

$$f_1 = 1, \qquad g_2 = -\sqrt{2M/r} \tag{14.63}$$

and

$$G = 0, \qquad F = -\frac{M}{g_2 r^2} = \left(\frac{M}{2r^3}\right)^{1/2}. \tag{14.64}$$

In this gauge the $\bar{\mathsf{h}}$ function has the remarkably simple form

$$\bar{\mathsf{h}}(a) = a - \sqrt{2M/r}\, a \cdot e_r \, e_t. \tag{14.65}$$

This only differs from the identity through a single term. The line element obtained from this gauge choice is

$$ds^2 = dt^2 - \left(dr + \left(\frac{2M}{r}\right)^{1/2} dt\right)^2 - r^2(d\theta^2 + \sin^2(\theta)\, d\phi^2), \tag{14.66}$$

which is regular at the horizon ($r = 2M$) and covers all spacetime down to $r = 0$. This form of the line element was first derived by Painlevé and Gullstrand, not long after Schwarzschild's original work was published. Despite the many advantages of this form of the solution, it has not been routinely employed in solving physical problems.

The $\bar{\mathsf{h}}$ field of equation (14.65) is the form of the Schwarzschild solution we will use for studying the properties of spherically-symmetric black holes. Of course, all physical predictions must be independent of gauge, but this only reinforces the point that we should always endeavour to work in a gauge that simplifies the analysis as far as possible. The results for the extension to the action of $\bar{\mathsf{h}}$ to an arbitrary multivector A are useful in what follows. We find that

$$\bar{\mathsf{h}}(A) = A - \sqrt{2M/r}(A \cdot e_r) \wedge e_t. \tag{14.67}$$

It follows that $\det(\mathsf{h}) = 1$ and the inverse of the adjoint function, as defined by equation (4.152), is given by

$$\mathsf{h}^{-1}(A) = A + \sqrt{2M/r}(A \cdot e_t) \wedge e_r. \tag{14.68}$$

It is straightforward to verify that this function recovers the line element of equation (14.66).

14.3.1 Point particle trajectories

The motion of a classical point particle in free fall is governed by the geodesic equation

$$v \cdot \mathcal{D} v = \dot{v} + \omega(v) \cdot v = 0. \tag{14.69}$$

The mass m of the particle is unimportant (provided $m \ll M$), and is set to unity throughout this section. Since $G = 0$ in our chosen gauge, we immediately see that

$$\omega(e_t) = 0. \tag{14.70}$$

It follows that $v = e_t$ is a solution of the geodesic equation. The trajectory this defines has

$$\dot{x} = \mathsf{h}(v) = \mathsf{h}(e_t) = e_t + u e_r, \tag{14.71}$$

where

$$u = \dot{r} = -\sqrt{(2M/r)}. \tag{14.72}$$

Particles, or observers, following the geodesic defined by $v = e_t$ fall in radially with velocity $\dot{r}$ given by the familiar Newtonian formula. Furthermore, we see that $\dot{t} = 1$, so the time coordinate t is precisely the time measured by these infalling observers. This is, in part, why the gauge choice we have adopted turns out to simplify many calculations.

Now consider a more general trajectory, with covariant velocity

$$v = \dot{t}\, e_t + (\dot{t}\sqrt{2M/r} + \dot{r}) e_r + \dot{\theta} e_\theta + \dot{\phi} e_\phi. \tag{14.73}$$

Since the $\bar{\mathsf{h}}$ function is independent of t we have, from equation (13.272),

$$\mathsf{h}^{-1}(e_t) \cdot v = (1 - 2M/r)\dot{t} - \dot{r}\sqrt{2M/r} = \text{constant}. \tag{14.74}$$

So, for particles moving forwards in time ($\dot{t} > 0$ for $r \to \infty$), we can write

$$(1 - 2M/r)\dot{t} = \alpha + \dot{r}\sqrt{2M/r}, \tag{14.75}$$

where the constant α satisfies $\alpha > 0$. The radial equation is found from the constraint that $v^2 = 1$, which gives

$$\dot{r}^2 = \alpha^2 - (1 - 2M/r)\Big(1 + r^2(\dot{\theta}^2 + \sin^2(\theta)\, \dot{\phi}^2)\Big). \tag{14.76}$$

Spherical symmetry implies that the angular velocity J is also conserved, where

$$J^2 = r^4(\dot{\theta}^2 + \sin^2(\theta)\, \dot{\phi}^2). \tag{14.77}$$

The motion of a particle around a black hole is therefore determined by the single radial equation

$$\dot{r}^2 = \alpha^2 - (1 - 2M/r)\left(1 + \frac{J^2}{r^2}\right). \tag{14.78}$$

This equation is gauge-*invariant*, as it relates local quantities. The radial coordinate r is defined locally by the magnitude of the Riemann tensor, and the dots denote the derivative with respect to (local) proper time. This transition from global to local variables is in keeping with the gauging process. The motion of a particle in spacetime is obtained by integrating equations (14.78) and (14.75). At the horizon we have $\dot{r} = -\alpha$, so there is no pole in equation (14.75), and the equations can be integrated down to the singularity.

Differentiating equation (14.78) we obtain

$$\ddot{r} = -\frac{M}{r^2} + \frac{J^2}{r^3} - \frac{3MJ^2}{r^4}. \tag{14.79}$$

The equivalent three-dimensional vector equation is

$$\ddot{\boldsymbol{x}} = -\left(\frac{M}{r^2} + \frac{3MJ^2}{r^4}\right)\hat{\boldsymbol{x}}. \tag{14.80}$$

This equation was analysed perturbatively in section 3.3.1. For stable orbits the main new effect introduced by relativity is a small perturbation of the eccentricity vector. The content of equation (14.78) can similarly be summarised in the radial effective potential (per unit mass)

$$V_{\mathit{eff}} = -\frac{M}{r} + \frac{J^2}{2r^2}\left(1 - \frac{2M}{r}\right). \tag{14.81}$$

We then have

$$\frac{\alpha^2 - 1}{2} = \frac{\dot{r}^2}{2} + V_{\mathit{eff}} \tag{14.82}$$

which identifies $m\alpha$ as the conserved relativistic energy of the particle. Bound states have $\alpha < 1$ and scattering states have $\alpha > 1$.

The effective potential differs from the Newtonian expression in the factor of $(1 - 2M/r)$ multiplying the centrifugal term. This has little effect at large distances, but dramatically alters the small-r behaviour. Inside $r = 2M$ the centrifugal term in the effective potential changes sign and becomes *attractive*. There is no longer any term in the potential applying an effective outward force, and the particle must inexorably move towards the central singularity. One can see this clearly in equation (14.73). Inside the horizon the velocity $\dot{r}$ must be negative in order for $v^2 = 1$ to remain satisfied. Once inside the horizon, no particle can escape the singularity, no matter what force is applied to attempt to counteract the gravitational pull. Eventually, the tidal forces (defined by the Riemann tensor) become so large that all objects are pulled apart into their constituent particles.

14.3.2 Photon trajectories

A full treatment of the properties of electromagnetic waves in a gravitational background involves solving the gravitationally-coupled Maxwell equations of section 13.5.8. For a range of practical problems it is sufficient to ignore the detailed properties of the electromagnetic field, and work in the geometric optics limit. In this approach, photons are treated as massless (scalar) point particles. These particles follow *null* trajectories with

$$k = \mathsf{h}^{-1}(\dot{x}), \qquad k^2 = 0. \tag{14.83}$$

The trajectories are still specified by the equation $k\cdot\mathcal{D}k = 0$. For radial *infall* we must have

$$k = \nu(e_t - e_r), \tag{14.84}$$

where $\nu = k \cdot e_t$ is the frequency measured by radially free-falling observers (at rest at infinity). The photon trajectory is independent of the frequency, as demanded by the equivalence principle. The path defined by k is given by

$$\dot{x} = \mathsf{h}(k) = \nu\big(e_t - (1 + \sqrt{(2M/r)})e_r\big). \tag{14.85}$$

It follows that

$$\frac{dr}{dt} = -(1 + \sqrt{(2M/r)}). \tag{14.86}$$

This integrates straightforwardly to give the photon path. We have therefore found the path without employing the equation of motion. This is possible because we restricted to motion in a single spacetime plane.

The equations of motion tell us how the frequency changes along the path. To find this we need

$$\omega(k) = -\nu\left(\frac{M}{2r^3}\right)^{1/2}\sigma_r, \tag{14.87}$$

from which we see that

$$\dot{\nu} = \nu^2\left(\frac{M}{2r^3}\right)^{1/2}. \tag{14.88}$$

This equation is more usefully expressed in terms of the derivative with respect to r. We use

$$\dot{r} = -\nu\big(1 + \sqrt{(2M/r)}\big) \tag{14.89}$$

to arrive at

$$\frac{1}{\nu}\frac{d\nu}{dr} = \frac{M}{r}\frac{1}{2M + \sqrt{(2Mr)}} = \frac{1}{2r}\frac{1}{\sqrt{r/r_S} + 1}, \tag{14.90}$$

where $r_S = 2M$ is the Schwarzschild radius. This equation can again be integrated straightforwardly to tell us how frequency ν changes with radius. We see that nothing untoward happens until $r = 0$ is reached.

We can repeat the previous analysis for outgoing photons. For this case we have

$$k = \nu(e_t + e_r) \tag{14.91}$$

and the path is

$$\dot{x} = \mathsf{h}(v) = \nu\big(e_t + (1 - \surd(2M/r))e_r\big). \tag{14.92}$$

It follows that

$$\frac{dr}{dt} = 1 - \surd(2M/r). \tag{14.93}$$

But now, when $r < 2M$ the path is still *inwards*. Inside $r = 2M$, not even light can escape. The surface $r = 2M$ is called the *event horizon*. It marks the boundary between two regions, one of which (the interior in this case) cannot signal to the other. We also find that

$$\frac{1}{\nu}\frac{d\nu}{dr} = \frac{M}{r}\frac{1}{2M - \surd(2Mr)} = -\frac{1}{2r}\frac{1}{\sqrt{r/r_S} - 1}, \tag{14.94}$$

which is negative outside the horizon. So, as photons climb out of a gravitational field, they are *redshifted*. This is one of the best-tested predictions of general relativity. The redshift becomes increasingly large as the horizon is approached, so photons emitted from near the horizon are strongly redshifted as they climb out to infinity. The various features of radial motion in a black hole background are shown in figure 14.1. One conclusion from this plot is that, as seen by *external observers*, any object falling through the horizon appears to hover outside the horizon and just fade out of existence as the redshift increases.

If any object collapses to within its event horizon, it must carry on collapsing to form a central *singularity*. There is no possible force capable of preventing the collapse. This is because matter is always constrained to follow timelike paths, and if the entire future light-cone points inwards towards the singularity, no matter can escape. The object remaining at the end of this process is called a *black hole*. All paths for infalling matter terminate on the singularity. There has been much research into the properties of singularities, though their nature remains enigmatic. In one sense, gravitational singularities are no more difficult to deal with than singularities in the electromagnetic field due to point sources. They can also be analysed in much the same way using integral equations. But this (classical) treatment of singularities can only contain part of the story. Quantum mechanically, black holes have an associated entropy, implying the existence of a series of microstates consistent with the macroscopic properties of the hole. It is widely believed that a more complete understanding of quantum gravity should explain this phenomenon through a detailed quantum description of the singularity.

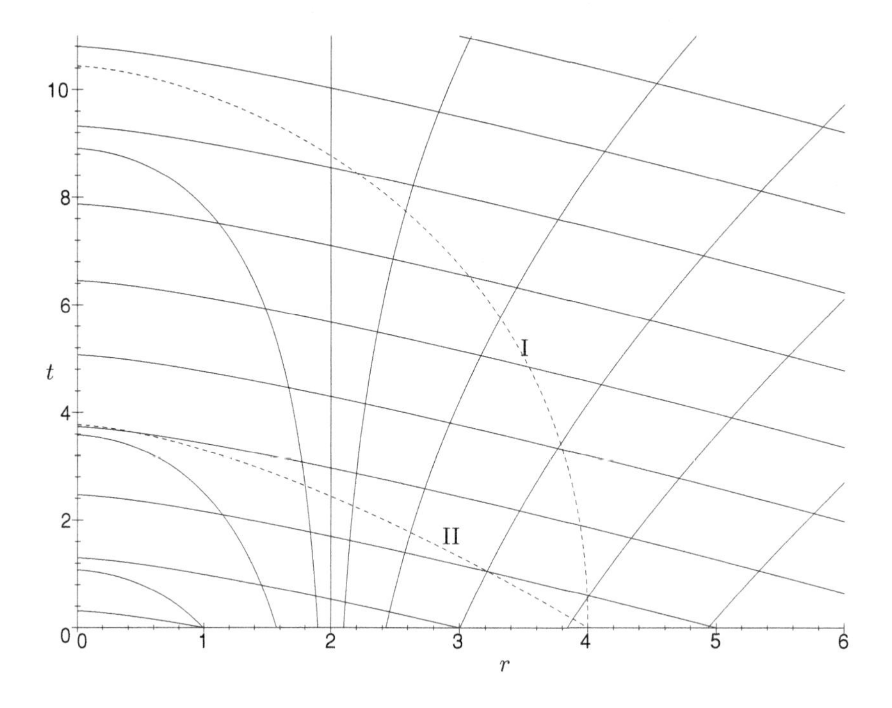

Figure 14.1 *Matter and photon trajectories in a black hole background.* The solid lines are photon trajectories, and the horizon lies at $r = 2$. Outside the horizon it is possible to send photons out to infinity, and hence communicate with the rest of the universe. As the emitter approaches the horizon, these photons are strongly redshifted and take a long time to escape. Once inside the horizon, all photon paths end on the singularity. The broken lines represent two possible trajectories for infalling matter. Trajectory I is for a particle released from rest at $r = 4$. Trajectory II is for a particle released from rest at $r = \infty$.

14.3.3 Stationary observers

It is instructive to see how physics appears from the point of view of stationary observers in a Schwarzschild background. These observers have constant r, θ, ϕ, so

$$\dot{x} = \dot{t} e_t. \tag{14.95}$$

It follows that

$$v = \dot{t}(e_t + \sqrt{(2M/r)} e_r). \tag{14.96}$$

But we require that $v^2 = 1$ for the path to be parameterised by the observer's proper time, so

$$\dot{t}^2(1 - 2M/r) = 1, \qquad \dot{t} = (1 - 2M/r)^{-1/2}. \tag{14.97}$$

This is a constant, since r is fixed for these observers. We can see immediately that it is only possible to remain at rest *outside* the horizon. This is reasonable given the preceding considerations, though the picture is not quite so clear if the black hole is rotating. For this case there is a region outside the horizon within which it is impossible to remain at rest (though it is still possible to escape).

The covariant acceleration bivector for a particle with velocity v is defined by

$$(v\cdot\mathcal{D}v)\wedge v = \dot{v}v + \omega(v)\cdot v\, v. \tag{14.98}$$

This gives the acceleration required to follow a given path. For stationary observers we have

$$(v\cdot\mathcal{D}v)\wedge v = \frac{M}{r^2(1-2M/r)^{1/2}}\sigma_r. \tag{14.99}$$

So an observer with mass m needs to apply force of $Mm/r^2 \times (1-2M/r)^{-1/2}$ to remain at rest. This is the Newtonian value multiplied by a relativistic correction term. This correction becomes increasingly large as the horizon is approached, as one would expect.

We can now look at physics from point of view of these observers, which can be viewed as both being stationary and having constant acceleration. For example, if a second observer has velocity γ_0 (so is in free fall), the relative velocity the two observers measure when their positions coincide is

$$\frac{v\wedge\gamma_0}{v\cdot\gamma_0} = \sqrt{(2M/r)}\sigma_r. \tag{14.100}$$

As we might expect, this is the Newtonian result. The only difference now lies in the interpretation of who is accelerating. The stationary observer is the one applying a force, so we now say that it is this observer that is accelerating. The observer in free fall is applying zero force, so is not accelerating. That is, we no longer view gravity as applying a force, as this would require a concept of what the particle would have done if the gravitational field were not present. Such a concept is not gauge-invariant, so is unphysical.

14.3.4 Absorption and scattering

The presence of the horizon implies that incident particles with total energy $E > mc^2$ can suffer two fates. Either they will be scattered by the gravitational fields, or they will be absorbed onto the central singularity. The crucial quantity that determines the fate of the particle is the angular velocity J. In figure 14.2 we plot the effective potential of equation (14.81) for a range of angular velocities. If J is too small there is nothing to prevent the particle hitting the singularity. As J increases, the effective potential develops a barrier. If this barrier is greater than the total (non-relativistic) energy, the particle is no longer absorbed, and instead is scattered by the black hole.

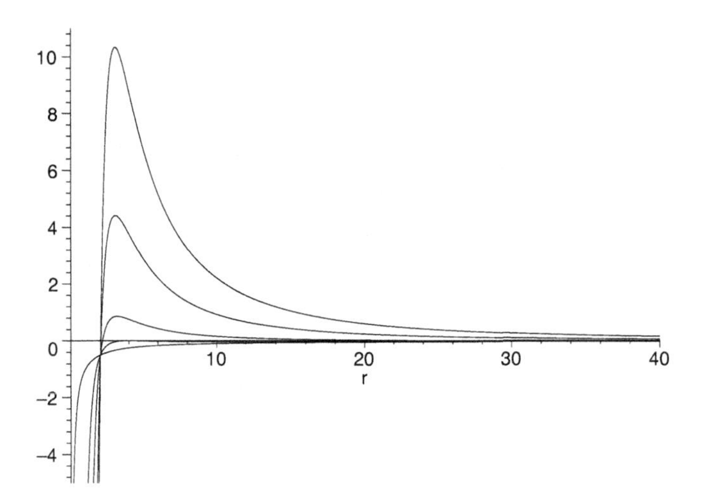

Figure 14.2 *The gravitational effective potential.* The potential for a unit mass particle is defined by equation (14.81), and units are chosen so that the horizon lies at $r = 2$. The plots are for J values of 0, 4, 8, 16 and 24. For small J nothing prevents the particle hitting the singularity. As J increases a barrier of increasing height is formed. If the particle has insufficient energy to surmount this barrier it is scattered.

For a given energy, we can determine the critical value of J that distinguishes between absorption and scattering. This is most usefully encoded in terms of an impact parameter b, as illustrated in figure 14.3. Asymptotically, the incoming particle has angular velocity

$$J = b\dot{r}(\infty). \tag{14.101}$$

But in this region the energy is determined entirely by $\dot{r}$, so the impact parameter is given by

$$b^2 = \frac{J^2}{\alpha^2 - 1}, \tag{14.102}$$

where α is the energy per unit mass of the incident particle, as defined in equation (14.75). For a fixed energy, the critical value of J therefore determines the critical value of the impact parameter. From the point of view of absorption, the black hole then appears as a disc of radius b, and the total absorption cross section is defined by

$$\sigma_{abs} = \pi b^2. \tag{14.103}$$

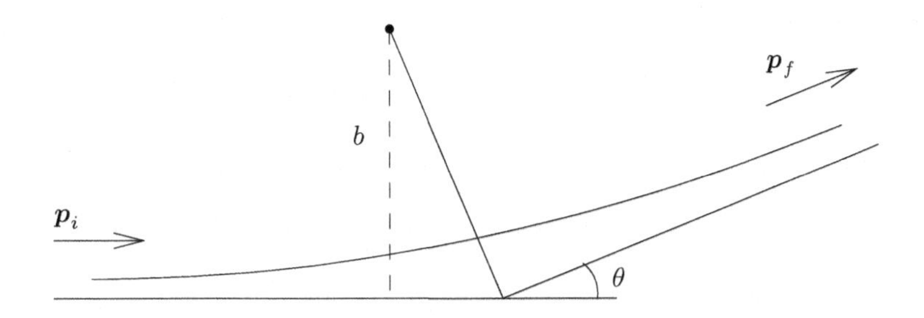

Figure 14.3 *The impact parameter.* In the asymptotic incoming region, the impact parameter b measures the distance between the incoming trajectory and a parallel radial trajectory. For a black hole there is a critical value of b inside which all geodesics terminate on the singularity. The diagram also defines the scattering angle θ.

This will be a decreasing function of energy — the faster the particle is travelling, the less likely it is to be absorbed.

The algebra needed to compute the absorption cross section is straightforward, if a little tedious. First we write $x = 1/r$, so that the effective potential becomes

$$V_{\mathit{eff}} = -Mx + \frac{b^2(\alpha^2-1)}{2}x^2(1-2Mx). \tag{14.104}$$

The turning point is at

$$x_c = \frac{1}{6M}\left(1 + \left(1 - \frac{12M^2}{(\alpha^2-1)b^2}\right)^{1/2}\right). \tag{14.105}$$

To find b the equation we need to solve is therefore

$$2V_{\mathit{eff}}(x_c) = \alpha^2 - 1. \tag{14.106}$$

The solution then returns the absorption cross section

$$\sigma_{abs} = \frac{\pi M^2}{2u^4}\left(8u^4 + 20u^2 - 1 + (1+8u^2)^{3/2}\right), \tag{14.107}$$

where we have expressed the result in terms of the velocity u:

$$u^2 = \frac{p^2}{E^2} = \frac{\alpha^2-1}{\alpha^2}. \tag{14.108}$$

The absorption cross section is plotted in figure 14.4. For small velocities we see that

$$\sigma_{abs} \mapsto \frac{16\pi M^2}{u^2}. \tag{14.109}$$

As the incident velocity decreases, the absorption cross section increases, as is to be expected. As the velocity increases the absorption cross section tends towards

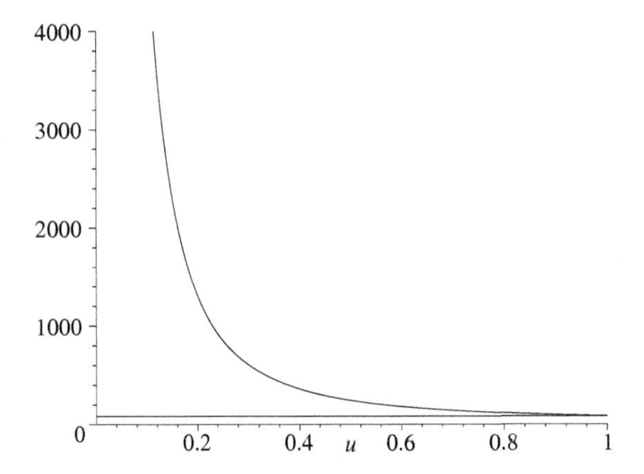

Figure 14.4 *The classical absorption cross section.* The cross section is a function of the incident velocity u (in units of c). As the velocity approaches the speed of light the cross section approaches the photon limit, as shown by the straight line. The vertical axis is in units of $(GM/c^2)^2$.

the limiting result for a massless particle. For these the effective potential is simply

$$V_{\textit{eff}} = \frac{J^2}{2r^2}\left(1 - \frac{2M}{r}\right). \tag{14.110}$$

The turning point occurs at $r = 3M$, at which the effective potential has the value $J^2/54M^2$. Equating this with asymptotic energy $J^2/2b^2$ we see that for photons $b^2 = 27M^2$, and the photon absorption cross section is

$$\sigma_{abs} = \pi b^2 = 27\pi M^2. \tag{14.111}$$

This is the limiting value of equation (14.107) as $u \mapsto 1$. In section 14.4.3 we study how these features are modified by a more complete, quantum treatment of the absorption process.

Scattering presents a more difficult problem. The differential scattering cross section for a Newtonian $1/r$ potential is determined by the Rutherford formula

$$\frac{d\sigma}{d\Omega} = \frac{M^2}{4u^2\sin^4(\theta/2)}, \tag{14.112}$$

where θ is the scattering angle and u is the velocity of the incident particle. This formula relates the incident cross sectional area σ to the solid angle $d\Omega$, where

$$d\sigma = 2\pi b\, db, \qquad d\Omega = 2\pi\sin(\theta)\, d\theta. \tag{14.113}$$

The Rutherford cross section formula is easily computed from the properties of hyperbolic trajectories. The relativistic corrections to the Rutherford formula are generated by the additional r^{-3} term in the potential. This term makes the problem considerably more difficult to solve, and no simple analytic formula exists for the classical scattering cross section. One problem is that it is now possible for particles to spiral around the centre before escaping. We could build up a perturbative picture of the scattering problem using the techniques described in section 3.3.1, though the resulting expressions are usually extremely complicated. A better approach to this problem is described in section 14.4.1, where the cross section is calculated using perturbative quantum theory.

14.3.5 Electromagnetism in a black hole background

Further insight into the nature and effects of a black hole is obtained by considering the electromagnetic fields surrounding charges held at rest outside the horizon. The relevant equations were obtained in section 13.5.8. We assume that the charge is placed at a distance $a > 2M$ along the z axis. The vector potential can be written in terms of a single scalar potential $V(r, \theta)$ as

$$A = V(r, \theta) \left(e_t + \frac{\sqrt{2Mr}}{r - 2M} e_r \right), \tag{14.114}$$

so that

$$\mathcal{F} = -\frac{\partial V}{\partial r} e_r e_t - \frac{1}{r - 2M} \frac{\partial V}{\partial \theta} \hat{\theta}(e_t + \sqrt{2M/r} e_r) \tag{14.115}$$

and

$$\boldsymbol{D} = -\frac{\partial V}{\partial r} e_r e_t - \frac{1}{r - 2M} \frac{\partial V}{\partial \theta} \hat{\theta} e_t. \tag{14.116}$$

The Maxwell equations now reduce to the single partial differential equation

$$\frac{1}{r^2} \frac{\partial}{\partial r} \left(r^2 \frac{\partial V}{\partial r} \right) + \frac{1}{r(r - 2M)} \frac{1}{\sin(\theta)} \frac{\partial}{\partial \theta} \left(\sin(\theta) \frac{\partial V}{\partial \theta} \right) = -\rho, \tag{14.117}$$

where $\rho = q\delta(\boldsymbol{x} - \boldsymbol{a})$ is a δ-function at $z = a$. The solution (originally found by Linet) is

$$V(r, \theta) = \frac{q}{ar} \frac{(r - M)(a - M) - M^2 \cos^2(\theta)}{d} + \frac{qM}{ar}, \tag{14.118}$$

where

$$d = \left(r(r - 2M) + (a - M)^2 - 2(r - M)(a - M) \cos(\theta) + M^2 \cos^2(\theta) \right)^{1/2}. \tag{14.119}$$

When this result is substituted back into equation (14.115) we see that the covariant field $\mathcal{F}$ is both finite and continuous at the horizon. It follows that we

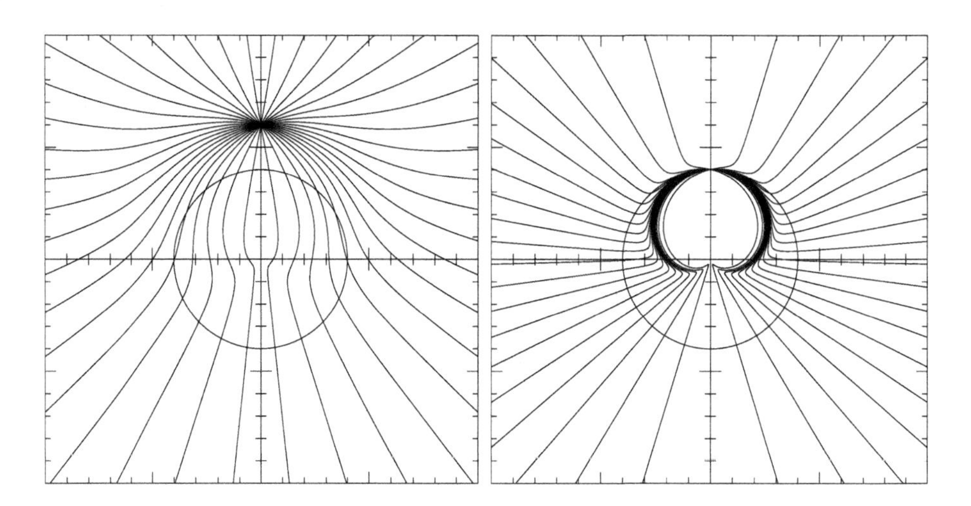

Figure 14.5 *Streamlines of the electric field in a black hole background.* The horizon lies at $r = 2$ and the charge is placed on the z axis. In the left-hand diagram the charge is held at $z = 3$, and in right-hand diagram it is at $z = 2.1$.

have found a *global* solution to the electromagnetic field equations, appropriate both inside and outside the horizon.

One way to illustrate the global properties of $\mathcal{F}$ is to plot the streamlines of $\boldsymbol{D}$. Equation (13.279) ensures that these streamlines begin and end on charges, so for our case of a single isolated charge they should therefore spread out from the charge and cover all space. Furthermore, since the distance scale r was chosen to agree with the gravitationally-defined distance, the streamlines of $\boldsymbol{D}$ convey genuine intrinsic information. The plots therefore encode gauge-invariant information about the electromagnetic field. Figure 14.5 shows streamline plots for charges held at different distances above the horizon. A polarisation charge is clearly visible at the origin, and streamlines are attracted towards this but never actually meet it. The effects of the polarisation charge can be felt outside the horizon as a repulsive force acting back on the charge. That is, less force is required to keep a charge at rest outside a black hole than is required for an uncharged particle. The fact that the origin of this effect lies inside the horizon reinforces the importance of constructing global solutions to the field equations.

14.3.6 Other gauges

Before proceeding it is useful to study the vacuum spherical equations in an arbitrary gauge. We return to the spherical equations before the $f_2 = 0$ gauge choice was made, and again impose that M is constant. The equations that remain are

$$g_1{}^2 - g_2{}^2 = 1 - 2M/r \tag{14.120}$$

and

$$\partial_r g_1 = G, \qquad \partial_r g_2 = F, \tag{14.121}$$

and all fields are functions of r only. The bracket relation of equation (14.35) gives

$$g_2 \partial_r f_2 - g_1 \partial_r f_1 = G f_1 - F f_2, \tag{14.122}$$

from which it follows that

$$\partial_r (f_1 g_1 - f_2 g_2) = \partial_r \det\,(\mathsf{h}) = 0. \tag{14.123}$$

The determinant of h is constant, and the value of this constant depends on the choice of gauge. Because the Riemann tensor falls off as r^{-3} we always choose to work in a gauge where $\bar{\mathsf{h}}$ tends to the identity as $r \mapsto \infty$. In this case we have $\det\,(\mathsf{h}) = 1$, so we can write

$$f_1 g_1 - f_2 g_2 = 1. \tag{14.124}$$

No other equations remain to fix the solution further. We therefore have two free functions in the choice of $\bar{\mathsf{h}}$ function.

A useful alternative to the Newtonian gauge chosen in this section is to write the solution in Kerr–Schild form. For this we set

$$\begin{aligned} g_1 &= 1 - M/r, & g_2 &= -M/r, \\ f_1 &= 1 + M/r, & f_2 &= M/r. \end{aligned} \tag{14.125}$$

In this case the $\bar{\mathsf{h}}$ function takes on the compact form

$$\bar{\mathsf{h}}(a) = a + \frac{M}{r} a \cdot e_- \, e_-, \qquad e_- = e_t - e_r. \tag{14.126}$$

This algebraic form has a number of convenient algebraic features. The first is that the solution is of the form of the identity plus an interaction term, as is also the case in the Newtonian gauge setup. The second is that this form of $\bar{\mathsf{h}}(a)$ is a symmetric function. Finally, e_- is a null vector that satisfies $\bar{\mathsf{h}}(e_-) = e_-$. All of these features can be employed to simplify calculations.

The line element generated by our general form of $\bar{\mathsf{h}}$ function is

$$\begin{aligned} ds^2 =& (1 - 2M/r)\, dt^2 + 2(f_1 g_2 - f_2 g_1)\, dt\, dr - (f_1{}^2 - f_2{}^2)\, dr^2 \\ & - r^2 (d\theta^2 + \sin^2(\theta)\, d\phi^2). \end{aligned} \tag{14.127}$$

This in effect contains one arbitrary function, because the constraint on the determinant fixes one of the two unknown coefficients. The remaining unspecified degree of freedom lies in the rotation gauge, which does not affect the metric. We can draw an important conclusion about the metric by considering its behaviour at the horizon. There we must have $g_1 = \pm g_2$, and we know that $f_1 g_1 - f_2 g_2 = 1$ globally. It follows that

$$f_1 g_2 - f_2 g_1 = \pm 1 \quad \text{at } r = 2M, \tag{14.128}$$

so the off-diagonal term must be either $+1$ or -1 at the horizon. The presence of the horizon must break time reversal symmetry. This is to be expected. For a black hole (corresponding to the negative solution), the horizon is the place where particles can fall in, but cannot escape. The opposite value at the horizon (corresponding to a positive value of g_2 in the Newtonian gauge) defines an object from which particles can escape, but no particle can cross the horizon. This is called a *white hole*, though it is unclear whether such a solution defines a physically relevant object.

14.4 Quantum mechanics in a black hole background

The gauge theory formulation of gravity is motivated by constructing gauge fields to ensure that the Dirac equation is covariant under local rotations and displacements. We now study the effects of the black hole gauge fields on a Dirac fermion. Assuming that no electromagnetic couplings are present, the minimally-coupled equation takes the familiar form

$$D\psi I\boldsymbol{\sigma}_3 = m\psi\gamma_0. \tag{14.129}$$

The simplicity of the $\bar{\mathsf{h}}$ field in the Newtonian gauge suggests that this will be the simplest gauge to work in. As always, we must ensure that the all physical predictions are gauge-invariant. With the gravitational fields as described in equations (14.64) and (14.65), the Dirac equation becomes

$$\nabla\psi I\boldsymbol{\sigma}_3 - \left(\frac{2M}{r}\right)^{1/2} \gamma_0 \left(\frac{\partial}{\partial r}\psi + \frac{3}{4r}\psi\right) I\boldsymbol{\sigma}_3 = m\psi\gamma_0. \tag{14.130}$$

If we pre-multiply by γ_0 and employ the i symbol to represent right-sided multiplication by $I\boldsymbol{\sigma}_3$, then equation (14.130) becomes

$$i\partial_t\psi = -i\boldsymbol{\nabla}\psi + i\left(\frac{2M}{r}\right)^{1/2} \frac{1}{r^{3/4}}\frac{\partial}{\partial r}(r^{3/4}\psi) + m\bar{\psi}, \tag{14.131}$$

where $\bar{\psi} = \gamma_0\psi\gamma_0$. We see that the Newtonian gauge has enabled us to write the Dirac equation in a very straightforward Hamiltonian form. One reason for this simplicity is that the spatial sections defined by the time coordinate t are flat.

The interaction Hamiltonian in equation (14.131), with all constants included, is

$$\hat{H}_I(\psi) = i\hbar\left(\frac{2GM}{r}\right)^{1/2} \left(\frac{\partial}{\partial r} + \frac{3}{4r}\right)\psi. \tag{14.132}$$

This single term incorporates all gravitational effects exerted by a black hole on a Dirac fermion. A number of observations can be made immediately. The first is that the interaction Hamiltonian does not depend on the mass of the particle, which is how the equivalence principle is embodied in the Dirac equation. The second point is that $\hat{H}_I$ does not depend on the speed of light. The non-relativistic approximation is therefore straightforward, following the technique of section 8.3.3. To lowest order in c^{-1} we obtain the Schrödinger equation with interaction determined by $\hat{H}_I$. For stationary states this equation is

$$-\frac{\hbar^2}{2m}\boldsymbol{\nabla}^2\psi + i\hbar\left(\frac{2GM}{r}\right)^{1/2}\frac{1}{r^{3/4}}\frac{\partial}{\partial r}(r^{3/4}\psi) = E\psi, \tag{14.133}$$

where ψ now denotes the Schrödinger wave function. This equation is simplified by introducing the phase-transformed variable

$$\Psi = \psi \exp\left(-i(8r/a_G)^{1/2}\right), \tag{14.134}$$

where

$$a_G = \frac{\hbar^2}{GMm^2}. \tag{14.135}$$

The distance a_G is the gravitational analogue of the Bohr radius for the hydrogen atom. The new variable Ψ satisfies the simple equation

$$-\frac{\hbar^2}{2m}\boldsymbol{\nabla}^2\Psi - \frac{Mm}{r}\Psi = E\Psi. \tag{14.136}$$

This is precisely the equation we would expect if we used the Newtonian gravitational potential. The solutions for Ψ are therefore Coulomb wavefunctions. The non-relativistic limit enables us to make two immediate predictions. The first is that a spectrum of bound states should exist, with similar properties to that of the hydrogen atom. The second is that, in the non-relativistic limit, the scattering cross section should be determined by the Rutherford formula. This latter prediction is confirmed in the following section.

The interaction Hamiltonian $\hat{H}_I$ hides a significant feature, which is that it is not Hermitian due to the presence of the singularity. To see this we form the difference between $\hat{H}_I$ and its adjoint. With ϕ and ψ both Dirac spinors we find that

$$\begin{aligned}\int d^3x\, \langle \phi^\dagger \hat{H}_I(\psi)\rangle_q &= \sqrt{2M}\int d\Omega \int_0^\infty dr\, \langle r^{3/4}\phi^\dagger \partial_r(r^{3/4}\psi) I\boldsymbol{\sigma}_3\rangle_q \\ &= \int d^3x\, \langle (\hat{H}_I(\phi))^\dagger \psi\rangle_q + \sqrt{2M}\int d\Omega \left[r^{3/2}\langle \phi^\dagger \psi I\boldsymbol{\sigma}_3\rangle_q\right]_0^\infty,\end{aligned} \tag{14.137}$$

where we follow the convention of section 8.1.2. We will see shortly that all wavefunctions approach the origin as $r^{-3/4}$. The boundary term at the origin therefore does not vanish, and the Hamiltonian is not Hermitian. It follows that any normalisable stationary state must have an imaginary component to its energy. This is sensible. For all states the covariant current vector is always timelike. Inside the horizon this vector must point inwards, towards the singularity, so current density is inevitably swept onto the singularity. This implies that bound states must necessarily decay, so we expect the energy to have an imaginary component.

14.4.1 Scattering

The Dirac equation (14.131) is ideally suited to a perturbative scattering calculation employing the methods of section 8.5. We seek an iterative solution to the Green's function equation

$$\left(i\hat{\nabla}_2 - \hat{B}(x_2) - m\right)S_G(x_2, x_1) = \delta^4(x_2 - x_1), \tag{14.138}$$

where

$$B(x) = i\hat{\gamma}_0 \left(\frac{2M}{r}\right)^{1/2} \left(\frac{\partial}{\partial r} + \frac{3}{4r}\right). \tag{14.139}$$

As usual, the hats denote operators which act on spinors, and in this section we retain the familiar i symbol to denote the complex structure.

The iterative solution to equation (14.138) is given by

$$\begin{aligned} S_G(x_f, x_i) &= S_F(x_f, x_i) + \int d^4x_1\, S_F(x_f, x_1)B(x_1)S_F(x_1, x_i) \\ &+ \iint d^4x_1\, d^4x_2\, S_F(x_f, x_1)B(x_1)S_F(x_1, x_2)B(x_2)S_F(x_2, x_i) + \cdots, \end{aligned} \tag{14.140}$$

where $S_F(x_2, x_1)$ is the free-field, position-space Feynman propagator. The interaction term $B(x)$ is independent of time so energy is conserved throughout the interaction. Converting to momentum space we find that the scattering multivector T_{fi}, as defined in equation (8.229), is given by

$$T_{fi} = (\hat{p}_f + m)\left(B(\boldsymbol{p}_f, \boldsymbol{p}_i) + \int \frac{d^3k}{(2\pi)^3} B(\boldsymbol{p}_f, \boldsymbol{k}) \frac{\hat{k} + m}{k^2 - m^2 + i\epsilon} B(\boldsymbol{k}, \boldsymbol{p}_i) + \cdots\right). \tag{14.141}$$

Here $B(\boldsymbol{p}_2, \boldsymbol{p}_1)$ denotes the spatial Fourier transform of the interaction term,

$$B(\boldsymbol{p}_2, \boldsymbol{p}_1) = (2M)^{1/2} i\hat{\gamma}_0 \int d^3x\, e^{-i\boldsymbol{p}_2\cdot\boldsymbol{x}} \frac{1}{r^{1/2}} \left(\frac{\partial}{\partial r} + \frac{3}{4r}\right) e^{i\boldsymbol{p}_1\cdot\boldsymbol{x}}, \tag{14.142}$$

where bold symbols refer to spatial components only. To evaluate this we first write

$$B(\boldsymbol{p}_2, \boldsymbol{p}_1) = (2M)^{1/2} i\hat{\gamma}_0 \left(\frac{3}{4} f(\boldsymbol{p}_1 - \boldsymbol{p}_2) + \left.\frac{\partial f(\lambda \boldsymbol{p}_1 - \boldsymbol{p}_2)}{\partial \lambda}\right|_{\lambda=1}\right), \tag{14.143}$$

where

$$f(\boldsymbol{p}) = \int d^3x \frac{e^{i\boldsymbol{p}\cdot\boldsymbol{x}}}{r^{3/2}} = \left(\frac{2\pi}{|\boldsymbol{p}|}\right)^{3/2}. \tag{14.144}$$

We therefore find that the momentum-space interaction is governed by the *vertex factor*

$$B(\boldsymbol{p}_2, \boldsymbol{p}_1) = 3\pi^{3/2} i(M)^{1/2} \frac{\boldsymbol{p}_2{}^2 - \boldsymbol{p}_1{}^2}{|\boldsymbol{p}_2 - \boldsymbol{p}_1|^{7/2}} \hat{\gamma}_0. \tag{14.145}$$

This factor has the unusual feature of vanishing if the ingoing and outgoing particles are on-shell, because energy is conserved throughout the process. It follows that the lowest order contribution to the scattering cross section vanishes. This is to be expected, as the vertex factor goes as $\sqrt{M}$, and we expect the amplitude to go as M to recover the Rutherford formula in the low velocity limit.

Working to the lowest non-zero order in M the scattering multivector becomes

$$T_{fi} = -9\pi^3 M(\hat{p}_f + m)\hat{\gamma}_0 I_1 \hat{\gamma}_0, \tag{14.146}$$

where

$$I_1 = \int \frac{d^3k}{(2\pi)^3} \frac{\boldsymbol{p}_f{}^2 - \boldsymbol{k}^2}{|\boldsymbol{p}_f - \boldsymbol{k}|^{7/2}} \frac{\hat{k} + m}{k^2 - m^2 + i\epsilon} \frac{\boldsymbol{k}^2 - \boldsymbol{p}_i{}^2}{|\boldsymbol{k} - \boldsymbol{p}_i|^{7/2}}. \tag{14.147}$$

Here we have explicitly included a factor of $i\epsilon$ to ensure that any poles in the complex plane are navigated in the correct manner. However, we have

$$k^2 - m^2 = E^2 - \boldsymbol{k}^2 - m^2 = \boldsymbol{p}^2 - \boldsymbol{k}^2, \tag{14.148}$$

where E is the particle energy and $\boldsymbol{p}^2 = \boldsymbol{p}_i{}^2 = \boldsymbol{p}_f{}^2$. The pole in the propagator is therefore cancelled by the vertex factors, so there is no need for the factor of $i\epsilon$ in the denominator. The integral we need to evaluate is therefore

$$I_1 = \int \frac{d^3k}{(2\pi)^3} \frac{\boldsymbol{k}^2 - \boldsymbol{p}^2}{|\boldsymbol{p}_f - \boldsymbol{k}|^{7/2}|\boldsymbol{k} - \boldsymbol{p}_i|^{7/2}} (\hat{k} + m), \tag{14.149}$$

and the result of this integral is

$$I_1 = \frac{1}{9\pi^2 \boldsymbol{q}^2}\big(2m + 3(\hat{p}_f + \hat{p}_i) - 4E\hat{\gamma}_0\big), \tag{14.150}$$

where $\boldsymbol{q} = \boldsymbol{p}_f - \boldsymbol{p}_i$. The scattering multivector is now given by

$$T_{fi} = -\frac{4\pi M}{\boldsymbol{q}^2}\big(E(2E + \boldsymbol{q}) + \boldsymbol{p}^2 + \boldsymbol{p}_f \boldsymbol{p}_i\big). \tag{14.151}$$

This should be contrasted with the equivalent expression for Coulomb scattering, given in equation (8.237). We see immediately that the coupling term goes with the particle energy, rather than its mass. This is because the interaction Hamiltonian is independent of m. The unpolarised cross section is given by

$$\begin{aligned}\frac{d\sigma}{d\Omega} &= \frac{|T_{fi}|^2}{16\pi^2} \\ &= \frac{2M^2}{\boldsymbol{q}^4}\big(m^2(E^2 - \boldsymbol{p}_f\cdot\boldsymbol{p}_i) + (2E^2 - m^2)^2 + 4E^2\boldsymbol{p}_f\cdot\boldsymbol{p}_i\big).\end{aligned} \tag{14.152}$$

If we now let $v = |\boldsymbol{p}|/E$ denote the particle velocity, and θ the scattering angle, we arrive at the simple expression

$$\frac{d\sigma}{d\Omega} = \frac{M^2}{4v^4\sin^4(\theta/2)}\big(1 + 2v^2 - 3v^2\sin^2(\theta/2) + v^4 - v^4\sin^2(\theta/2)\big). \tag{14.153}$$

As demanded by the equivalence principle, this formula depends only on the incident velocity, and not on the particle mass. This confirms that the equivalence principle is directly encoded in the Dirac equation as a consequence of minimal coupling. The final cross section formula is gauge-invariant. We can perform analogues of this calculation in a range of different gauges, and the same result is obtained in all cases. Furthermore, all terms in the result have local, gauge-invariant definitions. The mass M can be defined in terms of tidal forces, and the velocity v is that measured locally by observers in radial free fall from rest at infinity. The angle θ is the angle between asymptotic in and out states, measured locally in the asymptotic regime.

The cross section of equation (14.153) confirms that the low velocity limit recovers the Rutherford formula. The massless limit $m \mapsto 0$ is also well defined, and is obtained by setting $v = 1$. This produces the simple formula

$$\frac{d\sigma}{d\Omega} = \frac{M^2\cos^2(\theta/2)}{\sin^4(\theta/2)}. \tag{14.154}$$

The small angle limit to this gives a cross section going as $(4M)^2/\theta^4$. This recovers the classical formula for the bending of light by a massive source. While the calculation here has assumed a point mass source, the small angle limit is appropriate for any localised source of gravitational mass M. The massless limit contains a surprise in the backward direction, however. Simulations of scattering based on massless particles following null geodesics reveal a large 'glory' scattering in the backward direction. This is absent from the quantum treatment, and is a diffraction effect for massless spin-1/2 particles that is not evident at the classical level. The scheme described here can be modified to the case of a scalar field, and produces the differential cross section

$$\frac{d\sigma}{d\Omega} = \frac{M^2}{4v^4\sin^4(\theta/2)}(1 + v^2)^2. \tag{14.155}$$

Again, we see that the equivalence principle is obeyed, and the various small angle and low velocity approximations are retained. The classical cross section contains further structure, attributable to multiple orbits. In the quantum framework these effects should be present in the higher-order terms.

14.4.2 Stationary states and angular separation

The Dirac equation in the Newtonian gauge is immediately separable in space and time, and admits stationary state solutions of the form

$$\psi(x) = \psi(\boldsymbol{x}) \exp(-EtI\boldsymbol{\sigma}_3). \tag{14.156}$$

If the state is normalisable then E contains an imaginary component determined by

$$\mathrm{Im}(E) = -\frac{\sqrt{2M}}{2N} \lim_{r\to 0} r^{3/2} \int d\Omega \, \langle \psi^\dagger \psi \rangle, \tag{14.157}$$

where N is the normalisation constant

$$N = \int d^3x \, \langle \psi^\dagger \psi \rangle. \tag{14.158}$$

As expected, the sign of the imaginary component of E corresponds to a decaying wavefunction. This behaviour is independent of the sign of the real part of E, so both positive and negative energy states must decay. For scattering states we do not demand that ψ is normalisable, and can look for solutions where the energy is real, with $E > m$.

With the time dependence separated out, equation (14.131) reduces to

$$\boldsymbol{\nabla}\psi - (2M/r)^{1/2} r^{-3/4} \partial_r \big(r^{3/4}\psi\big) = iE\psi - im\bar{\psi}. \tag{14.159}$$

To solve this equation we follow the standard procedure for a central potential and separate out the angular dependence. This is achieved using the spherical monogenics, described in section 8.4.1. We assume that the wavefunction takes the standard form of

$$\psi(\boldsymbol{x}, \kappa) = \begin{cases} \psi_l^m u(r) + \sigma_r \psi_l^m v(r) I\boldsymbol{\sigma}_3 & \kappa = l+1, \\ \sigma_r \psi_l^m u(r) \boldsymbol{\sigma}_3 + \psi_l^m I v(r) & \kappa = -(l+1), \end{cases} \tag{14.160}$$

where κ is a non-zero integer and $u(r)$ and $v(r)$ are complex functions of r only. On substituting this wavefunction into the Dirac equation (14.159) we obtain the pair of coupled radial equations

$$\begin{pmatrix} 1 & -(2M/r)^{1/2} \\ -(2M/r)^{1/2} & 1 \end{pmatrix} \begin{pmatrix} u_1' \\ u_2' \end{pmatrix} = \boldsymbol{A} \begin{pmatrix} u_1 \\ u_2 \end{pmatrix}, \tag{14.161}$$

where

$$\boldsymbol{A} = \begin{pmatrix} \kappa/r & i(E+m) - (2M/r)^{1/2}(4r)^{-1} \\ i(E-m) - (2M/r)^{1/2}(4r)^{-1} & -\kappa/r \end{pmatrix}, \tag{14.162}$$

u_1 and u_2 are the reduced functions defined by

$$u_1 = ru, \qquad u_2 = irv \tag{14.163}$$

and the primes denote differentiation with respect to r. The form of this equation should be contrasted with the hydrogen atom of section 8.4.3.

To analyse equation (14.161) we first rewrite it in the equivalent form

$$(1 - 2M/r)\begin{pmatrix} u_1' \\ u_2' \end{pmatrix} = \begin{pmatrix} 1 & (2M/r)^{1/2} \\ (2M/r)^{1/2} & 1 \end{pmatrix} \boldsymbol{A} \begin{pmatrix} u_1 \\ u_2 \end{pmatrix}. \tag{14.164}$$

This makes it clear that the equations have regular singular points at the origin and horizon ($r = 2M$), as well as an irregular singular point at $r = \infty$. Unfortunately, the special function theory required to deal with such equations has not been developed. Hypergeometric functions are appropriate for differential equations with three regular singular points, or one regular and one irregular singular point. An attempt to generalise hypergeometric functions results in Heun's equation, but most techniques for handling this involve series solutions and numerical integration, so these are the techniques that must be applied here. The presence of the three singular points implies that any power series will have a limited radius of convergence, so typically these can only be used to define initial data for numerical integration routines.

A Frobenius series about the origin shows that both u_1 and u_2 approach the origin as $r^{1/4}$. It follows that the wavefunction goes as $r^{-3/4}$ near the origin, as was stated earlier. For normalisable states this behaviour ensures that the energy contains an imaginary decay factor. Next we construct a series about the horizon by writing

$$u_1 = \eta^s \sum_{k=0}^{\infty} a_k \eta^k, \qquad u_2 = \eta^s \sum_{k=0}^{\infty} b_k \eta^k, \tag{14.165}$$

where $\eta = r - 2M$. On substituting this series into equation (14.164), and setting $\eta = 0$, we obtain

$$\frac{s}{2M}\begin{pmatrix} a_0 \\ b_0 \end{pmatrix} = \begin{pmatrix} 1 & 1 \\ 1 & 1 \end{pmatrix} \begin{pmatrix} \kappa/(2M) & i(E+m) - (8M)^{-1} \\ i(E-m) - (8M)^{-1} & -\kappa/(2M) \end{pmatrix} \begin{pmatrix} a_0 \\ b_0 \end{pmatrix}. \tag{14.166}$$

The two values of the index s for which this has non-zero solutions are

$$s = 0 \quad \text{and} \quad s = -\tfrac{1}{2} + 4iME. \tag{14.167}$$

The $s = 0$ solution corresponds to an analytic power series with a well-defined wavefunction at the horizon. Such solutions are certainly physical. The second root gives rise to a wavefunction that is singular at the horizon, and as such is physically inadmissible. As a consequence, it is not possible to construct a complete set of outgoing modes at infinity and in any scattering process some of the wavefunction is lost. This is the quantum-mechanical description of absorption by a black hole.

Before proceeding, we should confirm that the two indicial roots at the horizon are gauge-invariant, and not an artifact of our various gauge choices. This is important because the singular index can be used to determine the Hawking temperature of the black hole. The method we use to confirm gauge invariance is quite general and can be applied to a range of situations. We start by keeping the gauge unspecified so that, after separating out the angular dependence, the Dirac equation reduces to

$$\begin{pmatrix} L_r & L_t \\ L_t & L_r \end{pmatrix} \begin{pmatrix} u_1 \\ u_2 \end{pmatrix} = \begin{pmatrix} \kappa/r - G/2 & im - F/2 \\ -im - F/2 & -\kappa/r - G/2 \end{pmatrix} \begin{pmatrix} u_1 \\ u_2 \end{pmatrix}. \tag{14.168}$$

We can still assume that the time dependence is of the form $\exp(-iEt)$, so that equation (14.168) becomes

$$\begin{pmatrix} g_1 & g_2 \\ g_2 & g_1 \end{pmatrix} \begin{pmatrix} u_1' \\ u_2' \end{pmatrix} = \boldsymbol{B} \begin{pmatrix} u_1 \\ u_2 \end{pmatrix}, \tag{14.169}$$

where

$$\boldsymbol{B} = \begin{pmatrix} \kappa/r - G/2 + if_2E & i(m + f_1E) - F/2 \\ -i(m - f_1E) - F/2 & -\kappa/r - G/2 + if_2E \end{pmatrix}. \tag{14.170}$$

The form of time dependence is gauge-invariant, since the time coordinate is defined by the requirement that the Riemann tensor is stationary. Given a time coordinate t, a general displacement consistent with this requirement takes the form

$$t \mapsto t' = t + \alpha(r), \tag{14.171}$$

where α is a differentiable function of r. This ensures that stationary states all go as $\exp(-iEt)$, regardless of the choice of time coordinate.

Now, since ${g_1}^2 - {g_2}^2 = 1 - 2M/r$ holds for vacuum solutions in all gauges, we obtain

$$(1 - 2M/r) \begin{pmatrix} u_1' \\ u_2' \end{pmatrix} = \begin{pmatrix} g_1 & -g_2 \\ -g_2 & g_1 \end{pmatrix} \boldsymbol{B} \begin{pmatrix} u_1 \\ u_2 \end{pmatrix}. \tag{14.172}$$

We again look for a power series solution of the form of equation (14.165), and setting $\eta = 0$ produces the indicial equation

$$\det \left[\begin{pmatrix} g_1 & -g_2 \\ -g_2 & g_1 \end{pmatrix} \boldsymbol{B} - \frac{s}{r} \mathsf{I} \right]_{r=2M} = 0, \tag{14.173}$$

where I is the identity matrix. For vacuum fields we know that

$$g_1 G - g_2 F = \tfrac{1}{2} \partial_r ({g_1}^2 - {g_2}^2) = M/r^2, \tag{14.174}$$

which is gauge-invariant. It follows that the solutions to the indicial equation are

$$s = 0 \quad \text{and} \quad s = -\tfrac{1}{2} + 4iME(g_1 f_2 - g_2 f_1). \tag{14.175}$$

But, as discussed in section 14.3.6, at the horizon we have

$$(g_1 f_2 - g_2 f_1) = \pm 1, \tag{14.176}$$

with the positive sign corresponding to the black hole case. The indices of the Dirac equation are therefore gauge-invariant. Similar arguments can be applied to scalar and higher-spin fields.

14.4.3 Quantum absorption

We are now in a position to give a full, quantum-mechanical description of absorption by a black hole. At the horizon the solutions of the Dirac equation separate into two branches, one regular and one singular. The singular branch is unphysical and cannot be excited by finite incoming waves. The regular branch is finite at the horizon, with an inward-pointing current. This gives rise to absorption. To understand this process in detail we need to study the asymptotic form of the regular solutions and determine their split into incoming and outgoing modes. We can then construct an arbitrary incoming mode (typically a plane wave) and study the amount of scattered radiation. Any radiation that is not scattered is absorbed.

In absorption and scattering problems we are interested in states with real energy E, $E > m$. For such states the spatial current $\boldsymbol{J}$ is conserved, and for angular eigenstates we obtain the conserved Wronskian W:

$$W = g_1(u_1 u_2^\dagger + u_1^\dagger u_2) + g_2(u_1 u_1^\dagger + u_2 u_2^\dagger). \tag{14.177}$$

This measures the total outward flux over a surface of radius r, and we have written W in an arbitrary gauge. At the horizon we see that

$$W = -g_1 |u_1 - u_2|^2, \tag{14.178}$$

and so the flux is *inwards* for all regular solutions. This is to be expected, as the current must point inwards at the horizon.

For explicit calculations we return to the Newtonian gauge. The radial equation (14.164) is straightforward to integrate numerically. We start with a power series expansion around the horizon of the regular solution. This allows us to find values of u_1 and u_2 a small distance either side of the horizon. These values are then used to initiate numerical integration of the equations, both inwards and outwards. To visualise the solutions it is convenient to plot the radial density function $P(r)$:

$$P(r) = |u_1|^2 + |u_2|^2. \tag{14.179}$$

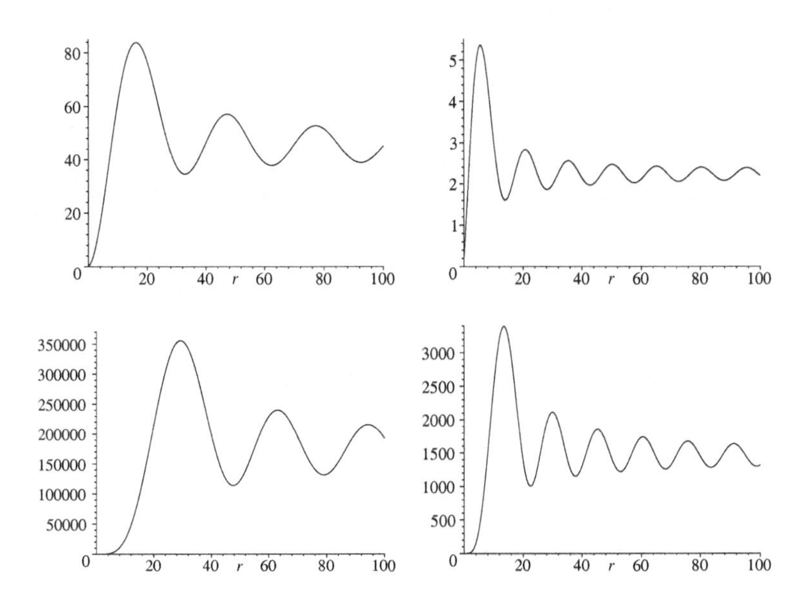

Figure 14.6 *The radial density for scattered states.* The plots show $P(r)$ as a function of radius. The horizon lies at $r = 2$, and the product mM is set to 0.01 in units of m_p^2, where m_p is the Planck mass. The modes are scaled so that the Wronskian is -1, and only the regular solution is plotted. The top two diagrams are for $\kappa = 1$, with $E = 10mc^2$ (left) and $E = 20mc^2$ (right). The bottom two diagrams are for $\kappa = 2$, with $E = 10mc^2$ (left) and $E = 20mc^2$ (right).

In physical terms $P(r)$ is r^2 times the timelike component of the Dirac current, as measured by observers in radial free fall from rest at infinity. It is only in the Newtonian gauge that this definition gives rise to the simple formula of equation (14.179).

In figure 14.6 we plot $P(r)$ for a range of energies and angula momenta. The plots are for scattering states, so the wavefunctions are unnormalised. For the sake of comparison the magnitude of each mode is fixed by setting the Wronskian to -1. The gravitational coupling is controlled by the dimensionless quantity

$$\frac{GMm}{\hbar c} = \frac{Mm}{m_p^2}, \tag{14.180}$$

where m_p is the Planck mass. In figure 14.6 we have used a dimensionless coupling of 0.01. The chosen energies of $10mc^2$ and $20mc^2$ imply that the modes are highly relativistic, and also ensure that the associated wavelengths are larger than the horizon size. To understand the asymptotic features of the plots we return to equation (14.161) and solve for the behaviour at large r. We find that the solutions behave asymptotically as

$$
\begin{aligned}
u_1 =& \beta \exp i \left(pr + \frac{M}{p}(m^2+2p^2)\ln(pr) \right) \mathrm{e}^{2iE(2Mr)^{1/2}} \\
&+ \alpha \exp -i \left(pr + \frac{M}{p}(m^2+2p^2)\ln(pr) \right) \mathrm{e}^{2iE(2Mr)^{1/2}}
\end{aligned} \tag{14.181}
$$

and

$$
\begin{aligned}
u_2 =& \frac{p\beta}{E+m} \exp i \left(pr + \frac{M}{p}(m^2+2p^2)\ln(pr) \right) \mathrm{e}^{2iE(2Mr)^{1/2}} \\
&- \frac{p\alpha}{E+m} \exp -i \left(pr + \frac{M}{p}(m^2+2p^2)\ln(pr) \right) \mathrm{e}^{2iE(2Mr)^{1/2}},
\end{aligned} \tag{14.182}
$$

where $p^2 = E^2 - m^2$. The Wronskian is therefore equal to

$$
W = -\frac{2p}{E+m}(|\alpha|^2 - |\beta|^2), \tag{14.183}
$$

and the radial probability $P(r)$ is given asymptotically by

$$
\begin{aligned}
|u_1|^2 + |u_2|^2 =& \frac{4m}{E+m}|\alpha|\,|\beta| \cos\left(2pr + \frac{2M(m^2+2p^2)}{p}\ln(pr) + \phi_0 \right) \\
&+ \frac{2E}{E+m}(|\alpha|^2 + |\beta|^2).
\end{aligned} \tag{14.184}
$$

The oscillations predicted by this formula are clearly visible in figure 14.6. The magnitudes of α and β determine the relative amounts of scattered and absorbed radiation present for a given mode. With the Wronskian held constant, all modes have a constant flux through the horizon onto the singularity. In the large r region $|\alpha|$ determines the amount of ingoing radiation, and $|\beta|$ the amount of outgoing radiation. As $|\alpha|$ increases, a smaller fraction of the radiation is absorbed and more is scattered. One effect that is clear in figure 14.6 is that as the angular momentum increases, for fixed energy, $|\alpha|$ also increases. That is, less radiation is absorbed for fixed energy as the angular momentum increases. This is precisely the behaviour we expect from classical considerations.

Given that each mode is normalised such that $W = -1$, then total absorption cross section is given by

$$
\sigma_{abs} = \frac{\pi}{2p(E-m)} \sum_{\kappa \neq 0} \frac{|\kappa|}{|\alpha_\kappa|^2}, \tag{14.185}
$$

where α_κ is the value of α for each angular eigenmode. The values of α_κ are determined numerically by integrating the radial equations out to a suitable distance from the horizon and matching to the asymptotic forms of equations (14.181) and (14.182). Typically, we need to sum over a range of κ values before the sum settles down to its final result. The result of this sum, for a massive fermion, is plotted in figure 14.7. For energies close to the rest energy the absorption

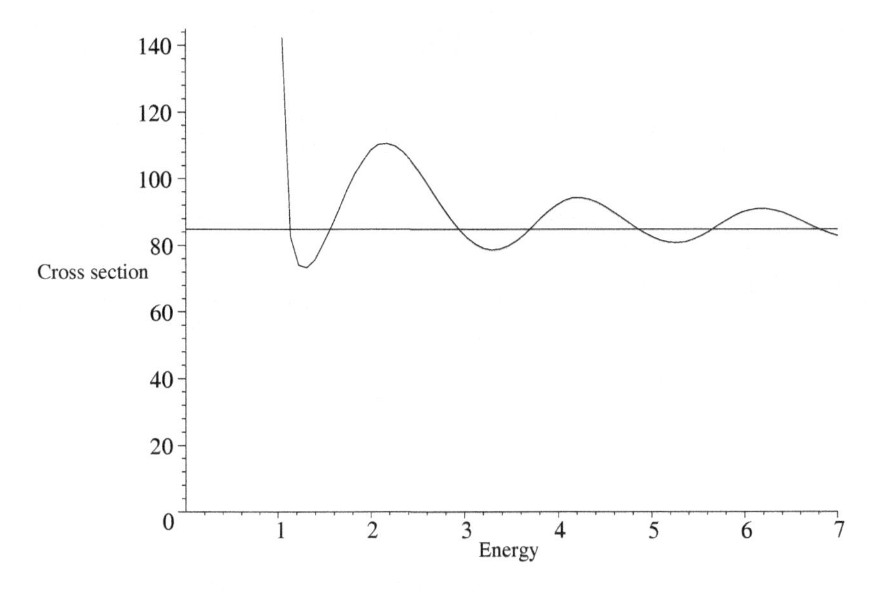

Figure 14.7 *The quantum absorption cross section.* The plot shows the total absorption cross section as a function of the incident energy. The dimensionless coupling Mm/m_p^2 is 0.1, and the energy is plotted in units of the rest energy mc^2. The horizontal line is the photon limit.

cross section follows the classical prediction. But at higher energies a series of oscillations are present as the wavelength becomes comparable with the horizon size. These oscillations take place around the photon limit of 27π, and are also present for massless particles. The precise form of these oscillations depends on the mass of the particle, so represents a quantum-mechanical violation of the equivalence principle.

14.5 Cosmology

The radial equations we have developed so far are easily adapted to the case of homogeneous, isotropic matter distributions. Such distributions provide a good model for the large scale distribution of matter in the observable universe. Before studying the field equations for such cosmological matter distributions, we must first introduce the cosmological constant. This was originally introduced by Einstein to allow the construction of static cosmological solutions, and for many years had been thought to be an unnecessary additional feature of general relativity. But experimental evidence, both from the cosmic microwave background and from distant supernovae, now favours models which do include a cosmological constant. There are also hints from quantum gravity that a cosmological constant should arise as a form of vacuum energy, though this is not well understood.

We start with the radial equations, as summarised in table 14.2. Inclusion of the cosmological constant Λ only modifies a handful of these equations. The mass function M becomes

$$M = \tfrac{1}{2}r\big({g_2}^2 - {g_1}^2 + 1 - \Lambda r^2/3\big), \tag{14.186}$$

and the derivatives of the g_1 and g_2 fields become

$$\begin{aligned} L_t g_2 &= G g_1 - M/r^2 + r\Lambda/3 - 4\pi r p, \\ L_r g_1 &= F g_2 + M/r^2 - r\Lambda/3 - 4\pi r \rho. \end{aligned} \tag{14.187}$$

The Riemann tensor is altered to

$$\begin{aligned} \mathcal{R}(B) =& 4\pi(\rho + p) B\cdot e_t\, e_t - \frac{1}{3}(8\pi\rho + \Lambda)B \\ & - \left(\frac{M}{2r^3} - \frac{2\pi}{3}\rho\right)(B + 3\sigma_r B \sigma_r) \end{aligned} \tag{14.188}$$

and we continue to assume that the matter distribution takes the form of an ideal fluid.

For cosmological models the matter distribution is assumed to be spatially homogeneous and isotropic, so that ρ and p are functions of time only. The mass function M is then given by

$$M(r,t) = \frac{4\pi}{3} r^3 \rho, \tag{14.189}$$

so the Riemann tensor also depends only on time. The equation for $L_r p$ tells us that G vanishes, and hence that

$$f_1 = 1. \tag{14.190}$$

The time coordinate t therefore measures the proper time for observers at rest with respect to the cosmological background. The derivatives of M and ρ similarly tell us that

$$F = \frac{g_2}{r} \tag{14.191}$$

and

$$\dot{\rho} = -\frac{3g_2}{r}(\rho + p). \tag{14.192}$$

For these to be consistent with the relation $L_r g_2 = F g_1$ we must have

$$F = H(t), \qquad g_2(r,t) = rH(t), \tag{14.193}$$

where $H(t)$ is a function of time only. The $L_t g_2$ equation now reduces to a simple equation for $H(t)$:

$$\dot{H} + H^2 - \frac{\Lambda}{3} = -\frac{4\pi}{3}(\rho + 3p). \tag{14.194}$$

The $\bar{\mathsf{h}}$ field	$\bar{\mathsf{h}}(a) = a + a\cdot e_r\big((g_1 - 1)e^r + H(t)re^t\big)$ ${g_1}^2 = 1 - kr^2 \exp\big(-2\int^t H(t')\,dt'\big)$
The ω field	$\omega(a) = H(t)a\wedge e_t - (g_1 - 1)/r\, a\wedge(e_r e_t)e_t$
Riemann tensor	$\mathcal{R}(B) = 4\pi(\rho + p)B\cdot e_t\, e_t - (8\pi\rho + \Lambda)/3\, B$
The density	$8\pi\rho = 3H(t)^2 - \Lambda + 3k\exp\big(-2\int^t H(t')\,dt'\big)$
Dynamical equations	$\dot{H} + H^2 - \Lambda/3 = -(4\pi/3)\,(\rho + 3p)$ $\dot{\rho} = -3H(t)(\rho + p)$

Table 14.3 *Equations governing a homogeneous, isotropic perfect fluid.* The covariant vector e_t defines the rest frame of the universe. This is determined experimentally from the cosmic microwave background radiation. No other direction is contained in $\mathcal{R}(B)$, and all physical fields are functions of time only.

Finally, we are left with a pair of equations for g_1,

$$\begin{aligned} L_t g_1 &= 0, \\ L_r g_1 &= ({g_1}^2 - 1)/r. \end{aligned} \tag{14.195}$$

The second equation tells us that g_1 is of the form

$${g_1}^2 = 1 + r^2\phi(t). \tag{14.196}$$

The equation for $L_t g_1$ then tells us that $\phi(t)$ satisfies

$$\dot{\phi} = -2H(t)\phi. \tag{14.197}$$

It follows that g_1 is given by

$${g_1}^2 = 1 - kr^2 \exp\left(-2\int^t H(t')\,dt'\right), \tag{14.198}$$

where k is an arbitrary constant of integration which turns out to define the spatial geometry. The full set of equations describing a homogeneous perfect fluid are summarised in table 14.3.

14.5.1 Comparison with standard approach

The derivation of the cosmological equations presented here, as a special case of a spherical solution, differs from most presentations. To recover a more familiar set of equations we first introduce the distance function S defined by

$$H(t) = \frac{\dot{S}(t)}{S(t)}. \tag{14.199}$$

With this substitution we find g_1 is now simply

$$ {g_1}^2 = 1 - kr^2/S^2. \tag{14.200} $$

Similarly, the $\dot{H}$ and density equations become

$$ \begin{aligned} \frac{\ddot{S}}{S} - \frac{\Lambda}{3} &= -\frac{4\pi}{3}(\rho + 3p), \\ \frac{\dot{S}^2 + k}{S^2} - \frac{\Lambda}{3} &= \frac{8\pi}{3}\rho. \end{aligned} \tag{14.201} $$

These are the *Friedmann equations* of cosmology. Our derivation has focused attention on the *Hubble function* $H(t)$, rather than the distance scale $S(t)$. This is natural, as $H(t)$ is a directly measurable (gauge-invariant) quantity, whereas $S(t)$ is only defined up to an arbitrary scaling.

The Friedmann equations are usually derived by starting with a diagonal line element. This is obtained from the radial setup by the displacement defined by

$$ f(x) = x \cdot e_t e_t + S x \wedge e_t e_t. \tag{14.202} $$

Under this displacement, $\bar{\mathsf{h}}(a)$ transforms to

$$ \bar{\mathsf{h}}'(a) = a \cdot e_t e_t + \frac{1}{S}\big((1 - kr^2)^{1/2} a \cdot e_r e^r + a \wedge \sigma_r \, \sigma_r\big), \tag{14.203} $$

and the line element this defines is

$$ ds^2 = dt^2 - \frac{S^2}{1 - kr^2} dr^2 - S^2 r^2 \big(d\theta^2 + \sin^2(\theta)\, d\phi^2\big). \tag{14.204} $$

In this gauge we can see clearly that S controls the distance scale, and k controls the spatial geometry. We can always choose the scale such that k is either zero or ± 1. A k of zero corresponds to a spatially flat universe, which is favoured on theoretical grounds and is consistent with observations. The non-zero values correspond to an open universe ($k < 0$, defining hyperbolic geometry) or a closed universe ($k > 0$, defining spherical geometries). These three spatial geometries are the only spatially homogeneous and isotropic models we can consider. These geometries are discussed in more detail in chapter 10. Which model is appropriate for the universe on its largest scales is determined by the present values of the density and Hubble function. Most experiments find that the universe is close to the critical density ($k = 0$), but no experiment can ever conclusively prove that k is zero. Any slight deviation in the density away from the critical value implies that k is non-zero. The fact that the universe is so close to its critical density has led theoreticians to propose a range of models which force the universe to have $k = 0$. The most popular of these is provided by *inflationary cosmology*, in which the universe passes through a stage of rapid inflation, so that all spatial sections are expanded dramatically and become essentially flat.

14.5.2 Density perturbations and cluster formation

We will not discuss the detailed solutions of the cosmological equations in this book. This is a large subject and is covered in detail in a range of modern textbooks. Here we discuss an application where the derivation from the radial equations is particularly helpful. The problem of interest is the growth of a perturbation in a cosmological background. The perturbation is assumed to be spherically-symmetric, and the coordinate system is centred on the perturbation. To simplify matters further, we ignore the cosmological constant and set the pressure to zero. We are therefore dealing with a simple model of a pressureless fluid collapsing under the influence of its own gravity.

Returning to the radial equations in table 14.2, we see that for zero pressure we have $G = 0$ and $f_1 = 1$. The matter therefore follows geodesics, and t measures the proper time for observers comoving with the matter. The mass satisfies

$$L_t M = 0, \tag{14.205}$$

which says that the mass M enclosed within radius r is conserved along the fluid streamlines. The operator L_t is clearly the comoving derivative along the fluid streamlines. The function g_1 is also conserved along a streamline, and the equations integrate straightforwardly to determine the streamlines (geodesics). The form of the geodesic depends on the value of g_1, and there are three cases to consider:

1. ${g_1}^2 < 1$. This case includes closed cosmologies, and the matter streamlines are defined by

$$\begin{aligned} r &= \frac{M}{1-{g_1}^2}(1-\cos(\eta)), \\ t - t_i &= \frac{M}{(1-{g_1}^2)^{3/2}}\big(\eta - \sin(\eta) - \eta_i + \sin(\eta_i)\big) \end{aligned} \tag{14.206}$$

where η parameterises the curve, and η_i is determined from the initial value of r at time t_i. The velocity g_2 is given by

$$g_2 = \frac{M}{r(1-{g_1}^2)^{1/2}} \sin(\eta) \tag{14.207}$$

and η_i is fixed in the range $0 < \eta_i < 2\pi$ by determining whether the initial velocity is inwards or outwards. Setting $\eta_i = \pi$ corresponds to starting from rest, and provides a simple model for black hole formation.

2. ${g_1}^2 = 1$. This case include flat cosmologies, and the equations integrate directly to give

$$t - t_i = \frac{2(r^{3/2} - r_i^{3/2})}{3(2M)^{1/2}}. \tag{14.208}$$

The velocity is chosen outwards to avoid a singularity forming instantaneously.

3. ${g_1}^2 > 1$. This case includes open cosmologies. The streamlines are parameterised by

$$\begin{aligned} r &= \frac{M}{{g_1}^2 - 1}(\cosh(\eta) - 1), \\ t - t_i &= \frac{M}{({g_1}^2 - 1)^{3/2}}(\sinh(\eta) - \eta - \sinh(\eta_i) + \eta_i) \end{aligned} \tag{14.209}$$

and the velocity is given by

$$g_2 = \frac{M}{r({g_1}^2 - 1)^{1/2}} \sinh(\eta). \tag{14.210}$$

For this case it is also necessary to start with an initial outward velocity, in order to avoid streamline crossing.

By working globally in the Newtonian gauge we keep simple control over the initial conditions. For these we wish to set up a small perturbation in a finite region, such that outside the perturbation the system evolves as a homogeneous cosmology. This will be the case provided the average density in the perturbation matches the external universe. Suppose that the perturbation initially has width r_i and the external cosmology has initial values ρ_i and H_i for the density and Hubble function respectively. We introduce the dimensionless variables

$$x = \frac{r}{r_i}, \qquad v(x) = \frac{g_2(r, t_i) - rH_i}{r_i H_i}, \qquad f(x) = \frac{\rho(r, t_i) - \rho_i}{\rho_i}. \tag{14.211}$$

The functions $f(x)$ and $v(x)$ are related by

$$x^2 f(x) = -\frac{d}{dx}(x^2 v(x)), \tag{14.212}$$

with both $f(x)$ and $v(x)$ vanishing at the boundary ($x = 1$). Equation (14.212) ensures that the model is correctly compensated, so that the perturbation has no effect on the external cosmology. (Equation (14.212) also ensures that no decaying modes are present in the perturbation left over from the linear regime.) To fix $f(x)$ and $v(x)$ we choose a parameter n, which controls the polynomial degree of the functions, and also fix the value of the velocity gradient at the origin. The function $v(x)$ is then a polynomial of degree $2n + 1$, formed as follows. At the centre we set $v = 0$, and the first derivative is determined by the velocity gradient. The remaining derivatives up to order n are set to zero. Similarly, at the boundary v is chosen such that g_2 matches the exterior value of rH_i up to the first n derivatives. The result is a simple function controlling the perturbation, and for each initial value of r the fluid streamlines can be plotted easily. An example of these streamlines is shown in figure 14.8.

If the system is allowed to evolve for a suitable amount of time, it provides

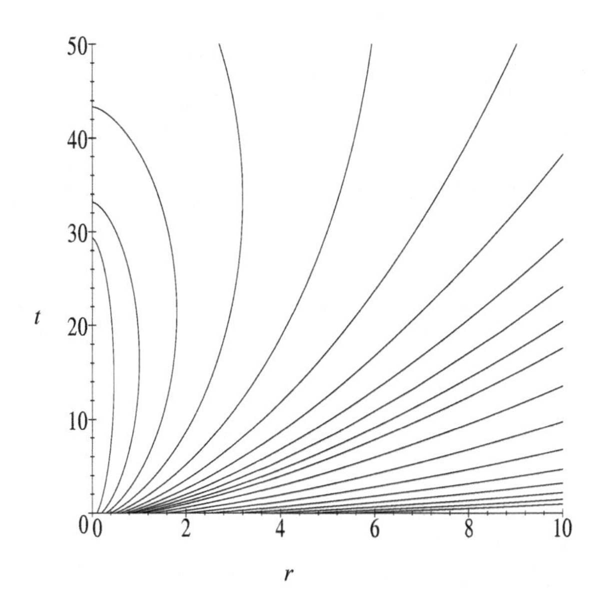

Figure 14.8 *Matter streamlines for an $n = 3$ model.* The perturbation has initial width 1, with $H_i = 1$ and $\rho_i = 3/8\pi$. The velocity gradient at the centre of the perturbation is 0.95. The central region is therefore moving inwards relative to the Hubble flow, so recollapses to a singularity after a finite time. All units are arbitrary.

a good model of a cluster of galaxies sitting inside a cosmological background. One can then study photon paths in this model, to look for lensing effects, or temperature perturbations in the cosmic microwave background. One weakness with these models is that no pressure is included, so the cluster has no means of supporting itself. This implies that a singularity forms after a finite term (determined by the central density and velocity gradient). The model then describes a black hole, sitting in an expanding universe.

14.5.3 The Dirac equation in a cosmological background

A good illustration of the full gravitational equations, with torsion included, is provided by the case of a Dirac field coupled self-consistently to gravity. The equations governing this system are

$$\begin{aligned} \mathcal{H}(a) &= 4\pi(\psi I\gamma_3\tilde{\psi})\cdot a, \\ \mathcal{G}(a) - \Lambda a &= 8\pi\langle a\cdot D\,\psi I\gamma_3\tilde{\psi}\rangle_1, \\ D\psi I\gamma_3 &= m\psi. \end{aligned} \tag{14.213}$$

This system of equations is highly non-linear and extremely difficult to analyse in all but the simplest of situations. Here we are interested in cosmological solutions, for which all fields are functions of time only. We also restrict our discussion to the spatially flat case ($k = 0$), so that we can write

$$\bar{\mathsf{h}}(a) = a + rH(t)a\cdot e_r\, e_t. \tag{14.214}$$

The ω function is given by

$$\omega(a) = H(t)a\wedge e_t + \tfrac{1}{2}\kappa a\cdot\mathcal{S}, \tag{14.215}$$

where $\kappa = 8\pi$ and $\mathcal{S}$ denotes the spin trivector:

$$\mathcal{S} = \tfrac{1}{2}\psi I\gamma_3\tilde{\psi}. \tag{14.216}$$

After a little work, the Einstein tensor evaluates to

$$\mathcal{G}(a) = 2\dot{H}a\wedge e_t\, e_t + 3H^2 a - \tfrac{1}{2}\kappa a\cdot(\mathcal{D}\cdot\mathcal{S}) + \tfrac{1}{2}\kappa^2 a\cdot\mathcal{S}\,\mathcal{S} - \tfrac{3}{4}\kappa^2\mathcal{S}^2 a, \tag{14.217}$$

and the matter energy-momentum tensor is

$$\mathcal{T}(a) = \langle a\cdot e_t\dot{\psi}I\gamma_3\tilde{\psi} + Ha\wedge e_t\,\mathcal{S} + \tfrac{1}{2}\kappa a\cdot\mathcal{S}\,\mathcal{S}\rangle_1. \tag{14.218}$$

Finally, the Dirac equation is now

$$(e_t\partial_t + \tfrac{3}{2}He_t + \tfrac{3}{4}\kappa\mathcal{S})\psi I\gamma_3 = m\psi, \tag{14.219}$$

which has the unusual feature of being nonlinear, due to the presence of the spin term.

We will construct the simplest solution to this system by setting the spinor ψ equal to a magnitude and phase only:

$$\psi = \rho(t)^{1/2}\mathrm{e}^{-I\boldsymbol{\sigma}_3\chi(t)}. \tag{14.220}$$

The Dirac equation therefore reduces to the pair of equations

$$\begin{aligned}\dot{\rho} &= -3\rho H,\\ \dot{\chi} &= 3\pi\rho + m.\end{aligned} \tag{14.221}$$

The Einstein equation yields the final pair of equations

$$\begin{aligned}3H^2 - 12\pi^2\rho^2 - 8\pi m\rho - \Lambda &= 0,\\ 2\dot{H} + 3H^2 + 12\pi^2\rho^2 - \Lambda &= 0.\end{aligned} \tag{14.222}$$

The second of these follows from the first and the equation for $\dot{\rho}$. These equations are solved by

$$\rho = \frac{\beta^2}{6\pi\sinh(\beta t)\big(m\sinh(\beta t) + \beta\cosh(\beta t)\big)}, \tag{14.223}$$

where

$$\beta = \frac{\sqrt{3\Lambda}}{2}. \tag{14.224}$$

The initial singularity is chosen to correspond to $t = 0$. The Hubble function is similarly given by

$$H(t) = \frac{\beta^2 + 2\beta\sinh^2(\beta t) + 2m\beta\sinh(\beta t)\cosh(\beta t)}{3\sinh(\beta t)\big(m\sinh(\beta t) + \beta\cosh(\beta t)\big)}. \tag{14.225}$$

The limit $\Lambda \mapsto 0$ is easily taken and gives rather simpler behaviour in the absence of a cosmological constant:

$$\rho(t) = \frac{1}{6\pi t(1+mt)}, \qquad H(t) = \frac{1+2mt}{3t(1+mt)}. \tag{14.226}$$

Antiparticle solutions can also be found, though these can have unusual properties. At large times the Hubble function tends to a constant value of $(\Lambda/3)^{1/2}$. This behaviour is typical of Λ cosmologies and leads to the surprising prediction that the universe will keep accelerating. The presence of a non-zero spin vector implies that these models break isotropy, but this fact is hidden from the line element, which remains isotropic. The spin direction is only seen by particles with non-zero spin, which interact directly with the torsion tensor.

14.6 Cylindrical systems

We now turn our attention to a different class of exact solutions — those exhibiting cylindrical symmetry. Such solutions can provide models for stringlike configurations, and some of the solutions are also appropriate for gravity in (2+1) dimensions. We first introduce cylindrical polar coordinates (t, ρ, ϕ, z), where

$$\rho = \big((x^1)^2 + (x^2)^2\big)^{1/2}, \qquad \tan(\phi) = \frac{x^2}{x^1} \tag{14.227}$$

and $x^\mu = \gamma^\mu \cdot x$. We use the symbol ρ for the cylindrical distance to avoid confusion with the radial coordinate r used throughout this chapter. When we come to describe the matter, the energy density is denoted ε in this section. The coordinate frame defined by cylindrical polar coordinate is

$$\begin{aligned} e_t &= \gamma_0, & e_\phi &= \rho(-\sin(\phi)\,\gamma_1 + \cos(\phi)\,\gamma_2), \\ e_\rho &= \cos(\phi)\,\gamma_1 + \sin(\phi)\,\gamma_2, & e_z &= \gamma_3, \end{aligned} \tag{14.228}$$

and we continue to write $\hat{\phi}$ for the unit vector e_ϕ/ρ. As a bivector basis we use the set $\{\sigma_\rho, \sigma_\phi, \sigma_3\}$, where

$$\sigma_\rho = e_\rho e_t, \qquad \sigma_\phi = \hat{\phi} e_t, \qquad \sigma_3 = e_z e_t. \tag{14.229}$$

We are interested in stationary fields that exhibit cylindrical symmetry. For these we can write a general $\bar{\mathsf{h}}$ function as

$$\begin{aligned} \bar{\mathsf{h}}(e^t) &= f_1 e^t + \rho f_2 e^\phi, & \bar{\mathsf{h}}(e^\rho) &= g_1 e^\rho, \\ \bar{\mathsf{h}}(e^\phi) &= \rho h_1 e^\phi + h_2 e^t, & \bar{\mathsf{h}}(e^z) &= e^z, \end{aligned} \tag{14.230}$$

where all of the arbitrary functions depend on ρ only. A suitable ω field consistent with this $\bar{\mathsf{h}}$ field is given by

$$\begin{aligned} \omega_t &= \omega(e_t) = -T\sigma_\rho + (K+h_2)I\sigma_3, \\ \omega_\rho &= \omega(e_\rho) = \bar{K}\sigma_\phi, \\ \omega_{\hat{\phi}} &= \omega(\hat{\phi}) = K\sigma_\rho + (h_1 - G)I\sigma_3, \\ \omega_z &= \omega(e_z) = 0. \end{aligned} \tag{14.231}$$

Again, the new scalar functions appearing here (T, K, $\bar{K}$, G) are functions of ρ alone. Since all expressions involving L_z must vanish, there are only three non-vanishing commutation relations to construct. These are

$$\begin{aligned} [L_\rho, L_t] &= TL_t + (K+\bar{K})L_{\hat{\phi}}, \\ [L_\rho, L_{\hat{\phi}}] &= -(K-\bar{K})L_t - GL_{\hat{\phi}}, \\ [L_t, L_{\hat{\phi}}] &= 0. \end{aligned} \tag{14.232}$$

Since neither L_t nor $L_{\hat{\phi}}$ contains derivatives with respect to ρ, the bracket relations immediately yield

$$\begin{aligned} L_\rho f_1 &= Tf_1 + (K+\bar{K})f_2, \\ L_\rho f_2 &= -Gf_2 - (K-\bar{K})f_1, \\ L_\rho h_1 &= -Gh_1 - (K-\bar{K})h_2, \\ L_\rho h_2 &= Th_2 + (K+\bar{K})h_1. \end{aligned} \tag{14.233}$$

The cylindrical derivative L_ρ is given by $L_\rho = g_1(\rho)\partial_\rho$. We can always make the position gauge choice $g_1 = 1$, though this is not always the simplest gauge to work with.

The Riemann tensor takes the general form

$$\begin{aligned} \mathcal{R}(\sigma_\rho) &= \alpha_1\sigma_\rho + \beta I\sigma_3, \\ \mathcal{R}(I\sigma_3) &= \alpha_2 I\sigma_3 - \beta\sigma_\rho, \\ \mathcal{R}(\sigma_\phi) &= \alpha_3\sigma_\phi, \end{aligned} \tag{14.234}$$

where the scalar functions are defined by

$$\begin{aligned} \alpha_1 &= -L_\rho T + T^2 - K(K+2\bar{K}), \\ \alpha_2 &= L_\rho G + G^2 - K(K-2\bar{K}), \\ \alpha_3 &= K^2 - GT, \\ \beta &= L_\rho K + G(K+\bar{K}) - T(K-\bar{K}). \end{aligned} \tag{14.235}$$

The same functions appear in the Einstein tensor,

$$\begin{aligned}
\mathcal{G}(e_t) &= -\alpha_2 e_t - \beta\hat{\phi}, \\
\mathcal{G}(e_\rho) &= -\alpha_3 e_\rho, \\
\mathcal{G}(\hat{\phi}) &= -\alpha_1\hat{\phi} + \beta e_t, \\
\mathcal{G}(e_z) &= -(\alpha_1 + \alpha_2 + \alpha_3)e_z.
\end{aligned} \tag{14.236}$$

It is a feature of gravity in $(2+1)$ dimensions that all of the information in the Riemann tensor is also contained in the Einstein tensor. That is, there is no Weyl tensor in three dimensions. It also turns out that no additional new information is obtained from the Bianchi identities, which are satisfied automatically from the equations we have already constructed.

The $\bar{\mathsf{h}}$ function of equation (14.230) contains a single rotational gauge freedom, which is the freedom to boost in the σ_ϕ plane. If we make the physical assumption that the matter energy-momentum tensor has a future-pointing timelike eigenvector, the gauge freedom can be used to set this eigenvector to the e_t direction. Once this is done all the rotational gauge freedom in the problem has been removed, and we are left with a complete set of field equations. These are

$$\begin{aligned}
-L_\rho G - G^2 + K(K - 2\bar{K}) &= 8\pi\varepsilon, \\
K^2 - GT &= 8\pi P_\rho, \\
-L_\rho T + T^2 - K(K + 2\bar{K}) &= 8\pi P_\phi, \\
L_\rho K + G(K + \bar{K}) - T(K - \bar{K}) &= 0,
\end{aligned} \tag{14.237}$$

where ε is the matter density, and P_ρ and P_ϕ are the radial and azimuthal pressures respectively. The coefficient of $\mathcal{G}(e_z)$ is determined algebraically by the other three coefficients, and the same must therefore be true of the matter energy-momentum tensor. It follows that the z-component of the Einstein equations contains no new information. Of course, if we were working in a genuine $(2+1)$ system, the e_z equation would not be present.

14.6.1 Vacuum solutions

In the vacuum region all of the scalars $\{\alpha_1, \alpha_2, \alpha_3, \beta\}$ are zero, so we are still free to perform an ρ-dependent boost in the σ_ϕ direction. This freedom can be employed to set $\bar{K}$ to zero. It is also useful in this region to work in a gauge where $g_1 = 1$. In this case the vacuum region is described by the simple pair of equations

$$\begin{aligned}
\partial_\rho G + G^2 - GT &= 0, \\
\partial_\rho T - T^2 + GT &= 0,
\end{aligned} \tag{14.238}$$

with K determined by $K^2 = GT$. On subtracting these equations and integrating we see that

$$G - T = 1/(\rho + \rho_0), \tag{14.239}$$

where ρ_0 is an arbitrary constant of integration. Similarly, adding the equations and integrating yields

$$G + T = c/(\rho + \rho_0), \tag{14.240}$$

where c is a second constant of integration.

The restriction that $GT = K^2 > 0$ means that $c^2 > 1$, and we can set

$$c = \pm \cosh(2\alpha). \tag{14.241}$$

There are two distinct vacuum configurations, depending on which sign is chosen for c. In either case, the constant α can be gauged to zero with a further constant boost in the σ_ϕ direction (which does not reintroduce a $\bar{K}$ term). The two vacuum sectors are therefore characterised by the solutions

$$\begin{aligned} &\text{type I:} \qquad && G = \frac{1}{\rho + \rho_0}, \quad T = K = \bar{K} = 0, \\ &\text{type II:} \qquad && T = -\frac{1}{\rho + \rho_0}, \quad G = K = \bar{K} = 0. \end{aligned} \tag{14.242}$$

All other vacuum solutions can be reached from this pair by ρ-dependent boosts in the σ_ϕ direction. No globally-defined gauge transformation exists between these solution classes. For both solutions the Riemann tensor vanishes, since there is no Weyl tensor for three-dimensional systems. It is therefore possible locally to gauge transform all of these fields to zero, but this is not possible globally. In this sense the solutions represent two distinct topological structures.

14.6.2 Physical properties of matter solutions

The key physical properties associated with matter solutions are the acceleration, vorticity, shear and angular momentum of the string. Given that we have chosen a gauge where the timelike eigenvector of the energy-momentum tensor is e_t, the acceleration vector w is defined by

$$w = e_t \cdot \mathcal{D} e_t = -T e_\rho. \tag{14.243}$$

This measures the extent to which particles comoving with the matter (with velocity e_t) depart from geodesic motion. The vorticity bivector ϖ is defined by

$$\varpi = \mathcal{D} \wedge e_t + w \wedge e_t = -(K - \bar{K}) I \sigma_3. \tag{14.244}$$

The definition ensures that ϖ satisfies $e_t \cdot \varpi = 0$. To define the shear tensor we require the linear function H that projects vectors into the 3-space orthogonal to e_t,

$$\mathsf{H}(a) = a - a \cdot e_t \, e_t. \tag{14.245}$$

In terms of this function the shear tensor $\sigma(a)$ is defined by

$$\begin{aligned}\sigma(a) &= \tfrac{1}{2}\big(\mathsf{H}(a)\cdot\mathcal{D}e_t + \mathsf{H}(\partial_b)(b\cdot\mathcal{D}e_t)\cdot a\big) - \tfrac{1}{3}\mathsf{H}(a)\mathcal{D}\cdot e_t \\ &= -\tfrac{1}{2}(K+\bar{K})(a\cdot e_\rho\,\hat{\phi} + a\cdot\hat{\phi}\,e_\rho).\end{aligned} \tag{14.246}$$

This is a symmetric, traceless linear function. We see that acceleration is controlled by T, the vorticity by $(K-\bar{K})$ and the shear by $(K+\bar{K})$. In the matter region all of these scalar quantities are physically measurable functions. The same is true of the fourth function, G, which can be determined from the radial pressure.

The remaining physical property of relevance is the angular momentum contained in the fields. The vector g_ϕ is a Killing vector for cylindrical solutions, so the vector $\mathcal{T}(g_\phi)$ is covariantly conserved. It follows that

$$\nabla\cdot\big(\mathsf{h}(\mathcal{T}(g_\phi))\det(\mathsf{h})^{-1}\big) = 0. \tag{14.247}$$

The total conserved angular momentum per unit length in the e_t frame is therefore given by the expression

$$J_S = \int_0^{\rho_s} d^2x\, g^t\cdot\mathcal{T}(g_\phi)\det(\mathsf{h})^{-1}, \tag{14.248}$$

where ρ_s is the string radius. In the $g_1 = 1$ gauge this expression evaluates to give

$$J_S = -2\pi\int_0^{\rho_s} d\rho\,(\varepsilon + P_\phi)f_1 f_2(f_1h_1 - f_2h_2)^{-2}, \tag{14.249}$$

which shows that a non-zero f_2 is required for angular momentum to be present.

14.6.3 Cosmic strings

Cosmic strings are an example of topological defects that can occur as a remnant of symmetry breaking processes in the early universe. They have zero radial and azimuthal pressures. It follows that there is a negative pressure along the length of the string — they are under *tension*. The energy-momentum tensor is

$$\mathcal{T}(a) = \tfrac{1}{2}\varepsilon(a - I\sigma_3\, a\, I\sigma_3). \tag{14.250}$$

From the Einstein equations we see that $\alpha_1 = \alpha_3 = \beta = 0$, and the Riemann tensor therefore has the compact form

$$\mathcal{R}(B) = 8\pi\varepsilon\langle BI\sigma_3\rangle I\sigma_3. \tag{14.251}$$

Tidal forces are only exerted in the $I\sigma_3$ plane and are controlled by the density.

The Einstein equations tell us that $T = K = \bar{K} = 0$, so all that remains is the single equation

$$L_\rho G + G^2 = -8\pi\varepsilon. \tag{14.252}$$

The full solution is then recovered by integrating the bracket equations (14.233). These imply that both f_1 and h_2 are constant. A global rotation can therefore be performed to transform to a gauge where $f_1 = 1$ and $h_2 = 0$. The remaining equations are

$$L_\rho h_1 = -G h_1, \qquad L_\rho f_2 = -G f_2. \tag{14.253}$$

It follows that $f_2 = \lambda h_1$, where λ is an arbitrary constant. But ρh_1 must tend to 1 as $\rho \mapsto 0$ so that $\bar{\mathsf{h}}(a)$ is well defined on the axis. It follows that h_1, and hence f_2, must diverge as ρ^{-1}. For f_2 this would imply that $\bar{\mathsf{h}}(e^t)$ is singular on the axis, which is not permitted. It follows that the constant λ must be zero, so the string has no angular momentum. This agrees with the fact that the shear and vorticity are both zero. Pressure is necessary for strings to have any angular momentum.

We have now restricted $\bar{\mathsf{h}}(a)$ to the simple form

$$\bar{\mathsf{h}}(a) = a + (g_1 - 1) a \cdot e_\rho \, e^\rho + (\rho h_1 - 1) a \cdot e_\phi \, e^\phi, \tag{14.254}$$

and the remaining equations are

$$L_\rho h_1 = -G h_1, \qquad L_\rho G = -8\pi\varepsilon - G^2, \tag{14.255}$$

with $L_\rho = g_1 \partial_\rho$. To complete the solution we must make a gauge choice for g_1. An obvious choice is to set $g_1 = 1$, so that ρ measures the proper radial distance from the string. A slightly simpler alternative is to choose a gauge such that $\bar{\mathsf{h}}(e^\phi) = e^\phi$. This requires that

$$h_1 = 1/\rho \tag{14.256}$$

and it follows that

$$G = g_1/\rho. \tag{14.257}$$

The equations now integrate to give

$$g_1{}^2 = 1 - \int_0^\rho 16\pi s \varepsilon(s) \, ds, \tag{14.258}$$

where the constant of integration is chosen so that $\bar{\mathsf{h}}(a)$ is well defined on the axis. On defining

$$M(\rho) = \int_0^\rho 2\pi s \varepsilon(s) \, ds, \tag{14.259}$$

the solution can be summarised neatly by

$$\bar{\mathsf{h}}(a) = a + \big((1 - 8M(\rho))^{1/2} - 1\big) a \cdot e_\rho \, e^\rho. \tag{14.260}$$

The choice of density function is arbitrary, provided $8M(\rho) < 1$. In the vacuum region outside the string we have $T = K = \bar{K} = 0$, so the vacuum region is described by a solution in the gauge class of type I. This can be described in terms of a flat spacetime, with a wedge of spacetime removed and the edges identified. This topological picture of a string defect can be used to provide a qualitative understanding of many of the string's properties.

14.6.4 Rigidly rotating strings

The simplest models that include pressure are those for a two-dimensional ideal fluid, with $P_\rho = P_\phi = P$. The two natural physical models to consider are those where the fluid is vorticity-free ($\bar{K} = K$) or shear-free ($\bar{K} = -K$). The latter case corresponds to a rigidly rotating string, and is the situation we analyse here. The equations governing this setup are (in the $g_1 = 1$ radial gauge)

$$\begin{aligned} \partial_\rho K - 2KT &= 0, \\ \partial_\rho G + G^2 &= -8\pi\varepsilon + 3K^2, \\ \partial_\rho T - T^2 &= -8\pi P + K^2, \\ K^2 - GT &= 8\pi P. \end{aligned} \tag{14.261}$$

These can be solved once the density distribution has been specified. A choice of density that produces a straightforward solution is

$$8\pi\varepsilon = 3K^2 + \lambda^2, \tag{14.262}$$

where λ is an arbitrary positive constant. This ansatz ensures that the density is always positive. The equations for G and T can be solved immediately to give

$$G = \frac{\lambda\cos(\lambda\rho)}{\sin(\lambda\rho)}, \qquad T = \frac{\lambda\sin(\lambda\rho)}{\cos(\lambda\rho) + A}, \tag{14.263}$$

where A is a constant satisfying $A < -1$.

We next solve for K to obtain

$$K = \frac{B}{(A + \cos(\lambda\rho))^2}, \tag{14.264}$$

where B is a further constant. The density and pressure can now be recovered from equations (14.261). The boundary of the string occurs where the pressure vanishes, and this must be reached before $\rho = \pi/\lambda$. Finally, we return to equations (14.233) to find a suitable form for the $\bar{\mathsf{h}}$ function. First we see that f_1/h_2 is a constant, so that a gauge transformation can be performed to set $h_2 = 0$.

The remaining functions are easily found by integration:

$$\begin{aligned} f_1 &= \frac{1+A}{\cos(\lambda\rho)+A}, \\ h_1 &= \frac{\lambda}{\sin(\lambda\rho)}, \\ f_2 &= \frac{-B({f_1}^2-1)}{\lambda(A+1)\sin(\lambda\rho)}. \end{aligned} \tag{14.265}$$

For f_1 the arbitrary time-scale factor has been used to set $f_1 = 1$ on the axis. It is simple to verify that this solution is well defined on the axis of the string. For completeness, the corresponding line element is

$$\begin{aligned} ds^2 &= \frac{\big(\cos(\lambda\rho)+A\big)^2}{(1+A)^2}\,dt^2 + \frac{2B}{\lambda^2(A+1)^3}\big(1-\cos(\lambda\rho)\big)\big(2A+1+\cos(\lambda\rho)\big)\,dt\,d\phi \\ &- \frac{\sin^2(\lambda\rho)}{\lambda^2}\left(1-\frac{B^2\big(1-\cos(\lambda\rho)\big)^2\big(2A+1+\cos(\lambda\rho)\big)^2}{\lambda^2\sin^2(\lambda\rho)(1+A)^4\big(A+\cos(\lambda\rho)\big)^2}\right)d\phi^2 - d\rho^2 - dz^2. \end{aligned} \tag{14.266}$$

The exterior vacuum fields can be found simply by returning to the vacuum equations, and solving these in the case where $K+\bar{K}=0$. The general form of vacuum fields outside a rigidly rotating string is then given by

$$\begin{aligned} G &= \frac{-\alpha^2}{(\rho+\rho_0)\big((\rho+\rho_0)^2-\alpha^2\big)}, \\ T &= -\frac{\rho+\rho_0}{(\rho+\rho_0)^2-\alpha^2}, \\ K &= \frac{\alpha}{(\rho+\rho_0)^2-\alpha^2}, \end{aligned} \tag{14.267}$$

where ρ_0 and α are constants to be determined by the fields at the boundary. This solution falls into the second class of vacuum solutions, as defined by equation (14.242). The h function is determined by

$$\begin{aligned} f_1 &= -(1+A)(\alpha/B)^{1/2}\big((\rho+\rho_0)^2-\alpha^2\big)^{-1/2}, \\ h_1 &= (\alpha/B)^{1/2}\lambda^2\frac{\big((\rho+\rho_0)^2-\alpha^2\big)^{1/2}}{(\rho+\rho_0)}, \\ f_2 &= \frac{\alpha}{f_1(\rho+\rho_0)}(f_1^2-1). \end{aligned} \tag{14.268}$$

These fields have an unusual property. At large distances, f_1 falls off as ρ^{-1}, whereas f_2 tends to a constant value. Beyond the point where the magnitude of f_2 overtakes that of f_1, a closed circular path orbiting the string becomes *timelike*. This solution admits closed timelike curves, even out at infinity. Such solutions are often thought of as unphysical, due to the bizarre acausal effects they would allow. But there is nothing outrageous in the matter distribution used to generate the solution, and it is difficult to pin down a precise statement of what constitutes a 'physically acceptable' matter distribution.

14.7 Axially-symmetric systems

As a further application of the gauge theory treatment of gravity, we now turn to the equations governing a stationary axisymmetric system. Such fields are produced by rotating stars, galaxies and black holes, and as such are of considerable importance in astrophysics. The prototype axisymmetric configuration is described by the *Kerr solution*, which uniquely describes the fields produced by an uncharged rotating black hole. The more complicated problem of finding the fields outside a rotating massive object such as a star or planet has yet to be fully solved. Here we discuss two forms of the Kerr solution. The first continues the solution strategy adopted in the cylindrical setup, and can be generalised to include matter fields. The second form generalises the Newtonian gauge for the Schwarzschild solution, and has a number of significant features.

14.7.1 Intrinsic form of the axisymmetric equations

We employ a standard spherical-polar coordinate system to describe axisymmetric fields, and the notation is precisely as defined at the start of section 14.2. A suitable form of the $\bar{\mathsf{h}}$ function consistent with axial symmetry is

$$\begin{aligned}
\bar{\mathsf{h}}(e^t) &= f_1 e^t + f_4 e^\phi, \\
\bar{\mathsf{h}}(e^r) &= g_1 e^r + g_3 e^\theta, \\
\bar{\mathsf{h}}(e^\theta) &= i_1 e^\theta + i_3 e^r, \\
\bar{\mathsf{h}}(e^\phi) &= h_1 e^\phi + h_2 e^t,
\end{aligned} \tag{14.269}$$

where all of the variables $\{f_1, \ldots, i_3\}$ are scalar functions of r and θ. The labelling convention for the $\{f_i, \ldots, i_i\}$ is chosen to allow for a more general parameterisation appropriate for time-dependent systems. We have ignored the possibility of any coupling between the e^t and e^r, so strictly speaking are looking for the fields outside an extended source with no horizon present. On solving the vacuum field equations we will construct a form of the Kerr solution, which will turn out to be ill defined at the horizon. As with the Schwarzschild solution, the singular nature of the fields is a consequence of a bad gauge choice, rather than an intrinsic property of the fields. In section 14.7.3 we give a form of the Kerr solution which avoids this problem.

A suitably general form of ω function consistent with the $\bar{\mathsf{h}}$ field of equation (14.269) is given by

$$\begin{aligned}
\omega(e_t) &= -(T+IJ)e_r e_t - (S+IK)\hat{\theta} e_t + h_2 I\sigma_3, \\
\omega(e_r) &= (S'+IK')e_r\hat{\theta} - i_3 e_r\hat{\theta}, \\
\omega(\hat{\theta}) &= (G'+IJ')e_r\hat{\theta} - (i_1/r)e_r\hat{\theta}, \\
\omega(\hat{\phi}) &= (H+IK)\hat{\theta}\hat{\phi} + (G+IJ)e_r\hat{\phi} + h_1/(r\sin(\theta))\, I\sigma_3.
\end{aligned} \tag{14.270}$$

The variables written in capitals are also functions of r and θ, except for the pseudoscalar I. The reason for the labelling scheme will become clearer when the final set of equations is derived. There are 40 independent scalar variables in gravity, so it is difficult to construct a labelling scheme that does not conflict with existing conventions somewhere. A significant feature of our scheme is that a complex structure naturally emerges, generated by the pseudoscalar I. It is a well-known feature of the Kerr solution that it is underpinned by a complex analytic structure. The origin of this lies in the natural complex structure of spacetime bivectors. Throughout this section we use *complex* to refer to a combination of scalar and pseudoscalar quantities.

The bracket structure defined by our choice of the ω function is

$$\begin{aligned}
[L_t, L_r] &= -TL_t - (K+K')L_{\hat{\phi}}, & [L_r, L_{\hat{\theta}}] &= -S'L_r - G'L_{\hat{\theta}}, \\
[L_t, L_{\hat{\theta}}] &= -SL_t + (J-J')L_{\hat{\phi}}, & [L_r, L_{\hat{\phi}}] &= -(K-K')L_t - GL_{\hat{\phi}}, \\
[L_t, L_{\hat{\phi}}] &= 0, & [L_{\hat{\theta}}, L_{\hat{\phi}}] &= (J+J')L_t - HL_{\hat{\phi}}.
\end{aligned} \tag{14.271}$$

The Riemann tensor generated by these fields is complicated and, rather than giving its full algebraic expression, it is simpler to consider the general form. This can be written as

$$\begin{aligned}
\mathcal{R}(\sigma_r) &= \alpha_1\sigma_r + \beta_1\sigma_\theta, & \mathcal{R}(I\sigma_r) &= \alpha_4 I\sigma_r + \beta_4 I\sigma_\theta, \\
\mathcal{R}(\sigma_\theta) &= \alpha_2\sigma_\theta + \beta_2\sigma_r, & \mathcal{R}(I\sigma_\theta) &= \alpha_5 I\sigma_\theta + \beta_5 I\sigma_r, \\
\mathcal{R}(\sigma_\phi) &= \alpha_3\sigma_\phi, & \mathcal{R}(I\sigma_\phi) &= \alpha_6 I\sigma_\phi,
\end{aligned} \tag{14.272}$$

where each of the α_i and β_i is a complex combination. If we now specialise to the case of vacuum solutions, so that the Riemann tensor is determined solely by the Weyl tensor, the duality relation $\mathcal{W}(IB) = I\mathcal{W}(B)$ immediately sets

$$\alpha_1 = \alpha_4, \quad \alpha_2 = \alpha_5, \quad \alpha_3 = \alpha_6, \quad \beta_1 = \beta_4 \quad \beta_2 = \beta_5. \tag{14.273}$$

In addition, for a vacuum solution $\mathcal{R}(B)$ must be symmetric and traceless. The most general form of tensor consistent with this requirement is

$$\begin{aligned}
\mathcal{R}(\sigma_r) &= \alpha_1\sigma_r + \beta\sigma_\theta, \\
\mathcal{R}(\sigma_\theta) &= \alpha_2\sigma_\theta + \beta\sigma_r, \\
\mathcal{R}(\sigma_\phi) &= -(\alpha_1+\alpha_2)\sigma_\phi,
\end{aligned} \tag{14.274}$$

with α_i and β complex combinations.

Next we consider the rotational gauge freedom in our choice of axisymmetric fields. We are free to perform a rotation in the $I\sigma_\phi$ plane, and a boost in the σ_ϕ direction. These can be summarised in the single rotor R:

$$R = \exp(wI\sigma_\phi/2), \tag{14.275}$$

where the scalar + pseudoscalar quantity w is an arbitrary function of (r, θ). This gauge freedom can be employed to diagonalise the Riemann tensor by setting $\beta = 0$. This removes all of the gauge freedom present, and enables us to write

$$\mathcal{R}(\sigma_r) = \alpha_1\sigma_r, \quad \mathcal{R}(\sigma_\theta) = \alpha_2\sigma_\theta, \quad \mathcal{R}(\sigma_\phi) = -(\alpha_1 + \alpha_2)\sigma_\phi. \tag{14.276}$$

The form of the Riemann tensor for the Schwarzschild solution is algebraically special, in that two of its eigenvalues are degenerate. This is referred to as having Petrov type D. There is no reason to expect the same to be true for axisymmetric fields, and the field outside a general rotating star is almost certainly not of type D. But it turns out that, if a horizon is present, the solution must be of type D. As we are interested here in deriving the Kerr solution, we therefore impose the additional condition that the Riemann tensor is degenerate, with the general algebraic form

$$\mathcal{R}(B) = \frac{\alpha}{2}(B + 3\sigma_r B\sigma_r), \tag{14.277}$$

with α a scalar + pseudoscalar quantity. This final restriction on the form of $\mathcal{R}(B)$ is *not* a gauge choice — it is a restriction on the form of solution we can construct.

Comparing the general form of equation (14.277) with the explicit Riemann tensor constructed from the ω field, we establish that

$$\alpha = (G + IJ)(T + IJ) + (S + IK)(H + IK). \tag{14.278}$$

The remaining identities reduce to a series of equations, an example of which is

$$\begin{aligned} L_r(G+IJ) =& (S' + IK' - S - IK)(H + IK) - I(K - K')(S + IK) \\ & - (G + T + IJ)(G + IJ). \end{aligned} \tag{14.279}$$

In all there are ten equations of this type. They all relate intrinsic derivatives of the variables in the ω field to quadratic combinations of the same variables. By forming suitable combinations of these equations we find that

$$L_r\alpha = -3\alpha(G + IJ), \qquad L_{\hat{\theta}}\alpha = -3\alpha(S + IK), \tag{14.280}$$

so the intrinsic derivatives of α are quite simple.

Next we must consider the Bianchi identities. These contain higher order consistency relations between the $\bar{\mathsf{h}}$ and ω fields. For the Schwarzschild and cylindrical cases these contained no new information, but this is not the case for the axisymmetric setup. If we consider the equation

$$\mathcal{D}\mathcal{R}(\sigma_r) - \partial_a \mathcal{R}(a \cdot \mathcal{D}\sigma_r) = 0 \tag{14.281}$$

we obtain the pair of equations

$$\begin{aligned} L_r \alpha &= -\frac{3}{2}\alpha(G + IJ + G' + IJ'), \\ L_{\hat{\theta}} \alpha &= -\frac{3}{2}\alpha(S + IK + S' + IK'), \end{aligned} \tag{14.282}$$

Comparing these with equation (14.280), we see that

$$G' + IJ' = G + IJ, \qquad S' + IK' = S + IK. \tag{14.283}$$

This simplification for type D fields explains our choice of notation of primed and unprimed variables.

With four of the variables now solved for, the remaining equations simplify to

$$\begin{aligned} L_r(G + IJ) &= -(G + IJ)^2 - T(G + IJ), \\ L_r(T + IJ) &= (S + IK)^2 - \big(2(G + IJ) - T\big)(T + IJ) \\ &\quad - 2S(H + IK), \\ L_r(S + IK) &= -IJ(S + IK) - 2IK(G + IJ), \\ L_r(H + IK) &= -(G + IJ)(S + IK) - G(H + IK) \end{aligned} \tag{14.284}$$

and

$$\begin{aligned} L_{\hat{\theta}}(S + IK) &= (S + IK)^2 + H(S + IK), \\ L_{\hat{\theta}}(H + IK) &= -(G + IJ)^2 + \big(2(S + IK) - H\big)(H + IK) \\ &\quad + 2G(T + IJ), \\ L_{\hat{\theta}}(G + IJ) &= IK(G + IJ) + 2IJ(S + IK), \\ L_{\hat{\theta}}(T + IJ) &= (G + IJ)(S + IK) + S(T + IJ). \end{aligned} \tag{14.285}$$

These equations are all consistent with the bracket structure, which now takes the form

$$[L_r, L_{\hat{\theta}}] = -SL_r - GL_{\hat{\theta}}. \tag{14.286}$$

Our set of equations is now complete. We have explicit forms for the intrinsic derivatives of all of our variables; these are all consistent with the bracket structure, and the full Bianchi identities are all satisfied. We have achieved the first main goal of the intrinsic method.

14.7.2 The Kerr solution

The vacuum equations summarised in equations (14.284) and (14.285) display a number of remarkable features. They are naturally complex, with the spacetime pseudoscalar as the unit imaginary, and there is a clear symmetry between the r and $\hat{\theta}$ equations. We now demonstrate that, subject to certain boundary conditions, these equations admit a unique, two-parameter family of solutions. This is the Kerr solution. The proof is constructive, but it is slightly involved and we will skip some of the details.

The first step in solving a set of intrinsic equations is the identification of suitable integrating factors. To find the first of these consider the function

$$Z = Z_0 \alpha^{-1/3}, \tag{14.287}$$

where Z_0 is an arbitrary complex constant. The function Z satisfies

$$L_r Z = (G + IJ)Z, \qquad L_{\hat{\theta}} Z = -(S + IK)Z. \tag{14.288}$$

On separating Z into modulus X and argument χ,

$$Z = X \mathrm{e}^{I\chi} \tag{14.289}$$

we find that

$$L_r X = GX, \qquad L_{\hat{\theta}} X = -SX. \tag{14.290}$$

It follows that X acts as an integrating factor for G and S. But if we recall the bracket of equation (14.286), we see that

$$[XL_r, XL_{\hat{\theta}}] = 0. \tag{14.291}$$

We have therefore constructed a pair of commuting derivations. This is sufficient to ensure that we can fix our displacement gauge freedom by setting $g_3 = i_3 = 0$. With this done, we can then write

$$XL_r = g(r)\partial_r, \quad XL_{\hat{\theta}} = i(\theta)\partial_\theta, \tag{14.292}$$

where $g(r)$ and $i(\theta)$ are arbitrary functions that we can choose with further gauge fixing.

More generally, if a pair of variables A and B satisfy the equation

$$L_{\hat{\theta}} A - L_r B = GB + SA \tag{14.293}$$

then an integrating factor C exists defined (up to an arbitrary magnitude) by

$$L_r C = AC, \qquad L_{\hat{\theta}} C = BC. \tag{14.294}$$

One such pair is T and $-H$. For these we define the integrating factor F, satisfying

$$L_r F = TF, \qquad L_{\hat{\theta}} F = -HF. \tag{14.295}$$

With the integrating factors X, Z and F at our disposal, we can considerably simplify our equations for $G+IJ$ and $S+IK$ to obtain

$$\begin{aligned} L_r\big(FZ(G+IJ)\big) &= 0, \qquad & L_{\hat{\theta}}\big(XZ(G+IJ)\big) &= -2XZ(SG+JK), \\ L_{\hat{\theta}}\big(FZ(S+IK)\big) &= 0, \qquad & L_r\big(XZ(S+IK)\big) &= 2XZ(SG+JK). \end{aligned} \tag{14.296}$$

These equations focus attention on the quantity $SG+JK$. On forming the derivatives of this quantity we see that

$$L_r\big(XF(SG+JK)\big) = L_{\hat{\theta}}\big(XF(SG+JK)\big) = 0, \tag{14.297}$$

and it follows that $XF(SG+JK)$ is a constant. For the Schwarzschild solution this constant is zero. We therefore expect that this term should also vanish for a rotating source since, at large distances, the fields should tend to the Schwarzschild case. It turns out that one can construct solutions with $XF(SG+JK) \neq 0$, but these are appropriate for an infinite disc of matter and not a localised source. As we are looking for the fields outside a localised rotating source, we can set

$$SG+JK = 0. \tag{14.298}$$

It follows that

$$XFZ^2(G+IJ)(S+IK) = C_1, \tag{14.299}$$

where C_1 is an arbitrary complex constant.

Remarkably, we are now close to a complete solution to the problem. Equation (14.296) tells us that we can set

$$FZ(G+IJ) = W(\theta), \qquad FZ(S+IK) = U(r), \tag{14.300}$$

where U and W are complex functions of r and θ respectively. If we now form

$$\frac{W(\theta)}{U(r)} = \frac{G+IJ}{S+IK} = I\frac{SJ-GK}{S^2+K^2}, \tag{14.301}$$

we see that the result is a pure imaginary quantity. It follows that W and U are $\pi/2$ out of phase and, since U and W are separately functions of r and θ, their phases must be constant. Next we construct the derivatives of Z to obtain

$$XL_rZ = g(r)\partial_rZ = XZ(G+IJ) = C_1/U(r) \tag{14.302}$$

and

$$XL_{\hat{\theta}}Z = i(\theta)\partial_\theta Z = XZ(S+IK) = C_1/W(\theta). \tag{14.303}$$

It follows that Z must be the sum of a function of r and a function of θ. Furthermore, these functions must also have constant phases, $\pi/2$ apart. Since the overall phase of Z is arbitrary (Z was defined up to an arbitrary complex scale factor), we can write

$$Z = R(r) + I\Psi(\theta), \tag{14.304}$$

where $R(r)$ and $\Psi(\theta)$ are real functions. These satisfy the equation

$$XL_r\big(XL_r \ln(Z)\big) + XL_{\hat{\theta}}\big(XL_{\hat{\theta}} \ln(Z)\big) = \alpha = \frac{Z_0^3}{Z^3}, \tag{14.305}$$

which is to be solved for R and Ψ.

There is considerable gauge freedom in equation (14.305), since we are free to choose the functions $g(r)$ and $i(\theta)$. The most convenient choice of gauge is to set

$$Z = r - Ia\cos(\theta). \tag{14.306}$$

The remaining functions are then found by integration. The end result, after a series of further gauge choices, is the Kerr solution in the form

$$\begin{aligned}
\bar{\mathsf{h}}(e^t) = g^t &= \frac{r^2+a^2}{\rho\Delta^{1/2}}e^t + \frac{ar\sin^2(\theta)}{\rho}e^\phi, \\
\bar{\mathsf{h}}(e^r) = g^r &= \frac{\Delta^{1/2}}{\rho}e^r, \\
\bar{\mathsf{h}}(e^\theta) = g^\theta &= \frac{r}{\rho}e^\theta, \\
\bar{\mathsf{h}}(e^\phi) = g^\phi &= \frac{r}{\rho}e^\phi + \frac{a}{\rho\Delta^{1/2}}e^t,
\end{aligned} \tag{14.307}$$

where

$$\rho^2 = X^2 = r^2 + a^2\cos^2(\theta) \tag{14.308}$$

and

$$\Delta = r^2 - 2Mr + a^2. \tag{14.309}$$

The mass is given by M, and the angular momentum by aMc. The quantity a is the angular momentum per unit mass, and has dimensions of distance. The limit $a \mapsto 0$ recovers the Schwarzschild solution in the form appropriate for the exterior of a non-rotating star. The reciprocal vectors are

$$\begin{aligned}
g_t &= \frac{\Delta^{1/2}}{\rho}e_t - \frac{a}{r\rho}e_\phi, \\
g_r &= \frac{\rho}{\Delta^{1/2}}e_r, \\
g_\theta &= \frac{\rho}{r}e_\theta, \\
g_\phi &= \frac{r^2+a^2}{r\rho}e_\phi - \frac{a\Delta^{1/2}\sin^2(\theta)}{\rho}e_t.
\end{aligned} \tag{14.310}$$

The variables controlling the ω field are given by

$$\begin{aligned} G+IJ &= \frac{\Delta^{1/2}}{\rho\big(r-Ia\cos(\theta)\big)}, \\ S+IK &= \frac{-Ia\sin(\theta)}{\rho\big(r-Ia\cos(\theta)\big)}, \\ T-G &= -\frac{r-M}{\rho\Delta^{1/2}}, \\ H-S &= \frac{\cos(\theta)}{\rho\sin(\theta)}. \end{aligned} \tag{14.311}$$

The equation for T shows that a horizon exists where $\Delta = 0$. The fact that the solution is singular there is a reflection of our choice of time coordinate. This measures the time for observers at a constant distance from the source. Such observers cannot exist inside the horizon, and the solution breaks down there. As with the Schwarzschild system, the resolution of this problem is to express the fields in terms of a different time coordinate.

The Riemann tensor for the Kerr solution can now be written in the compact form

$$\mathcal{R}(B) = -\frac{M}{2\big(r-Ia\cos(\theta)\big)^3}(B+3\sigma_r B\sigma_r). \tag{14.312}$$

This is obtained from the Schwarzschild solution by simply replacing r by the scalar + pseudoscalar combination $r - Ia\cos(\theta)$. Precisely such a replacement can be used to generate the Kerr solution using a 'complex coordinate transformation' in the Newman–Penrose formalism. This transformation does produce the Kerr solution, but there is no *a priori* reason to expect that such a transformation applied to a vacuum solution will generate a new vacuum solution. Our extremely compact form of the Riemann tensor for the Kerr solution is a significant advantage of the gauge theory approach to gravitation advocated in this book. The comparison with the standard tensor formulation of general relativity is dramatic — most textbooks devote nearly a page to listing all of the components of the Riemann tensor, if they list them at all.

14.7.3 A Newtonian gauge for the Kerr solution

The form of the Kerr solution developed in the preceding section gives rise to a metric that expresses the geometry in terms of Boyer–Lindquist coordinates. Such a form is only appropriate for the region outside an extended object. If a horizon has formed we must find an alternative gauge choice which covers the horizon smoothly. From our discussion of the Schwarzschild solution, we would like to find an analogue of the Newtonian gauge appropriate for rotating black holes. Such a gauge does exist, though it is not straightforwardly obtained from the Boyer–Lindquist setup.

The first step in expressing the Kerr solution in a Newtonian gauge is the introduction of spheroidal coordinates $(\bar{r}, \bar{\theta}, \phi)$, as described in section 6.2.2. The spheroidal coordinates are related to their spherical counterparts (r, θ, ϕ) as follows:

$$\begin{aligned} (\bar{r}^2 + a^2)^{1/2} \sin(\bar{\theta}) &= r \sin(\theta), \\ \bar{r} \cos(\bar{\theta}) &= r \cos(\theta). \end{aligned} \tag{14.313}$$

The scalar parameter a is the same as that controlling the angular momentum. In the limit $a \mapsto 0$, the barred coordinates reduce to their unbarred spherical-polar equivalents. Surfaces of constant $\bar{r}$ are ellipses in flat space, though a statement such as this relates to the properties of the coordinate system, and not necessarily to physically measurable features. It is convenient to introduce the hyperbolic coordinate u, defined by

$$a \sinh(u) = \bar{r}. \tag{14.314}$$

The coordinate frame vectors are given by

$$\begin{aligned} e_{\bar{r}} &= \tanh(u) \sin(\bar{\theta}) \big(\cos(\phi)\, \gamma_1 + \sin(\phi)\, \gamma_2\big) + \cos(\bar{\theta})\, \gamma_3, \\ e_{\bar{\theta}} &= a \cosh(u) \cos(\bar{\theta}) \big(\cos(\phi)\, \gamma_1 + \sin(\phi)\, \gamma_2\big) - a \sinh(u) \sin(\bar{\theta})\, \gamma_3 \end{aligned} \tag{14.315}$$

with e_ϕ unchanged from its spherical definition. We also define the unit vectors

$$\hat{e}_{\bar{r}} = \frac{a \cosh(u)}{\bar{\rho}} e_{\bar{r}}, \qquad \hat{e}_{\bar{\theta}} = \frac{1}{\bar{\rho}} e_{\bar{\theta}}, \tag{14.316}$$

where $\bar{\rho}$ is defined by

$$\bar{\rho}^2 = a^2 \sinh^2(u) + a^2 \cos^2(\bar{\theta}) = \bar{r}^2 + a^2 \cos^2(\bar{\theta}). \tag{14.317}$$

The unit frame vectors satisfy

$$e_t \hat{e}_{\bar{r}} \hat{e}_{\bar{\theta}} \hat{\phi} = I. \tag{14.318}$$

The Newtonian gauge form of the Schwarzschild solution, defined in equation (14.65), contains the unit vectors e_t and e_r. The generalisation of this function to the Kerr solution is given by

$$\bar{\mathsf{h}}(n) = n - \left(\frac{2M\bar{r}}{\bar{r}^2 + a^2}\right)^{1/2} n \cdot \hat{e}_{\bar{r}}\, v, \tag{14.319}$$

where the vector argument is denoted by n to avoid confusion with the scalar parameter a. The timelike velocity vector v is defined by

$$v = \cosh(\beta)\, e_t + \sinh(\beta)\, \hat{\phi} \tag{14.320}$$

where

$$\tanh(\beta) = \frac{\sin(\bar{\theta})}{\cosh(u)} = \frac{ar\sin(\theta)}{\bar{r}^2 + a^2}. \tag{14.321}$$

It follows that

$$\cosh(\beta) = \frac{a\cosh(u)}{\bar{\rho}}, \qquad \sinh(\beta) = \frac{a\sin(\bar{\theta})}{\bar{\rho}}. \tag{14.322}$$

Comparison with equation (14.65) shows how the various terms are generalised in moving from the Schwarzschild to the Kerr solution.

The $\omega(a)$ function generated by equation (14.319) has

$$\begin{aligned}
\omega(e_t) &= 0,\\
\omega(\hat{e}_{\bar{r}}) &= -\frac{M}{\alpha(\bar{r} - Ia\cos(\bar{\theta}))^2}\,\hat{e}_{\bar{r}}\wedge v,\\
\omega(\hat{e}_{\bar{\theta}}) &= \frac{\alpha}{\bar{r} - Ia\cos(\bar{\theta})}\,\hat{e}_{\bar{\theta}}\wedge v,\\
\omega(\hat{\phi}) &= \frac{\alpha}{\cosh(\beta)(\bar{r} - Ia\cos(\bar{\theta}))}\,\sigma_\phi,
\end{aligned} \tag{14.323}$$

where

$$\alpha = -\frac{(2M\bar{r})^{1/2}}{\bar{\rho}}. \tag{14.324}$$

The terms in the ω function also neatly generalise their counterparts in the Schwarzschild solution. In particular, the fact that $\omega(e_t)$ vanishes implies that e_t satisfies the geodesic equation. The trajectories defined by this velocity define a family of observers whose proper time is given by t.

The remaining covariant object to construct is the Riemann tensor. If we define the unit bivector

$$\hat{N} = \hat{e}_{\bar{r}}\wedge v, \tag{14.325}$$

then the Riemann tensor takes on the simple form

$$\mathcal{R}(B) = -\frac{M}{2\big(\bar{r} - Ia\cos(\bar{\theta})\big)^3}(B + 3\hat{N}B\hat{N}). \tag{14.326}$$

This is obtained from the form of equation (14.312) by a displacement (taking the unbarred to the barred coordinates) and a boost from e_t to v. Both are gauge transformations, so the intrinsic information in equations (14.312) and (14.326) is precisely the same. The same transformations are involved in taking the $\bar{\mathsf{h}}(a)$ function from the form of equation (14.307) to that of equation (14.319). In addition, further (singular) transformations are also required to convert t to the time measured by a set of infalling observers with covariant velocity e_t.

14.7.4 Geodesics and the horizon

The $\bar{\mathsf{h}}$ and ω fields for the Newtonian form of the Kerr solution are well defined over all spacetime, down to the ring $\bar{r} = \cos(\bar{\theta}) = 0$. There are no problems with motion through the horizon, and infalling observers reach the central singularity in a finite coordinate time. This is because the coordinate t now measures the proper time for a family of free-falling observers with covariant velocity e_t. The trajectories defined by this velocity have

$$x' = \mathsf{h}(e_t) = e_t - \alpha \hat{e}_{\bar{r}} = e_t - \dot{\bar{r}} e_{\bar{r}}. \tag{14.327}$$

This defines a family of observers all infalling along directions with constant $\bar{\theta}$ and ϕ, and with infall velocity

$$\dot{\bar{r}} = \left(\frac{2M\bar{r}}{\bar{r}^2 + a^2}\right)^{1/2}. \tag{14.328}$$

This family neatly generalises the observers in radial free fall from rest at infinity employed in the Schwarzschild solution. As in the spherical case, many physical phenomena are simplest to interpret when expressed in terms of observers with covariant velocity e_t. A curious feature of these observers is that they appear to 'slow down' as the singularity is approached, though they do reach $\bar{r} = 0$ in a finite proper time.

The next task is to locate the horizon in our new form of the Kerr solution. A horizon marks the boundary between regions where one cannot signal to the other. This occurs where it is no longer possible to send null photons outwards. If k denotes the covariant photon velocity, with $k^2 = 0$, a horizon will occur when it is no longer possible to satisfy

$$\hat{e}_{\bar{r}} \cdot \mathsf{h}(k) < 0. \tag{14.329}$$

The left-hand side of this inequality can also be written as

$$\bar{\mathsf{h}}(\hat{e}_{\bar{r}}) \cdot k = \left(\hat{e}_{\bar{r}} + \left(\frac{2M\bar{r}}{\bar{r}^2 + a^2}\right)^{1/2} v\right) \cdot k. \tag{14.330}$$

It is not possible for two future-pointing null vectors to have an inner product less than 0, so the horizon occurs at

$$\frac{2M\bar{r}}{\bar{r}^2 + a^2} = 1. \tag{14.331}$$

This defines a quadratic equation, with two solutions when $a < M$, one when $a = M$ and no solutions for $a > M$. In the case where $a < M$, the outer horizon defines an event horizon. Photons can cross this on an inward trajectory, but no photons can escape. The inner horizon is slightly different. On the inside of the inner horizon it is possible for photons to travel outwards, but they cannot cross the horizon. Instead, they pile up just inside the boundary, forming an unstable *Cauchy horizon.*

Instead of considering observers attempting to exit to infinity, suppose instead that we look for observers at rest with respect to the background $(\bar{r}, \bar{\theta}, \phi)$ coordinates. Such observers can be constructed from observations of distant stars, for example. These observers have covariant velocity

$$\mathsf{h}^{-1}(\dot{x}) = \dot{t}\mathsf{h}^{-1}(e_t) = \dot{t}\left(e_t + \frac{2M\bar{r}}{\bar{r}^2 + a^2}\cosh(\beta)\,\hat{\phi}\right), \tag{14.332}$$

and the condition that this is a unit timelike vector forces

$$\dot{t}^2\left(1 - \frac{2M\bar{r}}{\bar{r}^2 + a^2\cos^2(\bar{\theta})}\right) = 1. \tag{14.333}$$

The surface within which it is not possible to remain at rest is called the *ergosphere*. For non-rotating black holes the horizon and ergosphere coincide. But for rotating black holes the ergosphere is defined by

$$\bar{r}^2 + a^2\cos^2(\bar{\theta}) - 2M\bar{r} = 0. \tag{14.334}$$

This surface lies outside the horizon, and touches the horizon at the poles. In the intervening region it is impossible to remain at rest, but it is still possible to escape. One can think of this in terms of the angular momentum of the hole dragging observers around with it.

To gain some further insight into the properties of the Kerr solution, consider circular orbits in the equatorial plane ($\bar{\theta} = \pi/2$). For these we have

$$\left(\mathsf{h}^{-1}(\dot{x})\right)^2 = \dot{t}^2 - (\bar{r}^2 + a^2)\dot{\phi}^2 - \frac{2M}{\bar{r}}(\dot{t} - a\dot{\phi})^2 = 1. \tag{14.335}$$

The $\bar{r}$ derivative of this expression must vanish for a circular orbit, which tells us that

$$\bar{r}^3 = M\left(\frac{\dot{t}}{\dot{\phi}} - a\right)^2. \tag{14.336}$$

If we let Ω denote the angular momentum measured by our set of preferred infalling observers (which are at rest at infinity), we have

$$\Omega = \frac{\dot{\phi}}{\dot{t}}. \tag{14.337}$$

It follows that, for circular orbits,

$$\Omega = \frac{M^{1/2}}{aM^{1/2} \pm \bar{r}^{3/2}}. \tag{14.338}$$

For a given distance, there are two possible values of the angular velocity for circular orbits. The larger value of Ω is for a particle corotating with the black hole, and the smaller for a counterrotating orbit. Again, this effect can be understood in terms of the black hole dragging matter around with it. The larger angular velocity for corotating orbits means it is possible to form stable orbits much closer to the event horizon than for the Schwarzschild case.

14.7.5 The Dirac equation in a Kerr background

As a final illustration of the utility of the Newtonian gauge form of the Kerr solution, we return to the Dirac equation. We first form

$$\partial_b \omega(b) = \frac{M}{\alpha \bar{\rho}^2} v - \left(\frac{2M\bar{r}}{\bar{r}^2 + a^2}\right)^{1/2} e_t \frac{1}{\bar{r} - Ia\cos(\bar{\theta})}. \tag{14.339}$$

The Dirac equation in the Newtonian gauge can therefore be written

$$\nabla\psi - (2M\bar{r})^{1/2}\left(\frac{v}{\bar{\rho}}\frac{\partial}{\partial \bar{r}}\psi + \frac{v}{4\bar{r}\bar{\rho}}\psi + e_t \frac{1}{2(\bar{r}^2 + a^2)^{1/2}(\bar{r} - Ia\cos(\bar{\theta}))}\psi\right) = -m\psi I\gamma_3. \tag{14.340}$$

If we again multiply through by e_t, we arrive at an interaction Hamiltonian of the form

$$\hat{H}_K\psi = \frac{i(2M)^{1/2}}{\bar{\rho}^2}\Bigg((\bar{r}^3 + a^2\bar{r})^{1/4}\frac{\partial}{\partial \bar{r}}\left((\bar{r}^3 + a^2\bar{r})^{1/4}\psi\right) - a\cos(\bar{\theta})\,\bar{r}^{1/4}\sigma_\phi \frac{\partial}{\partial \bar{r}}(\bar{r}^{1/4}\psi) + \frac{a\bar{r}^{1/2}\cos(\bar{\theta})}{2(\bar{r}^2 + a^2)^{1/2}} I\psi\Bigg), \tag{14.341}$$

where we continue to use i for the quantum imaginary. This Hamiltonian is (almost) Hermitian when integrated over flat three-dimensional space, because the measure in oblate spheroidal coordinates is

$$d^3x = \bar{\rho}^2 \sin(\bar{\theta})\, d\bar{r}\, d\bar{\theta}\, d\phi. \tag{14.342}$$

Our form of the Kerr solution therefore does generalise the many attractive features of the Newtonian gauge for the Schwarzschild solution. As in the Schwarzschild case, the Hamiltonian is not self-adjoint when acting on normalised wavefunctions. For the Kerr case a boundary term arises at $\bar{r} = 0$, which now defines a disc of radius a.

The Dirac equation (14.340) is separable in spheroidal coordinates, though the details of this separation are quite complicated. One problem is that the angular separation constant depends on the energy. This makes scattering calculations far more difficult than in the spherical case, as the separation constant must be recalculated for each energy. A considerable amount of work remains to be done in extending the detailed understanding of quantum theory in a Schwarzschild background to the Kerr case.

14.8 Notes

Many of the applications discussed in this chapter are covered in greater detail in the papers by Doran, Lasenby, Gull and coworkers. The solution method described in this chapter was first proposed in the paper 'Gravity, gauge theories and geometric algebra' by Lasenby, Doran & Gull (1998). This method should be compared with the spin coefficient formalism of Newman & Penrose (1962). The advantages of the Newtonian gauge for spherically-symmetric systems have been promoted by a handful of authors, most notably in the papers by Gautreau (1984), Gautreau & Cohen (1995), and by Martel & Poisson (2001).

The problem of the electromagnetic fields created by a point charge at rest outside a Schwarzschild black hole was first tackled by Copson (1928), who obtained a solution that was valid locally in the vicinity of the charge, but contained an additional pole at the origin. Linet (1976) modified Copson's solution by removing the singularity at the origin to obtain the potential described in section 14.3.5. Similar plots to those presented in section 14.3.5 were first obtained by Hanni & Ruffini (1973), though these authors did not extend their plots through the horizon. A popular means of interpreting these plots in terms of effects entirely around the horizon is advanced in *The Membrane Paradigm* by Thorne, Price & Macdonald (1986). We believe that a better understanding is gained by considering the global properties of fields, both inside and outside the horizon.

Scattering and absorption processes by black holes have been widely discussed by many authors. Summaries of this work can be found in the books by Futterman, Handler & Matzner (1988) and Chandrasekhar (1983), or the article by Andersson and Jensen (2000). The first attempt at a quantum calculation of the scattering cross section was by Collins, Delbourgo & Williams (1973), though their derivation did not employ a consistent perturbation scheme. The calculation described in this chapter was first published in the paper 'Perturbation theory calculation of the black hole elastic scattering cross section' by Doran and Lasenby (2002). Classical and quantum absorption processes are discussed in detail by Sanchez (1977, 1978) and Unruh (1976).

Cylindrical systems are discussed by Deser, Jackiw & 't Hooft (1984) and Jensen & Soleng (1992). The properties of cosmic strings are described in *Cosmic Strings and Other Topological Defects* by Vilenkin & Shellard (1994). The solutions described in this chapter were developed in the paper 'Physics of rotating cylindrical strings' by Doran, Lasenby & Gull (1996). The form of the fields outside a rotating black hole was first discovered by Kerr (1963), and has been widely discussed since. A fairly complete summary of this work in contained in Chandrasekhar's *The Mathematical Theory of Black Holes* (1983). The complex coordinate transformation trick for deriving the Kerr solution was discovered by Newman & Janis (1965), and later explained by Schiffer et al.(1973). The uniqueness theorem for black holes was developed by Carter (1971) and Robinson (1975). The analogue of the Newtonian gauge for the Kerr solution was discovered by Doran (2000).

The applications of the gauge theory approach to gravity discussed in this chapter have concentrated on the simplest Einstein–Cartan theory. Modern developments in quantum gravity have suggested a number of modifications to this theory. Two of the most common ideas include the introduction of local scale invariance, and the inclusion of higher order terms in the Lagrangian. The geometric algebra gauge theory approach is equally applicable in these settings. Some preliminary work on this subject is described by Lewis, Doran & Lasenby (2000). This field is developing rapidly, driven in part by developments in inflationary theory and observations of the cosmic microwave background. These observations could well revolutionise our understanding of gravitation in future years.

14.9 Exercises

14.1 Spherical symmetry of the h function can be imposed by demanding that

$$R\bar{\mathsf{h}}_{x'}(\tilde{R}aR)\tilde{R} = \bar{\mathsf{h}}(a),$$

where R is a constant spatial rotor ($Re_t\tilde{R} = e_t$), and $x' = \tilde{R}xR$. Prove that this symmetry implies that the $\{e^r, e^t\}$ and $\{e^\theta, e^\phi\}$ pairs decouple from each other. Show further that we must have

$$\begin{aligned}\bar{\mathsf{h}}(\hat{\theta}) &= \alpha\hat{\theta} + \beta\hat{\phi},\\ \bar{\mathsf{h}}(\hat{\phi}) &= \alpha\hat{\phi} - \beta\hat{\theta},\end{aligned}$$

and explain why we can always set $\beta = 0$ with a suitable gauge choice.

14.2 The energy-momentum tensor for an ideal fluid is

$$\mathcal{T}(a) = (\rho + p)a\cdot vv - pa.$$

Show that covariant conservation of the energy-momentum tensor results in the pair of equations

$$\begin{aligned}\mathcal{D}\cdot(\rho v) + p\mathcal{D}\cdot v &= 0,\\ (\rho + p)(v\cdot\mathcal{D}v)\wedge v - (\mathcal{D}p)\wedge v &= 0.\end{aligned}$$

Give a physical interpretation of these equations.

14.3 The Schwarzschild line element is defined by

$$ds^2 = \Big(1 - \frac{2M}{r}\Big)dt^2 - \frac{r}{r-2M}dr^2 - r^2\,d\theta^2 - r^2\sin^2(\theta)\,d\phi^2.$$

Find the equation for the free-fall time as measured by radially-infalling observers, starting from rest at infinity. Express the line element in terms of this new time coordinate to obtain the Painlevé-Gullstrand form

$$ds^2 = dt^2 - \left(dr + \left(\frac{2M}{r}\right)^{1/2} dt\right)^2 - r^2(d\theta^2 + \sin^2(\theta)\, d\phi^2).$$

14.4 Prove that the total absorption cross section for a spherically-symmetric black hole of mass M is given by

$$\sigma_{abs} = \frac{\pi M^2}{2u^4}(8u^4 + 20u^2 - 1 + (1 + 8u^2)^{3/2})$$

where u is the incident velocity.

14.5 The covariant electromagnetic field generated by a charge at rest on the z axis outside a Schwarzschild black hole is defined by

$$\mathcal{F} = -\frac{\partial V}{\partial r} e_r e_t - \frac{1}{r - 2M}\frac{\partial V}{\partial \theta}\hat{\theta}(e_t + \sqrt{2M/r}e_r),$$

where

$$V(r,\theta) = \frac{q}{ar}\frac{(r-M)(a-M) - M^2\cos^2(\theta)}{D} + \frac{qM}{ar}$$

and

$$D = \big(r(r-2M) + (a-M)^2 - 2(r-M)(a-M)\cos(\theta) + M^2\cos^2(\theta)\big)^{1/2}.$$

Prove that $\mathcal{F}$ is finite and continuous at the horizon.

14.6 In calculating the scattering cross section from a black hole we need to compute the integral

$$I_1 = \int \frac{d^3k}{(2\pi)^3}\frac{\boldsymbol{k}^2 - \boldsymbol{p}^2}{|\boldsymbol{p}_f - \boldsymbol{k}|^{7/2}|\boldsymbol{k} - \boldsymbol{p}_i|^{7/2}}(\hat{k} + m).$$

Evaluate this integral by first displacing the origin in $\boldsymbol{k}$-space by the amount $(\boldsymbol{p}_f + \boldsymbol{p}_i)/2$, and then introducing spheroidal coordinates

$$\begin{aligned}k_1 &= \alpha\sinh(u)\,\sin(v)\,\cos(\phi),\\ k_2 &= \alpha\sinh(u)\,\sin(v)\,\sin(\phi),\\ k_2 &= \alpha\cosh(u)\,\cos(v),\end{aligned}$$

where $0 \le u < \infty$, $0 \le v \le \pi$, $0 \le \phi < 2\pi$ and $\alpha = |\boldsymbol{q}|/2$.

14.7 The Kerr–Schild form of the Schwarzschild solution is defined by

$$\bar{\mathsf{h}}(a) = a + \frac{M}{r}a\cdot e_-\, e_-, \qquad e_- = e_t - e_r.$$

Construct the Dirac equation in this gauge, and find the interaction vertex factor in momentum space. Calculate the differential scattering cross section for a fermion in this gauge, and verify that it is the same as found in the Newtonian gauge.

14.8 Prove that det (h) is constant for spherically-symmetric vacuum gravitational fields.

14.9 For a particle in a circular orbit around a Schwarzschild black hole, prove that the non-relativistic binding energy (as defined by the effective potential) is given by ($G = c = 1$)

$$E_b = -\frac{M}{2r}\frac{r - 4M}{r - 3M}.$$

14.10 Derive the full set of time-dependent radial equations with the cosmological constant Λ included.

14.11 A spherically-symmetric distribution of dust is released from rest, with the initial density distribution chosen so that streamlines do not cross. Prove that a singularity forms at the origin after a time

$$t_f = \left(\frac{3\pi}{32\rho_0}\right)^{1/2},$$

where ρ_0 is the central density.

14.12 Solve the Dirac equation in a cosmological background with $k \neq 0$. Is the Dirac field homogeneous? Can you construct self-consistent solutions to this system of equations?

14.13 Construct a matched set of interior and exterior gravitational fields around a rigidly-rotating cylindrical string. Do closed timelike curves exist in this geometry?

14.14 Verify that the Kerr solution defined by equation (14.307) satisfies the vacuum field equations.

14.15 The Riemann tensor for the Kerr solution can be written as

$$\mathcal{R}(B) = -\frac{M}{2\big(r - Ia\cos(\theta)\big)^3}(B + 3\sigma_r B \sigma_r).$$

Prove that this satisfies $\partial_b \mathcal{R}(b \wedge c) = 0$ and interpret both parts of this result.

14.16 The Newtonian gauge form of the Kerr solution involves the spheroidal coordinates $\bar{r}$ and $\bar{\theta}$. Prove that $\bar{r} = \cos(\bar{\theta}) = 0$ defines a ring.

Bibliography

S.L. Altmann. *Rotations, Quaternions, and Double Groups.* Clarendon Press, Oxford, 1986.

S.L. Altmann. Hamilton, Rodrigues, and the quaternion scandal. *Mathematics Magazine*, **62**(5):291, 1989.

N. Andersson and B.P. Jensen. Scattering by black holes. In R. Pike and P. Sabatier, editors, *Scattering.* Academic Press Ltd, London, 2000.

S.S. Antman. *Nonlinear Problems in Elasticity.* Springer–Verlag, Berlin, 1995.

L. Aramanovitch. Spacecraft orientation based on space object observations by means of quaternion algebra. *J. Guidance Control and Dynamics*, **18**:859–866, 1995.

A.O. Barut and I.H. Duru. Path integral formulation of quantum electrodynamics from classical particle trajectories. *Phys. Rep.*, **172**(1):1, 1989.

A.O. Barut and N. Zanghi. Classical models of the Dirac electron. *Phys. Rev. Lett.*, **52**(23):2009, 1984.

I.W. Benn and R.W. Tucker. *An Introduction to Spinors and Geometry.* Adam Hilger, Bristol, 1988.

F.A. Berezin. *The Method of Second Quantization.* Academic Press Ltd, London, 1966.

F.A. Berezin. *Introduction to Superanalysis.* Reidel, Dordrecht, 1987.

F.A. Berezin and M.S. Marinov. Particle spin dynamics as the Grassmann variant of classical mechanics. *Ann. Phys.*, **104**:336, 1977.

J.D. Bjorken and S.D. Drell. *Relativistic Quantum Mechanics.* McGraw-Hill, New York, 1964.

D.A. Brannan, M.F. Espleen and J.J. Gray. *Geometry.* Cambridge University Press, Cambridge, 1999.

G. Breit. The effect of retardation on the interaction of two electrons. *Phys. Rev.*, **34**(4):553, 1929.

B. Carter. Axisymmetric black hole has only two degrees of freedom. *Phys. Rev. Lett.*, **26**(6):331, 1971.

T.E. Cecil. *Lie Sphere Geometry.* Springer–Verlag, New York, 1992.

A.D. Challinor et al. A relativistic, causal account of a spin measurement. *Phys. Lett. A*, **218**:128, 1996.

S. Chandrasekhar. *The Mathematical Theory of Black Holes.* Oxford University Press, Oxford, 1983.

T. Cheng and L. Li. *Gauge Theory of Elementary Particle Physics.* Oxford University Press, Oxford, 1984.

M. Chisholm. *Such Silver Currents.* Lutterworth Press, Cambridge, 2002.

W.K. Clifford. *Mathematical Papers.* Macmillan, London, 1882.

S. Coleman. *Aspects of Symmetry.* Cambridge University Press, Cambridge, 1985.

P.A. Collins, R. Delbourgo and R. M. Williams. On the elastic Schwarzschild scattering cross section. *J. Phys. A*, **6**:161–169, 1973.

A. Connes and J. Lott. Particle models and non-commutative geometry. *Nucl. Phys. B*, **18**:29, 1990.

A.H. Cook. A Hamiltonian with linear kinetic energy for systems of many bodies. *Proc. R. Soc. Lond. A*, **415**:35, 1988.

E.T. Copson. On electrostatics in a gravitational field. *Proc. R. Soc. Lond. A*, **118**:184, 1928.

R. Coquereaux, A. Jadczyk and D. Kastler. Differential and integral geometry of Grassmann algebras. *Rev. Math. Phys.* **3**(1):63, 1991.

J.F. Cornwell. *Group Theory in Phsics I.* Academic Press Ltd, London, 1984a.

J.F. Cornwell. *Group Theory in Phsics II.* Academic Press Ltd, London, 1984b.

J.F. Cornwell. *Group Theory in Physics III.* Academic Press Ltd, London, 1989.

B. de Witt. *Supermanifolds.* Cambridge University Press, Cambridge, 1984.

S. Deser, R. Jackiw and G. 't Hooft. Three-dimensional Einstein gravity: Dynamics of flat space. *Ann. Phys.*, **152**:220, 1984.

R. d'Inverno. *Introducing Einstein's Relativity.* Oxford University Press, Oxford, 1992.

C.J.L. Doran. Geometric algebra and its application to mathematical physics. PhD thesis, Cambridge University, 1994.

C.J.L. Doran. New form of the Kerr solution. *Phys. Rev. D*, **61**(6):067503, 2000.

C.J.L. Doran and A.N. Lasenby. Perturbation theory calculation of the black hole elastic scattering cross section. *Phys. Rev. D*, **66**(2):024006, 2002.

C.J.L. Doran, A.N. Lasenby and S.F. Gull. States and operators in the spacetime algebra. *Found. Phys.*, **23**(9):1239, 1993a.

C.J.L. Doran, A.N. Lasenby and S.F. Gull. Grassmann mechanics, multivector derivatives and geometric algebra. In Z. Oziewicz, B. Jancewicz and A. Borowiec, editors, *Spinors, Twistors, Clifford Algebras and Quantum Deformations*, page 215. Kluwer Academic, Dordrecht, 1993b.

C.J.L. Doran, A.N. Lasenby and S.F. Gull. Physics of rotating cylindrical strings. *Phys. Rev. D*, **54**(10):6021, 1996.

C.J.L. Doran *et al.* Lie groups as spin groups. *J. Math. Phys.*, **34**(8):3642, 1993.

C.J.L. Doran *et al.* Lectures in geometric algebra. In W.E. Baylis, editor, *Clifford (Geometric) Algebras*, pages 65–236. Birkhäuser, Boston, 1996a.

C.J.L Doran et al. Spacetime algebra and electron physics. *Adv. Imag. & Elect. Phys.*, **95**:271, 1996b.

C.J.L Doran et al. Effects of spin-torsion in gauge theory gravity. *J. Math. Phys.*, **39**(6):3303, 1998.

L. Dorst, C. Doran and J. Lasenby, editors. *Applications of Geometric Algebra in Computer Science and Engineering.* Birkhäuser, Boston, 2002.

A.W.M. Dress and T.F. Havel. Distance geometry and geometric algebra. *Found. Phys.*, **23**(10):1357–1374, 1993.

A.S. Eddington. *Relativity Theory of Protons and Electrons.* Cambridge University Press, Cambridge, 1936.

R.P. Feynman. *Quantum Electrodynamics.* Addison–Wesley, Reading, MA, 1961.

R.P. Feynman, F.B. Morningo and W.G. Wagner. *Feynman Lectures on Gravitation.* Addison–Wesley, Reading, MA, 1995.

A.P. French. *Special Relativity.* Nelson, London, 1968.

J.A.H. Futterman, F.A. Handler and R.A. Matzner. *Scattering from Black Holes.* Cambridge University Press, Cambridge, 1988.

A.P. Galeao and P.L. Ferreira. General method for reducing the two-body Dirac equation. *J. Math. Phys.*, **33**(7):2618, 1992.

R. Gautreau. Curvature coordinates in cosmology. *Phys. Rev. D*, **29**(2):186, 1984.

R. Gautreau and J.M. Cohen. Gravitational collapse in a single coordinate system. *Am. J. Phys.*, **63**(11):991, 1995.

H. Georgi. *Lie Algebras in Particle Physics.* Benjamin/Cummings, Reading, MA, 1982.

J.W. Gibbs. *Collected Papers,* Volumes I and II. Longmans, Green and Co., London, 1906.

R. Gilmore. *Lie groups, Lie algebras and Some of Their Applications.* Wiley, New York, 1974.

M. Göckeler and T. Schucker. *Differential Geometry, Gauge Theories, and Gravity.* Cambridge University Press, Cambridge, 1987.

H. Goldstein. *Classical Mechanics.* Addison–Wesley, Reading, MA, 1950.

I.S. Gradshteyn and I.M. Ryzhik. *Table of Integrals, Series and Products,* fifth edition. Academic Press Ltd, London, 1994.

W.T. Grandy, Jr. *Relativistic Quantum Mechanics of Leptons and Fields.* Kluwer Academic, Dordrecht, 1991.

W. Greiner. *Relativistic Quantum Mechanics.* Springer–Verlag, Berlin, 1990.

S.F. Gull, A.N. Lasenby and C.J.L. Doran. Imaginary numbers are not real — the geometric algebra of spacetime. *Found. Phys.*, **23**(9):1175, 1993a.

S.F. Gull, A.N. Lasenby and C.J.L. Doran. Electron paths, tunnelling and diffraction in the spacetime algebra. *Found. Phys.*, **23**(10):1329, 1993b.

L.N. Hand and J.D. Finch. *Analytical Mechanics.* Cambridge University Press, Cambridge, 1998.

R. S. Hanni and R. Ruffini. Lines of force of a point charge near a Schwarzschild black hole. *Phys. Rev. D.*, **8**(10):3259, 1973.

T.F. Havel and C.J.L. Doran. Geometric algebra in quantum information processing. In S.J. Lomonaco and H.E. Brandt, editors, *Quantum Computation and Information*, pages 81–100. American Mathematical Society, Providence, RI, 2002a.

T.F. Havel and C.J.L. Doran. Interaction and entanglement in the multiparticle spacetime algebra. In L. Dorst, C.J.L. Doran and J. Lasenby, editors, *Applications of Geometric Algebra in Computer Science and Engineering*, page 229. Birkhäuser, Boston, 2002b.

T.F. Havel et al. Geometric algebra methods in quantum information processing by NMR spectroscopy. In E. Bayro-Corrochanno and G. Sobczyk, editors, *Geometric ALgebra with Applications in Science and Engineering*, page 281. Birkhäuser, Boston, 2001.

D. Hestenes. *Space-Time Algebra.* Gordon and Breach, New York, 1966.

D. Hestenes and R. Gurtler. Local observables in quantum theory. *Am. J. Phys.*, **39**:1028, 1971.

D. Hestenes. Proper particle mechanics. *J. Math. Phys.*, **15**(10):1768, 1974a.

D. Hestenes. Proper dynamics of a rigid point particle. *J. Math. Phys.*, **15**(10):1778, 1974b.

D. Hestenes. Geometry of the Dirac theory. In J. Keller, editor, *The Mathematics of Physical Spacetime*, page 67. UNAM, Mexico, 1982a.

D. Hestenes. Space-time structure of weak and electromagnetic interactions. *Found. Phys.*, **12**(2):153, 1982b.

D. Hestenes. Celestial mechanics with geometric algebra. *Celestial Mech.*, **30**:151–170, 1983.

D. Hestenes. Curvature calculations with spacetime algebra. *Int. J. Theor. Phys.*, **25**(6):581, 1986.

D. Hestenes. The zitterbewegung interpretation of quantum mechanics. *Found. Phys.*, **20**(10):1213, 1990.

D. Hestenes. The design of linear algebra and geometry. *Acta Appl. Math.*, **23**:65, 1991.

D. Hestenes. Invariant body kinematics: I. Saccadic and compensatory eye movements. *Neural Networks*, **7**(1):65, 1994a.

D. Hestenes. Invariant body kinematics: II. Reaching and neurogeometry. *Neural Networks*, **7**(1):79, 1994b.

D. Hestenes. *New Foundations for Classical Mechanics,* second edition. Kluwer Academic Publishers, Dordrecht, 1999.

D. Hestenes. Old wine in new bottles: a new algebraic framework for computational geometry. In E.D. Bayro-Corrochano and G. Sobcyzk, editors, *Geometric Algebra with Applications in Science and Engineering*, page 16. Birkhäuser Boston, 2001.

D. Hestenes and G. Sobczyk. *Clifford Algebra to Geometric Calculus.* Reidel, Dordrecht, 1984.

D. Hestenes and R. Ziegler. Projective geometry with Clifford algebra. *Acta. Appli. Math.*, **23**:25, 1991.

D. Hestenes, H. Li and A. Rockwood. Generalized homogeneous coordinates for computational geometry. In G. Sommer, editor, *Geometric Computing with Clifford Algebras.* Springer–Verlag, Berlin, 1999a.

D. Hestenes, H. Li and A. Rockwood. New algebraic tools for classical geometry. In G. Sommer, editor, *Geometric Computing with Clifford Algebras.* Springer–Verlag, Berlin, 1999b.

R. Heumann and N. S. Manton. Classical supersymmetric mechanics. *Ann. Phys.*, **284**:52–88, 2000.

C. Itzykson and J-B. Zuber. *Quantum Field Theory.* McGraw–Hill, New York, 1980.

J.D. Jackson. *Classical Electrodynamics,* third edition. Wiley, New York, 1999.

B. Jancewicz. *Multivectors and Clifford Algebras in Electrodynamics.* World Scientific, Singapore, 1989.

B. Jensen and H.H. Soleng. General-relativistic model of a spinning cosmic string. *Phys. Rev. D*, **45**(10):3528, 1992.

L. C. Kannenberg. *The Ausdehnungslehre of 1844 and Other Works.* Open Court Publ., Chicago, 1995.

R.P. Kerr. Gravitational field of a spinning mass as an example of algebraically special metrics. *Phys. Rev. Lett.*, **11**(5):237, 1963.

T.W.B. Kibble. Lorentz invariance and the gravitational field. *J. Math. Phys.*, **2**(3):212, 1961.

M. Kline. *Mathematical Thought from Ancient to Modern Times.* Oxford University Press, Oxford, 1972.

Y. Koide. Exactly solvable model of relativistic wave equations and meson spectra. *Il Nuovo Cim.*, **70A**(4):411, 1982.

B.A. Kupershmidt. *The Variational Principles of Dynamics.* World Scientific, Singapore, 1992.

A.N. Lasenby and J. Lasenby. Applications of geometric algebra in physics and links with engineering. In E. Bayro and G. Sobczyk, editors, *Geometric algebra: A Geometric Approach to Computer Vision, Neural and Quantum Computing, Robotics and Engineering*, pages 430–457. Birkhäuser, Boston, 2000a.

A.N. Lasenby and J. Lasenby. Surface evolution and representation using geometric algebra. In R. Cippola, editor, *The Mathematics of Surfaces IX*, pages 144–168. Institute of Mathematics and its Applications, London, 2000b.

A.N. Lasenby, C.J.L. Doran and S.F. Gull. A multivector derivative approach to Lagrangian field theory. *Found. Phys.*, **23**(10):1295, 1993a.

A.N. Lasenby, C.J.L. Doran and S.F. Gull. 2-spinors, twistors and supersymmetry in the spacetime algebra. In Z. Oziewicz, B. Jancewicz and A. Borowiec, editors, *Spinors, Twistors, Clifford Algebras and Quantum Deformations*, page 233. Kluwer Academic, Dordrecht, 1993b.

A.N. Lasenby, C.J.L. Doran and S.F. Gull. Grassmann calculus, pseudoclassical mechanics and geometric algebra. *J. Math. Phys.*, **34**(8):3683, 1993c.

A.N. Lasenby, C.J.L. Doran and S.F. Gull. Gravity, gauge theories and geometric algebra. *Phil. Trans. R. Soc. Lond. A*, **356**:487–582, 1998.

A.M. Lewis, C.J.L. Doran and A.N. Lasenby. Quadratic Lagrangians and topology in gauge theory gravity. *Gen. Rel. Grav.*, **32**(1):161, 2000.

A.M. Lewis, C.J.L. Doran and A.N. Lasenby. Electron scattering without spin sums. *Int. J. Theor. Phys.*, **40**(1), 2001.

H. Li. Hyperbolic conformal geometry with Clifford Algebra. *Int.J.Theor.Phys.*, **40**:81, 2001.

B. Linet. Electrostatics and magnetostatics in the Schwarzschild metric. *J. Phys. A*, **9**(7):1081, 1976

P. Lounesto. *Clifford Algebras and Spinors.* Cambridge University Press, Cambridge, 1997.

J.E. Marsden and T.J.R. Hughes. *Mathematical Foundations of Elasticity.* Dover Publications Inc., New York, 1994.

J.E. Marsden and T.S. Ratiu. *Introduction to Mechanics and Symmetry.* Springer–Verlag, Berlin, 1994.

K. Martel and E. Poisson. Regular coordinate systems for Schwarzschild and other spherical spacetimes. *Am. J. Phys.*, **69**(4):476, 2001.

C.W. Misner, K.S. Thorne and J.A. Wheeler. *Gravitation.* W.H. Freeman and Company, San Francisco, 1973.

M. Nakahara. *Geometry, Topology and Physics.* Adam Hilger, Bristol, 1990.

S. Nasar. *A Beautiful Mind.* Faber and Faber, London, 1998.

J. Nash. The imbedding problem for Riemannian manifolds. *Ann. Math.*, **63**(1):20, 1956.

E.T. Newman and A.I. Janis. Note on the Kerr spinning-particle metric. *J. Math. Phys.*, **6**(4):915, 1965.

E.T. Newman and R. Penrose. An approach to gravitational radiation by a method of spin coefficients. *J. Math. Phys.*, **3**(3):566–578, 1962.

R. Parker and C.J.L. Doran. Analysis of 1 and 2 particle quantum systems using geometric algebra. In L. Dorst, C.J.L. Doran and J. Lasenby, editors, *Applications of Geometric Algebra in Computer Science and Engineering*, page 215. Birkhäuser, Boston, 2002.

R. Penrose. The apparent shape of a relativistically moving sphere. *Proc. Cam. Phil. Soc.*, **55**:137–139, 1959.

R. Penrose and W. Rindler. *Spinors and Space-time,* Volume I: *Two-Spinor Calculus and Relativistic Fields.* Cambridge University Press, Cambridge, 1984.

R. Penrose and W. Rindler. *Spinors and Space-time,* Volume II: *Spinor and Twistor Methods in Space-time Geometry.* Cambridge University Press, Cambridge, 1986.

C.B.U. Perwass and J. Lasenby. A geometric analysis of the trifocal tensor. In R. Klette, G. Gimel'farb and R. Kakarala, editors, *Proceedings of Image and Vision Computing New Zealand*, pages 157–162. The University of Auckland, Auckland, 1998.

J. Preskill. Quantum computation. Lecture notes available at http://theory.caltech.edu/~preskill/ph229, 1998.

R. Rau, D. Weiskopf and H. Ruder. Special relativity in virtual reality. In H.-C. Hege and K. Polthier, editors, *Mathematical Visualization*, pages 269–279. Springer–Verlag, Berlin, 1998.

J. Richter-Gebert and U. Kortenkamp. *The Interactive Geometry Software Cinderella.* Springer–Verlag, Berlin, 1999.

W. Rindler. *Essential Relativity.* Springer–Verlag, Berlin, 1977.

D.C. Robinson. Uniqueness of the Kerr black hole. *Phys. Rev. Lett.*, **34**(14):905, 1975.

E.E. Salpeter. Mass corrections to the fine structure of Hydrogen-like atoms. *Phys. Rev.*, **87**(2):328–343, 1952.

E.E. Salpeter and H.A. Bethe. A relativistic equation for bound-state problems. *Phys. Rev.*, **84**(6):1232, 1951.

N. Sanchez. Wave scattering and the absorption problem for a black hole. *Phys. Rev. D*, **16**(4):937–945, 1977.

N. Sanchez. Elastic scattering of waves by a black hole. *Phys. Rev. D*, **18**(6):1798–1804, 1978.

M.M. Schiffer et al. Kerr geometry as complexified Schwarzschild geometry. *J. Math. Phys.*, **14**(1):52, 1973.

G. Schubring, editor. *Hermann Gunther Grassmann (1809–1877): Visionary Mathematician, Scientist and Neohumanist Scholar.* Kluwer Academic, Dordrecht, 1996.

B.F. Schutz. *Geometrical Methods of Mathematical Physics.* Cambridge University Press, Cambridge, 1980.

J. Schwinger et al. *Classical Electrodynamics.* Perseus Books, Reading, MA, 1998.

J.G. Semple and G.T. Kneebone. *Algebraic Projective Geometry.* Clarendon Press, Oxford, 1998.

S. Somaroo, D.G. Cory and T.F. Havel. Expressing the operations of quantum computing in multiparticle geometric algebra. *Phys. Lett. A*, **240**:1–7, 1998.

S.S. Somaroo, A.N. Lasenby and C.J.L. Doran. Geometric algebra and the causal approach to multiparticle quantum mechanics. *J. Math. Phys.*, **40**(7):3327–3340, 1999.

A.X.S Stevenson and J. Lasenby. Decomplexifying the absolute conic. In R. Klette, G. Gimel'farb and R. Kakarala, editors, *Proceedings of Image and Vision Computing New Zealand*, page 163. The University of Auckland, Auckland, 1998.

E.L. Stiefel and G. Scheifele. *Linear and Regular Celestial Mechanics.* Springer–Verlag, Berlin, 1971.

J.C. Taylor. *Gauge Theories in the Twentieth Century.* Imperial College Press, London, 2001.

J. Terrell. Invisibility of the Lorentz contraction. *Phys. Rev.*, **116**(4):1041–1045, 1959.

K.S. Thorne, R.H. Price and D.A. Macdonald. *Black Holes: The Membrane Paradigm.* Yale University Press, New Haven, CT, 1986.

H. Turnbull. *The Theory of Determinants, Matrices and Inverses.* Dover, New York, 1960.

W.G. Unruh. Absorption cross section of small black holes. *Phys. Rev. D.*, **14**(12):3251, 1976.

R. Utiyama. Invariant theoretical interpretation of interaction. *Phys. Rev.*, **101**(5):1597, 1956.

A. Vilenkin and E.P.S. Shellard. *Cosmic Strings and Other Topological Defects.* Cambridge University Press, Cambridge, 1994.

T.G. Vold. An introduction to geometric algebra with an application to rigid-body mechanics. *Am. J. Phys.*, **61**(6):491, 1993a.

T.G. Vold. An introduction to geometric calculus and its application to electrodynamics. *Am. J. Phys.*, **61**(6):505, 1993b.

J. Vrbik. Celestial mechanics via quaternions. *Can. J. Phys.*, **72**:141–146, 1994.

J. Vrbik. Perturbed Kepler problem in quaterionic form. *J. Phys. A*, **28**:6245–6252, 1995.

J.A. Wheeler and R.P. Feynman. Classical electrodynamics in terms of direct interparticle action. *Rev. Mod. Phys.*, **21**(3):425, 1949.

Index

Printed in the United States
By Bookmasters